Uncertainty, Modeling, and Decision Making in Geotechnics

Uncertainty, Modeling, and Decision Making in Geotechnics shows how uncertainty quantification and numerical modeling can complement each other to enhance decision-making in geotechnical practice, filling a critical gap in guiding practitioners to address uncertainties directly.

The book helps practitioners acquire a working knowledge of geotechnical risk and reliability methods and guides them to use these methods wisely in conjunction with data and numerical modeling. In particular, it provides guidance on the selection of realistic statistics and a cost-effective, accessible method to address different design objectives, and for different problem settings, and illustrates the value of this to decision-making using realistic examples.

Bringing together statistical characterization, reliability analysis, reliability-based design, probabilistic inverse analysis, and physical insights drawn from case studies, this reference guide from an international team of experts offers an excellent resource for state-of-the-practice uncertainty-informed geotechnical design for specialist practitioners and the research community.

Challenges in Geotechnical and Rock Engineering

This series offers advanced level books focusing on state-of-the-art methods for handling problems across geotechnical engineering.

Chief Editor *Kok-Kwang Phoon, Singapore University of Technology and Design*
Assistant Editor *Dongming Zhang, Tongji University*

Evolutionary Process of a Steep Rocky Reservoir Bank in a Dynamic Mechanical Environment
Luqi Wang and Wengang Zhang

Uncertainty, Modeling, and Decision Making in Geotechnics
Edited by Kok-Kwang Phoon, Takayuki Shuku, and Jianye Ching

Uncertainty, Modeling, and Decision Making in Geotechnics

Edited by
Kok-Kwang Phoon, Takayuki Shuku, and Jianye Ching

CRC Press
Taylor & Francis Group
Boca Raton London New York

CRC Press is an imprint of the
Taylor & Francis Group, an **informa** business

Cover image: Shutterstock ©

First edition published 2024
by CRC Press
2385 NW Executive Center Drive, Suite 320, Boca Raton FL 33431

and by CRC Press
4 Park Square, Milton Park, Abingdon, Oxon, OX14 4RN

CRC Press is an imprint of Taylor & Francis Group, LLC

Library of Congress Cataloging-in-Publication Data

Names: Phoon, Kok-Kwang, editor. | Shuku, Takayuki, editor. | Ching, Jianye, editor.
Title: Uncertainty, modelling, and decision making in geotechnics / edited by Kok-Kwang Phoon, Takayuki Shuku, Jianye Ching.
Description: Boca Raton : CRC Press, 2024. | Series: Challenges in geotechnical and rock engineering | Includes bibliographical references and index.
Identifiers: LCCN 2023026503 | ISBN 9781032367491 (hardback) | ISBN 9781032367507 (paperback) | ISBN 9781003333586 (ebook)
Subjects: LCSH: Geotechnical engineering--Decision making--Mathematical models. | Geotechnical engineering--Statistical methods. | Uncertainty--Mathematical models.
Classification: LCC TA705 .U53 2024 | DDC 624.1/51--dc23/eng/20230925
LC record available at https://lccn.loc.gov/2023026503

ISBN: 978-1-032-36749-1 (hbk)
ISBN: 978-1-032-36750-7 (pbk)
ISBN: 978-1-003-33358-6 (ebk)

DOI: 10.1201/9781003333586

Typeset in Times New Roman
by Deanta Global Publishing Services, Chennai, India

Access the Support Material: www.routledge.com/9781032367491

Contents

About the editors vii
List of contributors viii
Challenges in Geotechnical and Rock Engineering x
Foreword xi
Preface xii

Chapter 1 Uncertainty-Informed Decision-Making in Burland Triangle 1
Kok-Kwang Phoon

PART 1 Statistical characterization

Chapter 2 Soil and Rock Parametric Uncertainties 39
Kok-Kwang Phoon, Jianye Ching, and Yuanqin Tao

Chapter 3 Uncertainty in Constitutive Models 117
Zhen-Yu Yin, Pin Zhang, and Yin-Fu Jin

Chapter 4 Data-Driven Site Characterization 143
Takayuki Shuku

Chapter 5 Variability of Predictions in Geotechnics 176
Chong Tang, Kok-Kwang Phoon, and Jun Yuan

PART 2 Probabilistic methods

Chapter 6 Geotechnical Reliability Analysis for Practice 219
Jian Ji and Zijun Cao

Chapter 7 Reliability Analysis with Reduced Order Model 247
Yu Otake and Taiga Saito

Chapter 8 Stochastic Finite Element Methods for Slope Stability Analysis and Risk Assessment 280
Shui-Hua Jiang, Te Xiao and Dian-Qing Li

Chapter 9 Reliability-Based Design with Numerical Models 319

Shadi Najjar and Imad El-Chiti

Chapter 10 Probabilistic Inverse Analysis for Geotechnics 351

Ikumasa Yoshida

PART 3 Probabilistic software

Chapter 11 Use of Geotechnical Software for Probabilistic Analysis and Design 373

Sina Javankhoshdel, Brigid Cami, Terence Ma, Angela Li, Zhanyu Huang, Ellen Yeh, Seok Hyeon Chai, Thamer Yacoub

Chapter 12 Reliability-Based Decision-Making with FE Models for Real-Life Case Studies 401

Commend Stéphane, Minini Jocelyn, Groslambert Marc, and Jacot-Descombes Gil

PART 4 Probabilistic applications

Chapter 13 Soil-Structure Interaction in Spatially Variable Ground 427

Andy Y.F. Leung

Chapter 14 Reduction of Uncertainty through Piling Data within the Same Site in the Press-in Piling Method 441

Naoki Suzuki

Chapter 15 Slope Reliability Assessments for Linear Infrastructures 457

Michael A. Hicks

Chapter 16 Uncertainty, Modeling, and Decision-Making for Ground Improvement 473

Yutao Pan and Rui Tao

Index 493

About the editors

Kok-Kwang Phoon is Cheng Tsang Man Chair Professor and Provost at the Singapore University of Technology and Design. He has edited three books and authored one book: *Model Uncertainties in Foundation Design* (CRC Press, 2021). He was awarded the ASCE Norman Medal twice, in 2005 and in 2020, and the Humboldt Research Award in 2017. He is the Founding Editor of *Georisk* and appointed a Board Member of ISSMGE and elected as a Fellow of the Academy of Engineering Singapore.

Takayuki Shuku is Associate Professor of Okayama University in Japan. He received the Best Paper Award from Japan Society of Civil Engineering in 2020, and the ISSMGE Bright Spark Lecture Award in 2019.

Jianye Ching is Distinguished Professor at National Taiwan University and Convener of the Civil & Hydraulic Engineering Program of the Ministry of Science and Technology of Taiwan. He served as Chair of ISSMGE's TC304 (risk) and Chair of Geotechnical Safety Network (GEOSNet). He is Managing Editor of the journal *Georisk*.

List of contributors

Kok-Kwang Phoon
Singapore University of Technology and Design
Singapore

Takayuki Shuku
Okayama University
Okayama, Japan

Jianye Ching
National Taiwan University
Taipei, Taiwan

Yuanqin Tao
Zhejiang University of Technology
Hangzhou, China

Zhen-Yu Yin
The Hong Kong Polytechnic University
Hong Kong, China

Pin Zhang
University of Cambridge
Cambridge, UK

Yin-Fu Jin
Shenzhen University
Shenzhen, China

Chong Tang
Dalian University of Technology
Dalian, China

Jian Ji
Hohai University
Nanjing, China

Zijun Cao
Southwest Jiaotong University
Chengdu, China

Yu Otake
Tohoku University
Sendai, Japan

Taiga Saito
Tohoku University
Sendai, Japan

Shui-Hua Jiang
Nanchang University
Nanchang, China

Te Xiao
Hong Kong University of Science and Technology
Hong Kong, China

Dian-Qing Li
Wuhan University
Wuhan, China

Shadi Najjar
American University of Beirut
Beirut, Lebanon

Imad El-Chiti
American University of Beirut
Beirut, Lebanon

Ikumasa Yoshida
Tokyo City University
Tokyo, Japan

Sina Javankhoshdel
Rocscience
Toronto, Canada

Brigid Cami
Rocscience
Toronto, Canada

Terence Ma
Rocscience
Toronto, Canada

Angela Li
Rocscience
Toronto, Canada

Ellen Yeh
Rocscience
Toronto, Canada

Zhanyu Huang
Rocscience
Toronto, Canada

Seok Hyeon Chai
Rocscience
Toronto, Canada

Thamer Yacoub
Rocscience
Toronto, Canada

Stéphane Commend
HEIA-FR, HES-SO University of Applied Sciences and Arts Western Switzerland
Fribourg, Switzerland

Jocelyn Minini
HEIA-FR, HES-SO University of Applied Sciences and Arts Western Switzerland
Fribourg, Switzerland

Marc Groslambert
GeoMod Consulting Engineers SA
Lausanne, Switzerland

Gil Jacot-Descombes
GeoMod Consulting Engineers SA
Lausanne, Switzerland

Andy Yat Fai Leung
Hong Kong Polytechnic University
Hong Kong, China

Naoki Suzuki
GIKEN Ltd.
Tokyo, Japan

Michael A. Hicks
Delft University of Technology
Delft, Netherlands

Yutao Pan
Norwegian University of Science and Technology
Trondheim, Norway

Rui Tao
Norwegian University of Science and Technology
Trondheim, Norway

Challenges in Geotechnical and Rock Engineering

Geotechnical and rock engineering have made significant strides in response to different challenges and opportunities that include measuring and understanding material/structural behaviors, handling special environmental conditions, performing complex numerical simulations, design, construction, and circular role for low carbon materials, novel structures, green construction and operation, life cycle and risk/reliability informed management, data-driven algorithms and AI-based decision-making, autonomous systems, digital twin and smart infrastructure, resilience engineering, climate change and sustainability, among others. These challenges are interrelated. Although some of the challenges are common to all industries, it is not meaningful to engage them in an abstract manner outside of the practice context. One example is data-centric geotechnics that takes a "data first" approach to decision-making, but the data are actual "ugly" field observations (multi-source, sparse, incomplete, spatially variable, corrupted, etc.) rather than ideal abstract numbers. Data-centric geotechnics should deploy digital technologies in cognizance of the context of geotechnical and rock engineering that includes physics, empirical knowledge, experience, and engineering judgment. How decision-making in geotechnical and rock engineering can be revolutionized through human machine collaboration is one grand challenge that epitomize the motivation for launching this book series. This book series presents exciting emerging solutions in geotechnical and rock engineering that are expected to transform practice and to meet fast-evolving environmental/societal trends in the twenty-first century. It is a timely response to the changing technological, environmental, and societal landscape presented in the Institution of Civil Engineers (ICE) State of the Nation Report on "Digital Transformation" and the American Society of Civil Engineers (ASCE) "Future World Vision: Infrastructure Reimagined" paper.

Book series editors
Kok-Kwang Phoon
Dongming Zhang

Foreword

I have the privilege of working at Rocscience for more than 20 years. I have witnessed the growth of geomechanics software in our profession, and I have been fortunate to play a key role in the development of several software engines that are now widely used in practice. Chapter 1 cited a recommendation from a 2006 classic paper by Curran and Hammah: "Uncertainty is king; make room for it." From my personal experience, an explicit treatment of uncertainty can offer useful insights for decisions in practice. At times, such insights are critical. Rocscience, which is a world leader in developing innovative geotechnical software tools, began introducing random field modeling and reliability analysis as early as 2000. Another example of our innovation is evident in Settle3; which is capable of characterizing a site based on CPT soundings using the latest research in Bayesian machine learning.

"Uncertainty, Modeling, and Decision Making in Geotechnics" is edited by the foremost experts in geotechnical reliability who have assembled a team of world-class researchers and practitioners to guide engineers in the use of reliability methods grounded in actual data and established physics (geomechanics) to enhance decision-making. The uniqueness of this book is not in the exposition of sophisticated reliability methods, but in the compilation of necessary information (soil statistics, practical methods, software solutions, and real-life examples) that an engineer will need to refer to when probabilistic software is used in real projects. To my knowledge, Chapter 2 and 5 are respectively the most extensive compilations of parameter statistics and model uncertainties to date.

Perhaps my best compliment to the editors and authors of this book is that it is at least two decades overdue. It is safe to say engineers appreciated the significant impact of uncertainties in design and construction well before 2006 (I should hasten to add that I have the utmost respect for Dr John Curran who is the founder of Rocscience and an accomplished rock engineer), but many have struggled to find methods sufficiently realistic and tractable for practice. For engineers who would like to apply probabilistic analysis in their projects and are agonizing over "where is the data", "how to use the software", and "are there worked examples to refer to," I would simply say, "go read this book".

Dr. Thamer Yacoub
CEO and President
Rocscience

Preface

Uncertainty is not an abstract mathematical concept. It is commonly visualized as a histogram (or box and whisker) for one variable or a scatter plot for two variables in practice. It is very hard to visualize uncertainty in higher dimensions though. One option listed in the EXCEL Chart menu is the radar diagram. Another possibility is a ternary (triangular) chart to visualize the scatter of 3D data points. The databases CLAY/10/7490 (Ching and Phoon 2014) and ROCK/10/4025 (Muzamhindo and Ferentinou 2023) have ten dimensions (ten clay or rock parameters). It is difficult for an engineer or any human to make sense of ten-dimensional data. Mathematical formalization is clearly needed for *consistent* decision-making in the presence of uncertainty at *higher dimensions* where complicated multivariate dependencies proliferate beyond the sense-making ability of engineering judgment. As shown in this book, almost all real-world problems contain more than two random variables. It is important to note that the probabilistic model (or an alternate uncertainty quantification model) has not been introduced at the exploratory data analysis step involving one or two variables. Exploratory data analysis (EDA) refers to an approach of analyzing data sets to summarize their salient features and extract insights, often using statistical or other data visualization methods. All measurements, be it a property or a response, are intrinsically uncertain in this empirical sense. The complicated multivariate dependencies are present in CLAY/10/7490, ROCK/10/4025, and others. They are not related to any mathematical model. The applicability of a probabilistic or any alternate model to geotechnical data is a separate consideration. Phoon (2023a) argued that histograms, scatter plots, and dependencies (seen in scatter plots, ternary charts, or unseen in higher dimensions) are our "data ground truths". They are as real as our physical ground truths such as soil layers. Uncertainty as an intrinsic attribute of geotechnical data, probability as a mathematical model for this attribute, the difficulty to quantify epistemic uncertainty (e.g., statistical uncertainty), and the inability of the frequentist interpretation of probability to manage small sample sizes are conflated as one issue in the literature, resulting in "throwing the baby out with the bathwater" so to speak. Because there is no confidence to estimate frequentist statistics from sparse data, the entire probability theory is deemed not applicable and uncertainty-informed decision-making is deemed impractical.

From the prevailing deterministic perspective, uncertainty is a nuisance or noise. The ramification of this perspective on practice is extremely profound and broad in terms of mindset, norms, and methods. Uncertainty becomes an aspect of data analysis that needs to be circumvented, minimized, or mitigated through extraordinary efforts. For an inexperienced engineer or a student, a measurement can even be seen as infinitely precise due to a lack of adequate exposure to probability and statistics. Hence, one may find a measurement reported with a long string of numbers after the decimal place that is unrelated to the precision of the measurement process. A prediction may be made with a false sense of confidence in its precision as well (Hammah and Curran 2009; Kalenchuk 2022). In these cases, uncertainty is not deliberately ignored. The decision-maker may be oblivious to its existence. Risks are mostly likely mismanaged when determinism is taken to this extreme without a reality check. Hence, the role of engineering judgment is indispensable in geotechnical engineering (Peck 1969). One should be mindful that determinism is as much a paradigm as probabilistic thinking. Both can be useful under different circumstances when wielded in the right way. Both have the potential to mislead a decision-maker when they are not grounded in reality. The need to stay grounded and to apply decision support tools (any kind) wisely is a given.

In routine practice, the impact of the deterministic perspective is far ranging. When confronted with a set of values for one design parameter, say the undrained shear strength, the engineer is compelled to choose a single "characteristic" value because the deterministic paradigm only admits one value. Uncertainty is addressed indirectly by choosing a cautious value and/or by applying a partial factor of safety (CEN 2004). The variabilities of soil and rock parameters have been comprehensively documented since 1999 (Phoon and Kulhawy 1999) and they are further updated by ISSMGE-TC304 (2021) and Phoon et al. (2023). There is no debate that geotechnical variabilities exist. The debate is whether geotechnical variabilities should be addressed directly by probabilistic methods or indirectly by deterministic methods. The former methods are data-informed, objective, and the validity of the calculation process can be scrutinized. The latter methods are subjective and primarily reliant on human judgment. The quality of the judgment depends on the experience and reputation of the engineer. There is limited appreciation that even the act of choosing a single characteristic value can be better informed by data and a rational method to draw insights from data (e.g., probabilistic analysis) than direct application of judgment (Phoon et al. 2023). Because many geotechnical software are deterministic, there is limited support for engineers who want to take the direct approach.

An experienced engineer who feels uncertain about the inputs (a sensible reaction to data sparsity) will need to carry out multiple runs of the deterministic software as part of a parametric study. This "what if" exploration of the parametric space (space of possible combinations of the design parameters) can be argued as a crude implementation of Monte Carlo simulation (Phoon 2023b). Duncan (2000) proposed a simple reliability analysis that involves describing the uncertainty in each design parameter using a standard deviation that is computed based on an engineer's judgment of the lowest conceivable value (LCV) and the highest conceivable value (HCV). Different combinations of design parameters falling between these HCVs and LCVs can be obtained. However, absurd combinations are possible in the absence of a correlation structure, which is ignored in a typical parametric study (Simpson 2011). This is identical to the load combination problem – it is unrealistic to design for the simultaneous occurrence of extreme wind, earthquake, snow, and other environmental loadings (Turkstra and Madsen 1980). The strategy of a parametric study is actually a workaround to allow uncertainty to be included in a deterministic analysis. It can be made more rigorous using experimental design, but this is rarely done in geotechnical engineering practice.

Eurocode 7 Clause 2.4.1(6)P mandates "Any calculation model shall be either accurate or err on the side of safety" (CEN 2004). However, it is not possible to know if a model "err on the side of safety" for a particular problem at a specific site before a load test is conducted. The empirical evidence is overwhelming that one can only talk about the *likelihood* of a model being conservative (Tang and Phoon 2021). Hence, Clause 2.4.1(6) cannot be followed at the design stage if one retains a deterministic interpretation (Tang et al. 2023). In terms of predictions, Curran and Hammah (2006) opined that

> single-point predictions of quantities have therefore practically zero likelihood of ever being realized in such a world. If room is therefore not made in geomechanics software analysis to accommodate uncertainty, any conclusions reached will be open to question.

In fact, Lambe (1973) proposed Classes A, B, and C for predictions made before, during, or after construction, respectively, because many geotechnical models are biased and imprecise when they are not calibrated by observations (Class A prediction). The literature demonstrating "accurate" models is mainly based on B and C predictions. In the language of Bayesian probability, the model uncertainty in a Class A prediction is prior information. This uncertainty can be reduced when observations are made (Classes B and C).

This book advocates a different perspective that uncertainty is part of the geotechnical data estate. It has value because it can enhance decision-making. In short, any data (not just high quality and/or large quantity of data) is an asset as long as it is not fake or fully corrupted. This perspective is more in line with data-centric geotechnics (Phoon et al. 2022a) and in the opinion of the editors, it offers the only pathway to a digital future for geotechnical engineering (Phoon and Zhang 2023). Many practitioners struggle to address uncertainties directly, in part because there is an absence of suitable guidance and software. The purpose of this book is to fill this critical gap.

It helps practitioners acquire a working knowledge of geotechnical reliability methods and guides them to use these methods wisely in conjunction with data and numerical modeling. In particular, it provides guidance on the selection of realistic statistics and an appropriate method to address different design objectives and for different problem settings (how to use?) and illustrates the value to decision-making in a concrete way using realistic examples (how useful?). It goes without saying that an appropriate method must be reasonably cost-effective and accessible to practitioners. It significantly expands the range of topics and practical guidelines covered in the Soils and Foundations review paper on "Geotechnical Uncertainty, Modeling, and Decision Making" (Phoon et al. 2022b).

Phoon (2023b) provided an update to the National Research Council Report (1995) entitled "Probabilistic Methods in Geotechnical Engineering" and argued that reliability or other uncertainty quantification methods have become more rather than less important since their inception in the sixties, because they are needed to respond to digital transformation and to address complex new design goals such as sustainability and resilience. For a practitioner who finds value in uncertainty-informed decision-making and chooses to apply it, he/she is less interested in the theoretical or numerical implementation details of the methods. The practitioner is more interested to know how reliability calculations could relieve engineering judgment from the unsuitable task of performance verification in the presence of uncertainties so that he/she can focus on setting up the right lines of scientific investigation, selecting the appropriate models and parameters for calculations and verifying the reasonableness of the results (Peck 1980). The chapters in this book are organized under four broad headings with this practice-oriented goal in mind:

1. Statistical characterization – Chapters 2–5 provide practical guidance on "probable ranges for all pertinent quantities that enter into the solution of the problem" (Casagrande 1965) spanning soil/rock parameters, constitutive models, subsurface profiles, and calculation models. These chapters emphasize that uncertainty quantification must be grounded in reality, specifically on databases. The scenarios adopted in parametric studies can be much better informed by actual statistics such as the coefficient of variation, the cross correlation and/or the spatial correlation matrix, among others. Statistics has value to deterministic design. Uncertain inputs to numerical models can be represented as: (a) random variables, (b) random vectors, (c) scalar random fields, or (d) vector time varying random fields, depending on the problem and the level of sophistication. One practical gap of major interest to an engineer is how to characterize these theoretical constructs using limited site data. The latest research on data-driven site characterization that can produce conditional random field realizations from limited site data (Shuku 2023) and uncertainty-based machine learning algorithms to model soil properties and behaviors (Yin et al. 2023) is presented. The realism of numerical modeling is already enhanced through the simulation of more realistic spatially varying soil profiles.
2. Probabilistic methods – Chapters 6–10 explain the forward propagation of uncertainties from inputs to outputs (reliability analysis) and background propagation (inverse analysis). For forward propagation, this is a more theoretically consistent method to conduct parametric studies in the face of multivariate uncertainties, because a probability distribution function models the correlation structure in actual data (Ji and Cao 2023). It is also more relevant to design because reliability analysis explores the failure domain more thoroughly than what is possible by a few parametric runs. The propagation of uncertainties through a numerical model is more challenging than through a "toy problem". The random finite element method based on direct simulation is not practical for large problems, although it is easy to understand. There is a need to introduce alternate methods that can solve large 3D problems with multiple limit states (large system reliability problems) at reasonable cost (Jiang et al. 2023). The quantification of uncertainties in the outputs (limit states) can lead to better reliability-based design decisions that are more fully connected to data (Najjar and El-Chitib 2023). Reliability-based design is an important first step

to monetize data (Phoon and Ching 2013, Ching et al. 2014). For backward propagation (called inverse or back analysis), it is required for the risk management of complex and/or new projects using the observational approach (Yoshida 2023). Inverse analysis is much more challenging than reliability analysis (Walton and Sinha 2022). The following two requirements in Clause 2.7.2 "Observational method" of Eurocode 7 (CEN 2004) illustrate the usefulness of these probabilistic methods:

- "the range of possible behaviour shall be assessed and it shall be shown that there is an acceptable probability that the actual behaviour will be within the acceptable limits". This requirement refers to a probabilistic analysis.
- "a plan of monitoring shall be devised, which will reveal whether the actual behaviour lies within the acceptable limits". For staged construction, it requires inverse analysis, because the range of possible behaviour for the next stage can be reduced in the presence of monitoring data from the preceding stages (Finno and Calvello 2005).

Current probabilistic methods can also handle time and spatially varying data with potential to implement an autonomous control model for construction (Otake and Saito 2023). Referring to the value matrix proposed by Phoon and Zhang (2023), it is possible that "Type 3" methods, defined as disruptive methods that can "transform research and practice resulting in new products, systems, and services and may even redefine the role of geotechnical engineering", could be deployed in the near future.

3. Probabilistic software – Chapters 11 and 12 demonstrate probabilistic analysis of realistic real-world problems such as soil/rock slope, rockfall, deep excavation retained by a braced slurry wall, and the construction of a tunnel under a sensitive building. Both chapters illustrate that some research methods have been adopted by commercial software and as such, the probabilistic approach is already within reach of practice for a range of common 2D and 3D problems (Javankhoshdel et al. 2023; Commend et al. 2023).
4. Probabilistic applications – Classical deterministic solutions are based on homogeneous or layered soils. Chapter 13 shows that spatially variable soils can result in more critical design scenarios. This aspect is entirely missing from current deterministic numerical analysis. The value of a probabilistic analysis is not restricted to more effective uncertainty-informed decision-making but can offer new insights on soil-structure interactions in the presence of "almost real" spatially varying soil profiles and explain obvious differences between theoretical and observed behaviors (Leung 2023). The most obvious differences are why a footing only fails in one particular direction (classical bearing capacity theory cannot break this symmetry) and why is there differential settlement between two footings (not reproducible in a homogeneous or layered soil profile). Chapters 14–16 present three special applications: press-in piling (Suzuki 2023), linear infrastructures such as dykes, cuttings, and embankment (Hicks 2023), and cement-based ground improvement (Pan and Tao 2023).

From a numerical modeling perspective, the focus is calculation methods or constitutive models. There is no uncertainty, although it is widely recognized that the quality of the outputs depends on the quality of the inputs and the model. From the uncertainty perspective, the focus is on the probabilistic method – direct simulation is most widely adopted in conjunction with numerical modeling. The gaps are evident. Books covering numerical modeling are sophisticated in calculation methods, but silent on uncertainty. Direct simulation such as the random finite element method is only one class of probabilistic approaches that may or may not be computationally practical for large 3D problems. The problem setting is mainly reliability analysis in the existing plethora of books. However, there are many problem settings of interest to the practitioner beyond calculating the probability of failure. This book is intended to fill several gaps in the literature to provide more comprehensive guidance to the practitioner on how different aspects of uncertainty quantification can be applied to enhance decision-making under different parts of the Burland Triangle (soil behavior, soil profile, and modeling) (Phoon 2023a). Additional supplementary resources are made available at: www.routledge.com/Uncertainty-Modeling-and-Decision-Making-in-Geotechnics/Phoon-Shuku-Ching/p/book/9781032367491.

The contributions of the invited authors are deeply appreciated. The editors would also like to thank Mr Tony Moore and Ms Aimee Wragg from CRC Press & Routledge for their tireless support and patient guidance.

Editors
Kok-Kwang Phoon
Takayuki Shuku
Jianye Ching

REFERENCES

Casagrande, A. 1965. Role of the 'calculated risk' in earthwork and foundation engineering. *Journal of the Soil Mechanics and Foundations Division*, 91(SM4), 1–40.

CEN. 2004. *EN 1997–1: Geotechnical Design – Part 1: General Rules*, Comité Européen de Normalisation, Brussels, Belgium.

Ching, J. and Phoon, K. K. 2014. Transformations and correlations among some clay parameters – The global database. *Canadian Geotechnical Journal*, 51(6), 663–685.

Ching, J., Phoon, K. K., and Yu, J. W. 2014. Linking site investigation efforts to final design savings with simplified reliability-based design methods. *Journal of Geotechnical and Geoenvironmental Engineering, ASCE*, 140(3), 04013032.

Commend, S., Minini, J., Groslambert, M., and Jacot-Descombes, G. 2023. Reliability-based decision-making with FE models for real-life cases studies. Chapter 12, *Uncertainty, Modeling, and Decision Making in Geotechnics*, CRC Press, Boca Raton.

Curran, J. H. and Hammah, R. E. 2006. Seven lessons of geomechanics software development. *Proceedings of the 41st U.S. Symposium on Rock Mechanics (USRMS): "50 Years of Rock Mechanics—Landmarks and Future Challenges"*. Golden, Colorado.

Duncan, J. M. 2000. Factors of safety and reliability in geotechnical engineering. *Journal of Geotechnical and Geoenvironmental Engineering, ASCE*, 126(4), 307–316.

Finno, R. J. and Calvello, M. 2005. Supported excavations: Observational method and inverse modeling. *Journal of Geotechnical and Geoenvironmental Engineering*, 131(7), 826–836.

Hammah, R. E. and Curran, J. H. 2009. It is better to be approximately right than precisely wrong – why simple models work in mining geomechanics. *Proceedings of the 43rd US Rock Mechanics Symposium and 4th US Canada Rock Mechanics Symposium*, Asheville, North Carolina.

Hicks, M. 2023. Slope reliability assessments for linear infrastructures. Chapter 15, *Uncertainty, Modeling, and Decision Making in Geotechnics*, CRC Press, Boca Raton.

ISSMGE-TC304. 2021. State-of-the-art review of inherent variability and uncertainty in geotechnical properties and models. *International Society of Soil Mechanics and Geotechnical Engineering (ISSMGE)* - Technical Committee TC304 'Engineering Practice of Risk Assessment and Management', March 2, 2021. http://140.112.12.21/issmge/2021/SOA_Review_on_geotechnical_property_variablity_and_model_uncertainty.pdf.

Javankhoshdel, S., Cami, B., Ma, T., Li, A., Yeh, E., Huang, Z., Chai, S. H., and Yacoub, T. 2023. Use of geotechnical software for probabilistic analysis and design. Chapter 11, *Uncertainty, Modeling, and Decision Making in Geotechics*, CRC Press, Boca Raton.

Ji, J. and Cao, Z. 2023. Geotechnical reliability analysis for practice. Chapter 6, *Uncertainty, Modeling, and Decision Making in Geotechnics*, CRC Press, Boca Raton.

Jiang, S.-H., Xiao, T., and Li, D.-Q. 2023. Stochastic finite element methods for slope stability analysis and risk assessment. Chapter 8, *Uncertainty, Modeling, and Decision Making in Geotechnics*, CRC Press, Boca Raton.

Kalenchuk, K. S. 2022. 2019 Canadian Geotechnical Colloquium: Mitigating a fatal flaw in modern geomechanics: Understanding uncertainty, applying model calibration, and defying the hubris in numerical modelling. *Canadian Geotechnical Journal*, 59(3), 315–329.

Lambe, T. W. 1973. Predictions in soil engineering. *Géotechnique*, 23(2), 151–202.

Leung, Y. F. 2023. Soil-structure interaction in spatially variable ground. Chapter 13, *Uncertainty, Modeling, and Decision Making in Geotechnics*, CRC Press, Boca Raton.

Muzamhindo, H. and Ferentinou, M. 2023. Generic compressive strength prediction model applicable to multiple lithologies based on a broad global database. *Probabilistic Engineering Mechanics*, 71, 103400.

Najjar, S. and El-Chitib, I. 2023. Reliability-based design with numerical models. Chapter 9, *Uncertainty, Modeling, and Decision Making in Geotechnics*, CRC Press, Boca Raton.

National Research Council 1995. *Probabilistic Methods in Geotechnical Engineering*, National Academies Press, Washington, DC.

Otake, Y. and Saito, T. 2023. Reliability analysis with reduced order model. Chapter 7, *Uncertainty, Modeling, and Decision Making in Geotechnics*, CRC Press, Boca Raton.

Pan, Y. T. and Tao, R. 2023. Uncertainty, modeling, and decision making for ground improvement. Chapter 16, *Uncertainty, Modeling, and Decision Making in Geotechnics*, CRC Press, Boca Raton.

Peck, R. B. 1969. *A Man of Judgement. R. P. Davis Lecture on the Practice of Engineering*, West Virginia University.

Peck, R. B. 1980. 'Where has all the judgment gone?' The fifth Laurits Bjerrum memorial lecture. *Canadian Geotechnical Journal*, 17(4), 584–590.

Phoon, K. K. 2023a. Uncertainty-informed decision-making in Burland Triangle. Chapter 1, *Uncertainty, Modeling, and Decision Making in Geotechnics*, CRC Press, Boca Raton.

Phoon, K. K. 2023b. What geotechnical engineers want to know about reliability. *ASCE-ASME Journal of Risk and Uncertainty in Engineering Systems, Part A: Civil Engineering*, 9(2), 03123001.

Phoon, K. K., Cao, Z., Ji, J., Leung, Y. F., Najjar, S., Shuku, T., Tang, C., Yin, Z. Y., Yoshida, I., and Ching, J. 2022b. Geotechnical uncertainty, modeling, and decision making. *Soils and Foundations*, 62(5), 101189.

Phoon, K. K. and Ching, J. 2013. Is site investigation an investment or expense? – A reliability perspective. *Proceedings of the 18th Southeast Asian Geotechnical Conference (18SEAGC) & Inaugural AGSSEA Conference (1AGSSEA)*, 29–31 May 2013, Singapore, 25–43.

Phoon, K. K., Ching, J., and Cao, Z. 2022a. Unpacking data-centric geotechnics. *Underground Space*, 7(6), 967–989.

Phoon, K. K., Ching, J., and Tao, Y. 2023. Soil and rock parametric uncertainties. Chapter 2, *Uncertainty, Modeling, and Decision Making in Geotechnics*, CRC Press, Boca Raton.

Phoon, K. K. and Kulhawy, F. H. 1999. Characterization of geotechnical variability. *Canadian Geotechnical Journal*, 36(4), 612–624.

Phoon, K. K. and Zhang, W. G. 2023. Future of machine learning in geotechnics. *Georisk: Assessment and Management of Risk for Engineered Systems and Geohazards*, 17(1), 7–22.

Shuku, T. 2023. Data-driven site characterization. Chapter 4, *Uncertainty, Modeling, and Decision Making in Geotechnics*, CRC Press, Boca Raton.

Simpson, B. 2011. Reliability in geotechnical design – some fundamentals. *Proceedings of the Third International Symposium on Geotechnical Safety & Risk, Federal Waterways Engineering and Research Institute*, Germany, 393–399.

Suzuki, N. 2023. Reduction of uncertainty through piling data within the same site in the press-in piling method. Chapter 14, *Uncertainty, Modeling, and Decision Making in Geotechnics*, CRC Press, Boca Raton.

Tang, C. and Phoon, K. K. 2021. *Model Uncertainties in Foundation Design*, CRC Press.

Tang, C., Phoon, K. K., and Yuan, J. 2023. Variability of predictions in geotechnics. Chapter 5, *Uncertainty, Modeling, and Decision Making in Geotechnics*, CRC Press, Boca Raton.

Turkstra, C. J. and Madsen, H. O. 1980. Load combinations in codified structural design. *Journal of the Structural Division, ASCE*, 106(12), 2527–2543.

Walton, G. and Sinha, S. 2022. Challenges associated with numerical back analysis in rock mechanics. *Journal of Rock Mechanics and Geotechnical Engineering*, 14(6), 2058–2071.

Yin, Z.-Y., Zhang, P., and Jin, Y.-F. 2023. Uncertainty in constitutive models. Chapter 3, *Uncertainty, Modeling, and Decision Making in Geotechnics*, CRC Press, Boca Raton.

Yoshida, I. 2023. Probabilistic inverse analysis for geotechnics. Chapter 10, *Uncertainty, Modeling, and Decision Making in Geotechnics*, CRC Press, Boca Raton.

1 Uncertainty-Informed Decision-Making in Burland Triangle

Kok-Kwang Phoon

ABSTRACT

Histograms, scatter plots, and unseen dependencies at higher dimensions are part of our "ground truths" and they are as real as physical features such as soil layers. It is appropriate to call them "data ground truths". Probabilistic methods have advanced tremendously over the past five or more decades and they are now at the cusp of transitioning to machine learning methods that are expected to have a profound and wide-ranging impact on the future of geotechnical engineering under the "data first practice central" agenda of data-centric geotechnics. The current practice of reducing a range of values to a single characteristic value, simplifying a spatially varying profile to a homogeneous or layered profile, and grappling with uncertainties implicitly through parametric studies cannot take advantage of rapidly emerging digital technologies. In addition, it is difficult for current practices to address complex new design goals (sustainability, resilience) and to compensate for the diminishing value of precedents while also having to onboard new adaptation strategies as a result of climate change. It is often underappreciated that the effectiveness of practical experience in informed decision-making decreases with increasing novelty and the growing rate of change of solutions. This chapter makes the case to exploit data more intensively and more comprehensively for decision-making, starting with the application of well-established probabilistic methodologies to enhance various elements in the Burland Triangle. Data infrastructure is now as important as physical infrastructure. This distinction and the rapid advancement of cyber-physical systems have received limited attention thus far, although it is critical to digital transformation. Data-driven methods are being developed to show that the geotechnical data estate is an asset in its own right. Uncertainty-informed decision-making is an integral part of this value.

1.1 ROLE OF NUMERICAL MODELING

One distinctive feature of geotechnical engineering is that the engineer has to work with natural materials in an environment that is largely outside his/her control (historical and current conditions). Some degree of control is possible through a variety of ground improvement methods. The physical ground reality that is of interest to a geotechnical engineer is very complex and changing at different timescales (geologic [millions of years] to seismic [seconds]). Modeling is frequently understood as a mathematical abstraction of some aspects of reality relevant to the problem at hand. At the design stage, the geotechnical engineer is interested to know about the performance of a system, particularly in the context of exceeding one or more limit states. Numerical modeling seeks to draw answers from mathematical equations (typically differential in nature) for a specific problem scenario. In the context of decision-making in geotechnical engineering practice, the scenario includes being situated at a specific site. Hence, site investigation is necessary. It is natural to think of the purpose of numerical modeling as "predicting" an actual response measured in the field, i.e.,

DOI: 10.1201/9781003333586-1

to match one observed aspect of reality as precisely as possible. This is the main purpose for physics and many scientific disciplines. The prediction accuracy in the physical sciences is extraordinary.

Prediction exercises in geotechnical engineering should be understood in the context proposed by Lambe (1973):

1. Class A: Predictions are carried out before the construction event with the available site investigation data.
2. Class B: Predictions are made during the construction event, so that they can be influenced by initial field data.
3. Class C: Analyses are carried out after the construction event when the complete set of field data is made available; appropriate soil properties can be adjusted by back-calculation.

One of the purposes for proposing the above classification is to point out that Class A predictions are notoriously difficult in geotechnical engineering. This is because the physical reality (ground) is complex, available data to characterize this reality is scarce, and construction effects are difficult to quantify and frequently left out of the model. This is not to say that a numerical model should replicate every aspect of reality. In fact, it is well known that improving the sophistication of the method alone does not necessarily improve the quality of the prediction (Figure 1.1). There are other practical considerations discussed elsewhere (Hammah and Curran 2009).

Barbour and Kahn (2004) suggested that numerical models can be used in three general ways:

1. Interpretation: Numerical models are used to gain physical insights from field or laboratory data. An example is the development of a model to help back-analyze a suite of monitoring information.
2. Design: Numerical models are used to compare the relative performance of various design alternatives, with less emphasis on the final predicted performance.
3. Prediction: Numerical models are used to make a quantitative prediction of actual field behavior.

Barbour and Kahn (2004) opined that geotechnical engineers rarely focus on "prediction" as it is not unusual for computed and measured responses to differ by one order of magnitude. Referring to a braced excavation in dry sand, Herle (2003) remarked

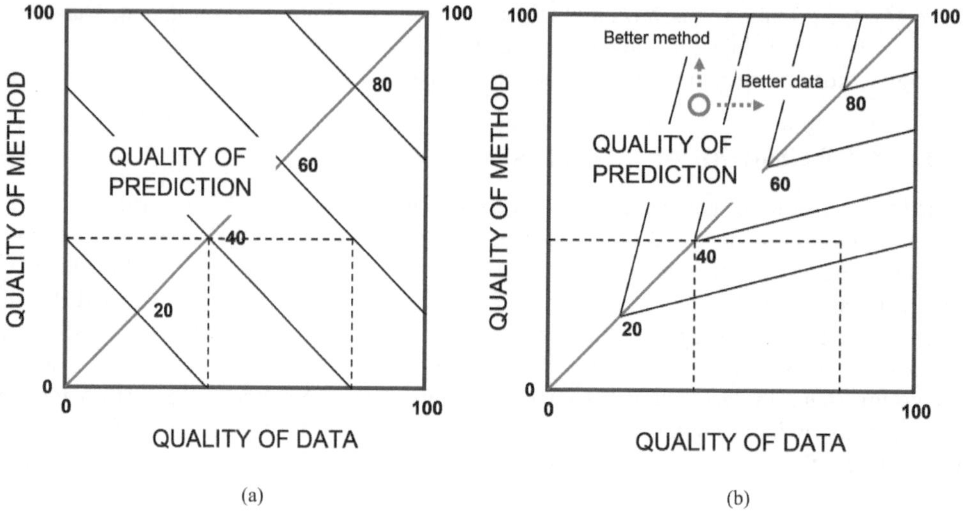

FIGURE 1.1 Quality of prediction as a function of quality of method and quality of data. (Source: Adapted from Lambe 1973 by Phoon and Ching 2017)

> Most calculation methods employed finite elements or subgrade modulus models. The comparison with measured values is very disappointing, see … final excavation stage with surface load, prior to the limit state. Especially worrying is that displacements have been predicted several times in the opposite direction than measured. However, even more depressing is the large scatter of predictions which becomes enormous in the limit state.

Tang and Phoon (2021) conducted the largest statistical model validation exercise to date using load test databases and found that geotechnical models are biased (typically on the conservative side) and imprecise. Curran and Hammah (2006) shared seven lessons learned from developing software for practice. One lesson listed in the paper is "uncertainty is king; make room for it". The authors elaborated that

> because geological materials are formed under a broad variety of complex, physical conditions, the history of which is not known, geomechanics involves large uncertainties. Single-point predictions of quantities have therefore practically zero likelihood of ever being realized in such a world. If room is therefore not made in geomechanics software analysis to accommodate uncertainty, any conclusions reached will be open to question.

Citing Starfield and Cundall's (1988) observation that "rock mechanics models fall into the class of 'data-limited problems'", Hammah and Curran (2009) opined that uncertainty is large and "it is dangerous to apply the methods of exactitude to mining geomechanics problems". In the 2019 Canadian Geotechnical Colloquium, Kalenchuk (2022) emphasized the

> need to have grounded conversations on numerical modelling regarding the reality that geomechanical designs are often data limited with high degrees of uncertainty. When data limits and uncertainty are overlooked, geomechanical engineers are at risk of introducing unforeseen fatal flaws into engineering design.

It is fair to say that numerical modeling is primarily targeted at solving the physics of the problem, assuming the right kind of data of sufficient precision exists to support the deterministic analysis. It goes without saying that the phenomenon must be sufficiently well understood to formulate the correct physics in the first place. It is not uncommon to calibrate incomplete physics with field data to achieve reasonable solutions within a finite range of conditions in geotechnical engineering. One can view the current emphasis on physics from a historical perspective. The state of practice has evolved from empirical or simple closed-form analytical solutions (Terzaghi 1943; Terzaghi and Peck 1948; Peck et al. 1953) to present-day numerical analyses as a result of growing computational and experimental capabilities. In a review of *Géotechnique* contributions to foundation engineering between 1948 and 2008, Salgado et al. (2008) opined that these *Géotechnique* papers

> have attempted to solve the problems faced by the foundation engineering industry, with a strong emphasis on the underlying science; as a result, these papers have played a key role in the advancement of both the science and its applications in our discipline.

The author argues that the geotechnical engineering industry is ready to advance to the next stage of development where the concept of a digital twinfrastructure is realizable (Phoon et al. 2022a). Physics alone is not sufficient, because it is not possible to simulate the behavior of an entire infrastructure system *at a specific site/region in real time* without reimagining the application of data, data-driven methodologies, connectivity, and other digital technologies. In an editorial for a special issue on "Probabilistic Site Characterization", Phoon (2018) opined that it is

> possible to envisage realizing 'precision construction', where characterization of 'site-specific' model factors and 'site-specific' soil parameters based on both site-specific and generic data can lead to further customization of design to a particular site and even a particular location in a site.

"Precision construction" is beyond the current state of the art, but it can serve as a talking point on what is meant by truly transformative in the next stage of development.

The objective of this chapter is to explain that (1) decision-making can be enhanced if current modeling achieves a more optimum balance between physics and data; (2) near-term development efforts should be invested in data-centric geotechnics to hasten this rebalance and to take advantage of emerging digital technologies; and (3) uncertainty quantification is necessary, regardless of the methodology, because geotechnical databases are relatively small, typically incomplete, and always heterogeneous (site specific, time dependent, multi-source, etc.). The rest of this book surveys a range of useful topics that a non-specialist can relate to in relation to familiar tasks in practice to support the objective of this chapter.

1.2 PREDICTION EXERCISES

Many prediction exercises have been conducted in the past (Table 1.1). The quantity of interest (QoI) in Table 1.1 refers to a response that affects the occurrence of a limit state. An engineer is rarely interested in all the responses that a calculation model can provide. As expected, the range of Class A predictions is very wide. Two predictions that are orders of magnitude apart are not uncommon. A sophisticated analysis does not necessarily lead to a more accurate prediction. One may argue that there is empirical support for Figure 1.1. One example is a shallow foundation prediction exercise conducted by the Centre of Excellence for Geotechnical Science and Engineering (Doherty et al. 2018). Fifty entries were submitted by 88 participants based in 13 countries that include Australia, Austria, Belgium, Canada, China, France, Germany, the Netherlands, Norway, Japan, Singapore, the United Kingdom, and the United States. The entries can be grouped based on the experience of the participants: 23 were from industry practitioners, 16 from academics, and 11 from undergraduate students. The participants were invited to predict the capacity and the settlements at 25%, 50%, and 100% of the capacity for two 1.8 m × 1.8 m square shallow foundations founded 1.5 m deep on estuarine silty clay (Figure 1.2). Hence, there are four quantities of interest. It can be seen that the predicted capacities can vary by more than one order of magnitude and the predicted settlements can vary by more than two orders of magnitude. In addition, the average of the predicted values is unconservative for the capacity and very conservative for the settlements. Note that the range of predicted values is usually much larger than the dispersion of the model factor, because there is a variety of choices for site characterization (stratification, characteristic profile, transformation model) and modeling (constitutive model, analysis method, boundary conditions), and the exercise of different judgment at all stages of the prediction process. Similar observations are made by Herle (2003).

The model factor is defined as the ratio of the measured to the calculated (predicted) response. The numerator and denominator are determined using a consistent methodology, as discussed in Chapter 5 of this book. However, a prediction exercise is conducted at a single site, while the model factor is evaluated from field tests conducted at many sites. Although the dispersion (as measured by the coefficient of variation [COV]) of the model factor for a shallow foundation is smaller, it is interesting that the prediction of capacity is also less dispersed compared to the prediction of settlement (Figure 1.3). Applying the simple classification scheme for the bias in Figure 1.3, the ratio of the measured value to the average predicted value in Figure 1.2 is "unconservative" for the capacity and "highly conservative" for the settlements. Prediction exercises have amply demonstrated that design decisions cannot be made solely based on numerical analyses alone. The outcomes of these exercises lent weight to Kalenchuk's (2022) remark that "Too many discussions of numerical methods in geomechanical engineering are centred on the impressive ability of numerical tools to conduct complex and sophisticated analyses with relative ease and efficiency". To be more specific, the deterministic manner in which numerical analysis is currently conducted is not sufficiently informative for design that requires risk management in some way. The global factor of safety or partial factors are introduced in design codes largely for this reason. Back-analysis is common in

TABLE 1.1
Review and Summary of Prediction Events in Geotechnics

	Subsurface			Geo-structure		Prediction exercise
No.	Site location	Soil	Geotechnical investigation	Type	Dimensions	Quantity of interest (QoI)
1	Massachusetts	Sand, clay, and glacial till		Embankment	$H = 12$ m $L = 90$ m	S_h, S_v, and u_w (10)
2	Ballina, Australia	Soft clay		Embankment	$L = 80$ m $B = 5$ m	S_h, S_v, and u_w (28)
3	NGES (Texas A&M Univ.)	Sand	In situ BST (3) CPT (5) SPT (6) PMT (4) DMT (3) Laboratory	Square footing	$B = 1–3$ m	Q and S_v (31)
4	Ballina, Australia	Soft clay		Shallow foundation	$B = 1.8$ m $t = 0.6$ m $D = 1.5$ m	Q_{tu} and S_v (50)
5		Silt, sand, clay, and sandstone		Steel H and pipe pile	$B = 0.25–0.41$ m $L = 21.6–57$ m	Q_{tu}, Q_{su}, and Q_{bu} (18)
6	Evanston, IL	Sand/clay	In situ CPT (4) SPT (2) PMT (1) DMT (1) Laboratory	Drilled pier, steel H, and pipe pile	$B = 0.46$ m $L = 15$ m	Q_{tu} and Q_t–w_t curve (24)
7	Dunkirk, France	Dense sand		Steel pipe pile (open-end)	$B = 0.46$ m $L = 10$ m	Q_t–w_t curve and Q_{tu} (16)
8	Sint-Katelijne-Waver, Belgium	Stiff clay	In situ CPT (43) SPT (1) PMT (1) DMT (2) Laboratory	DD pile	$B = 0.38–0.51$ m $L = 7.38–11.83$ m	Q_{tu}, Q_t–w_t, Q_s–w_t, and Q_b–w_t curves (90)
9	Limelette, Belgium	Sandy soil	In situ CPT (61) SCPT (3) SPT (3) PMT (2) DMT (11) SASW (3) Laboratory	DD pile	$B = 0.39–0.55$ m $L = 9.5$ m	Q_{tu}, Q_{bu}, Q_{su}, and Q_t–w_t curve (77)
10	Orlando, FL	Sand		Pipe pile (closed-end)	$B = 0.32$ m $L = 14$ m	Q_{tu} (33)

(Continued)

TABLE 1.1 (CONTINUED)
Review and Summary of Prediction Events in Geotechnics

	Subsurface			Geo-structure		Prediction exercise
No.	Site location	Soil	Geotechnical investigation	Type	Dimensions	Quantity of interest (QoI)
11	Merville, France	Silt/clay	In situ PMT (42) CPT (3) SPT (2) Laboratory	CFA pile	$B = 0.5$ m $L = 12$ m	Q_{tu} and Q_t–w_t curve (12)
12	CEFEU/ISC'2, Portugal	Sand	In situ SPT (4) CPT (9) DMT (9) PMT (3) Laboratory	Bored, CFA, and driven pile		Q_{tu} (33)
13	Edmonton, Alberta, Canada	Silt/clay	In situ SPT (2)	CFA pile	$B = 0.41$ m $L = 18.5$ m	Q_t–w_t curve (35) and Q_{tu} (41)
14	Santa Cruz, Bolivia	Sand	In situ SPT (3) CPT (3) Laboratory	Bored and FDP pile	$B = 0.36$–0.45 m $L = 9.6$–17.6 m	Q_t–w_t curve and Q_{tu} (50)
15	Santa Cruz, Bolivia	Sand		Bored pile	$B = 0.6$ m $L = 16.4$ m	Q_t–w_t curve and Q_{tu} (11)
16	B.E.S.T., Bolivia	Silt/sand	In situ SPT (8) SCPT (15) SDMT (9) PMT (5) SASW ReMi Laboratory	Bored, CFA, and FDP pile	$B = 0.3$–0.62 m $L = 9.5$ m	Q_t–w_t curve (72) and Q_{tu} (54)
17	Göteborg, Sweden	Soft marine clay		Precast concrete pile	$B = 0.28$ m $L = 50$ m	Q_t–w_t curve and Q_{tu} (22)
18		Clay/silt		Precast, prestressed concrete pile	$B = 0.41$ m $L = 25$ m	S_v (52)
19	Centrifuge	Silt		Flexible pile	$B = 0.96$ m $L = 21.6$ m	Monotonic and cyclic lateral load-displacement (P–y) curve, P_0 and M_0 (29)
20	Hunter's Point Naval Base, San Francisco	Sand		Steel pipe pile (closed-end) and five-pile group	$B = 0.27$ m $L = 15$ m	Q_{tu} and S_v (11)

(*Continued*)

TABLE 1.1 (CONTINUED)
Review and Summary of Prediction Events in Geotechnics

	Subsurface			Geo-structure		Prediction exercise
No.	Site location	Soil	Geotechnical investigation	Type	Dimensions	Quantity of interest (QoI)
21	Gulf of Mexico and North Sea	Single clay and layered soil		Spudcan	B= 10–20 m	Load-penetration (q–d) curve (30)

Source: Adapted from Tang et al. (2023a, 2023b).

Notes:

Embankment: 1 = Constructed Facilities Division (1975a, b) and 2 = Kelly et al. (2018).

Shallow foundation: 3 = Briaud and Gibbens (1997) and 4 = Doherty et al. (2018).

Pile foundation: 5 = Fellenius (1988), 6 = Finno (1989), 7 = Jardine et al. (2001), 8 = Holeyman et al. (2001), 9 = Holeyman and Charue (2003), 10 = Fellenius et al. (2004), 11 = Reiffsteck (2005, 2009), 12 = Viana da Fonseca and Santos (2008), 13 = Fellenius (2013), 14 = Fellenius and Terceros (2014), 15 = Fellenius (2015), 16 = Fellenius (2017), 17 = Fellenius et al. (2019), 18 = Fellenius (2021), and 19 = Guevara et al. (2022).

Pile group: 20 = DiMillio et al. (1988).

Spudcan penetration: 21 = van Dijk and Yetginer (2015).

H = height, L = length, B = width, t = thickness, D = embedment depth, S_h = horizontal movement, S_v = vertical settlement, u_w = pore water pressure, Q_{tu} = total capacity, Q_t and w_t = pile head load and displacement, Q_{tu} = bearing capacity, Q_s and w_s = pile shaft load and displacement, Q_b and w_b = pile tip load and displacement, and P and y = lateral load and deflection of pile head.

QoI = quantity of interest, DD = drilled displacement pile, CFA = continuous flight auger pile, FDP = full displacement pile, LDFE = large-deformation finite element analysis, BST = borehole shearing test, CPT = cone penetration test, SCPT = seismic cone penetration test, DMT = dilatometer test, SDMT = seismic dilatometer test, PMT = pressure meter test, ReMi = refraction microtremor, SASW = spectral analysis of surface wave test, SPT = standard penetration test.

landslides (particularly runout models), mining, and rock engineering because the premise that data is a secondary concern compared to physics is questionable.

One may argue that the diverse outcomes exhibited in prediction exercises are more representative of practice than the quality outcomes presented in most case studies, because of the number and diversity of the participants and the strict adherence to Class A protocol (predictions are submitted before the load tests). The difference in accuracy between a Class A prediction and a Class C prediction is well known (Indraratna et al. 2018). Case studies are very useful to illustrate various learning points (Focht 1994; Poulos 2004; Rogers 2008; Burland 2012), but in the opinion of the author, less so to demonstrate the accuracy of a prediction model when it is applied at different sites, in the presence of different data assets, and by different engineers. The accuracy of a prediction model is more correctly represented by a range or, to be more complete, a range with a likelihood associated with each value (histogram). Case studies, usually limited in number, do not provide sufficient statistical support to characterize this histogram (or even a range).

1.3 BURLAND TRIANGLE

In the opinion of the author, the purpose of numerical modeling is to *enhance decision-making in practice* rather than to enhance our prediction ability. Heyman (1996) made a similar point: "There is no correct solution to the equations, but one solution will lead to the greatest economy in material". As illustrated by the expanded Burland Triangle in Figure 1.4, modeling can only contribute meaningfully to decision-making in practice when it is linked to an understanding of the ground profile and soil behavior. However, it is crucial to appreciate the deterministic paradigm underlying

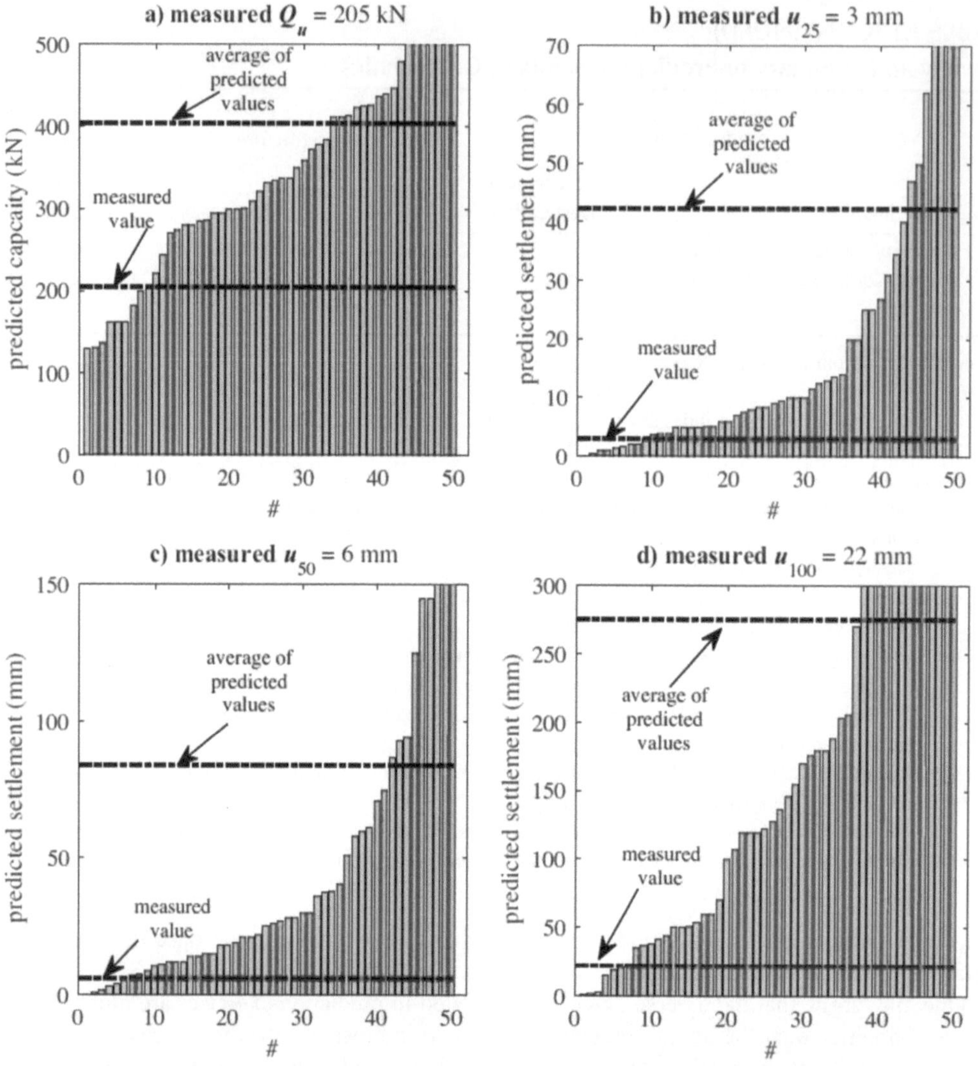

FIGURE 1.2 Range of predictions from 50 participants on the performance of two shallow foundations on estuarine silty clay: (a) capacity, (b) settlement at 25% of capacity, (c) settlement at 50% of capacity, and (d) settlement at 100% of capacity. (Source: Fig. 6, Doherty et al. 2018)

Figure 1.4. This deterministic paradigm continues to dominate our practice to this day (Burland 2012). All prediction exercises in Table 1.1 are deterministic in nature – they do not require the participants to submit a range of probable values or, more formally, a probability distribution of the desired response. To the knowledge of the author, no probabilistic prediction exercise involving a large number of practitioners has been conducted thus far. This is a significant gap in the literature, because it does not allow probabilistic models to be validated and the value of probabilistic analysis to be shared with practitioners. In fact, engineers do not rely on numerical modeling for "prediction", because they are aware that the deterministic answers ("single-point predictions" highlighted by Curran and Hammah 2006) are biased and imprecise. Even more importantly, engineers are aware that bias and uncertainty are not evaluated in a deterministic analysis. This means that it is fundamentally impossible to make a *quantitative* risk-informed decision, because being "cautious" (in a way linked to data rather than experience) requires a knowledge of the bias (how conservative?) and imprecision (how uncertain?). In the estimation of a 5% fractile characteristic value

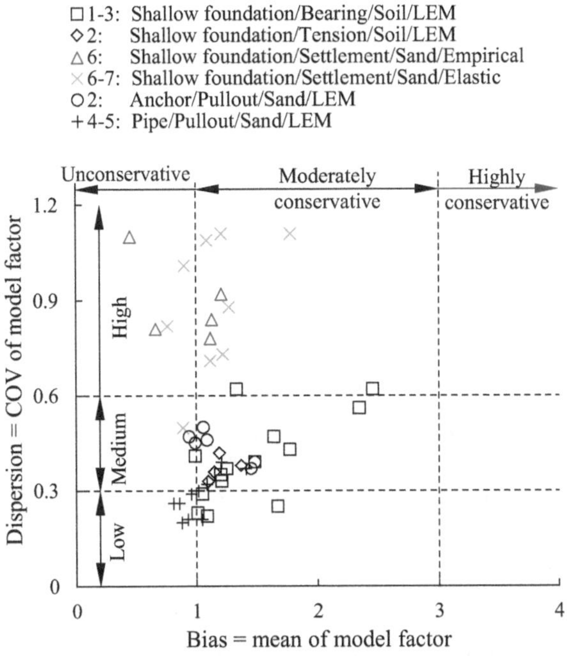

FIGURE 1.3 Classification of model uncertainty based on mean and coefficient of variation (COV) values of model factor, where LEM = limit equilibrium method. (Note: The capacity model factor is conservative on the average for a mean >1 and the settlement model factor is conservative on the average for a mean <1). (Source: Fig. 4.17, Tang and Phoon 2021) (Data sources: 1 = Strahler and Stuedlein 2014, 2 = Tang et al. 2020, 3 = Paikowsky et al. 2010, 4 = Ismail et al. 2018, 5 = Stuyts et al. 2016, 6 = Akbas 2007, and 7 = Samtani and Allen 2018)

(Clause 2.4.5.2(11), CEN 2004), the bias and imprecision are quantified by the mean and coefficient of variation, respectively (Länsivaara et al. 2022). Details are given in Chapter 2.

It is possible that engineers adopt the deterministic paradigm out of practical necessity (possibly from a legacy of having to make do with scarce data, costly data acquisition, and limited computing power) and/or in accordance with tradition (shaped by determinism in physics "God does not play dice", reliance on heuristics, and the evolutionary nature of codes that require changes to be made cautiously and deliberately). However, engineers do sensibly mitigate its significant limitations through clever work-arounds such as the observational method (refer to Clause 2.7.2, CEN 2004). At the same time, there are reservations concerning the practicality of a probabilistic paradigm. Perceptions abound that the cost of performing a probabilistic analysis on a realistic problem is excessive, there is insufficient (quality) site-specific data to characterize geotechnical uncertainties, the benefits do not appear to justify the elaborate computational efforts for conventional structures, and software is lacking or difficult to use. However, it is worthwhile balancing this debate by taking note of the following current trends: (1) rapid surge in computing power and digital technologies in general; (2) availability of more efficient and more powerful probabilistic and other data-driven methods (Phoon et al. 2022b); (3) sharing of large databases (termed big indirect data [BID] by Phoon et al. 2019) (e.g., 304dB at http://140.112.12.21/issmge/tc304.htm); (4) incorporation of probabilistic analysis in geotechnical software (Table 1.2); (5) increasing emphasis on risk-informed decision-making in design codes (Phoon et al. 2016); (6) innovations should be the norm in practice to address urgent challenges such as sustainability, but current technology acceptance criteria lean toward risk averse rather than risk informed (ASCE 2019); and (7) deterministic paradigm is at odds with digital transformation, because it pre-dates the personal computer era (Phoon et al. 2022a).

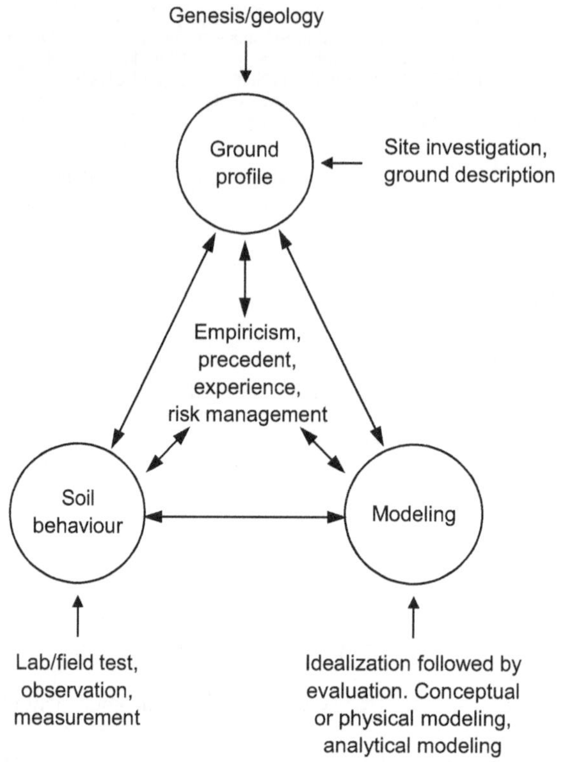

FIGURE 1.4 Expanded Burland Triangle (Source: Fig. 1, Barbour and Kahn 2004 updated from original idea presented by Burland 1987)

TABLE 1.2
List of Reliability Software with Geotechnical Features

Software	Applications	Reference
ABAQUS	Slopes, foundations, tunnels	https://www.3ds.com/products-services/simulia/disciplines/structures/
ANSYS	Slopes, tunnels	https://www.ansys.com/
COMSOL	Model uncertainty	https://www.comsol.com/uncertainty-quantification-module
COSSAN	Foundations	https://cossan.co.uk/
FLAC	Slopes, retaining structures, foundations, tunnels	https://www.itascacg.com/
GeoStudio	Slopes	https://www.geoslope.com/
GeoStru • CVSoil	• Characteristic value	https://www.geostru.eu
PLAXIS	Slopes, retaining structures, foundations	https://www.bentley.com/
Rocscience • RS2 and RS3 • Slide2 and Slide3	• Slopes, retaining structures, foundations, tunnels • 2D and 3D slopes	https://www.rocscience.com/
STRUREL	Tunnel, offshore structure	http://www.strurel.de/
UQLAB	Slopes, foundations, tunnels	https://www.uqlab.com/
ZSwalls+R	Retaining walls	https://geo-dev.ch

Source: Phoon (2023).

1.4 WHY UNCERTAINTY QUANTIFICATION?

The prediction debate in geotechnical engineering centers on the low likelihood for a single number to match a measured response. The author argues that this debate is largely an artifact of the deterministic paradigm that simplifies an output from a numerical analysis to a single number. The state of the art has advanced in tandem with computational power to the extent that a numerical model can produce a probability distribution for any output or, more generally, a multivariate probability distribution for a set of correlated outputs consistent with the physics and the parametric/model uncertainties (which are characterized from actual site data) and at a reasonable cost. This probability distribution can be easily visualized as a 95% confidence interval for a single output or a 95% confidence region for two or more outputs.

Two 95% confidence regions for the predicted undrained shear strength (s_u) and the preconsolidation stress (σ'_p) are illustrated in Figure 1.5 using actual site data from Onsøy (Norway) (Table 1.3). Note that the data is multivariate (ten soil parameters), sparse (nine records), and incomplete (there are three cells with missing values denoted by "NA"). They are obtained by training a multivariate site-specific probability distribution function (PDF) using Bayesian machine learning (Ching and Phoon 2019). The training dataset is Table 1.3 minus Row 4 and 8 (shaded in gray), which are ring-fenced for validation. Posterior (or predicted) samples of (s_u, σ'_p) at a validation depth equal to 5.2 m are obtained by updating the PDF using the record (LL = 56.8%, PI = 22.9%, LI = 1.07, σ'_v/P_a = 0.32, B_q = 0.35, q_{tl} = 7.70, q_{tu} = 6.11). Note that the soil parameters (σ'_p/P_a, s_u/σ'_v, S_t) are assumed to be unobserved. Because S_t includes s_u, it has to be assumed as unobserved for consistency. The 95% confidence region circumscribed by the dashed line is estimated from these posterior samples. The same procedure is repeated for the validation depth equal to 13.4 m. It can be seen that the actual values of (σ'_p/P_a, s_u/σ'_v) at both validation depths do fall in these 95% confidence regions. In addition, there is no unique "cautious" or "characteristic" value for (σ'_p/P_a, s_u/σ'_v). It depends on the limit state (Ching et al. 2020).

Note that the uncertainty of the prediction as represented by the size of the 95% confidence region is substantial, because there are only six training records (σ'_p/P_a is not observed at a depth equal to 9.5 m). There is an occasional perception that uncertainty quantification can improve the precision of the prediction. This is a misconception. Uncertainty quantification is meant to provide a correct representation of the precision (e.g., Figure 1.5), both size and shape, that is consistent with the data at hand (e.g., Table 1.3). It is correct that the precision is low when the data is limited. Uncertainty

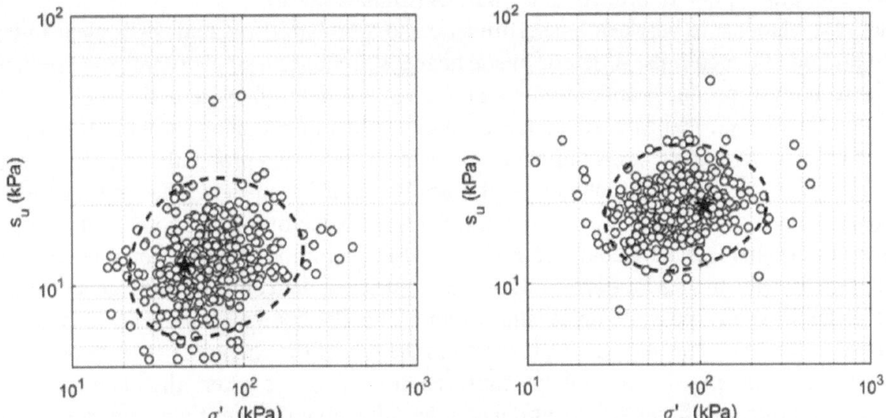

FIGURE 1.5 Open circles are posterior (or predicted) samples for σ'_p and s_u at two depths for a site in Onsøy (Norway). Dashed line is the 95% confidence boundary. Asterisk is the actual values of σ'_p and s_u for validation.

TABLE 1.3
Site Investigation Data for a Site in Onsøy (Norway)

Index	Depth (m)	LL (%)	PI (%)	LI	σ'_v/P_a	σ'_p/P_a	s_u/σ'_v	S_t	B_q	q_{t1}	q_{tu}
1	1.0	56.2	20.0	1.54	0.06	0.85	2.03	6	0.16	29.11	25.57
2	1.9	50.2	18.1	1.82	0.12	0.60	0.91	14	0.24	17.69	14.58
3	3.5	59.9	30.5	0.93	0.22	0.48	0.48	15	0.30	10.52	8.41
4	5.2	56.8	22.9	1.07	0.32	**0.45**	**0.37**	7	0.35	7.70	6.11
5	7.6	66.3	31.5	0.87	0.47	0.54	0.24	14	0.47	5.89	4.25
6	9.5	65.1	29.6	0.97	0.58	NA	0.25	12	0.41	6.19	4.74
7	10.8	74.4	36.1	0.81	0.65	0.84	0.25	9	0.46	5.93	4.31
8	13.4	71.4	35.8	0.87	0.81	**1.05**	**0.24**	NA	0.47	5.95	4.24
9	16.3	72.7	34.7	0.76	0.99	0.99	0.24	NA	0.55	6.13	3.88

Source: Table 1 (Ching and Phoon 2020); data extracted from Lacasse and Lunne (1982).

Note: LL = liquid limit; PI = plasticity index; LI = liquidity index; σ'_v = vertical effective stress; σ'_p = preconsolidation stress; P_a = atmospheric pressure = 101.3 kPa; s_u = undrained shear strength; s_u = the in situ undrained shear strength mobilized in embankment and slope failures (Mesri and Huvaj 2007); S_t = sensitivity; q_{t1} = normalized cone tip resistance = $(q_t - \sigma_v)/\sigma'_v$, where q_t = (corrected) cone tip resistance and σ_v = vertical total stress; q_{tu} = effective cone tip resistance = $(q_t - u_2)/\sigma'_v$, where u_2 = porewater pressure directly behind the cone; B_q = pore pressure ratio = $(u_2 - u_0)/(q_t - \sigma_v)$, where u_0 = hydrostatic pore pressure.

quantification provides a consistent valuation of data (Ching et al. 2014a). The choice of improving precision by gathering more data rests with the decision-maker. It has nothing to do with uncertainty quantification. The shape of the 95% confidence region depends in part on the correlation between σ'_p/P_a and s_u/σ'_v. It is safe to say that an engineer cannot visualize Figure 1.5 even in a broad approximate sense by staring at the numbers in Table 1.3 alone. No engineering judgment can lead to Figure 1.5. In particular, note that the lower left arc of the 95% confidence region can be regarded as cautious estimates of (σ'_p/P_a, s_u/σ'_v). Some estimates involve lower values of s_u/σ'_v but higher values of σ'_p/P_a and vice-versa. It is not possible for an engineer to infer a cautious *combination* of (σ'_p/P_a, s_u/σ'_v) correctly. The value of uncertainty quantification to the estimation of design parameters and in the monetization of data in general is clearly demonstrated in this concrete example. The posterior samples are natural inputs for probabilistic analysis (Chapters 6–8).

In fact, the relationship between a deterministic and a probabilistic analysis is clear in terms of this presentation of the outputs. A deterministic analysis produces a single point (one prediction). A probabilistic analysis produces an interval or a region (a set of possible predictions). A probabilistic analysis can thus be viewed as a collection of many deterministic analysis with different inputs. This is the basis for a Monte Carlo simulation. There is no doubt that probabilistic analysis enhances decision-making by revealing the outcomes of many "what if" scenarios. The identification of different combinations of inputs that lead to "failure" (exceedance of one or more limit states) is certainly more informative than a global factor of safety. In the opinion of the author, there are only two potential impediments to its adoption: (1) do we have sufficient data or the tools to generate plausible "what if" scenarios (e.g., possible input values)? and (2) is the computational cost reasonable? The short answer is yes. The long answer is presented in the rest of the chapters in this book.

If uncertainties and physics are suitably captured (not one or the other), the likelihood of a measured response falling within a 95% confidence interval is much higher than a measured response matching a single number. The latter deterministic match is commonly regarded as an "ideal" prediction, although extensive validation of many prediction models using large load test databases has shown that an "ideal" prediction is practically impossible (Tang and Phoon 2021). One may argue that an engineer is satisfied to achieve a prediction close enough to the measured response. However, there is no rational yardstick to measure "close enough". In fact, the 95% confidence interval or region is the rational yardstick to assess the adequacy of a numerical model. As Tukey (1962) pointed out, "far better an approximate answer to the right question, which is often vague, than an exact answer to the wrong question, which can always be made more precise". In the context of the 95% confidence interval, this means a larger interval that consistently captures the measured response is the preferred outcome from a numerical model. Sensitivity analysis can guide the collection of additional information (where and how much) to reduce the 95% confidence interval as well. In fact, the "value" of data mentioned above can be defined rationally by this reduction in the 95% confidence interval (Ching and Phoon 2012a; Ching et al. 2014a).

A knowledge of the probability distribution can directly benefit current practice by supporting a more rational and defensible choice of a "cautious" value (e.g., 5% fractile characteristic value specified in Clause 2.4.5.2(11), CEN 2004) (Länsivaara et al. 2022). It is more appropriate to apply engineering judgment as a "reality check" on the 5% fractile, rather than to select a "cautious" value from a single number (by imagining how it may vary from experience) or a few numbers from parametric studies. It has been pointed out above that it is unreasonable to expect an engineer to imagine a cautious combination of several soil parameters in the presence of a multivariate correlation structure that is not characterized in a deterministic analysis. If the 95% confidence interval fails to bracket the measured response, the engineer is alerted to the possibility of incomplete physics and/or an overly optimistic assessment of relevant uncertainties. In this way, the engineer is given an opportunity to explore the problem further. The body of research demonstrating how numerical analysis informed by uncertainty quantification can produce additional insights to enhance decision-making is significant (Phoon et al. 2022c).

As a result of representing information as a single number (a strong simplification imposed by the deterministic paradigm), it is not surprising that numerical models are more routinely used to gain physical insights through back-analyses (Walton and Sinha 2022) or to gain a rough sense of the range of possible answers through parametric studies. Hammah and Curran (2009) observed that

> a most powerful use of modelling tools is the proper formulation of questions … Models permit us to perform 'what if' analysis, which are experiments with different inputs, assumptions and conditions. Answers to these questions can often lead to the correct diagnosis of problems of key behaviours.

Note that conventional parametric studies entail an implicit recognition that inputs and/or scenarios are not precisely known. One can argue that it is a highly simplified deterministic version of a probabilistic analysis akin to carrying out a Monte Carlo simulation with a few samples. However, the alternate inputs/scenarios are rarely statistically informed by the actual spread of values exhibited by each input parameter, the dependencies between different input parameters measured at different depths, and statistical uncertainties arising from data scarcity ("curse of small sample size", Phoon 2017). Ironically, most engineers are more comfortable relying on their judgment to establish the range of plausible inputs/scenarios for parametric studies than relying on more consistent random sampling connected to measured data. For example, Simpson (2011) spoke of worst credible situations and parameter values as situations/values that could be imagined on the basis of a reasonable and well-informed engineering assessment. It is accurate to say that engineers are not familiar with probabilistic methods and/or not convinced they are applicable to practice. However, the potential for judgment to lead to inconsistent outcomes in the face of uncertainty (as explained below) is not emphasized. The partial preference for judgment is not surprising given the above circumstances.

Note that statistical uncertainties are significant, but unlike spatial variabilities, there are no physically intuitive methods to estimate them. Formal methods such as Bayesian analysis or bootstrapping are necessary. Even the development of such formal methods for spatially correlated data is technically challenging (Phoon and Fenton 2004; Ching and Phoon 2017, 2019). An example is shown in Figure 1.6. It is clear that statistical uncertainties decrease with data, but it is not possible to quantify this reduction using engineering judgment or even classical statistics. A large part of classical (frequentist) statistics assumes that the sample size is sufficiently large to treat the mean, coefficient of variation, and correlation coefficient as deterministic numbers. For geotechnical engineering, it is more reasonable to treat these statistics as random variables, as illustrated by Figure 1.6 for the case of the correlation coefficient. The probability distributions for these statistics become very narrow when the sample size is large, thus converging to the classical deterministic solutions as a special case. In short, the Bayesian approach of treating statistics as random variables is more general and more natural for the data-poor environment in geotechnical engineering (Baecher 2017). It can be shown that under appropriate conditions, the maximum a posteriori probability (MAP) estimates of the means and standard deviations converge to the maximum likelihood estimates (frequentist constants) when the sample size is large (Bernstein–von Mises theorem).

To the author's knowledge, parametric studies in practice are rarely preceded by a formal statistical analysis of the site data to quantify the sources of uncertainties and their dependencies. Hence, statistical uncertainties are not considered, which implies that routine parametric study underestimates the range of possible values for a given soil property. The degree of underestimation increases with data scarcity. Model uncertainties are not considered in parametric studies as well, although they are significant in geotechnical engineering (Phoon and Tang 2019a; Tang and Phoon 2021). If one is not interested in the *actual* range of variation (say the undrained shear strength at a specific site can vary between 50 and 100 kPa), a parametric study can provide useful information on the *sensitivity* of an output to some variation in one input. In a sensitivity analysis, the range of

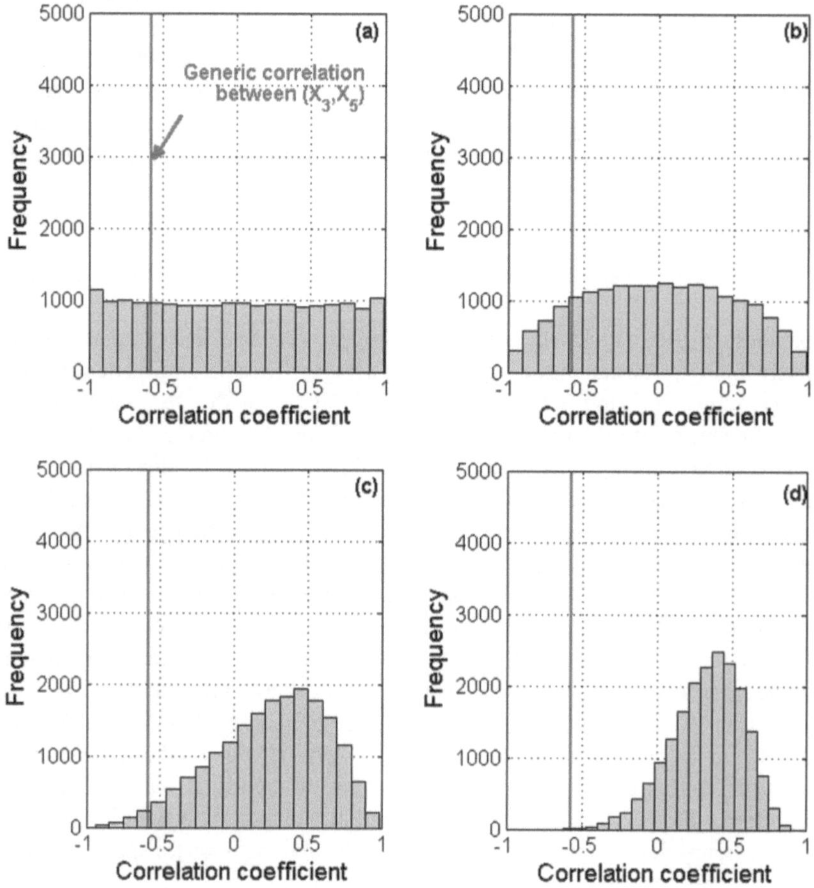

FIGURE 1.6 Statistical uncertainty of the correlation coefficient between X_3 (transformed liquidity index) and X_5 (transformed preconsolidation stress) at the Lilla Mellösa site, Sweden, from a sample containing: (a) no data; (b) 2 data points; (c) 5 data points; (d) abundant data. (Source: Fig. 5, Ching and Phoon 2019)

variation is typically regarded as a nominal perturbation to one input (say 10%), rather than related to statistics such as the coefficient of variation (say 30% for undrained shear strength, Phoon and Kulhawy 1999a). However, there is limited appreciation that this more modest sensitivity goal is also not easily achievable if more than one input is involved. The reason is that the variation in one input is related to the variation in a second input if they are dependent, as illustrated in Figure 1.7. In short, the relatively common practice of perturbing two inputs independently is incorrect and a violation of the dependency relationship between two inputs, even if the actual range of variation is not of primary concern in a sensitivity analysis. Duncan (2000) argued that simple reliability analyses involving neither complex theory nor unfamiliar terms and requiring only simple statistics (means, standard deviations, and correlations) can be an improvement. Another important complication but little known in geotechnical engineering is Simpson's paradox (Yule 1903; Simpson 1951), where a trend appears in several groups of data but disappears or reverses when the groups are combined. Simpson (2011) implicitly highlighted the practical importance of correlation:

> Many safety formats used in codes of practice generate extreme values of parameters by applying partial factors. In most cases it becomes incredible that several variables could attain very extreme values simultaneously. This underlies the principles of load combinations following the principles developed by Turkstra and Madsen (1980), and much of seismic design, in both of which the effect of one dominant variable is considered while others are given less extreme values.

When soil parameters are allowed to vary independently without restraint by a correlation structure, unrealistic "incredible" scenarios can appear. It should be apparent at this point that reasoning with uncertainties is much more complicated than what an engineer typically learned in basic statistics involving one random variable.

It is important to draw a distinction between the mathematical concept of a correlation and the empirical dependencies shown in scatter plots between two or more soil parameters. The latter

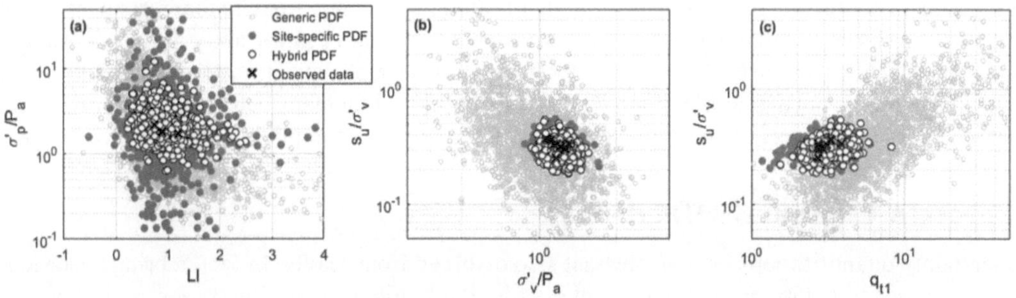

FIGURE 1.7 Correlations between some soil parameters at a Taipei site. (Source: updated Fig. 8, Ching and Phoon 2019) (Note: LI = liquidity index; σ'_v = vertical effective stress; σ_v = vertical total stress; σ'_p = preconsolidation stress; P_a = atmospheric pressure = 101.3 kPa; s_u = undrained shear strength; q_t = (corrected) cone tip resistance; undrained strength ratio (s_u/σ'_v); q_{t1} = normalized cone tip resistance = $(q_t - \sigma_v)/\sigma'_v$; X_3 = transformed (LI); X_5 = transformed (σ'_p/P_a); X_4 = transformed (σ'_v/P_a); X_6 = transformed (s_u/σ'_v); X_9 = transformed (q_{t1}).

dependencies are real regardless of how they are modeled theoretically (e.g., Pearson correlation, copula) and have been extensively documented in many soil/rock property databases (Ching et al. 2016). An example is shown in Figure 1.7. The values of a soil parameter measured in a small spatial neighborhood must be dependent in the sense that a large value must be preceded or followed by a large value if the measurement interval is small (using the depth direction as an example of a spatial dimension). Abrupt changes in a soil profile are not common and they are typically associated with transitions between soil types (layer boundaries). These spatial dependencies are well documented in the literature (Cami et al. 2020). There are also dependencies arising from curve fitting, such as fitting a hyperbolic model to a load-movement curve (Phoon et al. 2006, 2007; Tang and Phoon 2021) or fitting the van Genuchten model to a soil–water characteristic curve (Phoon et al. 2010). An engineer steeped in determinism is rarely conscious of these varied dependencies, how they are quantified, and their impact even on deterministic analysis. However, the engineer is certainly aware of important gaps between what has been calculated in a deterministic analysis and what has been measured in the field (Carter et al. 2000; Herle 2003; Poulos 2004). An extensive survey of predictions versus measurements for a variety of geotechnical structures is given in Chapter 5. The reasons postulated for the gaps are invariably grounded in physics, rather than thinking through whether the gaps can be explained partially by statistics before invoking missing physics. The absence of Bayesian thinking in practice is very apparent.

It is useful to point out a curious oddity between the sample size and a visual perception of uncertainty based on data scatter as another example of inconsistency. For the correlation between the liquidity index (LI) and the preconsolidation stress (σ'_p/P_a) in Figure 1.7a, there are only three observed data points at the Taipei site. For the correlation between the undrained strength ratio (s_u/σ'_v) and the normalized cone tip resistance (q_{t1}) in Figure 1.7c, there are nine observed data points. The data scatter in Figure 1.7a "looks" smaller than that in Figure 1.7c. This perception is, however, incorrect if one were to include statistical uncertainty as shown in the simulated samples of the site-specific probability density function (PDF). The above observations amply demonstrate the pitfalls in addressing uncertainties through intuition rather than through more formal Bayesian analysis. Engineering judgment based on this visual perception of uncertainty alone will lead to false conclusions. Contrary to popular belief, formal uncertainty quantification is actually more critical rather than not applicable to small datasets. It is likely that many deterministic analyses are carried out in violation of the uncharacterized statistical structure implicit in the measured data.

1.5 DATA-CENTRIC GEOTECHNICS

Uncertainty quantification is not an abstract step divorced from reality. In fact, it brings decision-making closer to reality beyond what is offered by the familiar deterministic approach, because it can model and draw insights from real-world data with greater fidelity. All geotechnical data appears as histograms, that is, there is a range of values and some values are more likely than others. Figure 1.8 presents the model factors for laterally loaded rigid drilled shafts in cohesionless soils in the form of histograms. Each data point in Figure 1.3 is produced by such a histogram. The bias (mean) and dispersion (COV) are second-moment statistics (analogous to the center of gravity and moment of inertia in solid mechanics) – it is apparent that they do not capture all the details in a histogram. Scatter plots of actual soil data show dependencies (Figure 1.7). There is no simple method to visualize more complicated dependency relationships in higher dimensions between more than two soil parameters. The ongoing digital transformation has altered our perspectives on the role and value of data. There is an increasing appreciation that histograms and scatter plots (dependencies) are part of our "ground truths" and they are as real as physical features such as soil layers. It is appropriate to call them "data ground truths". They should not be regarded as "noises" to be minimized or artificially eliminated by forcing a choice of a single cautious estimate. At one extreme, some may argue that no rational data analysis is possible, and decisions should rely primarily on engineering judgment informed by experience. Numerical modeling has been conflated

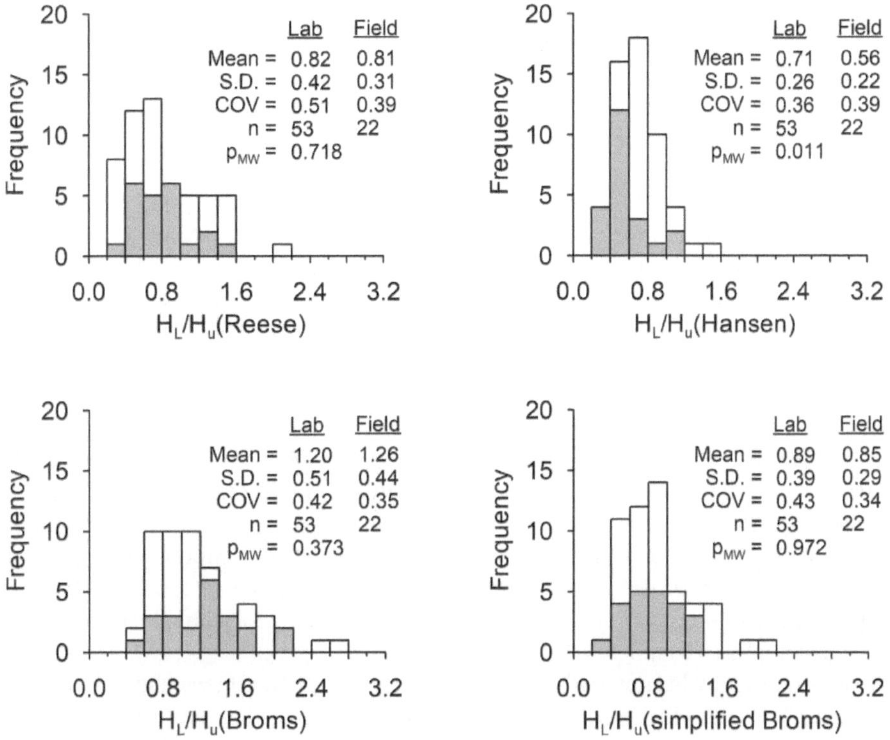

FIGURE 1.8 Model factors (M) for laterally loaded rigid drilled shafts in cohesionless soils. $M = H_L/H_u$, in which H_L = interpreted lateral or moment limit from load test data and H_u = calculated lateral capacity based on different lateral soil resistance models (Reese et al. 1974, Hansen 1961, Broms 1964, and Broms 1964 – simplified). (Source: Fig. 3, Phoon and Kulhawy 2005)

with the deterministic paradigm for several historical reasons. Some of the reasons are no longer applicable. In the Future of Geotechnical Engineering Report, Mitchell and Kopmann (2013) reflected that "information technology (IT) has had a very high impact on many facets of geotechnical engineering" since the New Millennium Report (National Research Council 2006) was published seven years ago. They spoke of the availability of big datasets and the challenge to make sense of this wealth of information. Hence, the future of practice is to develop methods to make sense of data, rather than to ignore data. When numerical modeling is combined with uncertainty quantification grounded on real-world data, its value to decision-making in practice is significantly enhanced, as demonstrated by Chapters 11–16 in this book.

More fundamentally, this section argues that uncertainty quantification plays a key role in data-centric geotechnics – an emerging field that attempts to prepare geotechnical engineering for digital transformation (Phoon et al. 2022b). The agenda is built on three core elements: "data centricity", "fit for (and transform) practice" and "geotechnical context". The first refers to the development of methods that make sense of all real-world data (discover insights). The second refers to actual value to critical real-world decisions for a specific project at a specific site. The third refers to the need to bear in mind the physical basis of the problem and the accumulated body of geotechnical and geological knowledge and experience that remains valuable for decision support. There is insufficient research to illuminate what type of methods are effective in the presence of geotechnical data associated with the different four Vs (volume, variety, velocity, veracity) collected for different problems, be it data driven, physics-informed data driven, explainable/interpretable artificial intelligence (AI), or others. A more detailed exploration of data-centric geotechnics is outside the scope of this book. It suffices to point out that uncertainty quantification is no longer "good to have" in decision-making. It should be regarded as "must have" when the construction industry is digitalized (Bilal et al. 2016; Gerbert et al. 2016; Munawar et al. 2022).

A large part of the geotechnical risk and reliability literature (which is the focus of the current book) remains classical in the sense that more advanced machine learning methods have yet to be extensively studied by researchers and practitioners (Phoon and Zhang 2023). There is significant room to improve the fairly simple statistical models that are widely used in geotechnical and rock engineering practice (Phoon 2017). One major limitation of the prevailing statistical models is the gap between the assumed and the actual attributes of geotechnical data. Many classical statistical models assume that data is homogeneous, abundant, independent, and normally distributed. For many years, practitioners have criticized these models as unrealistic (National Research Council 1995). Phoon et al. (2022b) opined that the "ugly data" challenge lies at the heart of any data-driven site characterization (DDSC). The authors defined DDSC as any site characterization methodology that relies solely on measured data, both site-specific data collected for the current project and existing data of any type collected from past stages of the same project or past projects at the same site, neighboring sites, or beyond. One example of ugly data is MUSIC-3X (multivariate, uncertain and unique, sparse, incomplete, and potentially corrupted with "3X" denoting three-dimensional spatial variations). An example is shown in Table 1.3. It is a useful mnemonic to highlight the attributes of real site data and to contrast with the highly idealized assumptions underlying classical statistics. Ideal data is "beautiful". Real-world data is "ugly". It is premature to say that ugly data cannot support decision-making. The central tenet in data-centric geotechnics is that data has value as long as it is not fake or completely corrupted.

It is an open question what data-driven site characterization can achieve and how useful are the outcomes for practice, but this "value of data" question (emphasized several times in this chapter) is of major interest given the rapid pace of digital transformation in many industries. The scientific aspects of this question are presented as three challenges by Phoon et al. (2022b): (1) ugly data, (2) site recognition, and (3) stratification. The site recognition challenge is as basic as the ugly data challenge. The objective is to quantify "site uniqueness", directly or indirectly, so that big indirect data (Tables 1.4 and 1.5) can be combined with sparse site-specific data in a manner sensitive to site differences. This idea is not new as geotechnical engineers have been relying on data from similar sites to supplement limited data collected at a given site. The engineer relies heavily on his/her experiences and knowledge of the local geology to identify similar sites, but this human-powered strategy cannot work when the amount of data becomes overwhelming large. The stratification challenge is partly covered in Chapter 4. Despite the apparent intractability of these challenges at first glance, promising solutions based on Bayesian machine learning have been found (Ching et al. 2021a, 2021b; Phoon and Ching 2021; Phoon and Ching 2022). Data-centric geotechnics, which includes DDSC, building information modeling (BIM), and digital twin, can be viewed as the next evolutionary leap in decision support that would succeed the deterministic expanded Burland Triangle (Figure 1.4). Its potential is likely to be tremendous and its impact far reaching. Managing uncertainty explicitly should be a cornerstone of any data-centric agenda. Details are given elsewhere (Phoon and Ching 2021; Phoon et al. 2022a; Phoon and Zhang 2023). It suffices to note that the geotechnical risk and reliability agenda is on the cusp of being changed to meet the challenging needs of practice such as ugly data and site recognition and to solve practical problems at a new level of realism beyond the fairly limited scope of solving realistic boundary-value problems.

1.6 UNCERTAINTY-INFORMED BURLAND TRIANGLE

Geotechnical engineering is widely perceived as an art as much as a science, in part because known unknowns and unknown unknowns are not fully addressed in its numerical models (Phoon 2017). Hence, the quantitative parts of the expanded Burland Triangle (Barbour and Kahn 2004) – ground profile, soil behavior, and modeling – must be supported qualitatively by "empiricism, precedent, experience, and risk management". The original Burland (2012) triangle considered

TABLE 1.4
Soil/Rock Property Databases

	Database	Reference	Soil/rock parameters	No. data points	No. sites/ studies	% complete
Univariate	CLAY/16	Phoon and Kulhawy (1999a)	γ, γ_d, w_n, PL, LL, PI, LI, ϕ', s_u, s_u^{FV}, q_c, q_t, SPT-N, DMT (A, B), PMT p_L		a	
	SAND/11	Phoon and Kulhawy (1999a)	ϕ', D_r, q_c, SPT-N, DMT (A, B, I_D, K_D, E_D), PMT (p_L, E_{PMT})		b	
	ROCK/8	Prakoso (2002)	γ (or γ_d), n, R, S_h, σ_{bt}, I_s, σ_c, E		c	
	ROCK/13	Aladejare and Wang (2017)	ρ, G_s, I_{d2}, n, w_c, γ, R_L, S_h, σ_{bt}, I_{s50}, σ_c, E, ν		d	
Multivariate	CLAY/5/345	Ching and Phoon (2012b)	LI, s_u, s_u^{re}, σ'_p, σ'_v	345	37 sites	100
	CLAY/7/6310	Ching and Phoon (2013)	s_u from seven different test procedures	6,310	164 studies	17.7
	CLAY/6/535	Ching et al. (2014b)	s_u/σ'_v, OCR, q_{tc}, q_{tu}, $(u_2 - u_0)/\sigma'_v$, B_q	535	40 sites	100
	CLAY/10/7490	Ching and Phoon (2014)	LL, PI, LI, σ'_v/P_a, σ'_p/P_a, s_u/σ'_v, S_t, q_{tc}, q_{tu}, B_q	7,490	251 studies	34.1
	FI-CLAY/7/216	D'Ignazio et al. (2016)	s_u^{FV}, σ'_v, σ'_p, w_n, LL, PL, S_t	216	24 sites	100
	JS-CLAY/5/124	Liu et al. (2016)	M_r, q_c, f_s, w_n, γ_d	124	16 sites	100
	JS-CLAY/7/372	Zou et al. (2017)	σ_v, σ'_v, q_{tc}, f_s/σ'_v, B_q, V_{s1}, s_u/σ'_v	372	25 sites	100
	SAND/7/2794	Ching et al. (2017)	D_{50}, C_u, D_r, σ'_v/P_a, ϕ', q_{t1}, $(N_1)_{60}$	2,794	176 studies	60.0
	EMI-ROCK/8/26000+	Kim and Hunt (2017)	σ_c, σ_{bt}, ρ, CAI, PPI, cohesion, direction shear, triaxial confining	26,000+	–	–
	FG/5/1000	Kootahi and Moradi (2017)	e, w_n, LL, PI, C_c	1,000	170 sites	100
	ROCK/9/4069	Ching et al. (2018)	γ, n, R_L, S_h, σ_{bt}, I_{s50}, V_p, σ_{ci}, E_i	4,069	184 studies	34.2
	FG-KSAT/6/1358	Feng and Vardanega (2019)	e, k, LL, PL, PI, G_s	1,358	33 studies	91.4
	SH-CLAY/11/4051	Zhang et al. (2020)	LL, PI, LI, e, K_0, σ'_v/P_a, s_u/σ'_v, S_t, q_c/σ'_v	4,051	50 sites	39.5
	CLAY/8/12225	Ching (2020)	LL, PI, w, e, σ'_v/P_a, C_c, C_{ur}, c_v	12,225	427 studies	–
	CLAY/12/3997	Ching (2020)	LL, PI, LI, σ'_v/P_a, σ'_p/P_a, s_u/σ'_v, K_0, E_u/σ'_v, B_q, q_{t1}, $N_{60}/(\sigma'_v/P_a)$	3,997	237 studies	–
	SAND/13/4113	Ching (2020)	e, D_r, σ'_v/P_a, σ'_p/P_a, K_0, E_{dn}, q_{c1n}, B_q, $(N_1)_{60}$, K_{DMT}, E_{DMTn}, E_{PMTn}, M_{dn}	4,113	172 studies	–
	ROCKMass/9/5876	Ching et al. (2021a)	RQD, RMR, Q, GSI, E_m, E_{em}, E_{dm}, E_i, σ_{ci}	5,876	225 studies	29.3

(Continued)

TABLE 1.4 (CONTINUED)
Soil/Rock Property Databases

Database	Reference	Soil/rock parameters	No. data points	No. sites/ studies	% complete
CLAY-C_u/6/6203	Ching et al. (2022)	LL, PI, w_n, e, C_c, C_{ur}	6,203	429 studies	61
Global-CPT/3/1196	Ching et al. (2023)	q_t, f_s, u_2	1,196	59 sites	100
Geo-Marine-CPT/3/398	Eslami et al. (2023)	q_t, f_s, u_2	398	58 sites	100
ROCK/10/4025	Muzamhindo and Ferentinou (2023)	γ, n, R_L, I_{s50}, V_p, σ_{bt}, BPI, m_i, E_i, σ_c	4,025	96 studies	50
SAND-Small/9/939	Lo et al. (2021)	D_{50}, C_u, e_{min}, e_{max}, σ'_3, e_c, G_{max}, σ'_{1p}, ϕ′	939	15 studies	79

Source: Phoon et al. (2023).

Notes:

Basic: C_u = coefficient of uniformity; D_{50} = median grain size; ρ = density; G_s = specific gravity; γ = unit weight; γ_d = dry unit weight; D_r = relative density; e = void ratio; e_{min} = minimum void ratio; e_{max} = maximum void ratio; e_c = void ratio under isotropic consolidation; n = porosity; w_n (or w_c) = water content; PL = plastic limit; LL = liquid limit; PI = plasticity index; LI = liquidity index; GSI = geological strength index.

Stress: σ_v = total vertical stress; σ'_v = effective vertical stress; σ'_p = preconsolidation stress; σ'_3 = effective confining stress under isotropic consolidation; σ'_{1p} = effective axial stress at peak state under drained triaxial compression; OCR = overconsolidation ratio; P_a = atmospheric pressure = 101.3 kPa; K_0 = at-rest lateral earth pressure coefficient.

Strength: ϕ′ = effective friction angle; s_u = undrained shear strength; s_u^{FV} = field vane s_u; s_u^{re} = remolded s_u; S_t = sensitivity; σ_{bt} = Brazilian tensile strength; σ_{ci} (or σ_c) = uniaxial compressive strength of intact rock; BPI = block punch index; m_i = Hoek–Brown constant.

Deformation: C_c = compression index; C_{ur} = unloading-reloading index; modulus; E_u = undrained modulus of clay; E_d = drained modulus of sand; $E_{dn} = (E_d/P_a)/(\sigma'_v/P_a)^{0.5}$; E_{dm} = dynamic modulus of rock mass; E_{em} = elasticity modulus of rock mass; E_i (or E) = Young's modulus of intact rock; E_m = deformation modulus of rock mass; ν = Poisson ratio; M_r = subgrade resilience modulus; M_d = effective constrained modulus determined by oedometer; M_{dn} = normalized $M_d = (M_d/P_a)/(\sigma'_v/P_a)^{0.5}$; G_{max} = small-strain shear modulus.

Permeability: k = hydraulic conductivity; c_v = coefficient of consolidation.

Dynamic: V_p = P-wave velocity; V_s = S-wave velocity; $V_{s1} = V_s(P_a/\sigma'_v)^{0.25}$.

Field test: SPT-N = standard penetration test blow count; N_{60} = corrected SPT-N; $(N_1)_{60} = N_{60}/(\sigma'_v/P_a)^{0.5}$; q_c = cone tip resistance; q_t = corrected cone tip resistance; f_s = sleeve frictional resistance; $q_{tc} = (q_t/P_a)/(\sigma'_v/P_a)^{0.5}$; $q_{t1} = (q_t - \sigma_v)/\sigma'_v$ = normalized cone tip resistance; $q_{tu} = (q_t - u_2)/\sigma'_v$ = effective cone tip resistance; $q_{c1n} = (q_c/P_a)/(\sigma'_v/P_a)^{0.5}$; B_q = pore pressure ratio = $(u_2-u_0)/(q_t-\sigma_v)$; $(u_2-u_0)/\sigma'_v$ = normalized excess pore pressure; u_2 = pore pressure behind cone tip; u_0 = hydrostatic pore pressure; PMT (p_L, E_{PMT}) = pressure meter limit stress, modulus; E_{PMTn} = normalized $E_{PMT} = (E_{PMT}/P_a)/(\sigma'_v/P_a)^{0.5}$; DMT (A, B, I_{DMT}, K_{DMT}, E_{DMT}) = dilatometer A and B readings, material index, horizontal stress index, modulus; E_{DMTn} = normalized $E_{DMT} = (E_{DMT}/P_a)/(\sigma'_v/P_a)^{0.5}$; CAI = Cerchar abrasivity index; PPI = punch penetration index; Q = Q-system; RMR = rock mass rating; RQD = rock quality designation; R = Schmidt hammer hardness (R_L = L-type Schmidt hammer hardness); S_h = shore scleroscope hardness; I_{d2} = slake durability index; I_s = point load strength index ($I_{s50} = I_s$ for diameter 50 mm).

a: The number of data groups varies between 2 and 42 depending on the clay parameter. Statistics are calculated at the data group level. The average number of data points/data groups varies between 16 and 564. Details given in Tables 1–3, Phoon and Kulhawy (1999b).

b: The number of data groups varies between 5 and 57 depending on the sand parameter. Statistics are calculated at the data group level. The average number of data points/data groups varies between 15 and 123. Details given in Tables 1–3, Phoon and Kulhawy (1999b).

c: The number of data groups varies between 30 and 174 depending on the rock parameter with no differentiation of rock type (igneous [intrusive, extrusive, pyroclastic], sedimentary [clastic, chemical], metamorphic [foliated, non-foliated]). Statistics are calculated at the data group level. The average number of data points/data groups varies between 3 and 161 for σ_c (Prakoso, 2017). Details given in Table 4.4, Prakoso (2002).

TABLE 1.4 (CONTINUED)

d: The number of data groups varies between 2 and 47 depending on the rock parameter and rock type (igneous, sedimentary, or metamorphic). Statistics are calculated at the data group level. The average number of data points/data groups varies between 7 and 92. Details given in Tables 2–4, Aladejare and Wang (2017).

"empiricism, case records, and well-winnowed experience" to be central to decision-making in practice. As argued above, the reason is that it is not possible to make a *safe decision* based on a single deterministic analysis alone unless the ground profile, soil behavior, physics, and construction effects are perfectly known, can be computed to perfect accuracy, and there are no unknown unknowns. Even for the rare occasion where this ideal state is approximately true, construction and other site-specific effects (e.g., spatial variability) are not considered in the quantitative parts of the expanded Burland Triangle. In this sense, the quantitative parts are fundamentally uncertain and likely incomplete. An experienced engineer is aware that the deterministic answers produced by numerical analyses cannot be applied directly to a real-world project at a specific site without moderation by an ad hoc combination of informal risk management strategies that include applying a global factor of safety (or partial factors of safety), selecting cautious input values and conservative calculation models, conducting parametric studies, learning from precedents, updating/validating designs and construction procedures based on prototype testing and observations, and keeping engineering judgment as an integral part of the decision-making loop (Marr 2006; Duncan and Sleep 2017; Phoon 2020). These strategies are effective, but their role in digital transformation is unclear. These strategies pre-date the current digital era and it would be sensible to review them.

Casagrande's (1965) classic paper on the "Role of the 'Calculated Risk' in Earthwork and Foundation Engineering" may be viewed as the beginning of more formal risk management strategies. Although his paper did not calculate risk in the quantitative sense, he did recommend adopting a more systematic approach to think about risk involving a deliberate assessment of all relevant uncertainties influencing the solution and the need to balance the likelihood and the consequences of failure:

1. The use of imperfect knowledge, guided by judgment and experience, to estimate the probable ranges for all pertinent quantities that enter into the solution to the problem.
2. The decision on an appropriate margin of safety, or degree of risk, taking into consideration economic factors and the magnitude of losses that would result from failure.

The adoption of classical statistics as a quantitative basis for the selection of a suitably cautious design value may arguably be traced to Lumb's classical *Canadian Geotechnical Journal* paper on "The Variability of Natural Soils" published in 1966 (Lumb 1966). The 2021 draft of the new EN 1997-1 defines the characteristic value using the same classical method. The larger role of probabilistic methods in design and decision-making was possibly first articulated in the National Research Council (1995) report:

> probabilistic methods, while not a substitute for traditional deterministic design methods, do offer a systematic and quantitative way of accounting for uncertainties encountered by geotechnical engineers, and they are most effective when used to organize and quantify these uncertainties for engineering designs and decisions.

An update is provided in Chapter 9 "Risk and Reliability" of the Future of Geotechnical Engineering Report (Mitchell and Kopmann 2013). In the Fifty-Fifth Rankine Lecture, Lacasse (2015) observed that

TABLE 1.5
Foundation Load Test Databases

Database/reference	Limit state	Soil type	*n*	Pile geometry B (m)	Pile geometry L/B	Soil parameters
NUS/ShalFound/919 (Tang et al. 2020)	Bearing	Clay	56	0.30–5.00	0–5.7	s_u = 9–200 kPa
		Sand	427	0.25–7.00	0–6.1	ϕ = 26–53°
	Tension	Clay	123	0.31–3.05	0.8–13.2	s_u = 15–300 kPa
		Sand	313	0.10–2.50	0.5–14.5	ϕ = 30–49°
NUS/ShalFound/Punch-Through/31 (Tang and Phoon 2019a)	Punch-through	Sand-over-clay	31	0.80–3.00	0.5–3.0	ϕ_{cv} = 32° D_r = 88% s_u = 8.7–85.9 kPa
NUS/Spudcan/Punch-Through/212 (Tang and Phoon 2019a)	Punch-through	Multi-layer clays with sand	140	3.00–20.00	0.2–1.2	ϕ_{cv} = 31–34° D_r = 44–99% s_u = 7.2–44.8 kPa
		Multi-layer clays with stiff layer	72	3.00–12.00		s_u = 3–50 kPa ρ = 0–2.6 kPa/m
NUS/DrilledShaft/542 (Tang et al. 2019)	Bearing	Clay	64	0.32–1.52	1.6–56.0	s_u = 41–256 kPa
		Sand	44	0.35–2.00	5.1–59.0	ϕ = 30–41°
		Gravel	41	0.59–1.50	6.2–30.0	ϕ = 37–47°
	Tension	Clay	32	0.36–1.80	3.4–55.0	s_u = 21–250 kPa
		Sand	30	0.30–1.31	2.5–43.0	ϕ = 30–45°
		Gravel	109	0.43–2.26	1.8–17.3	ϕ = 42–48°
NUS/DrivenPile/1243 H section (Tang and Phoon 2018a; Phoon and Tang 2019b)	Bearing	Clay	47	0.28–0.41	16.0–95.0	N_{SPT} = 5–50
		Sand	52	0.28–0.42	22.0–110.0	N_{SPT} = 7–40
		Mixed	50	0.28–0.42	17.0–85.0	N_{SPT} = 4–29
NUS/DrivenPile/1243 Tube/box section (Tang and Phoon 2019b)	Bearing	Clay	175	0.10–0.81	7.9–200.0	PI = 11–160% OCR = 1–43.2 S_t = 1–17
	Tension		64	0.10–0.81	12.0–110.0	PI = 12–110% OCR = 1–43.2 S_t = 1–8.3
NUS/DrivenPile/1243 Tube/box section (Tang and Phoon 2018b)	Bearing	Sand	134	0.14–0.76	13.0–251.0	ϕ = 30–42° D_r = 15–93%
	Tension		28	0.25–0.76	19.0–84.0	ϕ = 30–42° D_r = 31–97%
NUS/RockSocket/721 (Tang and Phoon 2021)	End bearing	Rock	270	0.10–2.50	1.0–31.3	σ_c = 0.5–99 MPa E_m = 7.82–75113 MPa GSI = 7.5–95 RQD = 20–100%
NUS/RockSocket/721 (Tang and Phoon 2021)	Shaft shearing	Rock	544	0.20–3.20	0–19.5	σ_c = 0.4–99 MPa E_m = 24–19844 MPa GSI = 50–70 RQD = 0–100%
NUS/HelicalPile/1113 (Tang and Phoon 2018c, 2020)	Bearing	Clay	270	0.21–1.02	6.0–74.0	$s_u \leq$ 305 kPa
		Sand	181	0.21–1.02	6.0–110.0	ϕ = 30–45°
	Tension	Clay	165	0.21–0.91	12.0–48.0	$s_u \leq$ 300 kPa
		Sand	121	0.21–0.91	10.0–62.0	ϕ = 30–45°

(*Continued*)

TABLE 1.5 (CONTINUED)
Foundation Load Test Databases

Database/reference	Limit state	Soil type	*n*	Pile geometry B (m)	Pile geometry L/B	Soil parameters
CYCU/ DrivenPCPile/152 Tube/box section (Marcos et al. 2013)	Bearing	Clay Sand	70 82	0.24–0.91 0.18–1.37	11.4–133.3 6.0–152.7	s_u = 13–261 kPa D_r = 8–81%
CYCU/ DrilledShaft/23 (Chou et al. 2022)	Lateral	Sand	23	0.30–1.58	12.8–59.0	ϕ = 28–47° D_r = 11–99%
CYCU/DrilledShaft/ TipGrouting/34 (Chen et al. 2021)	Bearing	Clay Sand	23 11	0.55–2.00 0.90–2.00	18.6–60.3 14.4–88.0	N_{SPT} = 8–41 N_{SPT} = 12–43
CYCU/RockSocket/50 (Topacio et al. 2022)	Bearing	Rock	50	0.6–2.00	4.0–55.8	σ_c = 0.2–79 MPa RQD = 0–100%
CYCU/ DrilledShaft/222 (Chen et al. 2011)	Shaft resistance (Bearing/uplift)	Clay Sand	148 74	0.18–1.80 0.14–2.00	1.6–64.2 2.5–70.5	s_u = 21–340 kPa ϕ = 30–45°
CYCU/ DrilledShaft/143 (Chen et al. 2023)	Bearing	Clay Sand	82 61	0.18–2.00 0.24–2.50	3.4–55.0 5.1–73.3	s_u = 41–505 kPa D_r = 28–92% ϕ = 29–41°

Source: Chen et al. (2023).

Note: B: foundation diameter; D: foundation embedment depth or thickness of sand layer; s_u: undrained shear strength of clay; ρ: strength gradient; ϕ: friction angle of sand; ϕ_{cv}: constant volume friction angle; D_r: relative density of sand; N_{SPT}: blow count in standard penetration test (SPT); PI: plasticity index; OCR: overconsolidation ratio; S_t: soil sensitivity index; σ_c: uniaxial compressive strength of rock; E_m: elasticity modulus of rock; GSI: geological strength index; RQD: rock quality designation.

> Reliability approaches do not remove uncertainty nor do they alleviate the need for judgment. They provide a way to quantify the uncertainties and to handle them consistently. Reliability approaches also provide the basis for comparing alternatives. Site investigations, laboratory test programmes, limit equilibrium and deformation analyses, instrumentation and monitoring and engineeringjudgment are necessary parts of the reliability approach.

Decision-making in geotechnical practice is expected to be informed by reliability assessment that enhances rather than replaces the deterministic toolbox (Figure 1.9).

The volume of literature on geotechnical risk and reliability since its inception in the 1960s is significant (Phoon 2017, 2020, 2023; Chwała et al. 2023). The purpose of this book is to review how uncertainty quantification and numerical modeling can complement each other to enhance decision-making in practice and possibly prepare the profession for the next digital stage of evolution. It is a significant expansion of the review conducted by Phoon et al. (2022c). The purpose is not to showcase the power of any methods, be it probabilistic or mechanical. The focus is to point out the value to decision-making (how useful?), to provide summary results for the statistical inputs from generic databases or methods to determine statistical inputs from site-specific data (how to use?), to explain the purposes of different methods and software (what to use?), and to share lessons learned from a variety of applications (what are useful?). The deterministic approach is easy to understand, simple to calculate, and arguably cheapest in computational cost compared to any probabilistic approach. However, Mitchell and Kopmann (2013) opined that

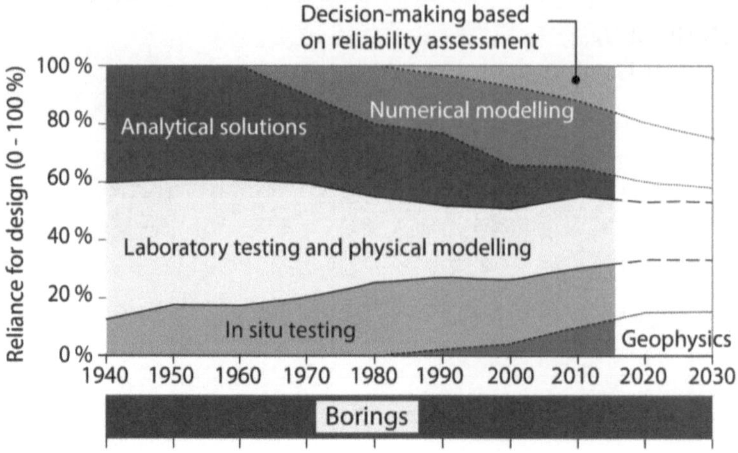

FIGURE 1.9 Evolution of geotechnical practice. (Source: Lacasse 2015)

> Methods of analysis that once were computationally prohibitive are now used routinely. On the other side of the coin, however, the digital age has presented engineers with new challenges in how to select, evaluate, and use all of the new resources wisely, as well as the need to acquire at least some proficiency in areas outside of the more classical geotechnical engineering discipline.

This book is intended to help practitioners acquire a working knowledge of geotechnical reliability methods and to guide them to use these methods wisely in conjunction with data and numerical modeling. The noteworthy distinction between a reliability and a deterministic method is that the former exploits data more intensively and more comprehensively. Data infrastructure is now as important as physical infrastructure. This distinction is not well appreciated by most engineers, although it is critical to digital transformation. For reliability methods, spatially varying data can inform decision-making directly without invoking ad hoc simplification to reduce the state of subsurface reality to a single deterministic value (e.g., average, cautious, or worse credible value). It does not make sense to abandon costly data or reduce high resolution features revealed by actual measurements to fit the input restriction of a deterministic method. In fact, some engineers justify doing this by regarding spatial variations as "noise" or "errors". The author disagrees that all variations are "errors". The more sensible approach is to develop methods that can handle spatially varying or more complicated real-world data as it stands. This "data first method second" philosophy is advocated for research and practice in the context of data-centric geotechnics (Phoon et al. 2022a). This shift from "method first data second" is necessary and urgently needed to accommodate the pace of digital transformation.

To bridge the gap between deterministic practice and less familiar probabilistic methods, the various chapters in this book can be situated as enhancements to an uncertainty-informed expanded Burland Triangle, as shown in Figure 1.10. The elements added to the original triangle are expanded highlighted in gray. Their relevance to decision-making is elaborated in various chapters as outlined below:

I. Statistical characterization

Although numerical models are highly sophisticated in the solution of boundary-value problems, the input ground profile is typically simple and deterministic. Homogeneous or horizontally layered soil profiles with constant properties assigned to each layer are commonly adopted in practice. Statistical guidelines for modeling soil properties as random variables and random fields are available. To emphasize "data first", Chapter 2 provides key results on the statistical characterization of more realistic spatially varying ground

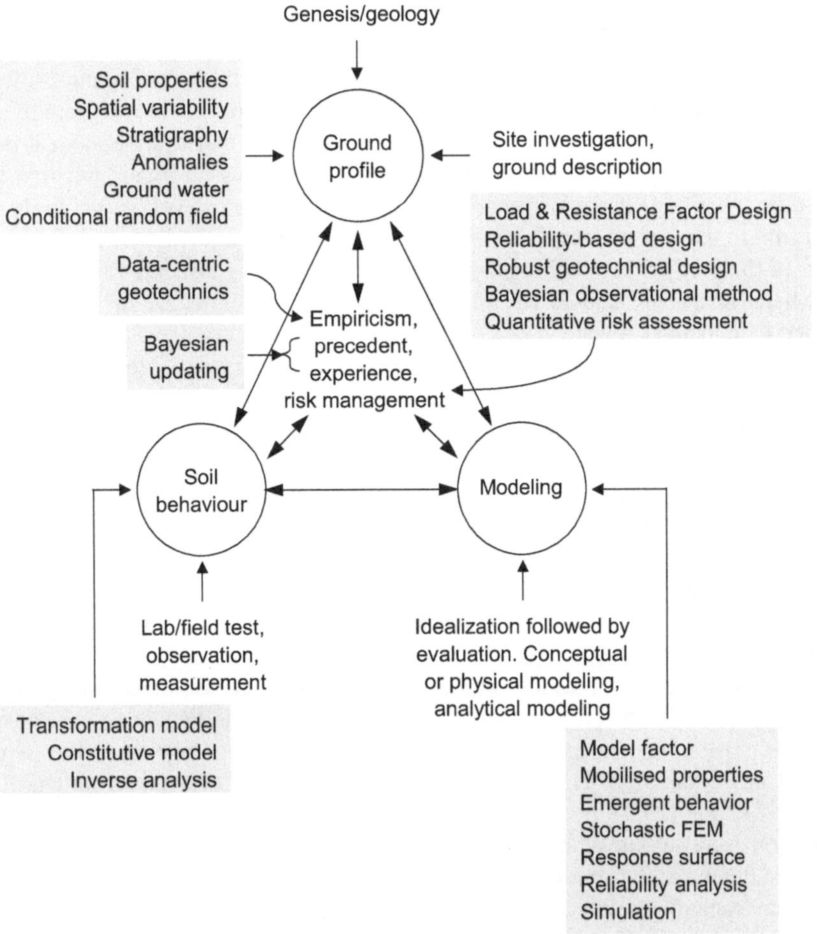

FIGURE 1.10 Uncertainty-informed decision-making in the expanded Burland Triangle (updated uncertainty elements are highlighted in gray). (Source: Fig. 1, Phoon et al. 2022c)

profiles, including the coefficient of variation, transformation uncertainty, and scale of fluctuation for soil/rock properties. Detailed statistics are given in Appendices A to D (Chapter 2), which are made available as soft copy at: www.routledge.com/Uncertainty-Modeling-and-Decision-Making-in-Geotechnics/Phoon-Shuku-Ching/p/book/9781032367491. Chapter 4 focuses on the uncertainty arising from structural and geometrical features in subsurface modeling (rather than the properties in Chapter 2), and this uncertainty is defined as "geologic uncertainty". The objective is to present some practical solutions to the stratification challenge that satisfy the following "fit for practice" requirements: (1) 3D stratification mapping of a realistic subsurface volume, (2) uncertainty quantification of the mapping, (3) tractable in the context of limited site data, and (4) reasonable computational cost. Chapters 2 and 4 are needed to define the "soil profile" element in Figure 1.10 as a probabilistic site model for stochastic/random finite element analysis. The ideal probabilistic model is one constructed from limited site data with or without data from "similar" sites (Phoon et al. 2022b). The former is a quasi-site-specific model. The latter is a local or site-specific model. Chapter 3 reviews the parametric uncertainties associated with constitutive models beyond linear elastic and elastoplastic models and presents a physics-informed machine learning methodology to predict real soil behaviors in practice. The main ideas are to exploit: (1) physics to improve the generality and interpretability of the predictions and (2) machine learning to reduce discrepancies between measurements and predictions in the presence of multi-fidelity data, some of which are sparse and costly. Chapter 3 can be regarded as a more advanced treatment of the transformation uncertainty presented in Chapter 2. Because advanced constitutive models are increasingly adopted in design, the identification of input parameters for these constitutive models from laboratory or field test data is necessary. Research on physics-informed machine learning methods in the "soil behavior" element in Figure 1.10 is, however, surprisingly limited, testifying to the current imbalance between physics and data. Another related method to improve the prediction of soil behaviors is inverse analysis discussed in Chapter 10 and elsewhere (Knabe et al. 2012). In addition, inverse analysis is needed for the observational method (Finno and Calvello 2005). It is rarely appreciated that the observational method needs to be situated in a probabilistic framework to be effective. However, Clause 2.7.2 "Observational method" of Eurocode 7 (CEN 2004) makes this explicit: "the range of possible behaviour shall be assessed and it shall be shown that there is an acceptable probability that the actual behaviour will be within the acceptable limits".

Chapter 5 covers the bias (mean) and dispersion (coefficient of variation) of predictions produced by common calculation models with emphasis on foundation engineering where abundant load test databases exist for the statistical characterization of the model factor. Full-scale tests are more limited for other geotechnical structures. Numerical analyses can serve as a proxy to full-scale tests under these circumstances. The calibration of closed-form calculation models using full-scale tests or numerical analyses is very useful for reliability analysis and reliability-based design (RBD), because these surrogate models, metamodels or emulators are typically inexpensive in computational cost. The cost saving can be very significant if many repeated calculations are necessary say in the context of a Monte Carlo simulation. Chapter 5 is a significant update of Phoon and Tang (2019a) and Tang and Phoon (2021), because it further covers the prediction dispersion exhibited in prediction exercises and numerical analyses. The dispersion of the model factor studied by Tang and Phoon (2021) arises primarily from model uncertainty (closed-form equations in design codes) and possibly some intra-site and inter-site soil variabilities. The load tests are conducted in different locations in a given site and a database is a compilation of load tests from many sites. The dispersion exhibited in a prediction exercise arises from the use of different models and different ways of exercising judgment. The predicted quantity refers to a single load test at a single location and the data provided to all participants is uniform. The dispersion observed in numerical analyses arises from the different methods of transforming measured parameters to input parameters that can be very sophisticated beyond the reach of commercial soil testing (transformation uncertainty) and the different problem scenarios. A systematic evaluation of the model factor for each numerical method is almost absent at this point, although some studies do exist (Zhang et al. 2015; Tang and Phoon 2017; Tang et al. 2017; Xu et al. 2021). Chapter 5 informs the "modeling" element in Figure 1.10. A significant database compilation effort is underway (Phoon and Tang 2024).

II. Probabilistic methods

Uncertainty quantification of soil properties, profiles, and constitutive models is not an end in itself. The engineer is ultimately interested in the occurrence of one or more limit states and possibly their inter-dependencies. In the context of the probabilistic paradigm, it is not meaningful to talk about avoiding the occurrence of a limit state completely. The more meaningful objective is to control the probability of occurrence to a tolerable value. The objective of reliability analysis is to calculate the probability of occurrence (called the "probability of failure"). This is covered in Chapter 6. The objective of reliability-based design is to achieve designs that are "safe" in the sense of capping the probability of failure below a target (tolerable) value (e.g., Table B2, CEN 2002). This is covered in Chapter 9. At first glance, the RBD *format* is identical to the allowable stress design (ASD) *format* of capping the factor of safety below a target value. The significant differences in value between RBD and ASD are, however, grossly under-appreciated in practice. First, RBD is sensitive to data. This means that there is more incentive to invest in a site investigation program (Ching et al. 2014a). Second, RBD can address multiple limit states or failure modes consistently. For example, Lacasse (2015) modeled the New Orleans levees as a series system of 560 reaches. A levee reach is a portion of a levee system (usually a length of a levee) that may be considered for analysis purposes to have approximately uniform representative properties. Each reach is assumed to be statistically independent and 1 km long. The salient point in this example is that the probability of failure for each reach (p_{fr}) must be very low to achieve a tolerable probability of failure for the entire system as a whole (p_{fs}), because

$$p_{fs} = 1 - (1 - p_{fr})^n \tag{1.1}$$

To provide a numerical sense of the significance of Eq. (1.1), p_{fs} is 0.43 even if p_{fr} is 0.001 for $n = 560$. This is not tolerable (e.g., target reliability index equals 3.8 or target probability of failure equals 7.2×10^{-5}, Table C2, CEN 2002). An engineer may "judge" that reducing p_{fr} by a factor of 10 to 0.0001 is sufficient. Simpson et al. (1981) suggested that the "worst credible" situations or parameter values "might be assumed to have a 1 in 1000 chance of occurrence, on the basis that designers would be unlikely to be able to believe that anything more remote might happen". Hence, a $p_{fr} = 0.0001$ may be perceived as "very safe". However, the resulting $p_{fs} = 0.054$ or system reliability index equal to 1.6 is still not tolerable. To achieve a tolerable $p_{fs} = 7.2\times10^{-5}$ (system reliability index equal to 3.8), p_{fr} must be less than 1.3×10^{-7} or the component reliability index for each reach must be larger than 5.2. It is not possible to consider system reliability under ASD. However, slope stability problems cannot be addressed correctly without system reliability, as pointed out in Ching et al. (2009) and elaborated in Chapters 8, 11, and 15. Third, there is no obvious pathway for ASD to take advantage of sophisticated numerical modeling to the extent of implementing real-time autonomous control in Chapter 7. One can speculate that the role of the engineer changes in a similar way to an autonomous vehicle driver from Level 0 (fully manual) to Level 5 (fully autonomous). Last, Phoon (2023) pointed out that a full probabilistic analysis is needed under the following scenarios:

a. Complicated physical model – complex ground conditions and boundaries, complete 3D representation, multiple failure modes, dynamic analysis, etc.
b. Complicated probabilistic model – MUSIC-3X.
c. Complicated design goals – life cycle management, resilience, adaptation to climate change.

The biggest hindrance to the application of probabilistic methods in practice is the cost of evaluating the performance function (G). The performance function demarcates the safe (G > 0) and fail (G < 0) domains based on a limit state (G = 0). The cost of evaluating G once is the same as the cost of one deterministic analysis. The Monte Carlo simulation is simple to implement, but it involves hundreds of thousands of G evaluations. Nonetheless, it is important to point out that more efficient methods are available, as covered in Chapters 6–8, 11, and 12. Probabilistic methods complement the risk management core of the expanded Burland Triangle in Figure 1.10.

III. Probabilistic software

Phoon (2023) recommends a design code to mandate a full probabilistic analysis for:

a. Scenarios entailing high consequences of failure such as the Consequences Class CC3 defined in Table B1 of EN1990 (CEN 2002). A more deliberate and detailed cost-benefit analysis is warranted.
b. Scenarios entailing complicated structures/ground conditions/loadings such as the Geotechnical Category 3 defined in Clause 2.1(21) of Eurocode 7 (CEN 2004)

> very large or unusual structures; structures involving abnormal risks, or unusual or exceptionally difficult ground or loading conditions; structures in highly seismic areas; structures in areas of probable site instability or persistent ground movements that require separate investigation or special measures.

A major part of an engineer or an industry's experience must be accrued from routine practice that may or may not be applicable to more unique and uncommon scenarios.

c. Scenarios entailing high social impact and invoking major public concerns. The public perception of risk is important and deemed critical to technology adoption.
d. Novel scenarios are not covered in design codes, because codification generally draws on a body of experience. Clause 1.5.2.2 of Eurocode 7 (CEN 2004) defines "comparable experience" as

> documented or other clearly established information related to the ground being considered in design, involving the same types of soil and rock and for which similar geotechnical behaviour is expected, and involving similar structures. Information gained locally is considered to be particularly relevant.

The adoption of precedent-based practice in the absence of comparable experience is difficult to justify in a court of law and/or public opinion. Novel scenarios are expected to grow in tandem with emerging external drivers such as climate change and sustainability.

Phoon (2023) argued that software is needed to bring data-driven site characterization and full probabilistic analysis into practice. The second biggest hindrance to the application of probabilistic methods in practice is arguably the lack of software. Nonetheless, Chapters 11 and 12 show that some geotechnical software have begun to incorporate probabilistic analysis tools and they are sufficiently powerful to solve a range of common 2D and 3D problems in a realistic way. Section 11.9 is particularly interesting – it describes the future development of Settle3 (Rocscience) to allow "almost real" spatially varying soil profiles to be automatically constructed from limited cone penetration test (CPT) soundings using the latest Bayesian machine learning. A three-dimensional probabilistic settlement analysis that does not require the engineer to agonize over the choice of statistics (coefficients of variation, scales of fluctuation, cross-correlation matrices) and probability distribution models is now within reach of practice.

IV. Probabilistic applications

The best that an engineer can do in the presence of sparse site investigation data is to simplify a soil profile as homogeneous or layered. A more realistic spatially varying profile cannot be constructed deterministically, because there are many plausible profiles that can fit the laboratory/field test data measured at limited locations/depths. All data-driven interpolation methods, including the most popular kriging method, are fundamentally probabilistic and rightfully so. They are capable of simulating many plausible realizations; some more likely than others. Note that the most likely realization is not necessarily the most critical. The most critical realization/profile contains the least favorable alignment of weak zones for the limit state of interest. Only a judgment-based soil profile can be deterministic. However, it is blind to more critical possibilities. Chapter 13 explains the ramifications of conducting numerical modeling in spatially varying soils. As expected, a specific spatially varying soil (associated with the "worst case" scale of fluctuation) produces the most critical failure mechanism. In other words, the classical homogeneous or layered soil profile assumption is unconservative. This is not surprising as the classical profile is a special case of the spatially varying profile. Despite this observation, failures are rare. The author hypothesized that the global factor of safety is sufficiently large to compensate for this limitation in current practice. Nonetheless, it is worthwhile to reflect on two questions: (1) can a factor of safety always compensate for a misidentification of the failure mechanism? and (2) can a factor of safety be increased without regard to sustainable development goals? Chapter 15 presents an application to linear infrastructures such as dykes. Although the design of linear infrastructures is not new, the loadings caused by climate change can be new. The reliance on precedent-based practice to construct and maintain infrastructures susceptible to climate change is questionable. The National Research Council Report (1995) observed in its executive summary that

> In other geotechnical problems, deterministic methods are so well developed and well understood that probabilistic methods do not offer any special benefit. For example, although probabilistic methods have the potential for use in conventional foundation design, traditional deterministic techniques are so widely accepted and effective for these problems that there has been little motive for change as only minimal benefits will be gained by using probabilistic methods in this case.

In the opinion of the author, this position needs to be updated. Chapters 14 and 16 cover two "non-traditional" applications on press-in piling and cement-based ground improvement for which probabilistic methods are known to be particularly useful since prevailing practice is not well established (National Research Council Report 1995).

1.7 CONCLUSIONS

The purpose of numerical modeling is to enhance decision-making in practice rather than to enhance our prediction ability. The expanded Burland Triangle places "soil behavior", "ground profile", and "modeling" at its three vertices, but situates "empiricism, precedent, experience, and risk management" at the core of decision-making because geotechnical engineers must cope with uncertainties arising from the variable nature of soil and rock, changeable environmental conditions, and imprecision in predicting field performance from models. Prevailing practice does not consider these uncertainties explicitly and as such, risk as recognized by the expanded Burland Triangle is managed informally through an ad hoc combination of strategies that include applying a global factor of safety (or partial factors of safety), selecting cautious input values and conservative calculation models, conducting parametric studies, learning from precedents, updating/validating designs and construction procedures based on prototype testing and observations, and keeping engineering judgment as an integral part of the decision-making loop. It is accurate to say that the value of applying probabilistic methods to manage risk more formally is not widely recognized.

The purpose of this chapter is to argue that probabilistic research conducted over the past several decades can complement numerical analysis to support increasingly complex decision-making in practice. Complexity is not related to the project details alone, but to system-level issues (e.g., resilience), to new design goals (e.g., sustainability), and to new loading conditions (e.g., climate change). It is not well appreciated that the effectiveness of experience in informing decision making in practice decreases with increasing novelty and rate of change of solutions. Currently, numerical analysis is conducted using deterministic inputs and simple soil profiles that are inconsistent with sparse site data. This deterministic mapping (one set of inputs to one set of outputs) does provide valuable physical insights, but cannot support risk-informed decision-making on its own without appealing to engineering judgment. Ironically, the usefulness of numerical analysis to decision-making and its role in digital transformation are diminished in the absence of explicit uncertainty quantification.

Statistical guidelines for modeling soil properties as random variables and random fields are available. The characterization of geologic uncertainties is more challenging than the characterization of geotechnical uncertainties mentioned above. However, the field of data-driven site characterization that addresses this "stratification" challenge among others is advancing rapidly using modern machine learning methods. Geotechnical models are biased and imprecise. It is possible to quantify the ratio of the measured response and the calculated response as a random variable called a model factor using extensive load test or other performance databases. The propagation of uncertainties from inputs to responses can be carried out more correctly than parametric studies (cannot address correlations between inputs, spatial variability, and statistical uncertainty) and more efficiently than a Monte Carlo simulation. For design, it is more important to calculate the probability of a response not satisfying a given performance criterion or a set of criteria rather than to calculate the probability of achieving a general response that is peripheral to the occurrence of one or more limit states. This is called the probability of failure.

Decision-making based on the probability of failure is more consistent than one based on the global factor of safety or partial factors of safety with respect to both mechanics and statistics. In addition, the probability of failure is sensitive to data (thus opening one potential pathway to digital transformation) and meaningful for both system and component failures. Resilience engineering requires system-level analysis. The impact of climate change requires explicit uncertainty quantification. Hence, to provide better decision support, geotechnical software should compute the probability of failure/reliability index as one basic output in addition to stresses, strains, forces, and displacements. However, for the engineer to understand why the probability of failure is high or low, the uncertainty of the explanatory variables/mechanisms should be provided in the form of a range of simulated outcomes. These simulated outcomes display a more complete and more consistent picture of "what if" design scenarios than prevailing parametric studies. Simplified reliability-based design approaches that can produce solutions meeting a target probability of failure for different degrees of understanding of the site and structures of different importance have been developed and are already adopted in new design codes.

This chapter further argues that probabilistic methods can lead to novel applications with the potential to lead to digital transformation (data-driven site characterization, physics-informed machine learning, inverse analysis, real-time autonomous control) and novel understanding of soil behaviors (non-classical failure mechanisms in spatially variable soils) outside the reach of deterministic methods. As probabilistic methods transition to machine learning methods that open new possibilities, the geotechnical data estate is increasingly being recognized as an asset in its own right. Uncertainty-informed decision-making is an integral part of this value.

ACKNOWLEDGMENTS

The author is grateful to Dr Yuanqin Tao for her kind editorial assistance. The author extends his appreciation to Professor Jianye Ching (Figure 1.5 and 1.7), Professor Chong Tang (Table 1.1), and Professor Yit-Jin Chen (Table 1.5) for their assistance to prepare some figures and tables. The author also gratefully acknowledges the contributions from Professor Zjiun Cao, Jianye Ching, Jian Ji, Yat Fai Leung, Shadi Najjar, Takayuki Shuku, Chong Tang, Zhen-Yu Yin, and Ikumasa Yoshida to Phoon et al. (2022c) that motivates the conceptualization of this chapter and this book.

REFERENCES

Akbas, S. O. 2007. *Deterministic and Probabilistic Assessment of Settlements of Shallow Foundations in Cohesionless Soils*. PhD thesis, Cornell University.

Aladejare, A. E. and Wang, Y. 2017. Evaluation of rock property variability. *Georisk: Assessment and Management of Risk for Engineered Systems and Geohazards*, 11(1), 22–41.

ASCE. 2019. *Future World Vision: Infrastructure Reimagined*. Reston, VA: ASCE.

Baecher, G. B. 2017. Bayesian thinking in geotechnics. In *Geo-Risk 2017*, edited by Jinsong Huang, Gordon A. Fenton, Limin Zhang, and D. V. Griffiths, 1–18. Reston, VA: ASCE.

Barbour, S. L. and Krahn, J. 2004. Numerical modelling–prediction or process? *Geotechnical News*, 22(4), 44–52.

Bilal, M., Oyedele, L. O., Qadir, J., Munir, K., Ajayi, S. O., Akinade, O. O., Owolabi, H. A., Alaka, H. A., and Pasha, M. 2016. Big Data in the construction industry: A review of present status, opportunities, and future trends. *Advanced Engineering Informatics*, 30(3), 500–521.

Briaud, J. L. and Gibbens, R. 1997. *Large-Scale Load Tests and Data Base of Spread Footings on Sand.* Rep. No. FHWA-RD-97-068. McLean, VA: Federal Highway Administration.

Broms, B. B., 1964. Lateral resistance of piles in cohesionless soils. *Journal of Soil Mechanics and Foundations Division*, 90(SM3), 123–156.

Burland, J. B. 1987. Nash lecture: The teaching of soil mechanics – A personal view. In *Proceedings of the 9th European Conference on Soil Mechanics and Foundation Engineering*, Vol. 3, 1427–1447. Rotterdam: A. A. Balkema.

Burland, J. B. 2012. *The geotechnical triangle. ICE Manual of Geotechnical Engineering: Geotechnical Engineering Principles, Problematic Soils and Site Investigation*, Eds. J. B. Burland, T. J. P. Chapman, H. Skinner, M. J Brown, Vol. 1, 17–26.

Cami, B., Javankhoshdel, S., Phoon, K. K., and Ching, J. 2020. Scale of fluctuation for spatially varying soils: Estimation methods and values. *ASCE-ASME Journal of Risk and Uncertainty in Engineering Systems, Part A: Civil Engineering*, 6(4), 03120002.

Carter, J. P., Desai, C. S., Potts, D. M., Schweiger, H. F., and Sloan, S. W. 2000. Computing and computer modeling in geotechnical engineering. In *Geo Eng2000*, Vol. 1, 1157–1252. Lancaster, PA: Technomic.

Casagrande, A. 1965. Role of the 'calculated risk' in earthwork and foundation engineering. *Journal of the Soil Mechanics and Foundations Division*, 91(4), 1–40.

CEN. 2002. *Eurocode–Basis of Structural Design.* EN 1990. Brussels, Belgium: European Committee for Standardization (CEN).

CEN. 2004. *Geotechnical Design—Part 1: General Rules.* EN 1997-1. Brussels, Belgium: European Committee for Standardization (CEN).

Chen, Y. J., Lin, S. S., Chang, H. W., and Marcos, M. C. 2011. Evaluation of side resistance capacity for drilled shafts. *Journal of Marine Science and Technology*, 19(2), 210–221.

Chen, Y. J., Lin, W. Y., Topacio, A., and Phoon, K. K. 2021. Evaluation of interpretation criteria for drilled shafts with tip post-grouting. *Soils and Foundations*, 61(5), 1354–1369.

Chen, Y. J., Phoon, K. K., Topacio, A., and Laveti, S. 2023. Uncertainty analysis for drilled shaft axial behavior using CYCU/DrilledShaft/143. *Soils and Foundations*, 63(4), 101337.

Ching, J. 2020. Unpublished databases.

Ching, J., Li, D. Q., and Phoon, K. K. 2016. Chapter 4. Statistical characterization of multivariate geotechnical data. In *Reliability of Geotechnical Structures in ISO2394*, edited by K. K. Phoon and J. V. Retief, 89–126. Leiden: CRC Press/Balkema.

Ching, J., Li, K. H., Phoon, K. K., and Weng, M. C. 2018. Generic transformation models for some intact rock properties. *Canadian Geotechnical Journal*, 55(12), 1702–1741.

Ching, J., Lin, G. H., Chen, J. R., and Phoon, K. K. 2017. Transformation models for effective friction angle and relative density calibrated based on a multivariate database of coarse-grained soils. *Canadian Geotechnical Journal*, 54(4), 481–501.

Ching, J. and Phoon, K. K. 2012a. Value of geotechnical site investigation in reliability-based design. *Advances in Structural Engineering*, 15(11), 1935–1945.

Ching, J. and Phoon, K. K. 2012b. Modeling parameters of structured clays as a multivariate normal distribution. *Canadian Geotechnical Journal*, 49(5), 522–545.

Ching, J. and Phoon, K. K. 2013. Multivariate distribution for undrained shear strengths under various test procedures. *Canadian Geotechnical Journal*, 50(9), 907–923.

Ching, J. and Phoon, K. K. 2014. Transformations and correlations among some clay parameters – The global database. *Canadian Geotechnical Journal*, 51(6), 663–685.

Ching, J. and Phoon, K. K. 2017. Characterizing uncertain site-specific trend function by sparse Bayesian learning. *Journal of Engineering Mechanics, ASCE*, 143(7), 04017028.

Ching, J. and Phoon, K. K. 2019. Constructing site-specific multivariate probability distribution model using Bayesian machine learning. *Journal of Engineering Mechanics, ASCE*, 145(1), 04018126.

Ching, J. and Phoon, K. K. 2020. Measuring similarity between site-specific data and records from other sites. *ASCE-ASME Journal of Risk and Uncertainty in Engineering Systems, Part A: Civil Engineering*, 6(2), 04020011.

Ching, J., Phoon, K. K., and Chen, C. H. 2014b. Modeling CPTU parameters of clays as a multivariate normal distribution. *Canadian Geotechnical Journal*, 51(1), 77–91.

Ching, J., Phoon, K. K., Chen, K. F., Orr, T. L. L., and Schneider, H. R. 2020. Statistical determination of multivariate characteristic values for Eurocode 7. *Structural Safety*, 82, 101893.

Ching, J., Phoon, K. K., Ho, Y. H., and Weng, M. C. 2021a. Quasi-site-specific prediction for deformation modulus of rock mass. *Canadian Geotechnical Journal*, 58(7), 936–951.

Ching, J., Phoon, K. K., and Hu, Y. G. 2009. Efficient evaluation of reliability for slopes with circular slip surfaces using importance sampling. *Journal of Geotechnical and Geoenvironmental Engineering, ASCE*, 135(6), 768–777.

Ching, J., Phoon, K. K., and Wu, C. T. 2022. Data-centric quasi-site-specific prediction for compressibility of clays. *Canadian Geotechnical Journal*, 59(12), 2033–2049.

Ching, J., Phoon, K. K., and Yu, J. W. 2014a. Linking site investigation efforts to final design savings with simplified reliability-based design methods. *Journal of Geotechnical and Geoenvironmental Engineering, ASCE*, 140(3), 04013032.

Ching, J., Uzielli, M., Phoon, K. K., and Xu, X. J. 2023. Characterization of autocovariance parameters of detrended cone tip resistance from a global CPT database. *Journal of Geotechnical and Geoenvironmental Engineering, ASCE*, 149(10), 04023090.

Ching, J., Wu, S., and Phoon, K. K. 2021b. Constructing quasi-site-specific multivariate probability distribution using hierarchical Bayesian model. *Journal of Engineering Mechanics, ASCE*, 147(10), 04021069.

Chou, S. A., Chen, Y. J., Chiou, J. S., Topacio, A., and Marcos, M. C. 2022. Evaluation of lateral capacity for flexible drilled shafts in cohesionless soils. *Science Progress*, 105, 1–30.

Chwała, M., Phoon, K. K., Uzielli, M., Zhang, J., Zhang, L., and Ching, J. 2023. Time capsule for geotechnical risk and reliability. In *Georisk: Assessment and Management of Risk for Engineered Systems and Geohazards*, 17(3), 439–466..

Constructed Facilities Division. 1975a. *Proceeding of the Foundation Deformation Prediction Symposium, Vol. 1: Symposium Summary*. Rep. No. FHWA-RD-75-515. Wellesley Hill, MA: Massachusetts Institute of Technology.

Constructed Facilities Division. 1975b. *Proceeding of the Foundation Deformation Prediction Symposium, Vol. 2: Appendix*. Rep. No. FHWA-RD-75-516. Wellesley Hill, MA: Massachusetts Institute of Technology.

Curran, J. H. and Hammah, R. E. 2006. Seven lessons of geomechanics software development. In *Proceedings of the 41st U.S. Symposium on Rock Mechanics (USRMS): "50 Years of Rock Mechanics – Landmarks and Future Challenges"*. Golden, CO: Red Hook. https://www.rocscience.com/assets/resources/learning/papers/Seven-Lessons-of-Geomechanics-Software-Development.pdf

DiMillio, A. F., Ng, E. S., Briaud, J. L., and O'Neill, M. W. 1988. *Pile Group Prediction Symposium: Summary, Vol. I: Sandy Soil*. Rep. No. FHWA-TS-87-221. McLean, VA: Federal Highway Administration.

Doherty, J. P., Gourvenec, S., and Gaone, F. M. 2018. Insights from a shallow foundation load-settlement prediction exercise. *Computers and Geotechnics*, 93, 269–279.

Duncan, J. M. 2000. Factors of safety and reliability in geotechnical engineering. *Journal of Geotechnical and Geoenvironmental Engineering, ASCE*, 126(4), 307–316.

Duncan, J. M. and Sleep, M. D. 2017. The need for judgement in geotechnical reliability studies. *Georisk: Assessment and Management of Risk for Engineered Systems and Geohazards*, 11(1), 70–74.

D'Ignazio, M., Phoon, K. K., Tan, S. A., and Lansivaara, T. 2016. Correlations for undrained shear strength of Finnish soft clays. *Canadian Geotechnical Journal*, 53(10), 1628–1645.

Eslami, A., Golafzani, S. H., and Naghibi, M. H. 2023. Developed triangular charts; deltaic CPTu-based soil behavior classification using AUT: CPTu-Geo-Marine Database. *Probabilistic Engineering Mechanics*, 71, 103380.

Fellenius, B. H. 1988. Variation of CAPWAP results as a function of the operator. In *Proceedings of the 3rd International Conference on the Application of Stress-Wave Theory to Piles*, 814–825. Vancouver: BiTech.

Fellenius, B. H. 2013. Capacity and load-movement of a CFA pile: A prediction event. In *Proceedings of the Geo-Congress 2013–Foundation Engineering in the Face of Uncertainty (Geotechnical Special Publication 229)*, 707–719. Reston, VA: ASCE.

Fellenius, B. H. 2015. Field test and predictions. In *Proceedings of the 2nd Bolivian International Conference on Deep Foundations*. Santa Cruz, Bolivia.

Fellenius, B. H. 2017. Report on the B.E.S.T. prediction survey of the 3rd CBFP event. In *Proceedings of the 3rd Bolivian International Conference on Deep Foundations*, Vol. 3, 7–25. Madison, WI: Omnipress.

Fellenius, B. H. 2021. Results of an instrumented static loading test. Application to design and compilation of an international survey. *Journal of Deep Foundations Institute*, 15(1), 71–87.

Fellenius, B. H., Edvardsson, F., Pettersson, J., Sabattini, M., and Wallgren, J. 2019. Prediction, testing, and analysis of a 50 m long pile in soft marine clay. *Journal of Deep Foundations Institute*, 13(2), 1–7.

Fellenius, B. H., Hussein, M., Mayne, P., and McGillivray, R. T. 2004. Murphy's law and the pile prediction at the 2002 ASCE Geo Institute's Deep Foundations Conference. In *Proceedings of the Deep Foundations Institute Meeting on Current Practice and Future Trends in Deep Foundations*, 29–43. Reston, VA: ASCE.

Fellenius, B. H. and Terceros, H. 2014. Response to load for four different bored piles. In *Proceedings of the DFI-EFFC International Conference on Piling and Deep Foundations*, 99–120. Hawthorne, NJ: Deep Foundations Institute.

Feng, S. and Vardanega, P. J. 2019. A database of saturated hydraulic conductivity of fine-grained soils: Probability density functions. *Georisk: Assessment and Management of Risk for Engineered Systems and Geohazards*, 13(4), 255–261.

Finno, R. J. 1989. *Predicted and Observed Axial Behavior of Piles: Results of a Pile Prediction Symposium.* Geotechnical Special Publication No. 23. New York: ASCE.

Finno, R. J. and Calvello, M. 2005. Supported excavations: Observational method and inverse modeling. *Journal of Geotechnical and Geoenvironmental Engineering, ASCE*, 131(7), 826–836.

Focht, J. A. 1994. Lessons learned from missed predictions. *Journal of Geotechnical Engineering, ASCE*, 120(10), 1653–1683.

Gerbert, P., Castagnino, S., Rothballer, C., Renz, A., and Filitz, R. 2016. *Digital in Engineering and Construction.* Boston, MA: Boston Consulting Group.

Guevara, M., Doherty, J., Gaudin, C., and Watson, P. 2022. Evaluating uncertainty associated with engineering judgement in predicting the lateral response of conductors. *Journal of Geotechnical and Geoenvironmental Engineering, ASCE*, 148(5), 05022001.

Hammah, R. E. and Curran, J. H. 2009. It is better to be approximately right than precisely wrong: Why simple models work in mining geomechanics. In *Proceedings of the 43rd US Rock Mechanics Symposium and 4th US Canada Rock Mechanics Symposium*. Asheville, NC: American Rock Mechanics Association (ARMA). https://www.rocscience.com/assets/resources/learning/papers/It-is-Better-to-be-Approximately-Right-than-Precisely-Wrong-Why-Simple-Models-Work-in-Mining-Geomechanics.pdf

Hansen, J. B. 1961. Ultimate resistance of rigid piles against transversal forces. *Bulletin 12 Danish Geotechnical Institute*, 5–9.

Herle, I. 2003. Numerical predictions and reality. *Advanced Mathematical and Computational Geomechanics.* Lecture Notes in Applied and Computational Mechanics, Vol. 13, 167–194. Berlin and Heidelberg: Springer.

Heyman, J. 1996. Hambly's paradox: Why design calculations do not reflect real behaviour. *Proceedings of the Institution of Civil Engineers - Civil Engineering*, 114(4), 161–166.

Holeyman, A. and Charue, N. 2003. International pile capacity prediction event at Limelette. In *Proceedings of the 2nd Symposium on Screw Piles*, 215–234. Lisse: Swets & Zeitlinger Publishers.

Holeyman, A., Couvreur, J. M., and Charue, N. 2001. Results of dynamic and kinetic pile load tests and outcome of an international prediction event. In *Proceedings of the Symposium on Screw Piles*, 247–273. Lisse: Swets & Zeitlinger Publishers.

Indraratna, B., Baral, P., Rujikiatkamjorn, C., and Perera, D. 2018. Class A and C predictions for Ballina trial embankment with vertical drains using standard test data from industry and large diameter test specimens. *Computers and Geotechnics*, 93, 232–246.

Ismail, S., Najjar, S. S., and Sadek, S. 2018. Reliability analysis of buried offshore pipelines in sand subjected to upheaval buckling. In *Proceedings of the Offshore Technology Conference (OTC).* Houston, TX: American Petroleum Institute. DOI: https://doi.org/10.4043/28882-MS

Jardine, R. J., Standing, J. R., Jardine, F. M., Bond, A. J., and Parker, E. 2001. A competition to assess the reliability of pile prediction methods. In *Proceedings of the 15 International Conference on Soil Mechanics and Foundation Engineering*, pp. 911–914. Rotterdam: A. A. Balkema.

Kalenchuk, K. S. 2022. 2019 Canadian Geotechnical Colloquium: Mitigating a fatal flaw in modern geomechanics: Understanding uncertainty, applying model calibration, and defying the hubris in numerical modelling. *Canadian Geotechnical Journal*, 59(3), 315–329.

Kelly, R. B., Sloan, S. W., Pineda, J. A., Kouretzis, G., and Huang, J. 2018. Outcomes of the Newcastle symposium for the prediction of embankment behaviour on soft soil. *Computers and Geotechnics*, 93, 9–41.

Kim, E. and Hunt, R. 2017. A public website of rock mechanics database from Earth Mechanics Institute (EMI) at Colorado School of Mines (CSM). *Rock Mechanics and Rock Engineering*, 50(12), 3245–3252.

Knabe, T., Schweiger, H. F., and Schanz, T. 2012. Calibration of constitutive parameters by inverse analysis for a geotechnical boundary problem. *Canadian Geotechnical Journal*, 49(2), 170–183.

Kootahi, K. and Moradi, G. 2017. Evaluation of compression index of marine fine-grained soils by the use of index tests. *Marine Georesources and Geotechnology*, 35(4), 548–570.

Lacasse, S. 2015. *55th Rankine Lecture: Hazard, Risk and Reliability in Geotechnical Practice.* London: Institution of Civil Engineers.

Lacasse, S. and Lunne, T. 1982. Penetration tests in two Norwegian clays. In *Proceedings of the 2nd European Symposium on Penetration Testing*, 661–670. Leiden: CRC Press/Balkema.

Lambe, T. W. 1973. Predictions in soil engineering. *Géotechnique*, 23(2), 151–202.

Länsivaara, T., Phoon, K. K., and Ching, J. 2022. What is a characteristic value for soils? *Georisk: Assessment and Management of Risk for Engineered Systems and Geohazards*, 16(2), 199–224.

Liu, S., Zou, H., Cai, G., Bheemasetti, T. V., Puppala, A. J., and Lin, J. 2016. Multivariate correlation among resilient modulus and cone penetration test parameters of cohesive subgrade soils. *Engineering Geology*, 209, 128–142.

Lumb, P. 1966. The Variability of natural soils. *Canadian Geotechnical Journal*, 3(2), 74–97.

Lo, M. K., Wei, X., Chian, S. C., and Ku, T. 2021. Bayesian network prediction of stiffness and shear strength of sand. *Journal of Geotechnical and Geoenvironmental Engineering, ASCE*, 147(5), 04021020.

Marcos, M. C., Chen, Y. J., and Kulhawy, F.H. 2013. Evaluation of compression load test interpretation criteria for driven precast concrete pile capacity. *KSCE Journal of Civil Engineering*, 17 (5), 1008–1022.

Marr, W. A. 2006. Geotechnical engineering and judgment in the information age. In *Geo-Congress 2006: Geotechnical Engineering in the Information Technology Age*, pp. 1–17. Reston, VA: ASCE.

Mesri, G. and Huvaj, N. 2007. Shear strength mobilized in undrained failure of soft clay and silt deposits. In *Advances in Measurement and Modeling of Soil Behavior (GSP 173)*, 1–22. Reston, VA: ASCE.

Mitchell, J. K. and Kopmann, J. 2013. *The Future of Geotechnical Engineering*. Report CGPR #70. Blacksburg, VA: Virginia Polytechnic Institute and State University.

Munawar, H. S., Ullah, F., Qayyum, S. and Shahzad, D. 2022. Big data in construction: Current applications and future opportunities. *Big Data and Cognitive Computing*, 6, 18.

Muzamhindo, H. and Ferentinou, M. 2023. Generic compressive strength prediction model applicable to multiple lithologies based on a broad global database. *Probabilistic Engineering Mechanics*, 71, 103400.

National Research Council. 1995. *Probabilistic Methods in Geotechnical Engineering*. Washington, DC: The National Academies Press.

National Research Council. 2006. *Geological and Geotechnical Engineering in the New Millennium – Opportunities for Research and Technological Innovation*. Washington, DC: The National Academies Press.

Paikowsky, S. G., Canniff, M. C., Lesny, K., Kisse, A., Amatya, S., and Muganga, R. 2010. *LRFD Design and Construction of Shallow Foundations for Highway Bridge Structures*. NCHRP Report No. 651. Washington, DC: National Academy of Sciences.

Peck, R. B., Hanson, W. E., and Thornburn, T. H. 1953. *Foundation Engineering*. New York: John Wiley and Sons.

Phoon, K. K. 2017. Role of reliability calculations in geotechnical design. *Georisk: Assessment and Management of Risk for Engineered Systems and Geohazards*, 11(1), 4–21.

Phoon, K. K. 2018. Editorial for special collection on probabilistic site characterization. *ASCE-ASME Journal of Risk and Uncertainty in Engineering Systems, Part A: Civil Engineering*, 4(4), 02018002.

Phoon, K. K. 2020. The story of statistics in geotechnical engineering. *Georisk: Assessment and Management of Risk for Engineered Systems and Geohazards*, 14(1), 3–25.

Phoon, K. K. 2023. What geotechnical engineers want to know about reliability. *ASCE-ASME Journal of Risk and Uncertainty in Engineering Systems, Part A: Civil Engineering*, 9(2), 03123001.

Phoon, K. K., Cao, Z., Ji, J., Leung, Y. F., Najjar, S., Shuku, T., Tang, C., Yin, Z. Y., Yoshida, I., and Ching, J. 2022c. Geotechnical uncertainty, modeling, and decision making. *Soils and Foundations*, 62(5), 101189.

Phoon, K. K. and Ching, J. 2017. Better correlations for geotechnical design. In *A Decade of Geotechnical Advances 2008–2017*, 73–102. Singapore: Geotechnical Society of Singapore.

Phoon, K. K. and Ching, J. 2021. Project DeepGeo - Data-driven 3D subsurface mapping. *Journal of GeoEngineering*, 16(2), 61–74.

Phoon, K. K. and Ching, J. 2022. Additional observations on the site recognition challenge. *Journal of GeoEngineering*, 17(4), 231–247.

Phoon, K. K., Chen, J. R., and Kulhawy, F. H. 2006. Characterization of model uncertainties for augered cast-in-place (ACIP) piles under axial compression. In *Foundation Analysis and Design: Innovative Methods (GSP 153)*, 82–89. Reston, VA: ASCE.

Phoon, K. K., Chen, J. R., and Kulhawy, F. H. 2007. Probabilistic hyperbolic models for foundation uplift movement. In *Probabilistic Applications in Geotechnical Engineering (GSP 170)*, CD-ROM, edited by K. K. Phoon, G. A. Fenton, E. F. Glynn, C. H. Juang, D. V. Griffiths, T. F. Wolff , and L. M. Zhang, pp. 1–12. Reston, VA: ASCE.

Phoon, K. K., Ching, J., and Cao, Z. 2022a. Unpacking data-centric geotechnics. *Underground Space*, 7(6), 967–989.

Phoon, K. K., Ching, J., and Shuku, T. 2022b. Challenges in data-driven site characterization. *Georisk: Assessment and Management of Risk for Engineered Systems and Geohazards*, 16(1), 114–126.

Phoon, K. K., Ching, J., and Tao, Y. 2023. Chapter 2. Soil and rock parametric uncertainties. In *Uncertainty, Modelling, and Decision Making in Geotechnics*. Boca Raton, FL: CRC Press.

Phoon, K. K., Ching, J., and Wang, Y. 2019. Managing risk in geotechnical engineering – from data to digitalization. In *Proceedings of the 7th International Symposium on Geotechnical Safety and Risk (ISGSR 2019)*, 13–34. Singapore: Research Publishing.

Phoon, K. K. and Fenton, G. A. 2004. Estimating sample autocorrelation functions using bootstrap. In *Proceedings of the 9th ASCE Specialty Conference on Probabilistic Mechanics and Structural Reliability*. Reston, VA: ASCE.

Phoon, K. K. and Kulhawy, F. H. 1999a. Characterization of geotechnical variability. *Canadian Geotechnical Journal*, 36(4), 612–624.

Phoon, K. K. and Kulhawy, F. H. 1999b. Evaluation of geotechnical property variability. *Canadian Geotechnical Journal*, 36(4), 625–639.

Phoon, K. K. and Kulhawy, F. H. 2005. Characterisation of model uncertainties for laterally loaded rigid drilled shafts. *Géotechnique*, 55(1), 45–54.

Phoon, K. K., Retief, J. V., Ching, J., Dithinde, M., Schweckendiek, T., Wang, Y., and Zhang, L. M. 2016. Some observations on ISO2394:2015 Annex D (reliability of geotechnical structures). *Structural Safety*, 62, 24–33.

Phoon, K. K., Santoso, A., and Quek, S. T. 2010. Probabilistic analysis of soil water characteristic curves. *Journal of Geotechnical and Geoenvironmental Engineering, ASCE*, 136(3), 445–455.

Phoon, K. K. and Tang, C. 2019a. Characterization of geotechnical model uncertainty. *Georisk: Assessment and Management of Risk for Engineered Systems and Geohazards*, 13(2), 101–130.

Phoon, K. K. and Tang, C. 2019b. Effect of extrapolation on interpreted capacity and model statistics of steel H-piles. *Georisk: Assessment and Management of Risk for Engineered Systems and Geohazards*, 13(4), 291–302.

Phoon, K. K. and Tang, C. 2024. *Database Approach for Data-Centric Geotechnics, Vol. 1: Site Characterization and Vol. 2: Geotechnical Structures*. Boca Raton, FL: CRC Press.

Phoon, K. K. and Zhang, W. 2023. Future of machine learning in geotechnics. *Georisk: Assessment and Management of Risk for Engineered Systems and Geohazards*, 17(1), 7–22.

Poulos, H. G. 2004. Success and failure in predicting pile performance. In *Proceedings of the Fifth International Conference on Case Histories in Geotechnical Engineering*. Paper No. SOAP 4. New York.

Prakoso, W. A. 2002. Reliability-based design of foundations on rock for transmission line and similar structure. PhD thesis, Cornell University.

Prakoso, W. A. 2017. Personal communication.

Reese, L. C., Cox, W. R., and Coop, F. D. 1974. Analysis of laterally loaded piles in sand. *Proceedings of the Sixth Offshore Technology Conference*, Vol. 2. Houston, 473–483.

Reiffsteck, P. 2005. Portance et tassements d'une fondation profonde - Présentation des résultats du concours de prévision. In *Proceedings of the Symp. Int ISP5/PRESSIO 2005: 50 ans de pressiomètres*, Vol. 2, 521–535. Paris: Presses de l'ENPC/LCPC.

Reiffsteck, P. 2009. ISP5 pile prediction revisited. *Proceedings of the Contemporary Topics in In Situ Testing, Analysis, and Reliability of Foundations*. Geotechnical Special Publication, Vol. 186, 50–57. Reston, VA: ASCE.

Rogers, J. D. 2008. A historical perspective on geotechnical case histories courses. In *Proceedings of the Sixth International Conference on Case Histories in Geotechnical Engineering*, pp. 1–18. Arlington, VA.

Salgado, R., Houlsby, G. T., and Cathie, D. N. 2008. Contributions to Géotechnique 1948–2008: Foundation engineering. *Géotechnique*, 58(5), 369–375.

Samtani, N. C. and Allen, T. M. 2018. Expanded database for service limit state calibration of immediate settlement of bridge foundations on soil. Report No. FHWA-HIF-18-008. Washington, DC: Federal Highway Administration.

Simpson, B. 2011. Reliability in geotechnical design – Some fundamentals. In *Proceedings of the Third International Symposium on Geotechnical Safety and Risk*, 393–399. Karlsruhe: Federal Waterways Engineering and Research Institute.

Simpson, B., Pappin, J. W., and Croft, D. D. 1981. An approach to limit state calculations in geotechnics. *Ground Engineering*, 14(6), 21–28.

Simpson, E. H. 1951. The interpretation of interaction in contingency tables. *Journal of the Royal Statistical Society, Series B*, 13(2), 238–241.

Starfield, A. M. and Cundall, P. A. 1988. Towards a methodology for rock mechanics modelling. *International Journal of Rock Mechanics and Mining Sciences and Geomechanics Abstracts*, 25(3), 99–106.

Strahler, A. W. and Stuedlein, A. W. 2014. Accuracy, uncertainty, and reliability of the bearing capacity equation for shallow foundations on saturated clay. In *GeoCongress 2014: Geo-Characterization and Modeling for Sustainability (GSP 234)*, 3262–3273. Reston, VA: ASCE.

Stuyts, B., Cathie, D., and Powell, T. 2016. Model uncertainty in uplift resistance calculations for sandy backfills. *Canadian Geotechnical Journal*, 53(11), 1831–1840.

Tang, C. and Phoon, K. K. 2017. Model uncertainty of Eurocode 7 approach for bearing capacity of circular footings on dense sand. *International Journal of Geomechanics*, 17(3), 04016069.

Tang, C. and Phoon, K. K. 2018a. Evaluation of model uncertainties in reliability-based design of steel H-piles in axial compression. *Canadian Geotechnical Journal*, 55(11), 1513–1532.

Tang, C. and Phoon, K. K. 2018b. Statistics of model factors in reliability-based design of axially loaded driven piles in sand. *Canadian Geotechnical Journal*, 55(11), 1592–1610.

Tang, C. and Phoon, K. K. 2018c. Statistics of model factors and consideration in reliability-based design of axially loaded helical piles. *Journal of Geotechnical and Geoenvironmental Engineering, ASCE*, 144(8), 04018050.

Tang, C. and Phoon, K. K. 2019a. Evaluation of stress-dependent methods for the punch-through capacity of foundations in clay with sand. *ASCE-ASME Journal of Risk and Uncertainty in Engineering Systems, Part A: Civil Engineering*, 5(3), 04019008.

Tang, C. and Phoon, K. K. 2019b. Characterization of model uncertainty in predicting axial resistance of piles driven into clay. *Canadian Geotechnical Journal*, 56(8), 1098–1118.

Tang, C. and Phoon, K. K. 2020. Statistical evaluation of model factors in reliability calibration of high-displacement helical piles under axial loading. *Canadian Geotechnical Journal*, 57(2), 246–262.

Tang, C. and Phoon, K. K. 2021. *Model Uncertainties in Foundation Design*. Boca Raton, FL: CRC Press.

Tang, C., Phoon, K. K., and Chen, Y. J. 2019. Statistical analyses of model factors in reliability-based limit state design of drilled shafts under axial loading. *Journal of Geotechnical and Geoenvironmental Engineering, ASCE*, 145(9), 05019042.

Tang, C., Phoon, K. K., Li, D. Q., and Akbas, S. O. 2020. Expanded database assessment of design methods for spread foundations under axial compression and uplift loading. *Journal of Geotechnical and Geoenvironmental Engineering, ASCE*, 146(11), 04020119.

Tang, C., Phoon, K. K., Zhang, L., and Li, D. Q. 2017. Model uncertainty for predicting the bearing capacity of sand overlying clay. *International Journal of Geomechanics*, 17(7), 04017015.

Tang, C., Phoon, K. K., Tao, Y., Feng, X., and Sun, H. 2023b. Development and use of databases of geotechnical information for data-centric geotechnics. *Computers and Geotechnics*, under review.

Tang, C., Yuan, J., Phoon, K. K., Feng, X., and Yu, X. 2023a. Variability of predictions in punch-through of spudcan penetration. *Computers and Geotechnics*, under review.

Terzaghi, K. 1943. *Theoretical Soil Mechanics*. New York: John Wiley and Sons.

Terzaghi, K. and Peck, R. B. 1948. *Soil Mechanics in Engineering Practice*. New York: John Wiley and Sons.

Topacio, A., Chen, Y. J., Phoon, K. K., and Tang, C. 2022. Evaluation of compression interpretation criteria for drilled shafts socketed into rocks. *Proceedings of the Institution of Civil Engineers – Geotechnical Engineering*, 1–17. https://doi.org/10.1680/jgeen.21.00120.

Tukey, J. W. 1962. The future of data analysis. *Annals of Mathematical Statistics*, 33(1), 1–67.

Turkstra, C. J. and Madsen, H. O. 1980. Load combinations in codified structural design. *Journal of the Structural Division, ASCE*, 106(12), 2527–2543.

van Dijk, B. F. J. and Yetginer, A. G. 2015. Findings of the ISSMGE jack-up leg penetration prediction event. In *Proceedings of the 3rd International Symposium on Frontiers in Offshore Geotechnics (ISFOG 2015): Frontiers in Offshore Geotechnics III*, Vol. 1, 1267–1274. London: Taylor & Francis.

Viana da Fonseca, A. and Santos, J. A. 2008. Behaviour of bored, CFA and driven piles in residual soil. In *International Prediction Event-Experimental Site-ISC'2*. Lisboa: University of Porto.

Walton, G. and Sinha, S. 2022. Challenges associated with numerical back analysis in rock mechanics. *Journal of Rock Mechanics and Geotechnical Engineering*, 14(6), 2058–2071.

Xu, S. J., Yi, J. T., Zhang, T. B., Wang, Z., and Yao, K. 2021. Characterising the model uncertainty of ISO methods for punch-through capacity prediction. *Proceedings of the Institution of Civil Engineers – Geotechnical Engineering*, 174(5), 549–562.

Yule, G. U. 1903. Notes on the Theory of association of attributes in statistics. *Biometrika*, 2(2), 121–134.

Zhang, D. M., Phoon, K. K., Huang, H. W., and Hu, Q. F. 2015. Characterization of model uncertainty for cantilever deflections in undrained clay. *Journal of Geotechnical and Geoenvironmental Engineering, ASCE*, 141(1), 04014088.

Zhang, D. M., Zhou, Y., Phoon, K. K., and Huang, H. W. 2020. Multivariate probability distribution of Shanghai clay properties. *Engineering Geology*, 273, 105675.

Zou, H., Liu, S., Cai, G., Puppala, A. J., and Bheemasetti, T. V. 2017. Multivariate correlation analysis of seismic piezocone penetration (SCPTU) parameters and design properties of Jiangsu quaternary cohesive soils. *Engineering Geology*, 228, 11–38.

Part 1

Statistical characterization

2 Soil and Rock Parametric Uncertainties

Kok-Kwang Phoon, Jianye Ching, and Yuanqin Tao

ABSTRACT

The purpose of Chapter 2 is to present practical tools and guidelines for the uncertainty quantification of soil and rock properties with a primary focus on enhancing decision-making in geotechnical design. The context underlying uncertainty quantification is emphasized in this chapter. A conceptual framework is presented to guide the engineer to consider uncertainty quantification in both the *absence* and the *presence* of a geotechnical structure (limit state) separately. This will provide conceptual clarity on the distinctive role of statistics (spatial variability), physics, and their interactions, which is somewhat lacking at present. The conceptual framework contains three building blocks: (1) point value, (2) spatial average, and (3) mobilized value. The first two blocks do not involve a geotechnical structure. The point value is the most basic building block. It refers to a value measured by one laboratory/field test (univariate case) at a given location/depth. It can also refer to a set of values from multiple tests conducted in close proximity (multivariate case). The classical model to describe spatial variability is to assume that two point values are independent and identically distributed, regardless of their measurement locations. Under this classical model, the engineer needs to characterize a single (univariate or multivariate) probability distribution function. The typical uncertainty quantification approach for the univariate case is to select a lognormal distribution and determine its mean and coefficient of variation (COV) based on site data. Extensive COV guidelines are provided in this chapter for clays, sands, rocks, and rock masses. These guidelines are immediately useful for the selection of a characteristic value based on the 5% fractile, first-order second-moment reliability analysis, and reliability-based design. The second building block is the spatial average. It is defined as the value affecting the occurrence of a slip along a *prescribed* trial line. To apply the spatial average, it may be sufficient to quantify a scale of fluctuation (roughly the characteristic length between strongly correlated measurements) in addition to the COV. The spatial correlation between two point values is not zero in this random field model. Extensive guidelines on the scale of fluctuation are provided in this chapter. The scale of fluctuation is typically measured in the vertical direction. Actual spatially varying soils are usually anisotropic in the sense that the vertical and horizontal scales of fluctuation are different. The latter is roughly 10 to 20 times longer than the former. Very little is known about the spatial variability of cement-admixed soils at present. When a limit state is governed by a spatial average, its uncertainty reduces with the characteristic length of the geotechnical structure. This important effect can be quantified using Vanmarcke's variance reduction function (VRF), which is available in simple analytical forms for the classical one-parameter autocorrelation models. This chapter provides a more general VRF (approximate) that is applicable to classical and non-classical Cosine Whittle–Matérn (CosWM) autocorrelation models. The CosWM model can address a broader class of spatial variability that exhibit roughness and pseudo-periodicity in addition to a scale of fluctuation. The new VRF is also applicable to inclined slip lines. The random field theory underlying the spatial average generalizes easily to a vector random field (at least theoretically) in the presence of multivariate data. The correlation between two soil properties produced by two different tests (called a cross correlation)

DOI: 10.1201/9781003333586-3

is distinct from spatial correlation which is the correlation between the same property measured at different locations. The most familiar application of cross correlation is the transformation model that relates a field or laboratory measurement to a design parameter. This chapter illustrates the reduction in the uncertainty of one or more design parameters when information in one or more tests is available using useful closed-form solutions from a conditional multivariate normal distribution. The key task for an engineer is to assemble a correlation matrix from the cross correlations between all pairs of soil parameters. It is not sufficiently emphasized that the matrix must be positive definite to be valid. Although classical transformation models constructed in this manner can be multivariate, they are unable to address the full set of MUSIC-3X (Multivariate, Uncertain and Unique, Sparse, Incomplete, potentially Corrupted, 3D spatially variations) attributes commonly found in geotechnical databases. The third building block is the mobilized value. It is closest to the "value affecting the occurrence of the limit state" in Clause 2.4.5.2(2) of Eurocode 7. The characteristic value is defined as a cautious estimate of this value. Note that the mobilized value used in this chapter refers to the probability distribution of this value. It is not the same as the spatial average. The reason is that the spatial average is defined over a prescribed trial surface that is not the critical surface affecting the occurrence of a limit state. The reliability-based characteristic value is fully compatible with physics and statistics. However, it is very costly to compute because it requires the random finite element method. An approximation called a mobilization-based characteristic value may be more applicable to practice. It has the same "look and feel" as the current spatial average-based characteristic and it is only slightly more costly. Extensive statistical guidelines are made available at: www.routledge.com/Uncertainty-Modelling-and-Decision-Making-in-Geotechnics/Phoon-Shuku-Ching/p/book/9781032367491.

2.1 INTRODUCTION

The most common question raised by geotechnical engineers is whether there is sufficient data for reliability applications (Phoon 2023). The answer depends on the application. For load and resistance factor design (LRFD) calibration, load test or other performance databases are needed to evaluate the bias (mean) and dispersion (coefficient of variation [COV]) of the model factor (Tang and Phoon 2021; Tang and Bathurst 2021; Phoon and Tang 2024). The model factor is defined as the ratio between the measured response and the predicted response. The characterization of the model factor is presented in Chapter 5. The short answer on whether there is sufficient data to support LRFD calibration is "yes" for foundations and "maybe" for other geotechnical structures. For first-order second-moment (FOSM) reliability analysis, second-moment statistics of the soil or rock parameters are needed. When it is implemented using an empirical "N-sigma rule", the standard deviation can be estimated as (HCV-LCV)/4, where HCV is the highest conceivable value and LCV is the lowest conceivable value (Duncan 2000). Only engineering judgment is needed, although it can be supplemented by a significant amount of published data on the coefficient of variation (Phoon and Kulhawy 1999a, 1999b; Prakoso 2002; Aladejare and Wang 2017; Phoon et al. 2016a; Tang and Phoon 2021; Guan et al. 2021; Pan et al. 2018; Cami et al. 2020; Phoon and Tang 2024). Details are given in Section 2.3. For first-order reliability analysis and random finite element analysis, a spatially variable vector random field may be needed. This will require the characterization of (1) marginal probability distributions of soil parameters at a point (typically non-normal) (Ching and Phoon 2015), (2) cross-correlations between different soil parameters (Zhou et al. 2021), and (3) spatial correlations between a soil parameter measured at different locations and depths (typically anisotropic in the sense that the spatial correlation length in the vertical direction is much shorter than the horizontal one) (Stuedlein et al. 2021). No practical characterization method existed until recently as a result of advancements in data-driven site characterization (DDSC; Phoon and Ching 2021, 2022; Phoon et al. 2022a, 2022b). There is ample scope and potential for data-driven site characterization to add significant or even transformative value to practice as shown in Table 2.1. Recent developments in DDSC are covered in Section 2.8.

TABLE 2.1
Tentative Literature Scan of Data-Driven Site Characterization Research (DDSC): What We Have and What Are the Gaps

State of DDSC research	Some literature	Limited literature	None?
Geomaterials	Soil	Rock	Improved ground
Data types	Geotechnical investigation	Geophysical	Emerging: wireless sensor network, laser scanning, drone photogrammetry, satellite imagery, etc.
Ugly data challenge	MUSIC-3X (multivariate, uncertain and unique, sparse, incomplete, potentially corrupted, 3D spatially varying)	Truncated/censored data, mixture of numerical and categorical data, image/video data, expert opinion, etc.	Labeled data, synthetic data, anonymized data, bad/ corrupted/low-quality data that diminishes rather than enhances decision support
Explainable site recognition challenge	Site recognition using geotechnical investigation data	Explainable site recognition using geotechnical investigation and other data (e.g., proof/load tests, monitoring, and remote sensing)	Explainable site recognition that learns from "human-in-the-loop"
Stratification/ mapping challenge	CPT-based soil behavior type	Discontinuities, surface cracks	Anomalies, voids, etc.
Value challenge	Type 1 (ML advantage)	Type 2a and 2b (ML advantage or supremacy)	Type 3 (disruptive value)
Learning how to learn challenge	Hierarchical Bayesian model	Transfer learning	Few-shot learning or other meta-learning
"Becoming smart" challenge	Data standardization, building information modeling (BIM)	Underground BIM or underground information modeling (UIM)	Digital shadow, digital twin

Source: Modified from Table 2, Phoon and Zhang (2023).
Note:
Type 1 refers to projects applying available machine learning (ML) methods to available data; Type 2a refers to projects applying new ML methods to available data; Type 2b refers to projects applying available ML methods to new types of data; Type 3 refers to projects applying new ML methods to new types of data.
"ML advantage" means that the ML method is effective but not indispensable. Value to practice is incremental. "ML supremacy" means that the ML method can solve problems where conventional methods are found to be impractical or ineffective. Value to practice is significant. The value of some Type 3 projects is considered to be disruptive when they transform practice or may even redefine the role of geotechnical engineering.

This chapter primarily focuses on presenting useful statistics of soil parameters for probabilistic methods that are *sufficiently mature for adoption in practice.* The statistics of some rock parameters are also included in this chapter for completeness. It is also worthwhile pointing out that statistical characterization should not be carried out in the abstract, but with full appreciation of the geotechnical context. Decision-making in every discipline is supported by its own data with unique attributes and a tradition of successful practice (investigation, design, construction, testing, monitoring, and risk management methodology) that has evolved to make the best use of this data and the prevailing technologies. Phoon and Ching (2014a) highlighted some unique features in the statistical characterization of geotechnical parameters because of this discipline specificity:

1. Coefficients of variation (COVs) for geotechnical design parameters can be potentially large, because geomaterials are naturally occurring and in situ variability cannot be reduced (in contrast, most structural materials are manufactured with quality control).
2. COVs for geotechnical design parameters are not unique and can vary over a wide range, depending on the test and the procedure in which they are derived.
3. Because geotechnical design parameter characteristics are different from one site to another, it is common to conduct a site investigation at each site. Because site-specific data is limited, statistical uncertainty must be handled with great care.
4. It is common to conduct both laboratory and field tests in a site investigation. A geotechnical design parameter is typically correlated with more than one laboratory and/or field test indices. It is important to consider this multivariate correlation structure where possible, because the COV of the design parameter reduces when consistent information increases.
5. Spatial variability of geotechnical design parameters cannot be readily dismissed, because the volume of geomaterial interacting with the structure is related to some multiple of the characteristic length of the structure and this characteristic length (e.g., height of slope, diameter of tunnel, depth of excavation) is typically larger than the scale of fluctuation of the design parameter, particularly in the vertical direction.
6. There are usually many different geotechnical calculation models for the same design problem. Hence, model calibration based on local field tests and local experience is important. The proliferation of model factors, possibly site-specific, is to be expected, because of the number of models and the number of calibration databases.
7. A geotechnical system, such as a pile group and a slope, is a system reliability problem containing multiple correlated failure modes. Some of these problems are further complicated by the fact that the failure surfaces are coupled to the spatial variability of the soil medium.

The last two features are not related to geotechnical parameters, but to the limit states. Simpson (2011) provided an elaboration of the third feature:

> In structural design, it is commonly the case that drafters of codes of practice have more knowledge about the parameters of strength and loads relevant to a particular design, and their variability, than does the designer. For example, code drafters may be more knowledgeable about wind loading, floor loading, variations in dimension of cast in situ concrete, or seismic loading than is the designer, and the same applies to the variability of steel and concrete. However, in geotechnical design, the designer knows the location of the site, something of its geology and ground water conditions and the results, or paucity of results, of the ground investigation, together with their likely reliability. This information varies considerably from one design to another and could not possibly be known by the code drafter.

The goal of Annex D, ISO2394:2015, is to emphasize the importance of this geotechnical context in risk-informed design, be it reliability-based design (RBD) or simplified formats such as the load and resistance factor design (Phoon et al. 2016b). Phoon (2017) made the same point in a review of the role of reliability calculations in geotechnical design:

> By introducing greater realism into reliability analysis that caters to the distinctive needs of geotechnical engineering practice, focusing on how it can add genuine value to practice, and be mindful of its limits (which exist for all methodologies), the author believes that reliability analysis 'has potential for wider application in geotechnical engineering, with attendant benefits for the profession, its clients, and the public' – the conclusion drawn by the National Research Council (1995) twenty years ago.

2.2 CHARACTERISTIC VALUE OF A SOIL PARAMETER

Although numerical models are highly sophisticated, the input ground profile is typically simple and deterministic. It is common to assume a homogeneous or horizontally layered soil profile in a finite element model. Only one "characteristic" set of soil parameters is selected in the analysis. Sometimes, several sets are selected when an additional parametric study is conducted. However, a formal statistical analysis of the site data is rarely conducted to inform this parametric study. Reasons include unfamiliarity with statistics and genuine difficulties in addressing complex data attributes such as site-specificity, multivariate statistics, spatial variability, and limited (or sparse) data. Simpson (2011) expressed a fairly common sentiment that a design decision should not be reduced to a computer activity in reference to Eurocode 7 (EC7): "EC7 attempts to do this by making the designer responsible for the selection of the characteristic values of materials, avoiding mathematical prescription of their derivation". A deterministic "select one value" approach is necessary, because a designer cannot reason with random variables/fields by judgment alone.

Three aspects of the input profile can be made more realistic: (1) spatially variable parameters; (2) "almost real" stratigraphy that are non-horizontal, variable layer thickness, and possibly discontinuous across the site (Shuku and Phoon 2023); and (3) range of "what if" ground scenarios (parameter profile and/or stratigraphy) consistent with limited site data. These three aspects are interlinked. It is also important to distinguish between probabilistic modeling (e.g., model spatial variability using a random field – Vanmarcke 1977a) and statistical characterization (e.g., estimate the scale of fluctuation in a random field model using cone penetration test data – Ching et al. 2016a; Ching et al. 2023b). This chapter takes the position that probabilistic and statistical methods enhance rather than diminish ownership in decision-making. It primarily focuses on the characterization of spatially variable soil parameters, because engineers require guidance on the selection of appropriate statistical values even for the most basic reliability applications (Sections 2.2–2.6). The second aspect is covered in Chapter 4, which presents the characterization of geologic uncertainty. Phoon et al. (2022a) posed this characterization as a "stratification challenge" in data-driven site characterization. The third aspect can be implemented through conditional random field simulations (Section 2.7), although other machine learning methods show significant potential (Shi and Wang 2021; Xiao et al. 2021; Xu et al. 2021; Yoshida et al. 2021).

Before discussing the sources of uncertainty and their impact on the evaluation of a soil parameter, it is useful to review the factors considered by an engineer in prevailing deterministic practice. Clause 2.4.5.2 "Characteristic values of geotechnical parameters" of Eurocode 7 (CEN 2004) provides some information on the selection of "characteristic values". First, Clause 2.4.5.2(2) defines the "characteristic value" as a "cautious estimate of the value affecting the occurrence of the limit state". It is immediately clear that uncertainty is recognized even in deterministic practice. There is *no need to be cautious if uncertainty does not exist*. The engineer feels that he/she does not have the tools to quantify uncertainty in a realistic way – hence, the appeal to engineering judgment to fill this gap. It is also clear that a soil parameter should be selected with a *limit state in mind*.

The characteristic value divided by an appropriate partial factor of safety is recommended as the design value in Eurocode 7. The design value is viewed as "a value sufficiently severe that a worse value is extremely unlikely to occur" (Simpson 2011). It is not easy to assess this design value directly, although Clause 2.4.6.2(3) allows this. The characteristic value is seen as "reasonably likely", "slightly cautious", "conservatively assessed mean", or "moderately conservative". The idea here is that the characteristic value is easier for the engineer to assess because it is a more common or less extreme value. The difference between the characteristic and design value is then accounted for using a codified partial factor of safety (usually a single prescribed number that does not depend on site-specific variability). Simpson (2011) argued that the characteristic value should not be a mean value, because there is no room to adjust for site-specific variability, unless a design code explicitly allows the designer to "vary the factors applied as a function of his perception of variability". It is apparent that uncertainty and site-specificity are difficult issues to address in a design code. The authors opine that probabilistic tools can be helpful – an avoidance of these tools does not simplify decision-making in the face of uncertainty.

Clause 2.4.5.2(4) presents a list of more detailed considerations in this important design decision (selection of a characteristic value):

- geological and other background information, such as data from previous projects;
- the variability of the measured property values and other relevant information, e.g., from existing knowledge;
- the extent of the field and laboratory investigation;
- the type and number of samples;
- the extent of the zone of ground governing the behavior of the geotechnical structure at the limit state being considered;
- the ability of the geotechnical structure to transfer loads from weak to strong zones in the ground.

There is a mixture of physical and statistical considerations in this list. There is a recognition that the limit state typically involves a geotechnical structure interacting with a volume of the ground, not with a single point in the ground ("extent of the zone of ground") and a load distribution effect ("transfer loads from weak to strong zones"). Uncertainty is recognized in the form of spatial variability ("variability of the measured property values") and statistical uncertainty ("number of samples"). While Clause 2.4.5.2 is sensible and consistent with what is generally regarded as good practice, there are no quantitative guidelines on how to calculate a characteristic value or profile given a set of data. For a 3D finite element analysis that requires specifying the property distribution in a volume of the ground, it is surely very difficult to rely on engineering judgment alone to select the characteristic *spatial distribution*. It is also obvious that a "reasonably likely" characteristic distribution is not related to a "sufficiently severe" design distribution by a simple partial factor of safety. One can imagine that both distributions are *qualitatively different* without appealing to any sophisticated random field analysis. It is also difficult to establish a set of characteristic values for multiple soil parameters by judgment alone, because an engineer needs to have a feel for the variabilities of the soil parameters and their *dependencies* (Ching et al. 2020c).

Clause 2.4.5.2(1) referred to the sources of data: "The selection of characteristic values for geotechnical parameters shall be based on results and derived values from laboratory and field tests, complemented by well-established experience". Given the qualitative nature of Clause 2.4.5.2, it is not surprising that different engineers select different characteristic values (Bond and Harris 2008). Simpson (2011) argued that "it is better to accept such subjectivity than to discard the valuable information it provides". The information refers to "subjective experience, knowledge and judgement of the designer". This objective versus subjective information divide calls for a reconciliation between big data and thick data that has not been pursued in geotechnical engineering (Phoon 2023). Nonetheless, recent advances in data-driven site characterization and digital technologies in general are contesting this argument.

This section provides a conceptual mobilization framework to quantify Clause 2.4.5.2 that is consistent with the limit state (physics) and the site data (uncertainty). It is based on a significant body of work on soil/rock parameter databases (Table 2.2) and the mobilization of a soil parameter in a *spatially varying field* (Ching and Phoon 2013b, c; Ching et al. 2014b, 2016b, c, d, 2017b, c, 2018b, 2023c; Hu and Ching 2015). It is important to note that spatial variability consists of two integral characteristics: (1) soil parameters are spatially heterogeneous and (2) limited observations (measurements) are not sufficient to describe such a spatially variable distribution uniquely (or with deterministic precision). As such, a spatially variable (or heterogeneous) soil mass model is irreducibly probabilistic in nature. Many plausible scenarios or "realizations" are consistent with the observations. This mobilization framework is a significant update on the current spatial average framework proposed by Vanmarcke (1977a, 1983) in two aspects. Vanmarcke's (1977a, 1983) framework is primarily probabilistic. It does not address practical questions on how random field parameters (e.g., coefficient of variation, scale of fluctuation) can be estimated from actual site data. These statistical characterization questions are non-trivial and are arguably more important to practice (Ching et al. 2016a, 2017d). Second, the spatial average is independent of the limit state. Figure 2.1 unpacks the characterization of a spatially variable soil parameter in three conceptual

TABLE 2.2
Generic Soil/Rock Parameter Databases

	Database	Reference	Soil/rock parameters	No. data points	No. sites/ studies	% complete
Univariate	CLAY/16	Phoon and Kulhawy (1999a)	γ, γ_d, w_n, PL, LL, PI, LI, ϕ', s_u, s_u^{FV}, q_c, q_t, SPT-N, DMT (A, B), PMT p_L		a	
	SAND/11	Phoon and Kulhawy (1999a)	ϕ', D_r, q_c, SPT-N, DMT (A, B, I_{DMT}, K_{DMT}, E_{DMT}), PMT (p_L, E_{PMT})		b	
	ROCK/8	Prakoso (2002)	γ (or γ_d), n, R, S_h, σ_{bt}, I_s, σ_c, E		c	
	ROCK/13	Aladejare and Wang (2017)	ρ, G_s, I_{d2}, n, w_c, γ, R_L, S_h, σ_{bt}, I_{s50}, σ_c, E, ν		d	
Multivariate	CLAY/5/345	Ching and Phoon (2012)	LI, s_u, s_u^{re}, σ'_p, σ'_v	345	37 sites	100
	CLAY/7/6310	Ching and Phoon (2013a)	s_u from seven different test procedures	6,310	164 studies	17.7
	CLAY/6/535	Ching et al. (2014a)	s_u/σ'_v, OCR, q_{tc}, q_{tu}, $(u_2-u_0)/\sigma'_v$, B_q	535	40 sites	100
	CLAY/10/7490	Ching and Phoon (2014a)	LL, PI, LI, σ'_v/P_a, σ'_p/P_a, s_u/σ'_v, S_t, q_{tc}, q_{tu}, B_q	7,490	251 studies	34.1
	FI-CLAY/7/216	D'Ignazio et al. (2016)	s_u^{FV}, σ'_v, σ'_p, w_n, LL, PL, S_t	216	24 sites	100
	JS-CLAY/5/124	Liu et al. (2016)	M_r, q_c, f_s, w_n, γ_d	124	16 sites	100
	JS-CLAY/7/372	Zou et al. (2017)	σ_v, σ'_v, q_{tc}, f_s/σ'_v, B_q, V_{s1}, s_u/σ'_v	372	25 sites	100
	SAND/7/2794	Ching et al. (2017a)	D_{50}, C_u, D_r, σ'_v/P_a, ϕ', q_{t1}, $(N_1)_{60}$	2,794	176 studies	60.0
	EMI-ROCK/8/26000+	Kim and Hunt (2017)	σ_c, σ_{bt}, ρ, CAI, PPI, cohesion, direction shear, triaxial confining	26,000+	–	–
	FG/5/1000	Kootahi and Moradi (2017)	e, w_n, LL, PI, C_c	1,000	170 sites	100
	ROCK/9/4069	Ching et al. (2018a)	γ, n, R_L, S_h, σ_{bt}, I_{s50}, V_p, σ_{ci}, E_i	4,069	184 studies	34.2
	FG-KSAT/6/1358	Feng and Vardanega (2019)	e, k, LL, PL, PI, G_s	1,358	33 studies	91.4
	SH-CLAY/11/4051	Zhang et al. (2020)	LL, PI, LI, e, K_0, σ'_v/P_a, s_u/σ'_v, S_t, q_c/σ'_v	4,051	50 sites	39.5
	CLAY/8/12225	Ching (2020)	LL, PI, w, e, σ'_v/P_a, C_c, C_{ur}, c_v	12,225	427 studies	–
	CLAY/12/3997	Ching (2020)	LL, PI, LI, σ'_v/P_a, σ'_p/P_a, s_u/σ'_v, K_0, E_u/σ'_v, B_q, q_{t1}, $N_{60}/(\sigma'_v/P_a)$	3,997	237 studies	–
	SAND/13/4113	Ching (2020)	e, D_r, σ'_v/P_a, σ'_p/P_a, K_0, E_{dn}, q_{c1n}, B_q, $(N_1)_{60}$, K_{DMT}, E_{DMTn}, E_{PMTn}, M_{dn}	4,113	172 studies	–
	ROCKMass/9/5876	Ching et al. (2021a)	RQD, RMR, Q, GSI, E_m, E_{em}, E_{dm}, E_i, σ_{ci}	5,876	225 studies	29.3
	CLAY-C_c/6/6203	Ching et al. (2023a)	LL, PI, w_n, e, C_c, C_{ur}	6,203	429 studies	61
	Global-CPT/3/1196	Ching et al. (2023b)	q_t, f_s, u_2	1,196	59 sites	100
	Geo-Marine-CPT/3/398	Eslami et al. (2023)	q_t, f_s, u_2	398	58 sites	100
	ROCK/10/4025	Muzamhindo and Ferentinou (2023)	γ, n, R_L, I_{s50}, V_p, σ_{bt}, BPI, m_i, E, σ_c	4,025	96 studies	50
	SAND-Small/9/939	Lo et al. (2021)	D_{50}, C_u, e_{min}, e_{max}, σ'_3, e_c, G_{max}, σ'_{1p}, ϕ'	939	15 studies	79

Source: Updated from Phoon and Ching (2022).

Note:

Basic: C_u = coefficient of uniformity; D_{50} = median grain size; ρ = density; G_s = specific gravity; γ = unit weight; γ_d = dry unit weight; D_r = relative density; e = void ratio; e_{min} = minimum void ratio; e_{max} = maximum void ratio; e_c = void ratio under isotropic consolidation; n = porosity; w_n (or w_c) = water content; PL = plastic limit; LL = liquid limit; PI = plasticity index; LI = liquidity index; GSI = geological strength index

Stress: σ_v = total vertical stress; σ'_v = effective vertical stress; σ'_p = preconsolidation stress; σ'_3 = effective confining stress

(*Continued*)

TABLE 2.2 (CONTINUED)
Generic Soil/Rock Parameter Databases

under isotropic consolidation; σ'_{1p} = effective axial stress at peak state under drained triaxial compression; OCR = overconsolidation ratio; P_a = atmospheric pressure = 101.3 kPa; K_0 = at-rest lateral earth pressure coefficient

Strength: ϕ' = effective friction angle; s_u = undrained shear strength; s_u^{FV} = field vane s_u; s_u^{re} = remoulded s_u; S_t = sensitivity; σ_{bt} = Brazilian tensile strength; σ_{ci} (or σ_c) = uniaxial compressive strength of intact rock; BPI = block punch index; m_i = Hoek–Brown constant

Deformation: C_c = compression index; C_{ur} = unloading–reloading index; modulus; E_u = undrained modulus of clay; E_d = drained modulus of sand; $E_{dn} = (E_d/P_a)/(\sigma'_v/P_a)^{0.5}$; E_{dm} = dynamic modulus of rock mass; E_{em} = elasticity modulus of rock mass; E_i (or E) = Young's modulus of intact rock; E_m = deformation modulus of rock mass; ν = Poisson ratio; M_r = subgrade resilience modulus; M_d = effective constrained modulus determined by oedometer; M_{dn} = normalized $M_d = (M_d/P_a)/(\sigma'_v/P_a)^{0.5}$; G_{max} = small-strain shear modulus

Permeability: k = hydraulic conductivity; c_v = coefficient of consolidation

Dynamic: V_p = P-wave velocity; V_s = S-wave velocity; $V_{s1} = V_s(P_a/\sigma'_v)^{0.25}$

Field test: SPT-N = standard penetration test blow count; N_{60} = corrected SPT-N; $(N_1)_{60} = N_{60}/(\sigma'_v/P_a)^{0.5}$; q_c = cone tip resistance; q_t = corrected cone tip resistance; f_s = sleeve frictional resistance; $q_{tc} = (q_t/P_a)/(\sigma'_v/P_a)^{0.5}$; $q_{t1} = (q_t - \sigma_v)/\sigma'_v$ = normalized cone tip resistance; $q_{tu} = (q_t - u_2)/\sigma'_v$ = effective cone tip resistance; $q_{c1n} = (q_c/P_a)/(\sigma'_v/P_a)^{0.5}$; B_q = pore pressure ratio = $(u_2 - u_0)/(q_t - \sigma_v)$; $(u_2 - u_0)/\sigma'_v$ = normalized excess pore pressure; u_2 = pore pressure behind cone tip; u_0 = hydrostatic pore pressure; PMT (p_L, E_{PMT}) = pressuremeter limit stress, modulus; E_{PMTn} = normalized $E_{PMT} = (E_{PMT}/P_a)/(\sigma'_v/P_a)^{0.5}$; DMT (A, B, I_{DMT}, K_{DMT}, E_{DMT}) = dilatometer A and B readings, material index, horizontal stress index, modulus; E_{DMTn} = normalized $E_{DMT} = (E_{DMT}/P_a)/(\sigma'_v/P_a)^{0.5}$; CAI = Cerchar abrasivity index; PPI = punch penetration index; Q = Q-system; RMR = rock mass rating; RQD = rock quality designation; R = Schmidt hammer hardness (R_L = L-type Schmidt hammer hardness); S_h = shore scleroscope hardness; I_{d2} = slake durability index; I_s = point load strength index ($I_{s50} = I_s$ for diameter 50 mm)

a: The number of data groups varies between 2 and 42 depending on the clay parameter. Statistics are calculated at the data group level. The average number of data points/data groups varies between 16 and 564. Details given in Tables 1–3, Phoon and Kulhawy (1999b).

b: The number of data groups varies between 5 and 57 depending on the sand parameter. Statistics are calculated at the data group level. The average number of data points/data groups varies between 15 and 123. Details given in Tables 1–3, Phoon and Kulhawy (1999b).

c: The number of data groups varies between 30 and 174 depending on the rock parameter with no differentiation of rock type (igneous [intrusive, extrusive, pyroclastic], sedimentary [clastic, chemical], metamorphic [foliated, non-foliated]). Statistics are calculated at the data group level. The average number of data points/data groups varies between 3 and 161 for σ_c (Prakoso 2017). Details given in Table 4.4, Prakoso (2002).

d: The number of data groups varies between 2 and 47 depending on the rock parameter and rock type (igneous, sedimentary, or metamorphic). Statistics are calculated at the data group level. The average number of data points/data groups varies between 7 and 92. Details given in Tables 2–4, Aladejare and Wang (2017).

building blocks detailed below. Quantitative guidelines for each building block are covered in the following sections. These guidelines should be followed with a more holistic appreciation of the uncertainty, physics, and their interactions in the geotechnical context elaborated in this section.

2.2.1 Point Value

Physically (or in each scenario/realization), point value refers to the value of a soil parameter at a fixed location and depth. A "point" is a mathematical idealization, but it is also close to reality because a field test/sampling mobilizes a soil mass much smaller than that activated by a realistic geotechnical structure. Although it is known at observation points where soil sampling or testing is conducted, this "measured reality" is sparse. At unobserved points, this point value is uncertain. An intuitive way to understand this uncertainty is to visualize the soil parameter at any location and depth falling within a range (e.g., 95% confidence interval) rather than taking a single value. Experienced engineers would accept that a range is more realistic than a single value, but the general perception is the lack of practical tools to deal with more complex information representation that can provide insights for decision-making. It is understandable that engineers feel ill-equipped to handle uncertainties at a spatial scale, as the majority is not familiar with random field theory. Note that the realistic ranges of many soil parameters are bounded even in the absence of observations. For example, the residual/critical values of strength parameters can be regarded as reasonable lower bounds. The 4 standard deviation range spanned by the lowest and highest conceivable value can be regarded as a judgment-based 95% confidence interval if the parameter is approximately normal.

Although there are multiple points in a volume, the probability model for a "point value" assumes that every point value is statistically independent, i.e., knowledge of other point values does not reduce the uncertainty at a particular unobserved point. This is generally untrue. Interpolation methods such as kriging are not effective if this assumption is true. The concept of a "point value" is illustrated in Figure 2.1a. The "point value" is customarily modeled as a random variable described by a probability distribution function (PDF). The Johnson system that includes the well-known lognormal distribution is a good choice, because it is quite flexible and there exists an analytical expression to obtain a Johnson distribution from a standard normal distribution (Section 2.6.1). The

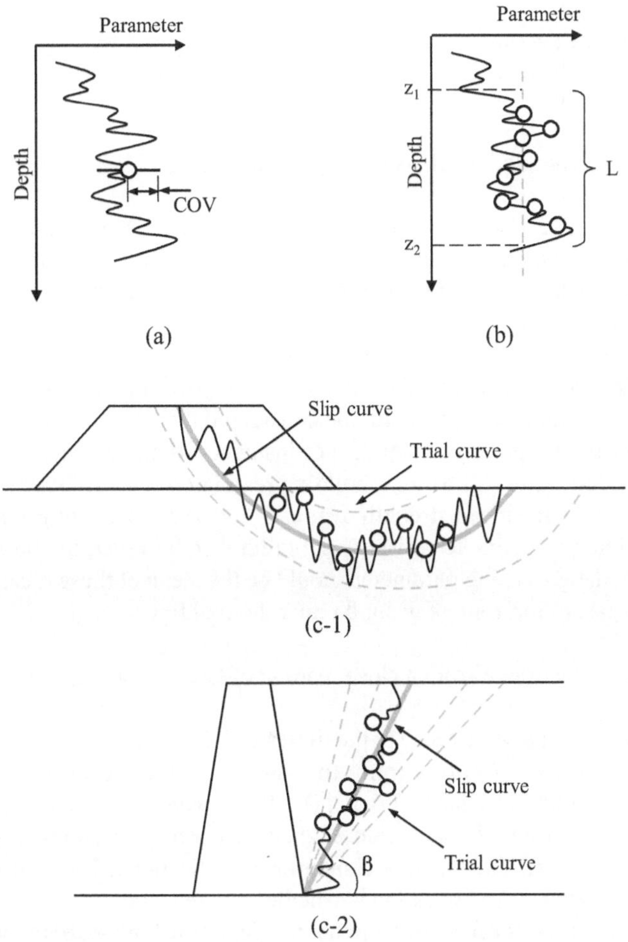

FIGURE 2.1 Conceptual mobilization framework to quantify Clause 2.4.5.2 of Eurocode 7 "Characteristic values of geotechnical parameters" in a *spatially varying* soil: (a) point value, (b) spatial average, (c-1) mobilized value (slope), and (c-2) mobilized value (retaining structure).

latter reason is important, because computation cost can be excessive in the absence of analytical expressions, particularly for a large number of points in a volume. For practice, it suffices to note that the basic statistics to describe the uncertainty in a point value is the *coefficient of variation*. Extensive information on COV has been compiled as presented in Section 2.3. The complete probability model for a "point value" must cover all points in a volume. The single point model naturally extends to a set of independent identically distributed (i.i.d) random variables for all points. Its characterization only requires fitting a PDF and calculating its parameters (e.g., COV) using site data. It is not uncommon for an engineer to assume a lognormal distribution and only calculate its COV. This univariate analysis is familiar to engineers, because it is covered in a basic probability and statistics course.

2.2.2 Spatial Average

Clause 2.4.5.2 of Eurocode 7 (CEN 2004) emphasizes the need to select a soil parameter that is relevant to the limit state under consideration. It distinguishes between "local failure" and "non-local failure", because of this consideration. "Local failure" is described in Clause 2.4.5.2(8):

> If the behaviour of the geotechnical structure at the limit state considered is governed by the lowest or highest value of the ground property, the characteristic value should be a cautious estimate of the lowest or highest value occurring in the zone governing the behaviour.

"Non-local failure" is elaborated in Clause 2.4.5.2(7) and Clause 2.4.5.2(9):

- Clause 2.4.5.2(7): The zone of ground governing the behavior of a geotechnical structure at a limit state is usually much larger than a test sample or the zone of ground affected in an in situ test. Consequently, the value of the governing parameter is often the mean of a range of values covering a large surface or volume of the ground. The characteristic value should be a cautious estimate of this mean value.
- Clause 2.4.5.2(9): When selecting the zone of ground governing the behavior of a geotechnical structure at a limit state, it should be considered that this limit state may depend on the behavior of the supported structure. For instance, when considering a bearing resistance ultimate limit state (ULS) for a building resting on several footings, the governing parameter should be the mean strength over each individual zone of ground under a footing, if the building is unable to resist a local failure. If, however, the building is stiff and strong enough, the governing parameter should be the mean of these mean values over the entire zone or part of the zone of ground under the building.

A few observations can be drawn from a close reading of Clause 2.4.5.2(7), (8), (9):

1. The "value affecting the occurrence of the limit state" involves the *mobilization of multiple point values for non-local failure* (refer to "range of values covering a large surface or volume of the ground" in Clause 2.4.5.2[7]). This means that the complete probability model describing a group of "point values" distributed in space and their potential spatial dependencies is critical. The i.i.d model mentioned in Section 2.2.1 is the simplest – it has no spatial dependencies. The question is whether it is realistic.
2. For non-local failure, the "surface" or "volume" mobilized in a limit state is a special domain that produces the *lowest factor of safety*. An arbitrarily *prescribed* surface or volume is unlikely to coincide with this special domain, as it is the solution of a boundary value problem. In fact, for finite element analysis, it is typically produced by applying a strength reduction method (Dawson et al. 1999). For slope stability analysis using the ordinary method of slices (Fellenius

1936), trial circles are prescribed, but a search is needed to look for the slip circle with the lowest factor of safety. Clause 2.4.5.2(7) and (9) did not make this crucial point explicit.
3. Clause 2.4.5.2(7) and (8) refer to a "cautious estimate", but it is clear from Clause 2.4.5.2(11) that statistics is an *optional* basis for this decision. Clause 2.4.5.2(11) suggested a 5% fractile as sufficiently cautious: "If statistical methods are used, the characteristic value should be derived such that the calculated probability of a worse value governing the occurrence of the limit state under consideration is not greater than 5%". This definition draws on Clause 4.2(3) of Eurocode 1990 (CEN 2002) "Material and product properties":
   Unless otherwise stated in EN 1991 to EN 1999:
   - where a low value of material or product property is unfavourable, the characteristic value should be defined as the 5% fractile value;
   - where a high value of material or product property is unfavourable, the characteristic value should be defined as the 95% fractile value.
4. Clause 2.4.5.2(11) has been interpreted by engineers based on standard assumptions in classical statistics (Frank et al. 2004; Orr 2015; Länsivaara et al. 2022; JRC 2024a, b):
   a. The "point value" denoted as X is normally distributed with mean X_{mean} and the coefficient of variation COV_X.
   b. Multiple point values over a surface or volume, denoted as $X_1, X_2, \ldots, X_{\mathrm{m}}$, are i.i.d.
   c. For local failure, the characteristic value is the 5% fractile of a "point value":

$$X_{\mathrm{kpoint}} = X_{\mathrm{mean}}\left[1+\Phi^{-1}(0.05)\mathrm{COV}_X\right] = X_{\mathrm{mean}}\left[1-1.645\mathrm{COV}_X\right] \tag{2.1}$$

   where Φ is the cumulative distribution function of a standard normal variable. This value is fairly extreme. It is not "reasonably likely", "slightly cautious", or "moderately conservative".
   d. For non-local failure, the characteristic value is the 5% fractile of a "mean value":

$$X_{\mathrm{kmean}} = X_{\mathrm{mean}}\left[1-1.645\mathrm{COV}_X\sqrt{\frac{1}{\mathrm{m}}}\right] \tag{2.2}$$

   where m is the number of point values. This value can be regarded as a "conservatively assessed mean". Schneider (1997) recommended a simplified version of Eq. (2.2) that effectively corresponds to m = 10:

$$X_{\mathrm{kmean}} = X_{\mathrm{mean}}\left[1-0.5\,\mathrm{COV}_X\right] \tag{2.3}$$

   e. When m is small, it is possible to correct the 95% fractile for a standard normal distribution (= 1.645) using a t distribution of (m − 1) degrees of freedom.

The application of classical statistics to interpret Clause 2.4.5.2(11) is incomplete, because it does not frame statistics (spatial correlation rather than statistical independence between point values), physics (slip surface rather than prescribed trial surface), or the interaction between statistics and physics (mobilized value rather than point value for local failure or spatial average for non-local failure) with sufficient realism:

1. Multiple point values over a surface or volume, say $X_1, X_2, \ldots, X_{\mathrm{m}}$, are spatially correlated. This is the core premise underlying spatial variability. For practice, it suffices to note that the basic statistics to describe spatial correlation is the scale of fluctuation (δ). The scale of fluctuation can be roughly understood as a characteristic length over which point values

are strongly correlated. Visually, the scale of fluctuation can be seen as equivalent to the dominant wavelength of the soil parameter fluctuations in Figure 2.1. When δ is large, there are less fluctuations over a given length. Extensive information on δ has been compiled as presented in Section 2.4.

2. For non-local failure, the mean value or "spatial average" (following Vanmarcke's [1983] nomenclature) is distributed with mean $= X_{mean}$ and coefficient of variation $= COV_X \times \Gamma(L)$, where $\Gamma^2(L)$ = Vanmarcke's variance reduction function ≈ 1 if $L < \delta$ and δ/L if $L > \delta$ and L is the prescribed averaging length. The concept of a "spatial average" is illustrated in Figure 2.1b. When spatial correlation is considered, Eq. (2.2) is modified as (Schneider and Schneider 2013):

$$X_{kmean} = X_{mean}\left[1 - 1.645 COV_X \sqrt{\frac{\delta}{L}}\right] \tag{2.4}$$

Because the scale of fluctuation is defined as the interval over which point values are strongly correlated, it is possible to argue that $m = L/\delta$ represents the *equivalent* number of statistically independent point values. Hence, the effect of a large δ is to reduce m. When δ = sampling interval, the value of m is the same as the actual number of observed point values. From this perspective, Eq. (2.2) is a special case of Eq. (2.4).

3. Eq. (2.4) accounts for spatial correlation, but it does not account for a slip surface or a mobilized volume arising from soil–structure interaction. For example, the length of a slip curve in a slope stability analysis is unknown, much less knowledge of its detailed trajectory. Even if Eq. (2.4) were to be correct (it is not), L is unknown. This complication is discussed in Sections 2.2.3 and 2.5.

4. It is plausible to interpret Clause 2.4.5.2(8) as referring to a limit state governed by classical extreme value statistics:

> If the behaviour of the geotechnical structure at the limit state considered is governed by the lowest or highest value of the ground property, the characteristic value should be a cautious estimate of the lowest or highest value occurring in the zone governing the behaviour.

If this "lowest value occurring in the zone governing the behaviour" is $X_{low} = \min(X_1, X_2, \ldots, X_m)$, Eq. (2.1) is incorrect. Let $Z = (X - \mu)/\sigma$ be a standard normal random variable. Then, the cumulative distribution function of X_{low} is

$$F(X_{low}) = 1 - \left[1 - \Phi(Z)\right]^m \tag{2.5}$$

The 5% fractile of X_{low} is

$$X_{klow} = X_{mean}\left[1 + \Phi^{-1}\left(1 - \sqrt[m]{0.95}\right) \cdot COV_X\right] \tag{2.6}$$

Eq. (2.6) reduces to Eq. (2.1) when m = 1. It is easy to verify numerically that $X_{klow} \leq X_{kpoint}$, which is logical because X_{low} is the *smallest* point value while X_{point} refers to *any* point value.

Note that Eqs. (2.1)–(2.4) are based on statistics (or data). The limit state (physics) does not play a role in these equations. Hence, Eqs. (2.1)–(2.3) are called statistics-based characteristic values as shown in Figure 2.2. Equation (2.4) is called a spatial average–based characteristic value to highlight the consideration of spatial correlation.

The effect of a limit state can be readily illustrated using an infinite slope. For a dry cohesionless soil, the factor of safety (FS_{slope}) is given by $\tan(\phi)/\tan(\beta)$, where β = slope angle and ϕ = friction angle. FS_{slope} is independent of the depth of the slip surface. If the friction angle is spatially variable in the depth direction and perfectly correlated in the slope direction, FS_{slope} is now given by $\min\{\tan[\phi(z)]/\tan(\beta)\}$, where z = depth. Denoting $X_i = \tan[\phi(z_i)]$, the "value affecting the occurrence of the limit state" for cohesionless infinite slope can be viewed as $X_{low} = \min(X_1, X_2,\ldots, X_m)$. For a cohesive soil with a spatially undrained shear strength in the depth direction, the factor of safety (FS_{slope}) is given by $\min[s_u(z)/(z\gamma\sin\beta\cos\beta)]$, where γ = unit weight and $s_u(z)$ = undrained shear strength (Santoso et al. 2009). Denoting $X_i = s_u(z_i)$, the "value affecting the occurrence of the limit state" for a cohesive infinite slope can be viewed as $\min(X_1/z_1, X_2/z_2,\ldots, X_m/z_m)$. This simple example illustrates three important points: (1) a failure depth exists only if the spatial variability of the governing soil parameter is considered. The homogeneous soil assumption produces an outcome contrary to reality – all depths are possible (cohesionless) or failure depth is always at the soil–rock interface (cohesive); (2) the "value affecting the occurrence of the limit state" depends on the limit state – it may not be the mean value or the lowest value; and (3) the minimization operator in FS_{slope} is a concrete example of the interaction between physics and statistics. It does not exist for a homogeneous slope.

2.2.3 Mobilized Value

Figure 2.1c illustrates the concept of a "mobilized value" for two ultimate limit states. A cursory inspection may mislead an engineer to conclude that the mobilized value is identical to the spatial average (Figure 2.1b). It appears to be an average strength value over a line or curve in a spatially varying soil. The difference between the mobilized value on an *actual* slip curve (solid line) and the spatial average on a *prescribed trial* curve (dashed line) is evident when one looks beyond a single realization. The actual slip curve does not maintain the same trajectory in different realizations, because it seeks the least favorable alignment of weak zones and this alignment changes when the distribution of weak zones changes (an intrinsic feature of spatial variability). In short, the solid line changes from one realization to the next. But the dashed line remains unchanged. The statistics of the spatial average (fixed averaging domain across all realizations) must therefore be different

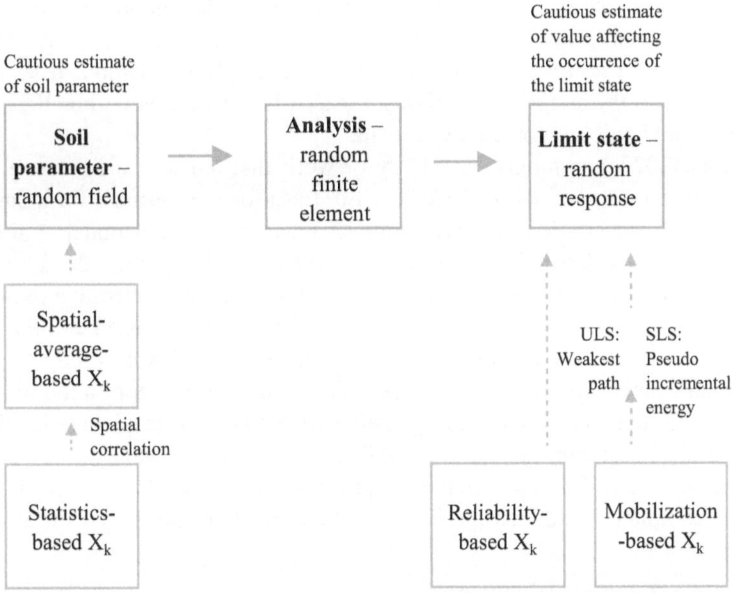

FIGURE 2.2 Relationship between statistics-based, spatial average–based, mobilization-based, and reliability-based characteristic value for soil–structure interaction in *spatially varying soil.*

from the statistics of the mobilized value (changing averaging domain across all realizations). In fact, the spatial average and the mobilized value are random variables following different probability distribution functions in general. Another way of appreciating this difference is to note that the spatial average along the dashed line can produce *any* factor of safety, but the mobilized value along the actual slip curve always produces the *lowest* factor of safety in each realization. Finally, it is worth pointing out that the slip curve is the solution of a boundary value problem. As such, the mobilized value must be problem (or limit state)-specific. In contrast, the spatial average depends on the random field parameters only (COV_X and δ). An important conclusion is that Eq. (2.4) does not apply to a mobilized value.

Clause 2.4.5.2(2) did not state that the "value affecting the occurrence of the limit state" is the mobilized value, but it is difficult to imagine a more relevant concept to match the intent of this phrase. Clause 2.4.5.2(4) did not guide an engineer to consider the interaction between physics and statistics. The code drafters may not have foreseen this complication, because it does not exist unless spatial variability is modeled. It is useful to point out Clause 4.2(3) of Eurocode 1990 (CEN 2002) (that Clause 2.4.5.2[11] of Eurocode 7 [CEN 2004] plausibly took reference from) does not refer to spatial variability, because it is primarily intended for structural materials. This example illustrates the need to apply statistical methods with the physical (geotechnical) context in mind.

As the actual slip curve is the solution of a boundary value problem in each realization, possibly determined through a complicated 3D finite element analysis, it is easy to see that the mobilized value (a random variable) and the statistics describing its probability distribution are not available in closed-form. It does not exist in closed-form even for a problem under simple stress states (Ching and Phoon 2013b, 2013c; Ching et al. 2014b). Hicks and Samy (2002) defined an "effective property" of a spatially variable soil mass as the equivalent homogeneous soil mass property that matches the response of the geotechnical structure being monitored in a limit state. The authors proposed a direct approach to obtain the probability distribution of this effective property using the random finite element method (RFEM) (Fenton and Griffiths 2008). This approach is rigorous but costly – it requires thousands of finite element analyses. The effective property is identical to the mobilized value. The reliability-based characteristic value is equal to the 5% fractile of this effective/mobilized property (Hicks 2013; Hicks et al. 2019). It is directly related to the random response, as shown in Figure 2.2. Varkey et al. (2020) used the 5% random response obtained from the RFEM as a benchmark, and compared the responses obtained by the finite element method using the characteristic soil properties obtained from six simplified methods (e.g., Eqs. [2.3] and [2.4]). In their problem, the simplified methods mostly give responses within 10% of the RFEM solution, but the results will be problem dependent.

Tabarroki et al. (2022a) summarized a body of work that led to a much more cost-effective approximation with the same "look and feel" as Eq. (2.4) for the ultimate limit state. This ULS mobilization-based characteristic value is calibrated from a weakest path model. A different approximation for the mobilized value at the serviceability limit state (SLS) is provided by Tabarroki et al. (2022b). This SLS mobilization-based characteristic value is calibrated using a pseudo incremental energy (PIE) method. As illustrated in Figure 2.2, the reliability-based and the mobilization-based characteristic values consider both physics and spatial variability correctly, in contrast to the statistics-based and spatial average–based values. The characterization of a mobilized value in a spatially varying soil remains an outstanding challenge and general results are lacking. However, useful research findings for common geotechnical structures are available and are summarized in Section 2.5. One surprising finding is that an approximation similar to Eq. (2.6) is possible, notwithstanding the complicated interactions between the limit state and spatial variability.

2.3 COEFFICIENT OF VARIATION

The simplest measure of soil/rock property uncertainty is the coefficient of variation, which is defined as the ratio between the standard deviation and the mean. It provides a second-moment dimensionless characterization of the data scatter *at a point* (Figure 2.1a). An observation or measurement practically takes place at a "point", because the "zone of ground governing the behaviour of a geotechnical structure at a limit state is usually much larger than a test sample or the zone of ground affected in an in situ test" (Clause 2.4.5.2[7], Eurocode 7 [CEN 2004]). The COV of a design parameter is not an intrinsic statistical property. It depends on the site condition, the measurement method, the transformation (correlation) model, and other reasons explained below. Hence, the COV takes a range of values rather than a unique value. Section D.1 of ISO2394:2015 (ISO 2015) made the same point: "COVs for geotechnical design parameters are not unique and can vary over a wide range, depending on the procedure in which they are derived". This section provides numerical guidelines for ISO2394:2015, Section D.2 "Uncertainty representation of geotechnical design parameters".

A classical statistical study on the uncertainty of soil design parameters was conducted by Phoon and Kulhawy (1999a, 1999b). The *total (or overall)* uncertainty in a design soil parameter arises from several sources of uncertainty, as illustrated in Figure 2.3. The three primary sources of total uncertainty are (1) spatial (or inherent) variability (site-specific); (2) measurement error (process-specific); and (3) transformation uncertainty (model-specific). The first results primarily from the natural geologic processes that produced and continually modify the soil mass in situ. The second is caused by equipment, procedure/operator, and random testing effects. The third source of uncertainty is introduced when field or laboratory measurements are transformed into design soil properties using empirical, semi-empirical, or theoretical models. An empirical model is: undrained shear strength/atmospheric pressure = 0.06 × SPT-N (Kulhawy and Mayne 1990). A semi-empirical model based on cavity expansion theory is undrained shear strength = $(q_t - \sigma_v)/N_k$, where q_t is the corrected cone tip resistance, σ_v is the total vertical stress, and N_k is the cone factor determined through calibration with site data (Ching et al. 2014a). Lim et al. (2020) adopted a more advanced press-replace numerical method and the modified Cam-clay soil model to develop more realistic physics-based correlation models assuming rough cone–soil interactions. The relative contribution of these three sources to the total uncertainty in the design soil parameter clearly depends on the site conditions, degree of equipment and procedural standardization and control, and quality of the correlation model. Therefore, design soil parameter statistics that are determined from total uncertainty analyses only can be applied to the specific set of circumstances (site conditions, measurement techniques, correlation models) for which the design soil parameters were derived. Some guidelines emerging from this classic study are reproduced in Table 2.3.

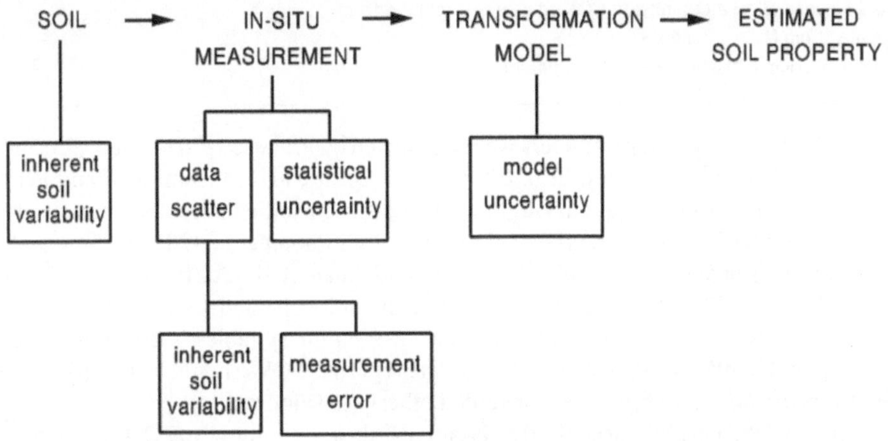

FIGURE 2.3 Sources of uncertainties contributing to total uncertainty in design soil parameter. (Source: Phoon and Kulhawy 1999a)

TABLE 2.3
Approximate Guidelines for Coefficients of Variation of some Design Soil Parameters Derived from *Total Uncertainty* Analyses.

Design Property[a]	Test[b]	Soil type	Point COV (%)	Spatial Avg. COV[c] (%)	Correlation Equation[f]
s_u(UC)	Direct (lab)	Clay	20-55	10-40	-
s_u (UU)	Direct (lab)	Clay	10-35	7-25	-
s_u (CIUC)	Direct (lab)	Clay	20-45	10-30	-
s_u (field)	VST	Clay	15-50	15-50	14
s_u (UU)	q_T	Clay	30-40[d]	30-35[d]	18
s_u (CIUC)	q_T	Clay	35-50[d]	35-40[d]	18
s_u (UU)	N	Clay	40-60	40-55	23
s_u[e]	K_D	Clay	30-55	30-55	29
s_u(field)	PI	Clay	30-55[d]	-	32
ϕ	Direct (lab)	Clay, sand	7-20	6-20	-
ϕ(TC)	q_T	Sand	10-15[d]	10[d]	38
ϕ_{cv}	PI	Clay	15-20[d]	15-20[d]	43
K_o	Direct (SBPMT)	Clay	20-45	15-45	-
K_o	Direct (SBPMT)	Sand	25-55	20-55	-
K_o	K_D	Clay	35-50[d]	35-50[d]	49
K_o	N	Clay	40-75[d]	-	54
E_{PMT}	Direct (PMT)	Sand	20-70	15-70	-
E_D	Direct (DMT)	Sand	15-70	10-70	-
E_{PMT}	N	Clay	85-95	85-95	61
E_D	N	Silt	40-60	35-55	64

Source: Table 5, Phoon and Kulhawy (1999b)

a - s_u = undrained shear strength; UU = unconsolidated-undrained triaxial compression test; UC = unconfined compression test; CIUC = consolidated isotropic undrained triaxial compression test; s_u(field) = corrected s_u from vane shear test; ϕ = effective stress friction angle; TC = triaxial compression; ϕ_{cv} = constant volume ϕ; K_o = in-situ horizontal stress coefficient; E_{PMT} = pressure-meter modulus; E_D = dilatometer modulus

b - VST = vane shear test; q_T = corrected cone tip resistance; N = standard penetration test blow count; K_D = dilatometer horizontal stress index; PI = plasticity index

c - averaging over 5 m using Vanmarcke (2010)'s variance reduction function

d - COV is a function of the mean; refer to COV equations in text for details

e - mixture of s_u from UU, UC, and VST

f - Equation numbering in Phoon and Kulhawy (1999b)

Comparable studies for rock parameters were conducted by Prakoso (2002), Aladejare and Wang (2017), Bozorgzadeh and Harrison (2019), Ching et al. (2021a), and Muzamhindo and Ferentinou (2023). Useful statistical tables quantifying the uncertainty in soil and rock parameters are given in Kulhawy et al. (2000: table 1) (reproduced as Table 2.4); Phoon et al. (2016b); Tang and Phoon (2021: chapter 2); Stuedlein et al. (2021: table 3.3), and Guan et al. (2021: tables 1.2–1.4) (reproduced as Tables 2.5–2.7). Tables 2.5–2.7 summarize the most updated statistics for common soil and rock parameters. The detailed source data for these summary tables is given in Appendix A and EPRI Report TR-105000 (Phoon et al. 1995). The reported COVs should be regarded as arising primarily from *spatial variability* unless specified otherwise (Figure 2.1a).

The range of COVs (spatial variability) is broad, as illustrated in Figures 2.4–2.6, demonstrating the difficulty of selecting a "representative" COV for each soil/rock parameter for design code calibration. This is not the case for structural materials such as concrete and steel that are manufactured to meet quality specifications (Phoon and Kulhawy 1999b). Phoon et al. (1995) further

TABLE 2.4
Coefficient of variation for spatial variability for some soil and rock parameters.

Test type	Property	Material type	Coefficient of variation (%) #group	mean	std dev	range
Index	γ, γ_d	fine-grained	14	7.8	5.8	2-20
	w_n	fine-grained	40	18.1	7.9	7-46
	w_P	fine-grained	23	15.7	6.0	6-34
	w_L	fine-grained	38	18.1	7.1	7-39
	PI - all data	fine-grained	33	29.5	10.8	9-57
	- ≤ 20%	fine-grained	13	35.0	11.4	16-57
	- > 20%	fine-grained	20	26.0	9.0	9-40
	γ, γ_d	rock	42	0.9	0.7	0.1-3
	n	rock	25	25.9	19.4	3-71
Strength	ϕ, tan ϕ	sand, clay	48	13.9	10.4	4-50
		sand	32	9.0	3.0	4-15
		clay	16	23.5	13.0	10-50
	s_u	clay	100	31.5	14.2	6-80
	q_u	rock	184	14.2	11.7	0.3-61
	$q_{t\text{-brazilian}}$	rock	74	16.6	10.4	2-58
Stiffness	$E_{t\text{-}50}$	rock	32	30.7	15.0	7-63
CPT	q_c	sand, clay	65	36.6	15.5	10-81
		sand	54	38.2	16.3	10-81
		clay	11	28.4	6.8	16-40
	q_T	clay	9	7.9	4.9	2-17
VST	s_u(VST)	clay	26	25.3	6.5	13-36
SPT	N	sand, clay	23	38.0	10.8	19-62
DMT	A, B	sand, clay	56	27.9	11.9	12-59
		sand	30	34.8	11.3	13-59
		clay	26	19.9	6.2	12-38
	I_D - all data	sand	30	40.7	21.6	8-130
	- w/o outliers		29	37.7	14.2	8-66
	K_D - all data	sand	31	41.2	19.2	15-99
	- w/o outliers		29	37.6	13.5	15-67
	E_D - all data	sand	31	42.7	19.6	7-92
	- w/o outliers		30	41.1	17.6	7-69

Source: Table 1, Kulhawy et al. (2000)

Note: γ = total unit weight; n = apparent porosity; γ_d = dry unit weight; w_n = natural water content; w_P = plastic limit; w_L = liquid limit, PI = plasticity index; ϕ = effective stress friction angle; s_u = undrained shear strength; q_u = uniaxial compressive strength; $q_{t\text{-brazilian}}$ = Brazilian indirect tensile strength; $E_{t\text{-}50}$ = tangent modulus at $0.5q_u$; q_c = cone penetration test (CPT) tip resistance; q_T = corrected tip resistance; s_u(VST) = undrained shear strength from vane shear test; N = standard penetration test (SPT) blow count; A, B = dilatometer test (DMT) A and B readings; I_D = dilatometer material index; K_D = dilatometer horizontal stress index; E_D = dilatometer modulus

observed that the COV is related to the mean for some soil parameters (Figure 2.7). For example, Figure 2.7a and b show that the standard deviation falls within a narrow range rather than the COV for the plasticity index (PI), liquidity index, and effective friction angle. The range of COVs is broad, because the range of the mean values is broad. Specifically, a small mean value results in a large COV and vice versa if the standard deviation stays constant. The COV appears to decrease with the mean dilatometer material index for sandy silt to silty sand in Figure 2.7c and to increase with the mean SPT-N value for sand in Figure 2.7d. However, these correlations are much weaker

TABLE 2.5
Summary of site-specific spatial variability statistics for clay.

Property	# groups	# cases/group		Site-specific mean			Site-specific COV		
		Range	Mean	Range	Mean	95% CI	Range	Mean	95% CI
LL (%)	103	10-2229	69	19.3-158.6	55.6	24.7-95.1	3.4-39	15.6	4.8-35.1
PL (%)	87	10-299	41	13.9-112.7	29.1	17.2-76.2	2.9-38.1	13.5	3.9-35.0
PI (%)	94	10-4044	93	6.2-60.8	29.0	10.5-56.2	6.5-57	23.5	6.8-47
w (%)	111	10-439	76	13.1-120.2	43.5	13.7-104.9	3.5-46	15.3	4.9-30
LI	49	10-2067	68	0.09-2.47	0.93	0.09-2.31	5.8-88	24.5	5.8-70.5
OCR	24	10-56	17	0.90-3.15	1.69	0.90-3.11	1.2-39	17.8	1.5-38.8
C_c	18	17-136	53	0.19-2.15	0.63	0.19-2.15	18.1-47.3	35.6	18.1-47.3
C_{ur}	9	17-115	44	0.03-0.21	0.10	0.03-0.21	22.6-50.5	42.4	22.6-50.5
ϕ' (°)	13	5-51	19	3-33.3	15.3	3-33.3	10-50	21.3	10-50
s_u (kPa)	91	9-393	59	6.3-712.8	148.0	7.2-558.4	6-56	28.2	9.9-53.5
s_u/σ'_v	45	10-352	27	0.05-1.14	0.39	0.06-1.07	3.2-39.4	20.8	5.0-39.3
S_t	17	10-384	51	2.2-38.6	8.8	2.2-38.6	12.4-63.4	30.8	12.4-63.4
q_c	11	47-53	50	1.2-2.1	1.65	1.2-2.1	16-40	28.4	16-40
q_t	9	-	-	0.4-2.7	1.54	0.4-2.7	2-17	7.9	2-17
q_{t1}	21	12-42	17	2.04-13.2	5.99	2.04-13.13	5.8-39.8	17.5	5.8-39.7
B_q	26	11-47	20	0.17-0.99	0.57	0.18-0.96	6.5-58.3	20.3	6.6-55.8
SPT-N	11	12-61	27	1.75-75.3	33.0	1.75-75.3	15.9-57	30.7	15.9-57
E_{DMT} (MPa)	25	10-32	17	0.71-33.7	7.2	0.76-32.36	4.6-45.8	24.0	5.3-56.6
E_{PMT} (MPa)	4	10-22	15	22.1-160.6	68.0	22.1-160.6	19.8-39.1	29.3	19.8-39.1
M_d (MPa)	5	10-13	11	0.49-4.60	2.66	0.49-4.60	20.8-46.8	34.6	20.8-46.8
K_0	8	10-264	45	0.48-2.88	1.28	0.48-2.88	2.4-22	13.5	2.4-22.0
K_{DMT}	47	10-50	18	1.34-15.12	3.91	1.70-12.69	6.2-49.4	18.2	6.3-40.6

Source: Table 1.2, Guan et al. (2021)

Note: LL = liquid limit; PL = plastic limit; PI = plasticity index; w = water content; LI = liquidity index; OCR = overconsolidation ratio; C_c = compression index; C_{ur} = unload/reload index; ϕ' = effective friction angle; s_u = undrained shear strength for clay; S_t = sensitivity; q_c = cone tip resistance; q_t = corrected cone tip resistance; $q_{t1} = (q_t - \sigma_v)/\sigma'_v$ = normalized cone tip resistance; B_q = CPT pore pressure ratio = $(u_2 - u_0)/(q_t - \sigma_v)$; u_0 = hydrostatic pore pressure; u_2 = pore pressure behind cone tip; SPT-N = standard penetration test blow count; E_{DMT} = soil modulus determined by dilatometer (DMT); E_{PMT} = soil modulus determined by pressuremeter (PMT); M_d = effective constrained modulus determined by oedometer; K_0 = at-rest lateral earth pressure coefficient; K_{DMT} = dilatometer horizontal stress index

and may not hold true in general. Guan et al. (2021) extended EPRI-105000 (Phoon et al. 1995) and showed that the dependency between the mean and COV exists for some rock parameters as well (Figure 2.8). The number of statistical studies on cement-mixed soils is limited. A summary is provided in Pan et al. (2018: table 4) (reproduced as Table 2.8).

The overall conclusions are (1) a simple total uncertainty analysis framework exists to evaluate the COV of any design soil/rock parameter; (2) considerable information on the COV of spatial variability is available (Tables 2.5–2.7 and Appendix A); (3) the COV of spatial variability covers a broad range in contrast to the COV for structural materials – reasons include the broad range of geo-materials and the dependency between the mean and COV; and (4) the COV of a design

TABLE 2.6
Summary of site-specific spatial variability statistics for sand.

Property	# groups	# cases/group Range	# cases/group Mean	Site-specific mean Range	Site-specific mean Mean	Site-specific mean 95% CI	Site-specific COV Range	Site-specific COV Mean	Site-specific COV 95% CI
e	6	11-17	14	0.47-0.63	0.55	0.47-0.63	7-19.9	11.1	7-19.9
ϕ' (°)	23	10-136	32	32.3-52	38.4	32.4-51.5	4.2-12.5	7.9	4.3-12.4
q_c	49	10-2039	125	0.7-26	3.3	0.85-13.17	17-81	39.7	17.0-77.4
q_{c1n}	25	10-28	15	14.1-254.6	90.4	14.2-247.4	11.5-68	36.9	11.9-68
SPT-N	26	10-300	62	6.8-74	32.9	6.8-73.3	18.4-62	34.3	18.5-61.0
$(N_1)_{60}$	9	11-35	21	5.7-28.6	15.3	5.7-28.6	16.5-38.8	32.2	16.5-38.8
E_{DMT} (MPa)	53	10-25	14	2.21-71.4	26.2	5.63-62.0	7-92	37.0	8.7-73.0
E_{PMT} (MPa)	7	10-53	26	5.24-26.1	12.6	5.24-26.1	15.7-68	34.3	15.7-68
K_0	4	13-15	15	0.64-2.20	1.16	0.64-2.20	25.8-36.9	33.1	25.8-36.9
K_{DMT}	15	10-25	15	1.9-28.3	15.1	1.9-28.3	20-99	44.3	20-99

Source: Table 1.3, Guan et al. (2021)

Note: e = void ratio; ϕ' = effective friction angle; σ'_v = vertical effective stress; P_a = atmospheric pressure = 101.3 kPa; q_c = cone tip resistance; $q_{c1n} = (q_c/P_a)/(\sigma'_v/P_a)^{0.5}$; SPT-N = standard penetration test blow count; N_{60} = corrected SPT-N; $(N_1)_{60} = N_{60}/(\sigma'_v/P_a)^{0.5}$; E_{DMT} = soil modulus determined by dilatometer (DMT); E_{PMT} = soil modulus determined by pressuremeter (PMT); K_0 = at-rest lateral earth pressure coefficient; K_{DMT} = dilatometer horizontal stress index

TABLE 2.7
Summary of site-specific spatial variability statistics for rock and rock mass.

Property	# groups	# cases/group Range	# cases/group Mean	Site-specific mean Range	Site-specific mean Mean	Site-specific mean 95% CI	Site-specific COV Range	Site-specific COV Mean	Site-specific COV 95% CI
n (%)	31	10-262	38	0.2-36.2	6.9	0.2-33.1	1.5-115.1	50.1	2.7-114.7
γ (kN/m^3)	56	10-778	44	5.4-30.1	24.6	18.0-28.1	0.4-21.5	5.2	0.6-18.5
V_P (km/s)	32	10-27	15	0.81-6.03	3.90	1.20-5.97	1.47-44.7	14.1	2.1-40.7
R_L	23	10-355	53	26.3-62.6	39.9	26.3-62.2	3.0-37.4	19.1	3.2-37.1
S_h	9	11-31	22	13.4-76.1	47.0	13.4-76.1	8.1-35.3	19.1	8.1-35.3
I_{s50} (MPa)	58	10-1305	63	0.17-9.04	3.69	1.21-9.02	5.1-91.5	34.4	5.1-91.4
σ_{bt} (MPa)	31	10-43	18	2.35-19.4	9.23	3.2-19.4	6.6-64.5	25.8	6.6-61.7
σ_{ci} (MPa)	116	10-470	29	1.9-226.9	66.6	8.7-151.2	5.7-108.4	33.8	6.6-84.1
E_i (GPa)	53	10-99	26	0.13-85.9	24.37	0.53-77.49	3.8-73.7	33.4	3.8-67.6
RQD	43	10-80	21	25.6-95.8	65.6	26.3-92.8	4.8-114.8	29.9	5.5-108.9
RMR	55	10-330	31	20.3-81.2	53.7	25.2-81.2	4.7-46.8	21.3	6.2-39.1
GSI	22	10-111	23	13.6-64.5	44.4	14.0-64.2	3.0-57.0	19.9	3.1-56.4
Q	26	10-28	18	0.13-74.28	11.7	0.16-70.17	17.6-303.5	104.7	19.4-289.4
E_m (GPa)	16	10-28	19	0.11-35.1	13.6	0.11-35.1	14.7-103.0	55.6	14.7-103.0

Source: Table 1.4, Guan et al. (2021)

Note: n = porosity; γ = unit weight; V_p = P-wave velocity; R_L = L-type Schmidt hammer hardness; S_h = shore scleroscope hardness; I_{s50} = point load strength index for diameter 50 mm; σ_{bt} = Brazilian tensile strength; σ_{ci} = uniaxial compressive strength of intact rock; E_i = Young's modulus of intact rock; GSI = geological strength index; I_{d2} = slake durability index; RQD = rock quality designation; RMR = rock mass rating; GSI = geological strength index; Q = Q-system; E_m = deformation modulus of rock mass

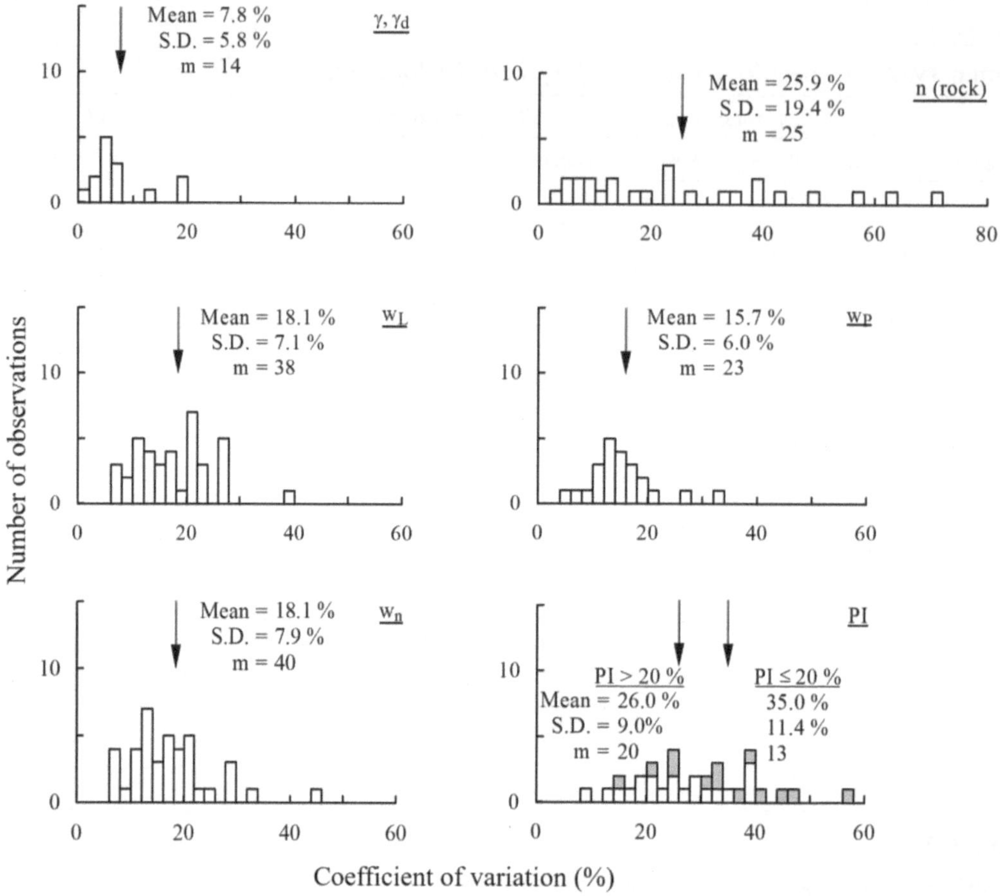

FIGURE 2.4 Coefficients of variation of index parameters for soil and rock (spatial variability). (Source: Figure 1, Kulhawy et al. 2000). Note: γ = total unit weight; n = apparent porosity; and γ_d = dry unit weight; w_n =natural water content; w_P = plastic limit; w_L = liquid limit; PI = plasticity index.

parameter also covers a broad range, because it depends on the site condition, measurement process, and quality of the correlation model.

The COV is a key input for reliability analysis and reliability-based design. A simplified first-order second-moment reliability analysis is presented below to illustrate the application of COVs in practical decision-making (Duncan 2000):

a. Estimate the standard deviation (σ_X) of each uncertain soil parameter using the "N-sigma rule":

$$\sigma_X = \frac{\text{HCV} - \text{LCV}}{4} \tag{2.7}$$

where HCV is the highest conceivable value and LCV is the lowest conceivable value. For a normal distribution, the exact denominator is 3.92 if (LCV, HCV) is the 95% confidence interval. This is a rough guide for engineers who are expected to estimate (LCV, HCV) based on judgment. However, even in the absence of site-specific data, σ_X can also be obtained from $X_{\text{mean}} \times \text{COV}_X$ with the help of Tables 2.3–2.7. There are frequentist and Bayesian methods to compute σ_X directly from site-specific data when it is available.

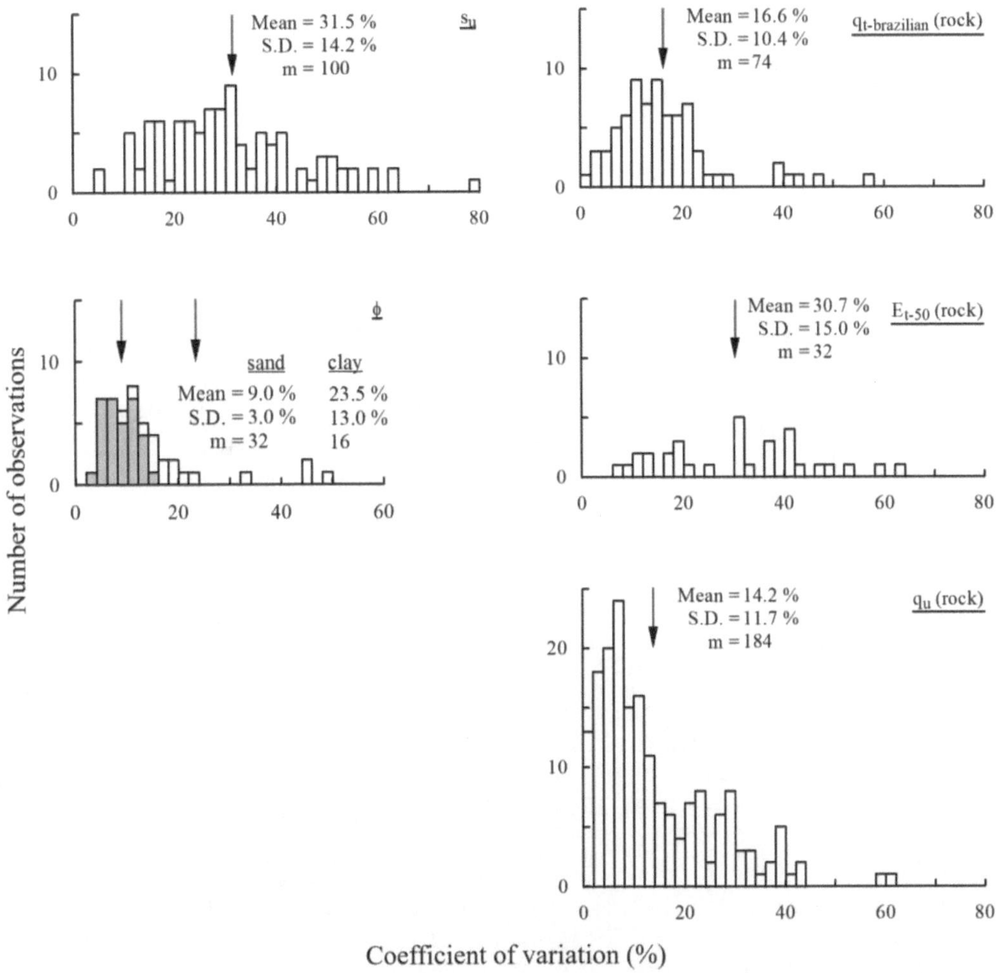

FIGURE 2.5 Coefficients of variation of laboratory parameters for soil and rock (spatial variability) (Source: Figure 2, Kulhawy et al. 2000). Note: ϕ = effective stress friction angle; s_u = undrained shear strength; q_u = uniaxial compressive strength; $q_{t\text{-brazilian}}$ = Brazilian indirect tensile strength; $E_{t\text{-}50}$ = tangent modulus at $0.5q_u$.

b. Estimate the standard deviation of the factor of safety (σ_{FS}) using:

$$\sigma_{FS} = \sqrt{\left(\frac{\Delta F_1}{2}\right)^2 + \left(\frac{\Delta F_2}{2}\right)^2 + \cdots} \tag{2.8}$$

where $\Delta F_1 = (F_1^+ - F_1^-)$. F_1^+ is the factor of safety calculated with the value of the first uncertain soil parameter increased by one standard deviation from its best estimate value and F_1^- is the factor of safety calculated with the value of the first parameter decreased by one standard deviation. In calculating F_1^+ and F_1^-, the values of all the other soil parameters are kept at their most likely values. Note that this seemingly logical approach of changing one parameter while keeping the rest unchanged violates the cross correlations between soil parameters. This problem can be solved by transforming correlated physical soil parameters to uncorrelated standard normal variables (Ching and Phoon 2015).

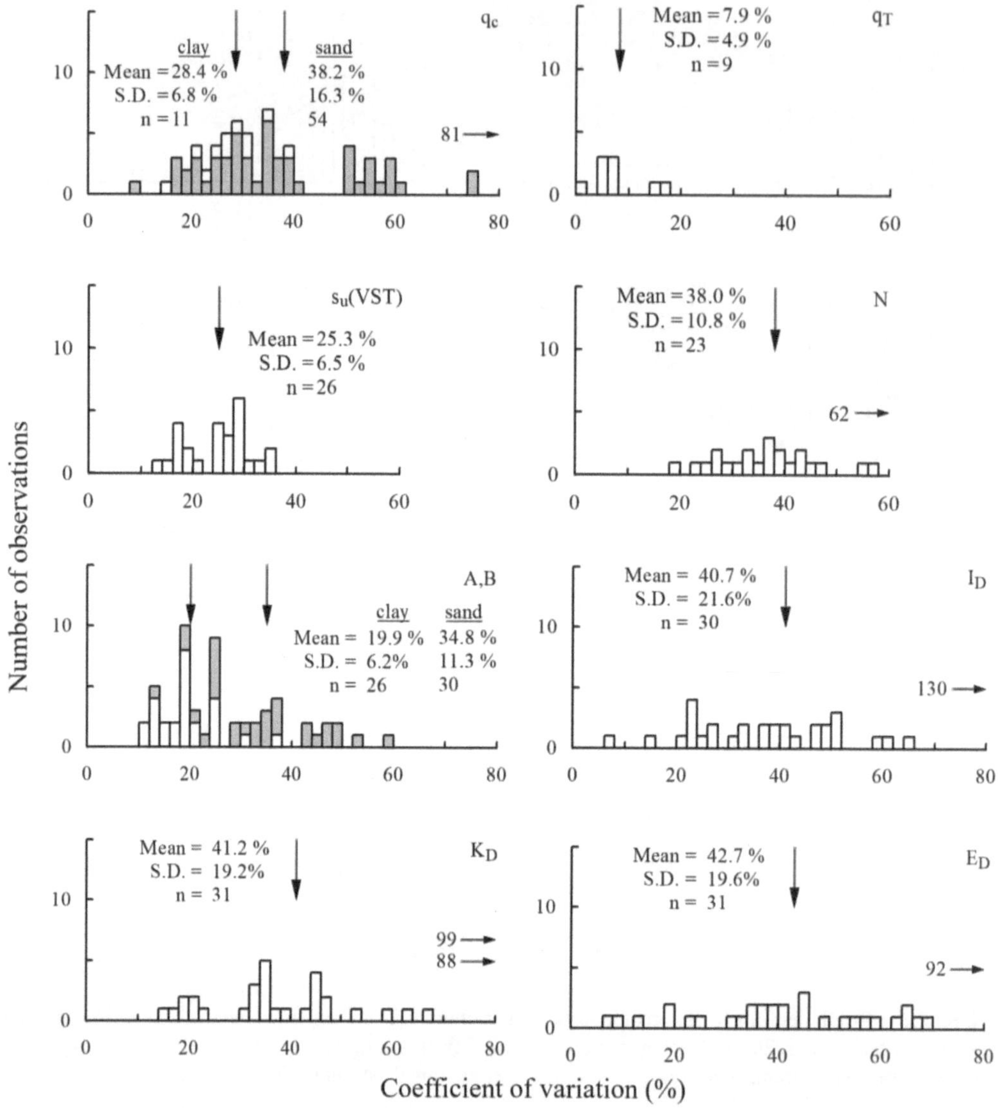

FIGURE 2.6 Coefficients of variation of field test parameters for soil (spatial variability) (Source: Figure 3, Kulhawy et al. 2000). Note: q_c = cone penetration test (CPT) tip resistance; q_T = corrected tip resistance; s_u(VST) = undrained shear strength from vane shear test; N = standard penetration test (SPT) blow count; A, B = dilatometer test (DMT) A and B readings; I_D = dilatometer material index; K_D = dilatometer horizontal stress index; E_D = dilatometer modulus.

c. Calculate the reliability index assuming the factor of safety is normally distributed:

$$\beta_{FOSM} = \frac{F_{MLV} - 1}{\sigma_{FS}} \tag{2.9}$$

where F_{MLV} is the most likely value of the factor of safety calculated using the most likely values of all soil parameters. A closed-form equation exists if FS is lognormally distributed as well. In reliability-based design, the goal is to achieve a reliability index larger than a target value (e.g., target reliability index = 3.8, Table C2, CEN 2002).

Limited knowledge/expertise in reliability is needed for FOSM. The values of F_{MLV}, F_1^+, F_1^-, are obtained by repeating the standard factor of safety calculation using different values of the soil

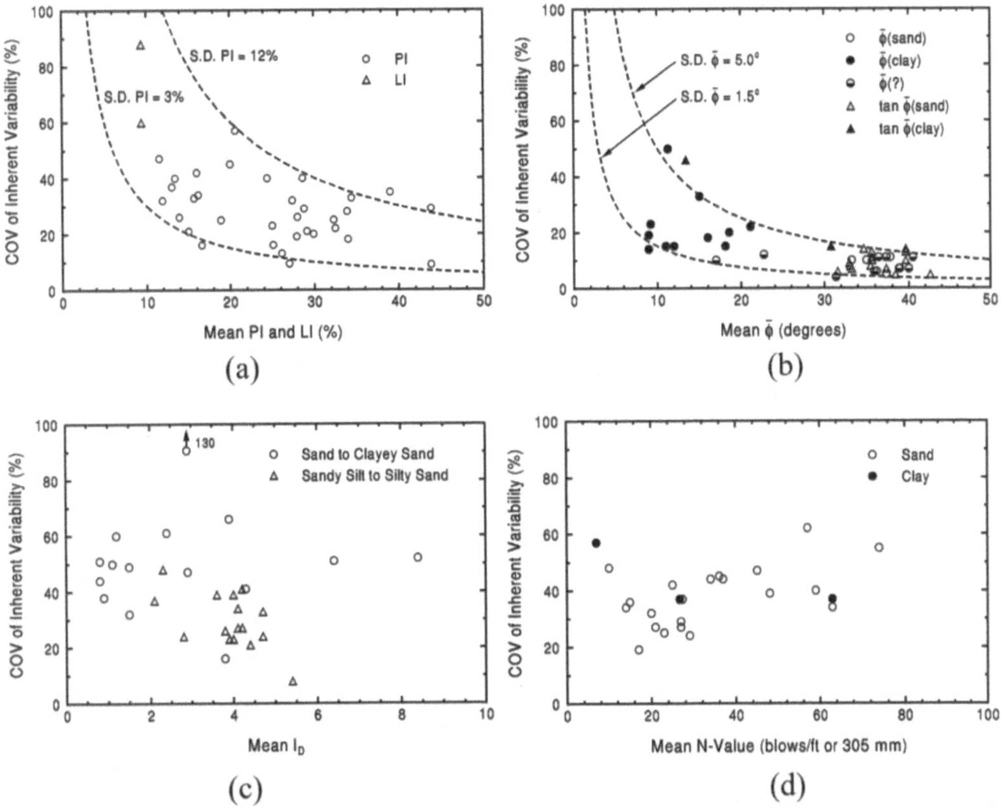

FIGURE 2.7 Dependency between mean and coefficient of variation (COV) for spatial variability: (a) plasticity index (PI) and liquidity index (LI), (b) effective friction angle ($\bar{\phi}$), (c) dilatometer material index (I_D), and (d) SPT blow count (N). (Source: Figures 4-5, 4-8, 4-11, 4-16, Phoon et al. 1995)

parameters. This is identical to a conventional parametric study. Hence, it is possible to obtain a useful probabilistic result for design in the form of a reliability index without using a probabilistic software. The standard deviation of each uncertain soil parameter can be estimated based on engineering judgment alone in the absence of site-specific data (Eq. 2.7), but it is better practice to refer to Tables 2.3–2.7 as a "reality check". Details are given in Duncan and Sleep (2015). The key take away is that results from a parametric study can be re-used to support reliability-informed design without incurring additional cost and efforts.

For design codes, reliability-based design is typically implemented in the form of a load and resistance factor design. It is important to emphasize that these resistance factors can be calibrated based on a range of COV (Table 2.9). It is not necessary to estimate the COV to decimal point precision for LRFD (Phoon et al. 2003a, 2003b). A similar approach was adopted by Paikowsky et al. (2004) in their reliability calibration of resistance factors for deep foundations. It appears that site variability is divided into low (COV < 25%), medium (25% < COV < 40%), and high (COV > 40%) in this National Cooperative Highway Research Program (NCHRP) study. In reference to the Canadian Highway Bridge Design Code (CAN/CSAS614:2014), Fenton et al. (2016) noted that "there is a real desire amongst the geotechnical community to have their designs reflect the degree of their site and modeling understanding". The value of a resistance factor in CAN/CSAS614:2014 depends on the "degree of understanding" (low, typical, high), but there is no quantitative guideline on how to assess the degree of understanding. The partial factors in Eurocode 7 do not vary according to an engineer's understanding of the variability at the site of interest. This site-specific knowledge is incorporated in the selection of a "cautious" characteristic value. Engineering judgment is

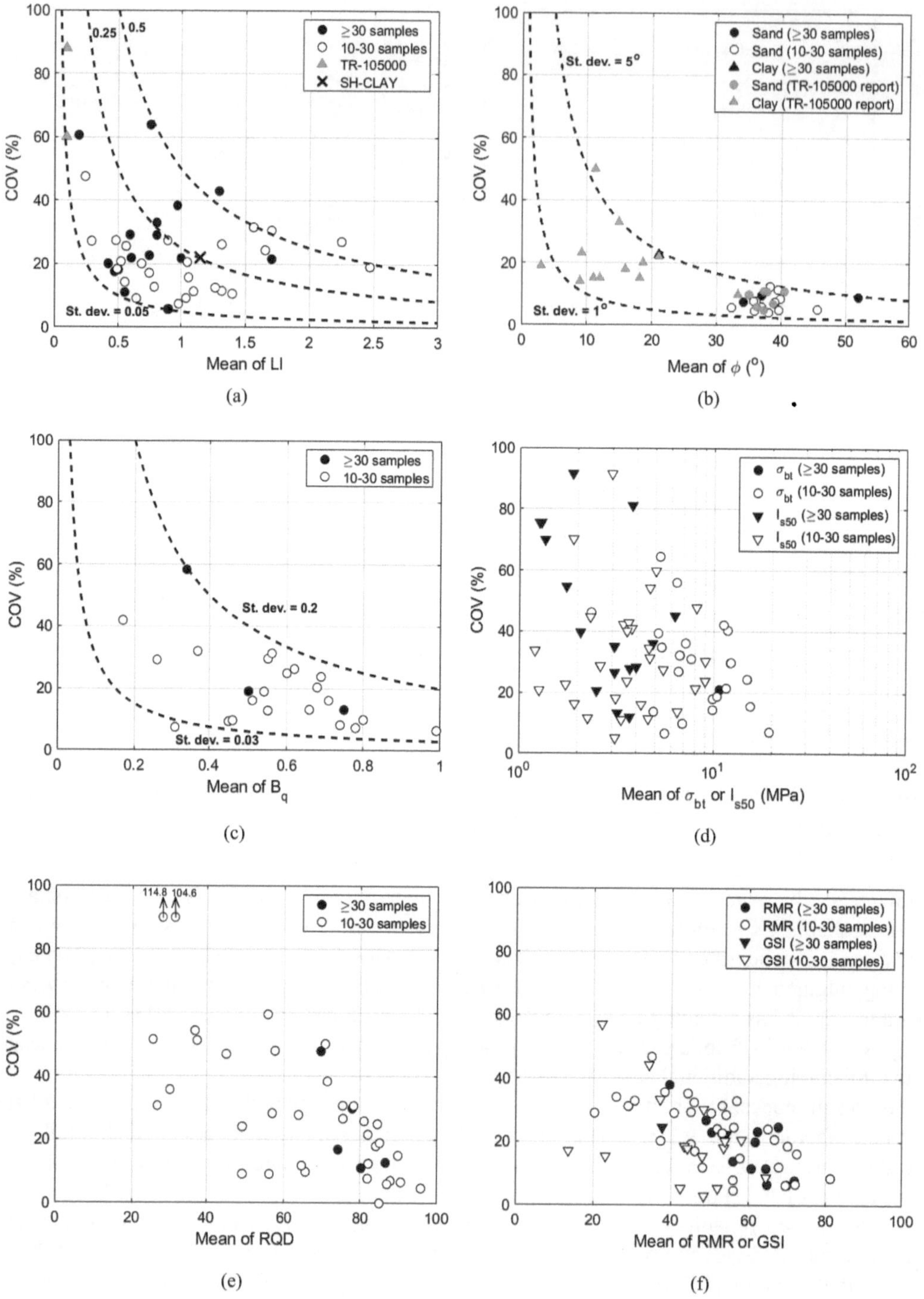

FIGURE 2.8 Dependency between mean and coefficient of variation (COV) for spatial variability: (a) liquidity index (LI), (b) effective friction angle (ϕ), (c) CPT pore pressure ratio (B_q), (d) Brazilian tensile strength (σ_{bt}) or point load strength index for diameter 50 mm (I_{s50}), (e) rock quality designation (RQD), and (f) rock mass rating (RMR) or geological strength index (GSI). (Source: Figures 1.2, 1.9, 1,14, 1.20, 1.23, 1.24, Guan et al. 2021)

TABLE 2.8
Statistical characteristics of cement-admixed soils

References		Test (Result)	Mean Value	COV	Scale of fluctuation[*] (m)		Skewness	Kurtosis	Marginal Distribution
					Vertical	Horizontal			
Honjo (1982)		Unconfined Compressive Test (UCS)	0.6-8.0 MPa	0.21-0.36(clay) 0.32-0.4(sandy soils)	0.8-8.0	-	-1.19 ~ 2.55	2.7-4.4	Normal
Babasaki et al. (1996)		Unconfined Compressive Test (UCS)	-	0.22-0.27	-	-	-	-	-
Hedman and Kuokkanen (2003)		Hand-operated penetrometer test (c_u)	-	-	0.38-1.12	0.07-0.33[‡]			
Navin and Filz (2005)		Unconfined Compressive Test (UCS)	1.0-4.7 MPa	0.34-0.79	-	Approximate 24.0	-	-	Lognormal
Larsson et al.(2005) [‡]		Hand-operated penetrometer test (c_u)	-	<0.60	-	Radial:<0.13 Orthogonal:<0.32[‡]	-	-	-
Larsson and Nilsson (2009)		Cone penetration test (Tip resistance)	-	0.20-0.60	-	1.8-3.6	-	-	-
Chen et al.(2011)	(MFBC)	Unconfined Compressive Test (UCS)	2.0-2.7 MPa	0.29-0.46	-	-	0.48 ~ 1.34	-	-
	(NCHS)	Unconfined Compressive Test (UCS)	3.2-4.5 MPa	0.29	-	-	-1.4 ~ -0.7	-	-
Al-Naqshabandy et al.(2012)		Cone penetration test (Tip resistance)	-	0.22-0.67	0.2-0.7	2.0-3.0	-	-	
Bergman et al. (2013)		Cone penetration test (Tip resistance)			0.08-0.77 m	<3.5 m			
Namikawa and Koseki (2013)		Unconfined compressive test (UCS)	1.7 MPa	0.2-0.4	-	-	-	-	Normal
Bruce et al. (2013)		Unconfined Compressive Test (UCS)	0.7-2.1 MPa	0.34-0.79	-	-		-	-
Chen et al. (2016)		Binder concentration	29%	0.19	-	-	-	-	Normal
Liu et al. (2017)[†]	(MFBC)	Unconfined Compressive Test (UCS)	1.7 MPa	0.42	-	-	1.10	4.67	Beta-distribution
	(Marina One)	Unconfined Compressive Test (UCS)	2.1 MPa	0.44	-	-	1.31	4.78	Beta-distribution

(Continued)

TABLE 2.8 (CONTINUED)
Statistical characteristics of cement-admixed soils

References	Test (Result)	Mean Value	COV	Scale of fluctuation* (m) Vertical	Horizontal	Skewness	Kurtosis	Marginal Distribution
Liu et al. (2019)	Centrifuge test (Binder Concentration)			1.0-3.33	**Small Scale[c] SOF:** Intracolumn: Radial: 0.12D-0.28D Circumferential: 67°-133° Intercolumn: 0.12D-0.28D **Large Scale[d] SOF:** 25 m			

(*Source:* updated from Pan et al. 2018)
Notes:
*The concept "auto-correlation distance" used in some studies (e.g. Namikawa and Koseki 2013) is converted to "scale of fluctuation" Vanmarcke (1983) by multiplying 2.0;
†Liu et al. (2017) normalized the strength to 28-day equivalent strength to eliminate the effect of curing period.
‡SOF within the column cross-section
c due to uneven distribution of binder during mixing
d due to effect of in-situ water content

TABLE 2.9
Three-tier classification scheme of soil property variability for reliability calibration.

Geotechnical parameter	Property variability	COV (%)
Undrained shear strength	Low[a]	10 – 30
	Medium[b]	30 – 50
	High[c]	50 – 70
Effective stress friction angle	Low[a]	5 – 10
	Medium[b]	10 – 15
	High[c]	15 – 20
Horizontal stress coefficient	Low[a]	30 – 50
	Medium[b]	50 – 70
	High[c]	70 – 90

(*Source*: Table 9.7, Phoon and Kulhawy 2008)
a – typical of good quality direct lab or field measurements
b – typical of indirect correlations with good field data, except for the standard penetration test (SPT)
c – typical of indirect correlations with SPT field data and with strictly empirical correlations

crucial, but there is limited guidance on how data can inform judgment. If the characteristic value is defined as the mean value, then site-specific variability is given due consideration in a variable factor approach (e.g., Boverket 1995; Phoon et al. 2003b; Paikowsky et al. 2004; Fenton et al. 2016).

2.4 STATISTICS OF SPATIAL AVERAGE

Spatial variability exists in natural deposits or formations for two reasons: (1) properties and/or geometric features (such as stratification discussed below, voids, and discontinuities) are spatially heterogeneous and (2) data is too limited to produce a single deterministic solution regardless of the site characterization method used (e.g., volume fraction for logging at a Brent Field site in the North Sea is 1×10^{-6}; Chilès and Delfiner 1999). A large part of the literature on spatial variability is founded on random field theory. As noted above, the *practice* of random field theory is more limited because statistical characterization is largely missing. This section attempts to fill this gap and provides numerical guidance for ISO2394:2015 (ISO 2015), Section D.2.3 "Inherent variability" and Section D.2.7 "Scale of fluctuation".

Vanmarcke's (1977a) classic paper on "Probabilistic Modeling of Soil Profiles" is arguably the first to introduce random field theory to geotechnical engineering. Random field theory provides a mathematically tractable framework to model spatial variability (Vanmarcke and Fenton 2003). Vanmarcke's (1983) key observations in his book *Random Fields: Analysis and Synthesis* are

1. Spatially averaged soil properties are more relevant to geotechnical engineering problems because soil–structure interaction mobilizes a finite volume of the ground (cf. Clause 2.4.5.2[7], Eurocode 7 [CEN 2004]). For example, Figure 2.1a illustrates the calculation of a spatial average (X_A) between depth z_1 and depth z_2:

$$X_A = \frac{1}{L}\int_{z1}^{z2} X(z)dz \tag{2.10}$$

where $L = |z_2 - z_1|$. Eq. (2.10) is a stochastic line integral. By definition, the integration path is *prescribed* with constant limits of integration.
2. The mean of X_A is the same as the mean of the point value X_{mean}. This follows by taking expectation on both sides of Eq. (2.10). In contrast, the mean of the mobilized value is always smaller than X_{mean} as discussed in Section 2.5.
3. The COV of this spatial average, COV_A, can be much smaller than the COV of the soil property at a point, COV_X. This uncertainty reduction can be calculated analytically using a variance reduction function, $\Gamma^2(L)$:

$$COV_A^2(L) = COV_X^2 \times \Gamma^2(L) \tag{2.11}$$

An approximate variance reduction function $\Gamma^2(L) \approx 1$ if $L < \delta$ and δ/L if $L > \delta$ was adopted in Eq. (2.4). Some indicative results over an averaging distance of 5 m are shown in Table 2.3.
4. This variance reduction function is dependent on a key random field parameter called the scale of fluctuation (δ), which can be regarded as a characteristic length parameter that elegantly unifies various classical one-parameter autocorrelation models, as shown in Table 2.10. Physically, a soil property measured at two spatial points can be regarded as strongly correlated if the distance apart is less than δ. Otherwise, the spatial correlation is weak.

There are different frequentist and Bayesian methods to estimate the scale of fluctuation from site data. The Bayesian methods are arguably more natural because statistical uncertainties can be significant when the sampling interval is too large, particularly in the horizontal direction (Ching et al. 2016a, 2017d, 2020a, 2021b; Ching and Phoon 2017). However, for practitioners, the graphical method shown in Figure 2.9 is likely to be most intuitive. This method is provided in Section D.2.7, ISO2394:2015 (ISO 2015). The equation in Figure 2.9 can be proven mathematically under the assumptions that the (1) random process is normal, (2) the autocorrelation function is a squared exponential model (QExp), and (3) the record length is sufficiently long (Zhu et al. 2019). Unfortunately, there are no analytical results for the other autocorrelation models shown in Table 2.10. The method of moments and the maximum likelihood estimation are commonly used

TABLE 2.10
Common autocorrelation models in geotechnical engineering

Autocorrelation model	Correlation as a function of lag τ	Smoothness ν	Frequency of usage
Single exponential (SExp)	$\rho(\tau)=\exp\left\{-2\frac{\lvert\tau\rvert}{\delta}\right\}$	0.5	48%
Second-order Markov (SMK)	$\rho(\tau)=\left(1+4\frac{\lvert\tau\rvert}{\delta}\right)\exp\left\{-4\frac{\lvert\tau\rvert}{\delta}\right\}$	1.5	5%
Third-order Markov (TMK)	$\rho(\tau)=\left(1+\frac{16}{3}\frac{\lvert\tau\rvert}{\delta}+\frac{256}{27}\left(\frac{\lvert\tau\rvert}{\delta}\right)^2\right)\exp\left\{-\frac{16}{3}\frac{\lvert\tau\rvert}{\delta}\right\}$	2.5	New to geotechnical practice
Squared exponential (QExp)	$\rho(\tau)=\exp\left\{-\pi\left(\frac{\lvert\tau\rvert}{\delta}\right)^2\right\}$	∞ (≈ WM with ν > 3.5)	19%
Spherical (Sph)	$\rho(\tau)=\begin{cases}1-\frac{9}{8}\left\lvert\frac{\tau}{\delta}\right\rvert+\frac{27}{128}\left\lvert\frac{\tau}{\delta}\right\rvert^3, & \text{if } \lvert\tau\rvert\le\frac{4}{3}\delta\\ 0, & \text{otherwise}\end{cases}$	Outside WM family	7%
Cosine exponential (CosExp)	$\rho(\tau)=\exp\left\{-\frac{\lvert\tau\rvert}{\delta}\right\}\cos\left\{\frac{\lvert\tau\rvert}{\delta}\right\}$	0.5	8%

(Continued)

TABLE 2.10 (CONTINUED)
Common autocorrelation models in geotechnical engineering

Autocorrelation model	Correlation as a function of lag τ	Smoothness ν	Frequency of usage
Binary noise (BN)	$\rho(\tau)=\begin{cases}1-\lvert\tau\rvert/\delta, & \text{if } \lvert\tau\rvert\le\delta \\ 0, & \text{otherwise}\end{cases}$	Outside WM family	12%
Whittle–Matérn (WM)	$\rho(\tau)=\frac{2}{\Gamma(\nu)}\left\{\frac{\sqrt{\pi}\,\Gamma(\nu+0.5)\lvert\tau\rvert}{\Gamma(\nu)\delta}\right\}^{\nu} K_{\nu}\left\{\frac{2\sqrt{\pi}\,\Gamma(\nu+0.5)\lvert\tau\rvert}{\Gamma(\nu)\delta}\right\}$	All ν	New to geotechnical practice
Cosine Whittle–Matérn (CosWM)	$\rho(\tau)=\frac{2^{1-\nu}}{\Gamma(\nu)}\left(\frac{\sqrt{2\nu}\,\lvert\tau\rvert}{s}\right)^{\nu} K_{\nu}\left(\frac{\sqrt{2\nu}\,\lvert\tau\rvert}{s}\right)\cos\left(\frac{\tau}{b}\right)$	All ν	New to geotechnical practice

Source: Table 2, Phoon et al. (2022c) citing Cami et al. (2020); Cami et al. (2021)

Note: δ = scale of fluctuation; ν = smoothness parameter that reduces the Whittle-Matérn model to a specific one-parameter autocorrelation model (e.g. $\nu = 0.5$ produces the Markovian exponential model); Γ = gamma function; and K_ν = modified Bessel function of second kind with order ν; δ is implicit function of s (scale parameter), ν, and b (hole parameter) for cosine Whittle–Matérn model; cosine exponential model can be regarded as a special case of cosine Whittle–Matérn model with $\nu = 0.5$ and a certain fixed relationship between s and b.

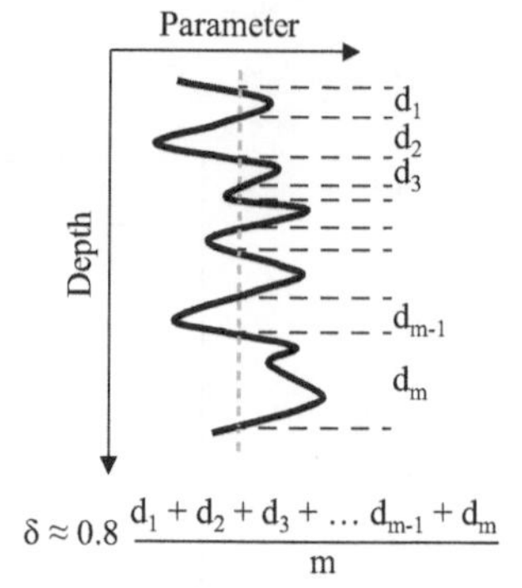

FIGURE 2.9 Estimation of vertical scale of fluctuation (δ) based on the average interval that a process fluctuates above or below the mean trend.

as well (Cami et al. 2020; 2021). A less common method is to back-calculate the scale of fluctuation from the variance reduction function (Fei et al. 2022).

To determine the COV for the spatial average, it is necessary to have four pieces of information:

1. COV of the point value, COV_X
   Extensive guidelines for selecting COV_X have been presented in Section 2.3.
2. Variance reduction function, $\Gamma^2(L)$
   The available analytical expressions for $\Gamma^2(L)$ are given in Table 2.11. To facilitate ease of calculation in practice, three approximate variance reduction functions are also listed. The approximate variance reduction function I (AVRF-I, first row in Table 2.11) is the classical one that found application in Eq. (2.4). AVRF-I and the approximate variance reduction function II (AVRF-II, second row in Table 2.11) are suitable approximations for the classical one-parameter autocorrelations models and the two-parameter WM model with a smoothness parameter $0.5 \le \nu < \infty$. The maximum errors are at most 0.43 for AVRF-I and 0.10 for AVRF-II, respectively. Note that the maximum value of a VRF is 1.0. Hence, an error of 0.43 is significant. For the most general three-parameter cosine Whittle–Matérn (CosWM) model proposed to date (Chang et al. 2021), an approximation variance reduction function is (1) AVRF-III (last row in Table 2.11) for $0.1 \le b/s < 1.0$ (relatively strong hole effect) and (2) AVRF-II for $1.0 \le b/s < \infty$ (relatively weak hole effect), where b is the hole (pseudo-periodicity) parameter. Details are given elsewhere (Tao et al. 2023).
3. Scale of fluctuation (δ)
   The scale of fluctuation in the vertical direction (δ_v) is generally much shorter than that in the horizontal direction (δ_h). Hence, many soil profiles look "layered", which is an indication that soil types and parameters tend to persist over a longer distance in the horizontal direction. Typical values of the scale of fluctuation are given in Tables 2.12 and 2.13. The detailed source data for Table 2.13 is given in Appendix B. Stuedlein et al. (2021) observed that δ_h/δ_v is between 3 and 500 with the ratio 10 to 20 as the most typical (Figure 2.10). For cement-mixed soils, typical values are provided by Liu et al. (2015) and Pan et al. (2018; 2019) (Table 2.8).
4. Spatial orientation of the averaging domain for an anisotropic field ($\delta_v \neq \delta_h$).
   For an averaging domain, say a line, that is vertical or horizontal, the COV of the spatial average can be obtained from Eq. (2.11) and substituting the relevant scale of fluctuation in the variance reduction function (Table 2.11). However, a trial slip line is rarely vertical or horizontal, as shown in Figure 2.1(c-2). In general, the scale of fluctuation along a line or curve in an anisotropic field is a function of δ_v, δ_h, and the trajectory of the averaging path. For a line oriented at an angle β to the horizontal as shown in Figure 2.1(c-2), Orr (2015) proposed the following equation:

TABLE 2.11
Variance reduction functions for common autocorrelation models in geotechnical engineering

Autocorrelation model	Variance reduction function, $\Gamma^2(L)$	Maximum error for AVRF-I[a]	(L/δ)[b]	Maximum error for AVRF-II[a]	(L/δ)[b]
Autocorrelation models for AVRF-I[c]	$\Gamma^2(L)=\begin{cases}1, \text{ if } L\le\delta\\ \delta/L, \text{ otherwise}\end{cases}$				
Autocorrelation models for AVRF-II[c]	$\Gamma^2(L)=\begin{cases}1-0.35\dfrac{L}{\delta}, \text{ if } L\le\delta\\ \dfrac{\delta}{L}-0.35\left(\dfrac{\delta}{L}\right)^2, \text{ otherwise}\end{cases}$				
Single exponential (SExp)	$\Gamma^2(L)=\frac{1}{2}(\frac{\delta}{L})^2\left(\exp\left(-\frac{2L}{\delta}\right)+\frac{2L}{\delta}-1\right)$	0.43	1	0.10	0.69
Second-order Markov (SMK)	$\Gamma^2(L)=\frac{\delta}{2L}\left(\exp\left(-\frac{4L}{\delta}\right)-\frac{3}{4}\frac{\delta}{L}\left(1-\exp\left(-\frac{4L}{\delta}\right)\right)+2\right)$	0.36	1	0.03	0.22
Third-order Markov (TMK)	$\Gamma^2(L)=\frac{\delta}{L}-\frac{45}{128}\left(\frac{\delta}{L}\right)^2+\exp\left(-\frac{16}{3}\frac{L}{\delta}\right)\left(\frac{45}{128}\left(\frac{\delta}{L}\right)^2+\frac{7}{8}\frac{\delta}{L}+\frac{2}{3}\right)$	0.34	1	0.04	0.29
Squared exponential (QExp)	$\Gamma^2(L)=\frac{1}{\pi}\left(\frac{\delta}{L}\right)^2\left(\exp\left(-\pi\left(\frac{L}{\delta}\right)^2\right)+\frac{\pi L}{\delta}\mathrm{Erf}\left(\frac{\sqrt{\pi}L}{\delta}\right)-1\right)$	0.32	1	0.06	0.41

(Continued)

TABLE 2.11 CONTINUED
Variance reduction functions for common autocorrelation models in geotechnical engineering

Autocorrelation model	Variance reduction function, $\Gamma^2(L)$	Maximum error for AVRF-I[a]	(L/δ)[b]	Maximum error for AVRF-II[a]	(L/δ)[b]
Spherical (Sph)	$\Gamma^2(L)=\begin{cases}1-\frac{3}{8}\frac{L}{\delta}+\frac{27}{1280}\left(\frac{L}{\delta}\right)^3, & \text{if } L\le\frac{4}{3}\delta\\ \frac{\delta}{L}-\frac{16}{45}\left(\frac{\delta}{L}\right)^2, & \text{otherwise}\end{cases}$	0.35	1	0.01	0.63
Cosine exponential (CosExp)	$\Gamma^2(L)=\frac{\delta}{L}-\left(\frac{\delta}{L}\right)^2\sin\left(\frac{L}{\delta}\right)\exp\left(-\frac{L}{\delta}\right)$	0.31	1	0.06	1.74
Binary noise (BN)	$\Gamma^2(L)=\begin{cases}1-\frac{1}{3}\frac{L}{\delta}, & \text{if } L\le\delta\\ \frac{\delta}{L}-\frac{1}{3}\left(\frac{\delta}{L}\right)^2, & \text{otherwise}\end{cases}$	0.33	1	0.02	1.00
Whittle–Matérn (WM)	$\Gamma^2(L)=\begin{cases}\Gamma^2(L) \text{ for SExp if } \nu=0.5\\ \Gamma^2(L) \text{ for SMK if } \nu=1.5\\ \Gamma^2(L) \text{ for TMK if } \nu=2.5\\ \Gamma^2(L) \text{ for QExp if } \nu\to\infty\end{cases}$	0.43 for $\nu \ge 0.5$	1	0.10 for $\nu \ge 0.5$	0.69

(Continued)

TABLE 2.11 CONTINUED
Variance reduction functions for common autocorrelation models in geotechnical engineering

Autocorrelation model	Variance reduction function, $\Gamma^2(L)$	Maximum error for AVRF-I[a]	(L/δ)[b]	Maximum error for AVRF-II[a]	(L/δ)[b]
Cosine Whittle–Matérn (CosWM)	$\Gamma^2(L)=\frac{2\left(\frac{b}{s}\right)^2}{\left[\left(\frac{b}{s}\right)^2+1\right]^2\left(\frac{L}{s}\right)^2}\left\{\begin{array}{c}\left(\frac{b}{s}\right)^2\left(\frac{L}{s}-1\right)+\frac{L}{s}+1+\exp\left(-\frac{L}{s}\right)\\ \left[\left(\left(\frac{b}{s}\right)^2-1\right)\cos\left(\frac{L}{s}\cdot\frac{s}{b}\right)-2\frac{b}{s}\sin\left(\frac{L}{s}\cdot\frac{s}{b}\right)\right]\end{array}\right\}$ if $\nu=0.5$[d]	–[e]	–	–	–
CosWM with $0.1 \le b/s < 1.0$	$\Gamma^2(L)=\begin{cases}\left(1-\lambda_1\frac{L}{s}\right)^{\lambda_2}, & \text{if } \frac{L}{s}\le\lambda_3\\ \left[\frac{s}{L}-\lambda_1\left(\frac{s}{L}\right)^2\right]^{\lambda_2}, & \text{otherwise}\end{cases}$ $\lambda_1=\frac{1}{\lambda_3+\frac{1}{\lambda_3}+1}$, $\lambda_2=\exp\left[0.13\left(\ln\frac{b}{s}\right)^2-0.67\ln\frac{b}{s}-0.14\right]$ $\lambda_3=-1.35\left(\frac{b}{s}\right)^2+3.26\frac{b}{s}+0.63$	–	–	–	–

Source: modified from Table 3, Tao et al. (2023)

a – maximum error between approximate and true variance reduction function over $L/\delta < 5$

b – value of L/δ corresponding to maximum error

c – Autocorrelation models for AVRF-I include the classical one-parameter models, the WM model ($0.5 \le \nu < \infty$), and the CosWM model ($0.5 \le \nu < \infty$, $1.0 \le b/s < \infty$)

d – for CosWM, a simple closed-form variance reduction function is only available for $\nu = 0.5$

e – Neither AVRF-I nor AVRF-II is suitable for CosWM with a strong hole effect

Erf – error function

TABLE 2.12
Scales of fluctuation for some soil parameters.

Direction	Parameter	Soil type	#studies	Scale of fluctuation (m) Range	Mean
Vertical	s_u	Clay	5	0.8–6.1	2.5
	q_c	Sand, clay	7	0.1–2.2	0.9
	q_T	Clay	10	0.2–0.5	0.3
	s_u(VST)	Clay	6	2.0–6.2	3.8
	N	Sand	1	–	2.4
	w_n	Clay, loam	3	1.6–12.7	5.7
	w_L	Clay, loam	2	1.6–8.7	5.2
	$\overline{\gamma}$	Clay	1	–	1.6
	γ	Clay, loam	2	2.4–7.9	5.2
Horizontal	q_c	Sand, clay	11	3.0–80.0	47.9
	q_T	Clay	2	23.0–66.0	44.5
	s_u(VST)	Clay	3	46.0–60.0	50.7
	w_n	Clay	1	–	170.0

Source: Table 4-4, Phoon et al. (1995)

s_u = undrained shear strength from laboratory tests; s_u(VST) = s_u from vane shear test; q_c = cone tip resistance; q_T = corrected cone tip resistance; N = standard penetration test blow count; w_n = natural water content; w_L = liquid limit; $\overline{\gamma}$ = effective unit weight; γ = total unit weight

TABLE 2.13
Typical values for the vertical and horizontal scale of fluctuation.

	Scale of fluctuation (m) Horizontal				Vertical			
Soil type	No. studies	Min	Max	Average	No. studies	Min	Max	Average
Alluvial	9	1.07	49	14.2	13	0.07	1.1	0.36
Ankara Clay	-	-	-	-	4	1	6.2	3.63
Chicago Clay	-	-	-	-	2	0.79	1.25	0.91
Clay	9	0.14	163.8	31.9	16	0.05	3.62	1.29
Clay, Sand, Silt mix	13	1.2	1000	201.5	28	0.06	21	1.58
Hangzhou Clay	2	40.4	45.4	42.9	4	0.49	0.77	0.63
Marine Clay	8	8.37	66	30.9	9	0.11	6.1	1.55
Marine Sand	1	15	15	15	5	0.07	7.2	1.43
Offshore Soil	1	24.6	66.5	45.6	2	0.48	1.62	1.04
Over Consolidated Clay	1	0.14	0.14	0.14	2	0.063	0.255	0.15
Sand	9	1.69	80	24.5	14	0.1	4	1.17
Sensitive Clay	-	-	-	-	2	1.1	2.0	1.55
Silt	3	12.7	45.5	33.2	5	0.14	7.19	2.08
Silty Clay	7	9.65	45.4	29.8	14	0.095	6.47	1.40
Soft Clay	3	22.2	80	47.6	8	0.14	6.2	1.70
Undrained Engineered soil	-	-	-	-	22	0.3	2.7	1.42
Water Content	9	2.8	22.2	12.9	8	0.05	6.2	1.70

Source: Table 8, Cami et al. (2020)

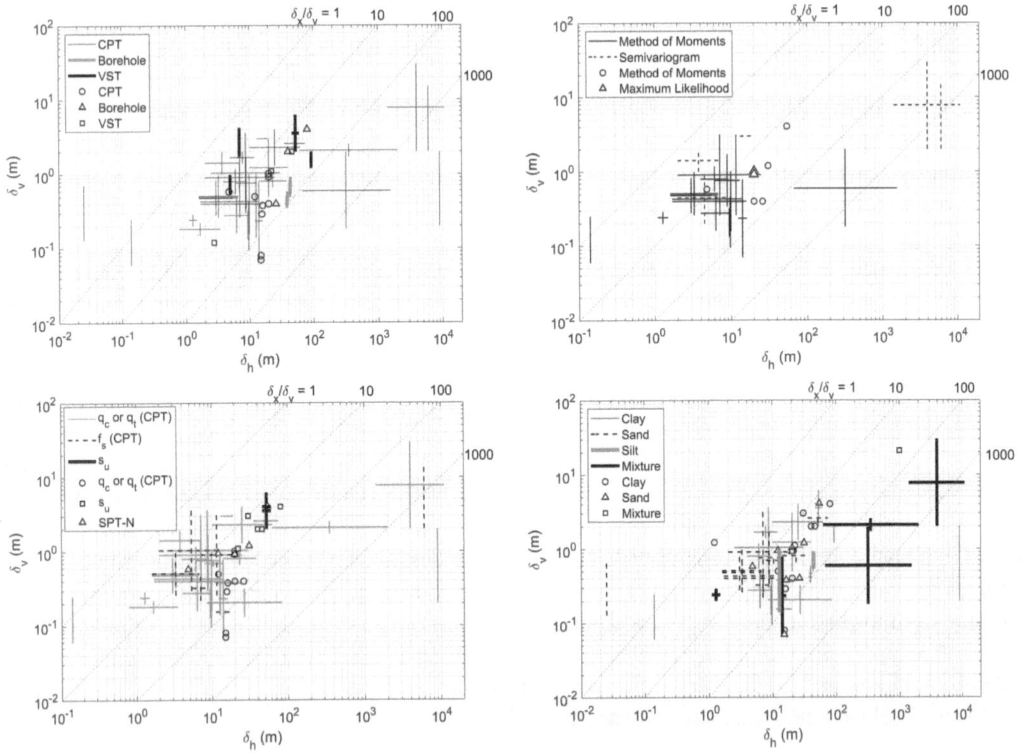

FIGURE 2.10 Variation of vertical scale of fluctuation, δ_v, with horizontal scale of fluctuation, δ_h, for various field tests, soil types, and methods of estimation. (Source: Figure 3.2, Stuedlein et al. 2021)

$$\delta_\beta = \frac{1}{\cos(\beta)/\delta_h + \sin(\beta)/\delta_v} \tag{2.12}$$

where δ_β is the scale of fluctuation along the potential inclined slip line direction. El-Ramly et al. (2006) adopted an isotropic equivalent δ_E for this inclined slip curve, which is calculated as the radius of a circle with the area equal to an ellipse with major axis δ_h and minor axis δ_v:

$$\delta_E = \sqrt{\delta_h \delta_v} \tag{2.13}$$

Tabarroki et al. (2022a) used an equivalent δ_E for a classical slip curve for a footing example (citing Vanmarcke 1977b). Their strategy is to box up the curve and assume the equivalent scale of fluctuation is a weighted average of δ_h and δ_v based on the horizontal and vertical sides of the box. Specifically, for the inclined line shown in Figure 2.1(c-2), the equivalent δ_E is as follows:

$$\delta_E = \frac{1}{\frac{\Delta h}{\Delta h + \Delta v} \cdot \frac{1}{\delta_h} + \frac{\Delta v}{\Delta h + \Delta v} \cdot \frac{1}{\delta_v}} \tag{2.14}$$

where Δh and Δv are the lengths of the horizontal and vertical sides of the box that contains the inclined line.

For a potential inclined slip line, the theoretical scales of fluctuation δ_β of five commonly used 2D separable autocorrelation models are summarized in Table 2.14. Detailed design charts for δ_β are provided in Appendix B. A comparison between the theoretical δ_β and the equivalent δ_E is shown in Figure 2.11. The equivalent scales of fluctuation δ_E are only applicable under special conditions and can lead to considerable errors in the general case. Tao et al. (2023) proposed an approximate variance reduction function for an inclined line, which is based on (1) the calculation of a theoretical scale of fluctuation δ_β and (2) a substitutability assumption that the variance reduction function for an inclined line is equal to the variance reduction function for a vertical line when the vertical δ_v is replaced by an inclined δ_β. This substitutability assumption is theoretically incorrect but found to be numerically accurate in all 2D autocorrelation models shown in Table 2.14.

The overall conclusions are (1) more realistic spatially varying soils can be modeled theoretically using an anisotropic random field; (2) there are at least two ramifications material to practice: (a) the COV of the spatial average along a prescribed curve can be much smaller than the COV for the point value and (b) more critical failure mechanisms and multiple failure modes are possible in a spatially heterogeneous soil mass (they do not exist in a conventional homogeneous or layered soil mass); (3) an anisotropic random field described by two parameters (vertical and horizontal scale of fluctuation, δ_v and δ_h) is sufficiently realistic for practice; (4) guidelines for the estimation of the scale of fluctuation are available (Cami et al. 2020); (5) some information on the vertical scale of fluctuation is available (Table 2.13 and Appendix B), the typical ratio of δ_h/δ_v is between 10 and 20, and very little is known about the spatial variability of cement-mixed soils (Table 2.8); (6) a broader class of spatial variability exhibiting three features (a) roughness (or conversely smoothness), (b) pseudo-periodicity (or hole effect), and (c) scale of fluctuation (related to correlation length) can be modeled using the cosine Whittle–Matérn autocorrelation model; (7) variance reduction functions are needed to quantify the reduction in uncertainty resulting from averaging along a potential slip line, which may be vertical, horizontal, or inclined; (8) Vanmarcke's approximate variance reduction function (AVRF-I) is not sufficiently accurate under more general conditions; and (9) there are more general anisotropy patterns such as rotated anisotropy (Huang and Leung 2021), but no statistical characterization has been reported.

2.5 STATISTICS OF MOBILIZED VALUE

The mobilized (or effective) property of the spatially variable soil mass is defined as the equivalent homogeneous soil mass property that matches the response of the geotechnical structure being considered in a limit state. In geotechnical design, there are two main types of limit states of concern: (a) the ultimate limit state (ULS) and (b) the serviceability limit state (SLS). The ULS usually concerns stability, which is affected by the soil shear strength. The SLS usually concerns deformation, which is affected by the soil modulus. This section summarizes some useful research findings for the characterization of the mobilized (or effective) property in a *spatially variable* soil for both ULS and SLS as well as for hydraulic conductivity in steady-state seepage. It will be clear that the mobilized properties for ULS and SLS exhibit different mechanisms:

1. For ULS, the mobilized shear strength exhibits a weak-zone seeking behavior, which can be characterized by the weakest path model.
2. For SLS, the effective Young's modulus can be approximated as a mobilization-based geometric spatial average, which can be characterized by the pseudo incremental energy model.
3. The effective hydraulic conductivity can also be approximated as a mobilization-based geometric spatial average.

It is important to emphasize that the mobilized value is not the same as the spatial average. Hence, the mean of the mobilized value may not be equal to the mean of the point value. Its variance may not be equal to the product of the variance of the point value and the classical variance reduction function (AVRF-I). In general, the probability distribution of the mobilized value depends on the limit state and spatial variability. The challenge is to estimate this probability distribution or its characteristic value (5% fractile of this distribution) while maintaining the simplicity and cost of Eq. (2.4). As shown in Figure 2.12, it is possible for the 5% fractile of the mobilized value to be the lowest and hence most critical for design among the 5% fractiles of the point value, spatial average, and mobilized value.

TABLE 2.14
Scale of fluctuation and variance reduction function for an inclined line

Autocorrelation model	Correlation as a function of lag τ_h and τ_v	Scale of fluctuation δ_β along an inclined line	Variance reduction function, $\Gamma^2_\beta(L)$
Single exponential (SExp)	$\rho(\tau_h,\tau_v)=\exp\left[-2\left(\frac{\tau_h}{\delta_h}+\frac{\tau_v}{\delta_v}\right)\right]$	$\delta_\beta=\frac{1}{\cos(\beta)/\delta_h+\sin(\beta)/\delta_v}$	$\Gamma^2_\beta(L)=\frac{1}{2}\left(\frac{\delta_\beta}{L}\right)^2\left(\exp\left(-\frac{2L}{\delta_\beta}\right)+\frac{2L}{\delta_\beta}-1\right)$
Second-order Markov (SMK)	$\rho(\tau_h,\tau_v)=\exp\left[-4\left(\frac{\tau_h}{\delta_h}+\frac{\tau_v}{\delta_v}\right)\right]\left(1+\frac{4\tau_h}{\delta_h}\right)\left(1+\frac{4\tau_v}{\delta_v}\right)$	$\delta_\beta=\frac{\delta_h\delta_v\left(\sin(\beta)^2\delta_h^2+3\cos(\beta)\sin(\beta)\delta_h\delta_v+\cos(\beta)^2\delta_v^2\right)}{\left(\sin(\beta)\delta_h+\cos(\beta)\delta_v\right)^3}$	—[a]
Squared exponential (QExp)	$\rho(\tau_h,\tau_v)=\exp\left[-\pi\left(\frac{\tau_h^2}{\delta_h^2}+\frac{\tau_v^2}{\delta_v^2}\right)\right]$	$\delta_\beta=\sqrt{\frac{1}{\cos^2(\beta)/\delta_h^2+\sin^2(\beta)/\delta_v^2}}$	$\Gamma^2_\beta(L)=\frac{1}{\pi}\left(\frac{\delta_\beta}{L}\right)^2\left(\exp\left(-\pi\left(\frac{L}{\delta_\beta}\right)^2\right)+\frac{\pi L}{\delta_\beta}\mathrm{Erf}\left(\frac{\sqrt{\pi}L}{\delta_\beta}\right)-1\right)$
Cosine exponential (CosExp)	$\rho(\tau_h,\tau_v)=\exp\left[-\left(\frac{\tau_h}{\delta_h}+\frac{\tau_v}{\delta_v}\right)\right]\cos\left(\frac{\tau_h}{\delta_h}\right)\cos\left(\frac{\tau_v}{\delta_v}\right)$	$\delta_\beta=\frac{\delta_h\delta_v\left(\sin(\beta)^2\delta_h^2+\cos(\beta)\sin(\beta)\delta_h\delta_v+\cos(\beta)^2\delta_v^2\right)}{\left(\sin(\beta)\delta_h+\cos(\beta)\delta_v\right)\left(\sin(\beta)^2\delta_h^2+\cos(\beta)^2\delta_v^2\right)}$	—[a]

(Continued)

TABLE 2.14 (CONTINUED)

Scale of fluctuation and variance reduction function for an inclined line

Autocorrelation model	Correlation as a function of lag τ_h and τ_v	Scale of fluctuation δ_β along an inclined line	Variance reduction function, $\Gamma^2_\beta(L)$
Binary noise (BN)	$\rho(\tau_h, \tau_v) = \begin{cases} \left(1 - \frac{\tau_h}{\delta_h}\right)\left(1 - \frac{\tau_v}{\delta_v}\right), & \tau_h \le \delta_h \text{ and } \tau_v \le \delta_v \\ 0, & \text{else} \end{cases}$	$\delta_\beta = \begin{cases} 3\delta_h\delta_v \sec(\beta) \frac{-\delta_h^2 \sec(\beta)\tan(\beta)}{3\delta_v}, & \text{if } \frac{\delta_h}{\cos(\beta)} \le \frac{\delta_v}{\sin(\beta)} \\ 3\delta_h\delta_v \csc(\beta) \frac{-\delta_v^2 \csc(\beta)\cot(\beta)}{3\delta_h}, & \text{else} \end{cases}$	—[a]

Source: Table 4, Tao et al. (2023)

a – The substitutability assumption of variance reduction function is theoretically correct for the 2D SExp and QExp, but theoretically incorrect for the 2D separable SMK, CosExp, and BN.

2.5.1 Ultimate Limit State

The weakest path model originates from the observations made by Ching and Phoon (2013b). They simulated the mobilized shear strength of a spatial variable soil subjected to a simple stress state (e.g., uniaxial compression) using the random finite element method. A key observation was that the mobilized shear strength is equal to the spatial average along the *actual* slip curve. Note that the actual slip curve is not a prescribed curve but an emergent curve that is the solution of the boundary value problem in RFEM. Based on this observation, Ching and Phoon (2013c) proposed the following simple model to characterize the mobilized shear strength:

$$X^{\text{mob}} = \min\left(X_1^{\text{ave}}\ X_2^{\text{ave}}, \ldots, X_m^{ave}\right) \tag{2.15}$$

where X^{mob} is the mobilized shear strength, the spatial average along the actual slip curve; X_i^{ave} is the spatial average along the *i*th potential slip curve; m is the number of equivalent independent potential slip curves. Note that X_i^{ave} is identical to Vanmarcke's spatial average, but the minimization operator in Eq. (2.15) results in a new random variable that is distinct from Vanmarcke's spatial average. For ULS design in clay, X is the undrained shear strength (s_u), and for ULS design in sand, X is the tangent friction angle [tan(ϕ)]. Equation (2.15) states that the actual slip curve is the potential

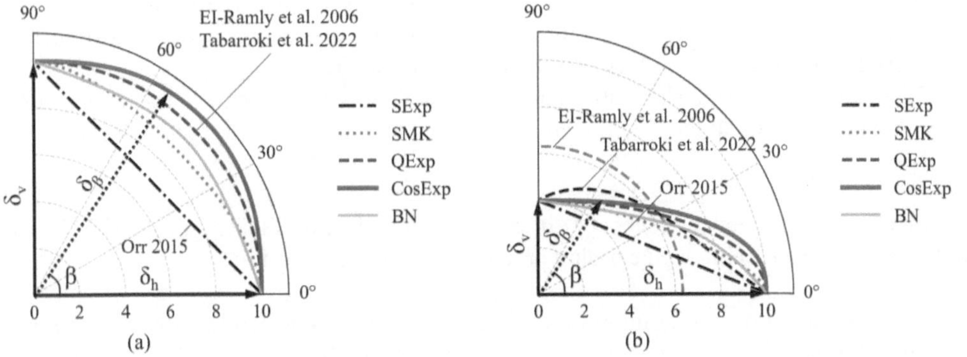

FIGURE 2.11 Scale of fluctuation along an inclined line in (a) isotropic ($\delta_h = \delta_v = 10$ m) and (b) anisotropic ($\delta_h = 10$ m, $\delta_v = 4$ m) random fields. (Source: Figure 9, Tao et al. 2023)

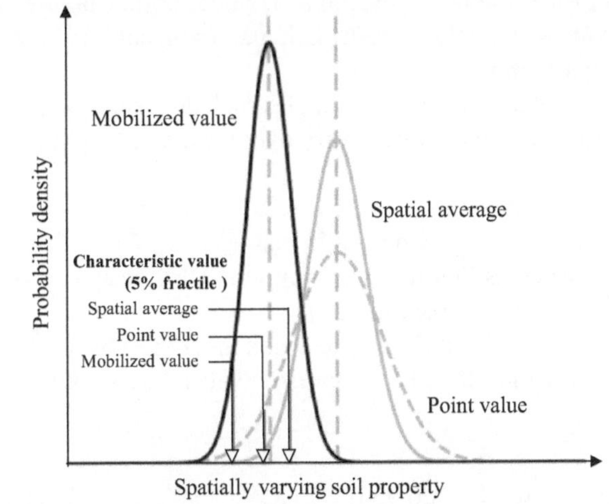

FIGURE 2.12 Probability distributions of point, spatial average, and mobilized values.

slip curve with the lowest resistance, i.e., the actual slip curve seeks the weakest path. The parameter "m" governs the tendency of weak-zone seeking (tendency is strong when m is large). The length and orientation for each potential slip curve can be approximated as those of the classical slip curve for a homogeneous problem. For instance, the classical slip curve for the active failure wedge behind a retaining wall is a line with an inclination angle $\beta = 45° + \phi/2$. The model in Eq. (2.15) was formally named as the "weakest path model" in Ching et al. (2017c) and Tabarroki et al. (2022a).

The parameter "m" in the weakest path model governs the tendency of weak-zone seeking. It can be calibrated by the mobilized shear strength data simulated by RFEM. Tabarroki et al. (2022a) showed that the weakest path model with the calibrated m can satisfactorily reproduce the statistical bahaviors of the mobilized shear strength simulated by RFEM for different ULS design examples, including friction pile, retaining wall, footing, and basal heave in excavation. Note that the simulation of the mobilized shear strength is computationally costly because it requires RFEM. Nonetheless, the calibrated weakest path model can simulate approximate mobilized shear strength with minimal computational cost. Tabarroki et al. (2022a) found that the calibrated m is mainly correlated to three factors:

1. m decreases with increasing δ_β/L (δ_β is the SOF along the classical slip curve; L is the length of the classical slip curve). When δ_β/L approaches infinity (homogeneous), m approaches 1 (no weak-zone seeking).
2. m decreases with increasing constraint on weak-zone seeking. When there is no weak-zone seeking (e.g., a friction pile subjected to axial compression; the actual failure surface is fixed at the pile-soil interface on pile shaft), m equals 1.
3. COV_X and δ_h/δ_v of the shear strength (point) random field also have a secondary effect on m: m increases slightly with increasing COV_X and decreases slightly with increasing δ_h/δ_v.

Based on the calibration m values for different design problems, Tabarroki et al. (2022a) classified the degree of constraint for the design problems into four categories:

1. Full constraint (m = 1): examples include a friction pile subjected to axial compression, where the actual slip curve is fully constrained at the shaft surface.
2. High constraint (m is relatively small): examples include the active failure of a retaining wall, where the actual slip curve must pass through the toe of the wall and the slip curve is usually close to a straight line.
3. Medium constraint (m is medium): examples include footing failure and basal heave in excavation, where the actual slip curve usually passes through the corners of the footing or the tip of the diaphragm wall.
4. Low constraint (m is relatively large): examples include a soil column subjected to axial loading, where the actual slip curve can vary over the entire height of the column.

Figure 2.13 shows how the calibrated m varies with δ_β/L for problems with different degrees of constraint. Figure 2.13a shows the ULS design in clay ($X = s_u$, COV_X is assumed to be 0.3, and δ_h/δ_v is assumed to be 1), whereas Figure 2.13b shows the ULS design in sand ($X = \tan(\phi)$, COV_X is assumed to be 0.1, and δ_h/δ_v is assumed to be 1).

Tabarroki et al. (2022a) showed that the weakest path model can be used to derive the following equation for the ULS mobilization-based characteristic value that has the same "look and feel" as Eq. (2.4):

$$X_k^{mob} = X_{mean}\left[1 + \Phi^{-1}\left(1 - \sqrt[m]{0.95}\right)\Gamma(L)COV_X\right] = X_{mean}\left[1 - k\Gamma(L)COV_X\right] \quad (2.16)$$

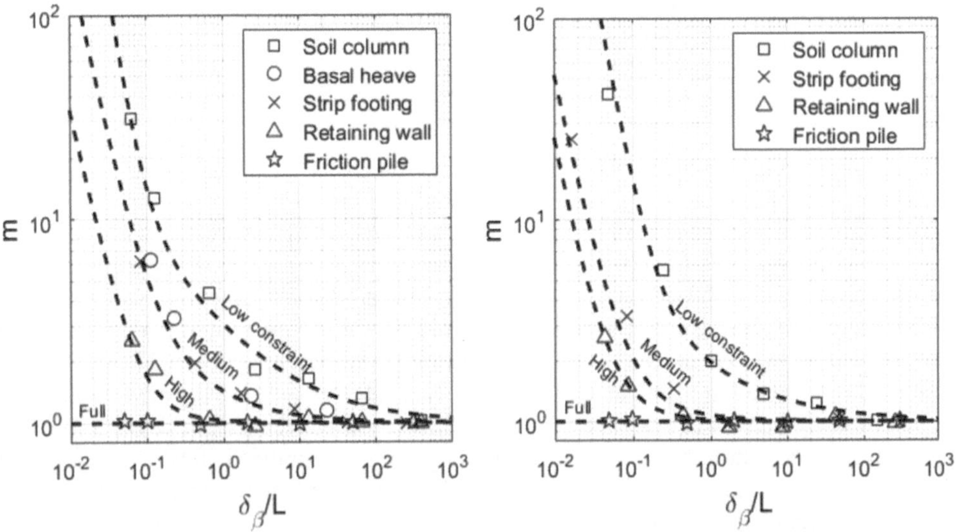

(a) Clay ($X = s_u$) (b) Sand ($X = \tan(\phi)$)

FIGURE 2.13 Variation of m with respect to δ_β/L for problems with different degrees of constraint: (a) ULS design in clay ($X = s_u$, $COV_X = 0.3$, and $\delta_h/\delta_v = 1$); (b) ULS design in sand ($X = \tan(\phi)$, $COV_X = 0.1$, and $\delta_h/\delta_v = 1$).

where the term $\Gamma(L)\ COV_X$ is exactly the COV of Vanmarcke's spatial average along the classical slip curve:

$$k = -\Phi^{-1}\left(1 - \sqrt[m]{0.95}\right) \tag{2.17}$$

where k is equal to the negative of the 5% fractile of min(Z_1, Z_2,…, Z_m) (it is a positive number). Figure 2.14 shows how k varies with δ_β/L for problems with different degrees of constraint. The parameter k is called the "mobilization factor" by Länsivaara et al. (2022). Note that if m = 1

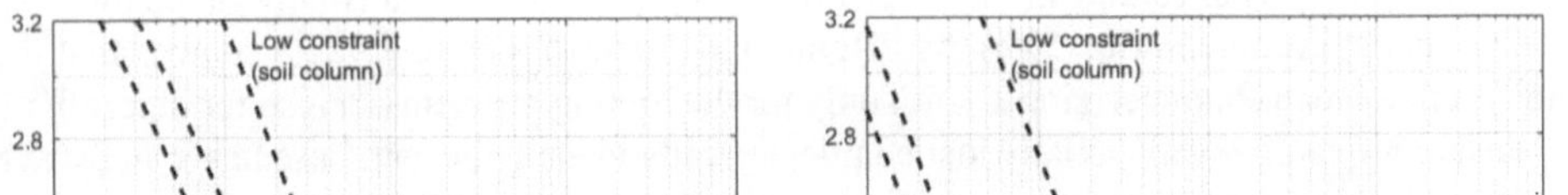

(a) Clay ($X = s_u$) (b) Sand ($X = \tan(\phi)$)

FIGURE 2.14 Variation of k with respect to δ_β/L for problems with different degrees of constraint: (a) ULS design in clay ($X = s_u$, $COV_X = 0.3$, and $\delta_h/\delta_v = 1$); (b) ULS design in sand ($X = \tan(\phi)$, $COV_X = 0.1$, and $\delta_h/\delta_v = 1$). (Source: Figure 4, Länsivaara et al. 2022) (no weak-zone seeking), $k = -\Phi^{-1}(1 - 0.95) = 1.645$, and if $\Gamma(L)$ is further approximated as $(\delta/L)^{0.5}$, Eq. (2.16) reduces to Eq. (2.4).

2.5.2 Serviceability Limit State

The mechanism for SLS is different from that for ULS. There is no clear weak-zone seeking behavior. Instead, the mobilized property for SLS (called the effective Young's modulus, denoted by E^{eff}) can be approximated as the mobilization-based geometric spatial average, as described in this section. For a spatially variable soil mass subjected to a simple stress state (e.g., a cube subjected to uniaxial compression), Ching et al. (2016d, 2017b) showed that E^{eff} simulated by RFEM can be characterized by Vanmarcke's spatial averaging, which is simply the spatial average of Young's modulus (E) values over all elements with uniform weights. Vanmarcke's spatial averaging works well for a cube subjected to a simple stress state because all elements are equally mobilized. For a footing problem, the elements are unequally mobilized: the elements right below a footing are highly mobilized, but those remote from the footing are not mobilized. Based on the E^{eff} data simulated by RFEM, Ching et al. (2018b) found that E^{eff} can be well approximated as the geometric spatial average with *non-uniform* weights:

$$\ln(E^{\mathrm{eff}}) \approx \sum_{i=1}^{N} w_i \ln(E_i) \Big/ \left[\sum_{i=1}^{N} w_i\right] \tag{2.18}$$

where E_i is Young's modulus of the ith element in RFEM; w_i is its weight; and N is the total number of elements. The weights ($w_1, w_2, \ldots, w_N$) quantify the degree of mobilization. For a footing problem, the elements right below a footing have larger weights because they are highly mobilized, and those remote from the footing have negligible weights because they are not mobilized. Ching et al. (2018b) further found that the weights ($w_1, w_2, \ldots, w_N$) can be approximately determined by a single deterministic finite element analysis using the pseudo incremental energy (PIE):

$$w_i = (\Delta\sigma_{x,i}\Delta\varepsilon_{x,i} + \Delta\sigma_{y,i}\Delta\varepsilon_{y,i} + \Delta\sigma_{z,i}\Delta\varepsilon_{z,i} + 2\Delta\tau_{xy,i}\Delta\varepsilon_{xy,i} + 2\Delta\tau_{yz,i}\Delta\varepsilon_{yz,i} + 2\Delta\tau_{xz,i}\Delta\varepsilon_{xz,i}) \times V_i \tag{2.19}$$

where $(\Delta\sigma_{x,i}, \Delta\sigma_{y,i}, \Delta\sigma_{z,i}, \Delta\tau_{xy,i}, \Delta\tau_{yz,i}, \Delta\tau_{xz,i})$ and $(\Delta\varepsilon_{x,i}, \Delta\varepsilon_{y,i}, \Delta\varepsilon_{z,i}, \Delta\varepsilon_{xy,i}, \Delta\varepsilon_{yz,i}, \Delta\varepsilon_{xz,i})$ are the stress and strain increments for the ith element due to the footing loading in a deterministic finite element (FE) analysis and V_i is the volume of the ith element. Ching et al. (2018b) formally called the weighted geometric spatial average in Eq. (2.18) the PIE method. Tabarroki et al. (2022b) showed that the PIE method is effective in characterizing E_{eff} not only for the footing problem but also for other SLS problems, such as a retaining wall, axially loaded pile, laterally loaded pile, and base heave in excavation.

Equation (2.18) can be used to derive the SLS mobilization-based characteristic value if Young's modulus of the spatially variable soil follows a lognormal random field with mean = E_{mean} and COV = COV_E, or equivalently $\ln(E)$ follows a normal random field with mean = $\ln[E_{\mathrm{mean}}/(1 + \mathrm{COV}_E^2)^{0.5}]$ and variance = $\ln(1 + \mathrm{COV}_E^2)$. Equation (2.18) suggests that the mean value of $\ln(E^{\mathrm{eff}})$ is still $\ln[E_{\mathrm{mean}}/(1 + \mathrm{COV}_E^2)^{0.5}]$, but the variance of $\ln(E^{\mathrm{eff}})$ has the following expression:

$$\mathrm{Var}[\ln(E^{\mathrm{eff}})] \approx \frac{\sum_{i=1}^{N}\sum_{j=1}^{N} w_i w_j \rho_{ij}}{\left(\sum_{i=1}^{N} w_i\right)^2} \ln(1 + \mathrm{COV}_E^2) \tag{2.20}$$

where ρ_{ij} is the autocorrelation between the centroids of the ith and jth elements and the term $\sum_{i=1}^{N}\sum_{j=1}^{N} w_i w_j \rho_{ij} \Big/ \left(\sum_{i=1}^{N} w_i\right)^2$ is the variance reduction factor for the weighted geometric average in Eq. (2.18). We denote this variance reduction by Γ^2_{WG}. One possible approximation for Γ^2_{WG} is to adopt an approximation similar to AVRF-I in Table 2.11:

$$\Gamma^2_{WG} \approx \begin{cases} 1, & \text{if } L_I \le \delta \\ \delta/L_I, & \text{otherwise} \end{cases} \tag{2.21}$$

where L_I is the size (e.g., depth) of the influence zone where the weight w_i is significant. A more accurate expression of Γ^2_{WG} in terms of the problem dimension and the scale of fluctuation is an ongoing research topic. The SLS mobilization-based characteristic value can then be derived as the 5% fractile of E^{eff}:

$$\ln(E_k^{eff}) = \ln\left(\frac{E_{mean}}{\sqrt{1+COV_E^2}}\right)\left[1-1.645\times\Gamma_{WG}\sqrt{\ln(1+COV_E^2)}\right] \tag{2.22}$$

2.5.3 Hydraulic Conductivity

Ching et al. (2023c) investigated the effective hydraulic conductivity (k^{eff}) of a spatially variable soil mass for steady state seepage in the presence of a geotechnical structure (e.g., steady seepage induced by dewatering during excavation); k^{eff} is defined as the equivalent homogeneous hydraulic conductivity (k) that matches the total steady state seepage flow rate. Ching et al. (2023c) noted that the mechanisms for E^{eff} and k^{eff} are similar. For instance, for a block domain with a vertically variable E (or k), E^{eff} (or k^{eff}) is equal to the harmonic average if the load (or seepage) is vertical, and equal to the arithmetic average if the load (or seepage) is horizontal. They further found that the weighted geometric average in Eq. (2.18) is also applicable to k^{eff}:

$$\ln(k^{eff}) \approx \sum_{i=1}^{N} w_i \ln(k_i) / \left[\sum_{i=1}^{N} w_i\right] \tag{2.23}$$

where k_i is the hydraulic conductivity of the ith element in RFEM; w_i is its weight; and N is the total number of elements. However, the weight is calculated differently (Ching et al. 2023c):

$$w_i = (\Delta v_{x,i}\Delta i_{x,i} + \Delta v_{y,i}\Delta i_{y,i} + \Delta v_{z,i}\Delta i_{z,i}) \times V_i \tag{2.24}$$

where $(\Delta v_{x,i}, \Delta v_{y,i}, \Delta v_{z,i})$ and $(\Delta i_{x,i}, \Delta i_{y,i}, \Delta i_{z,i})$ are the discharge velocity and hydraulic gradient increments for the ith element in a deterministic finite element analysis due to the steady state seepage induced by a geotechnical structure (e.g., dewatering during excavation). The idea of the mobilization-based characteristic value is also applicable to hydraulic conductivity: the characteristic hydraulic conductivity (k^{eff}_k) can be determined as the 95% fractile of k^{eff}. If the hydraulic conductivity of the spatially variable soil follows a lognormal random field with mean = k_{mean} and COV = COV_k:

$$\ln(k_k^{eff}) = \ln\left(\frac{k_{mean}}{\sqrt{1+COV_k^2}}\right)\left[1+1.645\times\Gamma_{WG}\sqrt{\ln(1+COV_k^2)}\right] \tag{2.25}$$

2.6 CROSS-CORRELATIONS

In a site investigation programme, some tests are conducted in close proximity. For example, a piezocone test (CPTU) is conducted close to a borehole where disturbed and undisturbed samples are extracted for laboratory testing. Multiple laboratory tests can be conducted on these soil samples. ISO2394:2015 (ISO 2015), Section D.3 explains the need to go beyond classical univariate statistical analysis:

> The practical significance of considering multivariate data is that the COV of a design parameter typically decreases when it is estimated from more than one parameter. Hence, site investigation is not a cost item but an investment item because reduction of uncertainties through multivariate tests can translate directly to design savings through RBD. This important link between the quality/quantity of site investigation and design savings cannot be addressed systematically in a deterministic design approach.

Simpson (2011) implicitly highlighted the importance of cross-correlation in the selection of design values for multiple soil parameters:

> Many safety formats used in codes of practice generate extreme values of parameters by applying partial factors. In most cases it becomes incredible that several variables could attain very extreme values simultaneously. This underlies the principles of load combinations following the principles developed by Turkstra and Madsen (1980), and much of seismic design, in both of which the effect of one dominant variable is considered while others are given less extreme values.

When multiple soil parameters are allowed to vary independently without restraint by a cross-correlation structure, unrealistic "incredible" scenarios can appear. Ching et al. (2020c) proposed the only quantitative and practical guideline thus far for multivariate characteristic values.

If a corrected cone tip resistance (q_t) is measured "close" to an undisturbed sample that produces an undrained shear strength (s_u) in a triaxial test (e.g., unconsolidated undrained compression or UU test), say within one scale of fluctuation vertically and horizontally, these soil parameters can be assumed to be observed at practically the "same" spatial point. A natural approach is to extend the random variable model for a point value at a single location to a *random vector* model:

$$\boldsymbol{X} = (X_1, X_2)^{\mathrm{T}} \tag{2.26}$$

where X_1 and X_2 are standard normal random variables related to q_t and s_u using suitable transformations, respectively (Section 2.6.2). Multiple point values over a surface or volume, say $\boldsymbol{X}_1, \boldsymbol{X}_2, \ldots, \boldsymbol{X}_n$, are spatially correlated. For concreteness, the ith point value is denoted as

$$\boldsymbol{X}_i = (X_{i1}, X_{i2})^{\mathrm{T}} \tag{2.27}$$

Figure 2.15 illustrates the concept of a vector random field. The first subscript indicates the location, 1, 2,..., m. The second subscript indicates the soil parameter, 1, 2,..., n. The point values at two locations can be assembled as a single vector by stacking Eq. (2.27) as follows:

$$\boldsymbol{X} = (\underbrace{X_{11}, X_{12}}_{\boldsymbol{X}_1}, \underbrace{X_{21}, X_{22}}_{\boldsymbol{X}_2})^{\mathrm{T}} \tag{2.28}$$

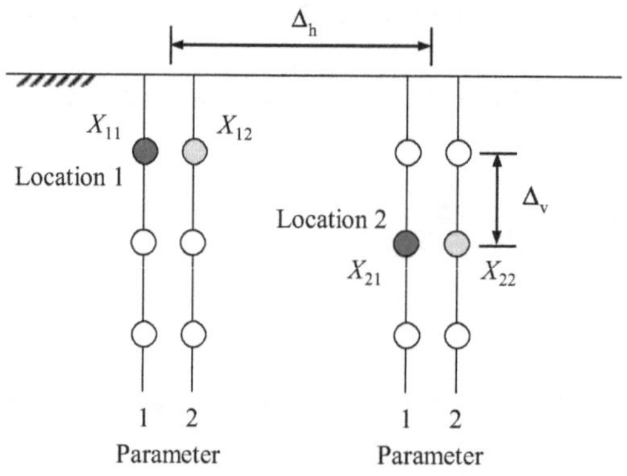

FIGURE 2.15 Definition of a vector random field containing two soil parameters at each observation (measurement) location.

Because $\boldsymbol{X}$ consists of standard normal variables, the mean vector for $\boldsymbol{X} = \boldsymbol{X}_{mean} = 0$. The correlation matrix for $\boldsymbol{X}$ is

$$C = \mathrm{E}\left[\begin{pmatrix} X_{11} \\ X_{12} \\ X_{21} \\ X_{22} \end{pmatrix} \begin{pmatrix} X_{11} & X_{12} & X_{21} & X_{22} \end{pmatrix}\right] = \begin{bmatrix} 1 & \rho_c & \rho_s & \rho_s\rho_c \\ \rho_c & 1 & \rho_s\rho_c & \rho_s \\ \rho_s & \rho_s\rho_c & 1 & \rho_c \\ \rho_s\rho_c & \rho_s & \rho_c & 1 \end{bmatrix} = \begin{bmatrix} 1 & \rho_s \\ \rho_s & 1 \end{bmatrix} \otimes \begin{bmatrix} 1 & \rho_c \\ \rho_c & 1 \end{bmatrix} = \Sigma_s \otimes \Sigma_c \tag{2.29}$$

where ρ_c = cross-correlation between parameter 1 (e.g., q_t) and parameter 2 (e.g., s_u); ρ_s = spatial correlation between location 1 and location 2 = $\rho_s(\Delta_v, \Delta_h)$ (a 2D autocorrelation model that depends on lag distances Δ_v and Δ_h); Σ_c = cross-correlation matrix (size $n \times n$); Σ_s = spatial correlation matrix (size $m \times m$); and $\otimes$ = Kronecker product. The assumptions in Eq. (2.29) are (1) the cross-correlation is independent of location (Ching et al. 2016e) and (2) the spatial correlation is independent of the soil parameter. The outcome of (1) and (2) is that the correlation between two soil parameters located in two locations is equal to the product of the cross-correlation and the spatial correlation. It is easy to generalize Eq. (2.29) for n soil parameters and m observed locations.

If we assume that $\boldsymbol{X}$ follows a multivariate normal distribution (Phoon 2006; Ching and Phoon 2015), the probability density function (PDF) is available in closed-form as a function of C and $\boldsymbol{X}_{mean}$ (mean vector):

$$\begin{aligned} f(\boldsymbol{X}) &= |C|^{-\frac{1}{2}} (2\pi)^{-\frac{nm}{2}} \exp\left[-\frac{1}{2}(\boldsymbol{X} - \boldsymbol{X}_{mean})^{\mathrm{T}} C^{-1} (\boldsymbol{X} - \boldsymbol{X}_{mean})\right] \\ &= |C|^{-\frac{1}{2}} (2\pi)^{-\frac{nm}{2}} \exp\left[-\frac{1}{2}\boldsymbol{X}^{\mathrm{T}} C^{-1} \boldsymbol{X}\right] \end{aligned} \tag{2.30}$$

For a general multivariate normal distribution, $\boldsymbol{X}_{mean} \neq 0$ and C is the covariance matrix with the leading diagonal variance of $\boldsymbol{X} \neq 1$. C is equal to the correlation matrix (Eq. 2.29) for the special case of a standard normal distribution only. The practical usefulness of Eq. (2.30) with C defined by Eq. (2.29) is that the dependency structure between n soil parameters located in m locations can be simplified into the (Kronecker) product of two correlation matrices. This dependency structure can be large scale in terms of the random dimension (or number of random variables) $n \times m$. For example, m = 1,000 (number of spatial points) and n = 10 (number of soil parameters), the random dimension is 10,000. The first matrix Σ_s can be populated once the autocorrelation model is selected (Table 2.10) and the scales of fluctuation (vertical and horizontal) are estimated (Table 2.13). The second matrix Σ_c contains only *bivariate* cross-correlations between any two soil parameters, although the correlations between *all possible pairs* of soil parameters are needed. The number of possible pairs is $n(n-1)/2$. As an example, for n = 10 soil parameters, information on 45 cross-correlations is needed. Both matrices must be positive definite. The Σ_s matrix is always positive definite because it is calculated from a positive definite autocorrelation model. But there is no guarantee that Σ_c is positive definite if each correlation entry in the matrix is estimated *separately* from a *bivariate* dataset. The estimation approach is practical, but this entry-by-entry assembly of Σ_c is inconsistent. Ching and Phoon (2014b) proposed a bootstrapping method to address this significant issue by (1) observing that each entry takes a range of values due to statistical uncertainty; (2) generating many matrices by varying each entry within its admissible statistical uncertainty range; (3) selecting positive definite versions; and (4) taking the average of these positive definite versions as the representative solution. However, there is no theoretical basis for this approach. The positive definite property is only guaranteed theoretically if the entries are estimated from a *single complete multivariate* dataset for the frequentist approach or they are estimated using a Bayesian machine learning approach (Ching and Phoon 2019). The following sections illustrate the estimation of cross-correlation matrices for three clay databases in Table 2.2: CLAY/10/7490 (Ching and Phoon 2014a), CLAY/6/535 (Ching et al. 2014a), and CLAY/5/345 (Ching and Phoon 2012). Spatial variability is discussed in Section 2.4.

2.6.1 Incomplete versus Complete Multivariate Soil Databases

The main practical challenge in the estimation of Σ_c is that complete multivariate data is rarely collected in a site investigation program because it is not cost-effective to conduct multiple tests in close proximity. There is an obvious tradeoff between conducting different tests in different locations and conducting different tests in the same location. The former strategy collects more information on the spatial variability of the site. The latter strategy collects information on the cross-correlations between soil parameters. Hence, the tradeoff is effectively seeking a balance between Σ_s and Σ_c in terms of accuracy. It is not possible to correlate two soil parameters spaced more than one scale of fluctuation apart vertically and horizontally, because of spatial variability. Referring to Figure 2.15, the correlation between X_{11} and X_{12} is ρ_c, because q_t (soil parameter 1) and s_u (soil parameter 2) are measured in close proximity. On the other hand, the correlation between two point values spaced apart, X_{11} and X_{22}, is $\rho_c\rho_s$. When these point values are far apart, no information on ρ_c is available because $\rho_s = 0$.

In practice, it is common to adopt an intermediate strategy involving collecting multiple sets of *bivariate* data in different locations, say take piezocone (CPTU) soundings next to one borehole and conduct vane shear tests (VSTs) next to a second borehole. Engineers rarely take CPTU soundings, conduct VSTs, or collect undisturbed samples for triaxial tests in three separate locations. It is not possible to produce a site-specific cross-correlation between say the undrained shear strength from a laboratory test and the CPTU data in this case. Engineers also do not collect CPTU, VST, or undrained shear strength data at a single location because soil data from a single location is unlikely to be sufficiently representative of the variability over the entire site. It is not surprising that many of the generic databases in Table 2.2 are incomplete (refer to the "% completeness" column). The construction of multivariate probability models based on the estimation of Σ_c from multiple sets of bivariate data possibly from different databases (the most realistic geotechnical information scenario) is explained below. More sophisticated Bayesian machine learning approaches are given elsewhere (Ching et al. 2020b).

2.6.2 Multivariate Probability Model

Multivariate information is usually gathered in a typical site investigation. An engineer will perceive "multivariate information" as data produced by multiple test types. This common understanding will be sharpened below in relation to a multivariate probability model. For instance, when undisturbed samples are extracted for oedometer and triaxial tests, SPT and/or piezocone test (CPTU) may be conducted in close proximity. Moreover, index parameters such as the unit weight, natural water content, plastic limit, liquid limit, and liquidity index are commonly determined from relatively simple laboratory tests on disturbed samples. A detailed list of soil/rock parameters is provided in the footnote of Table 2.2. It is generally known that data from these varied sources is not independent if they are measured in close spatial proximity. As noted above, the definition of "close" is related to the spatial variability of the site. These data sources are typically correlated to a design parameter, e.g., the undrained shear strength (s_u). In fact, this is one of the main purposes for conducting a site investigation. However, current deterministic practice is less cognizant that these correlations can be exploited to reduce the COV of a design parameter (Section 2.7). The impact on reliability-based design is obvious. For example, the COV of s_u may reduce from "high" to "medium" based on the classification in Table 2.9 in the presence of CPTU data and a larger resistance factor could be justified for design as a result of this COV reduction (Phoon et al. 2003b). Ching et al. (2014c) discussed the impact of site investigation efforts on design savings in the context of a simplified RBD.

Because bivariate correlations between soil parameters are more commonly available (e.g, s_u versus q_t), the multivariate normal distribution is a sensible and practical choice to capture the *multivariate* dependency among soil parameters in the form of a correlation matrix C (Phoon et al. 2012). The probability density function for the multivariate normal distribution covering both spatial correlation (Σ_s) and cross-correlation (Σ_c) is given in Eq. (2.30). If one assumes that the records (or point values) $\boldsymbol{X}_i$ in the soil databases are measured sufficiently far apart that they are statistically independent, $\rho_s = 0$ and C reduces to Σ_c. For a general record containing n soil parameters, $\boldsymbol{X}_i = (X_{i1}, X_{i2},\ldots, X_{in})$, the correlation matrix ($n \times n$) is

$$C = \Sigma_c = \begin{bmatrix} 1 & \rho_{12} & \cdots & \rho_{1n} \\ \rho_{12} & 1 & \cdots & \rho_{2n} \\ \vdots & \vdots & \ddots & \vdots \\ \rho_{1n} & \rho_{2n} & \cdots & 1 \end{bmatrix} \tag{2.31}$$

For conciseness, ρ_{cij} is denoted as ρ_{ij} in Eq. (2.31) and from hereon. It is necessary to estimate $n(n - 1)/2$ entries from soil databases. Engineers may think that there are too many unknown model parameters relative to the limited information they have on hand, but this multivariate normal probability model is already the most efficient in capturing high-dimensional dependencies with the least number of model parameters.

Some practical guidelines for the construction of a multivariate probability distribution for soil parameters are

1. The mean, standard deviation, and coefficient of variation are second-moment statistics derived from univariate data (Section 2.3). A Johnson distribution requires more statistics because it is described by four model parameters. However, only univariate data is needed. The key task in the construction of a multivariate probability model is to estimate the correlation matrix C that requires multivariate data.
2. It is worth emphasizing that a set of *multivariate* data and *multiple* sets of *bivariate* data are not the same in information content, even if they cover the same set of soil parameters. For example, one can measure the undrained shear strength (Y_1), overconsolidation ratio (OCR) (Y_2), and corrected cone tip resistance (Y_3) at the same depth in two different ways. The first way is to measure (Y_1, Y_2, Y_3) in the vicinity of one borehole. The data collected at various depths $[y_1(z_i), y_2(z_i), y_3(z_i)]$ is a proper multivariate dataset. The second way is to measure (Y_1, Y_2), (Y_1, Y_3), and (Y_2, Y_3) at three separate borehole locations. The data collected at various depths $[y_1(z_i), y_2(z_i)]$ is one bivariate dataset. There are three sets of bivariate data in this example. It is possible to evaluate the correlation matrix using one set of multivariate data or three sets of bivariate data. The critical distinction here is that the correlation coefficients (ρ_{12}, ρ_{13}, ρ_{23}) obtained from the first method are *related* in some sense because data are derived from a single borehole, whereas the correlation coefficients (ρ_{12}, ρ_{13}, ρ_{23}) obtained from the second method are *unrelated* because data is derived from three different boreholes possibly spaced far apart. This subtle relationship between the correlation coefficients is measured by a matrix property called positive definiteness. For example, if $\rho_{13} = 0.7$ and $\rho_{23} = 0.4$, ρ_{12} must be larger than -0.37 to preserve this property. The second method may produce $\rho_{12} = -0.5$, but this matrix is not valid. On the other hand, the second method is more practical because bivariate data is more common. Ching and Phoon (2014b) proposed an expedient approach to produce a positive definite correlation matrix from bivariate data. It is critical to appreciate that proper multivariate information does not refer to data produced by different tests *regardless of test locations*.
3. It is evident that most soil parameters do not follow the normal distribution. The normal distribution is unbounded from below. It is theoretically possible to generate negative values which will be absurd for many soil/rock parameters such as the undrained

shear strength. The popular solution is to assume that $\ln(Y)$ is normal, in which case the soil parameter Y is lognormally distributed. Phoon and Ching (2013) highlighted that the Johnson system of distributions enjoys the same analytical simplicity as the lognormal distribution, but provides a broader range of distributions for data fitting:

$$X = \begin{cases} b_X + a_X \ln\left(\dfrac{Y-b_Y}{a_Y} + \sqrt{1+\left(\dfrac{Y-b_Y}{a_Y}\right)^2}\right) & \text{SU} \\ b_X + a_X \ln\left[\dfrac{(Y-b_Y)/a_Y}{1-(Y-b_Y)/a_Y}\right] & \text{SB} \\ b_X^* + a_X \ln(Y-b_Y) & \text{SL} \end{cases} \quad (2.32)$$

where X = standard normal random variable with mean = 0 and standard deviation = 1; Y = Johnson random variable; (a_X, b_X, a_Y, b_Y) = Johnson distribution model parameters; and $b_X^* = b_X - a_X\ln(a_Y)$.

4. Based on the Johnson distribution given by Eq. (2.32), it is possible to convert each physical random variable (Y_i) to a standard normal random variable (X_i). It is *not necessary* that this component-by-component transformation procedure produces a random vector $\boldsymbol{X} = (X_1, X_2,\ldots, X_n)^T$ that is *jointly* normal, i.e., $\boldsymbol{X}$ follows the multivariate standard normal PDF given by Eq. (2.30). There is a simple way to check this: the linear sum of any subset of components drawn from a jointly normal vector is a normal random variable. Nonetheless, based on the authors' experience, this assumption is reasonable for all the soil databases studied thus far. Note that the correlation matrix (Eq. 2.31) is estimated from the transformed data, $(X_1, X_2,\ldots, X_n)$, not the original physical data, $(Y_1, Y_2,\ldots, Y_n)$.
5. It is simple to obtain realizations of *independent* standard normal random variables $\boldsymbol{Z}$ = $(Z_1, Z_2,\ldots, Z_n)^T$ using library functions in many softwares. Realizations of *correlated* standard normal random variables $\boldsymbol{X} = (X_1, X_2,\ldots, X_n)^T$ can be obtained using $\boldsymbol{X} = L\boldsymbol{Z}$, where L is the lower triangular Cholesky factor satisfying $C = LL^T$. Finally, each soil parameter is obtained using the inverse transform of Eq. (2.32):

$$Y = \begin{cases} b_Y + a_Y \sinh\left(\dfrac{X-b_X}{a_X}\right) & \text{SU} \\ \left[b_Y + (a_Y+b_Y)\exp\left(\dfrac{X-b_X}{a_X}\right)\right] \Big/ \left[1+\exp\left(\dfrac{X-b_X}{a_X}\right)\right] & \text{SB} \\ b_Y + \exp\left(\dfrac{X-b_X^*}{a_X}\right) & \text{SL} \end{cases} \quad (2.33)$$

The Johnson probability density function $f(y)$ and the effect of the model parameters (a_X, b_X, a_Y, b_Y) on its shape/location are illustrated in Figure 2.16. The closed-form transformations (Eq. 2.32) and inverse transformations (Eq. 2.33) are important to keep computational cost within reach of practice. The following subsection shows the construction of multivariate probability distributions using actual soil databases.

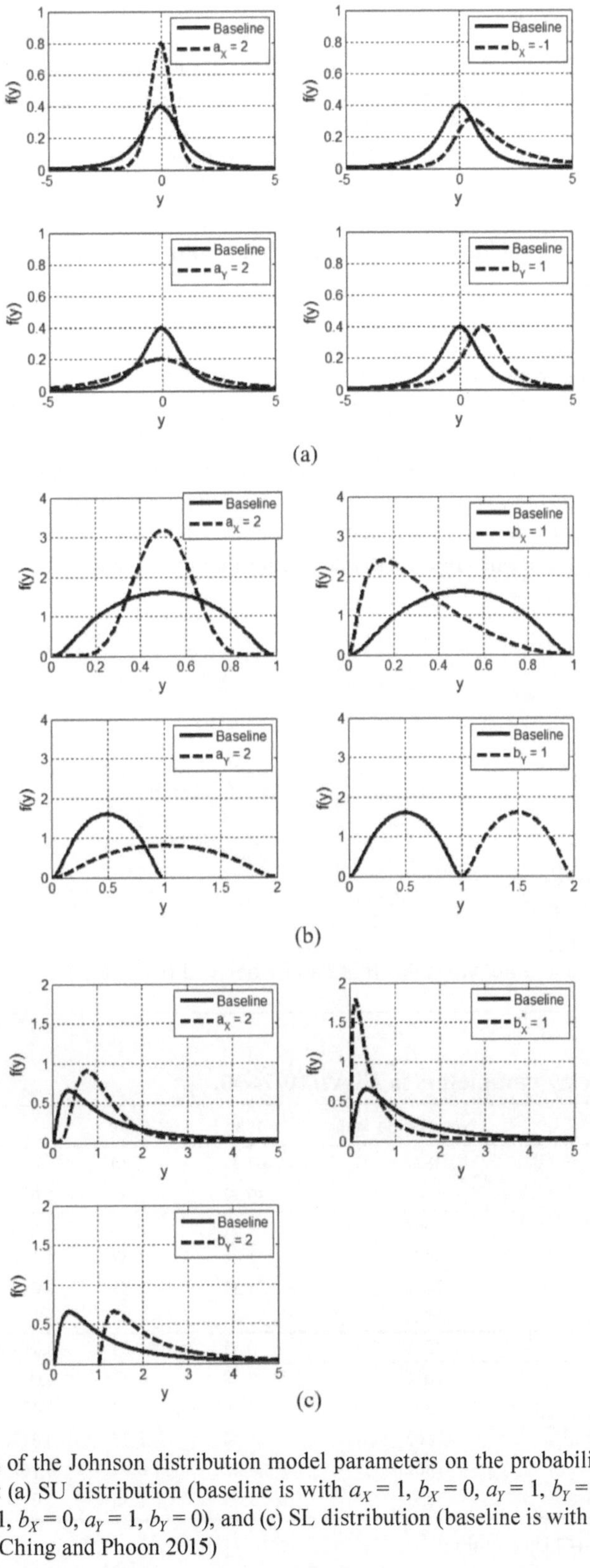

FIGURE 2.16 Effect of the Johnson distribution model parameters on the probability density function of the soil parameter (Y): (a) SU distribution (baseline is with $a_X = 1$, $b_X = 0$, $a_Y = 1$, $b_Y = 0$), (b) SB distribution (baseline is with $a_X = 1$, $b_X = 0$, $a_Y = 1$, $b_Y = 0$), and (c) SL distribution (baseline is with $a_X = 1$, $b_X^* = 0$, $b_Y = 0$). (Source: Figure 1.20, Ching and Phoon 2015)

2.6.3 Multivariate Probability Models for Clay Databases

Several multivariate probability models have been constructed for a variety of clay parameters (Ching and Phoon 2012, 2013a, 2014b; Ching et al. 2014a, 2023a; Zhang et al. 2020), sand parameters (Ching et al. 2017a), and rock parameters (Ching et al. 2018a, 2021a). These models are calibrated from generic databases, which are labeled in accordance to the template: (soil type)/(number of parameters of interest)/(number of data points). Details for these databases are given in Table 2.2. This section presents the construction of multivariate probability distributions for CLAY/10/7490, CLAY/6/535, and CLAY/5/345 to demonstrate that it is within reach of an engineer.

2.6.3.1 CLAY/10/7490

The CLAY/10/7490 database compiles data from 251 studies. The number of data points associated with each study varies from 1 to 419 with an average 30 data points per study. The clay parameters cover a wide range of overconsolidation ratio (but mostly 1–10), a wide range of sensitivity (S_t) (insensitive to quick clays; sites with S_t = 1 to tens or hundreds are fairly typical), and a wide range of plasticity index (but mostly 8–100). There are a few data points for fissured clays as well as organic clays. Details are given elsewhere (Ching and Phoon 2014a).

Ten dimensionless clay parameters are compiled in this database. The statistics are summarized in Table 2.15. To keep notations concise, the physical random variables (or natural logarithm transform) are denoted by

1. $Y_1 = \ln(\mathrm{LL})$
2. $Y_2 = \ln(\mathrm{PI})$
3. $Y_3 = \mathrm{LI}$
4. $Y_4 = \ln(\sigma'_v/P_a)$
5. $Y_5 = \ln(\sigma'_p/P_a)$
6. $Y_6 = \ln(S_t)$
7. $Y_7 = \ln(s_u/\sigma'_v)$
8. $Y_8 = B_q = (u_2-u_0)/(q_t-\sigma_v)$
9. $Y_9 = \ln[(q_t-\sigma_v)/\sigma'_v]$
10. $Y_{10} = \ln[(q_t-u_2)/\sigma'_v]$

There are two *dependent* variables that can be derived from $(Y_1, Y_2,\ldots, Y_{10})$:

TABLE 2.15
Statistics of the 10 clay parameters in CLAY/10/7490.

Parameter	No. of data (m)	Mean	COV	Min	Max
LL	3822	67.7	0.80	18.1	515
PI	4265	39.7	1.08	1.9	363
LI	3661	1.01	0.78	−0.75	6.45
σ'_v/P_a	3370	1.80	1.47	4.13E-3	38.74
σ'_p/P_a	2028	4.37	2.31	0.094	193.30
S_t	1589	35.0	2.88	1	1467
s_u/σ'_v	3538	0.51	1.25	3.7E-3	7.78
B_q	1016	0.58	0.35	0.01	1.17
$(q_t-\sigma_v)/\sigma'_v$	862	8.90	1.17	0.48	95.98
$(q_t-u_2)/\sigma'_v$	668	5.34	1.37	0.61	108.20
$(u_2-u_0)/\sigma'_v$	862	4.28	0.90	0.11	48.79
OCR	3531	3.85	1.56	1.0	60.23

Source: Table 2, Phoon and Ching (2014b)

11. $Y_{11} = (u_2 - u_0)/\sigma'_v = Y_8 \times \exp(Y_9)$
12. $Y_{12} = \ln(\text{OCR}) = Y_5 - Y_4$

The multivariate non-normal distribution for ($Y_1, Y_2, \ldots, Y_{10}$) is constructed using the approach in Section 2.6.2. The Johnson distribution type and model parameters (a_X, b_X, a_Y, b_Y) are summarized in Table 2.16.

Figure 2.17 presents the bivariate correlation structure underlying the ten soil parameters *after* they have been transformed into standard normal random variables using Eq. (2.32). There are 45 possible bivariate correlations for a database containing $n = 10$ parameters. Due to the absence of a genuine multivariate dataset, these 45 correlation coefficients were estimated from *independent* bivariate datasets taken from diverse geographic locations. The resulting correlation matrix is not positive definite as explained previously, but it is possible to identify a positive definite matrix by sampling within the 95% confidence interval for each sample correlation coefficient (Ching and Phoon 2014b; Ching et al. 2023a). Table 2.17 presents one such example constituted from the average of 1,000 positive definite matrices of size 10 × 10. It is obvious that the correlation matrix is not unique, but this limitation applies to all statistical inferences in the presence of finite sample sizes. For completeness, the correlation coefficients between (X_{11}, X_{12}) and ($X_1, X_2, \ldots, X_{10}$) are also listed in Table 2.17.

2.6.3.2 CLAY/6/535

The CLAY/6/535 database consists of six dimensionless parameters simultaneously measured in close proximity (Ching et al. 2014a). In other words, this database consists of genuine multivariate data. This is the critical difference between CLAY/6/535 and CLAY/10/7490. The original dataset is in the form [s_u/σ'_v, OCR, $(q_t - \sigma_v)/\sigma'_v$, $(q_t - u_2)/\sigma'_v$, $(u_2 - u_0)/\sigma'_v$, B_q]. To compare with the results presented in Section 2.6.3.1 for the CLAY/10/7490 database, logarithms are applied to the non-negative variables to obtain the following soil parameters:

1. $Y_1 = \ln(s_u/\sigma'_v)$
2. $Y_2 = B_q$
3. $Y_3 = \ln[(q_t - \sigma_v)/\sigma'_v]$

TABLE 2.16
Johnson distribution type and model parameters for ($Y_1, Y_2, \ldots, Y_{12}$) in CLAY/10/7490.

	Parameter	Johnson type	Model parameters			
			a_X	b_X	a_Y	b_Y
Y_1	ln(LL)	SU	1.636	−1.166	0.616	3.479
Y_2	ln(PI)	SU	1.433	−0.265	0.918	3.178
Y_3	LI	SU	1.434	−1.068	0.629	0.358
Y_4	$\ln(\sigma'_v/P_a)$	SB	3.150	0.256	14.458	−7.010
Y_5	$\ln(\sigma'_p/P_a)$	SB	4.600	21.548	576.785	−4.793
Y_6	$\ln(S_t)$	SU	2.393	−2.080	1.885	0.461
Y_7	$\ln(s_u/\sigma'_v)$	SU	2.039	−0.517	1.427	−1.461
Y_8	B_q	SU	2.676	0.161	0.513	0.615
Y_9	$\ln[(q_t-\sigma_v)/\sigma'_v]$	SU	1.340	−0.572	0.659	1.476
Y_{10}	$\ln[(q_t-u_2)/\sigma'_v]$	SU	2.134	−1.102	1.154	0.657
Y_{11}	$(u_2-u_0)/\sigma'_v$	SU	0.873	−1.095	0.790	1.869
Y_{12}	ln(OCR)	SB	0.307	0.825	3.290	0.000

Source: Table 3, Phoon and Ching (2014b)

4. $Y_4 = \ln[(q_t - u_2)/\sigma'_v]$
5. $Y_5 = (u_2 - u_0)/\sigma'_v$
6. $Y_6 = \ln(\text{OCR})$

The statistics of (Y_1, Y_2,…, Y_6) are summarized in Table 2.18. The database contains 535 clay data points from 40 sites with complete measurement of (Y_1, Y_2,…, Y_6) at close proximity. The clay

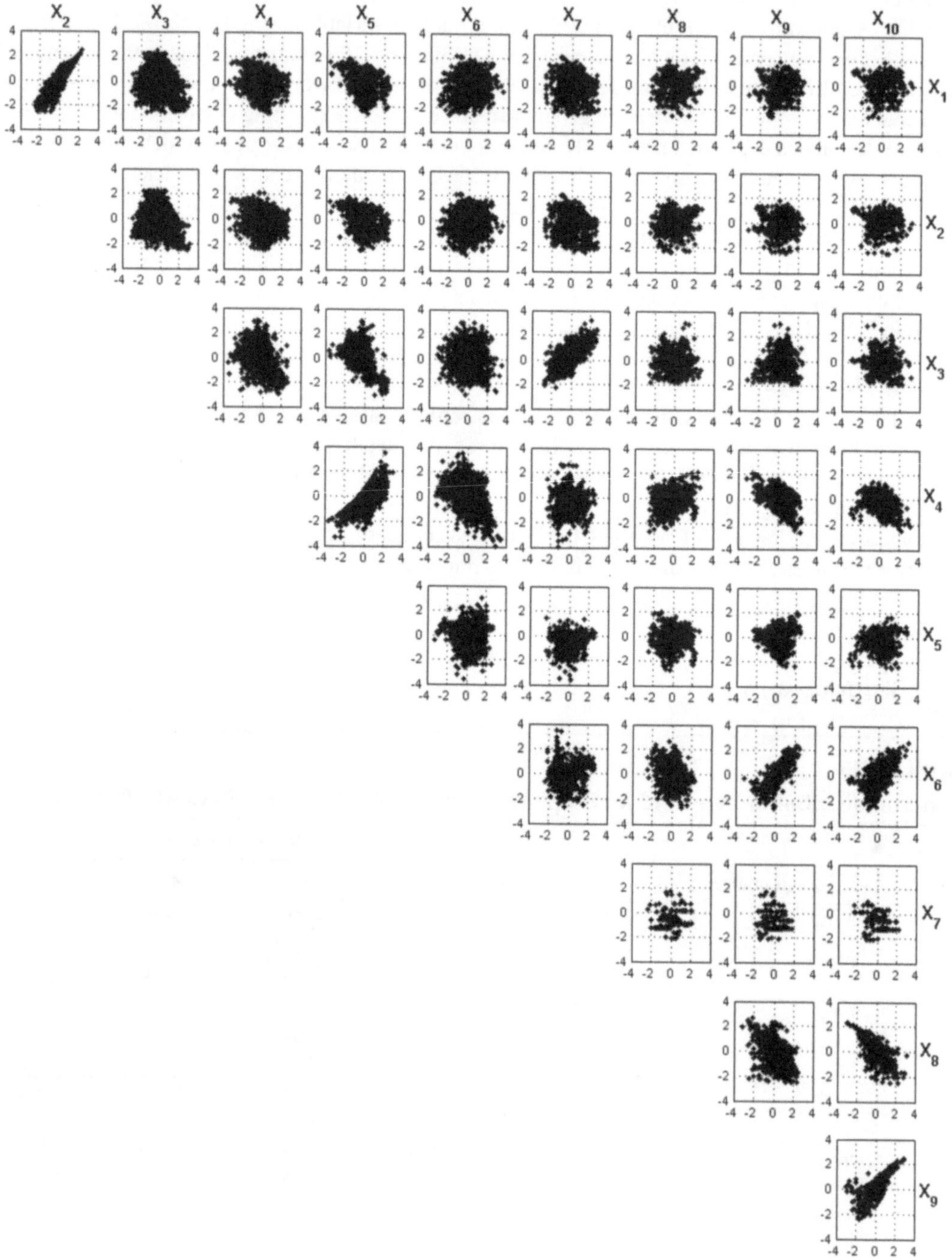

FIGURE 2.17 Scatter plots between X_i and X_j for CLAY/10/7490 (horizontal axis = column variable, vertical axis = row variable). (Source: Figure 5, Ching and Phoon 2014b)

TABLE 2.17
Average of 1000 positive definite matrices obtained from bootstrap samples.

	X_1	X_2	X_3	X_4	X_5	X_6	X_7	X_8	X_9	X_{10}	X_{11}	X_{12}
X_1	1.00	0.92	–0.28	–0.26	–0.31	–0.21	0.13	0.06	0.09	0.01	0.12	-0.03
X_2		1.00	–0.33	–0.17	–0.31	–0.23	0.06	0.10	0.02	–0.08	0.06	-0.01
X_3			1.00	–0.47	–0.54	0.55	0.02	–0.01	–0.01	–0.15	0.04	0.03
X_4				1.00	0.68	–0.02	–0.51	0.14	–0.39	–0.32	–0.26	–0.29
X_5					1.00	0.03	0.00	–0.02	0.08	–0.03	0.12	0.18
X_6						1.00	0.21	0.17	0.07	–0.09	0.28	0.14
X_7							1.00	–0.20	0.75	0.65	0.56	0.69
X_8								1.00	–0.41	–0.55	0.17	–0.24
X_9									1.00	0.75	0.74	0.51
X_{10}			Symmetry							1.00	0.40	0.39
X_{11}											1.00	0.45
X_{12}												1.00

Source: Table 4, Phoon and Ching (2014b)

Note: X_1, X_2, X_3, X_4, X_5, X_6, X_7, X_8, X_9, X_{10}, X_{11}, X_{12} are standard normal variables corresponding to: ln(LL), ln(PI), LI, $\ln(\sigma'_v/P_a)$, $\ln(\sigma'_p/P_a)$, S_t, $\ln(s_u/\sigma'_v)$, B_q, $\ln[(q_t\text{-}\sigma_v)/\sigma'_v]$, $\ln[(q_t\text{-}u_2)/\sigma'_v]$, $(u_2\text{-}u_0)/\sigma'_v$, ln(OCR).

parameters cover a wide range of OCR (but mostly 1–6), a wide range of S_t (insensitive to quick clays), and a wide range of PI (but mostly 10–100). Highly OC (fissured) and organic clays are nearly absent in this database. Details are given elsewhere (Ching et al. 2014a).

Note that (Y_1, Y_2,..., Y_6) in this database refer to the same physical parameters indexed as (Y_7, Y_8,..., Y_{12}) in the CLAY/10/7490 database (entries in the double-lined box in Tables 2.15 and 2.16). But, now a smaller database CLAY/6/535 is considered. The statistics in Table 2.18 can be compared to the entries in the double-lined box in Table 2.15. It is apparent that their mean values are similar, but the COVs for CLAY/10/7490 are significantly larger than those for CLAY/6/535. Figure 2.18 shows the s_u/σ'_v versus OCR transformation model, which is Y_1 versus Y_6 in the CLAY/6/535 database and Y_7 versus Y_{12} in the CLAY/10/7490 database. There are more s_u/σ'_v versus OCR data points ($m = 2020$) in the CLAY/10/7490 database. These 2020 data points cover a wider range than the 535 data points in CLAY/6/535. Nonetheless, the s_u/σ'_v versus OCR trends are very similar (Figure 2.18).

The multivariate non-normal distribution for (Y_1, Y_2,..., Y_6) is constructed using the approach in Section 2.6.2. The distribution type and model parameters are summarized in Table 2.19. There are

TABLE 2.18
Statistics of the 6 clay parameters in Clay/6/535.

Parameter	No. of data (m)	Mean	COV	Min	Max
s_u/σ'_v	535	0.64	0.60	0.11	3.04
B_q	535	0.56	0.34	–0.09	1.07
$(q_t\text{-}\sigma_v)/\sigma'_v$	535	9.35	0.68	2.55	58.88
$(q_t\text{-}u_2)/\sigma'_v$	535	5.28	0.88	0.61	43.69
$(u_2\text{-}u_0)/\sigma'_v$	535	4.70	0.58	-1.51	21.72
OCR	535	2.35	0.66	0.94	9.69

Source: Table 5, Phoon and Ching (2014b)

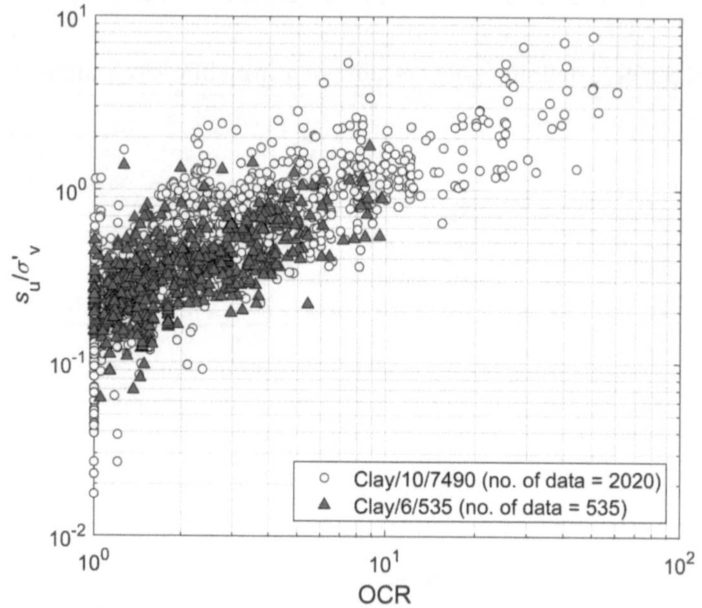

FIGURE 2.18 Comparison between CLAY/6/535 and CLAY/10/7490 in the s_u/σ'_v versus OCR transformation model. (Source: Figure 2, Phoon and Ching 2014b)

TABLE 2.19
Johnson distribution type and model parameters for (Y_1, Y_2, ..., Y_6) in Clay/6/535.

	Parameter	Johnson type	Model parameters			
			a_X	b_X	a_Y	b_Y
Y_1	$\ln(s_u/\sigma'_v)$	SU	2.889	−1.299	1.261	−1.218
Y_2	B_q	SU	2.961	0.049	0.544	0.570
Y_3	$\ln[(q_t-\sigma_v)/\sigma'_v]$	SU	2.211	−1.090	0.944	1.537
Y_4	$\ln[(q_t-u_2)/\sigma'_v]$	SU	2.708	−1.323	1.471	0.587
Y_5	$(u_2-u_0)/\sigma'_v$	SU	0.966	−0.733	1.126	3.155
Y_6	ln(OCR)	SB	0.787	0.920	2.652	−0.056

Source: Table 6, Phoon and Ching (2014b)

15 possible bivariate correlations for a database containing $n = 6$ parameters. Because these 15 correlation coefficients were estimated from a genuine multivariate dataset, the resulting correlation matrix is always positive definite. Table 2.20 presents the resulting 6 × 6 correlation matrix. This correlation matrix should be compared to the double-lined matrix in Table 2.17. Figure 2.19 shows the relationship between the 15 correlation coefficients (ρ_{ij}) obtained from the CLAY/10/7490 database and those obtained from the CLAY/6/535 database. The horizontal and vertical line segments represent the 95% confidence intervals of the ρ_{ij} estimates. It is evident that the two sets of correlation coefficients are fairly similar. This reinforces the observation made in Figure 2.18: although the scatter of the univariate data in the two databases is not the same (CLAY/10/7490 scatters more), the bivariate behaviors are fairly similar. Figure 2.18 shows the similarity of the linear trend and Figure 2.19 shows the similarity of the correlation coefficients.

TABLE 2.20
Product-moment (Pearson) correlations among ($X_1, X_2, \ldots, X_6$) for CLAY/6/535 (corresponding correlation from CLAY/10/7490 in parenthesis).

	X_1 (X_7)	X_2 (X_8)	X_3 (X_9)	X_4 (X_{10})	X_5 (X_{11})	X_6 (X_{12})
X_1	1.00	−0.28	0.67	0.60	0.49	0.61
(X_7)		(−0.20)	(0.75)	(0.65)	(0.56)	(0.69)
X_2		1.00	−0.45	−0.77	0.28	−0.15
(X_8)			(−0.41)	(−0.55)	(0.17)	(−0.24)
X_3			1.00	0.82	0.69	0.60
(X_9)				(0.75)	(0.74)	(0.51)
X_4				1.00	0.30	0.51
(X_{10})					(0.40)	(0.39)
X_5		Symmetry			1.00	0.53
(X_{11})						(0.45)
X_6						1.00
(X_{12})						

Source: Table 7, Phoon and Ching (2014b)
Note: $X_1, X_2, X_3, X_4, X_5, X_6$ are standard normal variables corresponding to: $\ln(s_u/\sigma'_v)$, B_q, $\ln[(q_t-\sigma_v)/\sigma'_v]$, $\ln[(q_t-u_2)/\sigma'_v]$, $(u_2-u_0)/\sigma'_v$, ln(OCR).

2.6.3.3 CLAY/5/345

The CLAY/5/345 database consists of five dimensionless parameters simultaneously measured in close proximity (Ching and Phoon 2012). In other words, this database also consists of genuine multivariate data. The original dataset is in the form of [LI, s_u, s_u^{re}, σ'_p, σ'_v]. To compare with the results presented in Section 2.6.3.1 for the CLAY/10/7490 database, normalizations (and also logarithms) are applied to some variables to obtain the following soil parameters:

1. Y_1 = LI
2. $Y_2 = \ln(\sigma'_v/P_a)$
3. $Y_3 = \ln(\sigma'_p/P_a)$
4. $Y_4 = \ln(S_t)$
5. $Y_5 = \ln(s_u/\sigma'_v)$

The statistics of ($Y_1, Y_2, \ldots, Y_5$) are summarized in Table 2.21. The database contains 345 lightly overconsolidated and structured (*sensitive to quick*) clay data points from 37 sites with complete measurement of ($Y_1, Y_2, \ldots, Y_5$) at close proximity. The clay parameters cover a narrower range of OCR (mostly 1–4) and a wide range of S_t (sensitive to quick clays). Insensitive clays, fissured clays, and organic clays are nearly absent in this database. Details are given elsewhere (Ching and Phoon 2012).

Note that ($Y_1, Y_2, \ldots, Y_5$) in this database refer to the same physical parameters indexed as ($Y_3, Y_4, \ldots, Y_7$) in the CLAY/10/7490 database (entries in the dashed box in Tables 2.15 and 2.16). But, now a smaller database CLAY/5/345 is considered. The statistics in Table 2.21 can be compared to the entries in the dashed box in Table 2.15. Again, the COVs for CLAY/10/7490 are significantly larger than those for CLAY/6/535. Figure 2.20 shows the LI versus σ'_p/P_a transformation model for the CLAY/5/345 and CLAY/10/7490 databases. There are more LI versus σ'_p/P_a data points (m = 1314) in the CLAY/10/7490 database. More importantly, there are many insensitive clays in these 1314 data points (data points in the dashed circle). This is evident from the low LI values. On the

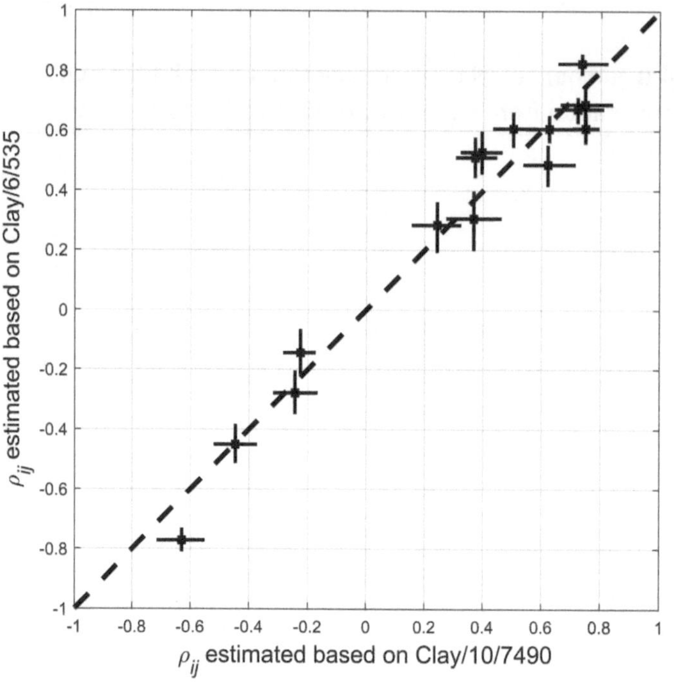

FIGURE 2.19 Relationship between the 15 correlation coefficients (ρ_{ij}) obtained from CLAY/10/7490 and the corresponding ones obtained from CLAY/6/535. (Source: Figure 3, Phoon and Ching 2014b)

TABLE 2.21
Statistics of the 5 clay parameters in CLAY/5/345.

Parameter	No. of data (m)	Mean	COV	Min	Max
LI	345	1.25	0.49	0.14	4.85
σ'_v/P_a	345	0.66	0.80	0.03	2.61
σ'_p/P_a	345	1.04	0.98	0.05	7.04
S_t	345	47.75	2.82	1.25	1443
s_u/σ'_v	345	0.51	0.61	0.10	2.31

Source: Table 8, Phoon and Ching (2014b)

other hand, the LI values for many data points in CLAY/5/345 are larger than 1, indicative of quick clays. As a result, the LI versus σ'_p/P_a trend in the CLAY/10/7490 database does not exist in the CLAY/5/345 database. The multivariate non-normal distribution for (Y_1, Y_2,..., Y_5) is constructed using the same approach discussed above. The distribution type and model parameters are summarized in Table 2.22.

There are ten possible bivariate correlations for a database containing n = 5 parameters. Because these ten correlation coefficients were estimated from a genuine multivariate dataset, the resulting correlation matrix is always positive definite. Table 2.23 presents the resulting 5 × 5 correlation matrix. This correlation matrix should be compared to the matrix in the dashed box in Table 2.17. Figure 2.21 shows the relationship between the ten correlation coefficients (ρ_{ij})

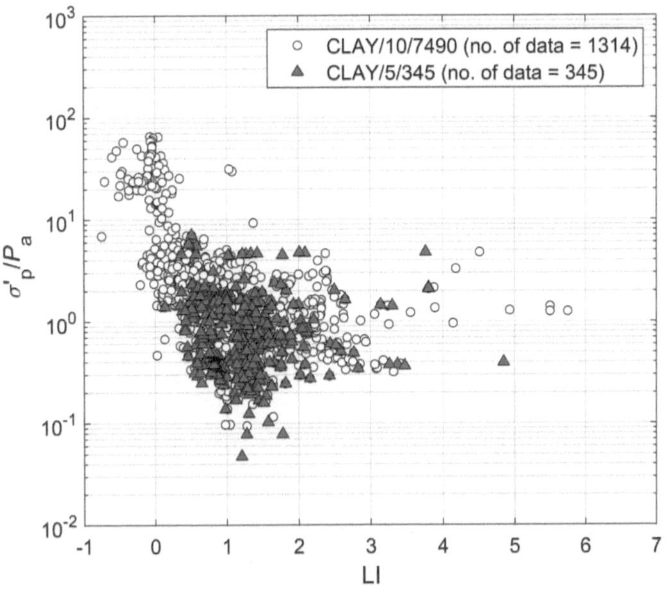

FIGURE 2.20 Comparison between CLAY/5/345 and CLAY/10/7490 in the LI versus σ'_p/P_a transformation model. (Source: Figure 4, Phoon and Ching 2014b)

TABLE 2.22
Johnson distribution type and model parameters for (Y_1, Y_2, ..., Y_5) in CLAY/5/345.

			Model parameters			
	Parameter	Johnson type	a_X	b_X	a_Y	b_Y
Y_1	LI	SU	1.536	−1.618	0.449	0.541
Y_2	$\ln(\sigma'_v/P_a)$	SB	6.226	−4.225	23.231	−16.113
Y_3	$\ln(\sigma'_p/P_a)$	SB	1.015	0.578	4.575	−2.018
Y_4	$\ln(S_t)$	SB	6.767	15.727	101.261	−6.271
Y_5	$\ln(s_u/\sigma'_v)$	SB	0.682	0.036	2.165	−1.877

Source: Table 9, Phoon and Ching (2014b)

obtained from the CLAY/10/7490 database and those obtained from the CLAY/5/345 database. It is evident that the two sets of correlation coefficients are less similar. The correlation coefficient for (LI, σ'_p/P_a) is different when computed using different databases. This difference can be explained from the transformation model shown in Figure 2.20 – the databases occupy different ranges of LI. The correlation coefficients for two pairs of variables, namely: (1) (S_t, LI) and (2) (σ'_p/P_a, σ'_v/P_a), are however comparable. The corresponding transformation models are presented in Figure 2.22a and b, respectively. In general, the differences shown in Figures 2.20 and 2.21 are not surprising given the soil types represented in the databases. Insensitive clays are prevalent in CLAY/10/7490 while CLAY/5/345 contains data for sensitive to quick clays only. This is also consistent with the well-known observation that correlations for unstructured soils cannot be applied to structured soils.

TABLE 2.23
Product–moment (Pearson) correlations among (X_1, X_2, …, X_5) for CLAY/5/345 (corresponding correlation from CLAY/10/7490 in parenthesis).

	X_1 (X_3)	X_2 (X_4)	X_3 (X_5)	X_4 (X_6)	X_5 (X_7)
X_1	1.00	−0.28	−0.15	0.68	0.19
(X_3)		(−0.47)	(−0.54)	(0.55)	(0.02)
X_2		1.00	0.77	0.14	−0.15
(X_4)			(0.68)	(−0.02)	(−0.51)
X_3			1.00	0.29	0.22
(X_5)				(0.03)	(0)
X_4	Symmetry			1.00	0.44
(X_6)					(0.21)
X_5					1.00
(X_7)					

Source: Table 10, Phoon and Ching (2014b)
Note: X_1, X_2, X_3, X_4, X_5 are standard normal variables corresponding to: LI, ln(σ'_v/P_a), ln(OCR), ln(σ'_p/P_a), S_t, s_u/σ'_v

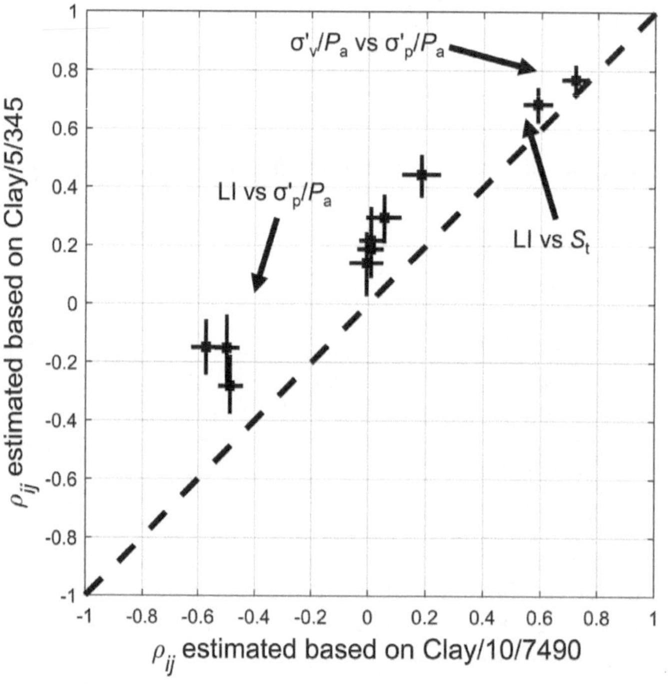

FIGURE 2.21 Relationship between the ten correlation coefficients (ρ_{ij}) obtained from CLAY/10/7490 and the corresponding ones obtained from CLAY/5/345. (Source: Figure 5, Phoon and Ching 2014b)

2.6.4 Site-Specific Correlations

The cross-correlations in Tables 2.17, 2.20, and 2.23 are generic in the sense that they are derived from generic soil databases that contain records from multiple sites. The cross-correlation matrix at the site level can vary from one site to the next. More sophisticated Bayesian machine learning models are needed to handle such "inter-site variability". Zhou et al. (2021) compiled site-specific correlations for sites found in the following clay/sand/rock databases:

1. CLAY/10/7490, CLAY/8/12225, and CLAY/12/3997
2. SH-CLAY/11/4051
3. SAND/7/2794 and SAND/10/4113
4. ROCK/9/4069
5. ROCKMASS/9/5876

The results of this study are reproduced in Appendix C. Engineers emphasize the importance of specific knowledge of the site in geotechnical design (Simpson 2011; Vardanega and Bolton 2016) usually in the context of highlighting this issue as a limitation of current probabilistic methods. The adoption of reliability-based design is impeded in part by this sentiment. It is worthwhile pointing out that the research agenda to address site specificity systematically has been formalized as the "site recognition challenge" in data-driven site characterization in recent years (Phoon et al. 2022a; Phoon and Ching 2021, 2022). It is no longer true to say that probabilistic methods are insensitive to inter-site variability (Section 2.8).

2.7 TRANSFORMATION MODEL

One major application of the multivariate probability model (and cross-correlations) is the estimation of pertinent soil parameters, particularly the values governing the behavior of a geotechnical structure at a limit state. Correlation models are prevalent in practice because an engineer can obtain an estimate of a soil parameter pertinent to design (called "design parameter") using more commonly available data that is indirectly related to this design parameter but cheaper to acquire, say data from a laboratory index test or a field test. In fact, these correlation models grew in popularity as a result of applying more rational soil mechanical principles to engineering practice (Kulhawy and Mayne 1990).

Many prevailing correlations are bivariate in the sense of estimating one desired design parameter (e.g., undrained shear strength) from one indirect source of data (e.g., corrected cone tip resistance). Examples are shown in Figures 2.18, 2.20, and 2.22. A minority is multivariate in the sense of estimating one design parameter from multiple sources of data. To the authors' knowledge, none has considered the simultaneous estimation of two or more design parameters from two or more sources of data. This is not the same as using two separate correlations. The reason is that the *correlation between two design parameters* cannot be obtained in this way (Phoon and Ching 2017). The general

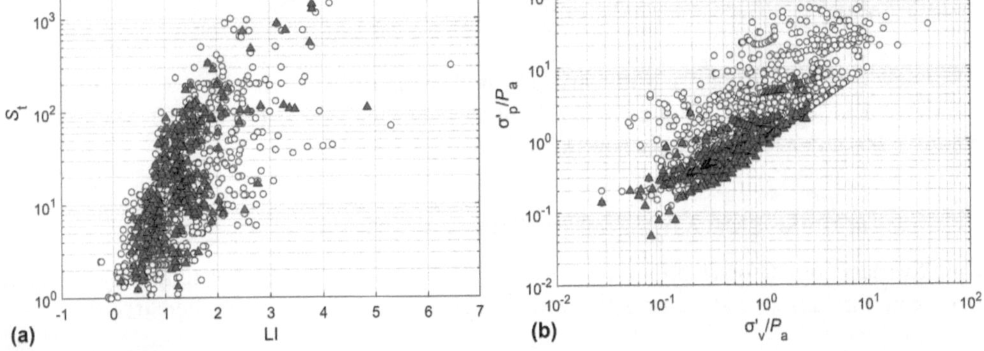

FIGURE 2.22 Comparison between CLAY/5/345 and CLAY/10/7490 in two transformation models: (a) S_t versus LI and (b) σ'_p/P_a versus σ'_v/P_a. (Source: Figures 6 and 7, Phoon and Ching 2014b)

approach is to derive a conditional distribution from a multivariate probability model so that (loosely speaking) the dependencies between all sources of data are accounted for (Section 2.7.1). The authors are of the opinion that our existing bivariate correlation models can be significantly improved by extending them using a multivariate probability model (Section 2.6). Although this model is more abstract and less familiar to most engineers, it can offer the following practical advantages:

1. Multivariate transformation models can be obtained. For example, one can derive a relation between the compression ratio and multiple sources of indirect data, say liquid limit, plasticity index, initial void ratio, natural water content, cone tip resistance, and dilatometer modulus. The natural generalization from the existing bivariate case to the multivariate case is called a "transformation model".
2. These transformation models can predict not only the mean value of the design parameter but also its coefficient of variation (Section 2.7.2). It is natural to expect all transformation models to contain transformation uncertainty (scatter of data about the mean regression line) and this uncertainty must be quantified explicitly as it has an impact on the choice of the characteristic value and design in general (Section 2.2.2).
3. The multivariate probability model can be used as a proper and empirically supported prior distribution to derive the posterior distribution of design parameters based on limited but site-specific laboratory/field data. This prior is clearly much preferred over a pure judgment-based or an uninformative prior. The multivariate probability model can be extended to a hierarchical Bayesian model (HBM) to combine site-specific data with a generic database in a manner sensitive to inter-site variability (refer to the "site recognition challenge" in data-driven site characterization discussed in Phoon et al. [2022a] and Phoon and Ching [2021]).
4. The entire multivariate distribution of multiple soil/rock parameters can be derived, not simply means and COVs (Section 2.6). When a multivariate distribution is available, multiple soil/rock parameters can be updated simultaneously from multiple laboratory/field measurements. This formal Bayesian updating approach is significantly more advantageous than relying on judgment alone to combine site-specific data with prior experience. Experience is usually applied to estimate a single characteristic value. It is exceedingly difficult to estimate a set of multiple characteristic values that will respect the correlation structure in Eq. (2.31) (Ching et al. 2020c).
5. The practical significance of updating is that the bias and imprecision are generally reduced in the presence of site-specific data and further reduced by incorporating *relevant* generic data (Ching et al. 2021c; Phoon and Ching 2022). It is possible to link this improvement to design savings explicitly in the context of reliability-based design discussed below (Ching et al. 2014c).

An overview of the evolution from univariate to multivariate characterization of soil uncertainty was presented in Phoon and Ching (2014a, 2014b, 2017). ISO2394:2015 (ISO 2015) covered the transformation of uncertainty in Section D.2.5.

2.7.1 Conditioning from a Multivariate Probability Model

The multivariate normal distribution shown in Eq. (2.30) is *unconditional*. For simplicity, we ignore spatial variability and consider a random vector consisting of four soil parameters only:

$$\boldsymbol{X} = (X_1, X_2, X_3, X_4)^{\mathrm{T}} \tag{2.34}$$

For conciseness, the first subscript in Eq. (2.27) is dropped. Physically, this means an *independent* record of Eq. (2.34) is measured in every location. One important result ensuing from a multivariate normality assumption is that the *conditional* distribution is available in closed-form. This explains why it is computationally advantageous to transform Y (soil parameter) to X (standard normal variable) in Eq. (2.32). Let $\boldsymbol{X}$ be partitioned into two parts ($\boldsymbol{X}_A$, $\boldsymbol{X}_B$), where $\boldsymbol{X}_A = (X_1, X_2)^T$ and $\boldsymbol{X}_B = (X_3, X_4)^T$. $\boldsymbol{X}_A$ can refer to a vector of design parameters such as s_u and σ'_p (transformed versions). $\boldsymbol{X}_B$ can refer to a vector of field test parameters such as q_t and N (transformed versions). It follows that the mean vector is partitioned as $\boldsymbol{X}_{mean} = (\boldsymbol{X}_{Amean}, \boldsymbol{X}_{Bmean})$, where $\boldsymbol{X}_{Amean} = (X_{1mean}, X_{2mean})^T = (0, 0)^T$ and $\boldsymbol{X}_B = (X_{3mean}, X_{4mean})^T = (0, 0)^T$. The mean vectors are zero, because $\boldsymbol{X}$ consists of standard normal variables. The correlation (covariance) matrix is similarly partitioned as

$$C = \begin{bmatrix} 1 & \rho_{12} & \rho_{13} & \rho_{14} \\ \rho_{21} & 1 & \rho_{23} & \rho_{24} \\ \rho_{31} & \rho_{32} & 1 & \rho_{34} \\ \rho_{41} & \rho_{42} & \rho_{43} & 1 \end{bmatrix} = \begin{bmatrix} C_{AA} & C_{AB} \\ C_{BA} & C_{BB} \end{bmatrix} \tag{2.35}$$

where $\rho_{ij} = \rho_{ji}$ and the sub-matrices C_{AA}, $C_{AB} = C_{BA}{}^T$, and C_{BB} are defined as

$$C_{AA} = \begin{bmatrix} 1 & \rho_{12} \\ \rho_{12} & 1 \end{bmatrix} \tag{2.36a}$$

$$C_{AB} = \begin{bmatrix} \rho_{13} & \rho_{14} \\ \rho_{23} & \rho_{24} \end{bmatrix} \tag{2.36b}$$

$$C_{BA} = \begin{bmatrix} \rho_{31} & \rho_{32} \\ \rho_{41} & \rho_{42} \end{bmatrix} \tag{2.36c}$$

$$C_{BB} = \begin{bmatrix} 1 & \rho_{34} \\ \rho_{43} & 1 \end{bmatrix} \tag{2.36d}$$

The conditional distribution of $\boldsymbol{X}_A$ given a *measured* $\boldsymbol{X}_B = \boldsymbol{x}_B = (x_3, x_4)^T$ is a bivariate normal distribution with

$$\text{mean vector} = \boldsymbol{X}_{Amean} + C_{AB}C_{BB}^{-1}(\boldsymbol{x}_B - \boldsymbol{X}_{Bmean}) = C_{AB}C_{BB}^{-1}\boldsymbol{x}_B \tag{2.37}$$

$$\text{covariance matrix} = C_{AA} - C_{AB}C_{BB}^{-1}C_{BA} \tag{2.38}$$

In general, the conditional distribution of $\boldsymbol{X}_A$ given $\boldsymbol{X}_B$ is a multivariate normal distribution fully defined by Eqs. (2.37) and (2.38), regardless of how $\boldsymbol{X}$ is partitioned. This theoretical result is the basis for many Bayesian methods. Equation (2.37) is the transformation model for predicting the design parameters (s_u, σ'_p) *jointly* when the field test parameters (q_t, N) are measured after suitable inverse transformations have been applied (Eq. 2.33). Equation (2.38) is the transformation uncertainty. One example is to update the distribution of $Y_3 = \ln(\sigma'_p/P_a)$ based on measurements from $(Y_1, Y_2, Y_4) = (\text{LI}, \ln[\sigma'_v/P_a], \ln[S_t])$. The estimation of a design property from multiple data sources is one important practical outcome of a site investigation program. The above "conditioning" procedure can be viewed as a rationalization of simpler empirical procedures widely adopted in practice, such as averaging estimates from different tests or choosing the most conservative estimate produced by all tests.

Consider an example involving updating the normalized preconsolidation stress (Y_3) at a given depth based on data from other sources measured at the same depth: liquidity index (Y_1), the normalized effective vertical stress (Y_2), and the sensitivity (Y_4). The following values measured at a depth of 9.33 m are extracted from a test fill in Gloucester (Canada) reported by Bozozuk and Leonards (1972):

$Y_1 = \text{LI} = 2.00$
$Y_2 = \ln(\sigma'_v/P_a) = \ln(0.679) = -0.387$
$Y_3 = \ln(\sigma'_p/P_a) = \ln(1.104) = 0.010$
$Y_4 = \ln(S_t) = \ln(40) = 3.689$

Based on the information on (Y_1, Y_2, Y_4), the probability distribution of Y_3 can be updated. The unconditioned and conditioned distribution for Y_3 given a subset of (Y_1, Y_2, Y_4) is shown in Figure 2.23. The unconditional distributions are the dashed lines, whereas the conditional distributions are the grey solid line. The actual measured value of Y_3 is 0.010 ($\sigma'_p/P_a = 1.104$) as highlighted by an arrow.

It is clear that conditioning on Y_1 = LI is not very helpful, because the conditional distribution is very similar to the unconditional distribution (upper left plot). This is correct, because the correlation between LI and $\ln(\sigma'_p/P_a)$ is weak (see the triangular data points in Figure 2.20). The correlation coefficient between X_1 and X_3 is only −0.15 (see Table 2.23). The presence of a second source of information, $Y_4 = \ln(S_t)$, is somewhat helpful, as the conditional distribution starts to look different (upper right plot). The addition of a third piece of information, $Y_2 = \ln(\sigma'_v/P_a)$ improves the estimate substantially, as the conditional distribution now looks quite different (lower left plot). Moreover, the center of the conditional distribution now moves toward the actual value 0.01. This is correct,

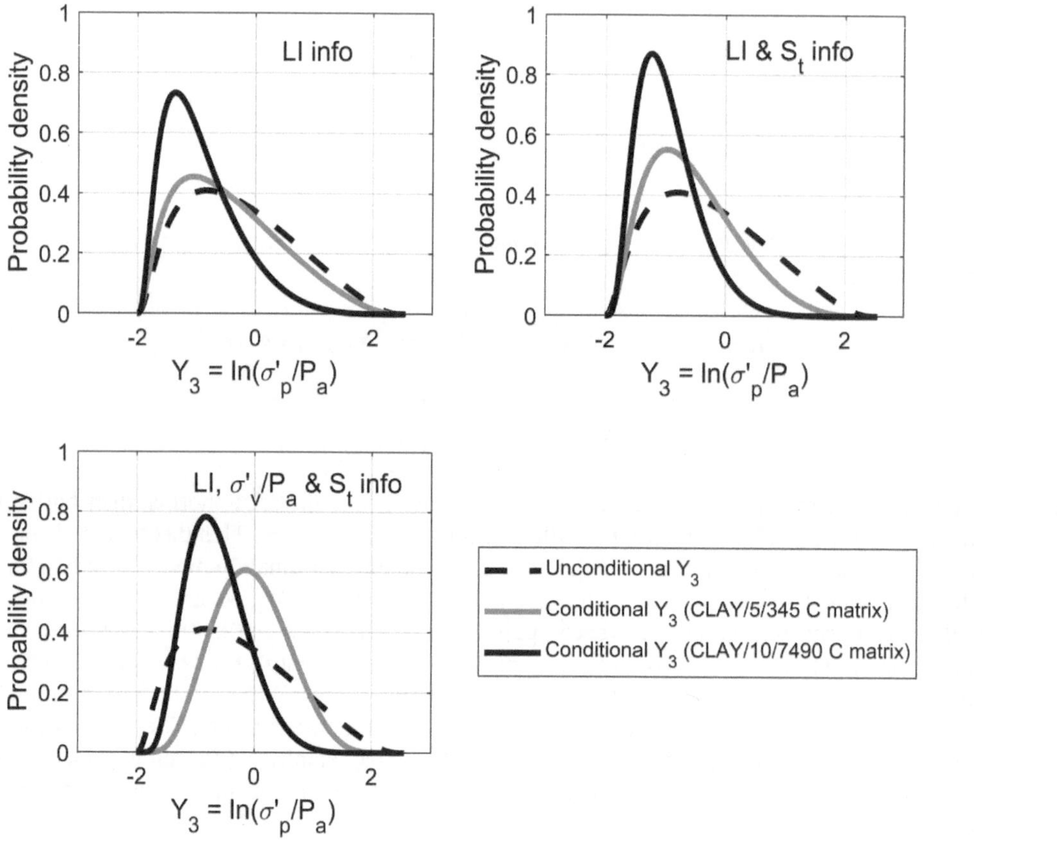

FIGURE 2.23 Conditional distributions for Y_3, given a subset of (Y_1, Y_2, Y_4).

because the correlation between ln(σ'_v/P_a) and ln(σ'_p/P_a) is strong (see the triangular data points in Figure 2.22b). The correlation coefficient between X_2 and X_3 is 0.77 (see Table 2.23).

As mentioned in Section 2.6.3.3, the correlation coefficients based on the CLAY/5/345 database are not quite the same as those based on the CLAY/10/7490 database. The former database was compiled specifically for structured (sensitive or quick) clays, whereas the latter is a larger and more generic database that is not restricted to structured clays. For the above example in the Gloucester test fill, the clay is a quick clay (S_t = 40). Hence, the correlation matrix constructed by the CLAY/5/345 database should be more appropriate for this clay. However, it will be interesting to see how the conditional distribution of Y_3 will change if the correlation matrix constructed by the CLAY/10/7490 database is used in the conditioning procedure. The black solid lines in Figure 2.23 are the conditional distributions of Y_3 based on this inappropriate correlation matrix. The conditional distribution of Y_3 based on the Y_1 information (black solid line in upper left plot) is worse than the unconditional distribution. It can be seen that the center of the conditional distribution of Y_3 moves away from the actual value of Y_3. This is because the correlation coefficient between LI and ln(σ'_p/P_a) is now −0.54 for the CLAY/10/7490 database, which is more negative than the corresponding value of −0.15 for the CLAY/5/345 database. This error actually increases when two pieces of information (Y_1, Y_4) are both given (black solid line in upper right plot). The center of the conditional distribution of Y_3 moves even further away from the actual value of Y_3. However, the error reduces when the third piece of information (Y_2) is considered (black solid line in lower left plot). The center of the conditional distribution of Y_3 now shifts back toward the actual value of Y_3. Note that the correlation coefficients for (Y_2, Y_3) based on the two databases are similar. Although it is strictly incorrect to adopt the CLAY/10/7490 correlation matrix to the quick clay example (because CLAY/10/7490 is not restricted to quick clays), the use of the CLAY/10/7490 correlation coefficient for (Y_2, Y_3) still produces results that are practically useful. This statement is verified in Figure 2.24, showing the conditional distributions of Y_3 based on the Y_2 information only. The two conditional distributions look quite similar.

2.7.2 Transformation Bias and Uncertainty

ISO2394:2015 (ISO 2015), Section D.4 defines a model factor as the "the ratio of the measured to the calculated capacity". The model factor is generally applied to characterize the bias and

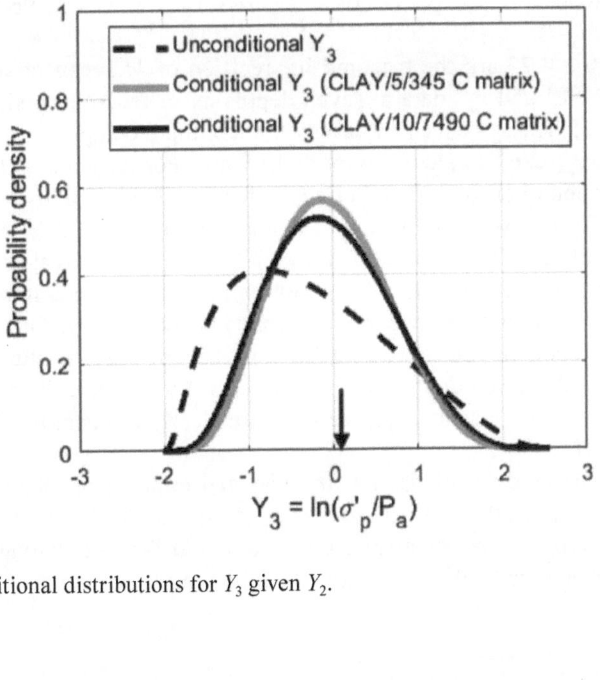

FIGURE 2.24 Conditional distributions for Y_3 given Y_2.

imprecision (uncertainty) of a calculation model for a response relevant to a limit state (e.g., capacity, settlement, deflection). Ching and Phoon (2014a) applied the same concept to the prediction of a soil/rock parameter from a transformation model. It is apparent from Figures 2.18, 2.20, 2.22 that any prediction produced by a linear or non-linear regression will be uncertain and this transformation uncertainty can be significant. It is straightforward to define the soil/rock parameter "model factor", M_T, as

$$M_T = \frac{\text{measured soil parameter}}{\text{predicted soil parameter from a transformation model}} \tag{2.39}$$

To illustrate the statistical characterization of M_T, consider the first transformation model in Table D.1 in Appendix D. The denominator of Eq. (2.39), the predicted soil parameter from a transformation model, is $s_u^{re}/P_a = 0.0144 \times LI^{-2.44}$, where s_u^{re} is the remolded undrained shear strength, P_a is one atmosphere pressure, and LI is the liquidity index. For each LI, the corresponding measured value of s_u^{re}/P_a is extracted from CLAY/10/7490. There are 899 values of M_T that can be calculated by dividing the measured value by the predicted value (Eq. 2.39). The mean and COV of these 899 M_T values are termed the "bias" and "imprecision" of the transformation model, respectively. Table D.1 shows that the bias = 1.92 and the imprecision = 1.25. This means the measured value is roughly two times the predicted value on average, i.e., the model underestimates s_u^{re}/P_a and it is conservative. If M_T is lognormal distributed, the probability of the transformation model producing an unconservative prediction is

$$\text{Prob}(M_T < 1) \approx \Phi\left[\frac{-\ln(\text{bias})}{\text{imprecision}}\right] \tag{2.40}$$

Note that ($M_T < 1$) means that the measured value is less than the predicted value. For the case of s_u^{re}/P_a, this overestimation is unconservative. Based on the bias = 1.92 and the imprecision = 1.25, this probability is 30%. Hence, despite the large bias on the conservative side, this model remains unconservative in roughly one out of three predictions. Another useful application of M_T is the estimation of the 95% confidence interval for the prediction. If M_T is lognormal distributed, this 95% confidence interval is roughly given by

$$0.0144\text{LI}^{-2.44}\lambda\exp(-1.96\theta) \le \frac{s_u^{re}}{P_a} \le 0.0144\text{LI}^{-2.44}\lambda\exp(1.96\theta) \tag{2.41}$$

where $\lambda = 1.92$ and $\theta = 1.25$ are the bias and imprecision of M_T, respectively. For illustration, Figure 2.25 shows the LI and s_u^{re} data at seven depths extracted from a site in Taipei, Taiwan (Chen and Hsieh 1996). The s_u^{re} data was based on unconfined compression (UC) tests. At each depth, the 95% CI of s_u^{re} can be calculated by its LI value. For instance, at the depth of 14.4 m, LI = 1.00. The lower bound of the 95% CI of s_u^{re}/P_a for this depth is $0.0144 \times 1.00^{-2.44} \times 1.92 \times \exp(-1.96 \times 1.25) = 0.0024$, whereas the upper bound is $0.0144 \times 1.00^{-2.44} \times 1.92 \times \exp(1.96 \times 1.25) = 0.32$. As a result, the 95% CI is $0.0024 \le s_u^{re}/P_a \le 0.32$, or equivalently 0.24 kPa $\le s_u^{re} \le$ 32.8 kPa. If the lower bounds of the 95% CI are unrealistically low, this implies that the lognormal distribution assumption with a lower bound of zero is not realistic. This does not imply that probabilistic analysis is not realistic (a widespread misconception among engineers). Figure 2.25b shows the 95% CIs for the seven depths. The measured s_u^{re} values are shown as circles for validation. The statistics for M_T (bias and imprecision) are clearly informative for decision-making (Phoon and Ching 2017).

Figure 2.26 summarizes the statistics for M_T (bias and imprecision) for some transformation models (Ching and Noorzad 2021). Details are given in Appendix D. The statistics are classified according to a system proposed by Phoon and Tang (2019) and Tang and Phoon (2021), as shown in Table 2.24. Ching and Noorzad (2021) presented the following observations:

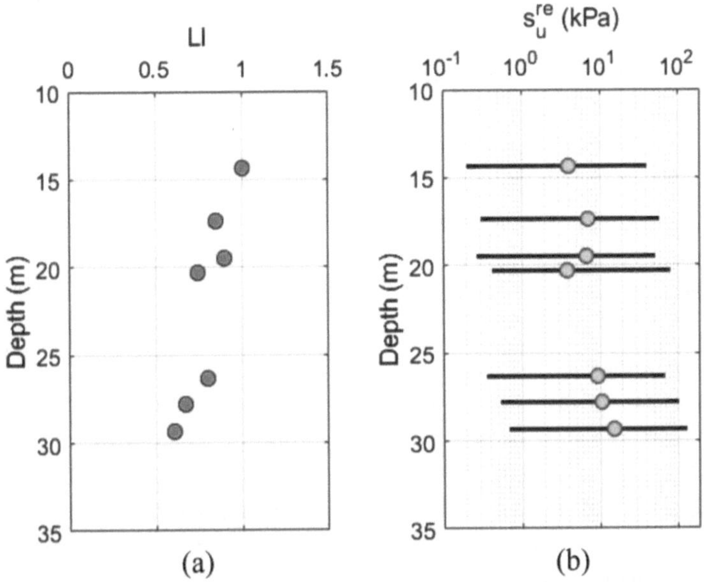

FIGURE 2.25 LI and s_u^{re} data extracted from a site in Taipei, Taiwan (data from Chen and Hsieh 1996): (a) LI versus depth and (b) s_u^{re} versus depth (the horizontal bars are the 95% CIs, whereas the circles are the measured s_u^{re} values for validation).

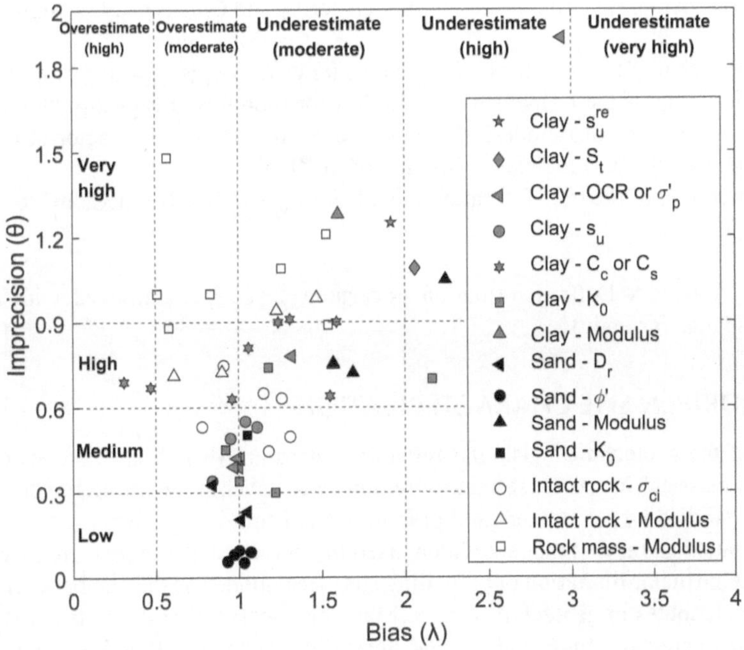

FIGURE 2.26 Model statistics (bias and imprecision) for some transformation models. The legend indicates the predicted parameter in clay, sand, intact rock, and rock mass. (Source: Figure 5.1, Ching and Noorzad 2021)

TABLE 2.24
Classification of model factor based on bias and imprecision.

Description	Imprecision (θ)	Bias (λ)	
		Overestimation	Underestimation
Low (imprecision)	$\theta < 0.3$		
Medium (imprecision) or Moderate (bias)	$0.3 \leq \theta < 0.6$	$0.5 \leq \lambda < 1$	$1 \leq \lambda < 2$
High (imprecision and bias)	$0.6 \leq \theta < 0.9$	$\lambda < 0.5$	$2 \leq \lambda < 3$
Very high (imprecision and bias)	$\theta \geq 0.9$		$\lambda \geq 3$

Source: Phoon and Tang (2019); Tang and Phoon (2021)

Note: bias > 1 is conservative on the average for the ultimate limit state; bias < 1 is conservative on the average for the serviceability limit state.

1. The common range of the bias is $0.5 \leq \lambda < 2$. This means that most transformation models underestimate or overestimate the measured values by a factor of 2 or less on the average (termed as "moderate bias").
2. The common range of imprecision is $0.3 < \theta < 0.9$ for clay, sand, and intact rock models. This is termed as "medium" to "high" imprecision. Rock mass models are "very highly" imprecise with $\theta > 0.9$.
3. The imprecision of the transformation models, s_u versus OCR (or σ'_p), is typically "medium": $0.3 < \theta < 0.6$.
4. The imprecision of the transformation models for C_c and C_s is typically "high": $0.6 < \theta < 0.9$.
5. The imprecision of the transformation models for ϕ' and D_r is typically "low": $\theta < 0.3$.
6. The imprecision of the transformation models for modulus is typically "high" to "very high": $\theta > 0.6$. They also underestimate the measured values by a factor of between 1.5 and 2.3 on average, i.e., the bias is "moderate" to "high".
7. The imprecision of the transformation models for K_0 is typically "medium" to "high": $0.3 < \theta < 0.9$.

The COVs in Appendix D (transformation uncertainty) can be contrasted with the COVs in Appendix A (spatial variability).

2.8 DATA-DRIVEN SITE CHARACTERIZATION

A large part of the geotechnical risk and reliability literature (which is the focus of the current book) remains classical in the sense that more advanced machine learning methods have yet to be extensively studied by researchers and practitioners. There is significant room to improve the fairly simple statistical models that are widely used in geotechnical and rock engineering practice (Phoon 2017). One major limitation of prevailing statistical models is the gap between the assumed and the actual attributes of geotechnical data. Many classical statistical models assume that data is homogeneous, abundant, independent, and normally distributed. Practitioners have criticized these models as unrealistic for many years. Phoon et al. (2022a) opined that the "ugly data" challenge lies at the heart of any data-driven site characterization. The authors defined DDSC as any site characterization methodology that relies solely on measured data, both site-specific data collected for the current project and existing data of any type collected from past stages of the same project or past projects at the same site, neighboring sites, or beyond. One example of ugly data is

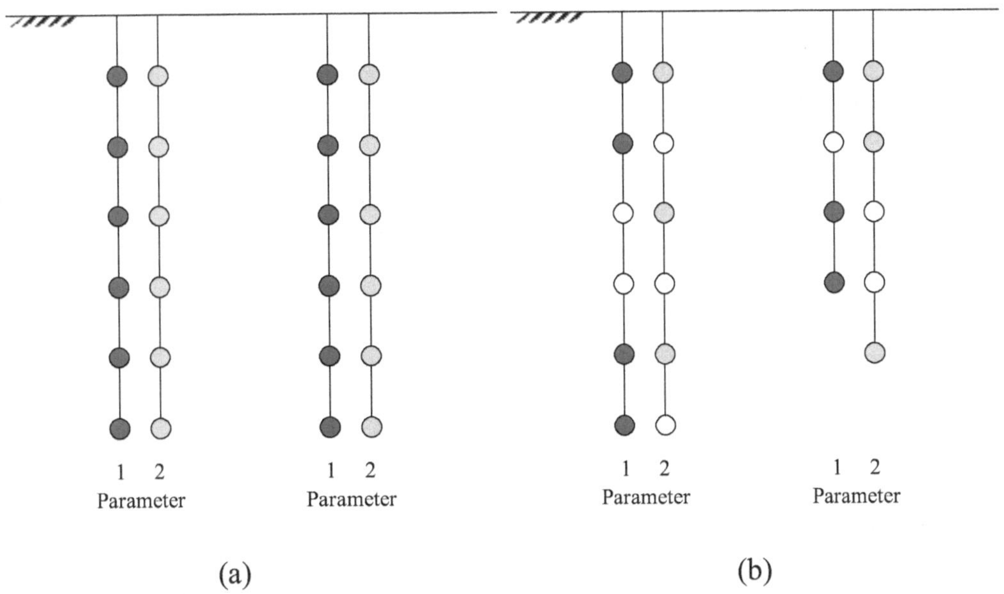

FIGURE 2.27 Bivariate site data: (a) complete (ideal case) and (b) incomplete (realistic case).

MUSIC-3X (multivariate, uncertain and unique, sparse, incomplete, and potentially corrupted with "3X" denoting three-dimensional spatial variations). An example of a complete and incomplete bivariate dataset is illustrated in Figure 2.27. It is a useful mnemonic to highlight the attributes of real site data and to contrast with the highly idealized assumptions underlying classical statistics. Ideal data is "beautiful". Real-world data is "ugly". It is premature to say that ugly data cannot support decision-making. The central tenet in data-centric geotechnics is that data has value as long as it is not fake or completely corrupted.

It is an open question what data-driven site characterization can achieve and how useful are the outcomes for practice, but this "value of data" question is of major interest given the rapid pace of digital transformation in many industries. The scientific aspects of this question are presented as three challenges by Phoon et al. (2022a): (1) ugly data, (2) site recognition, and (3) stratification. Additional challenges are given in Table 2.1. The site recognition challenge is as basic as the ugly data challenge. The objective is to quantify "site uniqueness", directly or indirectly, so that big indirect databases (BIDs) (Table 2.2) can be combined with sparse site-specific data in a manner sensitive to site differences. This idea is not new as geotechnical engineers have been relying on data from similar sites to supplement limited data collected at a given site. The engineer relies heavily on his/her experiences and knowledge of local geology to identify similar sites, but this human-powered strategy cannot work when the amount of data becomes overwhelmingly large or "fast" (real-time streaming). Despite the apparent intractability of these challenges at first glance, promising solutions based on Bayesian machine learning have been found (Ching et al. 2021a, 2021c). Data-centric geotechnics, which includes DDSC, building information modeling (BIM), and digital twin, can be viewed as the next evolutionary leap in decision support that would succeed the original deterministic Burland Triangle (Figure 1.4) and the enhanced uncertainty-informed Burland Triangle (Figure 1.10). Its potential is likely to be tremendous and its impact far reaching. Details are given elsewhere (Phoon and Ching 2021, 2022; Phoon et al. 2022d; Phoon and Zhang 2023). It suffices to note that the geotechnical risk and reliability agenda is on the cusp of being changed to meet the challenging needs of practice such as ugly data and site recognition and to solve practical problems at a new level of realism (Phoon 2023).

2.9 CONCLUSIONS

The purpose of Chapter 2 is to present practical tools and guidelines for the uncertainty quantification of soil and rock properties with a primary focus on enhancing decision-making in geotechnical design. The context underlying uncertainty quantification is important, but rarely emphasized in the geotechnical reliability literature. It should not be conducted as an abstract probabilistic exercise disconnected from the physical basis (geology and geotechnical engineering) that gives rise to the data at hand. The natural origin of the ground; spatial variability in properties, geomaterial types, and environmental factors; inter-site variability (sufficiently material for building regulations to mandate a site investigation at each site), and the possibility of extreme variations not detected during site investigation that may or may not be reasonably anticipated from experience; precedents; and local knowledge are examples of geotechnical contexts that shape current practice as exemplified by the typical language found in geotechnical design codes.

This chapter first presents a conceptual framework to guide the engineer to think about uncertainty quantification in the *absence* of a geotechnical structure (limit state) and in the *presence* of a geotechnical structure (limit state) separately. This provides conceptual clarity on the distinctive role of statistics, physics, and their interactions that is somewhat lacking at present. The conceptual framework contains three building blocks: (1) point value, (2) spatial average, and (3) mobilized value. The theoretical models for these building blocks are deeply related and in order of increasing complexity. The "data ground truths" being modeled are in order of increasing realism. The "data ground truths" should be distinguished from the more familiar "physical ground truths" although they emerge from the same reality. The former is the bedrock of a data-driven perspective. The latter is physics informed.

The point value is the most basic building block. It refers to a value measured by one laboratory/field test (univariate case). It can also refer to a set of values from multiple tests conducted in close proximity (multivariate case). It is a "point" value because the volume of ground mobilized in a test is much smaller than the volume mobilized by a typical geotechnical structure. The theoretical model to describe the uncertainty in the spatial distribution of a property is a set of point values that are independent and identically distributed. Under this simple model, the engineer needs to characterize a single (univariate or multivariate) probability distribution function. The most common univariate example is the lognormal distribution, although the Johnson system of distributions (that includes the lognormal distribution as one member) is more suitable because it can fit a wider range of empirical histograms and it can accommodate lower and/or upper bounds. The engineer is most familiar with the coefficient of variation or standard deviation that is a simple index of uncertainty in this point value model. The COVs of many structural materials are smaller than 10%–20%. For geomaterials that are not manufactured to a prescribed specification, COVs are much larger and they can vary between sites even for the same property. Site-specificity is an example of a geotechnical context that deserves more attention. Extensive COV guidelines are provided in this chapter. These guidelines are immediately useful for the selection of a characteristic value based on the 5% fractile, first-order second-moment reliability analysis, and reliability-based design. Besides soil properties, the precision of a calculation model can be quantified by a model factor that is also modeled by a probability distribution function. Extensive COV guidelines for the model factor are provided elsewhere (Tang and Phoon 2021, Phoon and Tang 2024).

The spatial average is the value affecting the occurrence of a slip along a *prescribed* trial line. To be specific, it affects the factor of safety for this trial slip line. Hence, the spatial average partially considers the presence of a geotechnical structure, because only plausible slip lines are of interest to the engineer. Traditionally, the spatial average is evaluated using Vanmarkce's random field theory. However, the spatial average and random field theory are separate concepts. The spatial average is simply the average of a set of random variables – it can be defined in discrete or continuous form. The averaging domain can be a line, an area, or a volume. The engineer is most interested in an averaging domain that is mobilized by a limit state, but this domain cannot be prescribed a priori because it is a solution to a boundary value problem. A random field is a generalization of the point value model. It is generally accepted that the correlation between a set of spatially distributed properties measured at different locations in the soil mass is not zero, particularly when the locations are sufficiently close. In short, the independence assumption between point values is not realistic. The introduction of a random field model increases the model complexity and substantially complicates the statistical characterization of this model. However, from a practical perspective, it may be sufficient to quantify the COV and a scale of fluctuation (roughly the characteristic length between strongly correlated measurements). Extensive guidelines on the scale of fluctuation are provided in this chapter. The scale of fluctuation is typically measured in the vertical direction. Actual spatially varying soils are usually anisotropic in the sense that the vertical and horizontal scales of fluctuation are different. The latter is roughly 10–20 times longer than the former. Very little is known about the spatial variability of cement-mixed soils at present. No statistical characterization of more general anisotropy such as the rotated anisotropy has been conducted.

When a limit state is governed by a spatial average (e.g., capacity of a floating pile is governed by the average soil strength along its shaft), its uncertainty reduces with the characteristic length of the geotechnical structure (e.g., pile length). Ignoring this "variance reduction" effect will result in an excessively conservative structure in the context of reliability-based design, because the characteristic value of the spatial average decreases with uncertainty. This important effect can be quantified using Vanmarcke's variance reduction function, which is available in simple analytical forms for the classical one-parameter autocorrelation models. However, there are two significant limitations. In recent years, it was found that some data exhibits roughness (or conversely smoothness) and pseudo-periodicity (or hole effect). The classical autocorrelation models are unable to capture these additional spatial variability features. Studies have shown that these features can affect the probability of failure of some limit states. The most general solution to date is the cosine Whittle–Matérn autocorrelation model that can account for three spatial variability features simultaneously: (1) scale of fluctuation, (2) roughness, and (3) pseudo-periodicity. Unfortunately, there are no guidelines for the new features, such as what soil types or geologic environments produce "rough" or "pseudo-periodic" data. However, it is possible to characterize the CosWM model if there is sufficient site-specific data. The second limitation is that the available variance reduction functions do not apply to inclined slip lines which are more common than vertical or horizontal lines. Practical but approximate solutions were published recently and reproduced in this chapter. The variance reduction function for averaging along a curve is not available.

It is important to point out that the application of random field theory is much more general than computing the 5% fractile characteristic value for the spatial average. When coupled with a finite element model, it can produce more critical failure mechanisms in two aspects: (1) a single realization of a random field is spatially heterogeneous and can produce a failure mechanism that is qualitatively different from the well-known "textbook" solution for a homogeneous soil; and (2) a random field can produce a range of plausible scenarios (or realizations) that are more consistent with the limited site investigation data. These scenarios are significantly more informative and some are certainly more severe than those selected for parametric studies based on the judgment of an engineer. A range of failure modes can appear when these plausible scenarios are analyzed using the random finite element method. In short, the current deterministic approach only produces one failure mode that may not be the most likely or the most critical one. Multiple failure modes and their relative likelihood of occurrence can only be obtained using a probabilistic analysis involving spatially varying soils. It is worth pointing out that the practice of reducing a "reasonably likely" characteristic value to a "sufficiently severe" design value by simple factoring does not work for spatially varying soils. The spatial distribution of values for a "reasonably likely" scenario and that of a "sufficiently severe" scenario are *qualitatively different*. It is not well appreciated that current practice is strongly restricted to homogeneous or layered soil profiles if it remains deterministic. When coupled with modern machine learning methods, a random field can produce realistic spatial distributions of one or more properties and a family of "near realistic" stratigraphy even in the presence of MUSIC-3X data. This data-driven site characterization research has attracted strong interests from the research community because of its obvious usefulness to practice.

The random field easily generalizes to a vector random field (at least theoretically) in the presence of multivariate data. All site investigations involve multiple tests. The correlation between two soil properties produced by two different tests (called a cross-correlation) is distinct from spatial correlation which is the correlation between the same property measured at different locations. The most familiar application of the cross-correlation is the transformation model that relates a field or laboratory measurement to a design parameter. Spatial correlation is not considered because the tests (most commonly two of them) are conducted in close proximity (within say one scale of fluctuation). Extensive guidelines are provided elsewhere (Phoon and Ching 2014a, 2014b, 2017). This chapter illustrates the reduction in the uncertainty of *one or more* design parameters when information on *one or more tests* is available using useful closed-form solutions from a conditional multivariate normal distribution. The conventional transformation models are "one to one", but it is relatively straightforward to construct more general "many to many" models. The key task for an engineer is to assemble a correlation matrix from the cross-correlations between all pairs of soil parameters. It is not sufficiently emphasized that the matrix must be positive definite to be valid and it is not well-known that an additional bootstrapping step is needed to fulfill this fundamental theoretical requirement (Ching and Phoon 2014b). While it is possible to characterize the multivariate normal distribution using familiar classical statistical tools for complete multivariate data (all tests measured in all locations), it is challenging to do this for incomplete data (some tests measured in some locations). Unfortunately, incompleteness is a characteristic feature of most geotechnical databases. In addition, the current transformation models widely in practice are generic in the sense that the data from different sites is simply pooled without regard to inter-site variability (Kulhawy and Mayne 1990). Much more sophisticated Bayesian machine learning methods are needed to tackle MUSIC-3X data and they have advanced in leaps and bounds in recent years. Software needs to be developed for these methods to impact decision-making in practice. This chapter does not cover this rapidly emerging body of work on data-driven site characterization that can be broadly categorized under three challenges: (1) ugly data, (2) site recognition, and (3) stratification (Phoon et al. 2022a). An engineer can readily appreciate that it is very difficult to address multiple correlated soil parameters in a consistent manner using judgment alone.

The third building block is mobilized value. It is closest to the "value affecting the occurrence of the limit state" in Clause 2.4.5.2(2) of Eurocode 7. The characteristic value is defined as a cautious estimate of this value. It is apparent that uncertainty is considered in this definition, although no quantitative methods are offered to compute this value from site data. While the role of engineering judgment is indispensable, this chapter argues that it can be enhanced by a data-driven approach. Sole reliance on experience, knowledge, and judgment impedes digital transformation. There are two aspects about the mobilized value that deserve more elaboration. First, the term "mobilized value" is already found in the geotechnical nomenclature. It is intended to make clear that the value on a critical slip line or surface is not always the ultimate value, depending on the movements mobilized by the geotechnical structure at a limit state. This is identical in meaning to Clause 2.4.5.2(2) of Eurocode 7. It is not identical to the reliability-based mobilized value used in this chapter (Hicks and Samy 2002), because there is no uncertainty associated with the current mechanical definition of the mobilized value. Specifically, the mechanical definition refers to the values (or an average value) at a critical slip surface computed by a deterministic finite element method. When uncertainty is considered, it is not meaningful to talk about "values at a critical slip surface", because this slip surface changes from realization to realization. Hence, the probabilistic approach to determine the mobilized value is to take an average at every realization. The ensemble of this average value is its probabilistic distribution. The 5% fractile of this distribution is the reliability-based characteristic value. This approach is clearly more informative in the sense that the cautious estimate is now quantified. The engineer can apply his/her judgment for a reality check. The second aspect is that this reliability-based characteristic value is widely misunderstood as the 5% fractile of the spatial average. This is theoretically incorrect, because the spatial average is defined over a prescribed trial surface that is not the critical surface affecting the occurrence of a limit state. This prescribed trial surface is independent of the different spatial distributions of soil properties found in different realizations. The critical slip surface crucially depends on the spatial distributions of soil properties by virtue of its definition. The reliability-based characteristic value is fully compatible with physics and statistics. Its physical limitations are identical to those in the adopted deterministic finite element model. This is a modeling choice that has nothing to do with uncertainty. Its statistical limitations (in its current stage of development) are the weak connection between the inputs of the random field and MUSIC-3X measured data and the lack of statistical data to characterize more advanced constitutive models. The first limitation has since been resolved by rapid advancements in data-driven site characterization, but it has not been adopted in the determination of a reliability-based characteristic value thus far. The second limitation is common to every application in geotechnical reliability. Advanced constitutive models are not widely used in practice. As such, data is not available to characterize the model parameters statistically. Finally, the reliability-based characteristic value is very costly to compute compared to a spatial average–based characteristic value. However, this chapter presents a practical approximation called a mobilization-based characteristic value. Tabarroki et al. (2022a) showed that the mobilization-based characteristic value is much more cost-effective and it can be presented with the same "look and feel" as the spatial average–based characteristic value. The ultimate limit state mobilization-based characteristic value is calibrated from a weakest path model. A different approximation for the mobilized value at the serviceability limit state is provided by Tabarroki et al. (2022b). This SLS mobilization-based characteristic value is calibrated using a pseudo incremental energy method. The constitutive model adopted thus far is the classic elastoplastic model. Research is in progress to bring this approach to practice.

Extensive statistical guidelines are made available at: www.routledge.com/Uncertainty-Modeling-and-Decision-Making-in-Geotechnics/Phoon-Shuku-Ching/p/book/9781032367491.

ACKNOWLEDGMENTS

Appendix A, B, C, and D are largely based on Chapter 1, 3, 2, and 5, respectively in the TC304 State-of-the-Art Review of Inherent Variability and Uncertainty in Geotechnical Properties and Models, edited by Jianye Ching and Timo Schweckendiek (http://140.112.12.21/issmge/2021/SOA_Review_on_geotechnical_property_variablity_and_model_uncertainty.pdf). The authors are grateful to the chapter authors and report editors for their assistance in preparing these appendices.

REFERENCES

Aladejare, A. E. and Wang, Y. 2017. Evaluation of rock property variability. *Georisk: Assessment and Management of Risk for Engineered Systems and Geohazards*, 11(1), 22–41.

Al-Naqshabandy, M. S., Bergman, N. S., and Larsson, S. 2012. Strength variability in lime-cement columns based on CPT data. *Proceedings of the Institution of Civil Engineers – Ground Improvement*, 165(1), 15–30.

Babasaki, R., Terashi, M., Suzuki, T., Maekawa, A., Kawamura, M., and Fukazawa, E. 1996. JGS TC report: Factors influencing the strength of improved soil. In *Proceedings of the 2nd International Conference on Ground Improvement Geosystems, Grouting and Deep Mixing*, Vol. 2, 913–918. London: Taylor & Francis.

Bergman, N., Al-Naqshabandy, M. S., and Larsson, S. 2013. Variability of strength and deformation properties in lime–cement columns evaluated from CPT and KPS measurements. *Georisk: Assessment and Management of Risk for Engineered Systems and Geohazards*, 7(1), 21–36.

Bond, A. and Harris, A. 2008. *Decoding Eurocode 7*. London: Taylor and Francis.

Boverket. 1995. *Design Regulations BKR 94 – Mandatory Provisions and General Recommendations*. Karlskrona: Swedish Board of Housing, Building and Planning.

Bozorgzadeh, N. and Harrison, J. P. 2019. Reliability-based design in rock engineering: Application of Bayesian regression methods to rock strength data. *Journal of Rock Mechanics and Geotechnical Engineering*, 11(3), 612–627.

Bozozuk, M. and Leonards, G. A. 1972. The Gloucester test fill. *Proceedings of the Performance of Earth and Earth-Supported Structures*, 1(1), 299–317.

Bruce, M. E. C., Berg, R. R., Collin, J. G., Filz, G. M., Terashi, M., and Yang, D. S. 2013. *Deep Mixing for Embankment and Foundation Support*. Report No. FHWA-HRT-13-046. Washington, DC: Federal Highway Administration (FHWA).

Cami, B., Javankhoshdel, S., Phoon, K. K., and Ching, J. 2020. Scale of fluctuation for spatially varying soils: Estimation methods and values. *ASCE-ASME Journal of Risk and Uncertainty in Engineering Systems, Part A: Civil Engineering*, 6(4), 03120002.

Cami, B., Javankhoshdel, S., Phoon, K. K., and Ching, J. 2021. Erratum for "scale of fluctuation for spatially varying soils: estimation methods and values". *ASCE-ASME Journal of Risk and Uncertainty in Engineering Systems, Part A Civil Engineering*, 7(4), 08221001.

CAN/CSAS614. 2014. *Canadian Highway Bridge Design Code*. Mississauga, ON: Canadian Standards Organization.

CEN. 2002. *Eurocode - Basis of Structural Design*. EN 1990. Brussels: European Committee for Standardization.

CEN. 2004. *Geotechnical Design – Part 1: General Rules*. EN 1997-1. Brussels: European Committee for Standardization.

Chang, Y. C., Ching, J., Phoon, K. K., and Yue, Q. X. 2021. On the hole effect in soil spatial variability. *ASCE-ASME Journal of Risk and Uncertainty in Engineering Systems, Part A: Civil Engineering*, 7(4), 04021039.

Chen, H. M. and Hsieh, B. J. 1996. Geotechnical properties of the clay deposits from in-situ tests on Keelung River reclaimed land. *Sino-Geotechnics*, 54, 55–66 (in Chinese).

Chen, J., Lee, F. H., and Ng, C. C. 2011. Statistical analysis for strength variation of deep mixing columns in Singapore. In *Geo-Frontiers 2011 Advances in Geotechnical Engineering*, edited by J. Han, D. E. Alzamora et al., 13–16. Reston, VA: ASCE.

Chen, J., Liu, Y., and Lee, F. H. 2016. A statistical model for the unconfined compressive strength of deep mixed columns. *Geotechnique*, 66(5), 351–365.

Chilès, J.-P. and Delfiner, P. 1999. *Geostatistics: Modeling Spatial Uncertainty*. New York: John Wiley & Sons.

Ching, J. 2020. Unpublished databases.

Ching, J. and Noorzad, A. 2021. Statistics for transformation uncertainties. In *TC304 State-of-the-Art Review of Inherent Variability and Uncertainty in Geotechnical Properties and Models*, 171–180. http://140.112.12.21/issmge/2021/SOA_Review_on_geotechnical_property_variablity_and_model_uncertainty.pdf

Ching, J. and Phoon, K. K. 2012. Modeling parameters of structured clays as a multivariate normal distribution. *Canadian Geotechnical Journal*, 49(5), 522–545.

Ching, J. and Phoon, K. K. 2013a. Multivariate distribution for undrained shear strengths under various test procedures. *Canadian Geotechnical Journal*, 50(9), 907–923.

Ching, J. and Phoon, K. K. 2013b. Mobilized shear strength of spatially variable soils under simple stress states. *Structural Safety*, 41, 20–28.

Ching, J. and Phoon, K. K. 2013c. Probability distribution for mobilized shear strengths of spatially variable soils under uniform stress states. *Georisk: Assessment and Management of Risk for Engineered Systems and Geohazards*, 7(3), 209–224.

Ching, J. and Phoon, K. K. 2014a. Transformations and correlations among some clay parameters – The global database. *Canadian Geotechnical Journal*, 51(6), 663–685.

Ching, J. and Phoon, K. K. 2014b. Correlations among some clay parameters – The multivariate distribution. *Canadian Geotechnical Journal*, 51(6), 686–704.

Ching, J. and Phoon, K. K. 2015. Chapter 1. Constructing multivariate distribution for soil parameters. In *Risk and Reliability in Geotechnical Engineering*, 3–76. Boca Raton, FL: CRC Press/Balkema.

Ching, J. and Phoon, K. K. 2017. Characterizing uncertain site-specific trend function by Sparse Bayesian Learning. *Journal of Engineering Mechanics*, 143(7), 04017028.

Ching, J. and Phoon, K. K. 2019. Constructing site-specific probabilistic transformation model by Bayesian machine learning. *Journal of Engineering Mechanics*, 145(1), 04018126.

Ching, J., Chen, Z. Y., and Phoon, K. K. 2023c. Homogenization of spatially variable hydraulic conductivity in the presence of a geotechnical structure. *Computers and Geotechnics*, 156, 105255.

Ching, J., Hu, Y. G., and Phoon, K. K. 2016c. On characterizing spatially variable shear strength using spatial average. *Probabilistic Engineering Mechanics*, 45, 31–43.

Ching, J., Hu, Y. G., and Phoon, K. K. 2018b. Effective Young's modulus of a spatially variable soil mass under a footing. *Structural Safety*, 73, 99–113.

Ching, J., Huang, W. H., and Phoon, K. K. 2020a. 3D probabilistic site characterization by Sparse Bayesian Learning. *Journal of Engineering Mechanics*, 146(12), 04020134.

Ching, J., Lee, S. W., and Phoon, K. K. 2016b. Undrained strength for a 3D spatially variable clay column subjected to compression or shear. *Probabilistic Engineering Mechanics*, 45, 127–139.

Ching, J., Phoon, K. K., and Chen, C. H. 2014a. Modeling CPTU parameters of clays as a multivariate normal distribution. *Canadian Geotechnical Journal*, 51(1), 77–91.

Ching, J., Phoon, K. K., and Kao, P. H. 2014b. Mean and variance of mobilized shear strength for spatially variable soils under uniform stress states. *Journal of Engineering Mechanics*, 140(3), 487–501.

Ching, J., Phoon, K. K., and Pan, Y. K. 2017b. On characterizing spatially variable soil Young's modulus using spatial average. *Structural Safety*, 66, 106–117.

Ching, J., Phoon, K. K., and Wu, C. T. 2023a. Data-centric quasi-site-specific prediction for compressibility of clays. *Canadian Geotechnical Journal*, 59(12), 2033–2049.

Ching, J., Phoon, K. K., and Yu, J. W. 2014c. Linking site investigation efforts to final design savings with simplified reliability-based design methods. *Journal of Geotechnical and Geoenvironmental Engineering*, 140(3), 04013032.

Ching, J., Sung, S. P., and Phoon, K. K. 2017c. Worst case scale of fluctuation in basal heave analysis involving spatially variable clays. *Structural Safety*, 68, 28–42.

Ching, J., Tong, X. W., and Hu, Y. G. 2016d. Effective Young's modulus for a spatially variable elementary soil mass subjected to a simple stress state. *Georisk: Assessment and Management of Risk for Engineered Systems and Geohazards*, 10(1), 11–26.

Ching, J., Wu, S., and Phoon, K. K. 2021c. Constructing quasi-site-specific multivariate probability distribution using hierarchical Bayesian model. *Journal of Engineering Mechanics*, 147(10), 04021069.

Ching, J., Wu, S. S., and Phoon, K. K. 2016a. Statistical characterization of random field parameters using frequentist and Bayesian approaches. *Canadian Geotechnical Journal*, 53(2), 285–298.

Ching, J., Wu, T. J., and Phoon, K. K. 2016e. Spatial correlation for transformation uncertainty and its applications. *Georisk: Assessment and Management of Risk for Engineered Systems and Geohazards*, 10(4), 294–311.

Ching, J., Yang, Z. Y., and Phoon, K. K. 2021b. Dealing with non-lattice data in three-dimensional probabilistic site characterization. *Journal of Engineering Mechanics*, 147(5), 0602100.

Ching, J., Li, K. H., Phoon, K. K., and Weng, M. C. 2018a. Generic transformation models for some intact rock properties. *Canadian Geotechnical Journal*, 55(12), 1702–1741.

Ching, J., Lin, G. H., Chen, J. R., and Phoon, K. K. 2017a. Transformation models for effective friction angle and relative density calibrated based on a multivariate database of coarse-grained soils. *Canadian Geotechnical Journal*, 54(4), 481–501.

Ching, J., Phoon, K. K., Beck, J. L., and Huang, Y. 2017d. Identifiability of geotechnical site-specific trend functions. *ASCE-ASME Journal of Risk and Uncertainty in Engineering Systems, Part A*, 3(4), 04017021.

Ching, J., Phoon, K. K., Ho, Y. H., and Weng, M. C. 2021a. Quasi-site-specific prediction for deformation modulus of rock mass. *Canadian Geotechnical Journal*, 58(7), 936–951.

Ching, J., Uzielli, M., Phoon, K. K., and Xu, X. J. 2023b. Characterization of autocovariance parameters of detrended cone tip resistance from a global CPT database. *Journal of Geotechnical and Geoenvironmental Engineering*, 149(10), 04023090.

Ching, J., Phoon, K. K., Chen, K. F., Orr, T. L. L., and Schneider, H. R. 2020c. Statistical determination of multivariate characteristic values for Eurocode 7. *Structural Safety*, 82, 101893.

Ching, J., Phoon, K. K., Khan, Z., Zhang, D. M., and Huang, H. W. 2020b. Role of municipal database in constructing site-specific multivariate probability distribution. *Computers and Geotechnics*, 124, 103623.

Dawson, E. M., Roth, W. H., and Drescher, A. 1999. Slope stability analysis by strength reduction. *Geotechnique*, 49(6), 835–840.

Duncan, J. M. 2000. Factors of safety and reliability in geotechnical engineering. *Journal of Geotechnical and Geoenvironmental Engineering*, 126(4), 307–316.

Duncan, J. M. and Sleep, M. D. 2015. Chapter 3. Evaluating reliability in geotechnical engineering. In *Risk and Reliability in Geotechnical Engineering*, 131–180. Boca Raton, FL: CRC Press.

D'Ignazio, M., Phoon, K. K., Tan, S. A., and Lansivaara, T. 2016. Correlations for undrained shear strength of Finnish soft clays. *Canadian Geotechnical Journal*, 53(10), 1628–1645.

El-Ramly, H., Morgenstern, N. R., and Cruden, D. M. 2006. Lodalen slide: A probabilistic assessment. *Canadian Geotechnical Journal*, 43(9), 956–968.

Eslami, A., Heidarie Golafzani, S., and Naghibi, M. H. 2023. Developed triangular charts; deltaic CPTu-based soil behavior classification using AUT: CPTu-Geo-Marine Database. *Probabilistic Engineering Mechanics*, 71, 103380.

Fei, S., Tan, X., Lin, X., Xiao, Y., Zha, F., and Xu, L. 2022. Evaluation of the scale of fluctuation based on variance reduction method. *Engineering Geology*, 308, 106804.

Fellenius, W. 1936. Calculation of the stability of earth dams. In *Transactions of the Second Congress on Large Dams*, 445–459. Washington, DC.

Feng, S. and Vardanega, P. J. 2019. A database of saturated hydraulic conductivity of fine-grained soils: Probability density functions. *Georisk: Assessment and Management of Risk for Engineered Systems and Geohazards*, 13(4), 255–261.

Fenton, G. A. and Griffiths, D. V. 2008. *Risk Assessment in Geotechnical Engineering*. New York: John Wiley & Sons.

Fenton, G. A., Naghibi, F., Dundas, D., Bathurst, R. J., and Griffiths, D. V. 2016. Reliability-based geotechnical design in the 2014 Canadian highway bridge design code. *Canadian Geotechnical Journal*, 53(2), 236–251.

Frank, R., Bauduin, C., Driscoll, R., Kawadas, M., Krebs Ovesen, N., Orr, T., and Schuppener, B. 2004. *Designers' Guide to EN 1997-1 Eurocode 7: Geotechnical Design - General Rules*. London: Thomas Telford.

Guan, Z., Chang, Y. C., Wang, Y., Aladejare, A., Zhang, D. M., and Ching, J. 2021. Site-specific statistics for geotechnical properties. In *TC304 State-of-the-Art Review of Inherent Variability and Uncertainty in Geotechnical Properties and Models*, 1–83. http://140.112.12.21/issmge/2021/SOA_Review_on_geotechnical_property_variablity_and_model_uncertainty.pdf

Hedman, P. and Kuokkanen, M. 2003. *Strength Distribution in Lime-Cement Columns – Field Tests at Strängnäs*. MSc thesis, 47–59. Slockholm: Royal Institute of Technology (in Swedish).

Hicks, M. A. 2013. An explanation of characteristic values of soil properties in Eurocode 7. In *Modern Geotechnical Design Codes of Practice*, edited by Patrick Arnold, Gordon A. Fenton, Michael A. Hicks, Timo Schweckendiek and Brian Simpson, 36–45. Amsterdam: IOS Press.

Hicks, M. A. and Samy, K. 2002. Reliability-based characteristic values: A stochastic approach to Eurocode 7. *Ground Engineering*, 35(12), 30–34.

Hicks, M. A., Varkey, D., van den Eijnden, A. P., de Gast, T., and Vardon, P. J. 2019. On characteristic values and the reliability-based assessment of dykes. *Georisk: Assessment and Management of Risk for Engineered Systems and Geohazards*, 13(4), 313–319.

Honjo, Y. 1982. A probabilistic approach to evaluate shear strength of heterogeneous stabilized ground by deep mixing method. *Soils and Foundations*, 22(1), 23–38.

Hu, Y. G. and Ching, J. 2015. Impact of spatial variability in soil shear strength on active lateral forces. *Structural Safety*, 52, 121–131.

Huang, L. and Leung, Y.F. 2021. Reliability assessment of slopes with three-dimensional rotated transverse anisotropy in soil properties. *Canadian Geotechnical Journal*, 58(9), 1365–1378.

ISO. 2015. *General Principles on Reliability of Structures*. ISO2394:2015. Geneva: ISO.

JRC. 2024a. Reliability-based verification of limit states for geotechnical structures – Guideline for the 2nd generation Eurocode 7. Developed by CEN/TC250/SC7, to be published by EU Joint Research Council (JRC); in progress, expected publication: 2024.

JRC. 2024b. Determination of representative values from derived values for verification of limit states with EN 1997 – Guideline for the 2nd generation Eurocode 7. Developed by CEN/TC250/SC7, to be published by EU Joint Research Council (JRC); in progress, expected publication: 2024.

Kim, E. and Hunt, R. 2017. A public website of rock mechanics database from Earth Mechanics Institute (EMI) at Colorado School of Mines (CSM). *Rock Mechanics and Rock Engineering*, 50(12), 3245–3252.

Kootahi, K. and Moradi, G. 2017. Evaluation of compression index of marine fine-grained soils by the use of index tests. *Marine Georesources and Geotechnology*, 35(4), 548–570.

Kulhawy, F. H. and Mayne, P. W. 1990. *Manual on Estimating Soil Properties for Foundation Design*. Report EL-6800. Palo Alto, CA: Electric Power Research Institute.

Kulhawy, F. H., Phoon, K. K., and Prakoso, W. A. 2000. Uncertainty in basic laboratory and field properties of geomaterials. In *Proceedings of the First International Conference on Geotechnical Engineering Education and Training*, 297–302. Rotterdam: Balkema.

Länsivaara, T., Phoon, K. K., and Ching, J. 2022. What is a characteristic value for soils? *Georisk: Assessment and Management of Risk for Engineered Systems and Geohazards*, 16(2), 199–224.

Larsson, S. and Nilsson, A. 2009. Horizontal strength variability in lime-cement columns. In *Deep Mixing 2009 Okinawa Symposium*, 629–634. Tokyo: Sanwa Company.

Larsson, S., Stille, H., and Olsson, L. 2005. On horizontal variability in lime-cement columns in deep mixing. *Geotechnique*, 55(1), 33–44.

Lim, Y. X., Tan, S. A., and Phoon, K. K. 2020. Friction angle and overconsolidation ratio of soft clays from cone penetration test. *Engineering Geology*, 274, 105730.

Liu, S., Zou, H., Cai, G., Bheemasetti, T. V., Puppala, A. J., and Lin, J. 2016. Multivariate correlation among resilient modulus and cone penetration test parameters of cohesive subgrade soils. *Engineering Geology*, 209, 128–142.

Liu, Y., Jiang, Y. J., Xiao, H., and Lee, F. H. 2017. Determination of representative strength of deep cement-mixed clay from core strength data. *Géotechnique*, 67(4), 350–364.

Liu, Y., Lee, F. H., Quek, S. T., Chen, E. J., and Yi, J. T. 2015. Effect of spatial variation of strength and modulus on the lateral compression response of cement-admixed clay slab. *Géotechnique*, 65(10), 851–865.

Liu, Y., He, L. Q., Jiang, Y. J., Sun, M. M., Chen, E. J., and Lee, F. H. 2019. Effect of in situ water content variation on the spatial variation of strength of deep cement-mixed clay. *Géotechnique*, 69(5), 391–405.

Lo, M. K., Wei, X., Chian, S. C., and Ku, T. 2021. Bayesian network prediction of stiffness and shear strength of sand. *Journal of Geotechnical and Geoenvironmental Engineering*, 147(5), 04021020.

Muzamhindo, H. and Ferentinou, M. 2023. Generic compressive strength prediction model applicable to multiple lithologies based on a broad global database. *Probabilistic Engineering Mechanics*, 71, 103400.

Namikawa, T. and Koseki, J. 2013. Effects of spatial correlation on the compression behaviour of a cement-treated column. *Journal of Geotechnical and Geoenvironmental Engineering*, 139(8), 1346–1359.

National Research Council (U.S.). 1995. *Probabilistic Methods in Geotechnical Engineering*. Washington, DC: National Academies Press.

Navin, M. P. and Filz, G. M. 2005. Statistical analysis of strength data from ground improved with DMM columns. In *Deep Mixing '05: International Conference on Deep Mixing Best Practice and Recent Advances*, 144–154. Linköping: Swedish Geotechnical Institute.

Orr, T. L. L. 2015. Chapter 10. Managing risk and achieving reliability geotechnical designs using Eurocode 7. In *Risk and Reliability in Geotechnical Engineering*, 395–433. Boca Raton, FL: CRC Press.

Paikowsky, S. G., Birgisson, B., McVay, M., Nguyen, T., Kuo, C., Baecher, G., Ayyub, B., Stenersen, K., O'Malley, K., Chernauskas, L., and O'Neill, M. 2004. *Load and Resistance Factors Design for Deep Foundations*. NCHRP Report 507. Washington, DC: Transportation Research Board of the National Academies.

Pan, Y., Yao, K., Phoon, K. K., and Lee, F. H. 2019. Analysis of tunnelling through spatially-variable improved surrounding–A simplified approach. *Tunnelling and Underground Space Technology*, 93, 103102.

Pan, Y., Liu, Y., Xiao, H., Lee, F. H., and Phoon, K. K. 2018. Effect of spatial variability on short-and long-term behaviour of axially-loaded cement-admixed marine clay column. *Computers and Geotechnics*, 94, 150–168.

Phoon, K. K. 2006. Modeling and simulation of stochastic data. In *GeoCongress 2006: Geotechnical Engineering in the Information Technology Age*, 1–17. Reston, VA: ASCE, 1–17.

Phoon, K. K. 2017. Role of reliability calculations in geotechnical design. *Georisk: Assessment and Management of Risk for Engineered Systems and Geohazards*, 11(1), 4–21.

Phoon, K. K. 2023. What geotechnical engineers want to know about reliability. *ASCE-ASME Journal of Risk and Uncertainty in Engineering Systems, Part A: Civil Engineering*, 9(2), 03123001.

Phoon, K. K. and Ching, J. 2013. Multivariate model for soil parameters based on Johnson distributions. In *Foundation Engineering in the Face of Uncertainty: Honoring Fred H. Kulhawy*, 337–353. Reston, VA: ASCE.

Phoon, K. K. and Ching, J. 2014a. Univariate to multivariate characterization of geotechnical variability. In *Proceedings of the International Symposium on Reliability Engineering and Risk Management (ISRERM 2014)*, 63–76. Taipei: National Taiwan University of Science and Technology.

Phoon, K. K. and Ching, J. 2014b. Characterization of geotechnical variability – A multivariate perspective. In *Proceedings of the 14th International Conference of International Association for Computer Methods and Advances in Geomechanics*, pp. 61–70. Oxfordshire, UK: Taylor and Francis.

Phoon, K. K. and Ching, J. 2017. Better correlations for geotechnical design. In *A Decade of Geotechnical Advances 2008–2017*, 73–102. Singapore: Geotechnical Society of Singapore.

Phoon, K. K. and Ching, J. 2021. Project DeepGeo - Data-driven 3D subsurface mapping. *Journal of GeoEngineering*, 16(2), 61–74.

Phoon, K. K. and Ching, J. 2022. Additional observations on the site recognition challenge. *Journal of GeoEngineering*, 17(4), 231–247.

Phoon, K. K. and Kulhawy, F. H. 1999a. Characterization of geotechnical variability. *Canadian Geotechnical Journal*, 36(4), 612–624.

Phoon, K. K. and Kulhawy, F. H. 1999b. Evaluation of geotechnical property variability. *Canadian Geotechnical Journal*, 36(4), 625–639.

Phoon, K. K. and Kulhawy, F. H. 2008. Chapter 9. Serviceability limit state reliability-based design. In *Reliability-Based Design in Geotechnical Engineering: Computations and Applications*, 344–383. London: CRC Press.

Phoon, K. K. and Tang, C. 2019. Characterisation of geotechnical model uncertainty. *Georisk: Assessment and Management of Risk for Engineered Systems and Geohazards*, 13(2), 101–130.

Phoon, K. K. and Tang, C. 2024. *Database Approach for Data-Centric Geotechnics, Vol. 1: Site Characterization and Vol. 2: Geotechnical Structures*. Boca Raton, FL: CRC Press.

Phoon, K. K. and Zhang, W. 2023. Future of machine learning in geotechnics. *Georisk: Assessment and Management of Risk for Engineered Systems and Geohazards*, 17(1), 7–22.

Phoon, K. K., Ching, J., and Cao, Z. 2022d. Unpacking data-centric geotechnics. *Underground Space*, 7(6), 967–989.

Phoon, K. K., Ching, J., and Huang, H. W. 2012. Examination of multivariate dependency structure in soil parameters. In *Geo Congress 2012 – State of the Art and Practice in Geotechnical Engineering*, 2952–2960. Reston, VA: ASCE.

Phoon, K. K., Ching, J., and Shuku, T. 2022a. Challenges in data-driven site characterization. *Georisk: Assessment and Management of Risk for Engineered Systems and Geohazards*, 16(1), 114–126.

Phoon, K. K., Kulhawy, F. H., and Grigoriu, M. D. 1995. *Reliability-Based Design of Foundations for Transmission Line Structures*. EPRI report TR-105000. Palo Altos, CA: Electric Power Research Institute (EPRI).

Phoon, K. K., Kulhawy, F. H., and Grigoriu, M. D. 2003a. Development of a reliability-based design framework for transmission line structure foundations. *Journal of Geotechnical and Geoenvironmental Engineering*, 129(9), 798–806.

Phoon, K. K., Kulhawy, F. H., and Grigoriu, M. D. 2003b. Multiple resistance factor design (MRFD) for spread foundations. *Journal of Geotechnical and Geoenvironmental Engineering*, 129(9), 807–818.

Phoon, K. K., Prakoso, W. A., Wang, Y., and Ching, J. 2016a. Chapter 3. Uncertainty representation of geotechnical design parameters. In *Reliability of Geotechnical Structures in ISO2394*, 49–87. Leiden: CRC Press/Balkema.

Phoon, K. K., Shuku, T., Ching, J., and Yoshida, I. 2022b. Benchmark examples for data-driven site characterization. *Georisk: Assessment and Management of Risk for Engineered Systems and Geohazards*, 16(4), 599–621.

Phoon, K. K., Retief, J. V., Ching, J., Dithinde, M., Schweckendiek, T., Wang, Y., and Zhang, L. M. 2016b. Some observations on ISO2394:2015 Annex D (reliability of geotechnical structures). *Structural Safety*, 62, 24–33.

Phoon, K. K., Cao, Z., Ji, J., Leung, Y. F., Najjar, S., Shuku, T., Tang, C., Yin, Z. Y., Yoshida, I., and Ching, J. 2022c. Geotechnical uncertainty, modeling, and decision making. *Soils and Foundations*, 62(5), 101189.

Prakoso, W. A. 2002. *Reliability-Based Design of Foundations on Rock for Transmission Line and Similar Structure*. PhD thesis, Cornell University.

Prakoso, W. A. 2017. Personal communication.

Santoso, A., Phoon, K. K., and Quek, S. T. 2009. Reliability analysis of infinite slope using subset simulation. In *Contemporary Topics in In Situ Testing, Analysis, and Reliability of Foundations*, edited by Magued Iskander, Debra F. Laefer, Mohamad H. Hussein, 278–285. Reston, VA: ASCE.

Schneider, H. R. 1997. Definition and characterization of soil properties. In *Proceedings of the Fourteen International Conference on Soil Mechanics and Geotechnical Engineering*. Hamburg, Balkema.

Schneider, H. R. and Schneider, M. A. 2013. Dealing with uncertainties in EC7 with emphasis on determination of characteristic soil properties. In *Modern Geotechnical Design Codes of Practice*, edited by H. Arnold, Gordon A. Fenton, Michael A. Hicks, Timo Schweckendiek, Brian Simpson, 87–101. Amsterdam: IOS Press.

Shi, C. and Wang, Y. 2021. Development of subsurface geological cross-section from limited site-specific boreholes and prior geological knowledge using iterative convolution XGBoost. *Journal of Geotechnical and Geoenvironmental Engineering*, 147(9), 04021082.

Shuku, T. and Phoon, K. K. 2023. Comparison of data-driven site characterization methods through benchmarking: Methodological and application aspects. *ASCE-ASME Journal of Risk and Uncertainty in Engineering Systems, Part A: Civil Engineering*, 9(2), 04023006.

Simpson, B. 2011. Reliability in geotechnical design – Some fundamentals. In *Proceedings of the Third International Symposium on Geotechnical Safety & Risk*, 393–399. Karlsruhe: Federal Waterways Engineering and Research Institute.

Stuedlein, A. W., Cami, B., Curzio, D. D., Javankhoshdel, S., Nishimura, S., Pula, W., Vessia, G., Wang, Y., and Ching, J. 2021. Summary of random field model parameters of geotechnical properties. In *TC304 State-of-the-Art Review of Inherent Variability and Uncertainty in Geotechnical Properties and Models*, 95–129. http://140.112.12.21/issmge/2021/SOA_Review_on_geotechnical_property_variablity_and_model_uncertainty.pdf

Tabarroki, M., Ching, J., Phoon, K. K., and Chen, Y. Z. 2022a. Mobilisation-based characteristic value of shear strength for ultimate limit states. *Georisk: Assessment and Management of Risk for Engineered Systems and Geohazards*, 16(3), 413–434.

Tabarroki, M., Ching, J., Lin, C. P., Liou, J. J., and Phoon, K. K. 2022b. Homogenizing spatially variable young modulus using pseudo incremental energy method. *Structural Safety*, 97, 102226.

Tang, C. and Bathurst, R. 2021. Statistics for geotechnical design model factors. In *TC304 State-of-the-Art Review of Inherent Variability and Uncertainty in Geotechnical Properties and Models*, 130–170. http://140.112.12.21/issmge/2021/SOA_Review_on_geotechnical_property_variablity_and_model_uncertainty.pdf

Tang, C. and Phoon, K. K. 2021. *Model Uncertainties in Foundation Design*. Boca Raton, FL: CRC Press.

Tao, Y. Q., Phoon, K. K., Sun, H. L., and Ching, J. 2023. Variance reduction function for a potential inclined slip line in a spatially variable soil, *Structural Safety*, in press.

Turkstra, C. J. and Madsen, H. O. 1980. Load combinations in codified structural design. *Journal of the Structural Division*, 106(12), 2527–2543.

Vanmarcke, E. H. 1977a. Probabilistic modeling of soil profiles. *Journal of the Geotechnical Engineering Division*, 103(11), 1227–1246.

Vanmarcke, E. H. 1977b. Reliability of earth slopes. *Journal of the Geotechnical Engineering Division*, 103(11), 1247–1265.

Vanmarcke, E. H. 1983. *Random Fields: Analysis and Synthesis*. Cambridge, MA: The MIT Press.

Vanmarcke, E. H. 2010. *Random Fields: Analysis and Synthesis*. Singapore: World Scientific.

Vanmarcke, E. H. and Fenton, G. A. 2003. *Probabilistic Site Characterization at the National Geotechnical Experimentation Sites (GSP 121)*. Reston, VA: ASCE.

Vardanega, P. J. and Bolton, M. D. 2016. Design of geostructural systems. *ASCE-ASME Journal of Risk and Uncertainty in Engineering Systems, Part A: Civil Engineering*, 2(1), 04015017.

Varkey, D., Hicks, M. A., van den Eijnden, A. P., and Vardon, P. J. 2020. On characteristic values for calculating factors of safety for dyke stability. *Géotechnique Letters*, 10(2), 353–359.

Xiao, T., Zou, H. F., Yin, K. S., Du, Y., and Zhang, L. M. 2021. Machine learning-enhanced soil classification by integrating borehole and CPTU data with noise filtering. *Bulletin of Engineering Geology and the Environment*, 80(12), 9157–9171.

Xu, J., Wang, Y., and Zhang, L. 2021. Interpolation of extremely sparse geo-data by data fusion and collaborative Bayesian compressive sampling. *Computers and Geotechnics*, 134, 104098.

Yoshida, I., Tomizawa, Y., and Otake, Y. 2021. Estimation of trend and random components of conditional random field using gaussian process regression. *Computers and Geotechnics*, 136, 104179.

Zhang, D. M., Zhou, Y., Phoon, K. K., and Huang, H. W. 2020. Multivariate probability distribution of Shanghai clay properties. *Engineering Geology*, 273, 105675.

Zhou, Y., Zhang, D., and Ching, J. 2021. Site-specific correlations between geotechnical properties. In *TC304 State-of-the-Art Review of Inherent Variability and Uncertainty in Geotechnical Properties and Models*, 84–94. http://140.112.12.21/issmge/2021/SOA_Review_on_geotechnical_property_variablity_and_model_uncertainty.pdf

Zhu, Y. X., Zheng, S., Cao, Z. J., and Li, D. Q. 2019. Revisiting the relationship between scale of fluctuation and mean cross distance. In *Proceedings of the 13th International Conference on Applications of Statistics and Probability in Civil Engineering*. Seoul, South Korea. https://s-space.snu.ac.kr/bitstream/10371/153405/1/217.pdf

Zou, H., Liu, S., Cai, G., Puppala, A. J., and Bheemasetti, T. V. 2017. Multivariate correlation analysis of seismic piezocone penetration (SCPTU) parameters and design properties of Jiangsu quaternary cohesive soils. *Engineering Geology*, 228, 11–38.

APPENDICES

Appendices A–D are available as soft copy only at: www.routledge.com/Uncertainty-Modeling-and-Decision-Making-in-Geotechnics/Phoon-Shuku-Ching/p/book/9781032367491

Appendix A: Site-specific spatial variability statistics for common clay, sand, and rock parameters

Table A.1 Clay parameter statistics from databases CLAY/10/7490, CLAY/8/12225, CLAY/12/3997, and SH-CLAY/11/4051 (Source: Table 1.A1, Guan et al. 2021).

Table A.2 Sand parameter statistics from databases SAND/7/2794 and SAND/13/4113 (Source: Table 1.A2, Guan et al. 2021).

Table A.3 Intact rock parameter statistics from databases ROCK/13 and ROCK/9/4069 (Source: Table 1.A3, Guan et al. 2021).

Table A.4 Rock mass parameter statistics from database ROCKMass/9/5876 (Source: Table 1.A4, Guan et al. 2021).

Appendix B: Scales of fluctuation

Table B.1 Summary of vertical and horizontal scales of fluctuation reported in the literature (Source: Table 3.2, Stuedlein et al. 2021 citing Cami et al. 2020 as main data source).

Figure B.1 Design charts for theoretical scales of fluctuation δ_β for an inclined line.

Appendix C: Site-specific correlations for common clay, sand, and rock parameters

Table C.1 Site-specific correlations for sites in CLAY/10/7490, CLAY/8/12225, and CLAY/12/3997 (Source: Table 2.2, Zhou et al. 2021).

Table C.2 Site-specific correlations for sites in SH-CLAY/11/4051 (Source: Table 2.3, Zhou et al. 2021).

Table C.3 Site-specific correlations for sites in SAND/7/2794 and SAND/10/4113 (Source: Table 2.4, Zhou et al. 2021).

Table C.4 Site-specific correlations for sites in ROCK/9/4069 (Source: Table 2.5, Zhou et al. 2021).

Table C.5 Site-specific correlations for sites in ROCKMASS/9/5876 (Source: Table 2.6, Zhou et al. 2021).

Appendix D: Transformation models for common clay, sand, and rock parameters

Table D.1 Transformation model for clay parameters and its model statistics, λ and θ (Source: Table 5.2, Ching and Noorzad 2021).

Table D.2 Transformation model for sand parameters and its model statistics, λ and θ (Source: Table 5.3, Ching and Noorzad 2021).

Table D.3 Transformation model for intact rock parameters and its model statistics, λ and θ (Source: Table 5.4, Ching and Noorzad 2021).

Table D.4 Transformation model for rock mass parameters and its model statistics, λ and θ (Source: Table 5.5, Ching and Noorzad 2021).

3 Uncertainty in Constitutive Models

Zhen-Yu Yin, Pin Zhang, and Yin-Fu Jin

ABSTRACT

Soils exhibit complex mechanical behaviors. Myriad phenomenological models have been proposed to describe these behaviors, and these models play an important role in geotechnical engineering. Model parameters obtained from any geotechnical site investigation or stress–strain curves obtained from any experimental test are subject to uncertainties due to the inherent spatial variability of ground, the limitations of the experimental techniques, and the limited number of soil samples used. Therefore, uncertainty should be considered in constitutive modeling and the application of constitutive models in geotechnical engineering. This chapter considers uncertainty in three ways: (1) how to develop correlation or surrogate models for soil properties considering uncertainty as key design and model parameters; (2) how to intelligently identify model parameters considering uncertainty from laboratory or in situ tests; and (3) how to integrate uncertainty into the framework of constitutive modeling.

3.1 INTRODUCTION

Soils exhibit complex mechanical behaviors such as state dependence (Su and Yang 2019; Kang et al. 2019a), stress dilatancy (Su et al. 2010), anisotropy (Yin et al. 2010c; Kang et al. 2019b), destructuration (Liu et al. 2013; Yin and Karstunen 2011), stress path dependence (Hu et al. 2018), time dependence (Yin et al. 2011a), and non-coaxiality (Tian and Yao 2017). Myriad phenomenological models have been proposed to describe these behaviors. Constitutive models play an important role in geotechnical engineering and can be classified as (1) linear elastic models; (2) elastic perfectly plastic models (such as the Mohr–Coulomb); (3) non-linear models (such as the hardening soil model [Vermeer 1978] and the non-linear Mohr–Coulomb [Jin et al. 2016b; Kolymbas 1991]); (4) critical state–based advanced models (such as the modified cam-clay model [Roscoe and Burland 1968]; the Nor-Sand model [Jefferies 1993]; the CSAM model [Yu 1998]; the Severn–Trent model [Gajo and Wood 1999]; UH models [Yao et al. 2004, 2008, 2009, 2014]; the SANISAND model [Taiebat and Dafalias 2008]; the SIMSAND model [Jin et al. 2016a, 2016b, 2017a]; the ANICREEP model [Yin et al. 2010b, 2011b]; hypoplasticity models [Mašín 2005; Wu et al. 1996; Wang et al. 2018; Kolymbas 1991, 1985; Von Wolffersdorff 1996]); and (5) micromechanical models (Chang and Hicher 2005; Yin et al. 2008, 2009, 2010a, 2014; Yin and Chang 2009; Xiong et al. 2017).

Uncertainty is an inherent characteristic of soil properties and behavior, thus this concept has been widely used in soil mechanics and geotechnical engineering (Phoon et al. 2021). Design parameters obtained from any geotechnical site investigation and stress–strain curves obtained from any experimental test are subject to uncertainties (Phoon 2008; Zhang et al. 2023). As reported by Mašín (2015), these are caused, in particular, by (a) an inherent spatial variability of soil properties, (b) experimental uncertainty (measurement scatter) due to the limitations of the experimental techniques, and (c) sampling uncertainty (statistical uncertainty) due to the limited number of soil samples used in the investigation (Phoon and Kulhawy 1999). In general, uncertainty includes

DOI: 10.1201/9781003333586-4

aleatoric and epistemic (or model) uncertainties. The former denotes the intrinsic randomness and cannot be mitigated, thereby epistemic uncertainty generated in the modeling process because of lack of knowledge has gained great attention (Kendall and Gal 2017).

To the authors' knowledge, uncertainty can be considered in constitutive modeling or the application of constitutive models in geotechnical engineering in three ways: (1) adopting some key mechanical properties with uncertainty as some of the parameters of constitutive models, for which we need to develop correlation or surrogate models for soil properties considering uncertainty; (2) identifying parameters of constitutive models considering uncertainty from laboratory or in situ tests; and (3) developing constitutive models with uncertainty, for which we need to invoke uncertainty into the framework of constitutive modeling.

3.2 CORRELATING SOIL PROPERTIES CONSIDERING UNCERTAINTY

3.2.1 Literature Survey

To identify the uncertainty of soil properties, Bayesian probability theory provides a mathematically solid method to explain the uncertainty of models or input parameters (Juang and Zhang 2017). This approach and its variants, e.g., multivariate probability distribution and hierarchical Bayesian learning, have been extensively used to predict the performance of geotechnical systems, such as soil properties (Houlsby and Houlsby 2013; Ching et al. 2016, 2020; Yan et al. 2009; Ching and Phoon 2019; Zhou et al. 2012, 2014, 2018; Zhang et al. 2017; Cao and Wang 2014a; Yang et al. 2021; Akeju et al. 2019; Kelly and Huang 2015; Zheng et al. 2021; Namikawa 2019), soil stratification (Wang et al. 2013, 2017, 2019), embankment (Huang et al. 2019; Zheng et al. 2018; Suzuki and Ishii 1994; Tian et al. 2021), slope stability (Zhang et al. 2009a, 2010), and braced excavation (Lo and Leung 2019; Juang et al. 2013). Bayesian inference is a commonly used method for obtaining the posterior distribution of a model and input parameters, thereafter updating them to guarantee further application in engineering practice (Qi and Zhou 2017; Jin et al. 2019d; Kelly and Huang 2015). The posterior distribution is always intractable in real-world engineering practice; hence, the Monte Carlo (MC) method is typically used to approximate a solution (Xiao et al. 2018; Cao and Wang 2013; Al-Bittar and Soubra 2014).

In practice, the empirical correlation for estimating soil properties (e.g., compressibility and undrained shear strength) is straightforward and convenient, which is more attractive in design. Most correlations whether based on simple algebraic, statistical methods or advanced machine learning (ML) methods neglect uncertainty mainly induced by measurement scatter (data from different laboratories/resources) and a limited number of samples. Actually, the uncertainties of the concerned parameters can be quantified by the Bayesian approach.

Zhang et al. (2021d, 2021a) proposed a novel framework to combine the empirical correlation represented by the ML surrogate model and uncertainty using the Bayesian neural network (BNN), which is applied to the creep index $C_{\alpha e}$ and permeability k (Zhang et al. 2021a), the compression index C_c, and the undrained shear strength s_u (Zhang et al. 2021d). The weights and biases of the BNN were treated as a multivariable distribution instead of scalar values, and the training aimed to obtain the posterior distribution of weights and biases. Two methods, variational inference (VI) (Blundell et al. 2015; Graves 2011b; Hinton and Van Camp 1993) and MC dropout (MCD) (Gal and Ghahramani 2015b), can be used to solve and approximate the BNN to represent model uncertainty within realistic time constraints, providing a basis for developing the BNN-based model. Overall, a BNN that integrates model uncertainty is a promising approach to overcome the challenge of ugly data and site recognition (Phoon et al. 2021) because of the strong non-linear fitting capacity of the neural network, and to leverage data for identifying potential uncertainty.

The following section demonstrates how to develop an uncertain model using a BNN.

3.2.2 Bayesian Neural Network

The ML algorithm without uncertainty means that its configuration and parameters are fixed. Figure 3.1a is an illustration of an artificial neural network (ANN) as a representative algorithm. Notably, the values of the weights are deterministic after the training process is completed, meaning that the predicted result of an ANN is deterministic. The models developed based on such ML algorithms without uncertainty may show high accuracy for datasets within the range of the training set, but these models still exhibit the potential to generate unreliable predictions on the space outside the original data. Moreover, because the predicted results do not involve uncertainties, i.e., reliability, the corresponding errors would inevitably be large and even induce huge risks if these predictions were directly applied in practice, inducing disasters in some fields, such as monitoring the deformation of a slope. For this reason, some fields have a negative attitude regarding the application of ML algorithms without uncertainty.

In a BNN, the weights and biases can be represented by a probability distribution, such as a Gaussian distribution instead of a single value, as shown in Figure 3.1b. Given that the weights or biases comply with a certain prior distribution $P(\omega)$, their posterior distribution can be obtained by Bayes' theorem, as follows:

$$P(\omega|D)=\frac{P(D|\omega)P(\omega)}{P(D)} \tag{3.1}$$

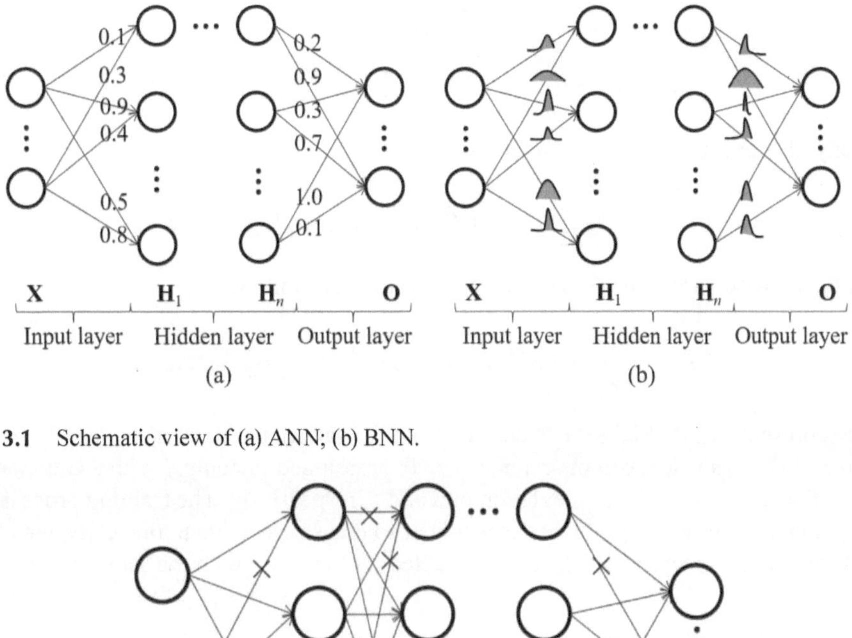

FIGURE 3.1 Schematic view of (a) ANN; (b) BNN.

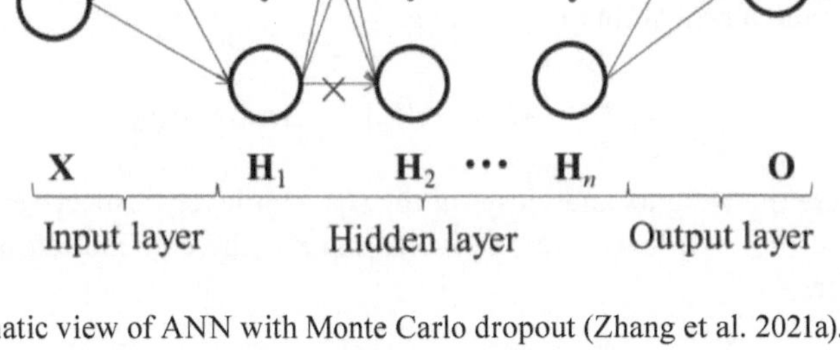

FIGURE 3.2 Schematic view of ANN with Monte Carlo dropout (Zhang et al. 2021a).

where $P(D)$ is model evidence, which can be treated as a constant, and $P(D|\omega)$ is the likelihood of observations. Generally, the maximum a posteriori (MAP) of weights and biases can be obtained based on MC, but the solution tends to be intractable because of the prohibitive computational cost and convergence issues. Thus, VI and MCD were proposed to solve this issue.

3.2.3 Variational Inference

VI is an approximation method to obtain the Bayesian posterior distribution of weights, which aims to find parameters θ of the distribution of weights $q_d(\omega|\theta)$ to approximate the $P(\omega|D)$ by minimizing the Kullback–Leibler (KL) divergence (Kullback and Leibler 1951) between them. Given two distributions p_d and q_d, the KL divergence between p_d and q_d can be defined as

$$\mathrm{KL}\left(p_d \| q_d\right) = \sum_i p_d\left(x_i\right)\left[\log \frac{p_d\left(x_i\right)}{q_d\left(x_i\right)}\right] \tag{3.2}$$

Therefore, the solution for θ can be defined as (Blundell et al. 2015):

$$\begin{aligned}\theta^* &= \arg\min_\theta \mathrm{KL}\left[q_d(\omega|\theta) \| P(\omega|D)\right] \\ &= \arg\min_\theta \int q_d(\omega|\theta)\log\frac{q_d(\omega|\theta)}{P(D|\omega)P(\omega)}d\omega \\ &= \arg\min_\theta \mathrm{KL}\left[q_d(\omega|\theta)\|P(\omega)\right] - \mathbb{E}_{q_d(\omega|\theta)}\left[\log P(D|\omega)\right]\end{aligned} \tag{3.3}$$

Accordingly, the loss function can be expressed by

$$F(D,\theta) = \mathrm{KL}\left[q_d(\omega|\theta)\|P(\omega)\right] - \mathbb{E}_{q_d(\omega|\theta)}\left[\log P(D|\omega)\right] \tag{3.4}$$

Based on the unbiased MC, Eq. (3.4) can be approximated as follows:

$$F(D,\theta) \approx \sum_i \log q_d\left(\omega^{(i)}|\theta\right) - \log P\left(\omega^{(i)}\right) - \log P\left(D|\omega^{(i)}\right) \tag{3.5}$$

where $\boldsymbol{\omega}^{(i)}$ represents the ith MC sample derived from the variational posterior $q_d(\boldsymbol{\omega}^{(i)}|\theta)$. In general, the variational posterior of a weight or a bias, i.e., θ, is assumed to comply with a Gaussian distribution with the mean and standard deviation of μ and ρ, respectively. The training process aims to optimize μ and ρ values using gradient descent instead of a single value in the ANN, which means that the number of parameters in the BNN is doubled for the ANN with the same architecture.

3.2.4 Monte Carlo Dropout

Dropout is a widely used overfitting prevention method (Srivastava et al. 2014), which inactivates each hidden neuron with a probability of p_h, using

$$y^{l+1} = f\left(\omega^{l+1} * r^l y^l + \mathrm{b}^{l+1}\right),\ r^l \cdot \mathrm{Bernoulli}(p_h) \tag{3.6}$$

where $\boldsymbol{\omega}^{l+1}$ and $\boldsymbol{b}^{l+1}$ are the weights and biases at the $(l+1)$th layer, respectively; $\boldsymbol{y}^l$ and $\boldsymbol{y}^{l+1}$ are the outputs at the lth and $(l+1)$th layers, respectively; and $\boldsymbol{r}$ is the distribution of the dropout probability at the lth layer.

Gal and Ghahramani (2015b) proposed a novel theoretical framework that casts dropout training in an ANN as an approximated BNN, in which the dropout operation also activates in the testing phase. This means that the weight and bias values are deterministic, but each hidden neuron has a fixed probability to be inactivated per implementation; therefore, the architecture of the ANN is not consistent. This characteristic is used to represent model uncertainty; moreover, Gal and Ghahramani (2015a) have proved that the application of dropout is mathematically equivalent to an approximate VI. By performing stochastic calculation T times, because the predicted result per time is different, the ultimate output and uncertainty can be obtained by

$$\mathbb{E}(y) = \frac{1}{T}\sum_{i=1}^{T} y\left(x, \omega^{t}\right)^{T} \tag{3.7}$$

$$\mathrm{Var}(y) = \frac{1}{T}\sum_{i=1}^{T} y(x)^{T} y(x) - \mathbb{E}(y)^{T}\,\mathbb{E}(y) \tag{3.8}$$

where the MC estimate is termed as MCD (Gal and Ghahramani 2015b). In comparison with VI, MCD represents model uncertainty without sacrificing in terms of a large computational cost.

3.2.5 Framework of the ML-Based Uncertainty Model

This study used both VI and MCD to approximate Bayesian inference. The models developed based on these two methods were labelled BNN_VI and ANN_MCD, respectively. Meanwhile, their performance was compared with the ANN model without uncertainty. A total of three models were ultimately developed.

For ANN and ANN_MCD, the weights and biases are scalar values. The mean square error, integrated with the k-fold cross-validation, was used as the loss function, which could be expressed as

$$L\left(y^{a}, y^{p}\right) = -\frac{1}{kn_{v}}\sum_{i}\left(y_{i}^{a} - y_{i}^{p}\right)^{2} \tag{3.9}$$

where y_i^a and y_i^p are the actual and predicted values of outputs, respectively; n_v is the number of datasets in the validation set; and $k = 10$ was used in this study.

For BNN_VI, the model can be represented by $P(y|x, \omega)$, and the output is the probability ranging from 0 to 1. Thus, the purpose of the training was to maximize the $P(y|x, \omega)$ that denotes the probability of the predicted result. It equals minimizing the negative log-likelihood (NLL), which can be used as the loss function in the model training, i.e., the last term in Eq. (3.5):

$$L\left(y^{a}, y^{p}\right) = -\frac{1}{k}\sum_{i}\log\left[P\left(y_{i}^{a} \middle| x, \omega\right)\right] \tag{3.10}$$

where NLL represents the logarithmic probability of y_i^a.

When considering uncertainty using the BNN_VI method, the KL divergence introduced in Eq. (3.5) was selected as the loss function. The training process is to find the optimal posterior distribution based on the training data for a given prior joint distribution.

For the above three methods, the *ReLU* was selected as the activation function, considering its fast convergence rate and saturation-avoiding properties (Goodfellow et al. 2016). A gradient-based adaptive algorithm *Adam* that integrated the advantages of AdaGrad, i.e., dealing with sparse gradients, and RMSProp, i.e., dealing with non-stationary objectives, was used to optimize the weights and biases of the BNN models. Based on the predefined loss function, the activation function, and the optimization algorithm, the framework of the BNN was roughly established.

Subsequently, a "trial-and-error" method was employed to adjust the hyperparameters of the BNN, such as the number of hidden layers and neurons, to determine its configuration details.

3.2.6 Examples and Validation

The compression index C_c with three influential variables was first employed to build a deterministic algorithm ANN-based model and two BNN-based models with different theoretical frameworks, i.e., VI and MCD. The performance of these three models was comprehensively compared in terms of accuracy, uncertainty, and monotonicity, in which the predictions on the sparse data area show a larger confidence interval, indicating the effectiveness of uncertainty-based models (Figure 3.3).

Recent research works have started to use these uncertainty-based ML algorithms to model soil properties and behaviors (Zhang et al. 2021d, 2021a), through which the prediction can be assigned with a confidence interval to explain the reliability of the prediction of the data-driven model.

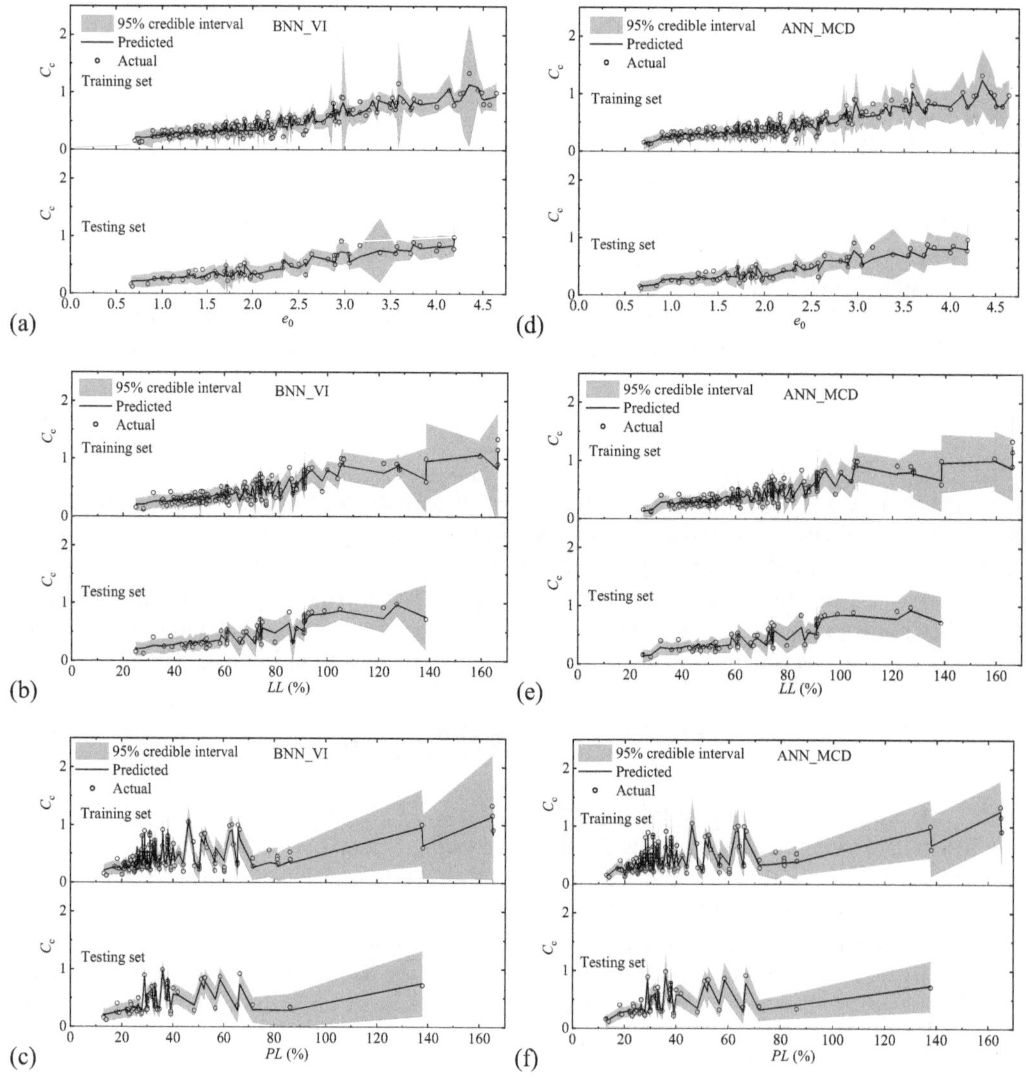

FIGURE 3.3 Evolution of predicted Cc along the input parameters using (a–c) BNN_VI; (d–e) ANN_MCD.

The interpretability of these data-driven models can thus be improved by integrating uncertainty, exhibiting a promising application.

3.3 IDENTIFYING SOIL MODEL PARAMETERS CONSIDERING UNCERTAINTY

3.3.1 Literature Survey

The accuracy of the parameters of a model significantly influences its modeling performance and can even result in inaccurate predictions. Direct use of the identified parameters without the consideration of uncertainty in engineering practice may thus pose risks.

Yin et al. (2018) distinguished three approaches to parameters determination from experimental data: analytical methods, empirical correlations, and inverse analysis methods. Among them, inverse analysis produces a relatively objective determination of the parameters for an adopted soil model, even for those without direct physical meaning, and thus has been widely adopted. The key methods of inverse analysis are divided into two categories: (1) deterministic methods and (2) probabilistic methods. Deterministic methods merely focus on finding a set of fixed values for the input parameters of concern (Jin et al. 2016a, 2016b, 2017a, 2017b, 2018, 2019b; Yin et al. 2017, 2018; Knabe et al. 2013; Levasseur et al. 2008; Papon et al. 2012) without taking into account the variability and uncertainty of soils. By contrast, probabilistic methods are competitive because of their consideration of uncertainty. Of these, the Bayesian approach to parameter identification has been applied in different fields (Zhang et al. 2009b, 2017; Cao and Wang 2014b; Miro et al. 2015; Ritto and Nunes 2015; Akeju et al. 2017; Cividini et al. 1983; Murakami et al. 2017; Hsiao et al. 2008; Juang et al. 2012; Most 2010; Honjo et al. 1994; Qi and Zhou 2017a; Ancey 2005; Eckert et al. 2007; Fischer et al. 2015; Gauer et al. 2009; Hellweger et al. 2016; Tan et al. 2018; Zhou et al. 2018) in which parameters of concern have been treated as random variables and expressed in terms of posterior distributions and statistics.

Up to now, the probabilistic methods for parameters identification of soil models have been applied to the linear elastic model (Honjo et al. 1994), one-dimensional elastoplastic model (Most 2010), a unified soil compression model (Jung et al. 2009), the hardening soil model (Miro et al. 2015), and advanced constitutive models (e.g., critical state–based models) (Jin et al. 2019a, 2019c).

This section presents how to develop an efficient approach to Bayesian parameters identification for advanced soil models through the enhancement of the transitional Markov chain Monte Carlo (TMCMC) approach. To this end, Bayesian parameter identification is first introduced in principle. To improve the performance of the TMCMC, a differential evolution–Markov chain algorithm is implemented in the process of new samples. To save on computational costs, a parallel computing implementation of the DE-TMCMC is achieved using the single program/multiple data (SPMD) technique in MATLAB®.

3.3.2 Framework of Bayesian Parameter Identification

According to Yuen (2010a), the Bayesian approach can update model parameters and characterize uncertainties using their posterior probability distribution functions (PDFs).

Following a Bayesian formulation (Beck 2010; Beck and Katafygiotis 1998; Yuen 2010a) and assuming that the observation data and the model predictions satisfy the prediction error equation, the observation can be expressed as

$$U_{\text{obs}} = c \cdot U_{\text{num}}(\mathbf{b}) \tag{3.11}$$

where $\mathbf{b}$ is a vector of the model parameters, and $c \sim N(1,\sigma_\varepsilon^2)$ is a one-mean Gaussian random variable with variance σ_ε^2 that represents the prediction error variance, and this σ_ε is an unknown parameter in addition to the soil model parameters $\mathbf{b}$.

Uncertainties of parameters can be evaluated using posterior PDFs, with the expression of the posterior PDF for data D written as

$$p(\theta|D) = \frac{p(\theta)p(D|\theta)}{p(D)} \tag{3.12}$$

where $\theta = [\mathbf{b}, \sigma_\varepsilon]$ is the uncertain parameters; $p(D)$ is the evidence; $p(\theta)$ is the prior PDF of the uncertain parameters θ, which is based on previous knowledge or the user's judgment; and $p(D|\theta)$ is the likelihood function expressing the level of data fitting. If the prediction errors in different measured data are statistically independent, then the likelihood function can be computed as follows (Yuen 2010b; Yuen and Mu 2015; Tan et al. 2016):

$$p(D|\theta) = (2\pi\sigma_\varepsilon^2)^{-\frac{N}{2}} \exp\left[-\frac{N}{2\sigma_\varepsilon^2} J_g(\mathbf{b}; D)\right] \tag{3.13}$$

where N is the number of measured data and $J_g(\mathbf{b}; D)$ is the goodness-of-fit function representing the degree of data fitting. For reasons of numerical stability and algebraic simplicity, it is often convenient to work with the log-likelihood. Accordingly, the log-likelihood $\ln p(D|\theta)$ is expressed as

$$\ln p(D|\theta) = -\frac{N}{2}\ln(2\pi\sigma_\varepsilon^2) - \frac{N}{2\sigma_\varepsilon^2} J_g(\mathbf{b}; D) \tag{3.14}$$

Generally, deformation and stress are two extremely important indicators of soil behaviors. The measurement produced by a laboratory test usually contains two curves, such as the curves ε_a–q and ε_a–e for the drained triaxial test or the curves ε_a–q and ε_a–u for the undrained triaxial test (where ε_a is axial strain, q is the deviatoric stress, e is the void ratio, and u is the excess pore water pressure). Accordingly, a goodness-of-fit function involving these two important indicators is reasonable. According to Jin et al. (2017a, 2016b), a normalized goodness-of-fit function is adopted due to the error independent of the magnitude of different variables (e.g., q and e or u), which is expressed as

$$J_g(\mathbf{b}; D) = \frac{1}{N_0 N} \sum_{j=1}^{N_0} \left[\sum_{i=1}^{N} \left(\frac{U_{\text{obs}}^i - U_{\text{num}}^i}{U_{\text{obs}}^i} \right)^2 \right]_j \tag{3.15}$$

where N is the number of measured values; N_0 is the number of curves for one test; U_{obs}^i is the value of measurement point I; and U_{num}^i is the value of the calculation at point i.

With multiple observations and types of observations, likelihood values for each observation must be combined into an overall value for each candidate parameter set (He et al. 2010). When the measured data D involves M tests during Bayesian parameter identification, the likelihood function is expressed as

$$\ln p(D|\theta) = \sum_{i=1}^{M} \text{w}_i \ln p(D_i|\theta) \tag{3.16}$$

where M is the number of involved tests; w_i is the weight of $p(D_i|\theta)$; and $p(D_i|\theta)$ is the likelihood corresponding to the test i. In this study, the weight of each likelihood for all involved tests is considered the same and thus equal to 1.

Because the soil model involves high-dimensional, non-linear functions, the posterior $p(\theta|D)$ must be evaluated numerically, such as by using the TMCMC method. The posterior PDF $p(\theta|D)$ represents the updated belief about the parameter vector θ after obtaining the evidence of D. An accurate estimator of the parameters θ for the adopted soil model is the maximum a posteriori estimation. The MAP parameter vector θ_{MAP} can be computed as follows:

$$\theta_{\text{MAP}} = \arg\max p(\theta|D) = \arg\max \frac{p(\theta)p(D|\theta)}{p(D)} \tag{3.17}$$

where the arg max is to find the points within a domain for a given function at which the function values are maximized.

3.3.3 Proposition of Enhanced DE-TMCMC

The TMCMC method was originally developed by Ching and Chen (2007) as a combination of the sequential particle filter method (Chopin 2002) and the MCMC. The method begins with the prior distribution $p(\theta)$ and makes a gradual transition to the posterior by optimization at each round of samplings. The key idea of the TMCMC is that of proposal density, which corresponds to the *j*th round of sampling $p(\theta)_j$, determined as

$$p(\theta)_j \propto p(\theta)\cdot L(\theta|D)^{q_j} \tag{3.18}$$

where $q_j \in [0, 1]$ is chosen following $q_0 = 0 < q_1 < \ldots < q_m = 1$ with $j = 0, 1, \ldots, m$ denoting the stage level. Consequently, $p(\theta)_0$ equals the prior distribution $p(\theta)$ for $j = 0$, and $p(\theta)_m$ is the posterior distribution $p(\theta|D)$ for $j = m$.

Details of the execution of the TMCMC algorithm can be found in Ching and Chen (2007). In this study, only certain key steps are summarized:

1. Calculate the q_j. If $q_j > 1$, then set $q_j = 1$.
2. For all samples $k = 1, 2, \ldots, N_s$, compute a weighting coefficient $w_{j,k}$:

$$w_{(j,k)} = \left[L\left(\theta_{(j-1,k)}|D\right)\right]^{q_j - q_{j-1}} \tag{3.19}$$

3. Compute the mean of the weighting coefficient:

$$S_j = \frac{1}{N_s}\sum_{k=1}^{N_s} w_{(j,k)} \tag{3.20}$$

4. Compute the covariance matrix of the Gaussian proposal distribution:

$$\Sigma_j = \beta^2 \cdot \sum_{k=1}^{N_s}\left[\frac{w_{(j,k)}}{S_j N_s}\left(\theta_{(j-1,k)} - \bar{\theta}_j\right)\cdot\left(\theta_{(j-1,k)} - \bar{\theta}_j\right)^T\right] \tag{3.21}$$

4. with

$$\bar{\theta}_j = \frac{\sum_{l=1}^{N_s} w_{(j,l)}\theta_{(j-1,l)}}{\sum_{l=1}^{N_s} w_{(j,l)}} \tag{3.22}$$

5. For each l in $[1, \ldots, N_s]$, set $\theta^c_{(j,l)} = \theta_{(j-1,l)}$. Then for $k = 1, 2, \ldots, N_s$, do the following MCMC sampling:
   5.1 *Select the index l from the set [1, 2, …, N_s] using the sequential importance sampling (SIS) method, where each l is assigned probability* $w_{(j,l)} \Big/ \sum_{n=1}^{N_s} w_{(j,n)}$.
   5.2 *Propose a new sample: Draw* θ^c *from the normal distribution* $N(\theta^c_{(j,l)}, \Sigma_j)$.

5.3 *The remaining steps are identical to those in the Metropolis algorithm.*
6. If $q_j = 1$, then stop the iteration; otherwise set $j = j + 1$ and continue the above process.

Details of the original TMCMC method, with its MATLAB code, can be found in Ching and Chen (2007).

To improve the performance of the original TMCMC, a differential evolution–Markov chain algorithm proposed by Vrugt (2016) was adopted in this study to replace the process that proposes a new sample in Step 5.2, which can be generated as

$$\theta_{(j,l)}^{\text{new}} = \theta_{(j,l)}^{c} + d\theta_{(j,l)} \tag{3.23}$$

with

$$d\theta_{(j,l)} = (1+\lambda)\cdot\gamma\cdot\left[\left(\theta_j^{best} - \theta_{(j,l)}^{c}\right) + \left(\theta_{(j,a)} - \theta_{(j,b)}\right)\right] + \zeta \tag{3.24}$$

where $\theta_{(j,l)}^{\text{new}}$ is the new sample; $\theta_{(j,l)}^{c}$ is the current sample; θ_j^{best} is the sample corresponding to the maximum weight in the current iteration; d is the dimension of θ; $\theta_{(j,a)}$ and $\theta_{(j,b)}$ are two vectors consisting of d variables, where the indices a and b are two integers drawn from $[1, \ldots, N_s]$; $\gamma = 2.38/\sqrt{2\delta d}$ is the jump rate; and δ denotes the number of chain pairs used to generate the jump with a default value of $\delta = 3$ according to Vrugt (Vrugt 2016). The values of λ and ζ are sampled independently from the uniform distribution $[-c, c]$ and the normal distribution $N(0, c^*)$, respectively. In this study, $c = 0.1$ and $c^* = 10^{-12}$ were employed, as recommended by Vrugt (Vrugt 2016).

After differential evolution, a binomial crossover operation forms the final sample:

$$\theta_{(j,l)}^{\text{new}} = \begin{cases} \theta_{(j,l)}^{\text{new}}, \text{ if rand}(0,1) \le CR \text{ or } l = l_{\text{rand}} \\ \theta_{(j,l)}^{c}, \text{ otherwise} \end{cases} \tag{3.25}$$

where rand(0, 1) is a uniform random number within [0, 1]; l_{rand} = randint (1, d) is an integer randomly chosen from 1 to d and is newly generated for each l; and the crossover probability $CR \in [0, 1]$ roughly corresponds to the average fraction of the vector components that are inherited from the mutation vector, with $CR = 0.9$ taken in this study.

For simplicity's sake, the original TMCMC is referred as "O-TMCMC" and the enhanced TMCMC as "DE-TMCMC" in the following sections.

3.3.4 Parallel Computing DE-TMCMC

Stochastic simulation algorithms, such as the TMCMC algorithm, used in Bayesian inverse modeling require numerous simulation runs when used for geotechnical engineering problems. For large-order computational models, the computational demands involved in the TMCMC sampling algorithm are excessive, sometimes to the point of being unacceptable. Accordingly, high-performance computing (HPC) techniques are vital for their ability to reduce computation time. Most MCMC algorithms involve a single Markov chain and are thus not parallelizable. Hence, parallel computing algorithms should be implemented for use with the enhanced DE-TMCMC algorithm, efficiently distributing computations on multicore central processing units (CPUs) (Angelikopoulos et al. 2012; Hadjidoukas et al. 2015).

In this study, the SPMD technique in MATLAB was adopted to achieve a parallel computing implementation of DE-TMCMC. Typical applications appropriate for SPMD are those that require simultaneous execution of a program on multiple datasets when communication or synchronization is required between workers. The SPMD efficiently distributes the computations involved in the

DE-TMCMC algorithm to available heterogeneous graphics processing units (GPUs) and multicore CPUs so that the number of log-likelihood evaluations is the same for each computer worker. The SPMD technique is computationally efficient when the computational time for a log-likelihood evaluation is the same independent of the location of the sample in the parameter space. More information about the SPMD technique can be found in the MATLAB manual.

Figure 3.4 shows the parallel computing strategy in DE-TMCMC using the SPMD technique, where log-likelihood represents the user-defined function for computing the log-likelihood; N_ workers is the total number of computer workers (e.g., eight for a four-core CPU); and "labindex" is the index of computer workers, automatically identified in MATLAB. The entire calculation will be distributed among different computer workers according to the value of labindex. Note that the number of samples N_s should be an integer multiple of the number of computer workers, N_workers, in parallel computation.

Not surprisingly, drastic reductions in time can be achieved using this parallel implementation of the DE-TMCMC algorithm. Furthermore, the time savings differ for the same case on different

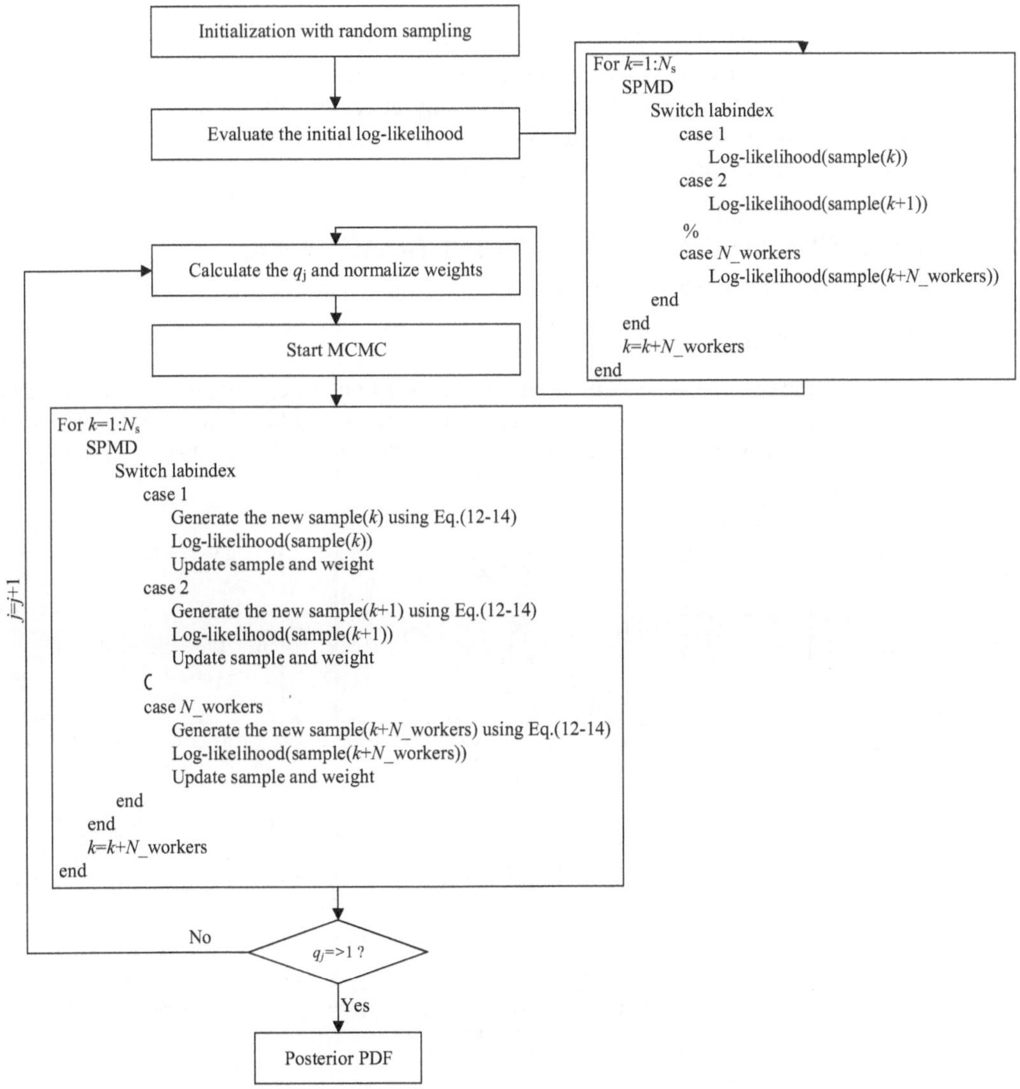

FIGURE 3.4 Parallel computing strategy in DE-TMCMC using the SPMD technique.

computers with different settings: powerful computers with many multicore CPUs can save more computational time.

3.3.5 Examples and Validation

In order to assess the efficiency of the parallel computing DE-TMCMC methodology for the identification of parameters of advanced constitutive models, several MCMC simulations were carried out using the O-TMCMC and DE-TMCMC, respectively, for a comparison on synthetic data and real experimental data (Jin et al. 2019a): (1) identifying the parameters of the SIMSAND model (a critical state–based model) from synthetic test data (three synthetic drained triaxial tests with/without noise) in terms of robustness and effectiveness for the numerical validation; and (2) identifying the parameters of the SIMSAND model from real experimental data (drained triaxial tests on Toyoura sand) for applicability. To avoid randomness, the calculations were independently conducted ten times. The results demonstrate that DE-TMCMC is highly robust and efficient, producing reasonably ranged posterior PDFs as well as a set of accurate parameters regardless of whether the used data is noisy.

Furthermore, the enhanced DE-TMCMC-based parameter identification approach was applied to two in situ pressuremeter tests on soft clays (Jin et al. 2019a). The elastic viscoplastic model ANICREEP was adopted to take into account the time effect for clays. The identified MAP parameters agree well with values measured in oedometer and triaxial tests, indicating a high level of efficiency for the proposed approach and its applicability to in situ testing (Figure 3.5).

3.4 INTEGRATING UNCERTAINTY INTO THE FRAMEWORK OF CONSTITUTIVE MODELING

3.4.1 Literature Survey

Phenomenological models involve certain assumptions based on experimental observations and tend to give rise to more parameters for modeling more complex soil behavior. The development of these models relies heavily on domain knowledge and the calibration of these parameters is

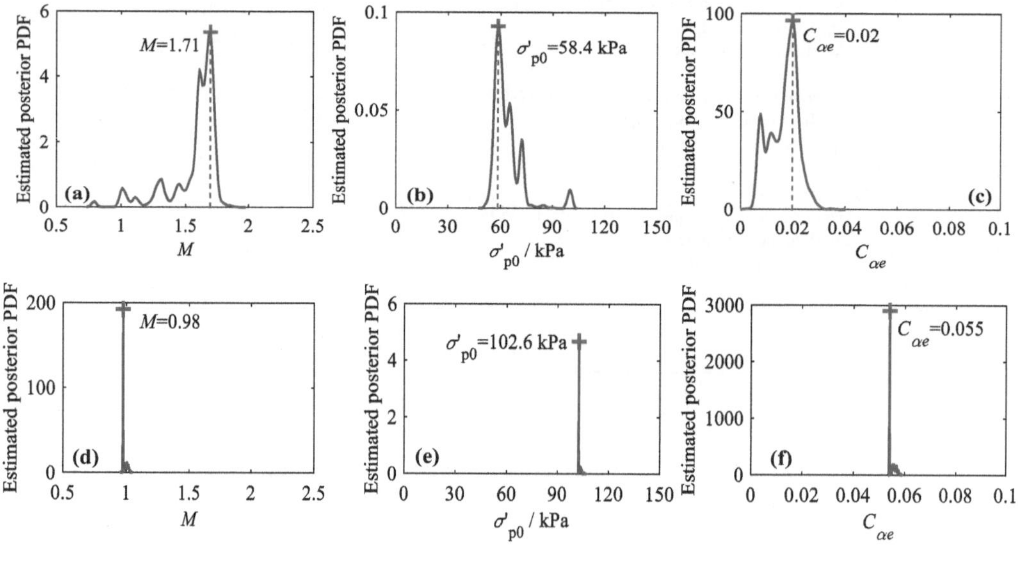

FIGURE 3.5 Posterior PDF of parameters of ANICREEP model for Burswood clay: (a–c) by DE-TMCMC; (d–f) by O-TMCMC.

case-specific and requires conducting experiments. To circumvent the limitations of these conventional constitutive models, increasing research works have recently resorted to various ML algorithms for constitutive modeling (Table 3.1).

Table 3.1 shows that artificial neural networks have dominated data-driven constitutive modeling particularly for recurrent neural networks (RNN) and their variants such as the long short-term memory neural network (Table 3.2), which has recently been elucidated to be appropriate for modeling path-dependent soil behavior (Zhang et al. 2020). ML-based data-driven constitutive modeling directly maps the stress–strain relationship without any assumptions or predefined parameters, thus providing a much more straightforward way (Ghaboussi et al. 1991; Basheer 2000; Zhang and Yin 2021b). Nevertheless, Zhang et al. (2021f) pointed out that most data-driven constitutive models were trained based on a large volume of datasets, but they were merely adopted to model soil behaviors under simple loading paths as the testing set or even the testing loading paths close to that used in the training set, thus they tend to generate plausible predictions. The generalization ability and these data-driven models and their applicability in engineering practice were not sufficiently examined. The superiority of data-driven models over conventional phenomenological models cannot be identified from this perspective.

Hence, new data-driven constitutive modeling paradigms have been proposed. One representative regime is to add known physical knowledge as additional constraints for rationalizing predictions, such as embodying the basic laws of thermodynamics into the architecture of ML algorithms (Masi et al. 2021), custom loss function (Vlassis and Sun 2021), and custom input–output pairs (Xu et al. 2021). One representative regime is to reduce the dependency of the data-driven model on expensive and time-consuming high-fidelity (HF) data such as multi-fidelity modeling (Zhang et al. 2022a). The other noteworthy regime is to invoke uncertainty into the modeling framework for enhancing the interpretability of data-driven models.

TABLE 3.1
Data-Driven Constitutive Model Developed Based on Myriad ML Algorithms

Machine learning algorithms	References
Evolutionary polynomial regression (EPR)	Javadi and Rezania (2009); Faramarzi et al. (2012); Javadi et al. (2012); Cuisinier et al. (2013); Nassr et al. (2018); Ahangar Asr et al. (2018); Zhang et al. (2021e)
Genetic programming (GP)	Cabalar and Cevik (2011)
Support vector machine (SVM)	Zhao et al. (2014); Kohestani and Hassanlourad (2016)
Extreme learning machine (ELM)	Zhang et al. (2021e)
Feedforward neural network (FNN)	Ellis et al. (1995); Sidarta and Ghaboussi (1998); Ghaboussi and Sidarta (1998); Penumadu and Zhao (1999); Basheer (2000, 2002); Habibagahi and Bamdad (2003); Banimahd et al. (2005); Shahin and Indraratna (2006); Fu et al. (2007); Hashash and Song (2008); He and Li (2009); Hashash et al. (2009); Johari et al. (2011); Sezer (2011); Lv et al. (2011); Araei (2014); Rashidian and Hassanlourad (2014); Stefanos and Gyan (2015); Kohestani and Hassanlourad (2016); Li et al. (2008, 2017); Zhang et al. (2021e)
Radial basis function neural network (RBFNN)	Peng et al. (2008)
Recurrent neural network (RNN)	Zhu et al. (1998a, 1998b); Romo et al. (2001); Wang and Sun (2018); Zhang et al. (2020, 2021c, 2022a); Zhang and Yin (2021b)

TABLE 3.2
Characteristics of ML-Based Constitutive Models of Soils

ML algorithms	Advantages	Limitations
EPR	Simple and explicit expression	Poor non-linear mapping ability No sequential prediction ability Single output prediction
GP	Simple and explicit expression	Numerous structure No sequential prediction ability Single output prediction
SVM	Structural risk minimization Adaptability for high-dimensional data	Poor readability and interpretability No sequential prediction ability Single output prediction
ELM	Low computational cost Multi-output prediction	No sequential prediction ability Poor generalization ability
FNN	Strong non-linear mapping ability Multiple output prediction	No sequential prediction ability Gradients exploding or vanishing
RBF	Low computational cost Multiple output prediction	No sequential prediction ability Poor generalization ability
RNN	Strong non-linear mapping ability Sequential prediction ability Multiple outputs prediction	Numerous weights, biases, and hyperparameters

Source: Zhang et al. (2021f).

For the first time, this section presents an interpretable data-driven model which is suitable for training on sparse datasets for the purpose of geotechnical constitutive modeling. A prior information-based neural network (PiNet) is used for incorporating any theoretical framework of constitutive modeling, to physically constrain the developed data-driven model. The PiNet-based model is subsequently applied to simulate the behavior of real soils in conjunction with a multi-fidelity residual (MR) framework to maximize the impact of (and therefore reduce the dependency on) sparse high-fidelity data. Finally, MC dropout is incorporated to achieve prediction uncertainty.

3.4.2 Physics-Informed Neural Network

A PiNet framework was first proposed for recovering and solving the partial differential equation for one-dimensional consolidation (Zhang et al. 2022b). In this study, a new PiNet approach is proposed to incorporate existing soil constitutive modeling theory. To that end, the architecture of the ANN constrained physically using the elasto-plastic (EP) theoretical concept is presented in Figure 3.6 (code on GitHub: https://github.com/PinZhang3/PiNet). With the incorporation of "tried-and-tested" theoretical knowledge, the architecture of the ANN becomes partially explicit, because key elements of the network are now computed and constrained using existing theoretical expressions. The computation mechanism of ANN can thus be partially interpreted in this way.

The EP-based PiNet includes four outputs: p', q, $\dot{\varepsilon}_v^e$, and $\dot{\varepsilon}_d^e$. Parameters p' and q are again predicted directly by neural network (*NN*) whereas $\dot{\varepsilon}_v^{e,t+1}$ and $\dot{\varepsilon}_d^{e,t+1}$ are obtained from p' and q using elastoplasticity theory. The pseudocode for this version of PiNet is as follows:

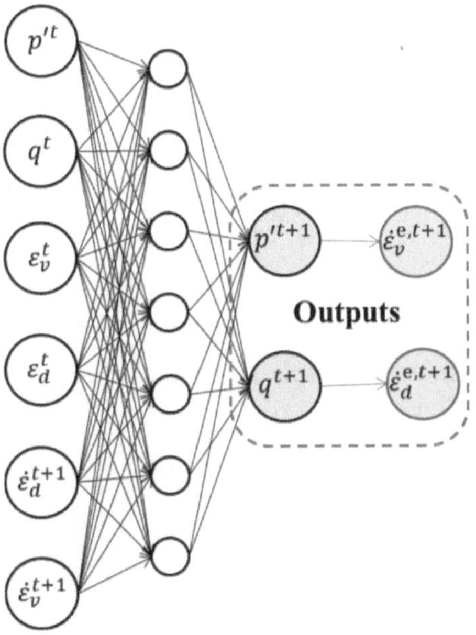

FIGURE 3.6 Adopted PiNet framework showing model inputs and interaction between neural network outputs and physics constraints for elastoplastic modeling.

Algorithm 1: Computation of EP stress–strain responses using PiNet

---

1. Initialize stress–strain status $p'^t, q^t, \varepsilon_v^t, \varepsilon_d^t$ and assign strain increment $\dot{\varepsilon}_v^{t+1}, \dot{\varepsilon}_d^{t+1}$
2. Predict p'^{t+1}, q^{t+1} using EP-based *NN* with $p'^t, q^t, \varepsilon_v^t, \varepsilon_d^t$, $\dot{\varepsilon}_v^{t+1}$, and $\dot{\varepsilon}_d^{t+1}$ as inputs
3. Compute $\dot{p}'^{t+1} = (p'^{t+1} - p'^t), \dot{q}^{t+1} = (q^{t+1} - q^t)$ and E^{t+1} (using constitutive equation)
4. Compute $\dot{\varepsilon}_v^{e,t+1}$ and $\dot{\varepsilon}_v^{e,t+1}$based on the predicted $\dot{p}'^{t+1}$, $\dot{q}^{t+1}$, and E^{t+1} using the constitutive equation
5. Update $\varepsilon_v^{t+1} = \varepsilon_v^t + \dot{\varepsilon}_v^{t+1}$; $\varepsilon_d^{t+1} = \varepsilon_d^t + \dot{\varepsilon}_d^{t+1}$; $t=t+1$
6. Repeat steps 1~5

The custom loss function for this case is expressed by

$$L = \gamma_p L_p + \gamma_q L_q + \gamma_{\varepsilon_v} L_{\varepsilon_v} + \gamma_{\varepsilon_d} L_{\varepsilon_d} \tag{3.26}$$

where $L_{\varepsilon_v} = \left(\dot{\varepsilon}_v^e - \dot{\varepsilon}_v^{e*}\right)^2 / n$; $L_{\varepsilon_d} = \left(\dot{\varepsilon}_d^e - \dot{\varepsilon}_d^{e*}\right)^2 / n$, and $\gamma_i = L_i / \left(L_p + L_q + L_{\varepsilon_v} + L_{\varepsilon_d}\right), i = p, q, \varepsilon_v, \varepsilon_d$.

Note that elastoplastic decomposition in EP are general assumptions in the respective theoretical framework. Thus, the proposed PiNet framework is generalizable for reproducing the responses of EP constitutive models.

For the sake of comparison, conventional feed-forward ANN is also used to model EP behavior. The corresponding pseudocode is as follows:

Algorithm 2: Computation of EP stress–strain responses using ANN

---

1. Initialize stress—strain status $p''^{t}, q'^{t}, \varepsilon_v^t, \varepsilon_d^t$ and assign strain increment $\dot{\varepsilon}_v^{t+1}, \dot{\varepsilon}_d^{t+1}$
2. Predict p''^{t+1}, q'^{t+1} using *NN* with $p''^{t}, q'^{t}, \varepsilon_v^t, \varepsilon_d^t$, $\dot{\varepsilon}_v^{t+1}$, and $\dot{\varepsilon}_d^{t+1}$ as inputs
3. Update $t = t+1$
4. Repeat steps 1—3

In this case, the training objective is to minimize the prediction errors of p' and q only such that the corresponding loss function is

$$L = \gamma_p L_p + \gamma_q L_q \tag{3.27}$$

where $\gamma_i = L_i / (L_p + L_q)$, $i = p$ and q.

3.4.3 Multi-Fidelity Residual Modeling

Practical applications of PiNet remain inhibited by big data requirements. This is incompatible with geotechnical modeling as the acquisition of real soil stress–strain is both time-consuming and expensive. MR modeling provides an option to build reliable data-driven models from sparse datasets (Zhang et al. 2022a). MR modeling with ANN as the low-fidelity (LF) model has exhibited competitive performance with advanced constitutive models in describing complex soil behaviors (Zhang 2022). To further refine this framework, the PiNet model is incorporated within an MR framework in four steps as follows (prediction of stress at the [t + 1]th step):

1. PiNet is first *trained* using an LF dataset to produce a trained LF version of the model NN_L. In this chapter, the low-fidelity dataset is generated using the respective constitutive model.

2. The trained NN_L is subsequently used to *predict* the outputs of a separate high-fidelity dataset $y_{L,p}^{t+1}$; the difference between model predictions $y_{L,p}^{t+1}$ and the HF measurements is the "residual error" $y_{E,a}^{t+1}$.

3. The residual error $y_{E,a}^{t+1}$ is combined with the input of the HF dataset in (2) to create a new dataset. A new ANN (NN_H) is subsequently trained on this dataset to capture the "residual error". The predictions of NN_H are given by $y_{E,p}^{t+1}$ (Figure 3.7a).

4. While $y_{L,p}^{t+1} + y_{E,p}^{t+1}$ could be taken as the final prediction, a further refinement for improving accuracy is applied here using a least-squares linear regression such that the final prediction, $y_{MR,p}^{t+1}$, is instead:

$$y_{MR,p}^{t+1} = c_1 \left(y_{L,p}^{t+1} + y_{E,p}^{t+1} \right) + c_2 \tag{3.28}$$

where c_1 and c_2 are the linear regression coefficients.

The MR training process can be completed based on these four steps. NN_L, NN_H, and regression coefficients are obtained and prepared for the applications to predict the stress–strain relationship of a studied soil. For a known initial stress state, the mechanical response of soils under random loading paths can be predicted using the following pseudocode:

Algorithm 3: Prediction of stress–strain responses of a studied soil using MR

---

1. Initialize stress–strain status $\boldsymbol{\sigma}^t$, $\boldsymbol{\varepsilon}^t$ and assign strain increment $\dot{\varepsilon}^{t+1}$
2. Predict preliminary stress $\sigma_{L,p}^{t+1}$ using NN_L with $\boldsymbol{\sigma}^t$, $\boldsymbol{\varepsilon}^t$, and $\dot{\varepsilon}^{t+1}$ as inputs
3. Predict residual $\sigma_{E,p}^{t+1}$ using NN_H with $\sigma_{L,p}^{t+1}$, $\boldsymbol{\varepsilon}^t$, and $\dot{\varepsilon}^{t+1}$ as inputs
4. Predict final stress $\sigma_{MR,p}^{t+1} = c_1\left(\sigma_{L,p}^{t+1} + \sigma_{E,p}^{t+1}\right) + c_2$
5. Update $\boldsymbol{\sigma}^{t+1} = \sigma_{MR,p}^{t+1}$; $\boldsymbol{\varepsilon}^{t+1} = \boldsymbol{\varepsilon}^t + \dot{\varepsilon}^{t+1}$; $t = t + 1$
6. Repeat steps 1–5

The MR training process resembles the search for parameters in conventional constitutive models. MR training aims to find an optimum NN_H to compensate for the difference between the predictions of NN_L and the real mechanical behaviors of the studied soil, while conventional constitutive modeling uses optimization algorithms or trial-and-error methods to calibrate their parameters (Jin et al. 2016b). Training of an NN_H can be completed within seconds based on an existing baseline NN_L, achieving competitive computational efficiency with the parameter calibration process applied in conventional constitutive models.

3.4.4 Incorporating Uncertainty

The principle of incorporating uncertainty is the same as that presented in Section 3.2.4, and thus not repeated here. MCD is selected and incorporated within NN_H (Figure 3.7b) because NN_H is trained on real experimental data that essentially includes uncertainty.

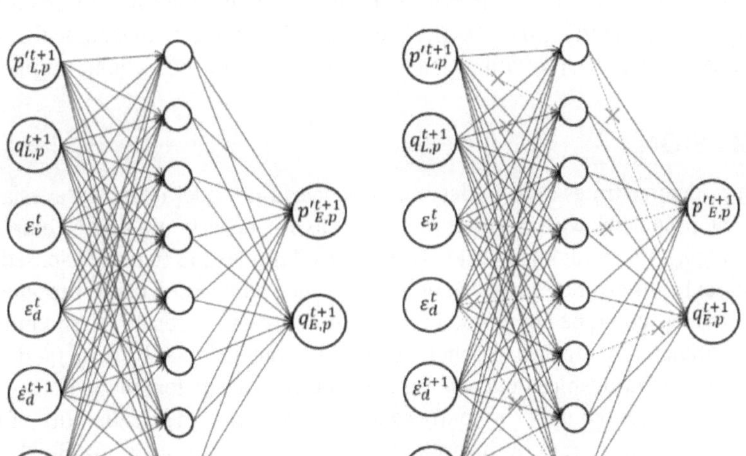

(a) (b)

FIGURE 3.7 Schematic illustration of the high-fidelity ANN model (a) without uncertainty; (b) with Monte Carlo dropout uncertainty showing one random initialization of the network (black crosses indicate dropped connections).

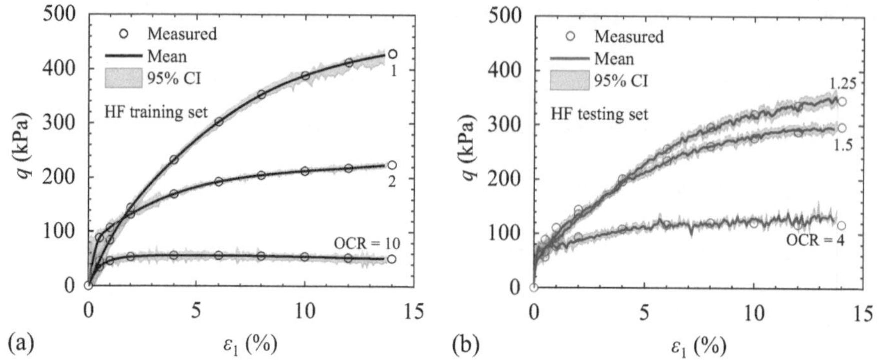

FIGURE 3.8 Elastoplastic model–based MR predictions of the mean and 95% CI predictions of the stress–strain responses in the HF experimental dataset: (a) training set; (b) testing set.

3.4.5 Examples and Validation

The elastoplastic model is used to generate separate sets of synthetic data based on strain-controlled loading (Zhang et al. 2023). The "pure" data-driven ANN is also trained on the LF dataset to benchmark predictions of the PiNet such that a total of two LF models are produced. The only requirement for the LF synthetic data is to ensure that various loading paths can be included so that the LF model can roughly identify the underlying path-dependent soil behavior.

Experimental data published elsewhere in the literature can be collated to form two separate datasets: the HF dataset to train the HF model and a "testing" dataset to validate the final MR framework. For the elastoplastic model–based LF model, six triaxial drained compression experiments on Lower Cromer till (LCT) clay considering this data have been extensively used to calibrate many constitutive models: experiments with over consolidation ratio (OCR) = 1, 2, and 10 are used as the HF training data while the remaining data (OCR = 1.25, 1.5, and 4) is reserved for testing (Figure 3.8).

The inclusion of the Monte Carlo dropout was subsequently examined to quantify the uncertainty of the data-driven model predictions. The final predictions of mechanical responses were assigned a confidence interval, providing a basis for interpretation and reliable engineering design.

3.5 CONCLUSION

This chapter first demonstrates how to develop an uncertain model using a Bayesian neural network. Mechanical properties with physical properties as variables could first be employed to build a deterministic algorithm artificial neural network–based model and two BNN-based models with different theoretical frameworks, i.e., variational inference and Monte Carlo dropout. The performance of these models could be compared in terms of accuracy, uncertainty, and monotonicity, in which a novel cumulative distribution function–based parametric analysis method could be proposed to examine the relationships between input and output parameters in the BNN-based models. The BNN generally increased uncertainty by sacrificing accuracy. The reliability of the predicted result using the BNN can be high in the area of dense datasets; however, the BNN gave low reliability to the predicted result in the area of sparse datasets. These factors were reliable and implied that the BNN knew what it did not know. Moreover, the BNN with the integration of VI showed a more conservative performance with a larger credible interval than that of the MCD, particularly in the case of sparse datasets, which was related to the weights and biases approximated by the VI. These factors indicate that the predicted result using a BNN with a reliability value is more suitable for applying in engineering practice in comparison with other ML or deep learning algorithms.

Next, to identify parameters considering soil uncertainty, a method of Bayesian parameter identification using an enhanced DE-TMCMC with parallel computing technique is presented. The DE-TMCMC, enhanced through implementing a differential evolution into TMCMC to replace the process of proposing a new sample, has been proposed. To save on computational costs, a parallel computing DE-TMCMC was achieved by implementing the SPMD technique. The performance of DE-TMCMC during Bayesian parameter identification was first validated through comparison with the original TMCMC for the case of identifying the parameters of a constitutive model from triaxial tests in terms of robustness and effectiveness, and was then successfully applied to in situ pressuremeter tests on soft clays.

Finally, an interpretable data-driven constitutive modeling framework considering uncertainty for use with sparse data was presented. A prior information-based neural network combined with known theoretical soil mechanics knowledge was first developed; this can include a theoretical concept from elastoplasticity which can be each encoded into separate PiNet models. Compared with conventional artificial neural networks, the PiNet showed greater accuracy in modeling mechanical responses without increasing the computational cost. The PiNet also showed a significantly improved ability to generalize outside the range of training data and demonstrated a strong capacity to extrapolate. Multi-fidelity residual modeling with PiNet was adopted to predict real soil behaviors in practice. Then, the inclusion of MCD was shown to be able to quantify the uncertainty of the data-driven model predictions. The final predictions of mechanical responses were assigned a confidence interval, providing a basis for interpretation and reliable engineering design.

ACKNOWLEDGMENTS

This research was financially supported by the Research Grants Council (RGC) of Hong Kong Special Administrative Region Government (HKSARG) of China (Grant No.: 15220221, 15227923). The authors are grateful to Mr Gengfu He for his kind editorial assistance.

REFERENCES

Ahangar Asr, A., Faramarzi, A., and Javadi, A. A. 2018. An evolutionary modelling approach to predicting stress-strain behaviour of saturated granular soils. *Engineering Computations*, 35(8), 2931–2952.

Akeju, O. V., Senetakis, K., and Wang, Y. 2019. Bayesian parameter identification and model selection for normalized modulus reduction curves of soils. *Journal of Earthquake Engineering*, 23(2), 305–333.

Al-Bittar, T. and Soubra, A.-H. 2014. Probabilistic analysis of strip footings resting on spatially varying soils and subjected to vertical or inclined loads. *Journal of Geotechnical and Geoenvironmental Engineering*, 140(4), 04013043.

Ancey, C. 2005. Monte Carlo calibration of avalanches described as Coulomb fluid flows. *Philosophical Transactions of the Royal Society A: Mathematical, Physical and Engineering Sciences*, 363(1832), 1529–1550.

Angelikopoulos, P., Papadimitriou, C., and Koumoutsakos, P. 2012. Bayesian uncertainty quantification and propagation in molecular dynamics simulations: A high performance computing framework. *The Journal of Chemical Physics*, 137(14), 144103.

Araei, A. A. 2014. Artificial neural networks for modeling drained monotonic behavior of rockfill materials. *International Journal of Geomechanics*, 14(3), 04014005.

Banimahd, M., Yasrobi, S. S., and Woodward, P. K. 2005. Artificial neural network for stress–strain behavior of sandy soils: Knowledge based verification. *Computers and Geotechnics*, 32(5), 377–386.

Basheer, I. A. 2000. Selection of methodology for neural network modeling of constitutive hystereses behavior of soils. *Computer-Aided Civil and Infrastructure Engineering*, 15(6), 440–458.

Basheer, I. A. 2002. Stress-strain behavior of geomaterials in loading reversal simulated by time-delay neural networks. *Journal of Materials in Civil Engineering*, 14(3), 270–273.

Baydin, A. G., Pearlmutter, B. A., Radul, A. A., and Siskind, J. M. 2018. Automatic differentiation in machine learning: A survey. *Journal of Machine Learning Research*, 18(1), 5595–5637.

Beck, J. L. 2010. Bayesian system identification based on probability logic. *Structural Control and Health Monitoring*, 17(7), 825–847.

Beck, J. L. and Katafygiotis, L. S. 1998. Updating models and their uncertainties. I: Bayesian statistical framework. *Journal of Engineering Mechanics*, 124(4), 455–461.

Blundell, C., Cornebise, J., Kavukcuoglu, K., and Wierstra, D. 2015a. Weight uncertainty in neural networks. arXiv:1505.05424v2.

Cabalar, A. F. and Cevik, A. 2011. Triaxial behavior of sand–mica mixtures using genetic programming. *Expert Systems with Applications*, 38(8), 10358–10367.

Cao, Z. and Wang, Y. 2014a. Bayesian model comparison and characterization of undrained shear strength. *Journal of Geotechnical and Geoenvironmental Engineering*, 140(6), 04014018.

Cao, Z. and Wang, Y. 2014b. Bayesian model comparison and selection of spatial correlation functions for soil parameters. *Structural Safety*, 49, 10–17.

Cao, Z. J. and Wang, Y. 2013. Bayesian approach for probabilistic site characterization using cone penetration tests. *Journal of Geotechnical and Geoenvironmental Engineering*, 139(2), 267–276.

Chang, C. S. and Hicher, P. Y. 2005. An elasto-plastic model for granular materials with microstructural consideration. *International Journal of Solids and Structures*, 42(14), 4258–4277.

Ching, J. and Chen, Y.-C. 2007. Transitional Markov chain Monte Carlo method for Bayesian model updating, model class selection, and model averaging. *Journal of Engineering Mechanics*, 133(7), 816–832.

Ching, J. and Phoon, K.-K. 2019. Constructing site-specific multivariate probability distribution model using Bayesian machine learning. *Journal of Engineering Mechanics*, 145(1), 04018126.

Ching, J., Phoon, K. K., Ho, Y. H., and Weng, M. C. 2020. Quasi-site-specific prediction for deformation modulus of rock mass. *Canadian Geotechnical Journal*, 58(7), 936–951.

Ching, J., Wu, S.-S., and Phoon, K.-K. 2016. Statistical characterization of random field parameters using frequentist and Bayesian approaches. *Canadian Geotechnical Journal*, 53(2), 285–298.

Chopin, N. 2002. A sequential particle filter method for static models. *Biometrika*, 89(3), 539–552.

Cividini, A., Maier, G., and Nappi, A. 1983. Parameter estimation of a static geotechnical model using a Bayes' approach. *International Journal of Rock Mechanics and Mining Sciences and Geomechanics Abstracts*, 20(5), 215–226.

Cuisinier, O., Javadi, A. A., Ahangar-Asr, A., and Masrouri, F. 2013. Identification of coupling parameters between shear strength behaviour of compacted soils and chemical's effects with an evolutionary-based data mining technique. *Computers and Geotechnics*, 48, 107–116.

Eckert, N., Parent, E., and Richard, D. 2007. Revisiting statistical–topographical methods for avalanche predetermination: Bayesian modelling for runout distance predictive distribution. *Cold Regions, Science and Technology*, 49(1), 88–107.

Ellis, G. W., Yao, C., Zhao, R., and Penumadu, D. 1995. Stress-strain modeling of sands using artificial neural networks. *Journal of Geotechnical Engineering*, 121(5), 429–435.

Faramarzi, A., Javadi, A. A., and Alani, A. M. 2012. EPR-based material modelling of soils considering volume changes. *Computers and Geosciences*, 48, 73–85.

Fischer, J.-T., Kofler, A., Fellin, W., Granig, M., and Kleemayr, K. 2015. Multivariate parameter optimization for computational snow avalanche simulation. *Journal of Glaciology*, 61(229), 875–888.

Fu, Q., Hashash, Y. M. A., Jung, S., and Ghaboussi, J. 2007. Integration of laboratory testing and constitutive modeling of soils. *Computers and Geotechnics*, 34(5), 330–345.

Gajo, A. and Wood, M. 1999. Severn–Trent sand: A kinematic-hardening constitutive model: The q–p formulation. *Géotechnique*, 49(5), 595–614.

Gal, Y. and Ghahramani, Z. 2015a. Dropout as a Bayesian approximation: Appendix. arXiv:1506.02157v5.

Gal, Y. and Ghahramani, Z. 2015b. Dropout as a Bayesian approximation: Representing model uncertainty in deep learning. arXiv:1506.02142.

Gal, Y. and Ghahramani, Z. 2016. Dropout as a Bayesian approximation: Representing model uncertainty in deep learning. *In the 33rd International Conference on Machine Learning*. New York, Vol. 48, pp. 1050–1059.

Gauer, P., Medina-Cetina, Z., Lied, K., and Kristensen, K. 2009. Optimization and probabilistic calibration of avalanche block models. *Cold Regions, Science and Technology*, 59(2–3), 251–258.

Ghaboussi, J., Garret, J., and Wu, X. 1991. Knowledge-based modeling of material behavior with neural networks. *Journal of Engineering Mechanics*, 117(1), 132–153.

Ghaboussi, J. and Sidarta, D. E. 1998. New nested adaptive neural networks (NANN) for constitutive modeling. *Computers and Geotechnics*, 22(1), 29–52.

Goodfellow, I., Bengio, Y., and Courville, A. 2016. *Deep Learning*. MIT Press. ISBN: 9780262035613.

Graves, A. 2011a. Practical variational inference for neural networks. In *Proceedings of the 24th International Conference on Neural Information Processing Systems*. Curran Associates Inc., Granada Spain, pp. 2348–2356.

Graves, A. 2011b. *Practical Variational Inference for Neural Networks*. Proceedings of the 24th International Conference on Neural Information Processing Systems, December 2011, pp. 2348–2356. NIPS.

Habibagahi, G. and Bamdad, A. 2003. A neural network framework for mechanical behavior of unsaturated soils. *Canadian Geotechnical Journal*, 40(3), 684–693.

Hadjidoukas, P. E., Angelikopoulos, P., Papadimitriou, C., and Koumoutsakos, P. 2015. Π4U: A high performance computing framework for Bayesian uncertainty quantification of complex models. *Journal of Computational Physics*, 284, 1–21.

Hashash, Y. M. A., Fu, Q., Ghaboussi, J., Lade, P. V., and Saucier, C. 2009. Inverse analysis–based interpretation of sand behavior from triaxial compression tests subjected to full end restraint. *Canadian Geotechnical Journal*, 46(7), 768–791.

Hashash, Y. M. A. and Song, H. 2008. The integration of numerical modeling and physical measurements through inverse analysis in geotechnical engineering. *KSCE Journal of Civil Engineering*, 12(3), 165–176.

He, J., Jones, J. W., Graham, W. D., and Dukes, M. D. 2010. Influence of likelihood function choice for estimating crop model parameters using the generalized likelihood uncertainty estimation method. *Agricultural Systems*, 103(5), 256–264.

He, S. and Li, J. 2009. Modeling nonlinear elastic behavior of reinforced soil using artificial neural networks. *Applied Soft Computing*, 9(3), 954–961.

Hellweger, V., Fischer, J.-T., Kofler, A., Huber, A., Fellin, W., and Oberguggenberger, M. 2016. Stochastic methods in operational avalanche simulation - From back calculation to prediction. In *Proceedings of International Snow Science Workshop 2016*. Montana State University Library, pp. 1357–1381.

Hinton, G. E. and Van Camp, D. 1993. Keeping the neural networks simple by minimizing the description length of the weights. In *Proceedings of the Sixth Annual Conference on Computational Learning Theory*, pp. 5–13. https://doi.org/10.1145/168304.168306

Honjo, Y., Wen-Tsung, L., and Guha, S. 1994. Inverse analysis of an embankment on soft clay by extended Bayesian method. *International Journal for Numerical and Analytical Methods in Geomechanics*, 18(10), 709–734.

Houlsby, N. M. T. and Houlsby, G. T. 2013. Statistical fitting of undrained strength data. *Géotechnique*, 63(14), 1253–1263.

Hsiao, E. C., Schuster, M., Juang, C. H., and Kung, G. T. 2008. Reliability analysis and updating of excavation-induced ground settlement for building serviceability assessment. *Journal of Geotechnical and Geoenvironmental Engineering*, 134(10), 1448–1458.

Hu, X., Zhang, Y., Guo, L., Wang, J., Cai, Y., Fu, H., and Cai, Y. 2018. Cyclic behavior of saturated soft clay under stress path with bidirectional shear stresses. *Soil Dynamics and Earthquake Engineering*, 104, 319–328.

Huang, J., Zeng, C., and Kelly, R. 2019. Back analysis of settlement of Teven Road trial embankment using Bayesian updating. *Georisk: Assessment and Management of Risk for Engineered Systems and Geohazards*, 13(4), 320–325.

Javadi, A. A., Faramarzi, A., and Ahangar-Asr, A. 2012. Analysis of behaviour of soils under cyclic loading using EPR-based finite element method. *Finite Elements in Analysis and Design*, 58, 53–65.

Javadi, A. A. and Rezania, M. 2009. Applications of artificial intelligence and data mining techniques in soil modeling. *Geomechanics and Engineering*, 1(1), 53–74.

Jefferies, M. 1993. Nor-Sand: A simple critical state model for sand. *Géotechnique*, 43(1), 91–103.

Jin, Y.-F., Wu, Z.-X., Yin, Z.-Y., and Shen, J. S. 2017a. Estimation of critical state-related formula in advanced constitutive modeling of granular material. *Acta Geotechnica*, 12(6), 1329–1351.

Jin, Y.-F., Yin, Z.-Y., Shen, S.-L., and Hicher, P.-Y. 2016a. Investigation into MOGA for identifying parameters of a critical-state-based sand model and parameters correlation by factor analysis. *Acta Geotechnica*, 11(5), 1131–1145.

Jin, Y.-F., Yin, Z.-Y., Shen, S.-L., and Hicher, P.-Y. 2016b. Selection of sand models and identification of parameters using an enhanced genetic algorithm. *International Journal for Numerical and Analytical Methods in Geomechanics*, 40(8), 1219–1240.

Jin, Y.-F., Yin, Z.-Y., Shen, S.-L., and Zhang, D.-M. 2017b. A new hybrid real-coded genetic algorithm and its application to parameters identification of soils. *Inverse Problems in Science and Engineering*, 25(9), 1343–1366.

Jin, Y.-F., Yin, Z.-Y., Wu, Z.-X., and Zhou, W.-H. 2018. Identifying parameters of easily crushable sand and application to offshore pile driving. *Ocean Engineering*, 154, 416–429.

Jin, Y.-F., Yin, Z.-Y., Zhou, W.-H., and Horpibulsuk, S. 2019a. Identifying parameters of advanced soil models using an enhanced transitional Markov chain Monte Carlo method. *Acta Geotechnica*, 14(6), 1925–1947.

Jin, Y.-F., Yin, Z.-Y., Zhou, W.-H., and Huang, H.-W. 2019b. Multi-objective optimization-based updating of predictions during excavation. *Engineering Applications of Artificial Intelligence*, 78, 102–123.

Jin, Y.-F., Yin, Z.-Y., Zhou, W.-H., and Shao, J.-F. 2019c. Bayesian model selection for sand with generalization ability evaluation. *International Journal for Numerical and Analytical Methods in Geomechanics*, 43(14), 2305–2327.

Jin, Y. F., Yin, Z. Y., Zhou, W. H., and Horpibulsuk, S. 2019d. Identifying parameters of advanced soil models using an enhanced transitional Markov chain Monte Carlo method. *Acta Geotechnica*, 14(6), 1925–1947.

Johari, A., Javadi, A. A., and Habibagahi, G. 2011. Modelling the mechanical behaviour of unsaturated soils using a genetic algorithm-based neural network. *Computers and Geotechnics*, 38(1), 2–13.

Juang, C. H., Luo, Z., Atamturktur, S., and Huang, H. 2012. Bayesian updating of soil parameters for braced excavations using field observations. *Journal of Geotechnical and Geoenvironmental Engineering*, 139(3), 395–406.

Juang, C. H., Luo, Z., Atamturktur, S., and Huang, H. W. 2013. Bayesian updating of soil parameters for braced excavations using field observations. *Journal of Geotechnical and Geoenvironmental Engineering*, 139(3), 395–406.

Juang, C. H. and Zhang, J. 2017. Bayesian methods for geotechnical applications—A practical guide. *Geotechnical Safety and Reliability*, 215–246.

Jung, B. C., Biscontin, G., and Gardoni, P. 2009. Bayesian updating of a unified soil compression model. *Georisk: Assessment and Management of Risk for Engineered Systems and Geohazards*, 3(2), 87–96.

Kang, X., Xia, Z., Chen, R., Ge, L., and Liu, X. 2019a. The critical state and steady state of sand: A literature review. *Marine Georesources and Geotechnology*, 37(9), 1105–1118.

Kang, X., Xia, Z., and Chen, R. P. 2019b. Measurement and correlations of K_0 and V_s anisotropy of granular soils. *Proceedings of the Institution of Civil Engineers - Geotechnical Engineering*, 173(6), 546–561.

Kelly, R. and Huang, J. 2015. Bayesian updating for one-dimensional consolidation measurements. *Canadian Geotechnical Journal*, 52(9), 1318–1330.

Kendall, A. and Gal, Y. 2017. What uncertainties do we need in Bayesian deep learning for computer vision? arXiv: 1703.04977.

Knabe, T., Datcheva, M., Lahmer, T., Cotecchia, F., and Schanz, T. 2013. Identification of constitutive parameters of soil using an optimization strategy and statistical analysis. *Computers and Geotechnics*, 49, 143–157.

Kohestani, V. R. and Hassanlourad, M. 2016. Modeling the mechanical behavior of carbonate sands using artificial neural networks and support vector machines. *International Journal of Geomechanics*, 16(1), 04015038.

Kolymbas, D. 1985. A generalized hypoelastic constitutive law in proceedings of proceedings XI international conference soil mechanics and foundation engineering. Balkema, Rotterdam, 5, 2626.

Kolymbas, D. 1991. An outline of hypoplasticity. *Archive of Applied Mechanics*, 61(3), 143–151.

Kullback, S. and Leibler, R. A. 1951. On information and sufficiency. *Annals of Mathematical Statistics*, 22(1), 79–86.

Levasseur, S., Malécot, Y., Boulon, M., and Flavigny, E. 2008. Soil parameter identification using a genetic algorithm. *International Journal for Numerical and Analytical Methods in Geomechanics*, 32(2), 189–213.

Li, X. D., Zhang, G. Y., Fang, X. P., Tao, W. J., and Hui, X. 2008. Normalization characteristic of sands under triaxial compression and numerical modeling method (in Chinese). *Chinese Journal of Rock Mechanics and Engineering*, 27(S1), 3082–3087.

Li, Z., Chow, J. K., and Wang, Y. H. 2017. Applying the artificial neural network to predict the soil responses in the DEM simulation. *IOP Conference Series: Materials Science and Engineering*, 216, 012040. DOI: 10.1088/1757-899X/216/1/012040

Liu, D. and Wang, Y. 2019. Multi-fidelity physics-constrained neural network and its application in materials modeling. *Journal of Mechanical Design*, 141(12), 121403.

Liu, W. Z., Shi, M. L., Miao, L. C., Xu, L. R., and Zhang, D. W. 2013. Constitutive modeling of the destructuration and anisotropy of natural soft clay. *Computers and Geotechnics*, 51, 24–41.

Lo, M. K. and Leung, Y. F. 2019. Bayesian updating of subsurface spatial variability for improved prediction of braced excavation response. *Canadian Geotechnical Journal*, 56(8), 1169–1183.

Lv, Y., Nie, L., and Xu, K. 2011. Study of the neural network constitutive models for turfy soil with different decomposition degree. In *Proceedings of 2011 Second International Conference on Mechanic Automation and Control Engineering*. IEEE, pp. 6111–6114.

Masi, F., Stefanou, I., Vannucci, P., and Maffi-Berthier, V. 2021. Thermodynamics-based artificial neural networks for constitutive modeling. *Journal of the Mechanics and Physics of Solids*, 147, 104277.

Mašín, D. 2005. A hypoplastic constitutive model for clays. *International Journal for Numerical and Analytical Methods in Geomechanics*, 29(4), 311–336.

Mašín, D. 2015. The influence of experimental and sampling uncertainties on the probability of unsatisfactory performance in geotechnical applications. *Géotechnique*, 65(11), 897–910.

Miro, S., König, M., Hartmann, D., and Schanz, T. 2015. A probabilistic analysis of subsoil parameters uncertainty impacts on tunnel-induced ground movements with a back-analysis study. *Computers and Geotechnics*, 68, 38–53.

Most, T. 2010. Identification of the parameters of complex constitutive models: Least squares minimization vs. Bayesian updating. *Reliability and Optimization of Structural Systems*, 119.

Murakami, A., Shinmura, H., Ohno, S., and Fujisawa, K. 2017. Model identification and parameter estimation of elastoplastic constitutive model by data assimilation using the particle filter. *International Journal for Numerical and Analytical Methods in Geomechanics*, 42(1), 110–131.

Namikawa, T. 2019. Evaluation of statistical uncertainty of cement-treated soil strength using Bayesian approach. *Soils and Foundations*, 59(5), 1228–1240.

Nassr, A., Esmaeili-Falak, M., Katebi, H., and Javadi, A. 2018. A new approach to modeling the behavior of frozen soils. *Engineering Geology*, 246, 82–90.

Papon, A., Riou, Y., Dano, C., and Hicher, P. Y. 2012. Single-and multi-objective genetic algorithm optimization for identifying soil parameters. *International Journal for Numerical and Analytical Methods in Geomechanics*, 36(5), 597–618.

Peng, X.-H., Wang, Z.-C., Luo, T., Yu, M., and Luo, Y.-S. 2008. An elasto-plastic constitutive model of moderate sandy clay based on BC-RBFNN. *Journal of Central South University of Technology*, 15(s1), 47–50.

Penumadu, D. and Zhao, R. D. 1999. Triaxial compression behavior of sand and gravel using artificial neural networks (ANN). *Computers and Geotechnics*, 24(3), 207–230.

Phoon, K. K. 2008. *Reliability-Based Design in Geotechnical Engineering: Computations and Applications*. Taylor and Francis, New York.

Phoon, K. K., Ching, J., and Shuku, T. 2021. Challenges in data-driven site characterization. *Georisk: Assessment and Management of Risk for Engineered Systems and Geohazards*, 16(1), 114–126.

Phoon, K. K. and Kulhawy, F. H. 1999. Characterization of geotechnical variability. *Canadian Geotechnical Journal*, 36(4), 612–624.

Qi, X.-H. and Zhou, W.-H. 2017a. An efficient probabilistic back-analysis method for braced excavations using wall deflection data at multiple points. *Computers and Geotechnics*, 85, 186–198.

Rashidian, V. and Hassanlourad, M. 2014. Application of an artificial neural network for modeling the mechanical behavior of carbonate soils. *International Journal of Geomechanics*, 14(1), 142–150.

Ritto, T. and Nunes, L. 2015. Bayesian model selection of hyperelastic models for simple and pure shear at large deformations. *Computers and Structures*, 156, 101–109.

Romo, M. P., García, S. R., Mendoza, M. J., and Taboada-Urtuzuástegui, V. 2001. Recurrent and constructive-algorithm networks for sand behavior modeling. *International Journal of Geomechanics*, 1(4), 371–387.

Roscoe, K. H. and Burland, J. 1968. On the generalized stress-strain behaviour of wet clay. In *Proceedings of Engineering Plasticity*. Cambridge University Press, pp. 535–609.

Seoh, R. 2019. Qualitative analysis of monte Carlo dropout. arXiv:2007.01720v1.

Sezer, A. 2011. Prediction of shear development in clean sands by use of particle shape information and artificial neural networks. *Expert Systems with Applications*, 38(5), 5603–5613.

Shahin, M. A. and Indraratna, B. 2006. Modeling the mechanical behavior of railway ballast using artificial neural networks. *Canadian Geotechnical Journal*, 43(11), 1144–1152.

Sidarta, D. E. and Ghaboussi, J. 1998. Constitutive modeling of geomaterials from non-uniform material tests. *Computers and Geotechnics*, 22(1), 53–71.

Srivastava, N., Hinton, G., Krizhevsky, A., Sutskever, I., and Salakhutdinov, R. 2014. Dropout: A simple way to prevent neural networks from overfitting. *Journal of Machine Learning Research*, 15, 1929–1958.

Stefanos, D. and Gyan, P. 2015. On neural network constitutive models for geomaterials. *Journal of Civil Engineering Research*, 5(5), 106–113.

Su, D. and Yang, Z. X. 2019. Drained analyses of cylindrical cavity expansion in sand incorporating a bounding-surface model with state-dependent dilatancy. *Applied Mathematical Modelling*, 68, 1–20.

Su, L.-J., Yin, J.-H., and Zhou, W.-H. 2010. Influences of overburden pressure and soil dilation on soil nail pull-out resistance. *Computers and Geotechnics*, 37(4), 555–564.

Suzuki, M. and Ishii, K. 1994. Parameter identification and probabilistic prediction of settlement of embankment. *Structural Safety*, 14(1), 47–59.

Taiebat, M. and Dafalias, Y. F. 2008. SANISAND: Simple anisotropic sand plasticity model. *International Journal for Numerical and Analytical Methods in Geomechanics*, 32(8), 915–948.

Tan, F., Zhou, W.-H., and Yuen, K.-V. 2016. Modeling the soil water retention properties of same-textured soils with different initial void ratios. *Journal of Hydrology*, 542, 731–743.

Tan, F., Zhou, W. H., and Yuen, K. V. 2018. Effect of loading duration on uncertainty in creep analysis of clay. *International Journal for Numerical and Analytical Methods in Geomechanics*, 42(11), 1235–1254.

Tian, H.-M., Cao, Z.-J., Li, D.-Q., Du, W., and Zhang, F.-P. 2021. Efficient and flexible Bayesian updating of embankment settlement on soft soils based on different monitoring datasets. *Acta Geotechnica*, 17(4), 1273–1294.

Tian, Y. and Yao, Y. P. 2017. Modelling the non-coaxiality of soils from the view of cross-anisotropy. *Computers and Geotechnics*, 86, 219–229.

Vermeer, P. 1978. A double hardening model for sand. *Géotechnique*, 28(4), 413–433.

Vlassis, N. N. and Sun, W. 2021. Sobolev training of thermodynamic-informed neural networks for interpretable elasto-plasticity models with level set hardening. *Computer Methods in Applied Mechanics and Engineering*, 377, 113695.

Von Wolffersdorff, P. A. 1996. A hypoplastic relation for granular materials with a predefined limit state surface. *Mechanics of Cohesive-Frictional Materials: An International Journal on Experiments, Modelling and Computation of Materials and Structures*, 1(3), 251–271.

Vrugt, J. A. 2016. Markov chain Monte Carlo simulation using the DREAM software package: Theory, concepts, and MATLAB implementation. *Environmental Modelling and Software*, 75, 273–316.

Wang, H., Wang, X., Wellmann, J. F., and Liang, R. Y. 2019. A Bayesian unsupervised learning approach for identifying soil stratification using cone penetration data. *Canadian Geotechnical Journal*, 56(8), 1184–1205.

Wang, K. and Sun, W. 2018. A multiscale multi-permeability poroplasticity model linked by recursive homogenizations and deep learning. *Computer Methods in Applied Mechanics and Engineering*, 334, 337–380.

Wang, S., Wu, W., Yin, Z.-Y., Peng, C., and He, X.-Z. 2018. Modelling time-dependent behaviour of granular material with hypoplasticity. *International Journal for Numerical and Analytical Methods in Geomechanics*, 42(12), 1331–1345.

Wang, X., Wang, H., and Liang, R. Y. 2017. A method for slope stability analysis considering subsurface stratigraphic uncertainty. *Landslides*, 15(5), 925–936.

Wang, Y., Huang, K., and Cao, Z. 2013. Probabilistic identification of underground soil stratification using cone penetration tests. *Canadian Geotechnical Journal*, 50(7), 766–776.

Wu, W., Bauer, E., and Kolymbas, D. 1996. Hypoplastic constitutive model with critical state for granular materials. *Mechanics of Materials*, 23(1), 45–69.

Xiao, T., Li, D. Q., Cao, Z. J., and Zhang, L. M. 2018. CPT-based probabilistic characterization of three-dimensional spatial variability using MLE. *Journal of Geotechnical and Geoenvironmental Engineering*, 144(5), 04018023.

Xiong, H., Nicot, F., and Yin, Z. 2017. A three-dimensional micromechanically based model. *International Journal for Numerical and Analytical Methods in Geomechanics*, 41(17), 1669–1686.

Xu, K., Huang, D. Z., and Darve, E. 2021. Learning constitutive relations using symmetric positive definite neural networks. *Journal of Computational Physics*, 428, 110072.

Yan, W. M., Yuen, K. V., and Yoon, G. L. 2009. Bayesian probabilistic approach for the correlations of compression index for marine clays. *Journal of Geotechnical and Geoenvironmental Engineering*, 135(12), 1932–1940.

Yang, H.-Q., Zhang, L., Pan, Q., Phoon, K.-K., and Shen, Z. 2021. Bayesian estimation of spatially varying soil parameters with spatiotemporal monitoring data. *Acta Geotechnica*, 16(1), 263–278.

Yao, Y., Hou, W., and Zhou, A. 2009. UH model: Three-dimensional unified hardening model for overconsolidated clays. *Géotechnique*, 59(5), 451–469.

Yao, Y., Sun, D., and Luo, T. 2004. A critical state model for sands dependent on stress and density. *International Journal for Numerical and Analytical Methods in Geomechanics*, 28(4), 323–337.

Yao, Y., Sun, D., and Matsuoka, H. 2008. A unified constitutive model for both clay and sand with hardening parameter independent on stress path. *Computers and Geotechnics*, 35(2), 210–222.

Yao, Y.-P., Kong, L.-M., Zhou, A.-N., and Yin, J.-H. 2014. Time-dependent unified hardening model: Three-dimensional elastoviscoplastic constitutive model for clays. *Journal of Engineering Mechanics*, 141(6), 04014162.

Yin, Z.-Y., Jin, Y.-F., Shen, J. S., and Hicher, P.-Y. 2018. Optimization techniques for identifying soil parameters in geotechnical engineering: Comparative study and enhancement. *International Journal for Numerical and Analytical Methods in Geomechanics*, 42(1), 70–94.

Yin, Z.-Y., Jin, Y.-F., Shen, S.-L., and Huang, H.-W. 2017. An efficient optimization method for identifying parameters of soft structured clay by an enhanced genetic algorithm and elastic–viscoplastic model. *Acta Geotechnica*, 12(4), 849–867.

Yin, Z.-Y., Karstunen, M., Chang, C. S., Koskinen, M., and Lojander, M. 2011a. Modeling time-dependent behavior of soft sensitive clay. *Journal of Geotechnical and Geoenvironmental Engineering*, 137(11), 1103–1113.

Yin, Z.-Y., Zhao, J., and Hicher, P.-Y. 2014. A micromechanics-based model for sand-silt mixtures. *International Journal of Solids and Structures*, 51(6), 1350–1363.

Yin, Z.-Y. and Chang, C. S. 2009. Microstructural modelling of stress-dependent behaviour of clay. *International Journal of Solids and Structures*, 46(6), 1373–1388.

Yin, Z.-Y., Chang, C. S., and Hicher, P. Y. 2010a. Micromechanical modelling for effect of inherent anisotropy on cyclic behaviour of sand. *International Journal of Solids and Structures*, 47(14–15), 1933–1951.

Yin, Z.-Y., Chang, C. S., Hicher, P. Y., and Karstunen, M. 2008. Microstructural modeling of rate-dependent behavior of soft soil. *The 12th International Conference of International Association for Computer Methods and Advances in Geomechanics (IACMAG)* 1–6 October, Goa, India, 862–868.

Yin, Z.-Y., Chang, C. S., Hicher, P. Y., and Karstunen, M. 2009. Micromechanical analysis of kinematic hardening in natural clay. *International Journal of Plasticity*, 25(8), 1413–1435.

Yin, Z.-Y., Chang, C. S., Karstunen, M., and Hicher, P. Y. 2010b. An anisotropic elastic-viscoplastic model for soft clays. *International Journal of Solids and Structures*, 47(5), 665–677.

Yin, Z.-Y., Chang, C. S., Karstunen, M., and Hicher, P. Y. 2010c. An anisotropic elastic–viscoplastic model for soft clays. *International Journal of Solids and Structures*, 47(5), 665–677.

Yin, Z.-Y. and Karstunen, M. 2011. Modelling strain-rate-dependency of natural soft clays combined with anisotropy and destructuration. *Acta Mechanica Solida Sinica*, 24(3), 216–230.

Yin, Z.-Y., Karstunen, M., Chang, C. S., Koskinen, M., and Lojander, M. 2011b. Modeling time-dependent behavior of soft sensitive clay. *Journal of Geotechnical and Geoenvironmental Engineering*, 137(11), 1103–1113.

Yu, H. 1998. CASM: A unified state parameter model for clay and sand. *International Journal for Numerical and Analytical Methods in Geomechanics*, 22(8), 621–653.

Yuen, K.-V. 2010a. *Bayesian Methods for Structural Dynamics and Civil Engineering*, pp. 11–50. John Wiley & Sons.

Yuen, K.-V. 2010b. Recent developments of Bayesian model class selection and applications in civil engineering. *Structural Safety*, 32(5), 338–346.

Yuen, K. V. and Mu, H. Q. 2015. Real-time system identification: An algorithm for simultaneous model class selection and parametric identification. *Computer-Aided Civil and Infrastructure Engineering*, 30(10), 785–801.

Zhang, J., Zhang, L. M., and Tang, W. H. 2009a. Bayesian framework for characterizing geotechnical model uncertainty. *Journal of Geotechnical and Geoenvironmental Engineering*, 135(7), 932–940.

Zhang, L., Li, D.-Q., Tang, X.-S., Cao, Z.-J., and Phoon, K.-K. 2017. Bayesian model comparison and characterization of bivariate distribution for shear strength parameters of soil. *Computers and Geotechnics*, 95, 110–118.

Zhang, L. L., Zhang, J., Zhang, L. M., and Tang, W. H. 2010. Back analysis of slope failure with Markov chain Monte Carlo simulation. *Computers and Geotechnics*, 37(7–8), 905–912.

Zhang, P. 2022. Data-driven modelling of soil properties and behaviours with geotechnical applications. The Hong Kong Polytechnic University, Hong Kong SAR, Ph.D thesis.

Zhang, P., Jin, Y.-F., and Yin, Z.-Y. 2021a. Machine learning–based uncertainty modelling of mechanical properties of soft clays relating to time-dependent behavior and its application. *International Journal for Numerical and Analytical Methods in Geomechanics*, 45(11), 1588–1602.

Zhang, P., Yang, Y., and Yin, Z.-Y. 2021c. BiLSTM-based soil–structure interface modeling. *International Journal of Geomechanics*, 21(7), 04021096.

Zhang, P. and Yin, Z.-Y. 2021b. A novel deep learning-based modelling strategy from image of particles to mechanical properties for granular materials with CNN and BiLSTM. *Computer Methods in Applied Mechanics and Engineering*, 382, 113858.

Zhang, P., Yin, Z.-Y., and Jin, Y.-F. 2021d. Bayesian neural network-based uncertainty modelling: Application to soil compressibility and undrained shear strength prediction. *Canadian Geotechnical Journal*, 59(4), 546–557.

Zhang, P., Yin, Z.-Y., Jin, Y.-F., and Liu, X.-F. 2021e. Modelling the mechanical behaviour of soils using machine learning algorithms with explicit formulations. *Acta Geotechnica*, 17(4), 1403–1422.

Zhang, P., Yin, Z.-Y., Jin, Y.-F., Yang, J., and Sheil, B. B. 2022a. Physics-informed multi-fidelity residual neural networks for hydromechanical modelling of granular soils and foundation considering internal erosion. *Journal of Engineering Mechanics*, 148(4), 04022015.

Zhang, P., Yin, Z.-Y., and Sheil, B. 2022b. A physics-informed data-driven approach for consolidation analysis. *Géotechnique*, 1–12.

Zhang, P., Yin, Z. Y., and Jin, Y. F. 2021f. State-of-the-art review of machine learning applications in constitutive modeling of soils. *Archives of Computational Methods in Engineering*, 28(5), 3661–3686.

Zhang, P., Yin, Z. Y., and Jin, Y. F. 2022c. Bayesian neural network-based uncertainty modelling: Application to soil compressibility and undrained shear strength prediction. *Canadian Geotechnical Journal*, 59(4), 546–557.

Zhang, P., Yin, Z. Y., Jin, Y. F., and Ye, G. L. 2020. An AI-based model for describing cyclic characteristics of granular materials. *International Journal for Numerical and Analytical Methods in Geomechanics*, 44(9), 1315–1335.

Zhang, P., Yin, Z. Y., and Sheil, B. 2023. Interpretable data-driven constitutive modelling of soils with sparse data. *Computers and Geotechnics*, 160, 105511.

Zhang, X., Srinivasan, R., and Bosch, D. 2009b. Calibration and uncertainty analysis of the SWAT model using genetic algorithms and Bayesian model averaging. *Journal of Hydrology*, 374(3), 307–317.

Zhao, H., Huang, Z., and Zou, Z. 2014. Simulating the stress-strain relationship of geomaterials by support vector machine. *Mathematical Problems in Engineering*, 2014, 1–7.

Zheng, D., Huang, J., Li, D.-Q., Kelly, R., and Sloan, S. W. 2018. Embankment prediction using testing data and monitored behaviour: A Bayesian updating approach. *Computers and Geotechnics*, 93, 150–162.

Zheng, S., Zhu, Y.-X., Li, D.-Q., Cao, Z.-J., Deng, Q.-X., and Phoon, K.-K. 2021. Probabilistic outlier detection for sparse multivariate geotechnical site investigation data using Bayesian learning. *Geoscience Frontiers*, 12(1), 425–439.

Zhou, W.-H., Tan, F., and Yuen, K.-V. 2018. Model updating and uncertainty analysis for creep behavior of soft soil. *Computers and Geotechnics*, 100, 135–143.

Zhou, W.-H., Yuen, K.-V., and Tan, F. 2012. Estimation of maximum pullout shear stress of grouted soil nails using Bayesian probabilistic approach. *International Journal of Geomechanics*, 13(5), 659–664.

Zhou, W.-H., Yuen, K.-V., and Tan, F. J. E. G. 2014. Estimation of soil–water characteristic curve and relative permeability for granular soils with different initial dry densities. *Engineering Geology*, 179, 1–9.

Zhu, J.-H., Zaman, M. M., and Anderson, S. A. 1998a. Modeling of soil behavior with a recurrent neural network. *Canadian Geotechnical Journal*, 35(5), 858–872.

Zhu, J.-H., Zaman, M. M., and Anderson, S. A. 1998b. Modelling of shearing behaviour of a residual soil with recurrent neural network. *International Journal for Numerical and Analytical Methods in Geomechanics*, 22(8), 671–687.

4 Data-Driven Site Characterization

Takayuki Shuku

ABSTRACT

This chapter overviews the current state of data-driven site characterization (DDSC) methods. A 2×2 matrix focusing on types of data (numerical or categorical) and uncertainties (geotechnical or geologic) is used to organize different DDSC methods. The strengths and limitations of DDSC methods are highlighted by comparing the methods from a Bayesian perspective. When reviewing the literature on DDSC in geotechnics, it is important to note that many studies and researchers evaluate the proposed methods based on their own validation methods or data, thereby making it challenging to fairly compare the performance of different models. This chapter also discusses benchmark examples or benchmarking for evaluating the strengths and weaknesses of DDSC methods in a balanced and unbiased way.

4.1 INTRODUCTION

Decision-making in geotechnical engineering is always related to a project carried out at a specific site. It is natural for data-driven site characterization (DDSC) to attract the most attention in data-centric geotechnics. Traditional site characterization mainly relies on engineering judgment and practitioners' experience to derive 3D stratigraphic profiles or 3D spatial distribution of soil properties, which can be time-consuming, labor-intensive, and subject to significant uncertainty. In recent years, the increasing availability of large-scale, high-resolution datasets and advancements in computational methods have paved the way for data-driven approaches in various engineering disciplines, including geotechnical engineering. The term "data-driven site characterization" refers to any site characterization methodology that relies solely on measured data, both site-specific data collected from past projects at the same site, neighboring sites, or beyond (Phoon et al. 2022a). In principle, the source of data can go beyond numerical such as categorical, text, images and expert opinion. One part of DDSC is pursued under Project DeepGeo (Phoon and Ching 2021), a research effort to bring data-driven 3D subsurface mapping into practice. The purpose of Project DeepGeo is to produce a 3D stratigraphic map of the subsurface volume below a full-scale project site and to estimate the relevant engineering properties at each spatial point based on actual site investigation data and other relevant big indirect data (BID). One key complication in DDSC is that real data is MUSIC-3X (multivariate, uncertain and unique, sparse, incomplete, and potentially corrupted with "3X" denoting three-dimensional spatial variation) (Phoon et al. 2022a). It is an open question whether DDSC can solve real-world subsurface mapping problems based on real-world MUSIC-3X data from routine projects with minimum ad hoc assumptions.

Extensive efforts have been made to answer the above open question, and many DDSC methods have been developed because of the advancement in machine learning (ML) algorithms and computers capable of handling the high computational loads that ML requires. There is a large volume of papers on ML in geotechnics, and more than 30% of ML applications in geotechnics are about site characterization (Phoon and Zhang 2023). Ultimately, the goal of DDSC is to provide a more complete picture of a site's conditions, so that stakeholders can make informed decisions about how best

DOI: 10.1201/9781003333586-5

to manage. Since this goal can be achieved through numerous routes, various DDSC methods with different features are being proposed for various purposes. For example, some methods are designed for numerical data, while others are intended for categorical data such as soil/rock types.

This chapter overviews the current state of DDSC methods. The strengths and limitations of DDSC methods are highlighted by comparing the methods from a Bayesian perspective. When reviewing the literature on ML or DDSC in geotechnics, it is important to note that many studies/researchers evaluate the proposed methods based on their own validation methods or data, making it challenging to compare the performance of different models across studies. This chapter also discusses benchmark examples or benchmarking for evaluating the strengths and weaknesses of DDSC methods in a balanced and unbiased way. Although DDSC refers to any site characterization methodology that relies on "data", which can be site-specific data collected for the current project or existing data of any type collected from past projects at the same site, neighboring site, or beyond, this chapter specifically deals with DDSC that relies solely on "site-specific" data. The DDSC methods that can integrate generic databases can be seen in Ching and Phoon (2019, 2020b) and Ching et al. (2021a, 2021b).

4.2 TYPES OF DATA AND THE CORRESPONDING UNCERTAINTY

In geotechnics, (site-specific) data can be numerical or categorical. For example, a standard penetration test (SPT), which is commonly used for geotechnical practice in Japan, gives a depth profile of penetration resistance (SPT-N value or N_{SPT}) and soil stratigraphy (or soil type). While the N_{SPT} value is numerical, soil stratigraphy is categorical (Figure 4.1). DDSC methods are basically intended for estimating the spatial distribution or spatial continuity of "numerical" soil properties because design calculation is based mainly on numerical data. However, some methods are designed for categorical data to identify the geological structures of the ground, which is called "soil stratification" or "stratigraphic profiling". In situ tests such as Swedish weight sounding (SWS) test measure continuous and numerical data, but no information on "soil stratigraphy" is obtained. The information on soil stratigraphy (soil types and layer boundaries) is important in the discussion of liquefaction susceptibility and the selection of constitutive models in numerical simulations.

When cone penetration test (CPT) data is available, the soil type can be estimated from the empirical transformation model proposed by Robertson (2016) and Robertson and Wride (1998), which is known as the "Robertson chart". Based on the tip resistance q_t and sleeve friction f_s measured through the CPT, the soil behavior type (SBT) index (I_c) is calculated. The physical significance of I_c is that it is related to the soil behavior type, as shown in Table 4.1. The Robertson chart transforms "numerical data" into "categorical data" and is widely used in stratigraphic delineation in practice because of its simplicity.

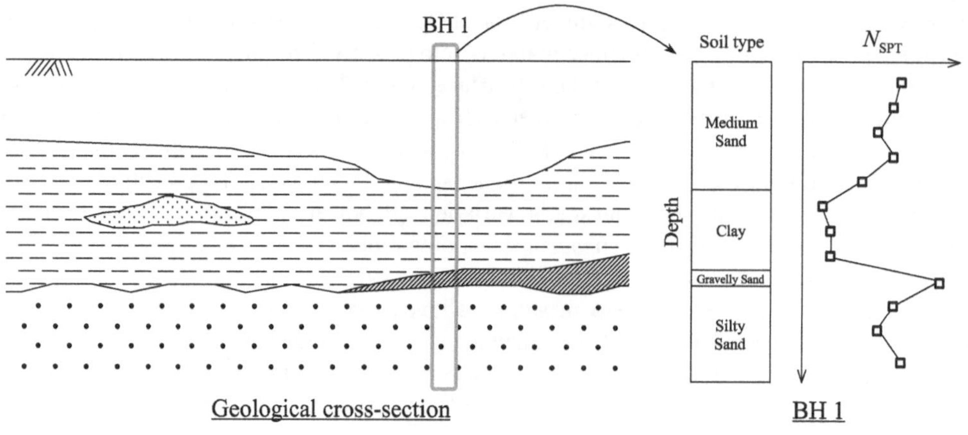

FIGURE 4.1 Numerical and categorical data in DDSC.

TABLE 4.1
Boundaries of Soil Behavior Type

Soil behavior type index (I_c)	Zone	Soil behavior type
$I_c < 1.31$	7	Gravelly sand to dense sand
$1.31 < I_c < 2.05$	6	Sands: clean sand to silty sand
$2.05 < I_c < 2.60$	5	Sand mixtures: silty sand to sandy silt
$2.60 < I_c < 2.95$	4	Silt mixtures: clayey silt to silty clay
$2.95 < I_c < 3.60$	3	Clays: silty clay to clay
$I_c > 3.60$	2	Organic soils: peats

Source: Robertson and Wride (1998).

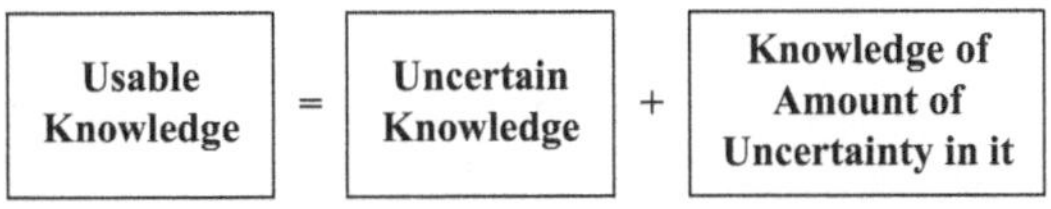

FIGURE 4.2 A logical equation stated by C.R. Rao.

Uncertainty quantification is necessary in DDSC, as current real-world data is insufficient, incomplete, and/or not directly relevant to derive a deterministic map. Renowned statistician C.R. Rao stated that the logical equation is important in decision-making under uncertainties (Figure 4.2). If we have to make decisions under uncertainty, we cannot avoid mistakes. If mistakes cannot be avoided, we had better know how often we make mistakes (Rao 1997). Therefore, it is natural to consider the probabilistic methods as suitable for geotechnical applications.

In subsurface modeling, uncertainty can be classified into two categories: geologic and geotechnical uncertainties. Bárdossy and Fodor (2004) used the terms "structured" and "unstructured" to specify the two types of uncertainties. The "structured" variability means more or less regular spatial changes that can be described by a trend surface analysis, and examples include gradual compositional transitions of one strata/rock into the other, or cyclic repetitions of sedimentary features in a sequence of layers. On the other hand, "unstructured" variability may occur unexpectedly in a stratum and their spatial position and/or magnitude cannot be exactly predicted. They appear in the trend surface analysis as residuals and outliers. The uncertainty in soil stratification is one form of geologic uncertainty that has recently attracted attention (Phoon et al. 2022a). Thus, the data type, numerical or categorical, usually corresponds to the type of uncertainty, geotechnical or geologic. By distinguishing between two uncertainties, engineers and geologists can develop appropriate mitigation measures and design solutions to address potential risks and uncertainties associated with a site. Deng et al. (2017) have studied the effects of geotechnical and geologic uncertainties on slope stability problems using numerical experiments. Yeh et al. (2021) studied the benefit of reducing geologic uncertainty in practice by reanalyzing an actual landslide that occurred at a freeway in Northern Taiwan. A wider dissemination of such case studies would encourage practitioners to consider both geotechnical and geologic uncertainties in subsurface modeling.

4.3 DDSC FROM A BAYESIAN PERSPECTIVE

DDSC can be derived in a unified way through a Bayesian statistical framework (e.g., Wang et al. 2016; Shuku and Phoon 2023a). Let us assume that we have site-specific data t and want to estimate a stratigraphic map x of the subsurface volume below a full-scale project site and/or relevant engineering properties at each spatial point. This typical DDSC challenge, estimating for unknown x based on data t, can be formulated based on Bayes' rule:

$$p(x \mid t) = \frac{p(t \mid x)p(x)}{p(t)} \tag{4.1}$$

where $p(x|t)$ is the probability (or probability density function, PDF) of x given t or *posterior* probability; $p(t|x)$ is the probability of t given x or *likelihood*; $p(x)$ is the probability of x or *prior* probability; and $p(t)$ is the probability of t or *marginal* probability. Since $p(t)$ is a normalization term and is often left out, Eq. (4.1) can be written as

$$p(x \mid t) \propto p(t \mid x)p(x) \tag{4.2}$$

The data t and unknown x can be numerical or categorical. The prior probability $p(x)$ is updated or inferred based on data t (or likelihood) through Bayes' rule: this is called "Bayesian updating" or "Bayesian inference".

The available site-specific data t is often spatially sparse in practice, and unknown data x is usually estimated or interpolated using some kind of "model f" to compensate for the data scarcity. The model f can be a linear function, advanced deep neural networks (DNNs), or a statistical model such as a Markov random field (MRF) model. With model f, Eq. (4.2) can be rewritten as

$$p(f(\theta) \mid t) \propto p(t \mid f(\theta))p(f(\theta)) \tag{4.3}$$

where $\boldsymbol{\theta}$ is a parameter (or hyperparameter vector). In case the model f is given by a linear function $f = w_0 + w_1 z$ (z is depth), the vector $\boldsymbol{\theta}$ is given by

$$\theta = \begin{bmatrix} w_0 & w_1 \end{bmatrix}^T \tag{4.4}$$

We believe that the majority of DDSC methods can be consistently derived from Eq. (4.3) in a unified way, and from this theoretical standpoint, the difference between DDSC methods stems from the difference between model f, prior $p(f)$, and likelihood $p(t|f)$.

4.4 CLASSIFICATION OF DATA-DRIVEN SITE CHARACTERIZATION METHODS

This section organizes DDSC methods using a 2 × 2 matrix, which is shown in Figure 4.3, based on the types of data (numerical or categorical) and uncertainties (geotechnical and geologic). The subsequent sections outline the methodologies in each category and highlight the characteristics from a Bayesian perspective. Comprehensive reviews of DDSC methods can be seen in Phoon et al. (2019, 2022a, 2022b) and Phoon and Ching (2021).

4.4.1 Category 1: Numerical Data – Geotechnical Uncertainty

Since geotechnical design is performed based on numerical data such as cohesion c and internal friction angle φ, most DDSC methods have been developed for estimating the 3D spatial distribution of numerical soil properties. Therefore, *geotechnical* uncertainty tends to be the main focus in DDSC. The DDSC methods for *numerical* data and *geotechnical* uncertainty are classified as Category 1, which is labeled DDSC-C1. As shown in Figure 4.3, DDSC-C1 includes sparse Bayesian learning (SBL; Ching and Phoon 2017; Ching et al. 2020a, 2021c; Ching and Yoshida 2023), Bayesian compressive sampling/sensing (BCS; Wang and Zhao 2016, 2017; Wang et al. 2019; Lyu et al. 2023), Gaussian process regression (GPR; Yoshida et al. 2021, Tomizawa and Yoshida 2023), and geotechnical lasso with basis functions (Glasso-BFs; Phoon and Shuku 2022; Shuku and Phoon 2023a).

Let us first consider DDSC-C1 for a depth profile of numerical data (Figure 4.4a). One of the main goals of DDSC is to estimate unknown or unseen data in space. The relationship between data **t** and model $f(\mathbf{z})$ can be formularized by

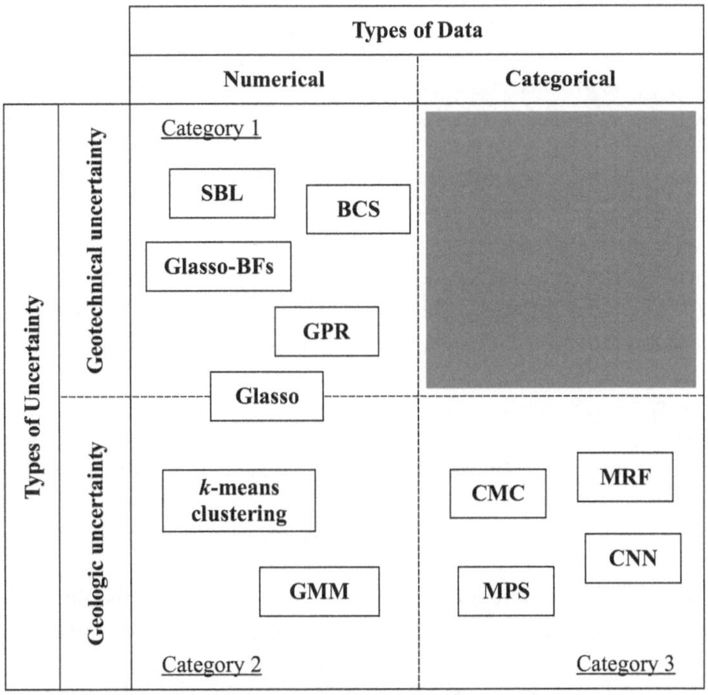

Note: BCS = Bayesian compressive sampling; SBL = Sparse Bayesian learning; Glasso = geotechnical least absolute shrinkage and selection operator; Glasso-BFs = geotechnical lasso with basis functions; GPR = Gaussian process regression; GMM = Gaussian mixture models; CNN = convolutional neural networks; CMC = coupled Markov chain; MRF = Markov random field; MPS = Multiple point statistics

FIGURE 4.3 2×2 matrix to organize DDSC methods.

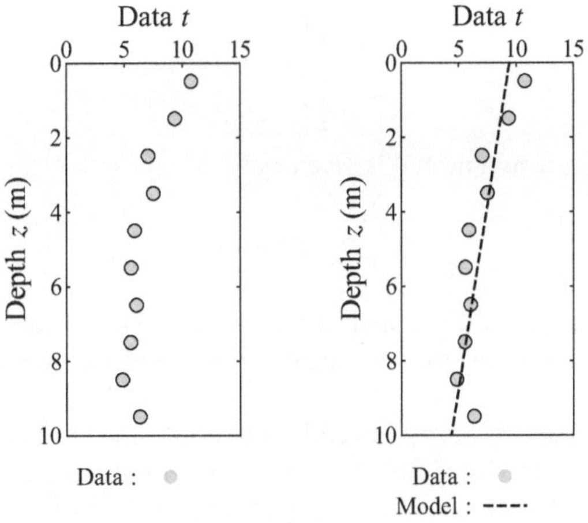

FIGURE 4.4 1D depth profile: (a) data t; (b) estimated linear model.

$$\mathbf{t} = f(\mathbf{z}) + \varepsilon \tag{4.5}$$

where $\mathbf{t} = [t_1, t_2, \ldots, t_m]^T$ is the observation data vector; $\mathbf{z} = [z_1, z_2, \ldots, z_m]^T$ is the depth vector; f is a prescribed model or function; and $\boldsymbol{\varepsilon} = [\varepsilon_1, \varepsilon_2, \ldots, \varepsilon_m]^T$ is the vector of (geotechnical) uncertainty.

As stated in Section 4.3, many DDSC methods can be derived from Bayes' rule defined by Eq. (4.3) in a unified way, and from this theoretical standpoint, methodological differences between different DDSC methods stem from how to model or assume $\boldsymbol{\varepsilon}$, $f(\mathbf{z})$, and $p(f(\mathbf{z}))$. In addition, different DDSC methods offer different levels of sophistication in uncertainty quantification, ranging from least-square estimation (LSE) to full Bayesian analysis, depending on the imposed assumptions. For example, a linear model fitting by the least-square method (LSM), which is widely used in DDSC practice, can be derived from Eq. (4.3) under the following assumptions.

1. $f(z_i) = w_0 + w_1 z_i$
2. ε_i is unbiased
3. ε_i is a set of zero mean i.i.d. Gaussian random variables with variance σ^2
4. $\sigma^2 = 1$
5. $p(f(\mathbf{w})) = p(w_0, w_1)$ follows a uniform distribution (or uninformative prior)

In this case, soil profile modeling reduces to a problem of estimating the deterministic unknown coefficients (w_0 and w_1) which are obtained by minimizing the following objective function J:

$$J_{\mathrm{LSM}} = \frac{1}{2}\sum_{j=1}^{m} \varepsilon_j^2 = \frac{1}{2}\sum_{j=1}^{m} \{t_j - (w_0 + w_1 z_j)\}^2 \tag{4.6}$$

$$J_{\mathrm{LSM}} = \frac{1}{2}\boldsymbol{\varepsilon}^T\boldsymbol{\varepsilon} = \frac{1}{2}\left\|\mathbf{t} - \boldsymbol{\Phi}\mathbf{w}\right\|_2^2 \tag{4.7}$$

where $\mathbf{w} = [w_0, w_1]$ and $\boldsymbol{\Phi}$ is an $m \times n$ matrix, often called a design matrix, whose elements are given by

$$\boldsymbol{\Phi} = \begin{bmatrix} 1 & z_1 \\ 1 & z_2 \\ \vdots & \vdots \\ 1 & z_m \end{bmatrix} \tag{4.8}$$

The closed-form solution is available and is given by

$$\hat{\mathbf{w}}_{\mathrm{LSM}} = (\boldsymbol{\Phi}^T\boldsymbol{\Phi})^{-1}\boldsymbol{\Phi}^T\mathbf{t} \tag{4.9}$$

The estimated model $f(\hat{\mathbf{w}}_{\mathrm{LSM}})$ is illustrated in Figure 4.4b. This model can be seen as the least sophisticated DDSC method in uncertainty quantification because no uncertainty information of the estimate is provided.

If we assume that $\boldsymbol{\varepsilon}$ is spatially correlated and is unknown, the maximum likelihood estimation (MLE) can be derived (e.g., DeGroot and Baecher 1993). A more sophisticated Bayesian method can also be derived with different assumptions (e.g., Ching et al. 2016). In general, numerical difficulty and computational cost increase with model complexity and the sophistication of uncertainty quantification. More sophisticated Bayesian methods tend to have more hyperparameters to train. For some methods, even training a few hyperparameters requires impractical computation time, and it can be more reasonable to focus only on point estimates from a practical point of view.

TABLE 4.2
DDSC-C1 Methods

Methods	Model f	Unknown parameter	Hyperparameter	Prior	Estimate
Linear regression	Linear function, e.g., $f = w_0 + w_1 z$	Coefficients of basis functions $\mathbf{w}$	–	Uninformative prior	Point estimation (LSE, ML)
BCS	Cosine, wavelet (DWT)	Coefficients of basis functions $\mathbf{w}$	α_i	Sparsity of $\mathbf{w}$	Probability estimation (interval estimation)
SBL	Shifted Legendre polynomial	Coefficients of basis functions $\mathbf{w}$	$\sigma, \delta_v, \delta_h$	Sparsity of $\mathbf{w}$	Probability estimation
Glasso-BFs	Shifted Legendre polynomial	Coefficients of basis functions $\mathbf{w}$	λ	Sparsity of $\mathbf{w}$	Point estimation (MAP)
GPR	Kernel functions	–	$\boldsymbol{\theta}=[\theta_0, \theta_1,\ldots, \theta_n]^T$	Gaussian $N(0, \mathbf{C}_0)$	Probability estimation
Glasso	Piecewise constant function	$\mathbf{f}=[f_0, f_1,\ldots,f_n]^T$	λ_v, λ_h	Sparsity of $\mathbf{Bf}$	Point estimation (MAP)

Note: DWT: discrete wavelet transform; LSE: least square estimation; ML: maximum likelihood estimation; MAP: maximum a posteriori.

Table 4.2 summarizes the model f, prior p, and the parameters and hyperparameters of five DDSC-C1 methods. Glasso can be categorized as DDSC-C1 and C3, and is explained in the following section. There are many possible choices for the basis functions $\mathbf{\Phi}$ such as the Gaussian basis function (or the radial basis function, RBF), the sigmoidal basis function, or the Fourier basis function. Although all types of polynomials can be used in the DDSC-C1 methods, they should be orthogonal basis functions because they have some desirable properties for DDSC (e.g., easing multicollinearity). Linear combinations of the fixed basis functions of the input variable $\mathbf{z}$ can be expressed by

$$f(\mathbf{z},\mathbf{w}) = \sum_{j=0}^{M-1} w_j \phi_j(\mathbf{z}) = \mathbf{w}^T \boldsymbol{\phi}(\mathbf{z}) \tag{4.10}$$

The design matrix $\mathbf{\Phi}$ is also defined by

$$\mathbf{\Phi} = \begin{bmatrix} \phi_0(z_1) & \phi_1(z_1) & \cdots & \phi_{M-1}(z_1) \\ \phi_0(z_2) & \phi_1(z_2) & \cdots & \phi_{M-1}(z_2) \\ \vdots & \vdots & \ddots & \vdots \\ \phi_0(z_N) & \phi_1(z_N) & \cdots & \phi_{M-1}(z_N) \end{bmatrix} \tag{4.11}$$

The model f in the DDSC-C1 methods can be generalized as follows:

$$f(\mathbf{w}) = \mathbf{\Phi}\mathbf{w} \tag{4.12}$$

4.4.1.1 Sparse Modeling

BCS, SBL, and Glasso-BFs utilize the "sparsity of the solution **w**" as the prior in common, and the data-driven modeling based on the *sparsity* are often called *sparse modeling.* The "sparsity of the solution" refers to a property of a solution or model that has a small number of non-zero coefficients or features relative to the total number of possible coefficients or features. In other words, a *sparse* solution is one where only a subset of the available variables or features is relevant for predicting or explaining the output. To find the best *sparse* solution is usually intractable when the number of parameters M is large because the computation is 2^M combinatorial, and approximate approaches need to be used.

BCS and SBL use the following prior probability of **w** to obtain the *sparse* subset:

$$p(\mathbf{w} \mid \alpha) = \prod_{k=1}^{M} \mathcal{N}(w_k \mid 0, \alpha_k^{-1}) \tag{4.13}$$

where α_k represents the precision of the corresponding w_k. When the evidence or marginal likelihood with respect to these hyperparameters is maximized, most of them go to infinity, and the corresponding posterior PDFs of these coefficients are concentrated at zero. This means that the basis functions associated with these practically zero coefficients play no role in the model, and a *sparse* model is automatically selected. This methodology to select the best subsets is known as the relevant vector machine (RVM) (Mackay 1992; Tipping 2001). To maximize the evidence, some algorithms such as the expectations and maximization (EM) algorithm and the Markov chain Monte Carlo (MCMC) methods are applicable (e.g., Tipping 2001; Bishop 2006). The *sparse* prior is based on a principle of sparsity which favors simple theories/models over complex ones, known as *Occam's razor.*

Glasso-BFs uses the regularization technique called the least absolute shrinkage and selection operator (lasso, Tibshirani 1996) to get the *sparse* subset of the coefficients of a model f. In lasso, the following prior probability distribution is used to impose sparsity on the coefficient from a Bayesian perspective:

$$p(\mathbf{w} \mid \kappa) \propto \exp\left(-\kappa \|\mathbf{w}\|_1\right) \tag{4.14}$$

where κ is the diversity parameter and $\| \, \|_1$ is an l_1 norm of **w**, which stands for the sum of the absolute values of w_k. If we focus only on the point that has the highest posterior probability (maximum a posteriori, MAP), the Bayesian updating reduces to the problem of minimizing the following objective function:

$$J_{\text{Glasso-BFs}} = \frac{1}{2}\|\mathbf{t} - \Phi\mathbf{w}\|_2^2 + \lambda\|\mathbf{w}\|_1 \tag{4.15}$$

where $\lambda = \kappa/\sigma^2$ is called the regularization parameter or hyperparameter that controls the relative intensity between the data (likelihood) term and the prior term. Higher values of λ lead to more sparsity, while lower values lead to less sparsity and potentially more overfitting. The role of *sparse* prior in the estimation of $f(\mathbf{z})$ for a depth profile **t** is illustrated in Figure 4.5. Synthetic profiles of **t** were generated using the following model:

$$t_i = 20 + 10 \times \cos(4\pi \cdot z_i) + \varepsilon(z_i) \tag{4.16}$$

where ε is the spatial variability with standard deviation $\sigma = 5$ and scale of fluctuation (SoF) $\delta = 0.001$. Shifted Legendre polynomials of order up to 20 ($M = 20$) are used for $f(\mathbf{z})$ in Glasso-BFs. Figure 4.5a compares a realization of **t** and the estimated model $f(\mathbf{z})$ (thick continuous line). Figure

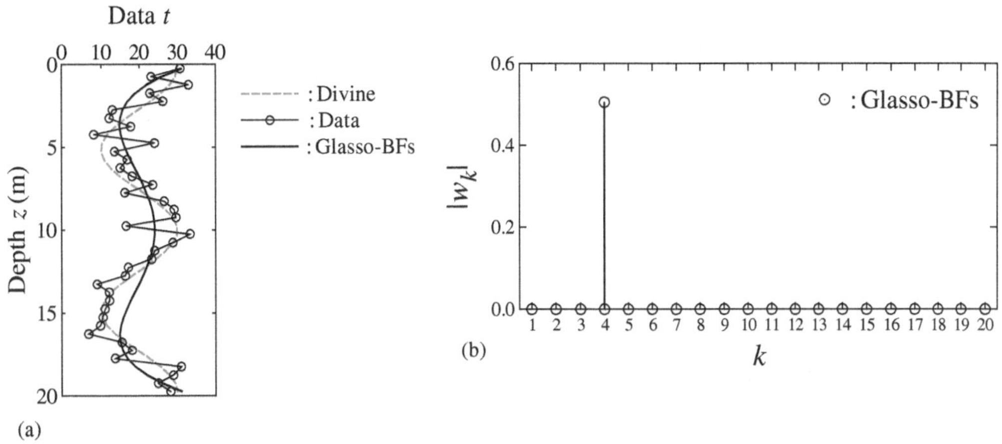

FIGURE 4.5 Results of Glasso-BFs for 1D depth profiles: (a) estimated trend model $f(\mathbf{z})$; (b) absolute values of estimated coefficients $|w_k|$.

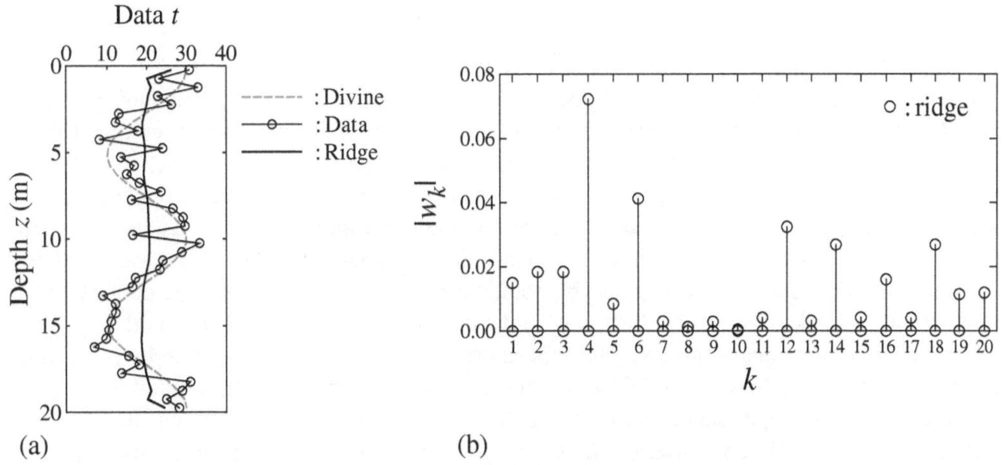

FIGURE 4.6 Results of ridge regression for 1D depth profiles: (a) estimated trend model $f(\mathbf{z})$; (b) absolute values of estimated coefficients $|w_k|$.

4.5b shows the absolute values of the estimated coefficients w_k. The *sparse* prior defined by Eq. (4.14) works well, and most of the coefficients w_k get to zero. Although only one coefficient ($k = 4$) is active, this simple model can capture the data trend well. Figure 4.6 shows the result by ridge regression which has been commonly used in curve fitting to limit overfitting. In ridge regression, underfitting occurs (Figure 4.6a), and most of the coefficients are active or non-zeros (Figure 4.6b). A comparison of the two models, *sparse* and *non-sparse*, tells us that (1) increasing the model complexity does not always contribute to an improvement in prediction accuracy and may even lead to overfitting and (2) the most important features for predicting data $\mathbf{t}$ can be identified using the sparse prior.

4.4.1.2 Gaussian Process Regression

GPR (e.g., Rasmussen and Williams 2006), also known as kriging (Matheron 1963, 1973), is a Bayesian non-parametric method for regression and interpolation. The key idea behind GPR is to model the unknown function or model f as a Gaussian process, which is a collection of random variables, any finite number of which have a joint Gaussian distribution. A Gaussian process is fully specified by a mean function (usually set to zero, **0**) and a covariance function **C**, also known as the *kernel*. The *kernel*, which is also called the autocorrelation function in random field simulation, determines the shape and smoothness of the function (e.g., Cami et al. 2020). Some common kernel functions used in GPR are

Radial basis function kernel:
$$k(z_i, z_j) = \exp\left(-\frac{\|z_i - z_j\|^2}{2\theta}\right) \tag{4.17}$$

Exponential kernel:
$$k(z_i, z_j) = \exp\left(-\frac{d}{\theta}\right) \tag{4.18}$$

Matérn kernel:
$$k(z_i, z_j) = \frac{2^{1-\nu}}{\Gamma(\nu)}\left(\frac{\sqrt{2\nu}d}{\theta}\right)^{\nu} K_\nu\left(\frac{\sqrt{2\nu}d}{\theta}\right) \tag{4.19}$$

White noise kernel:
$$k(z_i, z_j) = \sigma^2 \times \delta_{ij} = \theta_0 \times \delta_{ij} \tag{4.20}$$

where ν is the smoothness parameter; Γ is the gamma function; K_ν is the modified Bessel function of the second kind; d is $|z_i - z_j|$; θ is the hyperparameter or SoF; and σ (or θ_0) is the noise level.

GPR works by placing a prior distribution over the unknown function, and then updating this prior with observed data to obtain a posterior distribution. The posterior distribution reflects the updated belief about the function given the observed data, and it can be used to make predictions at new input points, along with uncertainty estimates. GRP is highly advantageous because it gives closed-form expressions for predictions and uncertainty estimates, making it a powerful tool for modeling complex data. However, it can be computationally expensive for large datasets, as the computational complexity grows cubically with the number of data points. Some methodologies to deal with the computational complexity have been proposed and can be seen in Quiñonero-Candela and Rasumusse (2005), Snelson and Ghahramani (2005), and Wilson and Nickisch (2015).

Figure 4.7 shows an application example of GPR with the superposition of exponential and white noise kernels for the 1D depth profile. The thick line and dashed lines indicate the mean function and 95% confidence interval (CI), respectively. The mean function captures the data trend well and all the data fall within the CI.

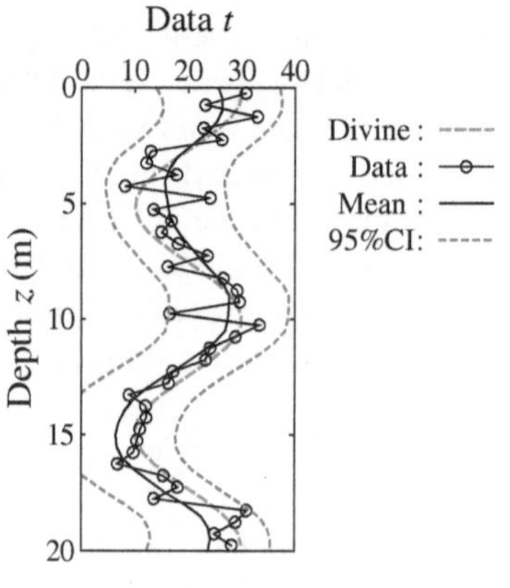

FIGURE 4.7 Result of GPR.

4.4.2 Category 2: Numerical Data – Geologic Uncertainty

The methods in Category 2 are designed for numerical data and geologic uncertainty, and are labeled DDSC-C2. DDSC-C2 includes cluster analysis or clustering-based methods which have been used for stratigraphic delineation (e.g., Hegazy and Mayne 2002; Facciorusso and Uzielli 2004; Liao and Mayne 2007). The goal of clustering is to partition the dataset into some number of clusters, and the "clusters" correspond to "soil types" or "strata" in the context of DDSC. A few representative methods are outlined below.

4.4.2.1 *K*-means Clustering

The most basic algorithm for clustering is the *K*-means algorithm (Lloyd 1982). A *K*-means algorithm defines a center of cluster *k*, as $\boldsymbol{\mu}_k$, and finds an assignment of data points to clusters such that the sum of the squares of the distances of each data point to its closest vector $\boldsymbol{\mu}_k$ is a minimum:

$$J = \sum_{i=1}^{m} \sum_{k=1}^{K} r_{ik} \left\| \mathbf{x}_i - \boldsymbol{\mu}_k \right\|^2 \tag{4.21}$$

where m is the number of data; K is the number of clusters; and r_{ik} is an indicator variable. The result of *K*-means clustering for a dataset is shown in Figure 4.8. This dataset is known as the "Old Faithful" dataset and comprises 272 measurements of the eruption of the Old Faithful geyser at Yellowstone National Park in the United States (e.g., Bishop 2006). The *K*-means algorithm reasonably divides the data into two clusters.

Clustering can be applied to soil stratification problem which is shown in Figure 4.9. The profiles were recorded in the Kobe Airport reclamation project in 1997. Although the depth profile of the soil types was already obtained from the site investigation, this real case history is used for methodological validation. In addition to two soil properties (natural water content w_n and compression index C_c), corresponding spatial coordination information, depth z (m) in this problem, should be used for reasonable clustering. Since the soil types are available, the true clusters can be visualized in 3D space (Figure 4.10). Simple *K*-means clustering was applied to the dataset, and whether the *K*-means algorithm can detect true clusters in real data was investigated.

Figure 4.11 compares the estimated clusters with the true ones on the $w_n - C_c$ plane. The number of clusters was determined using the elbow method. Even a simple *K*-means algorithm reasonably categorizes the data into five clusters, and the identified clusters are almost the same as the true clusters. Clustering methods are often used together with the DDSC-C1 methods for a more reasonable site characterization (e.g., Zhao and Wang 2020; Ching and Yoshida 2023).

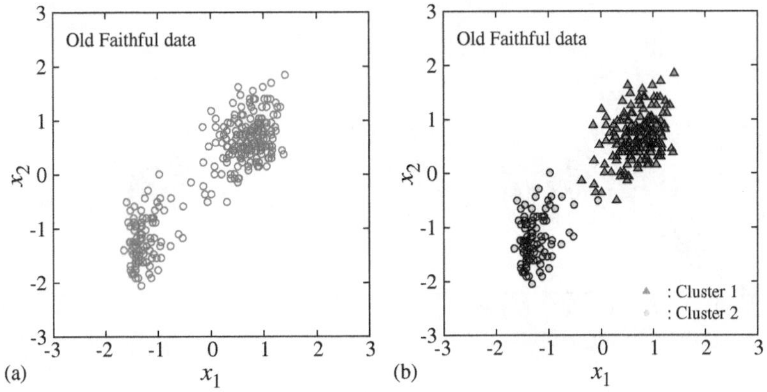

FIGURE 4.8 *K*-means clustering: (a) Old Faithful data; (b) clustered data.

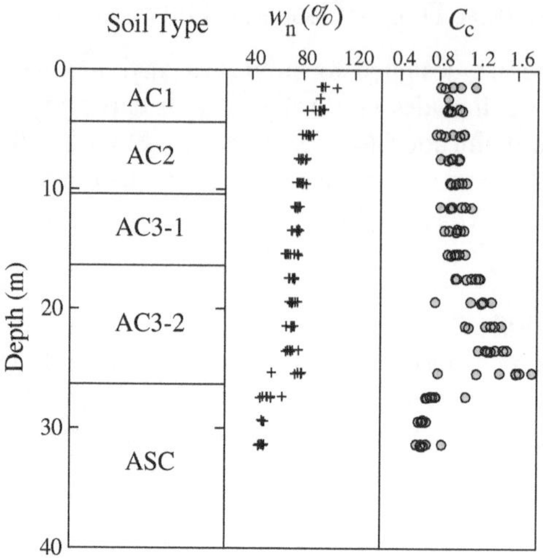

FIGURE 4.9 Soil profile recorded for Kobe Airport reclamation project site.

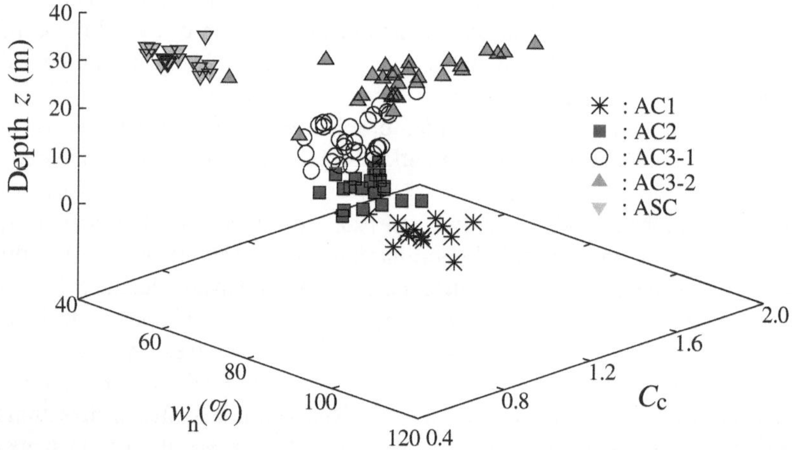

FIGURE 4.10 True clusters in 3D (w_n–C_c–z) space.

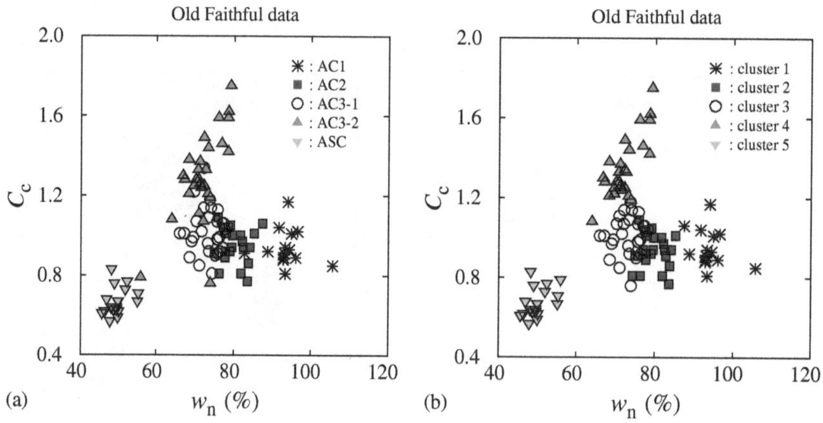

FIGURE 4.11 Comparison between true clusters and estimated clusters: (a) true clusters; (b) result of *K*-means clustering.

4.4.2.2 Gaussian Mixture Model (GMM)

One notable feature of the K-means algorithm is that every data point is uniquely assigned to only one of the clusters. This is called "hard assignment". Geotechnical data usually includes uncertainties, and it may not be reasonable to employ such hard assignment in soil stratification. By adopting a probabilistic approach, we can obtain "soft" assignments of data points to clusters in a way that reflects the level of uncertainty over the most appropriate assignment. The most common probabilistic clustering algorithm is the Gaussian mixture model (e.g., Bishop 2006), which models the data as a mixture of several K Gaussian distributions, each with its own mean $\boldsymbol{\mu}_k$ and covariance matrix $\mathbf{C}_k$. The mixture of Gaussian models is given by

$$p(\mathbf{x}) = \sum_{k=1}^{K} \pi_k \mathcal{N}(\mathbf{x} \mid \boldsymbol{\mu}_k, \mathbf{C}_k) \tag{4.22}$$

where the parameters π_k are called the mixing coefficients. The number of mixture components in a GMM can be determined using some criterion, such as the Bayesian information criterion (BIC) or the Akaike information criterion (AIC), or using a validation set. Figure 4.12 shows an application example of GMMs to the Old Faithful data.

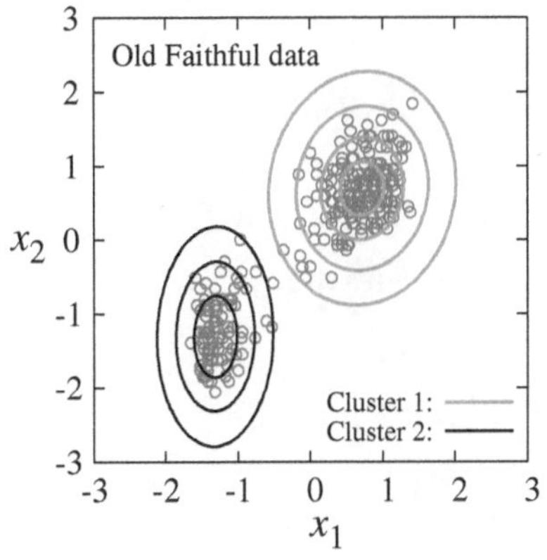

FIGURE 4.12 Example of GMM clustering for the Old Faithful data.

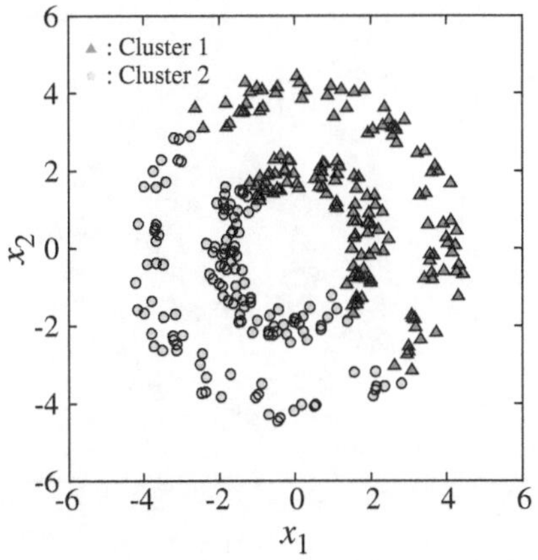

FIGURE 4.13 Example of linearly inseparable data.

4.4.2.3 Kernel Methods

K-means clustering finds the optimal "linear" separation of data points. It can only work if clusters can be linearly separated. It is clear that non-linear decision boundaries can be defined for the data shown in Figure 4.13. In the figure, two clusters detected by the *K*-means algorithm are shown. However, if the mapping $(x_1, x_2) \mapsto (z_1, z_2, z_3) = (x_1^2, x_2^2, \sqrt{2}x_1 x_2)$ is used, the data can be linearly separable in the 3D space (Figure 4.14). Although this approach seems promising, finding an appropriate mapping for future data is not easy. The *kernel trick* allows us to project our data into a higher-dimensional space to achieve linear separability and solve the *K*-means problem in a more efficient way. Figure 4.15 shows the results of kernel *K*-means clustering for the data in Figure 4.13. In the analysis, the RBF kernel (Eq. [4.17]) was used. An extensive review of kernel clustering can be found in Fillippone et al. (2008).

4.4.3 Categories 1 and 2: Numerical Data – Geotechnical and Geologic Uncertainty

The estimation of model *f* (also called trend estimation) and stratigraphic delineation are two different tasks and are usually solved independently using different ML algorithms designed for each task. As mentioned in Section 4.4.2, clustering is often used together with the methods of DDSC-C1 (e.g., Zhao and Wang 2020; Ching and Yoshida 2023) to accomplish both trend estimation and soil

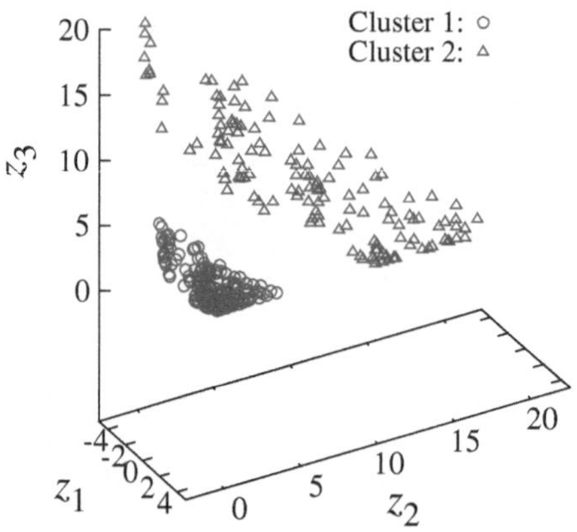

FIGURE 4.14 Result of the mapping $(x_1, x_2) \mapsto (z_1, z_2, z_3) = (x_1^2, x_2^2, \sqrt{2}x_1 x_2)$.

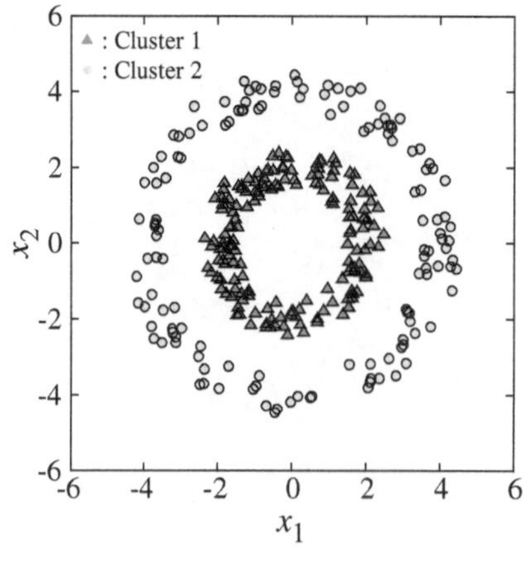

FIGURE 4.15 Result of kernel *K*-means clustering.

stratification. Geotechnical lasso (Glasso, Shuku and Phoon 2021), however, makes it possible to accomplish these two tasks simultaneously in a unified theoretical framework. Glasso is based on a machine learning method called l_1 trend filtering (Kim et al. 2009) which is another form of lasso. Glasso can be interpreted as the combined approach of DDSC-C1 and DDSC-C2 methods and can evaluate both geotechnical and geologic uncertainties.

Although both Glasso-BFs and Glasso are based on lasso, they employ a different model f: while Glasso-BFs use continuous basis functions, Glasso uses discrete basis functions which are defined by

$$f(\mathbf{z}) = \mathbf{f} = \left[f_1, f_2, \ \ldots \ , f_n\right]^{\mathrm{T}} \tag{4.23}$$

where f_i is a piecewise constant function, and subscript i corresponds to the depth z_i. In this method, the computational domain or target ground is discretized into cells or meshes, and the number of unknown parameters corresponds to the number of cells or meshes (Figure 4.16).

Glasso uses the following Laplace PDF as the prior:

$$p(\mathbf{f} \mid \kappa) \propto \exp\left(-\kappa\left\|\mathbf{Bf}\right\|_1\right) \tag{4.24}$$

where κ is the diversity parameter, and $\| \ \|_1$ is an l_1 norm of $\mathbf{Bf}$. The matrix $\mathbf{B}$ imposes some "structured sparsity" on the model $\mathbf{f}$, which seeks to explain $\mathbf{t}$ using the minimum number of "jumps" or "knots" when f is modeled as a spatial series of piecewise constant functions. Two types of matrices $\mathbf{D}_1$ and $\mathbf{D}_2$ are often used as the $\mathbf{B}$ matrix to induce the structured sparsity:

$$\mathbf{D}_1 = \begin{bmatrix} -1 & 1 & 0 & \cdots & 0 & 0 \\ 0 & -1 & 1 & \cdots & 0 & 0 \\ \vdots & \vdots & \vdots & \ddots & \vdots & \vdots \\ 0 & 0 & 0 & \cdots & 1 & 0 \\ 0 & 0 & 0 & \cdots & -1 & 1 \end{bmatrix} \tag{4.25}$$

$$\mathbf{D}_2 = \begin{bmatrix} -1 & 2 & -1 & \cdots & 0 & 0 & 0 \\ 0 & -1 & 2 & \cdots & 0 & 0 & 0 \\ \vdots & \vdots & \vdots & \ddots & \vdots & \vdots & \vdots \\ 0 & 0 & 0 & \cdots & 2 & -1 & 0 \\ 0 & 0 & 0 & \cdots & -1 & 2 & -1 \end{bmatrix} \tag{4.26}$$

The matrices $\mathbf{D}_1$ and $\mathbf{D}_2$ enforce structural sparsity associated with a constant trend and linear trend of the $\mathbf{f}$ vector, respectively. In image processing, $\mathbf{D}_1$ and $\mathbf{D}_2$ are known as a gradient operator and a Laplace operator. These matrices can be generalized to any higher-order difference (Tibshirani 2014).

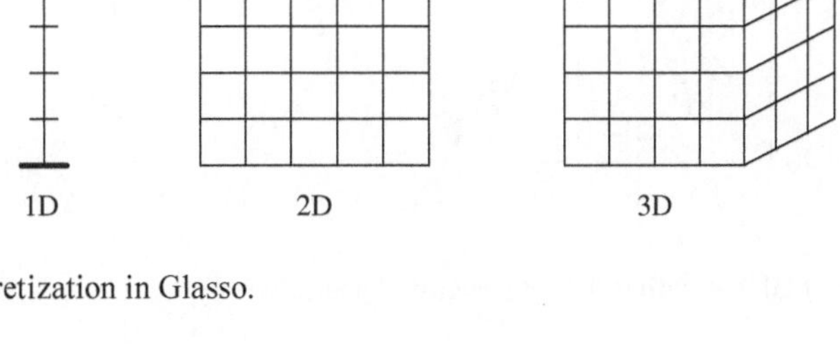

FIGURE 4.16 Discretization in Glasso.

If we focus solely on the **f** that has the highest posterior probability, i.e., MAP, Bayesian updating reduces to the problem of minimizing the following objective function:

$$J_{\text{Glasso}} = \frac{1}{2}\left\|\mathbf{t} - \mathbf{Af}\right\|_2^2 + \lambda\left\|\mathbf{Bf}\right\|_1 \quad (4.27)$$

where **A** is a matrix representing a linear operator with entries that are either 0 or 1, and $\lambda = \kappa/\sigma^2$ is the regularization parameter. Examples of Glasso with different matrices ($\mathbf{D}_1$, $\mathbf{D}_2$, and $\mathbf{D}_3$) are shown in Figure 4.17. In the figure, "Divine" shows the depth profile data **t** and the true trend (Eq. [4.16]). The abrupt changes in the trend correspond to *active* coefficients and can be interpreted as layer boundaries. By minimizing the objective function (Eq. [4.27]), the model *f* that reasonably captures the data trend and the depth of layer boundaries can be detected in a unified theoretical framework.

One of the main disadvantages of Glasso includes high computation cost and high memory consumption. Although there are some sophisticated algorithms to minimize the lasso-type objective functions (e.g., Boyd et al. 2010; Shuku and Phoon 2021), there might still be a difficulty in application for practical large-scale problems.

4.4.4 Category 3: Categorical Data – Geologic Uncertainty

Site characterization for categorical data often means estimating geological or stratigraphic cross sections/maps. Stratigraphic maps can help identify potential hazards such as faults, fractures, and water-bearing formations that could impact the stability of slopes, excavations, and foundations. DDSC methods for categorical data and geologic uncertainty are classified as Category 3 and are labeled DDSC-C3. Different types of statistical or machine learning models are used in the DDSC-C3 methods such as the coupled Markov chain model (CMC, e.g., Krumbein 1967; Elfeki and Dekking 2001; Qi et al. 2016; Li et al. 2016), the Markov random field model (e.g., Geman and Geman 1987 Besag 1986; Greig et al. 1989; Li et al. 2016; Wang et al. 2016, 2017), multiple point statistics (MPS, e.g., Guardiano and Srivasta 1993; Mariethoz and Caers 2014; Shi and Wang 2021a), and convolutional neural networks (CNN, e.g., Fukushima 1980; Lecun et al. 1998; Shi and Wang 2021b). This section mainly focuses on the first two statistical models, CMC and MRF, because they have been widely used in geological modeling due to their simple theory and ease of implementation.

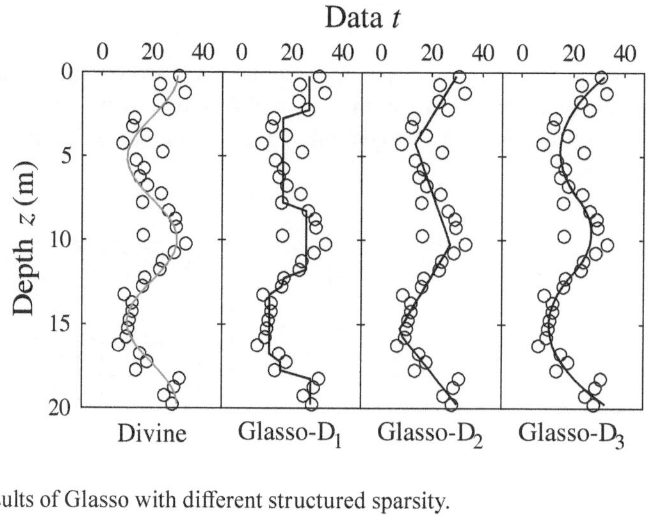

FIGURE 4.17 Results of Glasso with different structured sparsity.

4.4.4.1 Coupled Markov Chain Simulation

A Markov chain is a stochastic model describing a sequence of possible events/states in which the probability of each event/state depends only on the state attained in the previous event/state. The idea of using the CMC model for subsurface modeling was first introduced into geologic profiling by Krumbein (1967). After this pioneering work, Elfeki and Dekking (2001) extended the method to more practical 2D problems based on coupled CMC models.

An example of a graphical model for the Markov chain model with three states (gravel, sand, and clay) is shown in Figure 4.18. The numbers indicate the probabilities of changing from one state to another state and are called "transition probabilities". All the probabilities of transitioning between these states can be represented by a single matrix called a transition matrix:

$$\mathbf{T} = T(i,j) = \begin{bmatrix} 0.7 & 0.3 & 0.0 \\ 0.4 & 0.4 & 0.2 \\ 0.3 & 0.3 & 0.4 \end{bmatrix} \begin{matrix} \text{G} \\ \text{S} \\ \text{C} \end{matrix} \quad (4.28)$$
$$\begin{matrix} \text{G} & \text{S} & \text{C} \end{matrix}$$

With this setup and Figure 4.18, the probability of state of x_{i+1} = Gravel (G) given x_{i-1} = Silt (S), $p(x_{i+1}|x_{i-1})$ can be computed as follows:

$$\begin{aligned} & p(x_{i+1} = \text{G} \mid x_{i-1} = \text{S}) = p(x_i = \text{G} \mid x_{i-1} = \text{S}) \times p(x_{i+1} = \text{G} \mid x_i = \text{G}) + \\ & p(x_i = \text{S} \mid x_{i-1} = \text{S}) \times p(x_{i+1} = \text{G} \mid x_i = \text{S}) + \\ & p(x_i = \text{C} \mid x_{i-1} = \text{S}) \times p(x_{i+1} = \text{G} \mid x_i = \text{C}) \\ & \quad = T(2,1)T(1,1) + T(2,2)T(2,1) + T(2,3)T(3,1) \\ & \quad = 0.4 \times 0.7 + 0.4 \times 0.4 + 0.2 \times 0.3 \\ & \quad = 0.50 \end{aligned} \quad (4.29)$$

Once a transition probability matrix is given, the Markov chain model can be used to predict unknown states and analyze the behavior of the system, i.e., the spatial patterns of the soil types. The images simulated by CMC models are shown in Figure 4.19. Three states were considered in the CMC simulations, and the following two transition probability matrices were used:

$$\mathbf{T}_1 = \begin{bmatrix} 0.33 & 0.33 & 0.33 \\ 0.33 & 0.33 & 0.33 \\ 0.33 & 0.33 & 0.33 \end{bmatrix}, \mathbf{T}_2 = \begin{bmatrix} 0.7 & 0.3 & 0.0 \\ 0.0 & 0.7 & 0.3 \\ 0.3 & 0.0 & 0.7 \end{bmatrix} \quad (4.30)$$

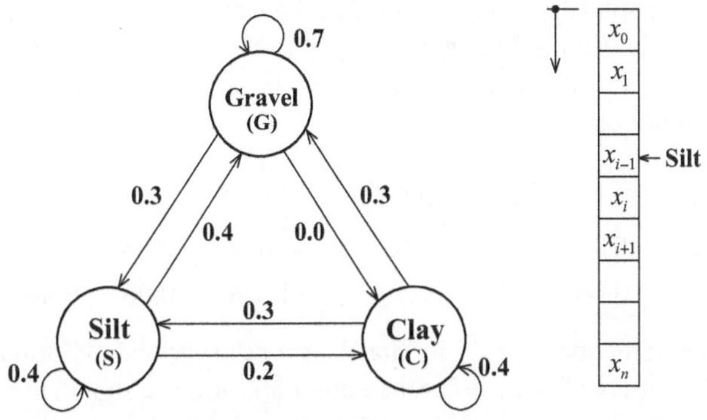

FIGURE 4.18 Schematic of 1D Markov chain model with three states.

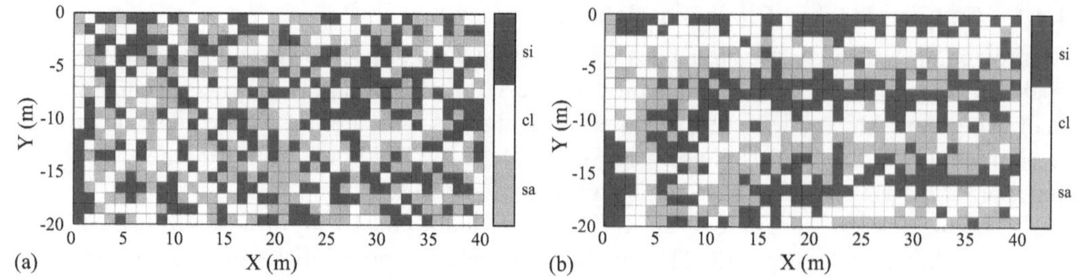

FIGURE 4.19 Images generated by three states of the CMC model: (a) $\mathbf{T}_1$; (b) $\mathbf{T}_2$.

Only the right-side cells have fixed states to initiate the CMC simulation. The states are randomly generated based on the transition probability (Eq. [4.30]). It is clear that different images are obtained depending on the transition probability.

In practice, a conditional Markov chain needs to be considered because site-specific data is available at some locations. The probability of state at x_i given the states at x_{i-1} and x_n, $p(x_i|x_{i-1}, x_n)$ can be defined by

$$p(x_i \mid x_{i-1}, x_n) = \frac{p(x_{i-1}, x_i, x_n)}{p(x_{i-1}, x_n)} = \frac{p(x_n \mid x_{i-1}, x_i)p(x_{i-1}, x_i)}{p(x_{i-1}, x_n)} \tag{4.31}$$

By applying the Markov property on the conditional probability in the numerator of Eq. (4.31), one obtains:

$$p(x_i \mid x_{i-1}, x_n) = \frac{p(x_n \mid x_i)p(x_{i-1}, x_i)}{p(x_{i-1}, x_n)} = \frac{p(x_n \mid x_i)p(x_i \mid x_{i-1})p(x_{i-1})}{p(x_n \mid x_{i-1})p(x_{i-1})} \tag{4.32}$$

Elfeki and Dekking (2001) extended a 1D Markov chain simulation to a coupled Markov chain on 2D lattice domains and proposed a very simple and efficient algorithm to approximately simulate a conditional CMC. For 2D (or 3D) problems, different transition probability matrices should be defined for the vertical and horizontal directions to consider the realistic heterogeneity of natural grounds. After this pioneering work, numerous researchers were inspired to apply CMC to DDSC, especially evaluating geologic uncertainty (e.g., Deng et al. 2017). Park (2010) discussed an extension of 2D CMC models to more practical 3D settings and applied the 3D CMC to a real case history. From a Bayesian perspective, conditioning data (x_{i-1} and x_n in Eq. [4.32]), the CMC model, and transition probability matrices $\mathbf{T}$ correspond to data vector $\mathbf{t}$ without observation noise, model f, and the (hyper)parameter of the model $\boldsymbol{\theta}$ in Eq. (4.3), respectively.

Stratigraphic cross sections estimated by the CMC model are shown in Figure 4.20. Four borehole logs (BH1–4) and the soil types of the ground surface are assumed to be data $\mathbf{t}$. The simulation is proceeding through cells from left to right, and top to bottom. The data $\mathbf{t}$ is assumed to have no observation error or "perfect information" for simplicity. The following transition matrices were used for the CMC simulations:

$$\mathbf{T}^{\mathrm{v}} = \begin{bmatrix} 0.83 & 0.17 & 0.00 \\ 0.11 & 0.87 & 0.02 \\ 0.00 & 0.11 & 0.89 \end{bmatrix}, \mathbf{T}^{\mathrm{h}} = \begin{bmatrix} 0.91 & 0.09 & 0.00 \\ 0.06 & 0.93 & 0.01 \\ 0.00 & 0.06 & 0.94 \end{bmatrix} \tag{4.33}$$

where $\mathbf{T}^{\mathrm{v}}$ and $\mathbf{T}^{\mathrm{h}}$ are the transition probability matrices for the vertical "v" and horizontal "h" directions, respectively. The results of three different simulation runs are shown to investigate the effects of random sampling on the estimated cross sections. Conditioned CMC simulations can generate realistic cross sections based on the available boring logs. All three results are not the same due to

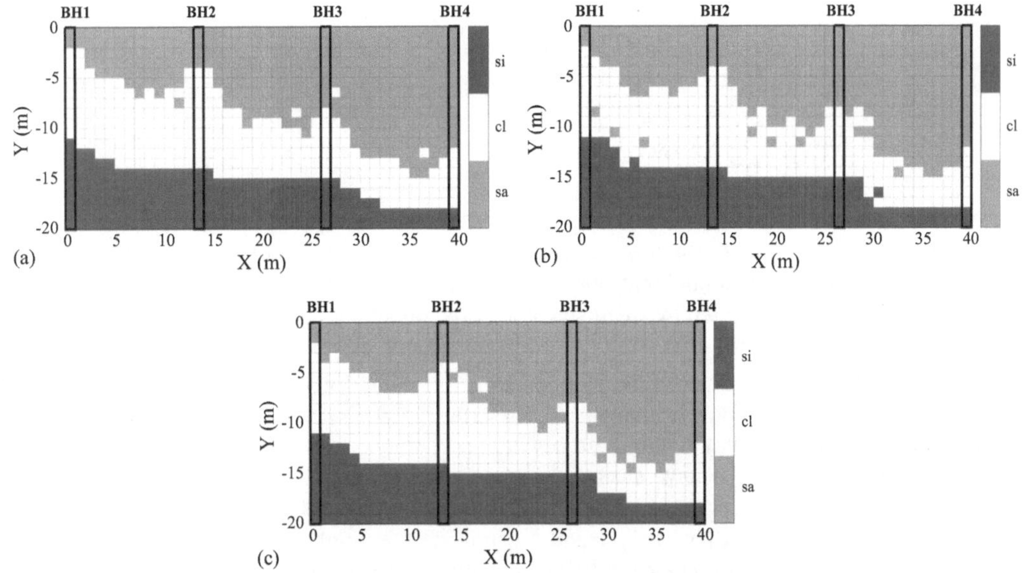

FIGURE 4.20 Simulated geological cross sections using a conditioned CMC.

random sampling. Geologic uncertainty and some statistical parameters of the estimation such as the mean, mode, and standard deviation can be quantified by combining the CMC simulation with the Monte Carlo simulation.

While the vertical transition probability matrix $\mathbf{T}^v$ can be easily estimated from borehole logs (e.g., Elfeki and Dekking 2001), the horizontal transition probability matrix $\mathbf{T}^h$ is difficult to determine because horizontal spaces between boreholes are usually wide, at least wider than 10 m in practice. To overcome the difficulty, some approaches were proposed to estimate a reasonable horizontal transition probability matrix with limited borehole data (e.g., Qi et al. 2016; Zhang et al. 2022).

4.4.4.2 Markov Random Field Model

An MRF is a class of graphical model that expresses the conditional dependence structure between random variables. There are two types of graphical models: MRFs (undirected graphical models) and Bayesian networks (directed graphical models). While directed models are used for causal inference and sequential data analysis, undirected models are used in cases where the data does not have any direction or causal relations among it, e.g., image processing for image restoration, image recovery, and noise removal (e.g., Geman and Geman 1987; Besag 1986; Greig et al. 1989). Since geological cross sections can be considered images, an MRF can also work in subsurface modeling. Geological modeling based on an MRF model has been studied by many researchers (e.g., Norberg et al. 2002; Li et al. 2016; Wang et al. 2016, 2017). This section outlines the MRF-based method called GPotts proposed by Shuku and Phoon (2023b) as an example. The GPotts is based on a Potts model, which is a form of MRF model, and the advantages of their method over existing methods include (1) only a few site-specific boreholes on soil types are required for the modeling. Other geological prior knowledge such as geological cross sections interpreted by geologists, the orientation of strata, and stratigraphic dips are not necessary. (2) Spatial correlation is implicitly considered in the Potts model, and the intensity of the correlation can be controlled by tuning two hyperparameters in the model. Additional models for spatial correlations are not necessary. (3) The proposed method or algorithm is directly applicable to 3D problems without any methodological changes.

In GPotts, the probability of state of $\mathbf{x}$ is defined by

$$p(\mathbf{x}) = \frac{1}{Z}\exp\{-E(\mathbf{x})\} \tag{4.34}$$

$$E(\mathbf{x}) = \theta\sum_{j=1}^{n}\delta(x_k, x_l), x_k = 0,1,\ldots,q-1, \delta(x_k, x_l) = \begin{cases} 1 & \text{for } x_k = x_l \\ 0 & \text{for } x_k \neq x_l \end{cases} \tag{4.35}$$

where Z is a normalizing constant called a partition function; θ is a hyperparameter called "inverse temperature", δ is Dirac's delta function; and x_k is a "state" or category at site k. The energy function Eq. (4.35) is called Potts model that can consider many numbers of states q. Figure 4.21 shows examples of 2D images generated by an anisotropic Potts model with three states. The "states" correspond to soil types, with "sa", "cl", and "si" indicating sand, clay, and silt, respectively. Clearly, the Potts model can generate a wide variety of complex patterns and structures of soil types by adjusting the hyperparameters. The concept of the hyperparameters θ_h and θ_v are identical to the horizontal and vertical correlation length (or SoF).

In subsurface modeling, we have an interest in estimating geological cross sections based on limited site-specific data $\mathbf{t}$. Let us consider an MRF model with random variable $\mathbf{x}$ on the lattice $L\{(i, j): 1 \leq i \leq N_X, 1 \leq j \leq N_Y\}$ where N_X and N_Y indicate the number of cells for horizontal and vertical directions respectively. The conditional probability of $\mathbf{x}$ given $\mathbf{t}$ is defined by

$$p(f(\mathbf{x})\,|\,\mathbf{t}) = \frac{1}{Z}\exp\{-E(f(\mathbf{x})\,|\,\mathbf{t})\} \tag{4.36}$$

$$\begin{aligned} E(f(\mathbf{x})\,|\,\mathbf{t}) &= h\sum_{i=1}^{m}\Phi(x_i, t_i) \\ &+ \sum_{i=1}^{N_X}\sum_{j=1}^{N_Y}\Big[\theta_v\{\delta(x_{i,j}, x_{i,j-1}) + \delta(x_{i,j}, x_{i,j+1})\} \\ &+ \theta_h\{\delta(x_{i,j}, x_{i-1,j}) + \delta(x_{i,j}, x_{i+1,j})\}\Big] \end{aligned} \tag{4.37}$$

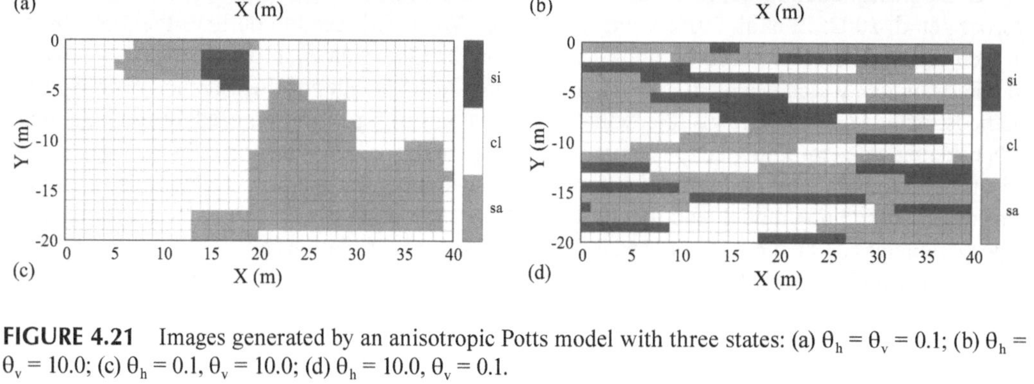

FIGURE 4.21 Images generated by an anisotropic Potts model with three states: (a) $\theta_h = \theta_v = 0.1$; (b) $\theta_h = \theta_v = 10.0$; (c) $\theta_h = 0.1, \theta_v = 10.0$; (d) $\theta_h = 10.0, \theta_v = 0.1$.

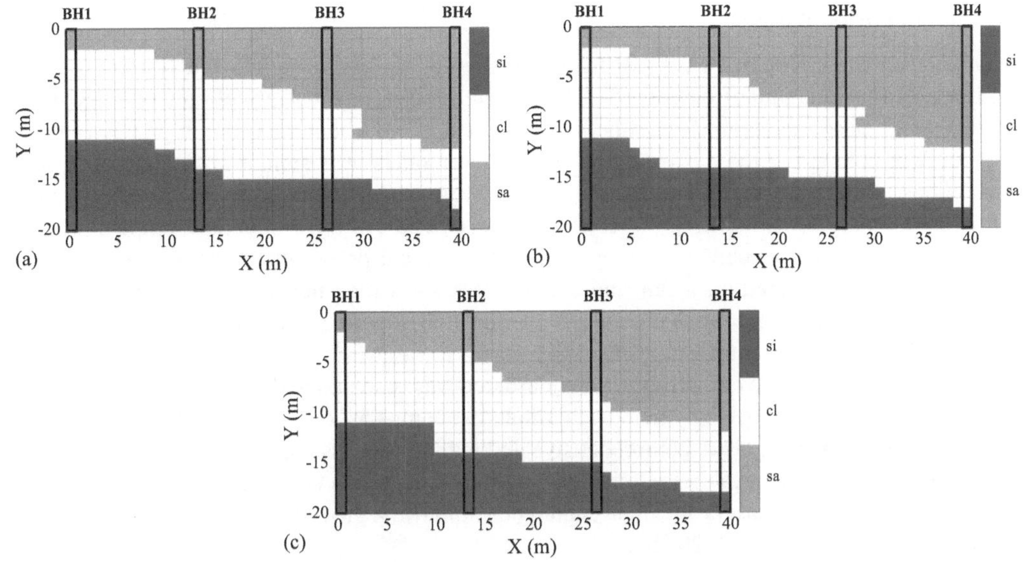

FIGURE 4.22 Geological cross sections simulated by conditioned CMC and MRF.

where h is a hyperparameter that controls the intensity of the likelihood term and Φ is a likelihood function.

Geological cross sections estimated using GPotts are shown in Figure 4.22. The numerical setup is the same as the simulation in Figure 4.20: we assume that only four borehole data and the soil types of the ground surface are available. The initial state of $\mathbf{x}$ is randomly sampled from the uniform distribution for each simulation and can impact on the estimation result. The results of three different simulation runs are shown in Figure 4.22. The hyperparameters $\theta_v = 0.1$ and $\theta_h = 10.0$ were used for simplicity. It is clear that GPotts can infer the 2D cross section well with only limited borehole data. The results are strongly influenced by the hyperparameters, and they need to be determined using an authorized method such as cross-validation (e.g., Bishop 2006; Shuku and Phoon 2023b).

4.4.4.3 Other DDSC-C3 Methods

Aside from CMC and MRF models, other statistical/machine learning models such as multiple-point statistics (MPS, Guardiano and Srivastava 1993; Mariethoz and Caers 2014) and convolutional neural networks (CNN, Fukushima 1980, Lecun et al. 1998) have been utilized in the DDSC-C3 methods. Geostatistical simulations based on conventional two-point statistics (variogram or covariance) are unable to evaluate complex and heterogeneous patterns. MPS simulation algorithms borrow high-order statistics from a visually and statistically explicit model, which is called a “training image” to overcome the limitation of two-point statistics. The training image can be seen as a prior model, and the prior is updated based on local observation data. In 1993, at the time when MPS was proposed, computational limitations made the MPS impossible to apply in real situations. This method, however, has recently attracted attention with the development of the computer and efficient algorithms (Renard and Mariethoz 2014). CNN is an emerging technology and has recently been actively utilized in geotechnical engineering (e.g., Phoon and Zhang 2023). Shi and Wang (2021b) proposed a method for underground stratification based on limited site-specific data and training images of geological cross sections using the eXtreme gradient boost (XGBoost, Chen and Guestrin 2016), which is an efficient algorithm for training CNN models. They demonstrated the method through numerical examples and real case histories.

4.5 BENCHMARKING

Benchmarking is the process of comparing and evaluating the performance of a system, product, or process against a standard or best practice. It can be used in a variety of contexts, including business, technology, and manufacturing. In the context of ML, benchmarking typically refers to the process of comparing the performance of different models on a standardized dataset or task, which allows researchers and practitioners to objectively evaluate and compare the effectiveness of different models. Benchmarking in ML is an important part of the model development process. By using standardized benchmark examples, researchers and practitioners can ensure that their models are being evaluated on a fair and consistent basis, which helps to drive innovation and progress in the field. Common benchmarks in machine learning include MNIST, CIFAR-10, ImageNet, COCO, and Stanford Question Answering Dataset (SQuAD). These benchmarks are widely used in ML research and provide a standardized way to evaluate the performance of models on specific tasks or datasets. They help to facilitate the comparison of different models and drive progress. DDSC, however, lacks comprehensive and publicly accessible benchmark examples that evaluate the strengths and weaknesses of DDSC methods in a balanced and unbiased way. The lack of common benchmark datasets and an accepted benchmarking procedure is impeding progress of the DDSC agenda (Phoon et al. 2022a).

Phoon et al. (2022c) proposed four standard benchmark examples for DDSC and a benchmarking procedure to measure the performance of the DDSC method in a balanced and unbiased way. The benchmark examples are based on the synthetic cone penetration test data (cone tip resistance q_t and sleeve friction f_s) and are plausible outcomes of some idealized stratigraphy (called virtual ground). Training and validation datasets are proposed to guide the demonstration of the performance to showcase: (1) ability to handle the attributes of real-world datasets such as MUSIC-3X directly, (2) generality over a range of ground conditions encountered in practice, (3) strengths and limitations compared to other methods using a common set of performance metrics, and (4) practicality in terms of computational resources needed to solve full-scale 3D problems. This section covers benchmark examples/benchmarking for DDSC methods.

4.5.1 Virtual Ground

In geotechnical engineering, benchmark datasets usually take one of three forms: actual data, experimental data, or synthetic data. Synthetic data has an advantage over actual data when it comes to validation because the "true" ground conditions are known everywhere. Four virtual ground benchmarking examples proposed by Phoon et al. (2022c) are summarized in Table 4.3 and Figure 4.23. The physical volume of the virtual ground for all the benchmark examples is defined by a 20 m long × 20 m wide × 10 m deep cuboid. This volume is discretized into 20 × 20 × 100 = 40,000 cells. The size of each cell is 1 m long × 1 m wide × 0.1 m deep. This geometrical

TABLE 4.3
Virtual Ground Benchmark Examples

	Description
S-VG1	Horizontal layers with constant property trend in each layer
S-VG2	Inclined layers with constant property trend in each layer
S-VG3	Inclined layers with linear property trend in each layer
S-VG4	Mixture of continuous and discontinuous layers with constant trend in each layer
S-VG5	Three-layer heterogeneous ground with constant property in each layer

Source: Phoon et al. (2022) and Shuku and Phoon (2023).

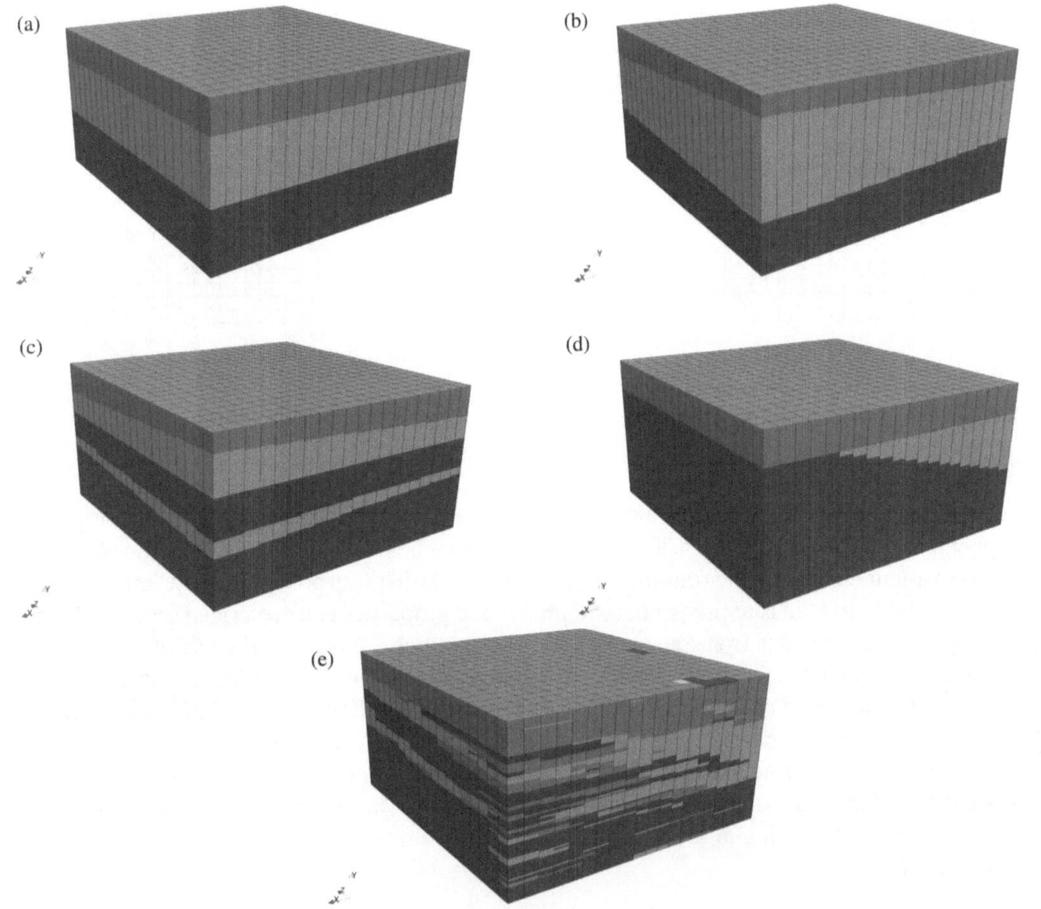

FIGURE 4.23 3D spatial distribution of SBT of five virtual grounds for benchmark examples: (a) S-VG1; (b) S-VG2; (c) S-VG3; (d) S-VG4; (e) S-VG5.

configuration is likely to be a lower band of the volume of ground encountered in practice. Hence, this virtual ground is labeled "small (S)".

All the virtual grounds consist of a top layer of sand, a middle layer of clay, and a base layer of silty sand. The boundary surfaces S are defined by

$$S(X_i, Y_j): aX_i + bY_j + cZ_k + d = 0 \tag{4.38}$$

where a, b, c, and d are scalar parameters that define the boundary surface, and (X_i, Y_j, Z_k) correspond to X, Y, and Z coordinates of the midpoint of the cells. In each layer, random samples of tip resistance q_t and sleeve friction f_s are generated from a bivariate normal distribution. The simulated q_t and f_s values are assigned to the midpoint of the cells and are assumed to be constant in each cell.

The four virtual grounds (S-VG1–4) are created based on a boundary-based approach in which stratigraphic boundaries are modeled by the explicit functions (Eq. [4.38]). This approach has limitations to simulate complex structures such as sand lenses and pockets in the ground. Shuku and Phoon (2023a) proposed a new benchmark example, S-VG5, using a category-based approach: CMC simulation. Clearly, S-VG5 has a more realistic looking heterogeneity than the S-VG1–4 models.

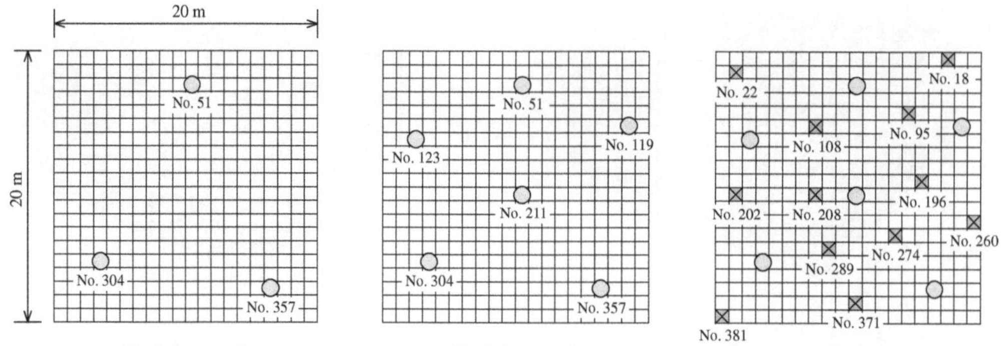

FIGURE 4.24 Locations of boreholes for training and validation sets in benchmarking.

4.5.2 Training and Validation Set

Everything is known within this virtual ground because it is generated by assuming a stratigraphy and populating the properties in each layer using a random field generator. However, the rule of the game is to assume that only the training dataset is available. The purpose of a DDSC model after it has been suitably trained is to predict the soil profile and properties at a prescribed set of validation soundings that are distinct from the set of training soundings. Note that the use of CPT data in the validation soundings is strictly restricted to cross-validation. The validation soundings are assumed to be "unobserved" and cannot be used for training. In Phoon et al. (2022c), the training and validation datasets consist of the depth profiles q_t, f_s, and I_c measured at various prescribed locations in the virtual ground (Figure 4.24). Both current and future benchmark examples (training and validation data sets) will be made available in Appendix 1 (available at https://www.routledge.com/Uncertainty-Modeling-and-Decision-Making-in-Geotechnics/Phoon-Shuku-Ching/p/book/9781032367491).

4.5.3 Performance Metrics

In benchmarking, it is important to use appropriate metrics that reflect the specific goals and requirements of the problem at hand. Different DDSC algorithms may perform better or worse on different metrics, and it is important to use multiple metrics and consider their relative importance in the context of the problem. Several performance metrics have been used to evaluate the performance of machine learning models, and some of them can be directly applicable to DDSC. The followings are some commonly used performance metrics:

Root mean square error (RMSE)/mean absolute error (MAE): The root mean square error and the mean absolute error are commonly used to measure the difference between predicted values $\hat{f}_i$ and actual values t_i in a regression analysis. They are calculated by

$$\text{RMSE} = \sqrt{\frac{1}{m}\sum_{i=1}^{m}\left(t_i - \hat{f}(z_i)\right)^2} \tag{4.39}$$

$$\text{MAE} = \frac{1}{m}\sum_{i=1}^{m}\left|t_i - \hat{f}(z_i)\right| \tag{4.40}$$

where m is the number of data. The RMSE is more commonly used in research and practice, particularly in situations where large errors have a more significant impact. However, the MAE has its own advantages, particularly when dealing with data with outliers, or in situations where the equal weighting of errors regardless of magnitude is preferred.

Accuracy/identification rate (IR): This measures the proportion of correctly classified estimations among all estimations. In other words, accuracy is the percentage of predictions made by a machine learning model that are correct. This metric is usually used to discuss the performance of soil stratification and is defined by

$$\text{Acc (or IR)} = \frac{1}{m}\sum_{i=1}^{m} I_i, I_i = \begin{cases} 1 & \text{for } t_i = \hat{f}_i \\ 0 & \text{for } t_i \neq \hat{f}_i \end{cases} \tag{4.41}$$

where m is the number of data and I is the indicator function. In DDSC, t_i and $\hat{f}_i$ are usually categorical data such as soil or rock types.

Coverage probability of confidence intervals: The coverage probability is a measure of how well the confidence intervals (CIs) cover the true data over multiple samples from the same population, which can be defined by

$$p(t_i \in \text{CI}) \tag{4.42}$$

It is usually expressed as a percentage, such as 95%, which indicates that, on average, 95% of the CIs computed from different samples will contain the data. The coverage probability gives an idea of how reliable these intervals are and helps practitioners to make more informed decisions based on the inference results.

Average width of a confidence interval: The width of a CI is the difference between its upper and lower limits, and it is a measure of the precision or uncertainty associated with the estimate or prediction. A narrower confidence interval indicates higher precision, while a wider confidence interval indicates more uncertainty. The average width of confidence intervals is the average of the widths of all confidence intervals across multiple samples, predictions, or estimates. In machine learning and statistical modeling, the average width of confidence intervals is often used to evaluate the performance and reliability of a model. When multiple models or parameter estimation methods are compared, a smaller width generally indicates that the method is more precise and provides more reliable inferences. It is important to note that the average width of confidence intervals should not be considered in isolation. It should be used in conjunction with other performance metrics, such as the coverage probability. Balancing the average width and coverage probability helps to ensure that the confidence intervals are both precise and reliable.

Information entropy: Information entropy, also known as Shannon entropy (Shannon 1948), is a measure of the average amount of information or uncertainty associated with a random variable. Information entropy quantifies the unpredictability or randomness of the possible outcomes, and it is used to assess the efficiency of communication systems, data compression algorithms, and various other applications in computer science, statistics, and machine learning. The information entropy is also widely used to quantify and visualize the uncertainties of inferred geological cross sections (e.g., Wellmanm et al. 2010, 2012), which can be defined as

$$H_l(x_l, i, j, k) = -\sum_{k=0}^{q-1} p_l(x_l)\log p_l(x_l) \tag{4.43}$$

where $p_l(x_l)$ is the occurrence probability of state x_l at (i, j, k)th location and q is the number of states or soil/rock types. A higher entropy value indicates a more uncertain or random variable, while a lower entropy value indicates a more predictable variable. This measure has been commonly used to check the performance of the DDSC-C3 methods.

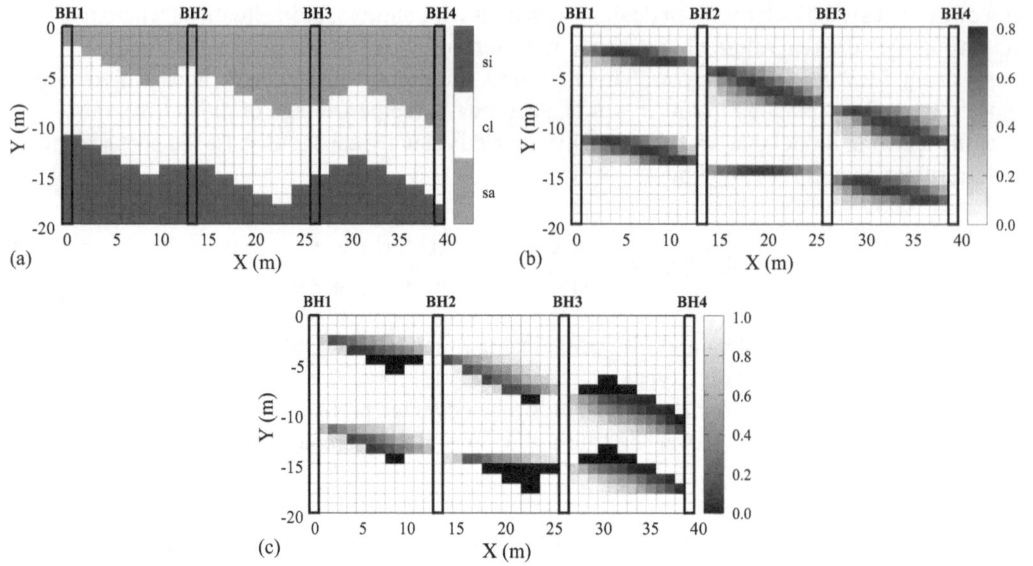

FIGURE 4.25 Information entropy map estimated using the GPotts model: (a) true geological cross section; (b) information entropy; (c) accuracy/identification rate.

An example of an information entropy map for a geological cross section is shown in Figure 4.25. Figure 4.25a shows the "true" cross section, and "BH" means available borehole data or training set for subsurface modeling. A geological cross section was estimated using the GPotts model with the training set, and the information entropy H_l and identification rate (or accuracy) were evaluated. Figures 4.25b and c show an information entropy map and IR map, respectively. Clearly, the layer boundaries tend to show high information entropy or low accuracy. This information on uncertainty can also be useful in practice such as planning site exploration programs and reliability-based design.

Runtime for training and validation (with PC specifications): This is not a direct performance metric for DDSC methods, it can be considered an indirect or practical performance. When deploying DDSC methods in real-world applications, practical aspects such as computational efficiency, resource usage, and runtime become important. A model that is highly accurate but requires a prohibitively long runtime or excessive computational resources may not be suitable for certain applications, especially those with real-time or near-real-time requirements. In such cases, comparing the runtime of different models on specific PC specs can provide valuable insights into their practical performance. This comparison can help practitioners make trade-offs between model accuracy and computational efficiency to find the best methods for their specific use case. It is important to note that runtime and computational efficiency are highly dependent on factors such as hardware, software, and the specific implementation of the algorithm. Therefore, when comparing the

TABLE 4.4
Time and Space Complexity

$O(1)$	Constant time complexity: Runtime/memory usage does not increase with the input size
$O(n)$	Linear time complexity: Runtime/memory usage grows linearly with the input size
$O(n^2)$	Quadratic time complexity: Runtime/memory usage grows quadratically with the input size

runtime of different models, it is essential to consider these factors and ensure that the comparison is fair and meaningful.

Computational complexity (Golub and van Loan 2012): Computational complexity is a measure of the amount of computational resources, such as time and memory, required to solve a problem or perform an algorithm. It is often expressed as a function of the input size, usually denoted by "n". There are two main ways to measure computational complexity: time complexity and space complexity.

Time complexity is a measure of the number of basic operations an algorithm takes to complete as a function of the input size. It represents how the runtime of the algorithm grows with the size of the input. The most common way to represent time complexity is using big-O notation, which provides an upper bound on the growth of the function. Space complexity is a measure of the amount of memory an algorithm uses as a function of the input size. It represents how the memory usage of the algorithm grows with the size of the input. Like time complexity, space complexity is often represented using big-O notation. A few examples of time and space complexity are summarized in Table 4.4.

4.5.4 Example of Benchmarking

As an example of benchmarking, the inference accuracy (RMSEs) and runtime of two DDSC methods, Glasso (with $\mathbf{D}_1$ model) and Glasso-BFs, were compared through a real ground benchmark example (RG1) that is based on actual CPT data collected at the South Parklands site in the city of Adelaide, South Australia (Jaksa 1995; Phoon et al. 2022c). Only the data of cone tip resistance q_t was used in benchmarking. Two types of training datasets, T1 and T2, were used for comparison. Details of the benchmarking setup can be seen in Phoon et al. (2022c).

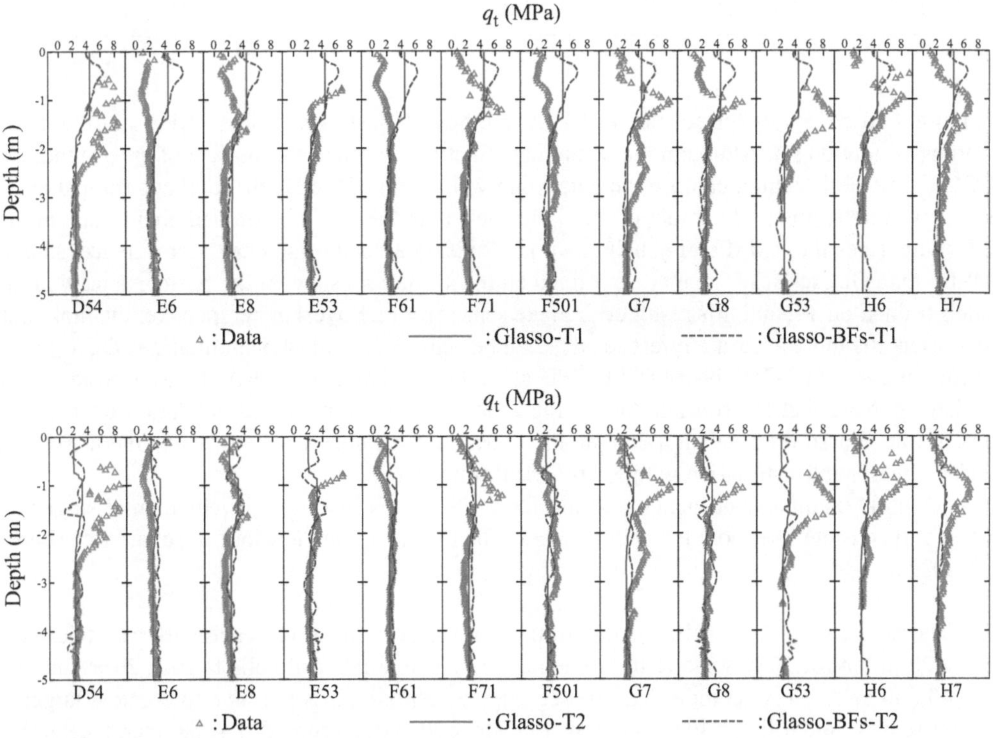

FIGURE 4.26 Results of benchmarking: (a) T1; (b) T2.

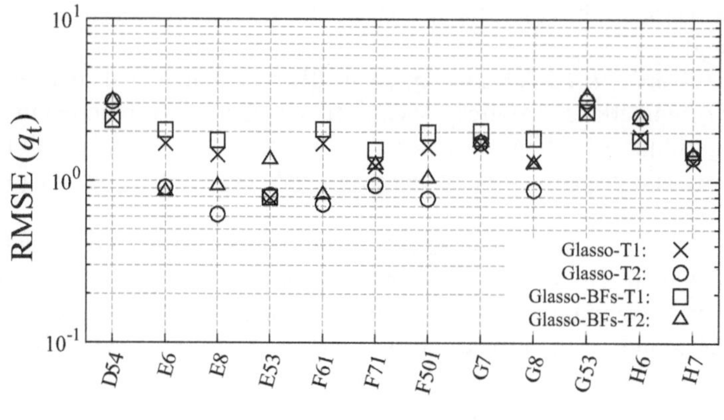

FIGURE 4.27 RMSE of q_t.

TABLE 4.5
Performance of Glasso and Glasso-BFs

Metric	Glasso-D$_1$-T2	Glasso-D$_1$-T2	Glasso-BFs-T1	Glasso-BFs-T2
Best RMSE	0.783	0.618	0.792	0.820
Median RMSE	1.629	0.922	1.909	1.314
Worst RMSE	2.662	3.136	2.683	3.338
Average RMSE	1.641	1.456	1.877	1.633
Runtime (s)	19,921	17,892	400	424

Note: PC specs: Intel Core i9-10980XE CPU at 3.00 GHz with 256GB RAM.

Figure 4.26 compares the results by DDSC methods (Glasso and Glasso-BFs) and the corresponding validation set. Although the actual data fluctuates around the surface of the ground, two DDSC-C1 methods cannot capture the data trend well. Both of the DDSC methods trained on T2 data show more simpler trends (almost straight lines) than the methods trained on T1 data. Figure 4.27 summarizes the RMSEs of q_t using two methods. Glasso shows better performance, smaller RMSEs, than Glasso-BFs. The models trained on the T2 dataset show better performance than the models trained on T1, and this result does make sense from a Bayesian perspective: the more data that is used, the more accurate inference that can be achieved. Table 4.5 summarizes the accuracy and runtime. An Intel Core i9-10980XE CPU at 3.00 GHz with 256GB RAM was used for all the simulations. Note that the runtime means the time for both training and validation or inference steps. While Glasso requires high computation cost (and memory consumption), Glasso-BFs can provide a reasonable estimation within a practical time frame.

Benchmark examples and benchmarking for DDSC are still in the developmental stage, and there is still considerable room for improvement. The followings are left for future research (Phoon et al. 2022c):

1. The physical volume of the virtual ground for the current set of benchmark examples is a 20 m long × 20 m wide × 10 m deep cuboid. The number of cells is 20 × 20 × 100 = 40,000 cells. This geometry can be regarded as "small". It is possible to create a larger virtual ground filling a 40 m long × 40 m wide × 20 m deep cuboid. The number of cells is 40 × 40 × 200 = 320,000 cells. For some DDSC methods, computation time and memory consumption are directly influenced by the number of cells and observation data. Scalable DDSC methods need to be developed in tandem with larger benchmark examples.

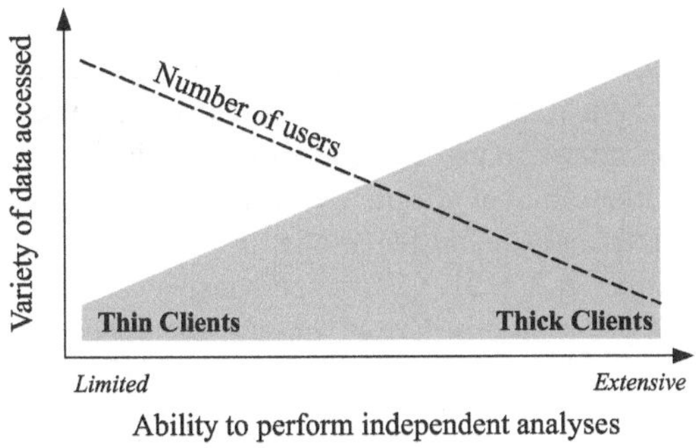

FIGURE 4.28 Concept of thick and thin clients.

2. The properties within each soil layer can be expanded to include properties from laboratory and other field tests. DDSC can potentially include data from similar sites. The reason is that although all sites are unique to some degree, they are not completely unique. Phoon (2018) posed the site challenge which is to quantify “site uniqueness”, directly or indirectly, so that big indirect data can be combined with sparse site-specific data. Phoon et al. (2022a) named this challenge a “site recognition” challenge.
3. The dataset proposed in the existing benchmark examples focused on only one attribute, namely CPT data. The dataset should be modified to be MUSIC-3X compliant.

4.6 SUMMARY

This chapter provided an overview of some DDSC methods by organizing them into three categories with the 2 × 2 DDSC matrix comprising two data rows (numerical or categorical) and two uncertainty columns (geotechnical and geologic). The strengths and limitations of the DDSC methods were highlighted by comparing the methods from a Bayesian perspective. This chapter also overviewed the recent advancement of benchmark examples for DDSC by showing an example of benchmarking DDSC methods.

As shown throughout this chapter, remarkable progress has been made in DDSC methods, and they can successfully provide useful subsurface information for geotechnical design. However, there are still several impediments to the broader use of advanced DDSC methods in design practice, such as lack of familiarity with probabilistic/stochastic methods including machine learning, lack of immediate and compelling value to real-world projects, and lack of user-friendly software to demonstrate concrete outcomes (Phoon et al. 2023). One of the biggest impediments is that the majority of practitioners may not be able to distinguish between theories and facts; in short, they clearly desire “solution, not data” and “information in understandable form” (Turner 2003). This point implies that we prefer a simple deterministic approach to a complicated approach. Turner (2006) classified users as “thick” or “thin” clients (or practitioners) in terms of their information acceptance capabilities (Figure 4.28). A “thick” client is one who can accept and interpret or evaluate a great deal of raw data. In contrast, relatively unsophisticated practitioners/users desire relatively simple, concise answers to their questions. Small volumes of carefully selected data or information usually suffice to meet their needs. Hence, such practitioners/users can be defined as “thin” clients. The “thin” clients are numerous – they represent the broad public – while the thick clients are much less numerous.

Utilizing advanced DDSC methods can surely improve site characterization and bring clear benefits to the planning, design, and construction of infrastructures (Loehr et al. 2016). Improved site characterization directly reduces the likelihood of encountering unforeseen ground conditions during construction, which often lead to change orders and cost overruns during construction, and may lead to unacceptable performance following construction (e.g., Poulos 2005). Improved site characterization also increases the reliability of estimates of important design parameters such as soil/rock properties (e.g., Loehr et al. 2015), which in turn can reduce the likelihood of unacceptable performance and/or allow designers to achieve some target reliability at less cost. Improved site characterization reduces the likelihood of failing to identify relevant geotechnical hazards that may negatively affect the construction or performance of a structure and lead to increased costs (e.g., Cheng and Huang 2003; Kruiver et al. 2017). These persuasive case studies will surely encourage the wider use of advanced DDSC methods in decision-making in geotechnical engineering.

REFERENCES

Bárdossy, G. and Fodor, J. (2004). *Evaluation of Uncertainties and Risks in Geology*. Springer, 221p.

Besag, J. (1986). On the statistical analysis of dirty pictures. *Journal of the Royal Statistical Society B*, 48(3), 259–302. https://www.jstor.org/stable/2345426

Bishop, C. M. (2006). *Pattern Recognition and Machine Learning*. Springer, 738p.

Boyd, S., Parikh, N., Chu, E., Peleato, B. and Eckstein, J. (2010). Distributed optimization and statistical learning via alternating direction method of multipliers. *Foundations and Trends in Machine Learning*, 3(1), 122. https://doi.org/10.1561/2200000016

Cami, B., Javankhoshdel, S., Phoon, K. K. and Ching, J. (2020). Scale of fluctuation for spatially varying soils: Estimation methods and values. *ASCE-ASME Journal of Risk and Uncertainty in Engineering Systems, Part A: Civil Engineering*, 6(4), 03120002. https://doi.org/10.1061/AJRUA6.0001083

Cheng, Z. and Huang, G. H. (2003). Integrated subsurface modeling and risk assessment of petroleum-contaminated sites in western Canada. *Journal of Environmental Engineering*, 129(9), 585–872. https://doi.org/10.1061/(ASCE)0733-9372(2003)129:9(858)

Chen, T. and Guestrin, C. (2016). XGBoost: A scalable tree boosting system. In *Proceedings of the 22nd ACM SIGKDD International Conference on Knowledge Discovery and Data Mining* (pp. 785–794). https://doi.org/10.1145/2939672.2939785

Ching, J., Wu, S. S. and Phoon, K. K. (2016). Statistical characterization of random field parameters using frequentist and Bayesian approaches. *Canadian Geotechnical Journal*, 53(2), 285–298. https://doi.org/10.1139/cgj-2015-0094

Ching, J. and Phoon, K. K. (2017). Characterizing uncertain site-specific trend function by sparse Bayesian learning. *Journal of Engineering Mechanics*, 143(7), 04017028.

Ching, J. and Phoon, K. K. (2019). Constructing site-specific multivariate probability distribution model using Bayesian machine learning. *Journal of Engineering Mechanics, ASCE*, 145(1), 04018126. https://doi.org/10.1061/(ASCE)EM.1943-7889.0001537

Ching, J. and Phoon, K. K. (2020a). 3D probabilistic site characterization by sparse Bayesian learning. *Journal of Engineering Mechanics*, 146(12), 04020134. https://doi.org/10.1061/(ASCE)EM.1943-7889.0001859

Ching, J. and Phoon, K. K. (2020b). Constructing a site-specific multivariate probability distribution using sparse, incomplete, and spatially variable (MUSIC-X) data. *Journal of Engineering Mechanics, ASCE*, 146(7), 04020061. https://doi.org/10.1061/(ASCE)EM.1943-7889.0001779

Ching, J., Phoon, K. K., Ho, Y. H. and Weng, M. C. (2021a). Quasi-site-specific prediction for deformation modulus of rock mass. *Canadian Geotechnical Journal*, 58(7), 936–951. https://doi.org/10.1139/cgj-2020-0168

Ching, J., Wu, S. and Phoon, K. K. (2021b). Constructing quasi-site-specific multivariate probability distribution using hierarchical Bayesian model. *Journal of Engineering Mechanics*, 147(10), 04021069. https://doi.org/10.1061/(ASCE)EM.1943-7889.0001964

Ching, J., Yang, Z. and Phoon, K. K. (2021c). Dealing with nonlattice data in three-dimensional probabilistic site characterization. *Journal of Engineering Mechanics*, 147(5), 06021003. https://doi.org/10.1061/(ASCE)EM.1943-7889.0001907

Ching, J. and Yoshida, I. (2023). Data-drive site characterization for benchmark examples: Sparse Bayesian learning versus Gaussian process regression. *ASCE-ASME Journal of Risk and Uncertainty in Engineering Systems, Part A: Civil Engineering*, 9(1). https://doi.org/10.1061/AJRUA6.RUENG-983

DeGroot, D. J. and Baecher, G. B. (1993). Estimating autocovariance of in-situ soil properties. *Journal of Geotechnical Engineering*, 119(1), 147–166. https://doi.org/10.1061/(ASCE)0733-9410(1993)119:1(147)

Deng, Z. P., Li, D. Q., Qi, X. H., Cao, Z. J. and Phoon, K. K. (2017). Reliability evaluation of slope considering geological uncertainty and inherent variability of soil parameters. *Computers and Geotechnics*, 92, 121–131. https://doi.org/10.1016/j.compgeo.2017.07.020

Elfeki, A. and Dekking, M. (2001). A Markov chain model for subsurface characterization: Theory and applications. *Mathematical Geology*, 33(5), 569–589. https://doi.org/10.1007/s11004-006-9037-9

Facciorusso, J. and Uzielli, M. (2004). Stratigraphic profiling by cluster analysis and fuzzy soil classification from mechanical cone penetration tests. In A. Viana da Fonseca and P. W. Mayne (Eds.), *Proceedings ISC-2 on Geotechnical and Geophysical Site Characterization* (pp. 905–912). IOS Press Inc.

Fillippone, M., Camastra, F., Masulli, F. and Rovetta, S. (2008). A survey of kernel and spectral methods for clustering. *Pattern Recognition*, 41(1), 176–190. https://doi.org/10.1016/j.patcog.2007.05.018

Fukushima, K. (1980). Neocognitron: A self-organizing neural network model for a mechanism of pattern recognition unaffected by shift in position. *Biological Cybernetics*, 36(4), 193–202. https://doi.org/10.1007/bf00344251

Geman, S. and Geman, D. (1987). Stochastic relaxation, Gibbs distributions, and the Bayesian restoration of images. *IEEE Transactions on Pattern Analysis and Machine Intelligence*, PAMI-6(6), 721–741. https://doi.org/10.1016/B978-0-08-051581-6.50057-X

Golub, G. H. and Van Loan, C. F. (2012). *Matrix Computations* (4th ed.). Johns Hopkins University Press, 756p.

Greig, D. M., Porteous, B. T. and Seheult, A. H. (1989). Exact maximum a posteriori estimation for binary images. *Journal of the Royal Statistical Society. Series B (Methodological)*, 51(2), 271–279. https://doi.org/10.1111/j.2517-6161.1989.tb01764.x

Guardiano, F. and Srivastava, M. (1993). Multivariate geostatistics: Beyond bivariate moments. In A. Soares (Ed.), *Geostatistics Troia* (pp. 133–144). Kluwer Academic Publications.

Hegazy, Y. A. and Mayne, P. W. (2002). Objective site characterization using clustering of piezocone data. *Journal of Geotechnical and Geoenvironmental Engineering, ASCE*, 128(12), 986–996. https://doi.org/10.1061/(ASCE)1090-0241(2002)128:12(986)

Jaksa, M. (1995). The influence of spatial variability on the geotechnical design properties of a stiff, over consolidated clay. Ph.D. Dissertation, University of Adelaide, 469p.

Kim, S. J., Koh, K., Boyd, S. and Gorinevsky, D. (2009). ℓ_1 trend filtering. *SIAM Review, Problems and Techniques Section*, 51(2), 339–360. https://doi.org/10.1137/070690274

Kruiver, P. P., Wiersma, A., Kloosterman, F. H., Lange, G. D., Korff, M., Stafleu, J., Busschers, F. S., Harting, R., Gunnink, J. L., Green, R. A., Elk, J. V. and Doornhof, D. (2017). Characterization of the Groningen subsurface for seismic hazard and risk modelling. *Netherlands Journal of Geosciences*, 96(5), s215–s233. https://doi.org/10.1017/njg.2017.11

Krumbein, W. C. (1967). *Fortran IV Computer Programs for Markov Chain Experiments in Geology, Computer Contribution 13*, State Geological Survey, The University of Kansas.

Lecun, Y., Bottou, L., Bengio, Y. and Haffner, P. (1998). Gradient-based learning applied to document recognition. *Proceedings of the IEEE*, 86(11), 2278–2324. https://doi.org/10.1109/5.726791

Li, Z., Wang, X., Wang, H. and Liang, R. Y. (2016). Quantifying stratigraphic uncertainties by stochastic simulation techniques based on Markov random field. *Engineering Geology*, 201, 106–122. https://doi.org/10.1016/j.enggeo.2015.12.017

Liao, T. and Mayne, P. W. (2007). Stratigraphic delineation by three-dimensional clustering of piezocone data. *Georisk: Assessment and Management of Risk for Engineered Systems and Geohazards*, 1(2), 102–119. https://doi.org/10.1080/17499510701345175

Lloyd, S. (1982). Least squares quantization in PCM. *IEEE Transactions on Information Theory*, 28(2), 129–137. https://doi.org/10.1109/TIT.1982.1056489

Loehr, J. E., Ding, D. and Likos, W. J. (2015). Effect of number of soil strength measurements on reliability of spread footing designs. *Transportation Research Record: Journal of the Transportation Research Board*, 2511(1), 37–44. https://doi.org/10.3141/2511-05

Loehr, J. E., Lutenegger, A., Rosenblad, B. and Boekmann, A. (2016). Geotechnical site characterization. Publication No. FHWA NHI-16-072.

Lyu, B., Hu, Y. and Wang, Y. (2023). Data-driven development of three-dimensional subsurface models from sparse measurements using Bayesian compressive sampling: A benchmarking study. *ASCE-ASME Journal of Risk and Uncertainty in Engineering Systems, Part A: Civil Engineering*, 9(2). https://doi.org/10.1061/AJRUA6.RUENG-935

MacKay, D. J. C. (1992). Bayesian interpolation. In C. R. Smith, G. J. Erickson and P. O. Neudorfer (Eds.), *Maximum Entropy and Bayesian Methods: Fundamental Theories of Physics* (Vol. 50). Springer. https://doi.org/10.1007/978-94-017-2219-3_3

Mariethoz, G. and Caers, J. (2014). *Multiple-Point Geostatistics: Stochastic Modeling with Training Images*. John Wiley & Sons.

Matheron, G. (1963). Principles of geostatistics. *Economic Geology*, 58(8), 1246–1266.

Matheron, G. (1973). The intrinsic random functions and their applications. *Advances in Applied Probability*, 5(3), 439–468. https://doi.org/10.2307/1425829

Nerberg, T., Rosen, L., Baran, A. and Baran, S. (2002). On modeling discrete geological structures as Markov Random Fields. *Mathematical Geology*, 34(1), 63–77.

Park, E. (2010). A multidimensional generalized coupled Markov chain model for surface and subsurface characterization. *Water Resources Research*, 46(11), W11509. https://doi.org/10.1029/2009WR008355

Phoon, K. K. (2018). Editorial for special collection on probabilistic site characterization. *ASCE-ASME Journal of Risk and Uncertainty in Engineering Systems, Part A: Civil Engineering*, 4(4), 02018002. https://doi.org/10.1061/AJRUA6.0000992

Phoon, K. K., Ching, J. and Wang, Y. (2019). Managing risk in geotechnical engineering – from data to digitalization. In *Proceedings 7th International Symposium on Geotechnical Safety and Risk* (pp. 13–34). ISGSR.

Phoon, K. K. and Ching, J. (2021). Project DeepGeo – Data-driven 3D subsurface mapping. *Journal of Geology Engineering*, 16(2), 61–73. http://doi.org/10.6310/jog.202106_16(2).2

Phoon, K. K. and Shuku, T. (2022). 3D data-driven site characterization using geotechnical lasso with basis functions. *Proceedings of the 8th International Symposium on Reliability Engineering and Risk Management*, 4–7 September 2022, Hannover, Germany, Published by Research Publishing, Singapore. DOI: 10.3850/978-981-18-5184-1_MS-13-044

Phoon, K. K., Ching, J. and Shuku, T. (2022a). Challenges in data-driven site characterization. *Georisk: Assessment and Management of Risk for Engineered Systems and Geohazards*, 16(1), 114–126. https://doi.org/10.1080/17499518.2021.1896005

Phoon, K. K., Ching, J. and Cao, J. (2022b). Unpacking data-centric geotechnics. *Underground Space*, 7(6), 967–989. https://doi.org/10.1016/j.undsp.2022.04.001

Phoon, K. K., Shuku, T., Ching, J. and Yoshida, I. (2022c). Benchmark examples for data-driven site characterization. *Georisk: Assessment and Management of Risk for Engineered Systems and Geohazards*, 16(4), 599–621. https://doi.org/10.1080/17499518.2022.2025541

Phoon, K. K. and Zhang, W. (2023). Future of machine learning in geotechnics. *Georisk: Assessment and Management of Risk for Engineered Systems and Geohazards*, 17(1), 7–22. https://doi.org/10.1080/17499518.2022.2087884

Phoon, K. K., Cao, Z., Liu, Z. and Ching, J. (2023). Report for ISSMGE TC309/TC304/TC222 third ML dialogue on "data-driven site characterization (DDSC)" 3 December 2021, Norwegian Geotechnical Institute, Oslo, Norway. *Georisk: Assessment and Management of Risk for Engineered Systems and Geohazards*, 17(1), 227–238. https://doi.org/10.1080/17499518.2022.2105366

Poulos, H. G. (2005). Pile behaviour – Consequences of geological and construction imperfections. *Journal of Geotechnical and Geoenvironmental Engineering*, 131(5), 538–563. https://doi.org/10.1061/(ASCE)1090-0241(2005)131:5(538)

Qi, X. H., Li, D. Q., Phoon, K. K., Cao, Z. J. and Tang, X. S. (2016). Simulation of geologic uncertainty using coupled Markov chain. *Engineering Geology*, 207, 129–140. https://doi.org/10.1016/j.enggeo.2016.04.017

Quiñonero-Candela, J. and Rasmussen, C. E. (2005). A unifying view of sparse approximate Gaussian process regression, 6(65), 1939–1959.

Rao, C. R. (1997). *Statistics and Truth* (2nd ed.). World Scientific Publishing, 212p. https://doi.org/10.1142/3454

Rasmussen, C. E. and Williams, C. K. I. (2006). *Gaussian Processes for Machine Learning.* MIT Press, 248p.

Renard, F. and Mariethoz, G. (2014). Special issue on 20 years of multiple-point statistics: Part 1. *Mathematical Geosciences*, 46, 129–131. https://doi.org/10.1007/s11004-014-9524-3

Robertson, P. K. (2016). Cone penetration test (CPT)-based soil behaviour type (SBT) classification system - An update. *Canadian Geotechnical Journal*, 53(12), 1910–1927. https://doi.org/10.1139/cgj-2016-0044

Robertson, P. K. and Wride, C. E. (1998). Evaluating cyclic liquefaction potential using the cone penetration test. *Canadian Geotechnical Journal*, 35(3), 442–459. https://doi.org/10.1139/t98-017

Shannon, C. E. (1948). A mathematical theory of communication. *Bell System Technical Journal*, 27(4), 623–666. https://doi.org/10.1002/j.1538-7305.1948.tb00917.x

Shi, C. and Wang, Y. (2021a). Nonparametric and data-driven interpolation of subsurface soil stratigraphy from limited data using multiple point statistics. *Canadian Geotechnical Journal*, 58(2). https://doi.org/10.1139/cgj-2019-0843

Shi, C. and Wang, Y. (2021b). Development of subsurface geological cross-section from limited site-specific boreholes and prior geological knowledge using iterative convolution XGBoost. *Journal of Geotechnical and Geoenvironmental Engineering*, 149(9), 04021082. https://doi.org/10.1061/(ASCE)GT.1943-5606.0002583

Shuku, T. and Phoon, K. K. (2021). Three-dimensional subsurface modelling using Geotechnical Lasso. *Computers and Geotechnics*, 133, 1034068. https://doi.org/10.1016/j.compgeo.2021.104068

Shuku, T. and Phoon, K. K. (2023a). Comparison of data-driven site characterization methods through benchmarking: Methodological and application aspects. *ASCE-ASME Journal of Risk and Uncertainty in Engineering Systems, Part A: Civil Engineering*, 9(2). https://doi.org/10.1061/AJRUA6.RUENG-977

Shuku, T. and Phoon, K. K. (2023b). Data-driven subsurface modelling using a Markov random field model. *Georisk: Assessment and Management of Risk for Engineered Systems and Geohazards*, 17(1), 41–63. https://doi.org/10.1080/17499518.2023.2181973

Snelson, E. and Ghahramani, Z. (2005). Sparse Gaussian processes using pseudo-inputs. *Advances in Neural Information Processing Systems*, 18(NIPS 2005), 1257–1264

Tibshirani, R. (1996). Regression shrinkage and selection via the lasso. *Journal of the Royal Statistical Society,. Series B*, 58(1), 267–288. https://www.jstor.org/stable/2346178

Tibshirani, R. (2014). Adaptive piecewise polynomial estimation via trend filtering. *The Annals of Statistics*, 42(1), 285–323. https://www.jstor.org/stable/43556281

Tipping, M. E. (2001). Sparse Bayesian learning and the relevance vector machine. *Journal of Machine Learning Research*, 1, 211–244. https://doi.org/10.1162/15324430152748236

Tomizawa, Y. and Yoshida, I. (2023). Benchmarking of Gaussian process regression with multiple random fields for spatial variability estimation. *ASCE-ASME Journal of Risk and Uncertainty in Engineering Systems, Part A: Civil Engineering*, 8(4). https://doi.org/10.1061/AJRUA6.0001277

Turner, A. K. (2003). Putting the user first: Implications for subsurface characterisation. In M. S. Rosenbaum and A. K. Turner (Eds.), *New Paradigms in Subsurface Prediction: Lecture Notes in Earth Sciences* (Vol. 99). Springer. https://doi.org/10.1007/3-540-48019-6_5

Turner, A. K. (2006). Challenges and trends for geological modelling and visualization. *Bulletin of Engineering Geology and the Environment*, 65(2), 109–127. https://doi.org/10.1007/s10064-005-0015-0

Wang, X., Li, Z., Wang, H., Rong, Q. and Liang, R. Y. (2016). Probabilistic analysis of shield-driven tunnel in multiple strata considering stratigraphic uncertainty. *Structural Safety*, 62, 88–100. https://doi.org/10.1016/j.strusafe.2016.06.007

Wang, H., Wellmann, J. F., Li, Z., Wang, X. and Liang, R. Y. (2017). A segmentation approach for stochastic geological modeling using hidden Markov random fields. *Mathematical Geology*, 49(2), 145–177. https://doi.org/10.1007/s11004-016-9663-9

Wang, Y., Cao, Z. and Li, D. (2016). Bayesian perspective on geotechnical variability and site characterization, 203, 117–125. https://doi.org/10.1016/j.enggeo.2015.08.017

Wang, Y. and Zhao, T. (2016). Interpretation of soil property profile from limited measurement data: A compressive sampling perspective. *Canadian Geotechnical Journal*, 53(9), 1547–1559. https://doi.org/10.1139/cgj-2015-0545

Wang, Y. and Zhao, T. (2017). Statistical interpolation of soil property profiles from sparse data using Bayesian compressive sampling. *Géotechnique*, 67(6), 523–536. https://doi.org/10.1680/jgeot.16.P.143

Wang, Y., Zhao, T., Hu, Y. and Phoon, K. K. (2019). Simulation of random fields with trend from sparse measurements without detrending. *Journal of Engineering Mechanics*, 145(2), 04018130. https://doi.org/10.1061/(ASCE)EM.1943-7889.0001560

Wellmann, J. F., Horowitz, F. G., Schill, E. and Regenauer-Lieb, K. (2010). Towards incorporating uncertainty of structural data in 3D geological inversion. *Tectonophysics*, 490(3–4), 141–151. https://doi.org/10.1016/j.tecto.2010.04.022

Wellmann, J. F. and Regenauer-Lieb, K. (2012). Uncertainties have a meaning: Information entropy as a quality measure for 3-D geological models. *Tectonophysics*, 526–529. https://doi.org/10.1016/j.tecto.2011.05.001

Wilson, A. and Nickisch, H. (2015). Kernel interpolation for scalable structured Gaussian process (KISS-GP). *International Conference on Machine Learning*, 37, 1775–1784.

Yeh, C. H., Dong, J. J., Khoshnevisan, S., Juang, C. H., Huang, W. C. and Lu, Y. C. (2021). The role of the geological uncertainty in a geotechnical design - A retrospective view of Freeway No. 3 Landslide in Northern Taiwan. *Engineering Geology*, 291, 106233. https://doi.org/10.1016/j.enggeo.2021.106233

Yoshida, I., Tomizawa, Y. and Otake, Y. (2021). Estimation of trend and random components of conditional random field using Gaussian process regression. *Computers and Geotechnics*, 136, 104179. https://doi.org/10.1016/j.compgeo.2021.104179

Zhao, T. and Wang, Y. (2020). Interpolation and stratification of multilayer soil property profile from sparse measurements using machine learning methods. *Engineering Geology*, 265, 105430. https://doi.org/10.1016/j.enggeo.2019.105430

Zhang, J. Z., Liu, Z. Q., Zhan, D. M., Huang, H. W., Phoon, K. K. and Xue, Y. D. (2022). Improved coupled Markov chain method for simulating geological uncertainty. *Engineering Geology*, 298, 106539. https://doi.org/10.1016/j.enggeo.2022.106539

5 Variability of Predictions in Geotechnics

Chong Tang, Kok-Kwang Phoon, and Jun Yuan

ABSTRACT

Predictions are at the heart of rational design methods in geotechnical engineering. Because of the limited understanding of site conditions, bias from imperfect models, measurement error, and construction effects, the predicted behavior is expected to deviate from the measured behavior. A quantification of this deviation is useful for decision-making in analysis and design, even if it is approximate. This chapter presents an extensive review of the variability of predictions in geotechnics found in three common sources: (1) statistical analysis of prediction variability from a performance database with a consistent definition for the measured and predicted response; (2) lessons from prediction events reflecting the variability of participants' judgment and experience in the selection of model and related parameter values; and (3) comparison of numerical modeling with field measurements that may be affected by site variability and error from the selection of model and parameter values. The variability arises from different causes in each data source. A comparison can provide deeper insights into how predictions are affected by the ground conditions, the installation of geotechnical structures, the modeling of ground–structure interactions, and the decisions made by the modeler (the engineer). The geotechnical structures covered in this chapter can be broadly divided into three groups based on the role of the soil in the predicted response: (1) foundation engineering (soil provides the resistance against external loading); (2) slope, excavation, and underground structure (soil acts as a load and a resistance); and (3) embankment constructed by soil (soil as a construction material). The model factor statistics presented in this chapter can be applied to satisfy Clause 2.4.16(P), EN 1997-1:2004 (CEN 2004) - "any calculation model shall be either accurate or err on the side of safety" in the probabilistic sense of ensuring Prob(model factor <1) < 0.05. The original deterministic intent of this clause cannot be fulfilled given the variability of predictions demonstrated in this chapter.

5.1 INTRODUCTION

5.1.1 Background

The ability to predict something has always been highly valued. Oracles or shaman were consulted by people for their decision-making in ancient times when predictions were mostly based on divination (Negro et al. 2009). Isaac Newton published his *Philosophiae Naturalis Principia Mathematica* (*Mathematical Principles of Natural Philosophy*) in 1687 and this work changed the world significantly. Since then, two major concepts of calculus (differential and integral) have found uses in all branches of science, engineering, and more. As for geotechnical engineering, differential equations are commonly used to model four requirements that predictions should approximately satisfy: equilibrium (stress or force), compatibility (strain or deformation), material constitutive behavior (stress–strain response of soil or rock possibly as a function of time), and boundary conditions (e.g., Dirichlet, Neumann, or mixed type). Through an understanding of the causes underlying a natural process, it is believed that its effects can be forecast and controlled by mankind.

 DOI: 10.1201/9781003333586-6

As in many other industries, civil engineers are expected to anticipate the performance of the structures they design and build. The successes in making predictions result largely from their ability to model the prototypes accurately through better understanding of the mechanics that governs the behavior and the associated influential factors. For structures that are built under stringent specifications for precast concrete or prefabricated steel (variability in material properties such as strength and modulus could be controlled within 10%) and quality assurance during construction (limited human error), predictions can be accurate. However, this is not so for geotechnical engineering in most cases because of the following (Mitchell 1986):

- Compared to man-made steel and concrete, geomaterial occurs naturally and exhibits an extremely wide range of mechanical behavior that is hard to capture precisely in a simplistic manner. Basically, geomaterials are classified into two broad groups: soil and rock. Some design manuals include a reference to intermediate geomaterial that lies between soil and rock on the basis of the strength of the geomaterial. According to the Unified Soil Classification System (USCS), soil is categorized into 15 groups and each group is assigned a symbol consisting of two capital letters that denote the main soil type (e.g., gravel, sand, silt, or clay) and gradation (e.g., well or poorly graded) or plasticity (e.g., high or low plastic). According to its origin and appearance, rock is usually divided into three groups: igneous (e.g., granite, basalt, or andesite), sedimentary (e.g., sandstone, mudstone, or shale), and metamorphic (transformation of existing rock to new types of rock as a result of extreme heat, intense pressure, or in some cases, both). Each group contains many different individual rock types, distinguished from one another by physical characteristics. An example of the wide range of the strength and small strain modulus is shown in Figure 5.1.

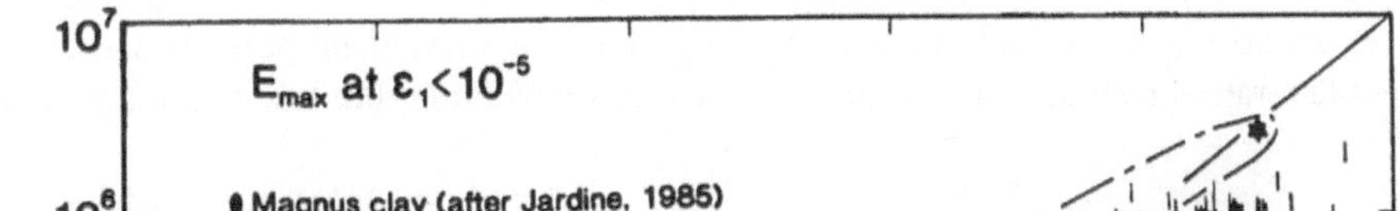

FIGURE 5.1 Generic transformation model between small strain elastic modulus E_{max} and compressive strength q_{max}. (Source: Figure 9 in Kim et al. 1991)

- Subsurface condition is highly uncertain, and the related uncertainties can be characterized as aleatory and epistemic (e.g., Baecher and Christian 2003; Tang and Phoon 2021).

Aleatory uncertainty is the inherent variability in geomaterial that manifests itself in both spatial and temporal scales (e.g., stratigraphy and index, strength and stiffness properties). In principle, aleatory uncertainty cannot be reduced; however, it can be characterized more accurately with more data (e.g., laboratory testing, in situ investigation, and geophysical survey). Epistemic uncertainty is often involved within site characterization (e.g., ground profile and stratification, classification of soil/rock type, and determination of geotechnical properties), largely due to limited knowledge and a lack of information and data to model the subsurface condition. Typically, it includes the uncertainty related to (1) the estimation of soil and rock parameters in Chapter 2, (2) data-driven constitutive modeling in Chapter 3, and (3) site stratification in Chapter 4. In practice, uncertainty also exists in the classification of a soil or rock type.

- The mechanism of interaction between a volume of soil (ground) and a structure built on it is complicated and affected by a variety of factors (e.g., mechanical properties of soil/rock that may be time, loading rate or direction, and stress or strain level dependent; and construction effects such as changes in ground conditions and soil stress states), especially in the presence of groundwater and spatial variability of soil properties as presented in Chapter 14.

The beginnings of modern predictive capability in geotechnics can be traced back to the final report of the Swedish Railways Geotechnical Commission in 1922 which investigated more than 300 embankment failures and land slips, and Terzaghi's seminal book *Erdbaumechanik: auf bodenphysikalischer Grundlage* in 1925. Laboratory and field investigation methods were introduced and rational approaches for geotechnical analysis were developed. This shift in practice to one supported in part by physics has led to a steady stream of evolutionary developments in the methods of geotechnical exploration (e.g., in situ investigation and geophysical survey), property measurement (e.g., a variety of laboratory testing systems), and analysis, as shown in Figure 5.2. The analysis methods

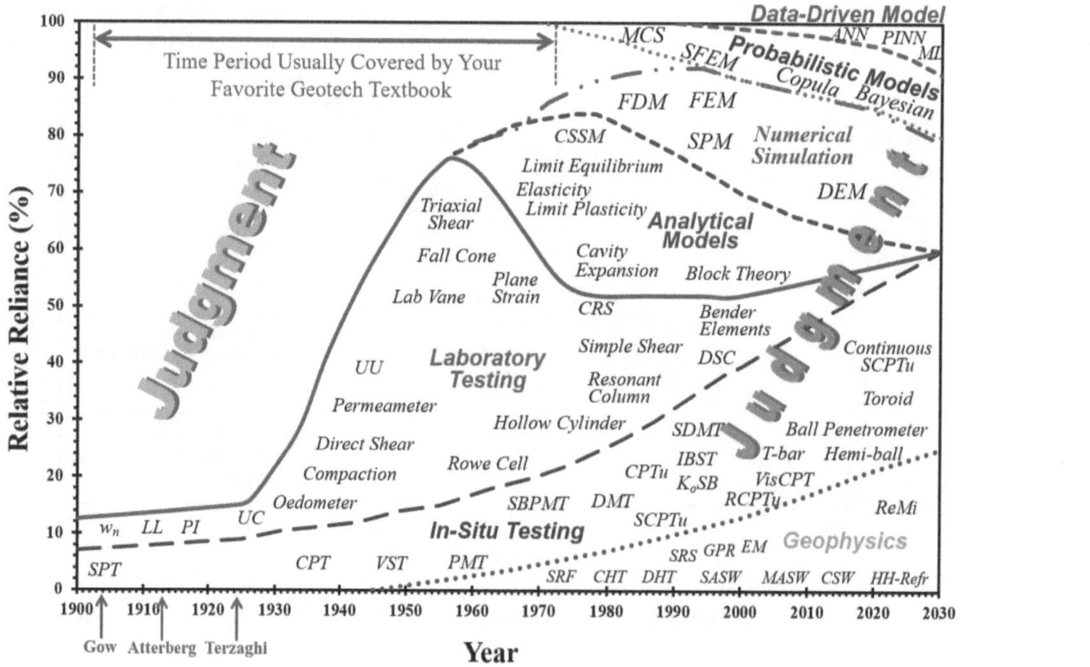

FIGURE 5.2 Evolution of geotechnical exploration, property measurement, and analysis. (Source: Adapted from Mayne 2015 and Tang and Phoon 2021 to include data-driven models)

can be identified as analytical (e.g., empirical, semi-empirical, or theoretical like elasticity or plasticity), numerical (e.g., finite element method [FEM], finite difference method [FDM], discrete element method [DEM], single particle method [SPM] or material point method [MPM]), probabilistic method based on the more "classical" statistics (e.g., Monte Carlo simulation [MCS], stochastic finite element method [SFEM]), and data science (a modernized version of statistics – Statistics 2.0) such as machine learning (ML) (e.g., supervised, unsupervised, or reinforcement learning).

As James Nicholas Gray (Turing awardee) opined, data science has emerged as a modern pillar of scientific research or as the "fourth paradigm of science" (besides empirical, theoretical, and numerical). Essentially, it is related to the analysis and interpretation of big data for knowledge extraction. A combination of data science and engineering led to the recently created field of data-centric engineering (Girolami 2020; Ley et al. 2020). For geotechnical applications, it is called data-centric geotechnics (Phoon and Ching 2021; Phoon et al. 2022a). A common theme across scientific domains is to think about how much data is available versus how much physics is known. The "physics–data" space is depicted in Figure 5.3 (e.g., Tartakovsky et al. 2020; Karniadakis et al. 2021).

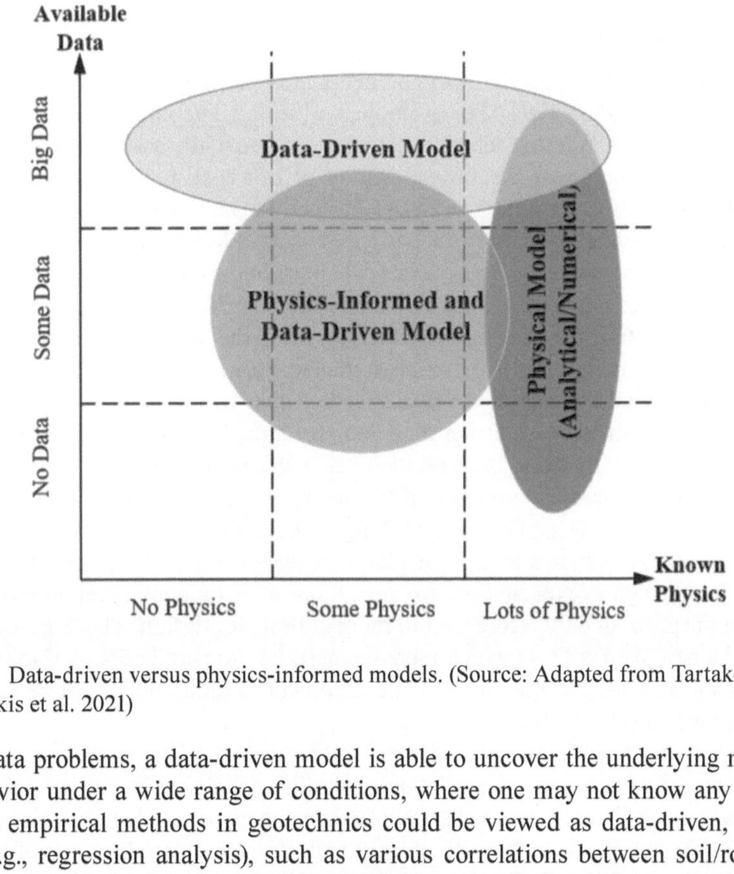

FIGURE 5.3 Data-driven versus physics-informed models. (Source: Adapted from Tartakovsky et al. 2020 and Karniadakis et al. 2021)

For big data problems, a data-driven model is able to uncover the underlying mechanism and predict behavior under a wide range of conditions, where one may not know any of the physics. Some of the empirical methods in geotechnics could be viewed as data-driven, albeit they are simplistic (e.g., regression analysis), such as various correlations between soil/rock parameters and in situ testing results in subsurface investigation manuals (e.g., Mayne et al. 2001; National Academies of Sciences, Engineering, and Medicine 2019). Data-driven quasi-site-specific predictions of soil/rock parameters (e.g., Ching et al. 2021a, 2021b, 2022a, 2022b) are more advanced and robust. At the other end, one may understand all the physics, which is effectively modeled by differential equations with known initial/boundary conditions and coefficients, referred to as a physical model. Theoretical (e.g., elasticity or plasticity) and numerical (e.g., FEM, FDM, or MPM) methods in geotechnics are largely physics based. In the real world, problems that lie between the two ideal scenarios present a significant challenge, where physics is partially known (e.g., possibly missing parameter values or even an entire term in the differential equation – imperfection of the physical model) and data is sparse. This intermediate scenario is ubiquitous in current geotechnical practice. Physics-informed learning machines (PhILMs) are preferable under this circumstance, as they seamlessly integrate data and the governing physical laws in a unified way and are more explainable than purely data-driven models (e.g., Tartakovsky et al. 2020; Karniadakis et al. 2021). This is important because an engineer needs to "understand" the predictions to incorporate judgment in decision-making. Some of the semi-empirical methods in geotechnics may be physics-informed and data-driven, as they are based on soil mechanics principles or experimental observations and calibrated against limited data. An example could be the mechanism-based approaches for spudcan penetration in stiff-over-soft soils, as discussed in Tang et al. (2023). Some recent studies have begun to explore the use of physics-informed probabilistic or data-driven approaches for geotechnical analyses (e.g., Bullock et al. 2019; Sheil 2021; Zhang et al. 2023). It is not clear if PhILMs can deal with geotechnical MUSIC-X (multivariate, uncertain and unique, sparse, incomplete, and potentially corrupted with "X" denoting spatial/temporal variability) data, in which sparsity is only one of the seven attributes (Phoon et al. 2019).

5.1.2 Objective and Scope

To make predictions tractable, geotechnical engineers have to make a variety of assumptions to simplify the problem (Vaughan 1994). For example, an idealized geomaterial layer profile will be developed for a design zone. Within each layer, geomaterial is usually assumed to be homogeneous and isotropic. Although the stress–strain response of soil or rock is non-linear, it is frequently idealized as being elastic (for settlement analysis), perfectly plastic (for stability analysis), or elastoplastic (for load-deformation analysis). In particular, soil–pile interaction under axial loading is commonly represented by a set of non-linear springs which are modeled by q_s–w_s (or t–z) and q_b–w_b (or q–w) curves, respectively, for load transfers in shaft shearing and tip end bearing. This approach is termed load transfer analysis, in which q_s or t is the shear stress developed along the pile shaft–soil interface and w_s or z is the relative movement between the shaft and geomaterial, while q_b or q is the tip end bearing pressure at displacement w_b or w. As a result, the predicted behavior of geotechnical structures (also termed geo-structures) will deviate from measurements. Tang and Pelletier (1990) stated that the deviation has a strong influence on the calculated probability of failure and thus on the estimation of the safety margin. It is particularly important for offshore geotechnical design (e.g., Gilbert and Tang 1995; Lacasse and Nadim 1996; Lacasse et al. 2013; Liu et al. 2019). Najjar and Gilbert (2009) revealed that the incorporation of a lower-bound capacity into design is expected to provide a more realistic quantification of reliability for decision-making purposes and therefore a more rational basis for design. The latest edition of ISO 2394 introduced a new Annex D, in which model uncertainty (or prediction variability) has been identified as one of the critical elements in the development of reliability-based design (RBD) for geotechnical engineering (ISO 2015).

The purpose of this chapter is to review the variability of predictions in geotechnics from three perspectives: (1) statistical analyses of prediction variability quantified by substantial "experimental" data (e.g., databases of load tests and observed settlements on full-scale foundations, pullout box tests of mechanically stabilized earth walls and soil nail walls, and case histories of slope failures and excavations); (2) lessons from some international prediction events (e.g., given a geo-structure like an embankment and pile foundation, a group of participants are provided with the same set of input data to predict one or more quantities of interest such as the capacity and the settlement for comparison purposes); and (3) comparison of numerical analyses with in situ measurements reported in diverse studies. Tang and Phoon (2021) conducted the largest model validation exercise to date using load test databases and found that geotechnical models are biased (typically on the conservative side) and imprecise. The predictions are obtained from analytical models commonly presented in textbooks or adopted in design codes. The load test databases are the largest and most diverse compiled to date, covering many foundation types (shallow, offshore spudcan, micropile, driven concrete/steel pile, steel helical pile, and drilled shaft) and a wide range of ground conditions (soft to stiff clay, silt, loose to dense sand, gravel, and soft rock). This chapter extends this model validation exercise to prediction events and predictions from numerical analyses. It is a significant update on the review paper by Phoon and Tang (2019) and the book by Tang and Phoon (2021) that only focus on prediction variability using the characterization of a model factor from a performance database. A second large database compilation effort is in progress (Phoon and Tang 2024).

5.2 PREDICTION METHOD CATEGORIES

Along with the historical development of the mechanics of deformable solids, the problems in geotechnical engineering are often categorized into two distinct groups, namely elasticity and stability (e.g., Terzaghi 1943; Terzaghi and Peck 1967; Chen 1975). The elasticity problems deal with the stress or deformation of soil or rock without failure, such as (1) point stress beneath a footing or behind an earth retaining wall, (2) deformation around a tunnel or an excavation, and (3) settlement analysis. One example is the analytical solution for tunneling-induced ground movements in clays (Loganathan and Poulos 1998). The stability problems are associated with the determination of a load that will cause the failure of soil, such as (1) bearing capacity, (2) passive and active earth pressure, and (3) slope stability. Analytical solutions such as the Terzaghi bearing capacity equation, the Rankine/Coulomb passive/active earth pressure, and the factor of safety (FS) for an infinite slope are widely known. In many design codes or manuals (e.g., CEN 2004; JRA 2017; CSA 2019; AASHTO 2020), elasticity is usually considered as a serviceability limit state, while stability is frequently considered as an ultimate limit state. Vardanega and Bolton (2016) observed that

> RBD is applied to the ultimate failure of the soil, rather than to the onset of disappointing deformations that later develop into serviceability issues, and then ultimately threaten structural collapse only if nothing has been done to interrupt the loading process or enhance the soil-foundation system. In that sense, the rigid demarcation between serviceability limit state (SLS) and ultimate limit state (ULS) failures in limit state design is unrealistic and unhelpful for a designer wishing to apply risk-based concepts.

Phoon (2017) clarified that reliability-based design can be applied to the ULS, SLS, or any other damage limit states. The limit state appears in a completely general way as a performance function in RBD. However, the large number of RBD applications that focused on the ULS have given practitioners the false impression that RBD is methodologically too limited to be useful in practice. Phoon (2023) explained this misconception:

> No reliability analysis can be conducted unless the relevant limit states are identified from a sufficient understanding of the physics. However, this step is physical modeling, not probabilistic modeling. A probabilistic model quantifies the uncertainties in a physical model – it does not make its prediction more accurate or more precise.

For analysis and design purposes, a large variety of useable and oftentimes analytically tractable models (link between theory and practice) have been developed for elasticity and stability analyses of geotechnical structures. Depending on the level of sophistication and rigor, the analysis and design procedures or methods in geotechnics can be categorized into three broad groups (Poulos 1989):

- Category 1: Empirical and not based on the theory of soil/rock mechanics. These methods were either rules of thumb or regression equations calibrated from experimental data (typically limited). Examples considered in this chapter include (1) Terzaghi and Peck (1967) and Alpan's (1964) methods for settlement analysis; (2) in situ test-based methods for pile axial capacity – standard penetration test (SPT) (e.g., Meyerhof 1976; Brown 2001) or cone penetration test (CPT) (e.g., Nottingham and Schmertmann 1975; Eslami and Fellenius 1997); and (3) empirical correlations of shaft shearing and tip end bearing resistances of a rock socket with uniaxial compressive strength (e.g., Horvath et al. 1980; Rowe and Armitage 1987; Kulhawy and Phoon 1993).

- Category 2: Semi-empirical and using simplified constitutive models such as a perfectly plastic model for stability and a linear elastic model for deformation to deliver analytical solutions or design charts/tables (usually based on the results from more sophisticated computations) (e.g., Poulos and Davis 1973; Gibson 1974; Chen 1975). Examples considered in this chapter include (1) classical bearing capacity theory (plasticity solutions) with empirical correction terms to account for the influence of footing shape and embedment, loading inclination and eccentricity, soil rigidity, and so on (e.g., Meyerhof 1965; Hansen 1970; Vesić 1973); (2) limit equilibrium solutions for slope stability (Duncan et al. 2014), punch-through resistance of spudcan penetrating in stiff-over-soft clay or sand-over-clay (Tang et al. 2023), and the stability of a mechanically stabilized earth wall (Leshchinsky et al. 2016) and a soil nail wall (Lazarte et al. 2015); (3) effective stress analysis of pile shaft shearing resistance as those presented in Brown et al. (2018) for a drilled shaft and Hannigan et al. (2016) for a driven pile; and (4) elastic solutions for the settlement of a single pile or piled raft (e.g., Poulos and Davis 1973; Vesić 1977; Randolph and Wroth 1978).
- Category 3: Relatively advanced methods incorporating numerical computations. For example, the implementation of load transfer analysis is becoming more common in pile design under axial loading (t–z/q–w curves for shaft shearing and tip end bearing) (e.g., Mosher and Dawkins 2000; API 2007; Abchir et al. 2016; Bohn et al. 2017; Ong et al. 2021) or lateral loading (p–y curve: the soil reaction per unit length of pile p occurs when the unit length of the pile is displaced a lateral deflection y into the soil) (e.g., Mosher and Dawkins 2000; Jeanjean et al. 2017; Lovera et al. 2021). More advanced numerical analyses (e.g., FEM, FDM, DEM, and MPM) also belong to this category. Categories 1 and 2 methods can only determine one or a few quantities of interest to the limit state (e.g., capacity, settlement). In contrast, Category 3 methods enable practitioners to obtain the complete behavior of geotechnical structures, such as foundation under loading, spudcan/pile during penetration, or excavation/tunnel at different stages of construction (e.g., Potts and Zdravković 1999, 2001).

The three categories of analysis and design methods outlined above can be regarded as corresponding to three classical paradigms of science: empirical, theoretical, and computational. EN 1997-1:2004 Clause 2.4.1(5) also allows three different calculation models for design: an analytical model, a semi-empirical model, and a numerical model (CEN 2004). Categories 1 and 2 methods are useful for the earlier stages of design and probably account for most geotechnical design throughout the world (Poulos 1989). At present, it is not uncommon to perform two-dimensional (2D) or three-dimensional (3D) finite element (or finite difference) analyses in the final detailed design stage, which can incorporate more advanced constitutive models to reflect such characteristics as non-linearity, shear dilatancy, and the dependency of geomaterial strength/stiffness on the strain level and/or stress path (Poulos 2017). A new set of rules to cover geotechnical design and the verification of limit states using numerical methods is being developed for the final project team draft of the next generation of Eurocode 7 Part 1 (EN 1997-1:202x) – Tomorrow's geotechnical toolbox (Lees 2019). If adopted, this would be the first code to reflect the increasing use of numerical methods in geotechnical engineering.

5.3 DATABASE ASSESSMENT OF PREDICTION VARIABILITY

Geotechnical design continues its gradual transition from allowable stress design (ASD) to reliability-based design worldwide (e.g., Fenton et al. 2016; Phoon and Retief 2016; Chwała et al. 2022; Phoon et al. 2022b). A global factor of safety is used by ASD to account for the uncertainties encountered by geotechnical engineers and to manage the associated risk arising from the decision-making process. RBD differs significantly from ASD in at least one key aspect, which is a rational and explicit characterization of geotechnical variability (e.g., Phoon 2017; Phoon et al. 2016a). Probabilistic methods in geotechnics are presented in Chapters 6–10. Prediction variability is an important element of the geotechnical RBD process as identified by Annex D of ISO2394:2015 (Phoon and Retief 2016). Section D.1(f) states that

> There are usually many different geotechnical calculation models for the same design problem. Hence, model calibration based on local field tests and local experience is important. The proliferation of model factors, possibly site-specific, is to be expected because of the number of models and the number of calibration databases.

Because of this core element, many studies have compiled test databases to quantify the prediction variability for the development of RBD, especially for highway bridge foundation design in the United States. Based on the work of Tang and Phoon (2021), an extensive review of test databases is presented in this section, together with the characterization of prediction variability by a model factor. It is worth mentioning that the model factor is also required in non-probabilistic design codes. For example, EN 1997-1:2004 Clause 2.4.1 refers to the application of a model factor to make a calculation model accurate or conservative for the purpose of design (CEN 2004):

> (6)P Any calculation model shall be either accurate or err on the side of safety.
> (7) A calculation model may include simplifications.
> (8) If needed, a modification of the results from the model may be used to ensure that the design calculation is either accurate or errs on the side of safety.
> (9) If the modification of the results makes use of a model factor, it should take account of the following:
> the range of uncertainty in the results of the method of analysis;
> any systematic errors known to be associated with the method of analysis.

Clause 2.4.1(6) is marked by "P", which refers to "Principles" that comprise "general statements and definitions for which there is no alternative". Despite its importance, it will be shown below that the model factor is a random variable and hence there is no practical way to ensure "any calculation model shall be either accurate or err on the side of safety" in a deterministic way. This clause can only be met probabilistically at the design stage where test results for model calibration are frequently not available, but there is no guidance from EN 1997-1:2004 (CEN 2004) on how to approach this in design.

5.3.1 Model Factor Approach

For a single response, the deviation between measurement (e.g., load test) and prediction can be directly captured by the model factor defined below (e.g., ISO 2015; Tang and Phoon 2021):

$$M = X_m / X_c \tag{5.1}$$

where M = model factor and X_m and X_c = measured and predicted (or calculated) quantity, respectively. This method is practical, familiar to engineers, and grounded realistically on a load test database. In the context of geotechnical engineering, the response X could be, e.g., a load, resistance, deformation, or displacement parameter. The simplest way to characterize the model factor M is to calculate the mean λ (bias) and the coefficient of variation θ (COV) (dispersion). Table 5.1 provides a classification of the bias and dispersion.

The mean of M would provide an engineer with a sense of the hidden factor of safety that either adds or subtracts from the nominal global FS, depending on whether the capacity calculation method is conservative ($\lambda > 1$) or unconservative ($\lambda < 1$) in the *average sense*. Note that a movement (settlement, deflection, rotation, etc.) calculation method is conservative in the average sense when $\lambda < 1$. It should not be inferred that a capacity calculation method is conservative or otherwise for a specific

TABLE 5.1
Classification of Model Factor Based on Bias and Imprecision

Description	Dispersion (COV)	Bias (λ)	
		Overestimation	Underestimation
Low (dispersion)	$\theta < 0.3$		
Medium (dispersion) or moderate (bias)	$0.3 \le \theta < 0.6$	$0.5 \le \lambda < 1$	$1 \le \lambda < 2$
High (dispersion and bias)	$0.6 \le \theta < 0.9$	$\lambda < 0.5$	$2 \le \lambda < 3$
Very high (dispersion and bias)	$\theta \ge 0.9$		$\lambda \ge 3$

Source: Phoon and Tang (2019) and Tang and Phoon (2021).
Note:
$\lambda > 1$ is conservative for the ultimate limit state; $\lambda < 1$ is conservative for the serviceability limit state.

case because M takes a range of values in actuality (hence it is random) that may depend on the scenarios covered in the database used in the calibration. This random nature is practically significant because it implies that a calculation method can be unconservative when applied to a specific case even though the method is conservative on the average. Figure 5.4 shows that the actual capacity model factor for a specific case can be unconservative (shaded area that lies below the reference line of mean λ = 1) when the COV = 0.6 and the mean = 1.5. However, 95% of the model factor cases exceeds the line of mean λ = 1 even when the COV = 0.6 provided the mean = 2.5. Basically, the mean has to be large enough to compensate for a large dispersion to achieve a conservative design

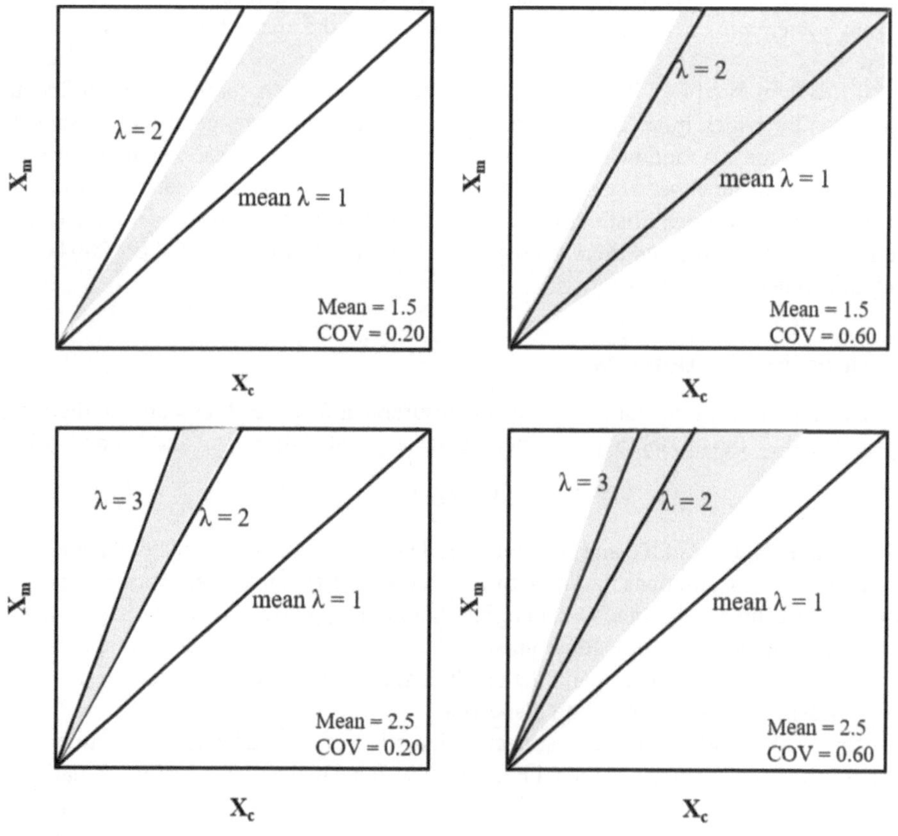

FIGURE 5.4 Effect of bias and dispersion on the 95% confidence interval of the capacity model factor (shaded area).

for 95% of the cases. In summary, it is necessary to consider the degree of scatter (dispersion) in M and to ensure that the probability of a measured value being lower than the calculated value is capped at a known value say p%. This idea is conceptually similar to EN 1997-1:2004 (CEN 2004), 2.4.5.2(11) which recommends that a cautious estimation (or characteristic value) for a geotechnical design parameter can be "derived such that the calculated probability of a worse value governing the occurrence of the limit state under consideration is not greater than 5%".

The authors argue that EN 1997-1:2004 Clause 2.4.16(P) "any calculation model shall be either accurate or err on the side of safety" must be interpreted in the same probabilistic sense of ensuring Prob($M < 1$) < 0.05 for the capacity model factor. If M is lognormally distributed, the following closed-form solution is available:

$$\text{Prob}(M<1)=\Phi\left(-\mu_{\ln M}/\sigma_{\ln M}\right) \tag{5.2}$$

where $\Phi(\cdot)$ is the standard normal cumulative distribution function, $\mu_{\ln M} = \ln(\lambda) - 0.5(\sigma_{\ln M})^2$ and $(\sigma_{\ln M})^2 = \ln(1 + \theta^2)$. The lognormal distribution is one member of the Johnson family of distributions (Ching and Phoon 2015). To ensure that a capacity calculation model is "sufficiently" conservative, it is necessary to introduce an additional partial factor of safety (Δ_p) on the model factor M:

$$M_p = \Delta_p \times M \tag{5.3}$$

The mean of $M_p = \Delta_p\lambda$. The COV of M_p is the same as the COV of $M = \theta$. If EN 1997-1:2004 Clause 2.4.16(P) is interpreted probabilistically as explained above, this additional partial factor of safety can be obtained:

$$\begin{gathered}\text{Prob}(M_p<1)<\frac{p}{100}\\ \text{Prob}(\Delta_p M<1)=\Phi\left(\frac{-\mu_{\ln\Delta M}}{\sigma_{\ln\Delta M}}\right)<\frac{p}{100}\end{gathered} \tag{5.4}$$

and

$$\begin{gathered}\frac{1}{\sqrt{\ln\left(1+\theta^2\right)}}\ln\left(\frac{\Delta_p\lambda}{\sqrt{1+\theta^2}}\right)=-\Phi^{-1}\left(\frac{p}{100}\right)\\ \Delta_p=\frac{\sqrt{1+\theta^2}\exp\left[-\Phi^{-1}(p/100)\sqrt{\ln\left(1+\theta^2\right)}\right]}{\lambda}\end{gathered} \tag{5.5}$$

For p = 5%, Eq. (5.5) is reduced to

$$\Delta_5=\frac{\sqrt{1+\theta^2}\,\exp\left[1.645\sqrt{\ln\left(1+\theta^2\right)}\right]}{\lambda} \tag{5.6}$$

Schneider (1997) recommended a characteristic value that is associated with a fractile of approximately p = 30%. Hence, p can take a value between 5% and 30%.

Unlike descriptive statistics that summarize the behavior of a population, inferential statistics allow engineers to make propositions about a population based on a set of sample data. Recently, Bayesian inference has become popular in geotechnical applications as indicated by some theme lectures (e.g., de Mello 1977; Whitman 1984; Christian 2004; Wu 2011; Juang and Zhang 2017; Baecher 2023). It should be noted that many statistical characterization studies pool all data into a single database (i.e., generic database) that could cover a variety of site locations, material types, test methods (e.g., scaled model test in 1g condition or centrifuge facility, and full-scale in situ test), construction sequence, and so on. The obtained model factor statistics may not be completely suitable for a specific site ("underfitting"). To address this shortcoming, Bozorgzadeh and Bathurst (2022) and Bozorgzadeh et al. (2023) introduced the hierarchical Bayesian approach in which data is parsed into smaller subgroups and the statistical parameters of different groups are assumed to be "similar but not identical".

Phoon et al. (2022c) framed this well-known problem of intra-site versus inter-site variability as a more general "site recognition" challenge. The challenge is to quantify "site uniqueness", directly or indirectly, so that sparse site-specific data can be supplemented by big indirect data (BID) to produce a quasi-site-specific model sensitive to site differences. Phoon et al. (2019) referred to big data as indirect to emphasize the point that big data exists in geotechnical engineering, but it is not directly relevant to one specific project at one specific site. Generic soil property databases (e.g., Phoon et al. 2016b; Ching et al. 2016) and load test databases (e.g., Phoon and Tang 2019; Tang and Phoon 2021) are examples of big indirect data. The value of a quasi-site-specific model is that it is less biased than a generic model (containing abundant data from many sites) and less imprecise than a site-specific model (containing sparse data from one site). This "site recognition" challenge is considered intractable because site-specific data is "ugly". "Ugly" refers to data attributes that depart significantly from the assumptions in classical statistics. One example is MUSIC-X (multivariate, uncertain and unique, sparse, incomplete, and potentially corrupted with "X" denoting the spatial/temporal variability). However, recent studies using Bayesian machine learning (Ching and Phoon 2019) have shown promising results (e.g., Ching and Phoon 2020; Ching et al. 2021a, 2021b, 2022a, 2022b; Phoon and Ching 2021, 2022; Sharma et al. 2022). More research studies on site-specific model calibration should be conducted in the future. This is an important aspect of a broad agenda of "precise" design and construction. EN 1997-1:2004 Clause 2.4.5.2(10) alludes to the site recognition challenge: "If statistical methods are employed in the selection of characteristic values for ground properties, such methods should differentiate between local and regional sampling and should allow the use of a priori knowledge of comparable ground properties" (CEN 2004). However, there is no statistical or quantitative guidance on how to differentiate between local sampling (site-specific data) and regional sampling (big indirect data).

5.3.2 Geotechnical Databases

It has been recognized that databases (soil/rock property, load test, case history, etc.) play an indispensable role in the advancement of geotechnical engineering (e.g., Abu-Hejleh et al. 2015; Tang and Phoon 2021), which can be used as (1) a research tool to increase our knowledge on the behavior of geotechnical structures (e.g., load transfers between pile shaft shearing and tip end bearing, failure mechanism of slope, interaction between soil and reinforcements in a mechanically stabilized earth wall and a soil nail wall, and load and deformation of a retaining structure and ground surface induced by excavation or tunnel); and (2) a practical tool to establish proper guidelines (e.g., estimation of geotechnical design parameters and calculation methods), develop site-specific design to reduce uncertainty, and assist in selecting an appropriate construction method and equipment to manage the risk (e.g., pile damage or collapse of support structures). Some notable databases are summarized in Table 5.2 covering in situ load testing on a great variety of foundations (e.g., shallow and deep – prefabricated or cast in situ), centrifuge modeling of spudcan penetration in stiff-over-soft clay and sand-over-clay, case histories of slope failure and basal heave stability in excavation, and laboratory model tests of a mechanically stabilized earth wall and a soil nail wall. Tang and Phoon (2021) suggested a more descriptive template to label these databases: <con tributor>/<structure>/<number of records>. Examples of databases labeled in this way are NUS/DrilledShaft/320 (Tang et al. 2019), NUS/HelicalPile/1113 (Tang and Phoon 2018a, 2020), CYCU/TipGrouting/34 (Chen et al. 2021), and CYCU/RockSocket/50 (Topacio et al. 2022).

5.3.3 Statistics of Model Factor and Implications

The model factor statistics (mean λ and COV θ) have been reviewed and summarized in the literature (e.g., Phoon and Tang 2019; Tang and Bathurst 2021; Tang and Phoon 2021). Recently, Otake

TABLE 5.2
Summary of Geotechnical Databases

Type	Database			
	References	ID	Geomaterial	Data records
Foundation	Kulhawy et al. (1983)	EPRI	Various	Load test (804)
Shallow foundation	Akbas (2007)	University of Cornell	Sand	Load test (167) Settlement (426)
	Paikowsky et al. (2010)	UML-GTR ShalFound07	Various	Load test (549)
		UML-GTR RockFound07	Rock	Load test (122)
	Tang et al. (2020)	NUS/ SpreadFound/919	Soil	Load test (919)
		NUS/RockFound/270	Soft rock	Load test (270)
	Bahmani and Briaud (2020)	TAMU-SHAL-SAND	Sand	Settlement (315)
	Bahmani and Briaud (2021)	TAMU-SHAL-CLAY	Clay	Settlement (103)
Spudcan	Tang et al. (2023)	DUT/Spudcan/242	Layered	Penetration (242)
Deep foundation	Paikowsky et al. (2004)	NCHRP Report 507	Various	Load test (804)
	Dithinde et al. (2011)	South Africa	Soil	Load test (174)
	Burlon et al. (2014)	IFSTTAR	Various	Load test (174)
	Chen et al. (2014)	WBPLT	Soil	Load test (673)
	Petek et al. (2016)	FHWA DFLTD v.2	Various	Load test (1341)
	Yu et al. (2017)	Caltrans	Soil	Load test (189)
	PWRI (2018)	PWRI	Various	Load test (696)
	Eslami et al. (2020)	AUT: GEO-CPT&Pile	Soil	Load test (600)
	Vardanega et al. (2021)	DINGO	Various	Load test (551)
Drilled shaft	Lin et al. (2012)	DSLT	Soil	Load test (351)
	AbdelSalam et al. (2015)	Egypt	Soil	Load test (318)
	Motamed et al. (2016)	NVDOT	Cemented	Load test (41)
	Briaud and Wang (2018)	TAMU-LATERAL	Soil	Load test (110)
	Asem (2018)	UIUC	Soft rock	Shaft shearing (317) End bearing (190)
	Kalmogo et al. (2019)	Iowa DOT	Various	Load test (51)
	Mullins et al. (2019)	FDOT (Tip grouted)	Soil	Load test (31)
	Tang et al. (2019)	NUS/ DrilledShaft/320	Soil	Load test (320)
	Chen et al. (2021)	CYCU/ TipGrouting/34	Soil	Load test (34)
	Ong et al. (2021)	Dr. Toh Associates	Soft rock	Load test (127)
	Chou et al. (2022)	CYCU/Lateral/23	Soil	Later test (23)
	Topacio et al. (2022)	CYCU/ RockSocket/50	Rock	Load test (50)
Auger cast pile	McVay et al. (2016)	FDOT	Various	Load test (78)
	Reddy and Stuedlein (2017)	Oregon State University	Sand	Load test (112)
	Figueroa et al. (2022)	UCI	Soil	Load test (130)
Helical pile	Tang and Phoon (2018a, 2020)	NUS/HelicalPile/1113	Soil	Load test (1113)
	Souissi et al. (2020)	CTL Thompson, Inc.	Various	Load test (799)
	Silva et al. (2023)	Brazil	Soil	Load test (107)
Driven pile	McVay et al. (2000)	PILEUF	Various	Load test (285)
	Long et al. (2009a)	IDOT	Various	Load test (250)
	Long et al. (2009b)	WisDOT	Various	Load test (316)
	Smith et al. (2011)	Full PSU Mater	Various	Load test (322)

(Continued)

TABLE 5.2 (CONTINUED)
Summary of Geotechnical Databases

Type	Database			
	References	ID	Geomaterial	Data records
	AbdelSalam et al. (2012)	PILOT	Soil	Load test (274)
	Long and Anderson (2014)	IDOT	Various	Load test (111)
	Tavera et al. (2016)	LODOTD	Various	Load test (1186)
	Lehane et al. (2017)	JIP Unified database	Soil	Load test (120)
	Yang et al. (2017)	ZJU-ICL	Sand	Load test (117)
	Tang and Phoon (2018b, c, 2019)	NUS/DrivenPile/550	Soil	Load test (550)
	Flynn and McCabe (2022)	ICP	Various	Load test (117)
	Ng et al. (2022)	MPC	IGM	Load test (223)
DCIS pile	Long (2013)	WisDOT	Various	Load test (182)
	Flynn (2014)	NUI Galway	Sand	Load test (105)
Micropile	Loehr et al. (2022)	NCHRP Report 989	Various	Load test (726)
Slope	Travis et al. (2011)	WEST Consulting, Inc.	Various	Slope failure (157)
	Timchenko and Briaud (2023)	TAMU-MineSlope	Various	Open pit mine slope failure (134)
MSE wall	Huang and Bathurst (2009)	Geogrid	Sandy soil	Pullout box test (478)
	Miyata and Bathurst (2012a)	Steel strip	Sand/gravel	Pullout box test (351) In situ pullout test (301)
	Miyata and Bathurst (2012b)	Geogrid	Soil	Pullout box test (503)
	Wood et al. (2012a)	TxDOT	Sandy soil	Pullout box test (367)
	Wood et al. (2012b)	TxDOT	Gravelly soil	Pullout box test (320)
	Yu and Bathurst (2015)	Steel Bart Mat	Coarse grained soil	Pullout box test (356)
	Miyata et al. (2014)	Geogrid	Soil	Creep test (362)
	Miyata and Bathurst (2015)	Geogrid	Soil	Installation damage test (520)
	Allen and Bathurst (2015)	Geosynthetic	Coarse grained soil	Tensile load (193)
		Steel		Tensile load (185)
	Miyata and Bathurst (2019)	Steel grid	Soil	Tensile load (113)
SNW	Lazarte (2011)	NCHRP Report 701	Various	Pullout resistance (153)
	Lin et al. (2017a)	Ryerson University	Soil	Nail load (123)
	Lin et al. (2017b)	Ryerson University	Soil	Pullout resistance (113)
	Yuan et al. (2019a)	Guangzhou University	Soil	Nail load (144)
	Yuan et al. (2019b)	Guangzhou University	Soil	Wall displacement (461)
	Liu et al. (2020)	Foshan University	Various	Facing tensile load (56)
Excavation	Long (2001)	UCD	Soil	296 case histories
	Moormann (2004)	Worldwide	Soft soil	530 case histories
	Wang et al. (2010)	Shanghai	Soft soil	300 case histories
	Tan et al. (2022)	Shanghai	Soft soil	592 case histories

Note: MSE = mechanically stabilized earth wall; SNW = soil nail wall; IGM = intermediate geomaterial; DCIS = driven cast-in-place; EPRI = Electrical Power Research Institute; IFSTTAR = Institut Français des Sciences et Technologies des Transports, de l'Aménagement et des Réseaux; WBPLT = web-based pile load test; Caltrans = California Department of Transportation; PWRI = Public Works Research Institute; DSLT = drilled shaft load test database; NVDOT = Nevada Department of Transportation; FDOT = Florida Department of Transportation; UCI = University of California, Irvine; WisDOT = Wisconsin Department of Transportation; IDOT = Illinois Department of Transportation; LODOTD = Louisiana Department of Transportation and Development; ICP = instrumented concrete pile; MPC = Mountain-Plains Consortium; NUI = National University of Ireland; TxDOT = Texas Department of Transportation; and UCD = University College Dublin.

and Honjo (2022) reported similar calibration studies with data from Japan. The results cover a variety of geotechnical structures (e.g., shallow foundation, offshore spudcan, anchor/pipe, drilled shaft, driven pile, helical pile, mechanically stabilized earth wall, soil nail wall, slope and deep excavation) in a wide range of ground conditions (e.g., soft to stiff clay, loose to dense sand, intermediate geomaterial, and soft rock). Analysis and design models for two typical limit states – ULS and SLS – are calibrated. The database used for calibration purposes includes measurements or observations from laboratory testing (scaled model or prototype in a centrifuge facility) (representing controlled soil and boundary conditions) and in situ investigation (representing natural soil and boundary conditions). The difference between the model factor statistics for laboratory and field tests was first studied by Phoon and Kulhawy (2005) using a database of laterally loaded rigid drilled shafts.

Some of the most recent studies are reviewed to update the summary of model factor statistics in Tang and Phoon (2021) and Tang and Bathurst (2021), including (1) settlement of shallow foundation in sand (Bahmani and Briaud 2020) and clay (Bahmani and Briaud 2021); (2) penetration behavior (resistance–depth profile and peak resistance) of offshore spudcan in stiff-over-soft clay and sand-over-clay (Tang et al. 2023); (3) geotechnical resistance of axially loaded micropile (Loehr et al. 2022); (4) axial capacity of large-diameter open-ended pile (LDOEP) (Rizk et al. 2022), auger cast pile (Figueroa et al. 2022), helical pile (Souissi et al. 2020; Silva et al. 2023), and driven pile on intermediate geomaterial (Ng et al. 2022); (5) pile settlement in fine-grained soil (Voyagaki et al. 2022); (6) resistance of laterally loaded pile (Otake and Honjo 2022); and (7) failure of open pit mine slope (Timchenko and Briaud 2023). The mean λ and COV θ values and the number of tests (N) are averaged over n-data groups for the same geotechnical structure, limit state, and geomaterial as plotted in Figure 5.5, from which it can be concluded (e.g., Phoon and Tang 2019; Tang and Bathurst 2021; Tang and Phoon 2021):

- The characterization of ULS model factors received most of the attention in the literature (foundation capacity is the most prevalent), while the characterization of the SLS model factor is relatively limited. This is because only strength parameters (e.g., cohesion,

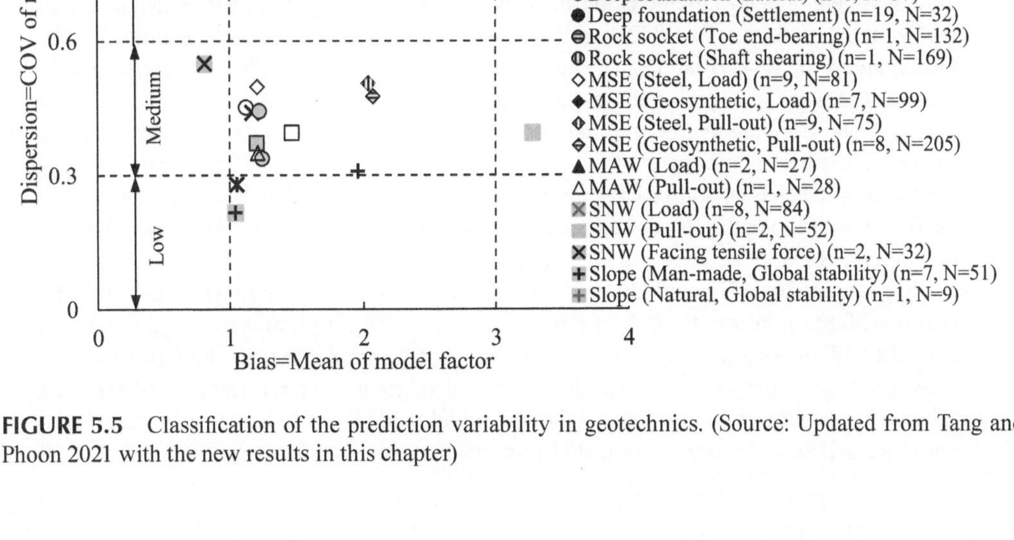

FIGURE 5.5 Classification of the prediction variability in geotechnics. (Source: Updated from Tang and Phoon 2021 with the new results in this chapter)

friction angle, or uniaxial compressive strength) are required in stability analysis that are familiar to engineers and are most often measured in laboratory testing or obtained from in situ investigation.

- The geotechnical resistance factors for ULS and SLS were recommended in Table 6.2 of the 2014 edition of the Canadian Highway Bridge Design Code with respect to the degree of (site and model) understanding (low, typical, and high) (Fenton et al. 2016). Site understanding refers to how well the ground providing the geotechnical resistance is known. Model understanding means the degree of confidence that a designer has in the model used to predict the geotechnical resistance. Model factor lumps the variability of ground conditions (site understanding) and predictions (model understanding). The variability of "ground conditions" is complex and not well studied. The reason is that the *value affecting the occurrence of a limit state* is not the point value or the spatial average but a mobilized value (Phoon et al. 2023a). To be more precise, it should be termed the variability of the "mobilized value" which is complex. To some extent, model factor statistics (mean λ and COV θ) can be an effective indicator of the degree of understanding.
- Most of the model factor statistics for the axial resistance of shallow and deep foundation in soil, pullout resistance, or the tensile load of steel-reinforced earth walls and soil nail walls vary between 1–3 (mean λ) and 0.3–0.6 (COV θ). Foundation analysis and design in soil is an important and classical problem in geotechnical engineering that has been studied since the early 1900's spanning over one century (e.g., load-displacement response and load transfers). For a reinforced soil wall, the interaction between steel reinforcements (strip or grid) and the soil is no more complex than the interaction observed for geosynthetics. The degree of site and model understanding is regarded as "typical".
- The θ values of a few calculation methods are around 0.3, such as (1) stability of soil slope, (2) pullout resistance of pipe (calibrated by scaled model testing in 1g condition), and (3) punch-through resistance of offshore spudcans in stiff-over-soft clays or clay with sand (evaluated by centrifuge tests on scaled model in ng condition). For these cases, soil samples are well prepared, corresponding to lower geotechnical variability. In addition, slope stability is an important and classical problem in geotechnical engineering that has been extensively studied since the 1930s, leading to a better understanding of the slope failure mechanism and improved analysis methods (Duncan et al. 2014). For the pullout capacity of pipes and the peak penetration resistance of offshore spudcans in stiff-over-soft clay or sand-over-clay, due to the increasing demand of offshore oil and gas, many laboratory tests were performed to study the underlying mechanism and improve the performance of the calculation models, as reviewed by Tang and Phoon (2021) and Tang et al. (2023). The degree of site and model understanding is regarded as "typical" to "high".
- The assessment of design methods for the resistance of large-diameter open-ended pile (e.g., Petek et al. 2020; Rizk et al. 2022) and laterally loaded pile (e.g., Phoon and Kulhawy 2005; Boeckmann et al. 2014; Briaud and Wang 2018) is relatively limited. This is because significant challenges are involved in addressing both problems, such as the complicated behavior (e.g., plug and development of internal friction) and difficulty in LDOEP installation and obvious deficiencies in the current limit state design method for laterally loaded pile that only cover a specific behavior and lack the ability to properly accommodate both ULS and SLS.
- The θ values of the methods for settlement analysis (e.g., Akbas 2007; Abchir et al. 2016; Bahmani and Briaud 2020; Voyagaki et al. 2022) and drilled or driven pile capacity in soft rock (e.g., Asem 2018; Ng et al. 2019, 2022) are greater than 0.6. The reasons may include: (1) soil stiffness is more difficult to determine than strength parameters – geotechnical variability; (2) only rock compressive strength is incorporated into the calculations, which is insufficient as rock mass is usually composed of joints, seams, faults, and bedding planes – incomplete physics in the predictive model; and (3) the interaction between geosynthetics and soil is more complicated than that between steel reinforcements and soil

– limited knowledge on the interaction behavior. They possibly correspond to a lower degree of site and model understanding.

- The studies of Phoon and Tang (2019), Tang and Bathurst (2021), and Tang and Phoon (2021) reported the model factor statistics for a variety of design conditions worldwide. On the other hand, the data from Japan is based on applying the same design method to each geotechnical structure (Otake and Honjo 2022). The design condition might be more similar. Nonetheless, the variability of predictions is comparable to those observed by Tang and Phoon (2021) and others.
- Based on the mean of the capacity model factor, the bias of the calculation model can be classified as (1) unconservative (mean $\lambda < 1$), (2) moderately conservative ($1 \leq \lambda \leq 3$), and (3) highly conservative ($\lambda > 3$). Note that the mean of the deformation model factor is conservative when $\lambda < 1$. Based on the COV of the model factor, the dispersion of the calculation model can be classified as (1) low (COV $\theta < 0.3$), (2) medium ($0.3 \leq \theta \leq 0.6$), and (3) high ($\theta > 0.6$). Note that "low dispersion" means "high precision" and vice versa. This three-tier classification scheme in Table 5.1 was suggested by Tang and Phoon (2021) based on extensive statistical analyses covering numerous geo-structures and soil types.

Section 5.3.1 showed that the model factor statistics are useful even for deterministic design codes such as EN 1997-1:2004 (CEN 2004). A simple additional partial factor of safety (Δ_p) can be introduced to the model factor (M) to satisfy Clause 2.4.16(P) "any calculation model shall be either accurate or err on the side of safety" in the probabilistic sense of ensuring Prob(model factor <1) < 0.05 for the capacity model factor. Engineers can divide the calculated capacity (X_c) by Δ_p based on the values summarized in Table 5.3 to achieve M_p: $M_p = \dfrac{X_m}{(X_c / \Delta_p)} = M \times \Delta_p$

- Note that M depends on the calculation method and it is not possible to achieve Prob(M < 1) < p/100 for any given p%. However, it is possible to achieve Prob(M_p < 1) < p/100 by reducing X_c to X_c/Δ_p. This is obvious because the fractile p% reduces when Δ_p increases (factored X_c becomes more conservative). Hence, Δ_p does act as an additional partial factor of safety. It can be seen that for conventional shallow and deep foundations in soil, most of the Δ_5 values (Δ_p values needed to satisfy a 5% fractile) are between 1.4 and 3, while those for a drilled shaft in soft rock and large-diameter open-ended pile can be larger than 5. One may argue that some calculation models are too imprecise for design, because Clause 2.4.16(P) cannot be complied with even probabilistically without introducing an extremely large factor of safety to the design. For a larger fractile = 25%, most of the Δ_{25} values are between 0.9 and 1.9 for shallow and deep foundations in soil and between 1 and 3 for a drilled shaft in soft rock and LDOEP (with the exception of the FHWA soil plug model). Table 5.3 demonstrates the value of characterizing the model factor as a random variable even for current deterministic practice. It is possible to derive a partial factor of safety consistently to ensure Prob(M_p < 1) < p/100 for any calculation method and any p% fractile (based on any probability distribution for M).

5.4 LESSONS FROM INTERNATIONAL PREDICTION EVENTS

A prediction event is occasionally organized as part of a conference, where geotechnical engineers are invited to predict the outcome of a field measurement or an in situ test on full-scale geotechnical structures. Some of the events can be very ambitious (e.g., Constructed Facilities Division 1975a, b; DiMillio et al. 1988; Finno 1989; Fellenius et al. 2017). They usually attract considerable interest at many international conferences. The participants are presented with information on the soil conditions at a specific site, details on the geotechnical structure to be measured or tested, and the testing program. Participation in a prediction event can be somewhat humbling, as it invariably reveals the limits of one's knowledge and the limits of prediction precision in the presence of uncertainties.

TABLE 5.3
Additional Partial Factor of Safety Δ_p to Ensure Prob(M_p < 1) < p/100 for Shallow and Deep Foundation with p = 5% and Revised as "25%"

Foundation	Design method	Case	Statistics of M		Additional partial factor of safety Δ_p	
			λ	θ	p = 5%	p = 25%
Shallow foundation	Vesic	Bearing (clay)	1.05	0.29	1.58	1.20
		Bearing (sand)	1.64	0.47	1.40	0.91*
	IEEE	Uplift (clay)	1.15	0.36	1.64	1.17
		Uplift (sand)	1.10	0.33	1.62	1.19
Driven steel pile (H)	API-α	Bearing (clay)	1.26	0.56	2.15	1.29
	Nordlund	Bearing (sand)	0.82	0.52	3.07	1.91
	Burland-β	Bearing (mixed)	0.81	0.40	2.51	1.72
Driven steel pile (pipe)	API-α	Bearing (clay)	1.02	0.32	1.72	1.27
	NGI-05	Bearing (clay)	1.10	0.29	1.51	1.15
	SHANSEP	Bearing (clay)	1.14	0.27	1.41	1.09
	ICP-05	Bearing (clay)	1.06	0.28	1.54	1.18
	Fugro-05	Bearing (sand)	0.95	0.36	1.99	1.42
	ICP-05	Bearing (sand)	1.13	0.30	1.50	1.13
	UWA-05	Bearing (sand)	1.08	0.37	1.78	1.26
Driven concrete pile	API-α	Bearing (clay)	1.09	0.34	1.67	1.21
	NGI-05	Bearing (clay)	0.95	0.26	1.66	1.29
	SHANSEP	Bearing (clay)	1.01	0.34	1.80	1.31
	ICP-05	Bearing (clay)	1.04	0.35	1.78	1.28
	API-β	Bearing (sand)	0.95	0.37	2.02	1.43
	NGI-05	Bearing (sand)	0.83	0.33	2.15	1.58
	Fugro-05	Bearing (sand)	0.87	0.41	2.38	1.62
	ICP-05	Bearing (sand)	1.13	0.29	1.47	1.12
	UWA-05	Bearing (sand)	1.00	0.33	1.79	1.31
Drilled shaft in soil	FHWA	Bearing (clay)	1.41	0.63	2.17	1.24
		Bearing (sand)	1.19	0.39	1.68	1.16
		Bearing (mixed)	1.69	0.47	1.36	0.88*
Drilled shaft in soft rock	Williams	Shaft shearing	1.15	1.09	5.51	2.34
	RA	Shaft shearing	2.58	1.09	2.46	1.04
	CK	Shaft shearing	1.38	1.20	5.35	2.14
LDOEP	FHWA	Soil (unplugged)	1.09	1.04	5.41	2.36
		Soil (plugged)	0.62	1.25	12.73	4.96
		Soil (internal plug)	0.82	1.00	6.78	3.02
	USACE	Soil (unplugged)	1.32	0.69	2.57	1.40
		Soil (plugged)	0.82	0.86	5.47	2.66
		Soil (internal plug)	1.03	0.66	3.13	1.74
	Revised lambda	Soil (unplugged)	1.47	0.61	2.01	1.16
		Soil (plugged)	0.83	0.68	4.02	2.21
		Soil (internal plug)	0.99	0.66	3.25	1.82
	API	Soil (unplugged)	1.52	0.73	2.39	1.27
		Soil (plugged)	0.86	0.80	4.74	2.39
		Soil (internal plug)	1.04	0.74	3.55	1.87
	NGI-05	Soil (unplugged)	1.16	0.69	2.92	1.59
		Soil (plugged)	0.92	0.71	3.81	2.05
		Soil (internal plug)	0.95	0.68	3.51	1.93

(*Continued*)

TABLE 5.3 (CONTINUED)
Additional Partial Factor of Safety Δ_p to Ensure Prob(M_p < 1) < p/100 for Shallow and Deep Foundation with p = 5% and Revised as "25%"

Foundation	Design method	Case	Statistics of M		Additional partial factor of safety Δ_p	
			λ	θ	p = 5%	p = 25%
	ICP-05	Soil (unplugged)	1.10	0.95	4.69	2.15
		Soil (plugged)	0.91	0.76	4.19	2.18
		Soil (internal plug)	0.93	0.82	4.52	2.25
	Fugro-05	Soil (unplugged)	1.23	0.87	3.70	1.79
		Soil (plugged)	0.86	0.81	4.81	2.42
		Soil (internal plug)	0.92	0.85	4.80	2.35
	UWA-05	Soil (unplugged)	1.19	0.87	3.83	1.85
		Soil (plugged)	0.87	0.82	4.84	2.41
		Soil (internal plug)	0.97	0.79	4.13	2.10

Source:
The statistics of model factor for LDOEP are taken from Rizk et al. (2022), while the other statistics are from Tang and Phoon (2021).

Note: *To be judicious, it may be better to impose $\Delta_p > 1$ in design. The use of $\Delta_p < 1$ implies adjusting the current calculation model to be less conservative. This is correct when λ is very large but an engineer may not be comfortable to do this.

Besides, it is also educational to demonstrate our current understanding – "what we know" and areas needing more work – "what we don't know", from which our future research can be guided to improve practice in a more purposeful way. Prediction exercises in geotechnical engineering should be understood in the context proposed by Lambe (1973):

- Class A: Predictions are carried out before the construction event only with the available site investigation (e.g., laboratory testing, in situ investigation, and geophysical survey) and geometry data.
- Class B: Predictions are made during the construction event so that they can be influenced by initial field data.
- Class C: Analyses are carried out after the construction event when the complete set of field data is made available and appropriate soil properties can be adjusted by back-calculation.

The prediction variability observed in these events is not the same as the COV of the model factor. This is because the predictions are made by different models and different choices of input parameters, and are subsequently modified in different ways by drawing on the experience of the participants. Another distinction is that the model factor statistics shown in Section 5.3 are generic in the sense that the records in the database are compiled from tests conducted at different sites. Prediction events are mainly carried out at a single site. Several international prediction events are revisited (Table 1.1): (1) embankment on soft soil (Constructed Facilities Division 1975a, b; Kelly et al. 2018), (2) shallow foundation (Briaud and Gibbens 1997; Doherty et al. 2018), (3) offshore spudcan (van Dijk and Yetginer 2015), (4) axially loaded single pile (Fellenius 1988, 2013, 2015, 2021; Finno 1989; Holeyman et al. 2001; Jardine et al. 2001; Holeyman and Charue 2003; Fellenius et al. 2004, 2019, 2017; Viana da Fonseca and Santos 2008; Reiffsteck 2005, 2009; Fellenius and Terceros 2014) and pile group (DiMillio et al. 1988), and (5) laterally loaded single pile (Guevara et al. 2022). The quantity of interest predicted includes capacity, settlement, pore water pressure, and load-deformation response. The following conclusions can be drawn from the outcomes of these prediction events:

- For the settlement and pore water pressure of embankment on soft soil, the COV values of the predicted/measured ratio are close to or larger than 1, indicating a high variability of predictions. The accurate prediction of embankment behavior on soft soil remains a challenge not only for practitioners, but also for academics.
- For shallow foundation, the stress dependency of the sand friction angle was not properly considered in the analysis. A large variety of methods were employed but it was difficult to identify the most accurate one, as most participants modified the methods further based on their personal experience. More significantly, predicted settlement values are very conservative on the safe side (mean of "model" factor < 0.5 or predicted value is two times larger than the measured value on the average) with a surprisingly high degree of variability (COV > 1), which is similar to the actual model factor results from database calibration (e.g., Akbas 2007; Bahmani and Briaud 2020). Possible reasons are that our predictive models remain disconnected from the data and the soil modulus is a complex function of strain, confining pressure, rate effects, soil structure, etc.
- For a single pile under axial loading, the "capacity" is defined in many different ways from the load–movement curve of a static loading test. There are more than 50 interpretation criteria, as reviewed and summarized in Niazi (2014). Interpretation criteria such as the L_1–L_2 (Hirany and Kulhawy 1988, 1989, 2002), the slope tangent method (O'Rourke and Kulhawy 1985), and Davisson (1972), DeBeer (1970), DIN 4026 (1978), Terzaghi and Peck (1967), Fuller and Hoy (1970), van der Veen (1953), and Chin's (1970) methods are widely adopted. Several prediction exercises showed that the COV of interpreted capacities ranged between 0.1 and 0.3. The prediction of the total capacity is of medium dispersion (COV = 0.3–0.6), which is similar to the results from database assessment (Tang and Phoon 2021). Interestingly, the variability of predicting shaft shearing resistance or tip end bearing pressure separately is almost always higher than that for the total capacity. One reason is that the soil mobilized by shaft shearing and tip end bearing does not occupy two distinct volumes. The uncertainty of the sum of two relatively independent quantities is less than the uncertainty of any one component quantity, because there is a variance reduction effect due to averaging. A simpler way to understand this variance reduction is that an overestimation of the shaft shearing can be compensated by an underestimation of the tip end bearing or vice versa. In addition to instrumented pile testing, the Osterberg cell test is another effective solution allowing the separation and direct measurement of shaft shearing and tip end bearing resistances. The prediction of displacement at serviceability loading (e.g., at one half of interpreted/calculated capacity or a factor of safety equal to 2) is of high dispersion (COV > 0.6). Also, this is similar to the results from database evaluation (e.g., Abchir et al. 2016; Voyagaki et al. 2022). The predicted and observed (pile head, shaft, and tip) load–displacement curves agree qualitatively only, which is also observed in the prediction of the penetration resistance–depth profile of an offshore spudcan by Tang et al. (2023). Moreover, experience drawn from the same/similar site and piling conditions can help the participant predict pile behavior more accurately. This supports the widely held view that engineering judgment is important. Engineering judgment when applied in this context can be regarded as an informal form of Bayesian updating (e.g., Christian 2004; Wu 2011; Baecher 2023).
- The absolute and differential settlements of pile group predicted by linear elastic, equivalent linear elastic, and non-linear linear methods were compared by Viggiani (1998) with 42 well-documented case histories. It was found that a simple linear elastic analysis led to rather satisfactory agreement with the observed values in most of the cases and may be expected to be adequate for engineering purposes, similar to those shown by the prediction results in DiMillio et al. (1988). This is because the linear elastic analysis was carried out based on a small strain moduli back-calculated from the initial stiffness of the load test on single piles.

- For centrifuge modeling of laterally loaded pile, the load–deflection curves and bending moment profiles predicted by the p–y curve and finite element/finite difference methods exhibit a large variation that qualitatively agrees with the database study of Briaud and Wang (2018) and Otake and Honjo (2022). The COV of the predicted pile head load could be as high as 0.7.
- For the case histories of offshore spudcan, there is a high scatter in the predicted vertical bearing pressure to penetration. As comprehensive soil data was not available for all of the cases, the more advanced large deformation finite element analyses did not necessarily provide better predictions than those using limit equilibrium methods. The observed wide range in predictions was mostly due to the variabilities in data interpretations and the selection of related soil parameter values.
- Many participants preferred to use their judgment to determine soil parameters, rather than exploit the high-quality laboratory and in situ test results provided. This could be one major limitation in making good predictions. As a consequence, a wide variety of soil parameters were selected, leading to a large prediction variability, even though the same method was adopted. Diaz-Segura (2013) assessed the range of variation of bearing capacity factor N_γ estimated by 60 (empirical or semi-empirical) methods for rough footing on sand. The analysis showed that N_γ can differ by up to 267% for the same ϕ value. Variability in the estimation of ϕ arising from the application of correlations with in situ tests leads to a range of variation for N_γ higher than that obtained by the 60 estimation methods. This confirmed the observation made by Lambe (1973) in the Thirteenth Rankine Lecture – the accuracy of a prediction depends both on the quality of the input data (also the quantity of data available) and the method used for prediction, as shown in Figure 5.6.

5.5 NUMERICAL ANALYSES VERSUS IN SITU MEASUREMENTS

5.5.1 Overview

Numerical analysis methods with relatively advanced and sophisticated constitutive models, which are mechanically consistent, provide very powerful tools to carry out parametric studies. They can account for a wider range of related parameters as an effective complement to laboratory or field tests (typically limited), increase our knowledge of the behavior of geotechnical structures, and

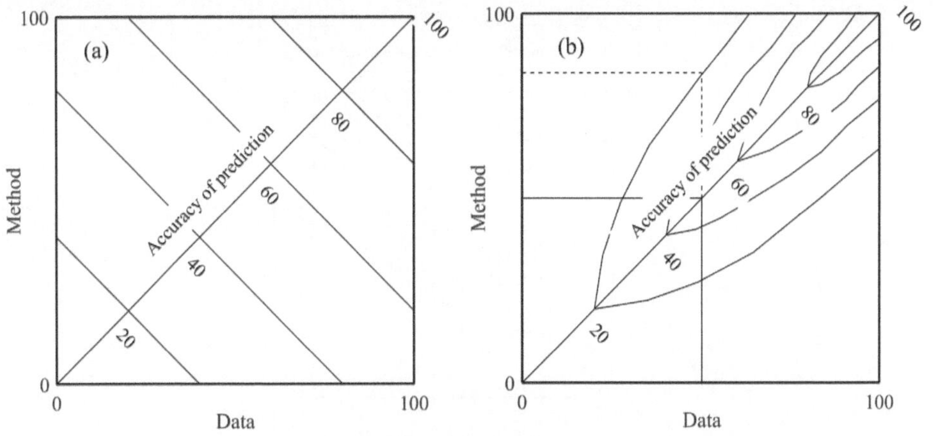

FIGURE 5.6 Accuracy of prediction as a function of the quality of model and input data. (Source: Adapted from Lambe 1973)

examine and improve the accuracy of predictions. It is tempting to say that advanced numerical analyses should outperform those simplified empirical and semi-empirical design methods. In her 2019 Canadian Geotechnical Colloquium paper, however, Kalenchuk (2022) provided an in-depth discussion of the practical limitations in the day-to-day application of numerical methods in geotechnical engineering. It is largely associated with practitioners' overconfidence of the ability of numerical tools to carry out sophisticated computations and their disregard (or lack of attention) of the uncertainty in the results obtained. In the KIVI lecture, Whyte (2018) pointed out three categories of constitutive models for FEM based on performance (speed, robustness, and calibration) versus predictive capability (Figure 5.7). The first is a simplistic model, which is much easier to use for prediction, but can be inaccurate (ease of calculation over performance). In contrast, a philosopher model is one constructed with significant theoretical rigor and sophistication (performance over ease of calculation). This type of model could be difficult to use in practice because the parameters may not be measured in a commercial soil testing laboratory and there may be computational convergence issues. The majority of the models published in the literature belong to this category. They overlook the fact that the degree of complexity in the adopted numerical model must be justified by the quality and quantity of the geomaterial and the performance data available. The third type is an engineer model that provides a more reasonable balance between performance and ease of calculation. Practitioners prefer to use the third type in analysis and design.

Model calibration is the process of correlating the observations of actual ground behavior to numerical output, and analysis/design models are then tailored to a site-specific problem (e.g., Lee et al. 2013; Taborda et al. 2020; Zdravković et al. 2020). This is a sensible and practical step to reduce numerical uncertainty. Depending on the type and quantity of data, model calibration ranges from qualitative to quantitative, with the increasing complexity of the model and the increasing quality and quantity of the data (Figure 5.8). Quantitative calibration is exceptionally difficult to achieve in

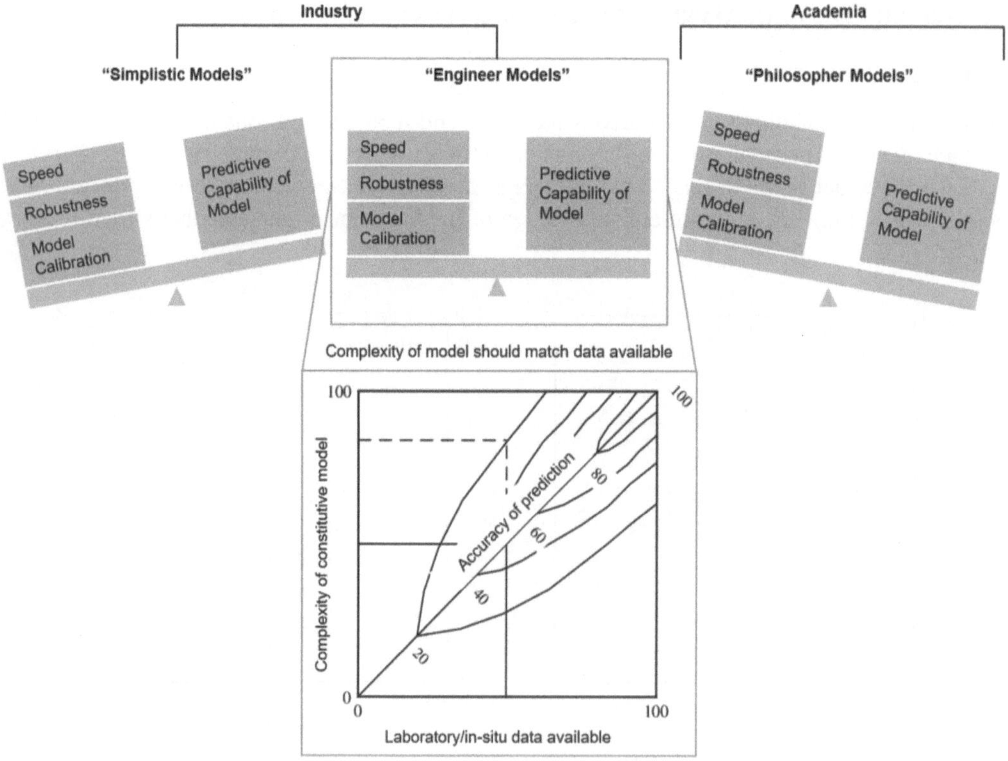

FIGURE 5.7 Three types of geotechnical analysis and design model. (Source: Adapted from Whyte 2018)

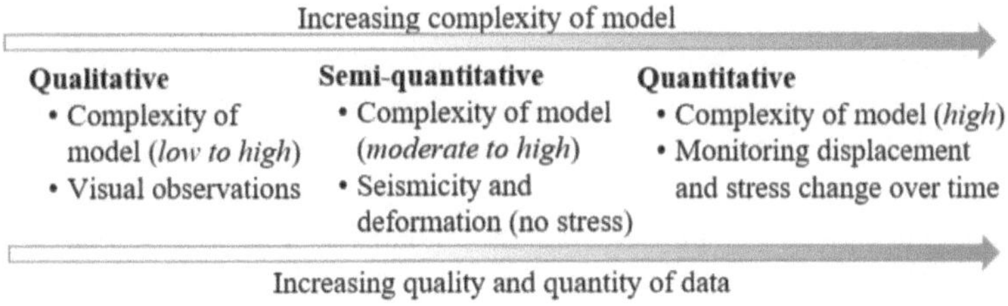

FIGURE 5.8 Qualitative versus quantitative model calibration.

a data-poor environment. Nevertheless, it is feasible nowadays with the advent of digital technologies. For example, wired and wireless networks allow nearly continuous monitoring of relevant parameters such as deformations and stress, while machine learning is very powerful in analyzing the collected big data that is beneficial for model calibration (Li et al. 2023; Yan et al. 2023). This is one of the problems that "data-centric geotechnics" aims to solve. Data from long-term monitoring should be integrated into the design process to improve site-specific knowledge and provide a basis for future decisions on the operation and maintenance of a facility (D'Appolonia 1990). Examples are given in Phoon et al. (2023b).

Compared to the use of numerical analysis methods in research and practice, systematic investigations on their prediction quality with sufficient measured or monitoring data are relatively limited in the literature. A few qualitative studies are available in the literature. However, they do not carry out statistical quantification of the discrepancies between numerical modeling results and in situ measurements. Duncan (1994) reviewed 100 publications on the use of finite element analyses in a variety of geotechnical problems (e.g., soil–structure interaction, slope, embankment, soil anchor, and tunnel) and discussed the role of advanced constitutive models of geomaterial in practical applications (e.g., elastoplastic and elastoviscoplastic). Sixty-five case histories were surveyed by Negro and de Queiroz (2000) to evaluate the capability of numerical modeling to predict soft ground tunneling performance. Comparisons between numerical predictions and field measurements were primarily made for deformation (e.g., surface settlement and lateral ground displacement), while another important aspect of tunneling performance (generation and dissipation of pore water pressure) was rarely investigated. An important observation is that three-dimensional modeling results are no better than two-dimensional analyses. When a simple 2D model is properly calibrated at the early stages of construction with actual field data, reasonable predictions of the performance are possible. Negro et al. (2009) presented an in-depth review of the prediction, monitoring, and evaluation of the performance of four types of geotechnical structures: deep foundation, earth fill, supported excavation, and tunnel. The authors concluded that (1) it is important to verify and calibrate numerical modeling with input from high-quality soil data against a well-documented case history; and (2) compilation of a well-documented database that covers the full performance of geotechnical structures is rare.

5.5.2 MOMIS Database

Since 1996, the Laboratoire Central des Ponts et Chaussées (LCPC, now called IFSTTAR) and the École Centrale de Nantes (ECN) have collaborated to develop a global database called Modélisation des Ouvrages et Mesures In Situ (MOMIS) (Numerical Structure Model and In Situ Measurements), consisting of Mestat (2001) for embankment, Mestat (2002) for tunnel, and Mestat et al. (2004a) for retaining structure in excavation. The dataset dated back to 1972. It was intended to compare the numerical modeling of geotechnical structures with in situ measurements. Laboratory-scaled model testing under a 1g or ng condition (centrifuge) was not considered. The current MOMIS database includes 518 case histories: shallow foundation (51), embankment (144), retaining structure (sheet pile: 66 and diaphragm wall: 102), and tunnel (155) (Mestat and Riou 2006). The case histories were collected from conference proceedings (dominant), journal papers, reports, and theses.

According to Mestat et al. (2004b), the information stored in the MOMIS database includes the type of analysis (drained versus undrained condition, consolidation, dynamic, cyclic), ground profile, construction technique employed, actual dimensions of geotechnical structure, constitutive models of geomaterial and construction materials, parameter values used in the computation, numerical model (e.g., dimensions, type of finite elements, mesh density, boundary conditions, loadings, time step, construction project phasing, and characteristics of interface), set of numerical data for the comparison between measured and calculated values, conclusions made from the comparison (maximum deviation, relative error, etc.), and references. The relative error for each quantity of interest was defined as the difference between the measured and calculated values relative to the measured value. Numerical results were provided by the finite element or finite difference methods. The numerical studies in MOMIS were mainly 2D finite element analyses. Only 9% of the studies were 3D finite element analyses. Constitutive models used in the analyses include elasticity (linear or non-linear), elastoplasticity with or without hardening, elastoviscoplasticity, and hypoplasticity. Based on the prediction classes proposed by Lambe (1973), numerical predictions in MOMIS are grouped as

- Class A (64 cases): Prediction before the construction event or loading test. It is an ideal type of prediction to evaluate the performance of numerical modeling. Class A prediction in MOMIS is limited.
- Class C (454 cases): Prediction after the construction event or loading test. It is often used as a back-calculation to determine the appropriate soil properties which is an essential part of the physics-informed data-driven predictions. Although Class C is not a true "prediction", useful lessons can be drawn from the comparisons to derive recommendations for subsequent numerical studies (e.g., selecting the constitutive model and the values of model parameters) and to quantify the prediction variability.

5.5.3 Results and Implications

The studies on embankment on soft and improved soil, retaining structures (sheet pile and diaphragm wall) in excavation and tunnel (Mestat and Bourgeois 2002; Mestat et al. 2004b; Mestat and Riou 2004, 2006) are reviewed herein. The breakdown of comparisons given in Table 5.4 is restricted to 2D finite element analyses only. For the same case, several numerical simulations may be conducted. This is the case for a prediction event or a comparison between several constitutive models. From these studies, the main results and conclusions are summarized as follows:

- *Embankment* (Mestat et al. 2004b): From a general standpoint, Class C predictions led to relatively satisfactory results for surface settlement along the embankment axis. For 67% of the computations, the associated relative error at the end of construction is less than 25%. Over the long term, the quality of predictions is better – 90% of the cases with a relative error of less than 25%. The best numerical modeling results were obtained by the elastoplastic constitutive relations with strain hardening. For the maximum excess pore water pressure along the embankment axis, the relative error is slightly higher – 62% of the predictions with a relative error of less than 25%. For the maximum lateral displacement at the toe of the embankment, the relative error at the end of construction is higher, where the maximum error could reach up to 250%. Only 37% of the models produced the predictions with a relative error below 25%.
- *Retaining structure* (Mestat and Bourgeois 2002; Mestat and Riou 2006): For the maximum horizontal displacement of a wall, 41% of the case studies had a relative error of less than 25%, and 61% provided a relative error within 50%. Predictions on the maximum surface settlement behind the wall were less accurate. In this situation, only 13.5% of the references led to a relative error below 25%. This may be explained by the fact that simple elastoplastic laws (e.g., Mohr–Coulomb or Drucker–Prager) without hardening

TABLE 5.4
Number of Comparisons for 2D FE Analyses Only

Geo-structure	Quantity of interest	Data entries	Relative error (%)	Percentile
Embankment on soft soil (end of construction)	Max. settlement on center line	39	<25	67
	Max. lateral displacement in depth (toe of slope)	24	>100 (max. of 250)	21
			<25	37
	Max. excess pore pressure	16	<25	62
Embankment on soft soil (long term)	Max. settlement on center line	38	<25 (max. of 80)	90
	Max. lateral displacement in depth (toe of slope)	26	<25	31
Embankment on improved soil (short term)	Max. settlement on center line	35		
	Max. lateral displacement in depth	21		
	Max. excess pore pressure	22		
Tunnel (end of construction)	Max. surface settlement (transversal model)	120	<25	60
			<50	80
	Point of inflexion of settlement trough	87	<25	54
	Crown settlement	30	<25	56
	Max. horizontal displacement	32	<25	43
Retaining structure (end of construction) (sheet pile/diaphragm wall)	Max. lateral displacement of the wall	69/98	<25	54
			<50	75
	Max. ground surface settlement behind the wall	37/28	<25	28
			<50	67
			>100	25
	Max. bending moment in the wall	24/16	<25	13.5

Note:
For retaining structure, data represents the number of sheet pile walls and diaphragm walls, respectively.
The last column "Percentile" should be understood in the following way using the first row as an example: relative error for maximum settlement on center line is less than 25% for approximately 67% of the FE analyses.

cannot accurately predict the plastic volume change and the soil–structure interaction is not modeled adequately. The quality of prediction on the horizontal distance of the point of maximum settlement was worse on the order of 100% with a maximum error rising to over 300%. The force in the struts (mean relative error <25%) was better estimated than the maximum bending moment (mean relative error around 125%).

- *Tunnel* (Mestat and Riou 2004; Mestat et al. 2004b): For the maximum surface settlement, the relative error is less than 25% for approximately 60% of the modeling setups. Numerical simulation of the maximum surface settlement using simple elastoplastic constitutive models was satisfactory, provided the model parameters were determined in a rational manner. Furthermore, 80% of numerical studies led to a relative error below 50%. A similar observation has also been made by Negro and de Queiroz (2000), where 80% of the reviewed cases were Class C predictions. It was found that numerical simulations are good enough for estimating the magnitude of the surface settlement, but in over half of the cases, the surface distortion was underpredicted. It was hypothesized that the shear strain concentration around the opening was not properly modeled in the numerical codes used.

> The extent of settlement trough was quite well represented where 54% of the numerical models had a relative error of less than 25%. Similar model errors were obtained for maximum horizontal displacement.

Section 5.5.2 mentioned that Class A prediction (benchmark) received much less attention in the literature. Mestat and Riou (2004) compiled a database containing 13 studies with 257 participants. It covered four types of geotechnical structure: embankment, retaining wall, spread footing, and tunnel. These benchmark studies include validation with in situ or laboratory measurements and a comparison of numerical models, from which several findings were obtained (Mestat and Riou 2004):

- Class C predictions were often performed to verify new sophisticated models such as hypoplasticity. Nevertheless, elastoplastic models without strain hardening were more used in the Class A predictions collected, possibly due to (1) benchmark participants generally prefer a simple model (like Mohr–Coulomb) especially for practical applications, and (2) some laboratory tests or in situ investigations were not provided in the report to determine the geotechnical parameters in the complex constitutive model.
- Young's modulus in the Mohr–Coulomb model is an interpreted value of soil stiffness, which considers the rate of deformation in the geotechnical work. It exhibits a large scatter. It is not representative of the intrinsic elastic modulus. The friction angle of sand also varies over a wide range from 31° to 42°.
- For two prediction competitions at the Haarajoki (Finland) and Mur Flats (Malaysia) embankment, the distributions of relative error were similar. The settlement relative error is less than 0.5 and so the calculation is slightly to highly conservative.

5.6 PHYSICS-INFORMED AND DATA-DRIVEN CHARACTERIZATION

The model factor itself is not constant but takes a range of values that may depend on the scenarios covered in the database used for evaluation. It is customary to model M as a random variable; however, the variation of M is sometimes explainable by other known variables (e.g., Huang and Bathurst 2009; Yu and Bathurst 2015; Zhang et al. 2015; Lin et al. 2017a; Tang and Phoon 2017; Tang et al. 2017; Asem 2018; Xu et al. 2021). This could be due in part to the oversimplification of complex real-world behavior and some important influential factors are not incorporated into the analysis. For example, the dependency of the sand friction angle on the stress level is not captured by the simplistic load spread and punch shear models for peak penetration resistance of spudcan foundation in sand-over-clay. The calculation of drilled shaft capacity in soft rock only considers the uniaxial compressive strength of rock. In this case, the model factor M cannot be directly treated as a random variable. Bathurst et al. (2017) showed that ignoring statistical dependency can lead to conservative (safe) resistance factor values. The resistance factor is part of the load and resistance factor design (LRFD) (Phoon et al. 2003).

The first way to remove statistical dependency is to conduct a regression analysis of M versus the influential parameters governing the quantities of interest. Albeit direct and simple, the influential parameters are not varied systematically in the database and the dependency cannot be completely represented by the regression equation. For the case beyond the calibration range, the model factor M might still depend on the parameters. The second way is called the generalized model factor approach in which the calculated quantity (e.g., load or resistance) is adopted as the predictor variable in the regression analysis (Dithinde et al. 2011). This method is purely empirical and does not provide physical insight into the sources of statistical dependency. The third way is presented in Figure 5.9 (e.g., Zhang et al. 2015; Tang and Phoon 2017; Tang et al. 2017; Xu et al. 2021). This physics-informed data-driven characterization of the model factor is explained below using the example of cantilever wall deflections in undrained clay (Zhang et al. 2015):

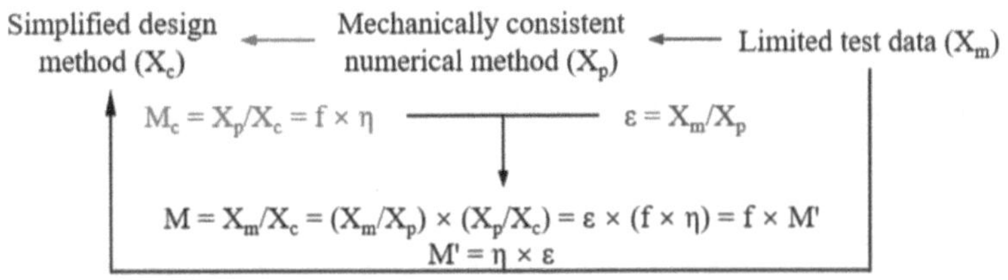

FIGURE 5.9 Physics-informed data-driven characterization of model factor.

- The model factor M is expressed as $M = M_c \times \varepsilon$, where M_c = correction factor to capture the variation of M with each influential parameter (e.g., excavation width and depth, wall thickness, coefficient of lateral earth pressure at rest, undrained shear strength, and soil stiffness) and ε = residual that is no longer dependent on these parameters. The correction term M_c is often used in engineering applications to improve the accuracy of simplified design methods. Unfortunately, it is not an easy task to establish M_c, as the test data only covers a limited range of influential parameters. It is insufficient to remove the statistical dependency adequately.
- An alternative way is to use a mechanically consistent numerical method for M_c that is defined as a ratio of the numerical prediction X_p (e.g., small-strain FEM with hardening soil model) to the solution from a simplified design model X_c (e.g., mobilized strength design [MSD]). It is reasonable to characterize M_c as the product of a trend term f (systematic variation) and error η (random). Parametric studies are implemented to account for a wide range of parameter values and to establish the trend term f. This is a popular method in research in particular for the development of a new analysis model (e.g., Al Hakeem and Aubeny 2019, 2021; Taborda et al. 2020; Zdravković et al. 2020). The behavior of a modeled geotechnical structure can be more fully understood in this way. Because all practical scenarios can be simulated, the numerical results would be a beneficial supplement to the limited test data.
- The residual part ε is presented as the ratio of X_m and X_p, that is the model factor or error of the numerical method adopted. It is likely to be random and can be evaluated by the testing data directly. Using this approach, the corrected model factor is finally characterized as $M' = \eta \times \varepsilon$.

This method is currently the best in terms of incorporating physics, correcting the bias in the original calculation model, handling problems with highly sensitive input parameters, and making efficient use of limited testing data. It is a combination of physics/mechanics (introduced through numerical modeling) and data, and therefore, can be termed as a physics-informed data-driven characterization method. Some examples are given in Table 5.5, covering the bearing capacity of strip footing on sand under combined loading (Phoon and Tang 2015a, 2015b), the uplift capacity of a helical anchor in undrained clay (Tang and Phoon 2016), spudcan penetration in sand-over-clay (Xu et al. 2021), and the lateral deflection of a cantilever wall in undrained clay (Zhang et al. 2015).

5.7 SUMMARY AND CONCLUSIONS

Prediction is one of the key activities in our geo-profession (e.g., bearing capacity and settlement of foundation, load exerted on the retaining structure, ground settlement induced by underground work such as excavation and tunneling, and the potential of sand liquefaction). It provides critical support to decision-making. Given the complexity of the real behavior of a geotechnical structure (e.g., inter-site variability and intra-site variability of soil/rock properties, limited knowledge on the governing physical laws, and construction effect or installation disturbance), it is very difficult to carry out a systematic analysis of the performance of geotechnical prediction models. For the

TABLE 5.5
Summary of Model Statistics with Response Modified by f

Problems	Numerical methods used	Variables		$\ln(f) = b_0 + \Sigma b_i x_i$		M′	
						Mean	COV
Strip footings on sand under positive combined loading (Phoon and Tang 2015a)	FELA (perfect plastic soil model)			b_0	0.28	1.02	0.09
		x_1	$\gamma D/p_a$	b_1	−5.05		
		x_2	ξ	b_2	11.4		
		x_3	$\tan\phi_a$	b_3	−0.26		
		x_4	d/B	b_4	−0.09		
		x_5	α/ϕ_a	b_5	0.21		
		x_6	e/B	b_6	−1.12		
		x_7	$(e/B)(\alpha/\phi_a)$	b_7	−0.98		
Strip footings on sand under negative combined loading (Phoon and Tang 2015b)	FELA (perfect plastic soil model)			b_0	0.1	1.03	0.10
		x_1	$\gamma D/p_a$	b_1	−4.5		
		x_2	ξ	b_2	10.4		
		x_3	$\tan\phi_a$	b_3	−0.25		
		x_4	d/B	b_4	−0.12		
		x_5	α/ϕ_a	b_5	−1.03		
		x_6	e/B	b_6	−0.45		
		x_7	$(e/B)(\alpha/\phi_a)$	b_7	−1.81		
Strip footings on sand under general combined loading (Phoon and Tang 2015b)	FELA (perfect plastic soil model)			b_0	0.1	1.02	0.11
		x_1	$\gamma D/p_a$	b_1	−4.5		
		x_2	ξ	b_2	10.25		
		x_3	$\tan\phi_a$	b_3	−0.15		
		x_4	d/B	b_4	0.05		
		x_5	α/ϕ_a	b_5	−0.93		
		x_6	e/B	b_6	−0.05		
		x_7	$(e/B)(\alpha/\phi_a)$	b_7	−2.53		
Helical anchors in clay under tension loading (Tang and Phoon 2016)	FELA (perfect plastic soil model)			b_0	0.78	0.92	0.16
		x_1	N	b_1	−0.07		
		x_2	S/D	b_2	−0.1		
		x_3	H/D	b_3	−0.02		
		x_4	$\gamma H/s_u$	b_4	0.04		
Cantilever retaining wall deflections in undrained clay (Zhang et al. 2015)	FEM in Plaxis 2.0 (hardening soil with small strain)			b_0	0.89	1.02	0.26
		x_1	2D/B	b_1	−0.13		
		x_2	$\ln(H_e/D)$	b_2	0.43		
		x_3	$\ln(\gamma D^4/EI)$	b_3	0.12		
		x_4	$1/K_0$	b_4	0.69		
		x_5	s_u/σ_v'	b_5	−0.74		
		x_6	E_{ur}/s_u	b_6	-7×10^{-4}		
Spudcan penetration in sand-over-clay (Xu et al. 2021)	CEL in Abaqus (clay: extended Tresca and sand: modified Mohr–Coulomb)			b_0	−0.04	0.97	0.18
		x_1	$\gamma_s'D/p_a$	b_1	0.34		
		x_2	D_R	b_2	0.40		
		x_3	H_s/D	b_3	0.71		
		x_4	$N_c s_u/p_a$	b_4	−0.39		

Source: Updated table 5.2 in Dithinde et al. (2016).

Note:

FELA = finite element limit analysis; CEL = coupled Eulerian–Lagrangian; D = foundation width, diameter of helix plate or wall depth; γ = unit weight of soil; p_a = atmospheric pressure; d = embedment depth; ϕ_a = repose angle of sand; e = load eccentricity; α = load inclination; ξ = empirical parameter ξ = 0.02–0.12; n = number of helix plates; S = plate spacing; H = embedment depth of top helix; s_u = undrained shear strength; H_e = excavation depth; EI = wall stiffness; B = wall width; E_{ur} = soil stiffness; K_0 = at-rest lateral earth pressure coefficient; σ_v' = effective vertical stress; γ_s' = effective unit weight of sand; D_R = relative density of sand; H_s = thickness of sand layer; N_c = bearing capacity factor; q_0 = surcharge; and ϕ_{cv} = critical-state friction angle of sand.

same geotechnical structure, the same prediction model, and comparable ground condition, the prediction performance may (1) vary between sites or even for different locations in a single site, (2) depend on the implementation of a predictive model and the input parameters adopted, and (3) be different because of the influence of engineering judgment at various stages of the prediction process. They could be referred to as the three characteristics of geotechnical analysis and design practice, respectively: site-specificity, imperfections in modeling and calculation, and diverse norms in practice (SID).

This chapter attempts to understand how predictions are affected by the ground conditions, the installation of geotechnical structures, the modeling of ground–structure interactions, and the decisions made by the modeler (engineer) by studying the prediction variability from three data sources: (1) statistical analyses based on geotechnical databases (e.g., loading tests and observed settlements on full-scale foundations, pullout box tests of mechanically stabilized earth walls and soil nail walls, and case histories of slope failures and excavations) performed by a variety of researchers or institutions (e.g., State Departments of Transportation in the United States) at different sites; (2) review of the prediction events organized at different international conferences typically for tests or measurements at a single site; and (3) comparisons between numerical modeling results and in situ measurements made by the LCPC. The results covered four types of geotechnical structure (e.g., foundation, slope, retaining structure, and tunnel), from which the following conclusions can be drawn:

- For most of the cases considered, database assessment of prediction variability indicates a medium degree of bias (mean of model factor λ = 1–3) and dispersion (COV of model factor θ = 0.3–0.6) (Figure 5.5). An exception is the prediction of (bored or driven) pile axial capacity in soft rock or intermediate geomaterial and settlement which produces high dispersion (θ = 0.6) (Figure 5.5). Except for the inter-site/intra-site variability and modeling/calculation imperfections, predictions also relied on the practitioners' knowledge, judgment, and experience to select the model and the values of related parameters. Different judgment made by a group of participants in a prediction event will significantly increase the variability of predictions. This is confirmed by the data in Sections 5.3 and 5.4. A database assessment is conducted by one researcher exercising consistent judgment. However, the prediction event takes place at a single site. The inter-site variability in a database assessment is not applicable to a prediction event. When the overall prediction variability in a prediction event (single site, many participants) is compared with that in a database assessment (multiple sites, one researcher), they may not be too different as a result of these contrasting factors.
- Model factor statistics can be applied to satisfy Clause 2.4.16(P), EN 1997-1:2004 (CEN 2004) - "any calculation model shall be either accurate or err on the side of safety" in the probabilistic sense of ensuring Prob(model factor <1) < 0.05 for the capacity model factor. Engineers can divide the calculated capacity (X_c) by Δp based on the values summarized in Table 5.3. The original deterministic intent of this clause cannot be fulfilled given the variability of predictions demonstrated in this chapter.
- Numerical methods are evolving rapidly and increasingly used in geotechnical engineering. It is essential to compare Class A predictions from a numerical model with in situ measurements using more case studies. We need to ascertain the degree of realism and the practicality of a given numerical analysis and establish the modeling guideline specific to each type of geotechnical structure, such as selecting a simple but robust model and determining the related model parameters to ensure the quality of prediction at an acceptable computation cost. This is consistent with the observation of Duncan (1994): wide use of advanced constitutive models in practice will require simple procedures to determine the values of the required parameters based on the results of conventional soil tests. Notwithstanding the rigor and sophistication of numerical methods, model calibration

with sufficient quality and quantity of data should be facilitated to reduce the prediction variability (Kalenchuk 2022).

- Current practice for foundation design is predominantly based on capacity analysis. A factor of safety or resistance factor is applied to reduce the capacity calculated or measured. This is aimed at controlling settlement indirectly by limiting the stress in the underlying soil. This practice is very conservative as indicated by a large mean of settlement model factor (>3) in Figure 5.4. Fellenius (2018) opined that foundation design should evolve toward settlement analysis which is no more complex than a conventional capacity approach. Thus, more studies should be conducted to develop accurate prediction methods, including the use of high-quality sampling techniques with high-quality laboratory and in situ test data, the adoption of an appropriate constitutive model, and the detailed calibration of the constitutive model by element test results. A good example is the development of monotonic p–y curves (e.g., Jeanjean et al. 2017; Zhang and Andersen 2017) and cyclic p–y curves (Zhang et al. 2017) for laterally loaded pile in cohesive soil. The shape of p–y curves was deduced from an extensive database of direct simple shear tests for soil stress–strain response combined with FE analysis results. Zhang et al. (2020) showed a noticeable improvement in calculating pile response in very soft to stiff clay, as compared to current industry practice.
- The prediction variability obtained from a database assessment is not the same as the one obtained from a prediction event. The former is generic, calibrating different design scenarios (e.g., ground conditions and prediction models). The latter is site-specific but it is based on a variety of prediction models moderated by engineering judgment. Nonetheless, they are useful benchmarks for machine learning methods that aim to predict site-specific quantities of interest to various limit states (cf. "precision construction" in Phoon 2018).
- Finally, engineers should bear in mind that the quality of the method (e.g., high degree of rigor and sophistication to precisely capture the underlying mechanisms and fully consider all of the influential parameters) and input data (e.g., determined by well-controlled laboratory and/or in situ testing) should be equally important. An advanced and sophisticated numerical model does not necessarily outperform a relatively simple and robust model. The results from some prediction events and comparisons between numerical modeling and in situ measurements demonstrated that accurate predictions with a bias close to 1 and COV <0.3 can be obtained even using a simple model with geotechnical parameters calibrated or back-calculated from performance data.

This chapter constitutes a significant update of the review paper by Phoon and Tang (2019) and the book by Tang and Phoon (2021) that only focus on prediction variability using the characterization of a model factor from a database. The prediction variability of Categories 1, 2, and 3 (Poulos 1989) and the effect of different modelers applying different judgment are notable updates in the chapter. A Category 4 (machine learning) is emerging in the literature but studies are too limited to assess the prediction variability of this new category.

REFERENCES

AASHTO (American Association of State Highway and Transportation Officials). 2020. *LRFD Bridge Design Specifications*. 9th ed. Washington, DC: AASHTO.

Abchir, Z., Burlon, S., Frank, R., Habert, J. and Legrand, S. 2016. t-z curves for piles from pressuremeter test results. *Géotechnique*, 66(2), 137–148. https://doi.org/10.1680/jgeot.15.P.097.

AbdelSalam, S.S., Ng, K.W., Sritharan, S., Suleiman, M. and Roling, M. 2012. Development of LRFD procedures for bridge pile foundations in Iowa – Volume III: Recommended resistance factors with consideration of construction control and setup. Report No. IHRB Project TR-584. Ames, IA: Iowa Department of Transportation.

AbdelSalam, S.S., Baligh, F.A. and El-Naggar, H.M. 2015. A database to ensure reliability of bored pile design in Egypt. *Proceedings of the Institution of Civil Engineers–Geotechnical Engineering*, 168(2), 131–143. https://doi.org/10.1680/geng.14.00051.

Abu-Hejleh, N.M., Abu-Farsakh, M.Y., Suleiman, M.T. and Tsai, C. 2015. Development and use of high-quality databases of deep foundation load tests. *Transportation Research Record*, 2511(1), 27–36. https://doi.org/10.3141/2511-04.

Akbas, S.O. 2007. Deterministic and probabilistic assessment of settlements of shallow foundations in cohesionless soils. PhD thesis, Department of Civil and Environmental Engineering, Cornell University.

Al Hakeem, N. and Aubeny, C. 2019. Numerical investigation of uplift behavior of circular plate anchors in uniform sand. *Journal of Geotechnical and Geoenvironmental Engineering, ASCE*, 145(9), 04019039. https://doi.org/10.1061/(ASCE)GT.1943-5606.0002083.

Al Hakeem, N. and Aubeny, C. 2021. Normally loaded inclined strip anchors in cohesionless soil. *Canadian Geotechnical Journal*, 58(10), 1478–1494. https://doi.org/10.1139/cgj-2019-0791.

Allen, T.M. and Bathurst, R.J. 2015. Improved simplified method for prediction of loads in reinforced soil walls. *Journal of Geotechnical and Geoenvironmental Engineering, ASCE*, 141(11), 04015049. https://doi.org/10.1061/(ASCE)GT.1943-5606.0001355.

Alpan, I. 1964. Estimating the settlements of foundations on sands. *Civil Engineering and Public Works Review*, 59(700), 1415–1418.

API (American Petroleum Institute). 2007. *Recommended Practice for Planning, Designing and Constructing Fixed Offshore Platforms–Working Stress Design*. 22nd ed. Washington, DC: API.

Asem, P. 2018. Axial behavior of drilled shafts in soft rock. PhD thesis, University of Illinois at Urbana-Champaign.

Baecher, G. 2023. Geotechnical systems, uncertainty, and risk. *Journal of Geotechnical and Geoenvironmental Engineering, ASCE*, 149(3), 03023001. https://doi.org/10.1061/JGGEFK.GTENG-10201.

Baecher, G.B. and Christian, J.T. 2003. *Reliability and Statistics in Geotechnical Engineering*. New York: John Wiley and Sons.

Bahmani, S.M. and Briaud, J.L. 2020. Settlement of shallow foundations on sand–a database study. *ISSMGE International Journal of Geoengineering Case Histories*, 6(2), 1–17. https://doi.org/10.4417/IJGCH-06-02-01.

Bahmani, S.M. and Briaud, J.L. 2021. Settlement of shallow foundations on clay–a database study. In *Proceedings of the International Foundations Conference and Equipment Expo 2021*, 326–335. Reston, VA: ASCE. https://doi.org/10.1061/9780784483435.032.

Bathurst, R.J., Javankhoshdel, S. and Allen, T.M. 2017. LRFD calibration of simple soil-structure limit states considering method bias and design parameter variability. *Journal of Geotechnical and Geoenvironmental Engineering, ASCE*, 143(9), 04017053. https://doi.org/10.1061/(ASCE)GT.1943-5606.0001735.

Boeckmann, A., Myers, S., Uong, M. and Loehr, J.E. 2014. Load and resistance factor design of drilled shafts in shale for lateral loading. Report No.: CMR 14-011. Jefferson City, MO: Missouri Department of Transportation.

Bohn, C., dos Santos, A.L. and Frank, R. 2017. Development of axial pile load transfer curves based on instrumented load tests. *Journal of Geotechnical and Geoenvironmental Engineering, ASCE*, 143(1), 04016081. https://doi.org/10.1061/(ASCE)GT.1943-5606.0001579.

Bozorgzadeh, N. and Bathurst, R.J. 2022. Hierarchical Bayesian approaches to statistical modelling of geotechnical data. *Georisk: Assessment and Management of Risk for Engineered Systems and Geohazards*, 16(3), 452–469. https://doi.org/10.1080/17499518.2020.1864411.

Bozorgzadeh, N., Liu, Z., Nadim, F. and Lacasse, S. 2023. Model calibration: A hierarchical Bayesian approach. *Probabilistic Engineering Mechanics*, 71, 103379. https://doi.org/10.1016/j.probengmech.2022.103379.

Briaud, J.L. and Gibbens, R. 1997. Large-scale load tests and data base of spread footings on sand. Report No. FHWA-RD-97-068. McLean, VA: Turner Fairbank Highway Research Center.

Briaud, J.L. and Wang, Y.C. 2018. Synthesis of load-deflection characteristics of laterally loaded large diameter drilled shafts. Report No. FHWA/TX-18/0-6956-R1. Austin, TX: Texas Department of Transportation.

Brown, D.A., Turner, J.P., Castelli, R.J. and Loehr, E.J. 2018. Drilled shafts: Construction procedures and design methods. Report No. FHWA NHI-18-024. Washington, DC: Federal Highway Administration.

Brown, R.P. 2001. Predicting the ultimate axial resistance of single driven piles. PhD thesis, Department of Civil, Architectural, and Environmental Engineering, University of Texas at Austin.

Bullock, Z., Karimi, Z., Dashti, S., Porter, K., Liel, A.B. and Franke, K.W. 2019. A physics-informed semi-empirical probabilistic model for the settlement of shallow-founded structures on liquefiable ground. *Géotechnique*, 69(5), 406–419. https://doi.org/10.1680/jgeot.17.P.174.

Burlon, S., Frank, R., Baguelin, F., Habert, J. and Legrand, S. 2014. Model factor for the bearing capacity of piles from pressuremeter test results–Eurocode 7 approach. *Géotechnique*, 64(7), 513–525. https://doi.org/10.1680/geot.13.P.061.

CEN (European Committee for Standardization). 2004. *Geotechnical Design. Part 1: General Rules*. EN 1997–1. Brussels, Belgium.

Chen, W.F. 1975. *Limit Analysis and Soil Plasticity*. Amsterdam: Elsevier.

Chen, Y.-J., Liao, M.-R., Lin, S.-S., Huang, J.-K. and Marcos, M.C.M. 2014. Development of an integrated web-based system with a pile load test database and pre-analyzed data. *Geomechanics and Engineering*, 7(1), 37–53. http://doi.org/10.12989/gae.2014.7.1.037.

Chen, Y.-J., Lin, W.Y., Topacio, A. and Phoon, K.K. 2021. Evaluation of interpretation criteria for drilled shafts with tip post-grouting. *Soils and Foundations*, 61(5), 1354–1369. https://doi.org/10.1016/j.sandf.2021.08.001.

Chin, F.K. 1970. Estimation of the ultimate load of piles not carried to failure. In *Proceedings of the 2nd Southeast Asian Conference on Soil Engineering*, 81–90, Singapore: Southeast Asian Society of Soil Engineering.

Ching, J., Li, D.Q. and Phoon, K.K. 2016. Chapter 4. Statistical characterization of multivariate geotechnical data. In *Reliability of Geotechnical Structures in ISO2394*, 89–126. London: CRC Press/Balkema.

Ching, J. and Phoon, K.K. 2015. Chapter 1. Constructing multivariate distribution for soil parameters. In *Risk and reliability in geotechnical Engineering*, 3–76. Boca Raton, FL: CRC Press/Balkema.

Ching, J. and Phoon, K.K. 2019. Constructing site-specific multivariate probabilistic distribution model by Bayesian machine learning. *Journal of Engineering Mechanics, ASCE*, 145(1), 04018126. https://doi.org/10.1061/(ASCE)EM.1943-7889.0001537.

Ching, J. and Phoon, K.K. 2020. Measuring similarity between site-specific data and records from other sites. *ASCE-ASME Journal of Risk and Uncertainty in Engineering Systems, Part A: Civil Engineering*, 6(2), 04020011. https://doi.org/10.1061/AJRUA6.0001046.

Ching, J., Phoon, K.K., Ho, Y.H. and Weng, M.C. 2021a. Quasi-site-specific prediction for deformation modulus of rock mass. *Canadian Geotechnical Journal*, 58(7), 936–951. https://doi.org/10.1139/cgj-2020-0168.

Ching, J., Phoon, K.K. and Wu, C.T. 2022a. Data-centric quasi-site-specific prediction for compressibility of clays. *Canadian Geotechnical Journal*, 59(12), 2033–2049. https://doi.org/10.1139/cgj-2021-0658.

Ching, J., Phoon, K.K., Yang, Z.Y. and Stuedlein, A.W. 2022b. Quasi-site-specific multivariate probability distribution model for sparse, incomplete, and three-dimensional spatially varying soil data. *Georisk: Assessment and Management of Risk for Engineered Systems and Geohazards*, 16(1), 53–76. https://doi.org/10.1080/17499518.2021.1971256.

Ching, J., Wu, S. and Phoon, K.K. 2021b. Constructing quasi-site-specific multivariate probability distribution using hierarchical Bayesian model. *Journal of Engineering Mechanics, ASCE*, 147(10), 04021069. https://doi.org/10.1061/(ASCE)EM.1943-7889.0001964.

Chou, S.A., Chen, Y.-J., Chiou, J.S., Topacio, A. and Marcos, M.C.M. 2022. Evaluation of lateral capacity for flexible drilled shafts in cohesionless soils. *Science Progress*, 105(3), 1–30. https://doi.org/10.1177/00368504221113.

Christian, J.T. 2004. Geotechnical engineering reliability: how well do we know what we are going? *Journal of Geotechnical and Geoenvironmental Engineering, ASCE*, 130(10), 985–1003. https://doi.org/10.1061/(ASCE)1090-0241(2004)130:10(985).

Constructed Facilities Division. 1975a. Proceeding of the foundation deformation prediction symposium. Volume 1. Symposium Summary. Report No. FHWA-RD-75-515. Wellesley Hill, MA: Massachusetts Institute of Technology.

Constructed Facilities Division. 1975b. Proceeding of the foundation deformation prediction symposium. Volume 2. Appendix. Report No. FHWA-RD-75-516. Wellesley Hill, MA: Massachusetts Institute of Technology.

CSA (Canadian Standards Association). 2019. Canadian highway bridge design code. CSA S6:19, Mississauga, Canada.

Chwała, M., Phoon, K.K., Uzielli, M., Zhang, J., Zhang, L. and Ching, J. 2023. Time capsule for geotechnical risk and reliability. *Georisk: Assessment and Management of Risk for Engineered Systems and Geohazards*, 17(3), 439–466. https://doi.org/10.1080/17499518.2022.2136717.

D'Appolonia, E. 1990. Monitored decisions. *Journal of Geotechnical Engineering, ASCE*, 116(1), 4–34. https://doi.org/10.1061/(ASCE)0733-9410(1990)116:1(4).

Davisson, M.T. 1972. High-capacity piles. In *Proceeding of Lecture Series on Innovations in Foundation Construction*, ASCE, Illinois Section, Chicago. https://ci.nii.ac.jp/naid/10007806422/.

DeBeer, E.E. 1970. Experimental determination of the shape factors of sand. *Géotechnique*, 20(4), 387–411. https://doi.org/10.1680/geot.1970.20.4.387.

De Mello, V.F.B. 1977. Reflections on design decisions of practical significance to embankment dams. *Géotechnique*, 27(3), 279–355. https://doi.org/10.1680/geot.1977.27.3.281.

Diaz-Segura, E.G. 2013. Assessment of the range of variation of N_γ from 60 estimation methods for footings on sand. *Canadian Geotechnical Journal*, 50(7), 793–800. https://doi.org/10.1139/cgj-2012-0426.

DiMillio, A.F., Ng, E.S., Briaud, J.-L. and O'Neill, M.W. 1988. Pile group prediction symposium: Summary. Volume I: Sandy soil. Report No. FHWA-TS-87-221. McLean, VA: Federal Highway Administration.

DIN 4026. 1978. Herstellung, Bemessung und zulässige Belastung Rammpfähle. Deutschen Gesellschaft für Geotechnik.

Dithinde, M., Phoon, K.K., De Wet, M. and Retief, J.V. 2011. Characterization of model uncertainty in the static pile design formula. *Journal of Geotechnical and Geoenvironmental Engineering, ASCE*, 137(1), 70–85. https://doi.org/10.1061/(ASCE)GT.1943-5606.0000401.

Dithinde, M., Phoon, K.K., Ching, J., Zhang, L.M. and Retief, J.V. 2016. Chapter 5. Statistical characterisation of model uncertainty. In *Reliability of Geotechnical Structures in ISO2394*, 127–158. Leiden: CRC Press/Balkema.

Doherty, J.P., Gourvenec, S. and Gaone, F.M. 2018. Insights from a shallow foundation load-settlement prediction exercise. *Computers and Geotechnics*, 93, 269–279. https://doi.org/10.1016/j.compgeo.2017.05.009.

Duncan, J.M. 1994. The role of advanced constitutive relations in practical applications. In *Proceedings of the 14th International Conference on Soil Mechanics and Foundation Engineering*, Vol. 5, 31–48. Rotterdam: A.A. Balkema.

Duncan, J.M., Wright, S.G. and Brandon, T.L. 2014. *Soil Strength and Slope Stability*. 2nd ed. Hoboken, NJ: John Wiley & Sons, Inc.

Eslami, A. and Fellenius, B.H. 1997. Pile capacity by direct CPT and CPTu methods applied to 102 case histories. *Canadian Geotechnical Journal*, 34(6), 886–904. https://doi.org/10.1139/t97-056.

Eslami, A., Moshfeghi, S., MolaAbasi, H. and Eslami, M.M. 2020. *Piezocone and Cone Penetration Test (CPTu and CPT) Applications in Foundation Engineering*. Oxford: Butterworth-Heinemann.

Fellenius, B.H. 1988. Variation of CAPWAP results as a function of the operator. In *Proceedings of the 3rd International Conference on the Application of Stress-Wave Theory to Piles*, 814–825. Vancouver, BC: BiTech Publishers.

Fellenius, B.H. 2013. Capacity and load-movement of a CFA pile: A prediction event. In *Proceedings of the Geo-Congress 2013–Foundation Engineering in the Face of Uncertainty*, 707–719. Reston, VA: ASCE. https://doi.org/10.1061/9780784412763.053.

Fellenius, B.H. 2015. Static loading test and prediction outcome. In *Invited Lecture at Proceedings of the 2nd Bolivian International Conference on Deep Foundations*. Santa Cruz, Bolivia.

Fellenius, B.H. 2018. Pitfalls and fallacies in foundation design. In *Proceedings of the Innovations in Geotechnical Engineering (IFCEE 2018)*. Geotechnical Special Publication 299, 299–316. Reston, VA: ASCE. https://doi.org/10.1061/9780784481639.020.

Fellenius, B.H. 2021. Results of an instrumented static loading test: Application to design and compilation of an international survey. *DFI Journal–The Journal of the Deep Foundations Institute*, 15(1), 71–87. https://doi.org/10.37308/DFIJnl.20200923.224.

Fellenius, B.H., Edvardsson, F., Pettersson, J., Sabattini, M. and Wallgren, J. 2019. Prediction, testing, and analysis of a 50 m long pile in soft marine clay. *DFI Journal–The Journal of the Deep Foundations Institute*, 13(2), 1–7. https://doi.org/10.37308/DFIJnl.201903219.201.

Fellenius, B.H., Hussein, M., Mayne, P. and McGillivray, R.T. 2004. Murphy's law and the pile prediction at the 2002 ASCE GeoInstitute's deep foundations conference. In *Proceedings of the Deep Foundations Institute Meeting on Current Practice and Future Trends in Deep Foundations*, 29–43. Reston, VA: ASCE.

Fellenius, B.H., Massarsch, K.R., Mandolini, A. and Terceros, H. 2017. The Bolivian experimental site for testing piles (B.E.S.T.): Predictions, comments and test results, and submitted papers. In *Proceedings of the 3rd Bolivian International Conference on Deep Foundations*, Vol. 3. Madison, WI: Omnipress.

Fellenius, B.H. and Terceros, H. 2014. Response to load for four different bored piles. In *Proceedings of the DFI-EFFC International Conference on Piling and Deep Foundations*, 99–120. Hawthorne, NJ: Deep Foundations Institute.

Fenton, G.A., Naghibi, F., Dundas, D., Bathurst, R.J. and Griffiths, D.V. 2016. Reliability-based geotechnical design in 2014 Canadian highway bridge design code. *Canadian Geotechnical Journal*, 53(2), 236–251. https://doi.org/10.1139/cgj-2015-0158.

Figueroa, G., Marinucci, A. and Lemnitzer, A. 2022. Axial load capacity predictions of drilled displacement piles with SPT- and CPT-based direct methods. *DFI Journal–The Journal of the Deep Foundations Institute*, 16(2), 1–22. https://doi.org/10.37308/DFIJnl.20220512.262.

Finno, R.J. 1989. *Predicted and Observed Axial Behavior of Piles: Results of a Pile Prediction Symposium*. New York: ASCE.

Flynn, K.N. 2014. Experimental investigations of driven cast-in-situ piles. PhD thesis, National University of Ireland, Galway.

Flynn, K.N. and McCabe, B.A. 2022. Instrumented concrete pile tests – Part 1: A review of instrumentation and procedures. *Proceedings of the Institution of Civil Engineers–Geotechnical Engineering*, 175(1), 86–111. https://doi.org/10.1680/jgeen.21.00126.

Fuller, F.M. and Hoy, H.E. 1970. Pile load tests including quick load test method, conventional methods, and interpretations. *Highway Research Record*, 333(8), 74–86.

Gibson, R.E. 1974. The analytical method in soil mechanics. *Géotechnique*, 24(2), 115–140. https://doi.org/10.1680/geot.1974.24.2.115.

Gilbert, R.B. and Tang, W.H. 1995. Model uncertainty in offshore geotechnical reliability. Offshore Technology Conference, Houston, Paper No. OTC-7757-MS. https://doi.org/10.4043/7757-MS.

Girolami, M. 2020. Introducing data-centric engineering: An open access journal dedicated to the transformation of engineering design and practice. *Data-Centric Engineering*, 1, e1. https://doi.org/10.1017/dce.2020.5.

Guevara, M., Doherty, J., Gaudin, C. and Watson, P. 2022. Evaluating uncertainty associated with engineering judgement in predicting the lateral response of conductors. *Journal of Geotechnical and Geoenvironmental Engineering, ASCE*, 148(5), 05022001. https://doi.org/10.1061/(ASCE)GT.1943-5606.0002759.

Hannigan, P.J., Rausche, F., Likins, G.E., Robinson, B.R. and Becker, M.L. 2016. *Design and Construction of Driven Pile Foundations, Volume I and II*. Report No. FHWANHI-16-009. Washington, DC: Federal Highway Administration.

Hansen, B.J. 1970. *A Revised and Extended Formula for Bearing Capacity*. Bulletin No. 28. Copenhagen: Danish Geotechnical Institute.

Hirany, A. and Kulhawy, F.H. 1988. Conduct and interpretation of load tests on drilled shaft foundations: Detailed guidelines. Report EL-5915 (1). Palo Alto, CA: Electrical Power Research Institute.

Hirany, A. and Kulhawy, F.H. 1989. Interpretation of load tests on drilled shafts – Part 1: Axial compression. In *Proceedings of the Foundation Engineering: Current Principles and Practices*, 1132–1149. New York: ASCE.

Hirany, A. and Kulhawy, F.H. 2002. On the interpretation of drilled foundation load test results. In *Proceedings of the Deep Foundations 2002*, Geotechnical Special Publication 116, 1018–1028. Reston, VA: ASCE. http://doi.org/10.1061/40601(256)71.

Holeyman, A. and Charue, N. 2003. International pile capacity prediction event at Limelette. In *Proceedings of the 2nd Symposium on Screw Piles*, 215–234. Lisse: Swets & Zeitlinger Publishers.

Holeyman, A., Couvreur, J.M. and Charue, N. 2001. Results of dynamic and kinetic pile load tests and outcome of an international prediction event. In *Proceedings of the Symposium on Screw Piles*, 247–273. Lisse: Swets & Zeitlinger Publishers.

Horvath, R.G., Kenney, T.C. and Trow, W.A. 1980. Results of tests to determine shaft resistance of rock socketed drilled piers. In *Proceedings of the International Conference on Structural Foundations on Rock*, Vol. 1, 349–361. Rotterdam: A.A. Balkema.

Huang, B. and Bathurst, R.J. 2009. Evaluation of soil-geogrid pullout models using a statistical approach. *Geotechnical Testing Journal, ASTM*, 32(6), 489–504. https://doi.org/10.1080/17499518.2016.1154160.

ISO (International Organization for Standardization). 2015. *General Principles on Reliability of Structures*. ISO2394:2015. Geneva: ISO.

Jardine, R.J. 1985. Investigations of pile-soil behavior with special reference to the foundation of offshore structures. PhD thesis, University of London.

Jardine, R.J., Standing, J.R., Jardine, F.M., Bond, A.J. and Parker, E. 2001. A competition to assess the reliability of pile prediction methods. In *Proceedings of the 15th International Conference on Soil Mechanics and Foundation Engineering*, Vols. 911–914. Rotterdam: A. A. Balkema.

Jeanjean, P., Zhang, Y., Zakeri, A., Andersen, K.H., Gilbert, R. and Senanayake, A.I.M.J. 2017. *A Framework for Monotonic p-y Curves in Clays. Keynote Lecture in Smarter Solutions for Future Offshore*

Developments, Vol. 1, 108–141. London: Society for Underwater Technology. https://doi.org/10.3723/OSIG17.108.
JRA (Japan Road Association). 2017. Specifications for highway bridges, part 4 substructures. Tokyo, Japan.
Juang, C.H. and Zhang, J. 2017. Bayesian methods for geotechnical applications–a practical guide. *Keynote Lecture in Proceedings of Geotechnical. Safety and Reliability*, 215–246. Reston, VA: ASCE. https://doi.org/10.1061/9780784480731.019.
Kalenchuk, K.S. 2022. Mitigating a fatal flaw in modern geomechanics: Understanding uncertainty, applying model calibration, and defying the hubris in numerical modelling. *Canadian Geotechnical Journal*, 59(3), 315–329. https://doi.org/10.1139/cgj-2020-0569.
Kalmogo, P., Sritharan, S. and Ashlock, J.C. 2019. Recommended resistance factors for load and resistance factor design of drilled shafts in Iowa. Report No. InTrans Project 14-512. Ames, IA: Iowa Department of Transportation.
Karniadakis, G.E., Kevrekidis, I.G., Lu, L., Perdikaris, P., Wang, S. and Yang, L. 2021. Physics-informed machine learning. *Nature Reviews Physics*, 3(6), 422–440. https://doi.org/10.1038/s42254-021-00314-5.
Kelly, R.B., Sloan, S.W., Pineda, J.A., Kouretzis, G. and Huang, J. 2018. Outcomes of the Newcastle symposium for the prediction of embankment behaviour on soft soil. *Computers and Geotechnics*, 93, 9–41. https://doi.org/10.1016/j.compgeo.2017.08.005.
Kim, Y., Ochi, K., Shibuya, S., Shi, D.M. and Tatsuoka, F. 1991. Deformation modulus and strength of artificial and natural soft rocks at low strain levels. In *Proceedings of the Symposium on Triaxial Test*, Tokyo, 265–272.
Kulhawy, F.H., O'Rourke, T.D., Steward, J.P. and Beech, J.F. 1983. Transmission line structure foundations for uplift-compression loading: Load test summaries. Report No. EL-3160-LD. Palo Alto, CA: Electrical Power Research Institute.
Kulhawy, F.H. and Phoon, K.K. 1993. Drilled shaft side resistance in clay soil to rock. In *Proceedings of the Design and Performance of Deep Foundations: Piles and Piers in Soil and Soft Rock*, 172–183. Reston, VA: ASCE.
Kuo, C., McVay, M. and Birgisson, B. 2002. Calibration of load and resistance factor design: Resistance factors for drilled shaft design. *Transportation Research Record*, 1808(1), 108–111. https://doi.org/10.3141/1808-12.
Lacasse, S. and Nadim, F. 1996. *Model Uncertainty in Pile Axial Capacity Calculations*. Houston: Offshore Technology Conference, Paper No. OTC-7996-MS. https://doi.org/10.4043/7996-MS.
Lacasse, S., Nadim, F., Langford, T. and Knudsen, S. 2013. Model uncertainty in axial pile capacity design methods. Offshore Technology Conference, Paper No. OTC 24066. https://doi.org/10.4043/24066-MS.
Lambe, T.W. 1973. Predictions in soil engineering. *Géotechnique*, 23(2), 151–202. https://doi.org/10.1680/geot.1973.23.2.151.
Lazarte, C.A. 2011. Proposed specifications for LRFD soil-nailing design and construction. NCHRP Rep. No. 701. Washington, DC: Transportation Research Board.
Lazarte, C.A., Robinson, H., Gómez, J.E., Baxter, A., Cadden, A. and Berg, R. 2015. Soil nail walls – Reference manual. Report No. FHWA-NHI-14-007. Washington, DC: Federal Highway Administration.
Lee, K.K., Cassidy, M.J. and Randolph, M.F. 2013. Bearing capacity on sand overlying clay soils: experimental and finite-element investigation of potential punch-through failure. *Géotechnique*, 63(15), 1271–1284. https://doi.org/10.1680/geot.12.P.175.
Lees, A.S. 2019. Tomorrow's geotechnical toolbox: EN 1997-1:202x–numerical methods. In *Proceedings of the XVII European Conference on Soil Mechanics and Geotechnical Engineering*. Reykjavik. Iceland Geotechnical Society. https://doi.org/10.32075/17ECSMGE-2019-1105.
Lehane, B.M., Lim, J.K., Carotenuto, P., Nadim, F., Lacasse, S., Jardine, R.J. and Van Dijk, B.F.J. 2017. Characteristics of unified databases for driven piles. In *Proceedings of the 8th International Conference of Offshore Site Investigation and Geotechnics (OSIG)*, 162–191. London: Society for Underwater Technology. https://doi.org/10.3723/OSIG17.162.
Leshchinsky, D., Leshchinsky, O., Zelenko, B., and Horne, J. 2016. Limit equilibrium design framework for MSE structures with extensible reinforcement. Report No. FHWA-HIF-17-004. Washington, DC: Federal Highway Administration.
Ley, C., Tibolt, M. and Fromme, D. 2020. Data-centric engineering in modern science from the perspective of a statistician an engineer, and a software developer. *Data-Centric Engineering*, 1, e2. https://doi.org/10.1017/dce.2020.2.
Li, X., Li, H., Du, S., Jing, J. and Li, P. 2023. Cross-engineering utilization of tunnel boring machines (TBM) construction data: A case study using big data from Yin-Song Diversion Project in China. *Georisk:*

Assessment and Management of Risk for Engineered Systems and Geohazards, 17(1), 127–147. https://doi.org/10.1080/17499518.2023.2184834.

Lin, P., Bathurst, R.J. and Liu, J. 2017a. Statistical evaluation of the FHWA simplified method and modifications for predicting soil nail loads. *Journal of Geotechnical and Geoenvironmental Engineering, ASCE*, 143(3), 04016107. https://doi.org/10.1061/(ASCE)GT.1943-5606.0001614.

Lin, P., Bathurst, R.J., Javankhoshdel, S. and Liu, J. 2017b. Statistical analysis of the effective stress method and modifications for prediction of ultimate bond strength of soil nails. *Acta Geotechnica*, 12(1), 171–182. https://doi.org/10.1007/s11440-016-0477-1.

Lin, S.-S., Marcos, M.C.M., Chang, H.-W. and Chen, Y.-J. 2012. Design and implementation of a drilled shaft load test database. *Computers and Geotechnics*, 41, 106–113. https://doi.org/10.1016/j.compgeo.2011.12.001.

Liu, H.F., Ma, H.H., Chang, D. and Lin, P.Y. 2020. Statistical calibration of federal highway administration simplified models for facing tensile forces of soil nail walls. *Acta Geotechnica*, 16(5), 1509–1526. https://doi.org/10.1007/s11440-020-01106-4.

Liu, Z., Nadim, F., Lacasse, S., Lehane, B. and Choi, Y.J. 2019. Method uncertainty for five axial pile capacity design methods. Offshore Technology Conference, Paper No. OTC-29514-MS. https://doi.org/10.4043/29514-MS.

Loehr, J.E., Boeckmann, A.Z. and Ding, D. 2022. *Reliability-Based Geotechnical Resistance Factors for Axially Loaded Micropiles*. Washington, DC: The National Academies of Science, Engineering and Medicine. https://doi.org/10.17226/26615.

Loganathan, N. and Poulos, H.G. 1998. Analytical prediction for tunneling-induced ground movements in clays. *Journal of Geotechnical and Geoenvironmental Engineering, ASCE*, 124(9), 846–856. https://doi.org/10.1061/(ASCE)1090-0241(1998)124:9(846).

Long, J.H. 2013. Improving agreement between static method and dynamic formula for driven cast-in-place piles in Wisconsin. Report No. 0092-10-09. Madison, WI: Wisconsin Department of Transportation.

Long, J.H. and Anderson, A. 2014. Improvement of driven pile installation and design in Illinois: Phase 2. Report No. FHWA-ICT-14-019. Springfield, IL: Illinois Department of Transportation.

Long, J.H., Hendrix, J. and Baratta, A.L. 2009a. Evaluation/modification of IDOT foundation piling design and construction policy. Report No. FHWA-ICT-09-037. Springfield, IL: Illinois Department of Transportation.

Long, J.H., Hendrix, J. and Jaromin, D. 2009b. Comparison of five different methods for determining pile bearing capacities. Final Report. Madison, WI: Wisconsin Department of Transportation.

Long, M. 2001. Database for retaining wall and ground movements due to deep excavations. *Journal of Geotechnical and Geoenvironmental Engineering, ASCE*, 127(3), 203–224. https://doi.org/10.1061/(ASCE)1090-0241(2001)127:3(203).

Lovera, A., Ghabezloo, S., Sulem, J., Randolph, M.F., Kham, M. and Palix, E. 2021. Pile response to multidirectional lateral loading using p-y curves approach. *Géotechnique*, 71(4), 288–298. https://doi.org/10.1680/jgeot.18.P.297.

Machairas, N., Highley, G.A. and Iskander, M.G. 2018. Evaluation of FHWA pile design method against the FHWA deep foundation load test database version 2.0. *Transportation Research Record*, 2672(52), 268–277. https://doi.org/10.1177/0361198118773196.

Mayne, P.W. 2015. In-situ geocharacterization of soils in the year 2016 and beyond. In *Proceedings of the 15th Pan-American Conference on Soil Mechanics and Geotechnical Engineering (Geotechnical Synergy in Buenos Aires)*, 139–161. Amsterdam: IOS Press.

Mayne, P.W., Christopher, B.R. and DeJong, J. 2001. Subsurface investigations (geotechnical site characterization). Report No. FHWA-NHI-01-031. Washington, DC: Federal Highway Administration.

McVay, M.C., Birgisson, B., Zhang, L., Perez, A. and, Putcha, S. 2000. Load and resistance factor design (LRFD) for driven piles using dynamic methods–a Florida perspective. *Geotechnical Testing Journal*, 23(1), 55–66. https://doi.org/10.1520/GTJ11123J.

McVay, M.C., Wasman, S., Huang, L. and Crawford, S. 2016. Load and resistance factor design resistance factors for Augercast in place piles. Final Report. Tallahassee, FL: Florida Department of Transportation.

Mestat, P. 2001. MOMIS: Une base de données sur la modélisation numérique des remblais sur sols compressibles et sur la confrontation calculs–mesures in situ. *Bulletin des Laboratoires des Ponts et Chaussées*, 232, 43–58.

Mestat, P. 2002. Applications de la base de données MOMIS à la validation du calcul des ouvrages souterrains. *Bulletin des Laboratoires des Ponts et Chaussées*, 236, 59–75.

Mestat, P. and Bourgeois, E. 2002. Prediction and performance: Numerical modeling of sheet pile walls and diaphragm walls. In *Proceedings of the 3rd International Symposium on Geotechnical Aspects of Underground Construction in Soft Ground*, 441–446. Lyon: Spécifique.

Mestat, P., Bourgeois, E. and Riou, Y. 2004a. MOMIS: Une base de données sur la confrontation modèles numériques d'ouvrages–mesures in situ. Applications aux rideaux de palplanches. *Bulletin des Laboratoires des Ponts et Chaussées*, 252–253, 49–76.

Mestat, P., Bourgeois, E. and Riou, Y. 2004b. Numerical modelling of embankments and underground works. *Computers and Geotechnics*, 31(3), 227–236. https://doi.org/10.1016/j.compgeo.2004.01.003.

Mestat, P. and Riou, Y. 2004. A database for case histories and numerical modelling. In *Proceedings of the 5th International Conference on Case Histories in Geotechnical Engineering*, 1–8. University of Missouri-Rolla.

Mestat, P. and Riou, Y. 2006. New developments of the MOMIS database applied to the performance of numerical modelling of underground excavations. In *Proceedings of the 5th International Conference of TC28 of the ISSMGE Amsterdam–the Netherlands*, 609–614. London: Taylor & Francis Group.

Meyerhof, G.G. 1965. Shallow foundations. *Journal of the Soil Mechanics and Foundations Division, ASCE*, 91(SM2), 21–31. https://doi.org/10.1061/JSFEAQ.0000719.

Meyerhof, G.G. 1976. Bearing capacity and settlement of pile foundations. *Journal of Geotechnical Engineering Division, ASCE*, 102(3), 197–228. https://doi.org/10.1061/AJGEB6.0000243.

Mitchell, J.K. 1986. Practical problems from surprising soil behavior. *Journal of Geotechnical Engineering, ASCE*, 112(3), 259–289. https://doi.org/10.1061/(ASCE)0733-9410(1986)112:3(255).

Miyata, Y. and Bathurst, R.J. 2012a. Analysis and calibration of default steel strip pullout models used in Japan. *Soils and Foundations*, 52(3), 481–497. http://doi.org/10.1016/j.sandf.2012.05.007.

Miyata, Y. and Bathurst, R.J. 2012b. Reliability analysis of soil-geogrid pullout models in Japan. *Soils and Foundations*, 52(4), 620–633. http://doi.org/10.1016/j.sandf.2012.07.004.

Miyata, Y. and Bathurst, R.J. 2015. Reliability analysis of geogrid installation damage test data in Japan. *Soils and Foundations*, 55(2), 393–403. https://doi.org/10.1016/j.sandf.2015.02.013.

Miyata, Y. and Bathurst, R.J. 2019. Statistical assessment of load model accuracy for steel grid-reinforced soil walls. *Acta Geotechnica*, 14(1), 57–70. https://doi.org/10.1007/s11440-018-0638-5.

Miyata, Y., Bathurst, R.J. and Allen, T.M. 2014. Reliability analysis of geogrid creep data in Japan. *Soils and Foundations*, 54(4), 608–620. http://doi.org/10.1016/j.sandf.2014.06.004.

Miyata, Y., Yu, Y. and Bathurst, R.J. 2018. Calibration of soil-steel grid pullout models using a statistical approach. *Journal of Geotechnical and Geoenvironmental Engineering, ASCE*, 144(2), 04017106. https://doi.org/10.1061/(ASCE)GT.1943-5606.0001811.

Moormann, C. 2004. Analysis of wall and ground movements due to deep excavations in soft soil based on a new worldwide database. *Soils and Foundations*, 44(1), 87–98. https://doi.org/10.3208/sandf.44.87.

Mosher, R.L. and Dawkins, W.P. 2000. Theoretical manual for pile foundation. Report No. ERDC/ITL TR-00-5. Washington, DC: U.S. Army Corps of Engineers.

Motamed, R., Elfass, S. and Stanton, K. 2016. LRFD resistance factor calibration for axially loaded drilled shafts in the Las Vegas Valley. Report No. 515-13-803. Carson City, NV: Nevada Department of Transportation.

Mullins, G., Gunaratne, M., Sarsour, A. and Mobley, S. 2019. Load and resistance factor design (LRFD) resistance factors for tip grouted drilled shafts. Final Report. Tallahassee, FL: Florida Department of Transportation.

Najjar, S.S. and Gilbert, R.B. 2009. Importance of lower-bound capacities in the design of deep foundations. *Journal of Geotechnical and Geoenvironmental Engineering, ASCE*, 135(7), 890–900. https://doi.org/10.1061/(ASCE)GT.1943-5606.0000044.

National Academies of Sciences, Engineering, and Medicine. 2019. *Manual on Subsurface Investigations*. Washington, DC: The National Academies Press.

Negro, A. and de Queiroz, P.I.B. 2000. Prediction and performance: A review of numerical analyses for tunnels. In *Proceedings of the 3rd International Symposium on Geotechnical Aspects of Underground Construction in Soft Ground*, 409–418. Lyon: Spécifique.

Negro, A., Karlsrud, K., Srithar, S., Ervin, M. and Vorster, E. 2009. Prediction, monitoring and evaluation of performance of geotechnical structures. In *Proceedings of the 17th International Conference on Soil Mechanics and Geotechnical Engineering*, 2930–3005. Amsterdam: IOS Press.

Ng, K.W., Adhikari, P. and Gebreslasie, Y.Z. 2019. Development of load and resistance factor design procedures for driven piles on soft rocks in Wyoming. Report No. WY-1902F. Cheyenne, WY: Wyoming Department of Transportation.

Ng, K.W., Masud, N., Oluwatuyi, O. and Wulff, S.S. 2022. Development of LRFD recommendations of driven piles on intermediate geomaterials. Report No. MPC-651. Fargo, ND: Mountain-Plains Consortium, North Dakota State University.

Niazi, F.S. 2014. Static axial pile foundation response using seismic piezocone data. PhD thesis, School of Civil and Environmental Engineering, Georgia Institute of Technology.
Nottingham, L.C. and Schmertmann, J.H. 1975. An investigation of pile capacity design procedures. Report No. D629. Gainesville, FL: University of Florida.
Ong, Y.H., Toh, C.T., Chee, S.K. and Mohamad, H. 2021. Bored piles in tropical soils and rocks: Shaft and base resistances, t–z and q–w models. *Proceedings of the Institution of Civil Engineers–Geotechnical Engineering*, 174(2), 193–224. https://doi.org/10.1680/jgeen.19.00106.
Otake, Y. and Honjo, Y. 2022. Challenges in geotechnical design revealed by reliability assessment: Review and future perspectives. *Soils and Foundations*, 62(3), 101129. https://doi.org/10.1016/j.sandf.2022.101129.
O'Rourke, T.D. and Kulhawy, F.H. 1985. Observations on load tests on drilled shafts. In *Proceedings of the Drilled Piers and Caissons II*, 113–128. New York: ASCE.
Paikowsky, S.G., Birgisson, B., McVay, M., Nguyen, T., Kuo, C., Baecher, G., Ayyub, B., Stenersen, K., O'Malley, K., Chernauskas, L. and O'Neill, M. 2004. Load and resistance factor design (LRFD) for deep foundations. NCHRP Report 507. Washington, DC: Transportation Research Board of the National Academies.
Paikowsky, S.G., Canniff, M.C., Lesny, K., Kisse, A., Amatya, S. and Muganga, R. 2010. LRFD design and construction of shallow foundations for highway bridge structures. NCHRP Report 651. Washington, DC: National Academy of Sciences.
Petek, K., McVay, M. and Mitchell, R. 2020. Development of guidelines for bearing resistance of large diameter open-end steel piles. Report No. FHWA-HRT-20-011. McLean, VA: U.S. Department of Transportation and Federal Highway Administration.
Petek, K., Mitchell, R. and Ellis, H. 2016. FHWA deep foundation load test database version 2.0 user manual. Report No. FHWA-HRT-17-034. McLean, VA: Federal Highway Administration.
Phoon, K.K. 2017. Role of reliability calculations in geotechnical design. *Georisk: Assessment and Management of Risk for Engineered Systems and Geohazards*, 11(1), 4–21. https://doi.org/10.1080/17499518.2016.1265653.
Phoon, K.K. 2018. Editorial for special collection on probabilistic site characterization. *ASCE-ASME Journal of Risk and Uncertainty in Engineering Systems, Part A: Civil Engineering*, 4(4), 02018002. https://doi.org/10.1061/AJRUA6.0000992.
Phoon, K.K. 2023. What geotechnical engineers want to know about reliability. *ASCE-ASME Journal of Risk and Uncertainty in Engineering Systems, Part A: Civil Engineering*, 9(2), 03123001. https://doi.org/10.1061/AJRUA6.RUENG-1002.
Phoon, K.K., Cao, Z., Ji, J., Leung, Y., Najjar, S., Shuku, T., Tang, C., Yin, Z., Ikumasa, Y. and Ching, J. 2022b. Geotechnical uncertainty, modeling, and decision making. *Soils and Foundations*, 62(5), 101189. https://doi.org/10.1016/j.sandf.2022.101189.
Phoon, K.K. and Ching, J. 2021. Project DeepGeo – Data-driven 3D subsurface mapping. *Journal of GeoEngineering*, 16(2), 47–59. https://doi.org/10.6310/jog.202106_16(2).2.
Phoon, K.K. and Ching, J. 2022. Additional observations on the site recognition challenge. *Journal of GeoEngineering*, 17(4), 231–247. https://doi.org/10.6310/jog.202212_17(4).6.
Phoon, K.K., Ching, J. and Cao, Z. 2022a. Unpacking data-centric geotechnics. *Underground Space*, 7(6), 967–989. https://doi.org/10.1016/j.undsp.2022.04.001.
Phoon, K.K., Ching, J. and Shuku, T. 2022c. Challenges in data-driven site characterization. *Georisk: Assessment and Management of Risk for Engineered Systems and Geohazards*, 16(1), 114–126. https://doi.org/10.1080/17499518.2021.1896005.
Phoon, K.K., Ching, J. and Tao, Y. 2023a. Chapter 2. Soil and rock parametric uncertainties. In *Uncertainty, modelling, and decision making in geotechnics*. Boca Raton, FL: CRC Press.
Phoon, K.K., Zhang, L.M. and Cao, Z.J. 2023b. Special Issue on "Machine Learning and AI in Geotechnics". *Georisk: Assessment & Management of Risk for Engineered Systems & Geohazards*, 17(1), 1–6.
Phoon, K.K., Ching, J. and Wang, Y. 2019. Managing risk in geotechnical engineering – from data to digitalization. In *Proceedings of the 7th International Symposium on Geotechnical Safety and Risk*, 13–34. Taipei, Taiwan: ISGSR 2019.
Phoon, K.K. and Kulhawy, F.H. 2005. Characterization of model uncertainties for laterally loaded rigid drilled shafts. *Géotechnique*, 55(1), 45–54. https://doi.org/10.1680/geot.2005.55.1.45.
Phoon, K.K., Kulhawy, F.H. and Grigoriu, M.D. 2003. Development of a reliability-based design framework for transmission line structure foundations. *Journal of Geotechnical and Geoenvironmental Engineering, ASCE*, 129(9), 798–806. https://doi.org/10.1061/(ASCE)1090-0241(2003)129:9(798).

Phoon, K.K., Prakoso, W.A., Wang, Y. and Ching, J. 2016b. Chapter 3. Uncertainty representation of geotechnical design parameters. In *Reliability of Geotechnical Structures in ISO2394*, 49–87. London: CRC Press/Balkema.

Phoon, K.K. and Retief, J.V. 2016. *Reliability of Geotechnical Structures in ISO2394*. London: CRC Press/Balkema.

Phoon, K.K., Retief, J.V., Ching, D., Dithinde, M., Schweckendiek, T., Wang, Y. and Zhang, L.M. 2016a. Some observations on ISO2394:2015 Annex D (reliability of geotechnical structures). *Structural Safety*, 62, 24–33. http://doi.org/10.1016/j.strusafe.2016.05.003.

Phoon, K.K. and Tang, C. 2015a. Model uncertainty for the capacity of strip footings under positive combined loading. In *Proceedings of the Geo-Risk 2017 – Geotechnical Safety and Reliability: Honoring Wilson H. Tang*, 40–60. Reston, VA: ASCE. https://doi.org/10.1061/9780784480731.005.

Phoon, K.K. and Tang, C. 2015b. Model uncertainty for the capacity of strip footings under negative and general combined loading. In *Proceedings of the 12th International Conference on Applications of Statistics and Probability in Civil Engineering (ICASP 12)*. Library of The University of British Columbia.

Phoon, K.K. and Tang, C. 2019. Characterisation of geotechnical model uncertainty. *Georisk: Assessment and Management of Risk for Engineered Systems and Geohazards*, 13(2), 101–130. https://doi.org/10.1080/17499518.2019.1585545.

Phoon, K. K. and Tang, C. 2024. *Database Approach for Data-Centric Geotechnics, Vol. 1: Site Characterization and Vol. 2: Geotechnical Structures*. Boca Raton, FL: CRC Press.

Potts, D.M. and Zdravković, L. 1999. *Finite Element Analysis in Geotechnical Engineering: Theory*. London: Thomas Telford.

Potts, D.M. and Zdravković, L. 2001. *Finite Element Analysis in Geotechnical Engineering: Application*. London: Thomas Telford.

Poulos, H.G. 1989. Pile behavior – Theory and application. *Géotechnique*, 39(3), 365–415. https://doi.org/10.1680/geot.1989.39.3.365.

Poulos, H.G. 2017. *Tall Building Foundation Design*. Boca Raton, FL: CRC Press.

Poulos, H.G. and Davis, E.H. 1973. *Elastic Solutions for Soil and Rock Mechanics*. New York: John Wiley & Sons, Inc.

PWRI (Public Works Research Institute). 2018. *Technical Note of PWRI (in Japanese)*. Ibaraki-Ken: PWRI.

Randolph, M.F. and Wroth, C.P. 1978. Analysis of deformation of vertically loaded piles. *Journal of the Geotechnical Engineering Division, ASCE*, 104(12), 1465–1488. https://doi.org/10.1061/AJGEB6.0000729.

Reddy, S.C. and Stuedlein, A.W. 2017. Ultimate limit state reliability-based design of augered cast-in-place piles considering lower-bound capacities. *Canadian Geotechnical Journal*, 54(12), 1693–1703. https://doi.org/10.1139/cgj-2016-0145.

Reiffsteck, P. 2005. Portance et tassements d'une fondation profonde: Présentation des résultats du concours de prévision. In *Proceedings of the Symposium on International ISP5/PRESSIO 2005: 50 ans de pressiomètres*, Vol. 2, 521–535. Paris: Presses de l'ENPC/LCPC.

Reiffsteck, P. 2009. ISP5 pile prediction revisited. In *Proceedings of the Contemporary Topics in In Situ Testing, Analysis, and Reliability of Foundations*, 50–57. Reston, VA: ASCE. https://doi.org/10.1061/41022(336)7.

Rizk, A., Kodsy, A., Iskander, M. and Machairas, N. 2022. Efficacy of design methods for predicting the capacity of large-diameter open-ended piles. *Journal of Geotechnical and Geoenvironmental Engineering, ASCE*, 148(10), 04022078. https://doi.org/10.1061/(ASCE)GT.1943-5606.0002824.

Rowe, R.K. and Armitage, H.H. 1987. A design method for drilled piers in soft rock. *Canadian Geotechnical Journal*, 24(1), 126–142. https://doi.org/10.1139/t87-011.

Schneider, H.R. 1997. Definition and characterization of characteristic soil properties. In *Proceedings of the Fourteen International Conference on Soil Mechanics and Geotechnical Engineering*, 2271–2274. Rotterdam: A.A. Balkema.

Sharma, A., Ching, J. and Phoon, K.K. 2022. A hierarchical Bayesian similarity measure for geotechnical site retrieval. *Journal of Engineering Mechanics, ASCE*, 148(10), 04022062. https://doi.org/10.1061/(ASCE)EM.1943-7889.0002145.

Sheil, B. 2021. Hybrid framework for forecasting circular excavation collapse: Combining physics-based and data-driven modeling. *Journal of Geotechnical and Geoenvironmental Engineering, ASCE*, 147(12), 04021140. https://doi.org/10.1061/(ASCE)GT.1943-5606.0002683.

Silva, B.C.D., Tsuha, C.H.C. and dos Santos Filho, J.M.S.M. 2023. A database of installation monitoring and uplift load tests of round-shaft helical anchors in Brazil. *Probabilistic Engineering Mechanics*, 71, 103378. https://doi.org/10.1016/j.probengmech.2022.103378.

Smith, T.D., Banas, A., Gummer, M. and Jin, J. 2011. Recalibration of the GRLWEAP resistance factor for Oregon DOT. Report No. FHWA-OR-RD-11-08. Salem, OR: Oregon Department of Transportation.

Souissi, M., Cherry, J.A. and Siller, T. 2020. Helical pile capacity-to-torque correlation: A more reliable capacity-to-torque factor based on full scale load tests. *DFI Journal–The Journal of the Deep Foundations Institute*, 14(2), 1–11. https://doi.org/10.37308/DFIJnl.20190716.208.

Suryasentana, S.K. and Lehane, B.M. 2014. Numerical derivation of CPT-based p-y curves for piles in sand. *Géotechnique*, 64(3), 186–194. https://doi.org/10.1680/geot.13.P.026.

Taborda, D.M.G., Zdravković, L., Potts, D.M., Abadias, D., Burd, H.J., Byrne, B.W., Gavin, K.G., Houlsby, G.T., Jardine, R.J., Liu, T., Martin, C.M. and Mcadam, R.A. 2020. Finite-element modelling of laterally loaded piles in a dense marine sand at Dunkirk. *Géotechnique*, 70(11), 1014–1029. https://doi.org/10.1680/jgeot.18.PISA.006.

Tan, Y., Fan, D. and Lu, Y. 2022. Statistical analyses on a database of deep excavations in Shanghai soft clays in China from 1995–2018. *Practice Periodical on Structural Design and Construction*, 27(1), 04021067. https://doi.org/10.1061/(ASCE)SC.1943-5576.0000646.

Tang, C. and Bathurst, R.J. 2021. Chapter 4. Statistics for geotechnical design model factors. In *State-of-the-Art Review of Inherent Variability and Uncertainty in Geotechnical Properties and Models*. International Society of Soil Mechanics and Geotechnical Engineering (ISSMGE)–Technical Committee TC304 'Engineering Practice of Risk Assessment and Management'. https://doi.org/10.53243/R0001.

Tang, C. and Phoon, K.K. 2016. Model uncertainty of cylindrical shear method for calculating the uplift capacity of helical anchors in clay. *Engineering Geology*, 207, 14–23. https://doi.org/10.1016/j.enggeo.2016.04.009.

Tang, C. and Phoon, K.K. 2017. Model uncertainty of Eurocode 7 approach for bearing capacity of circular footings on dense sand. *International Journal of Geomechanics, ASCE*, 17(3), 04016069. https://doi.org/10.1061/(ASCE)GM.1943-5622.0000737.

Tang, C. and Phoon, K.K. 2018a. Statistics of model factors and consideration in reliability-based design of axially loaded helical piles. *Journal of Geotechnical and Geoenvironmental Engineering, ASCE*, 144(8), 04018050. https://doi.org/10.1061/(ASCE)GT.1943-5606.0001894.

Tang, C. and Phoon, K.K. 2018b. Evaluation of model uncertainties in reliability-based design of steel H-piles in axial compression. *Canadian Geotechnical Journal*, 55(11), 1513–1532. https://doi.org/10.1139/cgj-2017-0170.

Tang, C. and Phoon, K.K. 2018c. Statistics of model factors in reliability-based design of axially loaded driven piles in sand. *Canadian Geotechnical Journal*, 55(11), 1592–1610. https://doi.org/10.1139/cgj-2017-0542.

Tang, C. and Phoon, K.K. 2019. Characterization of model uncertainty in predicting axial resistance of piles driven into clay. *Canadian Geotechnical Journal*, 56(8), 1098–1118. https://doi.org/10.1139/cgj-2018-0386.

Tang, C. and Phoon, K.K. 2020. Statistical evaluation of model factors in reliability calibration of high-displacement helical piles under axial loading. *Canadian Geotechnical Journal*, 57(2), 246–262. https://doi.org/10.1139/cgj-2018-0754.

Tang, C. and Phoon, K.K. 2021. *Model Uncertainties in Foundation Design*. Boca Raton, FL: CRC Press.

Tang, C., Phoon, K.K. and Chen, Y.-J. 2019. Statistical analyses of model factors in reliability-based limit-state design of drilled shafts under axial loading. *Journal of Geotechnical and Geoenvironmental Engineering, ASCE*, 145(9), 04019042. https://doi.org/10.1061/(ASCE)GT.1943-5606.0002087.

Tang, C., Phoon, K.K., Li, D.-Q. and Akbas, S.O. 2020. Expanded database assessment of design methods for spread foundations under axial compression and uplift loading. *Journal of Geotechnical and Geoenvironmental Engineering, ASCE*, 146(11), 04020119. https://doi.org/10.1061/(ASCE)GT.1943-5606.0002373.

Tang, C., Phoon, K.K., Zhang, L. and Li, D.-Q. 2017. Model uncertainty for predicting the bearing capacity of sand overlying clay. *International Journal of Geomechanics, ASCE*, 17(7), 04017015. https://doi.org/10.1061/(ASCE)GM.1943-5622.0000898.

Tang, C., Yuan, J., Phoon, K.K., Feng, X. and Yu, X. 2022. Variability in predictions of punch-through for spudcan penetration. *Computers and Geotechnics*, under review.

Tang, W.H. and Pelletier, J.H. 1990. Performance reliability of offshore piles. Offshore Technology Conference, Paper No. OTC 6379. https://doi.org/10.4043/6379-MS.

Tartakovsky, A.M., Marrero, C.O., Perdikaris, P., Tartakovsky, G.D. and Barajas-Solano, D. 2020. Physics-informed deep neural networks for learning parameters and constitutive relationships in subsurface flow problems. *Water Resources Research*, 56(5), 1–16. https://doi.org/10.1029/2019WR026731.

Tavera, E.A., Rix, G.J., Burnworth, G.H. and Jung, J. 2016. Calibration of region specific gates pile driving formula for LRFD. Report No. FHWA/LA.16/561. Baton Rouge, LA: Louisiana Department of Transportation and Development.

Terzaghi, K. 1943. *Theoretical Soil Mechanics*. New York: John Wiley and Sons, Inc.

Terzaghi, K. and Peck, R.B. 1967. *Soil Mechanics in Engineering Practice*. 2nd ed. New York: John Wiley and Sons, Inc.

Timchenko, A. and Briaud, J.-L. 2023. Analysis of a database of open-pit mine slope failures to predict travel distance, setback distance, and geometric properties. *Canadian Geotechnical Journal*, in press. https://doi.org/10.1139/cgj-2022-0117.

Topacio, A., Chen, Y.J., Phoon, K.K. and Tang, C. 2022. Evaluation of compression interpretation criteria for drilled shafts socketed into rocks. *Proceedings of the Institution of Civil Engineers–Geotechnical Engineering*, ahead of print. https://doi.org/10.1680/jgeen.21.00120.

Travis, Q.B., Schmeeckle, M.W. and Sebert, D.M. 2011. Meta-analysis of 301 slope failure calculations. I: database description. *Journal of Geotechnical and Geoenvironmental Engineering, ASCE*, 137(5), 453–470. https://doi.org/10.1061/(ASCE)GT.1943-5606.0000461.

Van Dijk, B.F.J. and Yetginer, A.G. 2015. Findings of the ISSMGE jack-up leg penetration prediction event. *Frontiers in Offshore Geotechnics*, III, 1267–1274. London, UK: Taylor & Francis Group.

Van der Veen, C. 1953. The bearing capacity of a pile. In *Proceedings of the 3rd International Conference on Soil Mechanics and Foundation Engineering*, Vol. 2, 85–90. London: ISSMGE (International Society for Soil Mechanics and Geotechnical Engineering.

Vardanega, P., Voyagaki, E., Crispin, J., Gilder, C. and Ntassiou, K. 2021. *The Dingo Database: Summary Report*. University of Bristol. https://doi.org/10.5523/bris.89r3npvewel2ea8ttb67ku4d.

Vardanega, P.J. and Bolton, M.D. 2016. Design of geostructural systems. *ASCE-ASME Journal of Risk and Uncertainty in Engineering Systems, Part A: Civil Engineering*, 2(1), 04015017. https://doi.org/10.1061/AJRUA6.0000849.

Vaughan, P.R. 1994. Assumption, prediction and reality in geotechnical engineering. *Géotechnique*, 44(4), 573–609.

Vesić, A.S. 1973. Analysis of ultimate loads of shallow foundations. *Journal of the Soil Mechanics and Foundations Division, ASCE*, 99(1), 45–73. https://doi.org/10.1061/JSFEAQ.0001846.

Vesić, A.S. 1977. *Design of Pile Foundations. NCHRP Synthesis 42*. Washington, DC: Transportation Research Board.

Viana da Fonseca, A. and Santos, J.A. 2008. *International Prediction Event: Behaviour of Bored, CFA and Driven Piles in Residual Soil, Experimental Site–ISC'2*. Lisboa: University of Porto.

Viggiani, C. 1998. Pile groups and piled rafts behaviour. In *Proceedings of the 3rd International Geotechnical Seminar on Deep Foundations on Bored and Auger Piles*, 77–94. Rotterdam: A. A. Balkema.

Voyagaki, E., Crispin, J., Gilder, C., Ntassiou, K., O'Riordan, N., Nowak, P., Sadek, T., Patel, D., Mylonakis, G. and Vardanega, P. 2022. The Dingo database of axial pile load tests for the UK: Settlement prediction in fine-grained soils. *Georisk: Assessment and Management of Risk for Engineered Systems and Geohazards*, 16(4), 640–661. https://doi.org/10.1080/17499518.2021.1971249.

Wang, J.H., Xu, Z.H. and Wang, W.D. 2010. Wall and ground movements due to deep excavations in Shanghai soft soils. *Journal of Geotechnical and Geoenvironmental Engineering, ASCE*, 136(7), 985–994 . https://doi.org/10.1061/(ASCE)GT.1943-5606.0000299.

Whitman, R.V. 1984. Evaluating calculated risk in geotechnical engineering. *Journal of Geotechnical Engineering, ASCE*, 110(2), 145–188. https://doi.org/10.1061/(ASCE)0733-9410(1984)110:2(143).

Whyte, S. 2018. Foundation optimization for ever larger offshore wind turbines: Geotechnical perspective. KIVI Lecture.

Wood, T.A., Jayawickrama, P.W., Surles, J.G. and Lawson, W.D. 2012a. Pullout resistance of MSE reinforcements in backfills typically used in Texas: Volume 2, test reports for MSE reinforcements in Type B (sandy) backfill. Report No. FHWA/TX-13/0-6493-1, Vol. 2. Austin, TX: Texas Department of Transportation.

Wood, T.A., Jayawickrama, P.W., Surles, J.G. and Lawson, W.D. 2012b. Pullout resistance of MSE reinforcements in backfills typically used in Texas: Volume 3, test reports for MSE reinforcements in Type A (gravelly) backfill. Report No. FHWA/TX-13/0-6493-1, Vol. 3. Austin, TX: Texas Department of Transportation.

Wu, T.H. 2011. The observational method: Case history and models. *Journal of Geotechnical and Geoenvironmental Engineering, ASCE*, 137(10), 862–873. https://doi.org/10.1061/(ASCE)GT.1943-5606.0000509.

Xu, S.J., Yi, J.T., Zhang, T.B., Wang, Z. and Yao, K. 2021. Characterising the model uncertainty of ISO methods for punch-through capacity prediction. *Proceedings of the Institution of Civil Engineers–Geotechnical Engineering*, 174(5), 549–562. https://doi.org/10.1680/jgeen.21.00003.

Yan, W., Yan, Y., Shen, P. and Zhou, W.H. 2023. A hybrid physical data informed DNN in axial displacement prediction of immersed tunnel joint. *Georisk: Assessment and Management of Risk for Engineered Systems and Geohazards*, 17(1), 169–180. https://doi.org/10.1080/17499518.2023.2169941.

Yang, Z.X., Guo, W.B., Jardine, R.J. and Chow, F. 2017. Design method reliability assessment form an extended database of axial load tests on piles driven in sand. *Canadian Geotechnical Journal*, 54(1), 59–74. https://doi.org/10.1139/cgj-2015-0518.

Yu, X., Abu-Farsakh, M., Hu, Y., Fortier, A.R. and Hasan, M.R. 2017. Calibration of LRFD geotechnical axial (tension and compression) resistance factors (ϕ) for California. Report No. CA18-2578. Sacramento, CA: California Department of Transportation.

Yu, Y. and Bathurst, R.J. 2015. Analysis of soil-steel bar mat pullout models using a statistical approach. *Journal of Geotechnical and Geoenvironmental Engineering, ASCE*, 141(5), 04015006. https://doi.org/10.1061/(ASCE)GT.1943-5606.0001281.

Yuan, J., Lin, P., Huang, R. and Que, Y. 2019a. Statistical evaluation and calibration of two methods for predicting nail loads of soil nail walls in China. *Computers and Geotechnics*, 108, 269–279. https://doi.org/10.1016/j.compgeo.2018.12.028.

Yuan, J., Lin, P., Mei, G. and Hu, Y. 2019b. Statistical prediction of deformations of soil nail walls. *Computers and Geotechnics*, 115, 103168. https://doi.org/10.1016/j.compgeo.2019.103168.

Zdravković, L., Taborda, D.M.G., Potts, D.M., Abadias, D., Burd, H.J., Byrne, B.W., Gavin, K.G., Houlsby, G.T., Jardine, R.J., Martin, C.M., Mcadam, R.A. and Ushev, E. 2020. Finite-element modelling of laterally loaded piles in a stiff glacial clay till at Cowden. *Géotechnique*, 70(11), 999–1013. https://doi.org/10.1680/jgeot.18.PISA.005.

Zhang, D.M., Phoon, K.K., Huang, H.W. and Hu, Q.F. 2015. Characterization of model uncertainty for cantilever deflections in undrained clay. *Journal of Geotechnical and Geoenvironmental Engineering, ASCE*, 141(1), 04014088. https://doi.org/10.1061/(ASCE)GT.1943-5606.0001205.

Zhang, P., Yin, Z.-Y. and Sheil, B. 2023. A physics-informed data-driven approach for consolidation analysis. *Géotechnique*, ahead of print. https://doi.org/10.1680/jgeot.22.00046.

Zhang, Y. and Andersen, K.H. 2017. Scaling of lateral pile p-y response in clay from laboratory stress-strain curves. *Marine Structures*, 53, 124–135. https://doi.org/10.1016/j.marstruc.2017.02.002.

Zhang, Y., Andersen, K.H., Jeanjean, P., Karlsrud, K. and Haugen, T. 2020. Validation of monotonic and cyclic p-y framework by lateral pile load tests in stiff, overconsolidated clay at the Haga site. *Journal of Geotechnical and Geoenvironmental Engineering, ASCE*, 146(9), 04020080. https://doi.org/10.1061/(ASCE)GT.1943-5606.0002318.

Zhang, Y., Andersen, K.H., Jeanjean, P., Mirdamadi, A., Gundersen, A.S. and Jostad, H.P. 2017. A framework for cyclic p-y curves in clay and application to pile design in GoM. In *Proceedings of the 8th International Conference on Offshore Site Investigation Geotechnics*, Vol. 1, 431–440. London: Society for Underwater Technology. https://doi.org/10.3723/OSIG17.431.

Part 2

Probabilistic methods

6 Geotechnical Reliability Analysis for Practice

Jian Ji and Zijun Cao

ABSTRACT

In recent years, the application of reliability analysis methods in geotechnical community has seen great prosperity. Geotechnical reliability analysis when targeted at practitioners requires simplified, straightforward guidelines for statistical modeling and uncertainty simulations. This chapter covers the fundamental concepts of the first-order reliability method (FORM) and Monte Carlo simulations (MCS) that are adapted to different variants for efficiently carrying out geotechnical reliability analysis. Given a geotechnical system with prescribed soil uncertainties, one may find this material helpful if he/she wants to answer these questions: (1) what is the probability of the system failure and how to compute it using the introduced reliability analysis methods; and (2) how to determine the desirable design parameters in the context of uncertain geotechnical factors. A short review of practical reliability analysis procedures illustrated to geotechnical problems via popular numerical packages is also provided.

6.1 INTRODUCTION: BACKGROUND AND GEOTECHNICAL RELIABILITY ANALYSIS

In the context of the uncertainty and variability of geological and geotechnical input parameters, the performance of a geotechnical system can hardly be predicted deterministically. In contrast, the reliability analysis is in a complementary role to quantify the combined effects of various uncertainties on a geotechnical system performance. The objective of geotechnical reliability analysis is to obtain the probability of failure of the geotechnical system given that all interested uncertain parameters (random variables) are properly characterized with their statistics. Mathematically, random variables related to a geotechnical system can be denoted by vector $\mathbf{x} = [x_1, x_2, \ldots, x_M]$, and the functional relationships corresponding to the system performance can be described by its performance function $z = g(\mathbf{x})$: $g(\mathbf{x}) < 0$ denoting failure, $g(\mathbf{x}) > 0$ denoting safety, and $g(\mathbf{x}) = 0$ which divides the design parameter space into safe and unsafe regions is called the limit state function (LSF). Figure 6.1 generalizes the concept of performance function in the space with two random variables x_1 and x_2 denoting the geotechnical system resistance and load, respectively.

The probability of failure P_f can be calculated by the integral:

$$P_f = \int \cdots \int_{g(\mathbf{x})<0} f(x_1, x_2, \cdots, x_M) \mathrm{d}x_1 \mathrm{d}x_2 \cdots \mathrm{d}x_M \tag{6.1}$$

where $f(x_1, x_2, \cdots, x_M)$ is the joint probability density function (PDF) of random variables $\mathbf{x}$, and $g(\mathbf{x}) < 0$ denotes the failure region for which the probability integration is conducted.

Equation (6.1) is the fundamental recipe of reliability analysis. In most cases, the analytical solution to Equation (6.1) can be very involved. Bearing this in mind, researchers have devised various reliability methods to replace the analytical solution. In the literature, those methods can be classified into two cohorts (Bjerager, 1990; Rackwitz, 2001; Shinozuka, 1983): (1) Monte Carlo simulations (MCS), and (2) numerical approximations in the perspective of the reliability index.

DOI: 10.1201/9781003333586-8

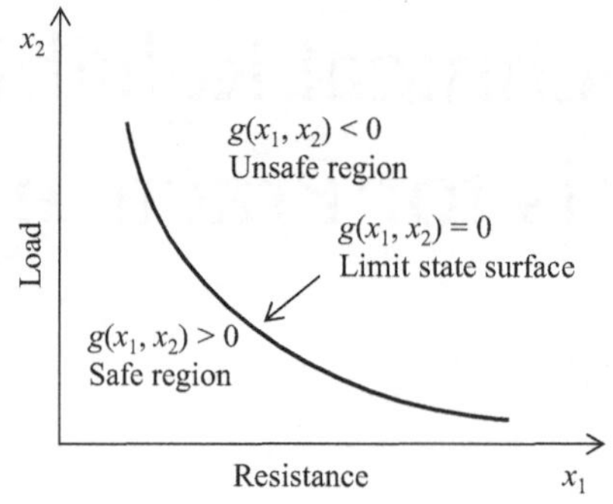

FIGURE 6.1 Concept of limit state surface between the safe and unsafe design regions.

6.2 PROBABILITY OF FAILURE AND RELIABILITY INDEX

When the performance function $z = g(\mathbf{x})$ is a linear combination of normally distributed random variables, it will also follow normal distribution. Let μ_z and σ_z denote the mean and the standard deviation of z, respectively. Then, the probability of failure can be calculated based on the PDF of a normal distribution as follows:

$$P_f = \mathrm{P}(z < 0) = \mathrm{P}\left[(z - \mu_z)/\sigma_z < (0 - \mu_z)/\sigma_z\right] \tag{6.2}$$

Note that $(z - \mu_z)/\sigma_z$ follows standard normal distribution. Hence,

$$P_f = \Phi\left[(0 - \mu_z)/\sigma_z\right] = 1 - \Phi(\mu_z/\sigma_z) \tag{6.3}$$

Introducing the reliability index

$$\beta = \mu_z/\sigma_z \tag{6.4}$$

The probability of failure is commonly denoted as

$$P_f = \Phi(-\beta) \tag{6.5}$$

In the literature, β defined above was originally proposed by Cornell (1969) for load and resistance reliability analysis, thus it is widely known as Cornell's reliability index β_c. A geometric illustration of β_c is given in Figure 6.2 as a one-dimensional (randomness) illustration.

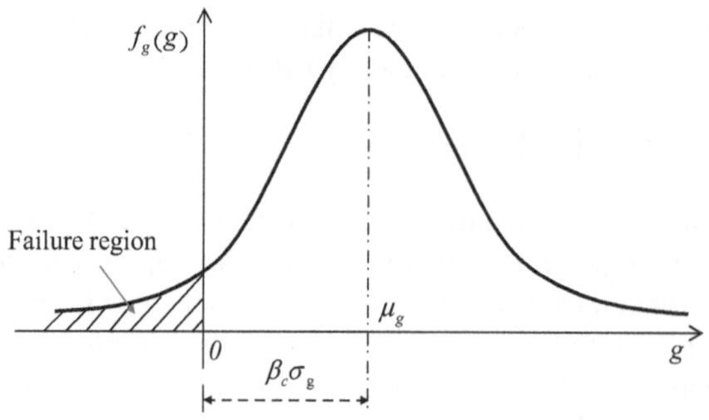

FIGURE 6.2 One-dimensional geometrical illustration of the Cornell reliability index.

6.3 FIRST-ORDER RELIABILITY METHOD (FORM)

6.3.1 Hasofer–Lind (HL) Reliability Index for Uncorrelated Normal Variables

The first-order reliability method proposed by Hasofer and Lind (Hasofer & Lind, 1974) has seen great success in approximating the reliability index over the last few decades. In the space (coordinate system) of uncorrelated standard normal variables $\mathbf{u} = \left[u_1, u_2, \ldots, u_M\right]^T$, the Hasofer–Lind reliability index β_{HL} is defined as the minimum distance from the origin of the axes to the limit state surface as follows:

$$\beta_{\text{HL}} = \sqrt{(\mathbf{u}_d)^T(\mathbf{u}_d)} \tag{6.6}$$

where $\mathbf{u}_d$ denotes the minimum distance point on limit state surface (LSS) in u-space, and is called the *design point* or *checking point* (Hasofer & Lind, 1974). It is also the most probable failure point (MPP) as it is the point in the failure domain with the maximum PDF (Shinozuka, 1983).

For a non-linear LSS, as shown in Figure 6.3, the design point $\mathbf{u}_d$ is the most probable failure point that represents the most likely combination of random variables which may lead to failure. By this definition, the Hasofer–Lind reliability index is invariant because regardless of the form in which the LSS equation is written, its geometric shape and the distance from the origin remain constant (Hasofer & Lind, 1974).

In general, the Hasofer–Lind reliability index β (for generality, subscript HL will now be omitted from β_{HL}) can be obtained through an optimization process of the distance D defined as

$$\beta = \min_{g(\mathbf{u})=0} D = \sqrt{\mathbf{u}^T\mathbf{u}} \tag{6.7}$$

6.3.2 Practical FORM: Constrained Optimization Algorithms

Due to the easy access of powerful optimization tools, the challenges for implementing FORM in geotechnical reliability analysis have been greatly relieved. In particular, Low and Tang (Low & Tang, 2004) showed that the Solver, i.e., the optimization tool embedded in Microsoft Excel, is a powerful tool for solving the aforementioned constraint optimization problem, as long as the performance function evaluation can be realized in Excel.

6.3.2.1 Constrained Optimization Algorithm-1 via Changing x_i

In x-space, the constrained optimization solution of β is much easier to follow, by directly using the statistical information of the original random variables. In this framework, the solution for FORM can be easily implemented via constrained optimization software.

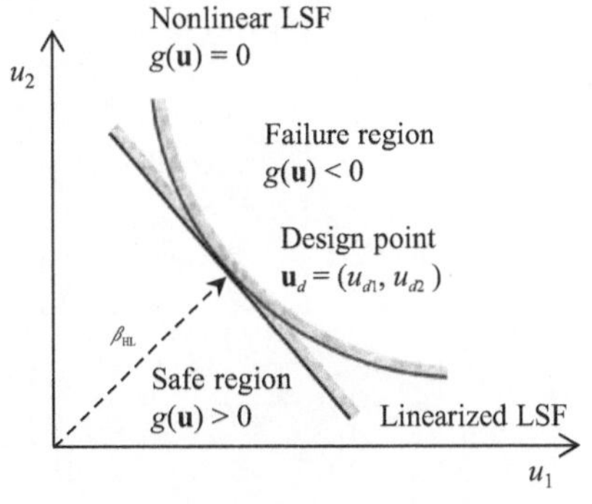

FIGURE 6.3 Hasofer–Lind reliability index: Non-linear performance function.

To ease the reliability analysis from unfamiliar concepts (for most practicing geotechnical engineers) of statistically-mathematical transformations, an alternative perspective of FORM is provided by Low and Tang (Low & Tang, 2004) by formulating the reliability index β in the space of original variables (x-space, Figure 6.4b in comparison to Figure 6.4a for β defined in u-space):

$$\beta = \underset{g(\mathbf{x})=0}{\text{Min}}\, D = \sqrt{\left[\frac{x_i - \mu_i^N}{\sigma_i^N}\right]^T \mathbf{R}^{-1}\left[\frac{x_i - \mu_i^N}{\sigma_i^N}\right]}, \quad \text{by changing } x_i \tag{6.8}$$

where x_i is the checking point value of the ith variable evaluated in x-space; μ_i^N and σ_i^N are the equivalent normal mean and standard deviation of the ith variable, respectively; and $\mathbf{R}$ is the correlation matrix. The Rackwitz–Fiessler transformation can be used to calculate μ_i^N and σ_i^N (Rackwitz & Fiessler, 1978).

The physical meaning of the FORM reliability index β can be comparatively explained in u-space and x-space, as illustrated in Figure 6.4. In u-space, β is the shortest distance from the coordinate origin to the limit state surface, or the radius of an expanding circle/sphere when it touches the limit state surface; in x-space, a one-sigma (1-σ) tilted ellipsoid is defined based on the mean, standard deviation, and correlation, and β is the dispersion ratio of the ellipsoid when it expands to touch the limit state surface, where the touch point is the design point $\mathbf{x}_d$ in x-space (Low & Tang, 2004).

6.3.2.2 Constrained Optimization Algorithm-2 via Changing Dimensionless n_i

Alternatively, the reliability index β can be solved in the space of correlated standard normal variables, i.e., the n-space with components $\mathbf{n} = \left[n_1, n_2, \ldots, n_M\right]^T$. The constrained optimization solution of β can be described by (Low & Tang, 2007)

$$\beta = \min_{g(\mathbf{n})=0} D = \sqrt{\mathbf{n}^T\mathbf{R}^{-1}\mathbf{n}}, \quad \text{by changing } n_i \tag{6.9}$$

where the dimensionless variable $n_i = \dfrac{x_i - \mu_i^N}{\sigma_i^N}$.

Low and Tang (Low & Tang, 2007) demonstrated that the constrained optimization algorithm-2 when implemented in the ubiquitous platform MS-Excel is more efficient than the algorithm-1. Note that transformations of checking point values n_i to x_i are needed to evaluate the LSF $g(\mathbf{n}) = g(\mathbf{x})$.

6.3.3 Practical FORM for Implicit Performance Function: Response Surface Method

To implement the geotechnical reliability analysis of implicit LSF using the constrained optimization approach, additional computations are required to construct surrogate models to replace the

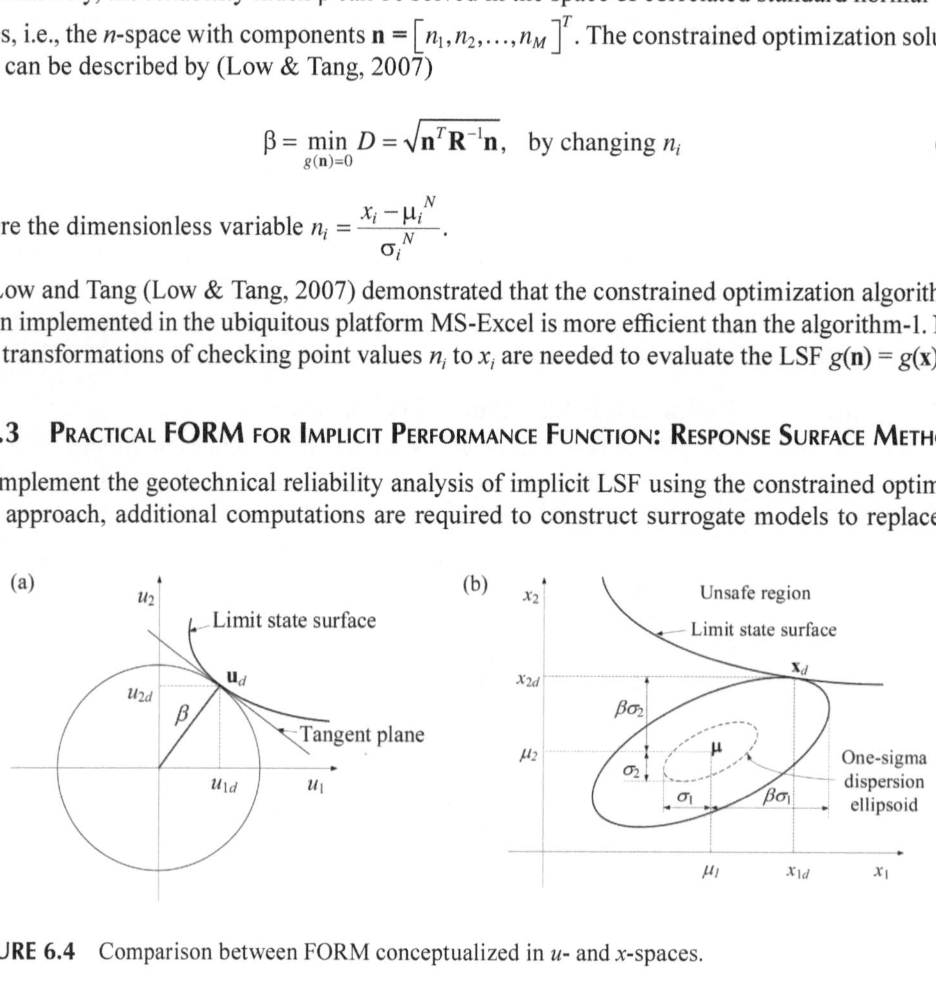

FIGURE 6.4 Comparison between FORM conceptualized in u- and x-spaces.

unknown LSF. In the context of FORM, the response surface method (RSM) is one of those surrogate model–based reliability analysis techniques that aim at constructing a functional LSF locally with experimental designs around the tentative checking point, for example, the polynomial-based response surface method (Wong, 1985; Ji & Low, 2012). In the recent literature, some kernel functions such as the support vector machine models have also been proposed to serve the RSM reliability analysis (Zhao et al., 2014; Zhao, 2008), but they bring additional computational efforts. A more cost-effective hybrid RSM procedure is to use only the first-order polynomial to serve the functional LSF to search in tandem for the design point, and after obtaining the final design point resort to other powerful non-linear regression tools such as the high-order polynomial or the kernel function–based model to capture the LSF curvature for a more accurate probability of failure estimation (Chan & Low, 2012). On the other hand, in the context of MCS reliability analysis, the global performance of surrogate models is more desirable, thus leading to many advanced RSM-MCS geotechnical reliability works such as the stochastic RSM (Li et al., 2011), the artificial neural network RSM (Cho, 2009), the kernel function–based models including the support vector machine-RSM (Ji et al., 2017), the Gaussian process-RSM (Kang et al., 2015), and the kriging-RSM (Zhang et al., 2013).

6.3.4 Practical FORM for Implicit Performance Function: HLRF Recursive Algorithms

Alternatively, the Hasofer–Lind–Rackwitz–Fiessler (HLRF) recursive algorithm for FORM (Liu & Der Kiureghian, 1991; Rackwitz & Fiessler, 1978; Haldar & Mahadevan, 2000) can be used without going through any surrogate models. In brief, the HLRF algorithm uses the gradients of LSF to iteratively compute the checking points, such that

$$\mathbf{u}_{k+1} = \frac{1}{\left|\nabla g(\mathbf{u}_k)\right|^2}\left[\nabla g(\mathbf{u}_k)^T\mathbf{u}_k - g(\mathbf{u}_k)\right]\nabla g(\mathbf{u}_k) \tag{6.10}$$

where $\mathbf{u}_k$ is the kth iteration (checking) point in u-space; and $g(\mathbf{u}_k)$ and $\nabla g(\mathbf{u}_k)$ are the performance function and the gradient vector of the performance function evaluated at $\mathbf{u}_k$, respectively. In general, the gradient vector is not constant when non-linear LSF is involved. As a result, an iterative evaluation of Eq. (6.10) is needed to obtain the design point $\mathbf{u}_d$.

To accommodate the HLRF algorithm in geotechnical applications that mostly incorporate stand-alone numerical package simulations with inputs of x-space variables, Ji and Kodikara (Ji & Kodikara, 2015) proposed a simplified version in x-space, leading to the HLRF-x algorithm:

$$\mathbf{x}_{k+1} = \boldsymbol{\mu}_k^N + \frac{1}{\nabla g(\mathbf{x}_k)^T\mathbf{T}_k\nabla g(\mathbf{x}_k)}\left[\nabla g(\mathbf{x}_k)^T\left(\mathbf{x}_k - \boldsymbol{\mu}_k^N\right) - g(\mathbf{x}_k)\right]\mathbf{T}_k\nabla g(\mathbf{x}_k) \tag{6.11}$$

where $\mathbf{T}_k = \left[\sigma_k^N\right]^T\mathbf{R}\left[\sigma_k^N\right]$ is a transformation matrix computed at the kth iteration point and the components of $\mathbf{T}_k$ are simply computed by $T_{k,ij} = \sigma_{k,i}^N R_{ij}\sigma_{k,j}^N$.

It is clear that for each iterative step, the computation for checking point values using HLRF-x only requires the gradient vector of the performance function in x-space. Hence, it is best suited for the FORM analysis of implicit performance functions, of which the gradient vector can be directly estimated by partial differentiations in x-space.

To stabilize the algorithm, Eq. (6.11) is recast by introducing a step size, yielding an improved version of iHLRF-x (Ji et al., 2018, 2019) as follows:

$$\mathbf{x}_{k+1} = \mathbf{x}_k + \lambda_k\mathbf{d}_k \tag{6.12}$$

$$\mathbf{d}_k = \boldsymbol{\mu}_k^N + \frac{1}{\nabla g(\mathbf{x}_k)^T\mathbf{T}_k\nabla g(\mathbf{x}_k)}\left[\nabla g(\mathbf{x}_k)^T(\mathbf{x}_k - \boldsymbol{\mu}_k^N) - g(\mathbf{x}_k)\right]\mathbf{T}_k\nabla g(\mathbf{x}_k) - \mathbf{x}_k \tag{6.13}$$

where λ_k and $\mathbf{d}_k$ are, respectively, the step size and the search direction defined in x-space. When the step size is fixed to be unity, the iHLRF-x reduces to HLRF-x (Ji & Kodikara, 2015). Likewise, the optimal step length can be determined by monitoring a merit function $m(\mathbf{x})$, which reads

$$m(\mathbf{x}) = \frac{1}{2}\left\| \left[\frac{x_i - \mu_i}{\sigma_i} \right] \right\|^2 + c\left| g(\mathbf{x}_k) \right| \tag{6.14}$$

where $c > 0$ is a penalty parameter, e.g., $c = 100$ suffices for most geotechnical reliability engineering problems.

As a backtracking line search technique in x-space, the corresponding Armijo rule for the optimal (maximum) step length is

$$\lambda_k = \max_j \left\{ b^j \left| m(\mathbf{x}_k + b^j \mathbf{d}_k) - m(\mathbf{x}_k) \le -ab^j \left\langle \nabla m(\mathbf{x}_k), \mathbf{d}_k \right\rangle \right. \right\} \tag{6.15}$$

where $a, b \in (0, 1)$ are prescribed parameters and j is an integer for the optimal solution.

In the literature, great efforts are being made to balance the robustness and efficiency of the FORM algorithms to solve practical reliability analysis problems. The HLRF is fundamental and can provide fast solutions for LSFs of low non-linearity. However, if the LSF is highly non-linear and there are multiple local minimums in terms of the reliability index, special care must be exercised and various enhanced algorithms have been proposed in ensuring the robustness of the HLRF algorithms (Periçaro et al., 2015; Ghohani Arab et al., 2019; Zhu et al., 2020; Keshtegar & Chakraborty, 2018; Ramesh et al., 2017; Liu & Der Kiureghian, 1991; Wang et al., 2016; Keshtegar & Meng, 2017). Nevertheless, the robustness of these improved HLRF algorithms is achieved at the cost of adding more performance evaluations.

The probability of failure calculated by $P_f = \Phi(-\beta)$ is indicative, and the accuracy can only be ensured when LSF is relatively linear. If more accuracy is desired, a second-order reliability method (SORM) or importance sampling (IS) method in conjunction with FORM's design point search can be employed to improve the P_f estimation (Huang & Griffiths, 2011; Zeng et al., 2016; Zhao & Ono, 1999; Koyluoglu & Nielsen, 1994). Also, the reliability index and the design point values are important information resulting from the FORM analysis, and they constitute a formula for conducting further reliability-based design (RBD) optimization (Phoon et al., 2003; Low & Phoon, 2015). In some recent geotechnical applications, reliability analysis has shown a tendency to deal with high-dimension random variables, e.g., for modeling random fields. In such cases, one may consider reducing the dimension of random variables prior to the use of FORM (Liao & Ji, 2021).

6.3.4.1 Application Example: Reliability Analysis of a Strut with Complex Supports

Coates et al. (Coates et al., 1994) presented a deterministic analysis of a strut with complex supports. The member is initially straight (Figure 6.5) with a pin support at 2, an elastic support at 1 (rotational stiffness λ_1) which provides a restoring moment $M_1 = \lambda_1\theta_1$, and a support at 3 (stiffness k_3) which provides a reaction force (k_3v_3) proportional to the vertical displacement v_3. The problem is to determine the smallest value of axial force P which will cause the strut to become elastically unstable. Coates et al. (Coates et al., 1994) showed that the problem reduces to that of finding the value of P, referred to as P_{crt}, which would make the determinant of a 7 × 7 matrix (functions of 7 physical parameters including P, L, a, E, I, λ_1, and k_3) vanish. It was also noted that a numerical procedure would be necessary to determine the determinant of the matrix by sequential variation of the axial load P until a zero determinant value is reached.

By imposing uncertainties to the 7 physical parameters, Low and Tang (Low & Tang, 2004, 2007) conducted reliability analysis of the strut using the spreadsheet's constrained optimization algorithm-2 for FORM (which needs to invoke the optimization toolbox SOLVER of MS-Excel). In this study, the reliability analysis is revisited using the HLRF-x recursive algorithm for FORM. A spreadsheet setup of the problem is shown in Figure 6.5. The statistical information for the 7 random

Procedure for Recursive Algorithm FORM in x-Space

ProbDist	Var.	Para1	Para2	Para3	Para4	μ_k^N	σ_k^N	$\mathbf{x}_k$	$\nabla g(\mathbf{x}_k)$	$\mathbf{x}_{k+1}$	n
Lognormal	P	700	140			633.3	192.8	973.5	-1	973.5	1.76
Triangular	L	800	1000	1200		1009.7	78.7	1110	-1.9	1110	1.27
Lognormal	a	500	50			495.7	53.9	540.4	0.25	540.4	0.83
BetaDist	E	3	3	2E+05	3E+05	2E+05	2E+04	2E+05	0.01	2E+05	-1.30
Lognormal	I	200	20			197.4	17.4	174.5	5.57	174.5	-1.32
PertDist	λ_1	350	500	650		500.0	63.8	499.9	0	499.9	0.00
Gamma	k_3	100	0.1			10.0	1.0	9.965	0.12	9.965	0.00

$P_{crit.}$ 974

$g(\mathbf{x}_k) = P_{cri.} - P$ 0.00

β 2.651

$P_{f\,(FORM)}$ 0.40%

Correlation matrix **R**

1	0	0	0	0	0	0
0	1	0.7	0	0	0	0
0	0.7	1	0	0	0	0
0	0	0	1	0.5	0	0
0	0	0	0.5	1	0	0
0	0	0	0	0	1	0.6
0	0	0	0	0	0.6	1

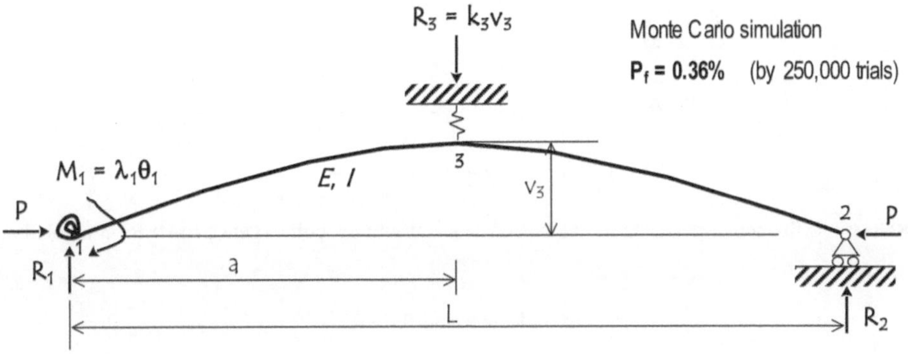

FIGURE 6.5 Reliability analysis of a strut with complex supports.

variables is denoted by Paral to Para4. For details, refer to figure 5 in Low and Tang (2007). The critical load P_{crt} can be obtained in a spreadsheet cell that calls the user-defined Excel function $P_{critical}$ (...), shown in figure 9 in Low and Tang (2004). The performance function $g(\mathbf{x}_k) = P_{cri.} - P$, where P refers to the first value of the $\mathbf{x}_k$ column. Hence, the single cell object "$g(\mathbf{x}_k) = P_{cri.} - P$" contains an implicit and iterative user-created program code.

As an implicit performance function, its gradient vector $\nabla g(\mathbf{x}_k)$ can be approximately obtained using the numerical differentiation method, i.e., each component of $\nabla g(\mathbf{x}_k)$ is approximated to be $\Delta g(\mathbf{x}_{k,i})/\Delta x_i$ (i = 1 to 7, denoting the ith random variable), where Δx_i is a small perturbation value. The automatic computation of $\Delta g(\mathbf{x}_{k,i})$ can be realized by the user-defined function $P_{critical}$. For simplicity, a forward difference scheme is adopted. Based on the proposed six-step procedure, the converged β value was found to be 2.651 (based on $\Delta x_i = 0.01\mu_{xi}$). The same result was reported in Low and Tang (2007). The failure probability P_f as obtained by $\Phi(-\beta)$ is 0.40%, which is comparable with the Monte Carlo simulations of P_f of about 0.36%. Note that the slight difference between the P_f of FORM and the P_f of Monte Carlo simulations could be improved by using a more advanced reliability method such as the second-order reliability method (Chan & Low, 2011), which is beyond the scope of this chapter.

Since the accuracy of approximation for a gradient vector using difference quotients depends mainly on the perturbation values Δx_i, it is interesting to investigate the sensitivity of β to Δx_i. In this regard, a parametric study was carried out by varying Δx_i from $0.01\mu_{xi}$ to $0.20\mu_{xi}$. The results are summarized in Figure 6.6. Obviously, β is slightly sensitive to Δx_i only for the first iteration. When additional iterations are carried out, β of each case converges very fast to almost the same solution. On the other hand, the use of relatively large perturbation values is not recommended. For example, the parametric study shows that $0.20\mu_{xi}$ consistently overestimates the β value, although by only a very small amount. It is the sole responsibility of the user to choose a proper perturbation value for each random variable and make cross-validations, as the accuracy of the numerical differentiation method depends on many factors.

In engineering reliability analysis involving implicit LSF, many studies have been carried out utilizing the polynomial response surface method (polynomial RSM). Thus, it is interesting to compare the widely used RSM-based FORM with the HLRF-*x* algorithm. For the strut with complex support, comparative results are presented in Figure 6.7.

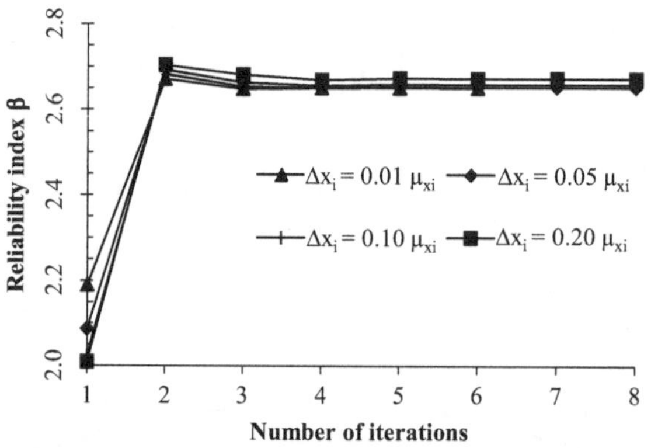

FIGURE 6.6 Influence of perturbation values on reliability index convergence in HLRF-*x* algorithm.

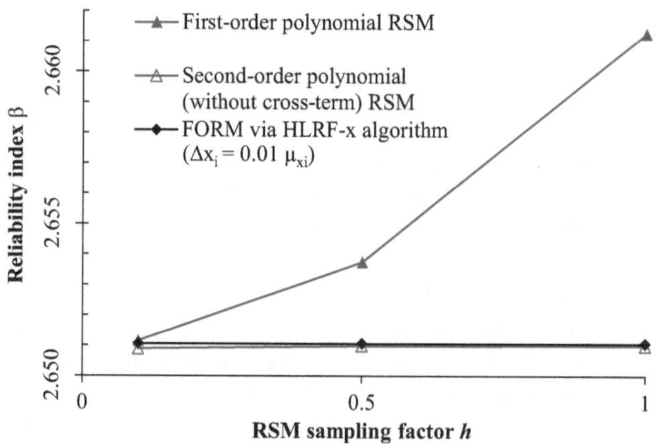

FIGURE 6.7 Comparison between RSM and FORM via HLRF-*x* algorithm.

The β value computed by the first-order polynomial RSM is somewhat sensitive to the sampling factor h (by which the sampling point for experimental design is determined at $x_i \pm h \times \sigma_{xi}^N$). In contrast, the second-order polynomial RSM is less sensitive to the sampling factor. In the case of sampling at $x_i \pm 0.1\sigma_{xi}^N$, both the first-order polynomial and the second-order polynomial (without cross terms) RSM yield a β value as good as that from the HLRF-x algorithm for FORM. However, considering the computational efficiency, the advantage of the HLRF-x is obvious. For a problem involving n basic random variables, $(n + 1)$ experimental evaluations of the performance function are required for each iteration when using a linear RSM; when using a simple second-order polynomial RSM, $2n + 1$ experimental evaluations are required for each iteration when cross terms are not considered, and $(n + 1)(n + 2)/2$ experimental evaluations when including cross terms. Note that the number of experimental evaluations increases exponentially with increasing the order of polynomial. In contrast, the HLRF-x algorithm requires only $n + 1$ (the same number as used in a linear RSM) runs of the performance function for each iteration when a forward or backward difference scheme is used, and $2n + 1$ runs when a central difference scheme is used. For most engineering problems, the forward or backward difference scheme is satisfactory. As a result, the number of experimental evaluations of the performance function is reduced significantly especially compared with non-linear RSM.

6.4 DITLEVSEN BOUNDS FOR SYSTEM RELIABILITY ANALYSIS

In geotechnical engineering, many subsystems denoting different failure modes may be mutually correlated. It is often difficult to accurately calculate the failure probability of a system. In practice, computing the possible range of system failure probability can still provide useful information for engineering decision-making. The Ditlevsen bounds method (Ditlevsen, 1979b) is often used in system reliability analysis of geotechnical problems. However, the bounds solution can only provide an accurate estimation of system failure probability when the LSFs of subsystems (or failure modes) are not highly non-linear because the reliability index of each failure mode is computed with FORM. For geotechnical problems with highly non-linear LSFs, the second-order reliability method could be used to refine the reliability index estimation (see e.g., Low et al., 2011; Zeng et al., 2018).

For practical reliability analysis purposes, the Ditlevsen bimodal bounds on the system failure probability P_F can be calculated by

$$P_{F_1} + \sum_{i=2}^{m} \max\left[\left\{P_{F_i} - \sum_{j=1}^{i-1} P(E_iE_j)\right\}; 0\right] \le P_F \le \min\left[\left\{\sum_{i=1}^{m} P_{F_i} - \sum_{i=2}^{m} \max_{j<i} P(E_iE_j)\right\}; 1\right] \quad (6.16)$$

where P_{F_1} is the largest failure probability among the potential failure modes and P_{F_i} is the failure probability of the ith failure mode; $P(E_iE_j)$ is the joint probability of events E_i and E_j; and P_F is the system failure probability. The joint probability $P(E_iE_j)$ can be reduced to $\max[P(A), P(B)] \le P(E_iE_j) \le P(A) + P(B)$ on condition that the two events are positively correlated, otherwise $0 \le P(E_iE_j) \le \min[P(A), P(B)]$, where the terms $P(A)$ and $P(B)$ are defined by

$$P(A) = \Phi(-\beta_i)\Phi\left(-\frac{\beta_j - \rho_{ij}\beta_i}{\sqrt{1-\rho_{ij}^2}}\right) \quad (6.17)$$

$$P(B) = \Phi(-\beta_j)\Phi\left(-\frac{\beta_i - \rho_{ij}\beta_j}{\sqrt{1-\rho_{ij}^2}}\right) \quad (6.18)$$

where β_i is the reliability index of the failure mode i and ρ_{ij} is the correlation coefficient between modes i and j as mentioned before. Low (Low, 2017) obtained the correlation coefficient ρ_{ij} of different failure modes by using the Cholesky decomposition of the correlation matrix **R** on n-space as follows:

$$\rho_{ij} = \frac{\mathbf{u}_i^{*T}}{\beta_i}\frac{\mathbf{u}_j^{*}}{\beta_j} = \frac{1}{\beta_i\beta_j}\mathbf{n}_i^{*T}\mathbf{R}^{-1}\mathbf{n}_j^{*} \tag{6.19}$$

where $\mathbf{n}_i^*$ and $\mathbf{n}_j^*$ are the vectors obtained by the equivalence normalization of design points which can be gained by the iHLRF-x algorithm. The reliability index β_i and the design point in x-space can be calculated according to the iHLRF-x algorithm (Ji et al., 2018; Ji and Kodikara, 2015).

6.5 MONTE CARLO SIMULATIONS FOR PRACTICAL RELIABILITY ANALYSIS

MCS is a simple, convenient, and relatively accurate method for probabilistic analysis. A routine MCS proceeds by first generating a number of random samples according to the input PDF $f(\mathbf{x})$, evaluating the performance function $g(\mathbf{x})$ for each sample, and then identifying failure samples that satisfy $g(\mathbf{x}) < 0$. Dividing the number of failure samples (e.g., N_f) over the total number of simulated samples (e.g., N), the estimated P_f by MCS is obtained, which can be mathematically expressed as

$$P_f = \int_{g(\mathbf{x})<0} f(\mathbf{x})\,d\mathbf{x} = \frac{1}{N}\sum_{k=1}^{N} I(\mathbf{x}^{(k)}) = \frac{N_f}{N} \tag{6.20}$$

where $I(\mathbf{x}^{(k)})$ is an indicator function with respect to the kth sample $\mathbf{x}^{(k)}$; if the performance function evaluated at $\mathbf{x}^{(k)}$ is less than zero (i.e., $g(\mathbf{x}^{(k)}) < 0$), then $I(\mathbf{x}^{(k)}) = 1$, or $I(\mathbf{x}^{(k)}) = 0$ for $g(\mathbf{x}^{(k)}) \geq 0$.

The estimator P_f in Equation (6.20) is statistically unbiased, and it approaches the accurate value of P_f as N becomes infinite. In general, the accuracy of the estimated P_f from MCS can be analytically measured, for example, by its coefficient of variation (COV):

$$\mathrm{COV}_{\mathrm{MCS}} = \sqrt{\frac{1-P_f}{NP_f}} \tag{6.21}$$

MCS is the most widely used simulation-based reliability analysis method due to its simplicity and robustness to the dimension of uncertain variables and the complexity of the reliability analysis problems concerned (e.g., high non-linearity of $g(\mathbf{x})$ without explicit expressions and the existence of multiple failure modes). These features explain the great popularity of MCS in addressing a large variety of reliability-based geotechnical problems.

At small probability levels, however, routine MCS suffers from a lack of resolution and efficiency. As indicated by Equation (6.21), the required number of samples in MCS for evaluating small probabilities of P_f is asymptotically estimated as

$$N = \frac{1-P_f}{\mathrm{COV}_{\mathrm{MCS}}^2 P_f} \approx \frac{1}{\mathrm{COV}_{\mathrm{MCS}}^2 P_f} \tag{6.22}$$

Equation (6.22) suggests that N increases dramatically as the probability level concerned (i.e., P_f) decreases and the desired accuracy (i.e., $\mathrm{COV}_{\mathrm{MCS}}$) improves. For example, given $P_f = 10^{-3}$ and $\mathrm{COV}_{\mathrm{MCS}} = 30\%$, a number of 10^4 samples are required to achieve practically acceptable accuracy. Such a significant number of MCS samples may require extensive computational efforts for solving complex geotechnical problems that involve computationally demanding models (e.g., finite element model [FEM]).

However, as most geotechnical reliability problems have to deal with a small probability of failure (e.g., $P_f < 0.001$) and high-dimensional random variables (in the case of a spatial variability simulation of geological properties), the amount of sampling calculations is usually large. In view of this shortcoming, the variance reducing MCS methods are found to be extremely useful for geotechnical reliability analysis. Among others, the adaptive Monte Carlo simulation (AMCS) (Liu et al., 2020), the Latin hypercube sampling (LHS) (Baecher and Christian, 2005), line sampling (LS) (Depina et al., 2016), importance sampling (Ching et al., 2009), subset simulations (SS) (Wang & Cao, 2013), generalized SS (GSS) (Gao et al., 2019), and weighted uniform simulations (WUS) (Ji & Wang, 2022) are found effective for simulating the failure of large and realistic geotechnical problems. Extensive reviews of the applications of these methods on geotechnical reliability problems are summarized in Phoon et al. (2022).

6.5.1 Advanced Simulation Techniques

6.5.1.1 Adaptive MCS

As a variant of direct MCS, the adaptive Monte Carlo simulation (AMCS) is suitable for the reliability analysis of geotechnical series systems such as slopes, pipelines, and levees of which the system failure may contain many (e.g., >1000) correlated components. The core idea of AMCS is to identify all failure samples of the number N_f among the total number N of independent identically distributed random samples by analyzing the series subsystems with a much smaller number of components (Liu et al., 2020, 2022). The P_f of a series system using the AMCS is approximated by

$$P_f \approx P_{f,k} = \frac{1}{N}\sum_{i=1}^{N} I\left[\min_{j=1}^{k} g_{t,j}\left(\mathbf{x}_i\right)\right] = \frac{N_{f,k}}{N} \tag{6.23}$$

where $P_{f,k}$ represents the failure probability of the series subsystem with k components; $g_{t,j}$ denotes the performance function of the jth trial component used to construct the subsystem; and $N_{f,k}$ is the number of failure samples identified by the series subsystem.

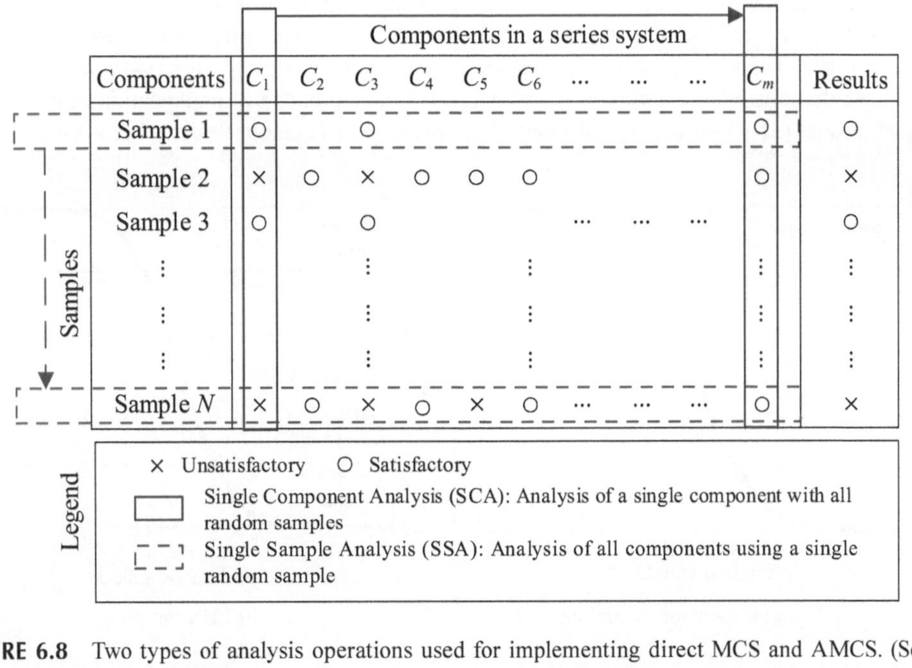

FIGURE 6.8 Two types of analysis operations used for implementing direct MCS and AMCS. (Source: Adapted from Liu et al., 2020)

The series subsystem is iteratively constructed by an adaptive procedure. As shown in Figure 6.8, direct MCS uses either single component analysis or single sample analysis to identify failure samples in an iterative manner. In contrast, AMCS utilizes both of the analysis operations to iteratively construct the series subsystem. AMCS can be conveniently implemented in the same manner as direct MCS, whereas the computational cost of AMCS can be two orders of magnitude lower. The computational efficiency of AMCS increases as the component correlation increases or the component number increases. AMCS has been applied to efficiently assess the system failure probability of slopes considering a large number of potential slip surfaces of circular or non-circular shapes (Liu et al., 2020). However, AMCS may be reduced to direct MCS for a series system with a component number less than 100.

6.5.1.2 Latin Hypercube Sampling

Latin hypercube sampling discretizes the sample space into a number of Latin hypercubes, each of which contains only one sample in each row and each column. When compared with the pure random sampling shown in Figure 6.9a, LHS inherits the idea from stratified sampling to partition the sample space into multiple subspaces with an equal probability and takes a random sample in each subspace (see Figure 6.9b) (McKay et al., 1979). Random samples generated from LHS are more evenly distributed in the sample space and their statistics (e.g., mean, standard deviation) converge faster to the inputs than those generated from pure random sampling (Stein, 1987), hopefully to reduce the variance of the estimated P_f. LHS is simple and easy to implement, and it has become a preferable alternative to MCS. Nonetheless, it might not guarantee the variance reduction for geotechnical engineering problems involving complex performance functions and/or a large number of random variables, and the improvement of computational efficiency for estimating a small failure probability is limited (Olsson et al., 2003, Li et al., 2015a).

6.5.1.3 Line Sampling

Line sampling divides an M-dimensional standard normal random vector into a one-dimension x_1 and the remaining $(M-1)$-dimensional random vector $\mathbf{x}_{-1}$, as illustrated in Figure 6.10. A random sample of $\mathbf{x}_{-1}$ becomes a line segment projected on the new sampling space, and the failure domain is also reshaped; x_1 is the so-called direction of line sampling. LS could significantly reduce the required number of samples for estimating a small P_f when the sampling direction is perpendicular to the failure domain. Mathematically, LS transforms the original M-dimensional integral for estimating P_f into an $(M-1)$-dimensional integral expressed as (Koutsourelakis et al., 2004):

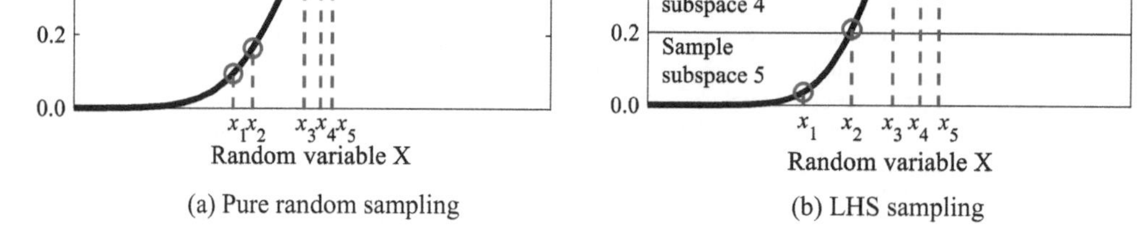

FIGURE 6.9 Comparison of pure random sampling and LHS.

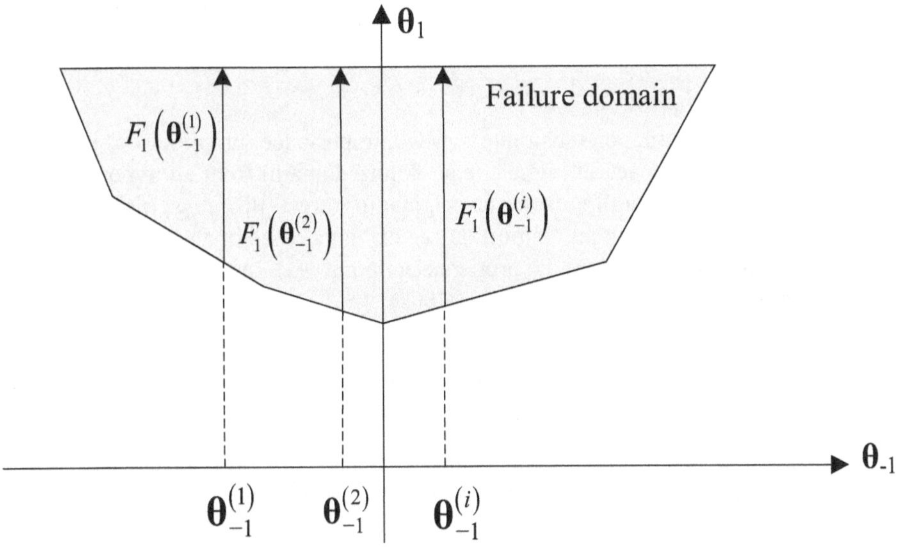

FIGURE 6.10 Illustration of line sampling. (Source: After Koutsourelakis et al., 2004)

$$
\begin{aligned}
P_f &= \underbrace{\int \cdots \int}_{M} I_F(\mathbf{x}) \prod_{i=1}^{M} \phi(x_i)\, d\mathbf{x} \\
&= \underbrace{\int \cdots \int}_{M-1} \left[\int I_{F_1}(\mathbf{x}_{-1}) \phi(x_1)\, dx_1 \right] \prod_{i=2}^{M} \phi(x_i)\, d\mathbf{x}_{-1} \\
&= \underbrace{\int \cdots \int}_{M-1} \Phi\left[F_1(\mathbf{x}_{-1}) \right] \prod_{i=2}^{M} \phi(x_i)\, d\mathbf{x}_{-1} \\
&= E_{\mathbf{x}_{-1}} \left\{ \Phi\left[F_1(\mathbf{x}_{-1}) \right] \right\}
\end{aligned}
\tag{6.24}
$$

where $\mathbf{x}$ is M-dimensional standard normal random vectors; $I_F(\mathbf{x})$ is the indicator function of the original failure event F associated with $\mathbf{x}$; $\phi(\cdot)$ is the probability density function of a univariate standard normal distribution; $\mathbf{x}_{-1}$ represents the standard normal random vectors removing x_1 from $\mathbf{x}$, and it has a dimension of $(M-1)$; $I_{F_1}(\mathbf{x}_{-1})$ is the indicator function of a failure event F_1 equivalent to F associated with $\mathbf{x}_{-1}$; and $\Phi\left[F_1(\mathbf{x}_{-1})\right] = \int I_{F_1}(\mathbf{x}_{-1})\phi(x_1)\,dx_1$ represents the Gaussian measure of subset $F_1(\mathbf{x}_{-1})$. For a performance function $g_1(\mathbf{x}_{-1}) - \theta_1 < 0$, $F_1(\mathbf{x}_{-1})$ is the half open interval $[g_1(\mathbf{x}_{-1}), \infty)$. Thus, the failure probability is estimated by

$$
P_f = \frac{1}{N} \sum_{i=1}^{N} \Phi\left[F_1\left(\mathbf{x}_{-1}^{(i)}\right) \right] \tag{6.25}
$$

where $\mathbf{x}_{-1}^{(i)}$ is the ith random sample of $\mathbf{x}_{-1}$.

LS provides an efficient tool to estimate the failure probability, even with high-dimensional problems. It is suitable for structural reliability problems with linear or weakly non-linear limit states in standard normal space. However, the performance of LS strongly depends on the quality of the chosen sampling direction and the non-linearity of reliability problems (Pradlwarter et al., 2007).

6.5.1.4 Importance Sampling

As a well-known variance reduction technique of MCS, importance sampling works by generating samples in a specific sample space of interest (e.g., failure domain) from an importance sampling PDF $h(\mathbf{x})$, instead of directly sampling from the original (or target) PDF (e.g., $q(\mathbf{x})$). By this means, samples will lie more frequently in an "important region" concerned in an analysis. Great computational savings can be achieved given the proper selection of $h(\mathbf{x})$. In the context of IS, the failure probability can be rewritten as (e.g., Ching et al., 2009)

$$P_f = \frac{1}{N}\sum_{i=1}^{N} I(\mathbf{x}_i)\frac{q(\mathbf{x}_i)}{h(\mathbf{x}_i)} \tag{6.26}$$

where $h(\mathbf{x}_i)$ is the value of the importance sampling density function corresponding to the ith sample $\mathbf{x}_i$. The variance of the estimator of P_f given by IS is estimated to be (Ibrahim, 1991)

$$\mathrm{Var}\left[P_f\right] = \frac{1}{N-1}\left[\frac{1}{N}\sum_{i=1}^{N} I(\mathbf{x}_i)\frac{q(\mathbf{x}_i)^2}{h(\mathbf{x}_i)} - P_f^2\right] \tag{6.27}$$

Theoretically, Equation (6.27) gives an unbiased estimate of P_f provided that the importance sampling PDF $h(\mathbf{x})$ is properly selected, and the determination of a suitable $h(\mathbf{x})$ can greatly improve the accuracy of the estimated P_f and computational efficiency. A simple and practical choice of $h(\mathbf{x})$ is to translate the center of the original PDF (e.g., $q(\mathbf{x})$) from its mean to the most probable failure point with a consistent covariance (Au & Wang, 2014). However, constructing the importance sampling PDF is problem-specific and is not a trivial task, particularly when high-dimensional random variables and highly non-linear performance functions are involved.

6.5.1.5 Subset Simulation

Subset simulation converts a small failure probability into a product of a sequence of relativity large conditional probabilities by introducing intermediate events adaptively, and employs specially designed Markov chains to generate conditional samples of these intermediate events until the target failure domain is achieved (Au & Beck, 2001; Au & Wang, 2014; Wang & Cao, 2013, 2014). SS inherits the advantages of MCS (e.g., robustness to the complexity of performance function $g(\mathbf{x})$ involving the high dimensionality of $\mathbf{x}$), and it is particularly designed for estimating the probability of rare events.

SS starts with direct MCS and proceeds level by level. It divides the sample space Ω of random variables $\mathbf{x}$ into $m + 1$ individual subsets $\{\Omega_i, i = 0, 1, 2,\ldots, m\}$ by the intermediate threshold values $\{y_i, i = 1, 2,\ldots, m\}$ of the driving variable Y, which is a key factor affecting the generation of conditional samples of interest in SS. The $\{y_i, i = 1, 2,\ldots, m\}$ in SS are adaptively determined, and samples in different subsets $\{\Omega_i, i = 0, 1, 2,\ldots, m\}$ are generated level by level and correspond to different probability weights, i.e., $P(\Omega_0) = 1 - p_0$; $P(\Omega_i) = p_0^{\,i} - p_0^{\,i+1}$ for $i = 1,\ldots, m - 1$; and $P(\Omega_m) = p_0^{\,m}$, where p_0 is a conditional probability (e.g., 0.1). Then, P_f is written as

$$P_f = P(F) = \sum_{\mathrm{i}=0}^{\mathrm{m}} \mathrm{P}(F \mid \Omega_i)\mathrm{P}(\Omega_i) \tag{6.28}$$

where $P(F)$ is the failure probability of failure event F; $P(F|\Omega_i)$ is the conditional failure probability given sampling in Ω_i, and it is estimated as the ratio of the failure sample number in Ω_i over the total sample number in Ω_i. Compared with direct MCS, SS efficiently generates a large number of failure samples to calculate P_f by constructing a driving variable Y that efficiently drives the sampling space to failure domains concerned in reliability analysis.

6.5.1.6 Generalized Subset Simulation

Generalized subset simulation is developed from SS to efficiently estimate the failure probabilities of multiple failure events (or limit states) by a single simulation run (Li et al., 2015b; Gao et al., 2019). In the context of GSS, each failure event $M^{(j)}$ concerned among a total of N_M failure events can be defined by its corresponding limit state function. The failure probabilities of the N_M failure events can be calculated by a single GSS. For this purpose, GSS defines a unified failure event F_U, i.e., the union of failure events of the N_M failure events concerned, i.e., $F_U = F^{(1)} \cup F^{(2)} \cup \cdots \cup F^{(j)} \cup \cdots \cup F^{(N_M)}$. Then, GSS turns to explore the unified failure domain (i.e., F_U) to simultaneously generate failure samples of different failure events. Using GSS, the failure domains (i.e., $F^{(j)}, j = 1, 2,\ldots, N_M$) corresponding to $M^{(j)}$ with different failure probability values, $P(F|M^{(j)})$, are reached at different simulation levels. Let $m^{(j)}$ denote the number of simulation levels needed to reach the failure domain $F^{(j)}$ of $M^{(j)}$. In general, $m^{(j)}$ increases with the decrease in $P(F|M^{(j)})$. Using GSS samples, $P(F|M^{(j)})$ is estimated as (Li et al., 2015b; Gao et al., 2019; Yang et al., 2021)

$$P\left(F|M^{(j)}\right) = P\left(F_{U,1}\right)P\left(F_{U,2}|F_{U,1}\right)\cdots P\left(F_{U,m^{(j)}-1}|F_{U,m^{(j)}-2}\right)P\left(F^{(j)}|F_{U,m^{(j)}-1}\right) = \prod_{i=1}^{m^{(j)}-1}\frac{N_i}{N}\times\frac{N^{(j)}}{N} \tag{6.29}$$

where $F_{U,i} = F_i^{(1)} \cup F_i^{(2)} \cup \cdots \cup F_i^{(j)} \cup \cdots \cup F_i^{(N_M)}$, $i = 1, 2,\ldots, m^{(j)}$ is the union of intermediate failure events (i.e., $F_i^{(j)}, j = 1, 2,\ldots, N_M$) of the N_M failure events in the ith simulation level; $P(F_{U,1}) = N_1/N$, and N_1 denotes the number of samples belonging to $F_{U,1}$ among the total number N of samples generated in each simulation level during GSS; $P(F_{U,i}|F_{U,i-1})$, $i = 2, 3\ldots, m^{(j)} - 1$ is the conditional probability of $F_{U,i}$ given sampling in $F_{U,i-1}$, and it is calculated as the ratio of the number N_i of "seed" samples selected among conditional samples in $F_{U,i-1}$ over N, i.e., N_i/N, for the ith level of Markov chain simulation; $P(F^{(j)}|F_{U,m^{(j)}-1})$ is the conditional probability of $F^{(j)}$ given sampling in the unified intermediate event $F_{U,m^{(j)}-1}$, and it is estimated as the ratio of the number $N^{(j)}$ of failure samples belonging to $F^{(j)}$ among the conditional samples generated in $F_{U,m^{(j)}-1}$ over N.

6.5.1.7 Weighted Uniform Simulations

Fundamentally, the weighted uniform simulation works in the manner of importance sampling, whereas it changes the sampling mode to a uniform distribution to calculate the probability of failure, and the joint PDF values can be regarded as weighting indices which when ordered in magnitude can be used to determine the design points of the most probable failure probabilities (Rashki et al., 2012). The WUS uses a uniform distribution for sampling instead of the original distributions of random variables, hence it can quickly cover the entire uncertainty variable space for the probability of failure analysis and locate the design points with a relatively small size of samples. The illustrative examples of Rashki et al. (2012) have shown that the WUS method can greatly improve the calculation efficiency for a probability of failure analysis while ensuring that the results are sufficiently accurate. At the same time, the design point information obtained by this simulation-based method is unique, which cannot be accomplished by ordinary MCS. Ji and Wang (2022) modified the WUS by accounting for random variable correlation in geotechnical applications. It is worth noting that the application of WUS in geotechnical reliability analysis could be challenged by the high-dimensional randomness simulations.

6.5.2 Sampling-Based Sensitivity Analysis

MCS-based reliability analysis can be viewed as a "black box" that takes samples of random variables as input and returns the failure probability (or the probability distribution of the design responses concerned) as output. Despite its simplicity in concept, it was often criticized for not providing reliability sensitivity information of random variables (Baecher & Christian, 2005; Wang & Cao, 2014). To overcome this shortcoming, the sampling-based probabilistic sensitivity analysis method was developed. This method can provide reliability sensitivity information with respect to either the statistical moment values of random variables or their probability distributions, as described below.

6.5.2.1 Sensitivity to the Statistical Moment Values of Random Variables

A probabilistic failure analysis method has been developed to prioritize the effects of various random variables and quantify their effects based on MCS samples. It contains two major components: hypothesis tests and Bayesian analysis. With hypothesis tests, the probabilistic failure analysis approach prioritizes the effects of random variables **x** on the failure probability by comparing, statistically, failure samples with the unconditional samples simulated by MCS. When the distribution of failure samples of a random variable deviates significantly from that of unconditional samples, the uncertainty in the parameter has a significant effect on P_f. The deviation between the distribution of failure samples and that of unconditional samples can be quantified as follows (Wang et al., 2010):

$$Z_{H,i} = \frac{\mu_i - \mu_{F,i}}{\sigma_i / \sqrt{N_F}} \tag{6.30}$$

where μ_i and σ_i are the mean value and standard deviation of the random variable x_i; N_F is the number of failure samples among a total of N unconditional samples; and $\mu_{F,i}$ is the mean value of the N_F failure samples of x_i. When $\mu_{F,i}$ deviates significantly from μ_i, the absolute value of $Z_{H,i}$ is relatively large. As the absolute value of $Z_{H,i}$ increases, the statistical difference between $\mu_{F,i}$ and μ_i becomes increasingly significant. The effect of the random variable on failure probability also becomes increasingly significant. The absolute value of $Z_{H,i}$ can therefore be used as an index to prioritize their relative effects on failure probability. By comparing the absolute values of $Z_{H,i}$ for various random variables, the important random variables that have significant effects on failure probability are identified.

The effects of random variables can be explicitly quantified through the subsequent Bayesian analysis (Wang et al., 2010; Wang, 2012):

$$P(F \mid x_i) = \frac{p(x_i \mid F)P(F)}{p(x_i)} \tag{6.31}$$

where $P(F|x_i)$ is the conditional failure probability for a given x_i value; $p(x_i|F)$ is the conditional PDF of x_i given that the failure occurs; and $p(x_i)$ is the PDF of x_i. Eq. (6.31) implies that a comparison between $p(x_i|F)$ and $p(x_i)$ can provide an indication of the effect of x_i on the failure probability. Generally speaking, $P(F|x_i)$ changes with x_i. However, when $p(x_i|F)$ is similar to $p(x_i)$, $P(F|x_i)$ remains almost constant regardless of x_i. This implies that the effect of x_i on the failure probability is minimal. Both hypothesis testing and Bayesian analysis results provide useful information on the reliability sensitivity of random variables.

6.5.2.2 Sensitivity to Probability Distributions of Random Variables

Based on the baseline distribution $f(\mathbf{x})$, the failure probability of a geotechnical system can be evaluated first. The corresponding failure samples can also be obtained. For geotechnical engineering, more information can be collected along with the site investigation, construction, and operation.

Such additional information can be used to update the uncertainties in random variables due to the wide range of variability in geotechnical parameters. Thus, the joint PDF of random variables can be updated, which will be further used to update the failure probability. Let $f(\mathbf{x})^U$ denote the updated joint PDFs of random variables. Consider, for example, that direct MCS-based reliability analysis has been performed for a given $f(\mathbf{x})$, providing a number of direct MCS samples of $\mathbf{x}$. As the information of the random variables changes, the failure probability will be updated as $P(F)^U$, which is written as

$$P(F)^U = \int I(F,\mathbf{x}) f(\mathbf{x})^U d\mathbf{x} \tag{6.32}$$

where $I(F, \mathbf{x})$ is the indicator function of the failure event given $\mathbf{x}$; $I(F, \mathbf{x})$ is equal to unity if the failure occurs, otherwise $I(F, \mathbf{x})$ is equal to zero.

Using Equation (6.32), $P(F)^U$ can be calculated by simulating random samples from $f(\mathbf{x})^U$ by rerunning the MCS. This, however, necessitates additional computational efforts for evaluating $I(F, \mathbf{x})$ for each sample and might render the updating a non-trivial task. Such computational difficulties can be avoided by a sample reweighting technique (Fonseca et al., 2007), which allows using the random samples obtained from $f(\mathbf{x})$ to calculate the $P(F)^U$ after the information of random variables changes. In the context of the sample reweighting technique, Equation (6.32) is rewritten as (Cao et al., 2019)

$$P(F)^U = \int I(F,\mathbf{x}) \frac{f(\mathbf{x})^U}{f(\mathbf{x})} f(\mathbf{x}) d\mathbf{x} \approx \frac{1}{N}\sum_{i=1}^{N} I(F,\mathbf{x}_i) \frac{f(\mathbf{x})^U}{f(\mathbf{x})} = \frac{1}{N}\sum_{i=1}^{N} I(F,\mathbf{x}_i)\omega_i \tag{6.33}$$

where $\mathbf{x}_i$, $i = 1, 2, \ldots, N$ are the N random samples of $\mathbf{x}$ generated from $f(\mathbf{x})$ (instead of $f(\mathbf{x})^U$), and their corresponding values of $I(F, \mathbf{x}_i)$ have been evaluated before updating the information of random variables; and ω_i is a weighting factor that is calculated as the ratio of $f(\mathbf{x})^U$ over $f(\mathbf{x})$ evaluated at $\mathbf{x} = \mathbf{x}_i$.

Based on Equation (6.33), the updated failure probabilities under different distributions of $\mathbf{x}$ specified by $f(\mathbf{x})^U$ are calculated using the same set of random samples simulated for the original distribution $f(\mathbf{x})$, avoiding resampling of $\mathbf{x}$ from $f(\mathbf{x})^U$ and re-evaluations of $I(F, \mathbf{x})$ for reliability updating under different information on random variables (i.e., distributions of $\mathbf{x}$). By this means, the computational efficiency of MCS-based reliability analysis updating can be substantially improved. This opens up the possibility of linking site investigation and monitoring efforts with reliability updating using direct MCS in a cost-effective manner (e.g., Ching & Phoon, 2012; Cao et al., 2019). In addition, only the failure samples influence the failure probability. Hence, Equation (6.33) can be further simplified as

$$P(F)^U \approx \frac{1}{N}\sum_{i=1}^{N_F} \omega_i \tag{6.34}$$

where N_F is the number of failure samples.

It is worth pointing out that Equation (6.33) is similar to Eq. (6.26) for estimating failure probability in IS. However, the equations in the two approaches serve different purposes. In Equation (6.33), $f(\mathbf{x})$ and $f(\mathbf{x})^U$ are analogues to $h(\mathbf{x})$ and $q(\mathbf{x})$ in Equation (6.26), respectively. However, the physical meanings of the two pairs of counterparts (i.e., $f(\mathbf{x})$ vs. $h(\mathbf{x})$, and $f(\mathbf{x})^U$ vs. $q(\mathbf{x})$) are different; $f(\mathbf{x})$ is determined according to the site information available prior to collecting new information from the site, and reflects the state of knowledge about $\mathbf{x}$ at the current stage. In contrary, $h(\mathbf{x})$ in IS contains information on the problem at hand and the target PDF. Following new site information, $f(\mathbf{x})^U$ reflects the updated knowledge about $\mathbf{x}$; however, its counterpart $q(\mathbf{x})$ in IS quantifies uncertainties in $\mathbf{x}$ either before or after new information is obtained.

6.6 RELIABILITY-BASED DESIGN OF GEOTECHNICAL SYSTEMS

In earthwork and geotechnical engineering design, due to the inherent uncertainty of the soil parameters as input of the design model, the method of risk-based partial safety factor (FS) is often used to meet the reliability design requirements. However, this often leads to over-design, that is, the calculation results obtained through the partial safety factor method cannot meet the economic requirements (Casagrande, 1965). To find a balance between safety and cost, a reliability-based design that can quantitatively incorporate those represented uncertainties into the design calculation process has been proposed and put into geotechnical application (Zhang & Ji, 2022; Aoues & Chateauneuf, 2010). Today, with the rapid development of computer science, it is possible to simulate complex real-world problems in the real world through methods such as the application of finite element software. RBD methods can be well combined with the finite element method and stand-alone numerical codes to risk-based optimize the design parameters for earthwork and geotechnical engineering practice (Zhang et al., 2009; Lü et al., 2017; Ji et al., 2019; Zhang & Ji, 2022).

6.6.1 RBD by Inverse Reliability Method

6.6.1.1 Single Design Parameters

For a target reliability index β, the inverse problem can be stated as

$$\begin{aligned} &\text{Given}\,\beta \\ &\text{Find}\,d \\ &\text{Subject to}: \min(\mathbf{u}^T\mathbf{u}) = \beta^2 \text{ and } G(\mathbf{x}) = g(\mathbf{u},d) = 0 \end{aligned} \tag{6.35}$$

where **u** are uncorrelated standard normal variables (i.e., in u-space); **x** are the corresponding non-normal random variables in x-space; $G(\cdot)$ and $g(\cdot)$ are performance functions in x-space and u-space; and d is the target design parameter.

The solution of the inverse reliability problem in u-space is given by Zhang and Kiureghian (1995):

$$\mathbf{u}_{k+1} = -\beta_t \frac{\nabla_u g(\mathbf{u}_k, \mathrm{d}_k)}{\left\|\nabla_u g(\mathbf{u}_k, \mathrm{d}_k)\right\|} \tag{6.36}$$

$$\mathrm{d}_{k+1} = \mathrm{d}_k + \frac{\left\langle \nabla_u g(\mathbf{u}_k, \mathrm{d}_k), \mathbf{u}_k \right\rangle - g(\mathbf{u}_k, \mathrm{d}_k) + \beta_t \left\|\nabla_u g(\mathbf{u}_k, \mathrm{d}_k)\right\|}{\partial g(\mathbf{u}_k, \mathrm{d}_k)/\partial \mathrm{d}} \tag{6.37}$$

Based on the Cholesky decomposition of the correlation matrix **R**, the gradient vector can be described as

$$\nabla_u g(\mathbf{u}_k) = \mathbf{L}^T \left[\sigma_k^N\right] \nabla_x g(\mathbf{x}_k) \tag{6.38}$$

Using the above transformation for a gradient vector, Equations (6.36) and (6.37) can be transformed into x-space:

$$\mathbf{x}_{k+1} = -\beta_t \frac{\mathbf{T}_k \nabla_x g(\mathbf{x}_k, \mathrm{d}_k)}{\left\|\nabla_x\right\|} + \mu_k^N \tag{6.39}$$

$$\mathrm{d}_{k+1} = \mathrm{d}_k + \frac{\left[\nabla_x g(\mathbf{x}_k, \mathrm{d}_k)\right]^T (\mathbf{x}_k - \mu_k^N) - g(\mathbf{x}_k, \mathrm{d}_k) + \beta_t \left\|\nabla_x\right\|}{\partial g(\mathbf{x}_k, \mathrm{d}_k)/\partial \mathrm{d}} \tag{6.40}$$

where $\|\nabla_x\| = \sqrt{\left[\nabla_x g(\mathbf{x}_k, \mathrm{d}_k)\right]^T \mathbf{T}_k \nabla_x g(\mathbf{x}_k, \mathrm{d}_k)}$, with revised correlation matrix $\mathbf{T}_k$ of $T_{k,ij} = \sigma_{k,i}^N R_{ij} \sigma_{k,j}^N$.

Introducing an iterative search step length λ and search direction $\mathbf{d}_k$, such that

$$\begin{pmatrix} \mathbf{x}_{k+1} \\ \mathrm{d}_{k+1} \end{pmatrix} = \begin{pmatrix} \mathbf{x}_k \\ \mathrm{d}_k \end{pmatrix} + \lambda \mathbf{d}_k^x \tag{6.41}$$

$$\mathbf{d}_k^x = \begin{pmatrix} -\beta_t \dfrac{\mathbf{T}_k \nabla_x g(\mathbf{x}_k, \mathrm{d}_k)}{\|\nabla_x\|} + \mu_k^N - \mathbf{x}_k \\ \dfrac{\left[\nabla_x g(\mathbf{x}_k, \mathrm{d}_k)\right]^T (\mathbf{x}_k - \mu_k^N) - g(\mathbf{x}_k, \mathrm{d}_k) + \beta_t \|\nabla_x\|}{\partial g(\mathbf{x}_k, \mathrm{d}_k) / \partial \mathrm{d}} \end{pmatrix} \tag{6.42}$$

For better convergence, the merit function defined below is used:

$$m(\mathbf{x}, \mathrm{d}) = \frac{1}{2} \left\| \left[\frac{x_i - \mu_i}{\sigma_i} \right] \right\|^2 + c \left| g(\mathbf{x}, \mathrm{d}) \right| \tag{6.43}$$

where $c > 0$ is a penalty parameter, and can be taken as $c = 100$ for most geotechnical reliability problems, as mentioned previously.

As a backtracking line search technique in x-space, the corresponding Armijo rule for the optimal (maximum) step length is

$$\lambda_k = \max_j \left\{ b^j \left| m(\mathbf{x}_k + b^j \mathbf{d}_k, \mathrm{d}_k) - m(\mathbf{x}_k, \mathrm{d}_k) \leq -ab^j \left\langle \nabla m(\mathbf{x}_k, \mathrm{d}_k), \mathbf{d}_k \right\rangle \right. \right\} \tag{6.44}$$

where $a, b \in (0, 1)$ are prescribed parameters; j is an integer for the optimal solution; and the gradient of the merit function is written as

$$\nabla m(\mathbf{x}, \mathrm{d}) = \begin{pmatrix} \left[\dfrac{x_i - \mu_i}{\sigma_i^2} \right]_k \\ 0 \end{pmatrix} + c \,\mathrm{sgn}\left[g(\mathbf{x}, \mathrm{d}) \right] \begin{pmatrix} \nabla g_x(\mathbf{x}, \mathrm{d}) \\ \partial g(\mathbf{x}, \mathrm{d}) / \partial \mathrm{d} \end{pmatrix} \tag{6.45}$$

Similar to the forward algorithm, the Armijo constants a and b are set to be 0.5.

6.6.1.2 Multiple Design Parameters

In the multiparameter RBD, a unique solution can be obtained when the number of reliability constraints is equal to the number of design parameters (Li & Foschi, 1998; Zhang & Ji, 2022). Suppose $\mathbf{d} = (d_1, d_2, \ldots, d_n)$ is a vector of n-dimensional design parameters including the LSF's deterministic model parameters or unknown statistics such as the mean and/or standard deviation of the random variables. The design parameter d_i can be computed using the inverse reliability algorithm by setting other d_i's as their tentative values and can be implicitly written as

$$d_i = f_i\left(d_1, d_2, \ldots, d_{i-1}, d_{i+1}, \ldots, d_n, \beta_i\right) \quad (i = 1, 2, \ldots, n) \tag{6.46}$$

where β_i is the target reliability index corresponding to the reliability constraint G_i.

To establish a coupling relationship between the design parameters and reliability constraints, the following function F_i is introduced:

$$\mathrm{F}_i = d_i - f_i\left(d_1, d_2, \ldots, d_{i-1}, d_{i+1}, \ldots, d_n, \beta_i\right) \quad (i = 1, 2, \ldots, n) \tag{6.47}$$

and applying the Newton–Raphson iterations to satisfy the constraint condition $F_i = 0$.

The function F_i can be expanded into a Taylor series:

$$F_i = d_i^* - f_i\left(d_1^*, d_2^*, \ldots, d_{i-1}^*, d_{i+1}^*, \ldots, d_n^*, \beta_i\right) + \sum_{j=1}^{n} \frac{\partial F_i}{\partial d_j}\bigg|_{\mathbf{d}^*}\left(d_j - d_j^*\right) \qquad (i = 1, 2, \ldots, n) \tag{6.48}$$

We can finally obtain a multiparameter iterative equation which allows updating of the design parameter d_i by setting $F_i = 0$, which will lead to the following solutions:

$$\mathbf{d}_{k+1} = \mathbf{d}_k - \begin{bmatrix} 1 & \frac{\partial F_1}{\partial d_2} & \cdots & \frac{\partial F_1}{\partial d_{n-1}} & \frac{\partial F_1}{\partial d_n} \\ \frac{\partial F_2}{\partial d_1} & 1 & \cdots & \frac{\partial F_2}{\partial d_{n-1}} & \frac{\partial F_2}{\partial d_n} \\ \vdots & \vdots & \ddots & \vdots & \vdots \\ \frac{\partial F_{n-1}}{\partial d_1} & \frac{\partial F_{n-1}}{\partial d_2} & \cdots & 1 & \frac{\partial F_{n-1}}{\partial d_{n-1}} \\ \frac{\partial F_n}{\partial d_1} & \frac{\partial F_n}{\partial d_2} & \cdots & \frac{\partial F_n}{\partial d_{n-1}} & 1 \end{bmatrix} \left(\mathbf{d}_k - \mathbf{f}\right) \tag{6.49}$$

where $\mathbf{d} = (d_1, d_2, \ldots, d_n)$ and $\mathbf{f} = (f_1, f_2, \ldots, f_n)$. It should be noted that the multiparameter RBD design may not be possible in some special cases, since the equation system represented by Equation (6.49) may not have a solution when reliability indices are specified.

6.6.2 RBD by Monte Carlo Simulations

6.6.2.1 MCS-Based Full Probabilistic Design

The core task of an MCS-based RBD is to determine feasible designs with $P(F|D^{(k)}) < P_T$, or equivalently $\beta^{(k)} > \beta_T$, among possible design samples $D^{(k)}$ ($k = 1, 2, \ldots, N_D$) in a design domain prescribed by users. Let $F^{(k)}$ denote the failure event of $D^{(k)}$, the corresponding occurrence probability is represented by $P(F|D^{(k)})$. The determination of feasible designs with $P(F|D^{(k)}) < P_T$ needs to calculate the probability $P(F|D^{(k)})$. A straightforward, but computationally expensive way to calculate the $P(F|D^{(k)})$ values of all possible designs is to repeatedly perform N_D runs of MCS, each of which will regenerate MCS samples and re-evaluate design calculation models for each sample. Alternatively, the $P(F|D^{(k)})$ values can be obtained by post-processing the MCS samples generated by a single MCS run without repeatedly running MCS for different possible designs.

An expanded RBD framework (Wang, 2011; Wang et al., 2011; Wang & Cao, 2014) was proposed for the MCS-based full probabilistic design. It can be viewed as an augmented reliability analysis of a geotechnical system, in which design parameters are artificially considered uncertain with probability distributions specified by users for design exploration purposes. For example, the design parameters $\mathbf{d}$ can be represented as discrete uniform random variables. Each possible design $D^{(k)}$ is specified by a possible combination of discrete values of design parameters and has a probability of $P(D^{(k)})$. Then, the RBD is formulated as a process of calculating the $P(F|D^{(k)})$ values of all possible designs and comparing them with P_T. This can be achieved by a single run of direct MCS with the aid of a Bayesian analysis of the MCS samples, in which $P(F|D^{(k)})$ is written as

$$P\left(F \mid D^{(k)}\right) = P\left(D^{(k)} \mid F\right) P(F) / P\left(D^{(k)}\right) \tag{6.50}$$

where $P(D^{(k)}|F)$ = the occurrence probability of $D^{(k)}$ given failure and $P(F)$ = the failure probability considering failure samples of all possible designs. Since design parameters are represented as discrete uniform random variables, $P(D^{(k)})$ is taken as equal to $1/N_D$, where N_D is the number of possible designs. Failure samples and those corresponding to each $D^{(k)}$ are identified from random samples generated by a single run of direct MCS to estimate the $P(F)$ and $P(D^{(k)}|F)$ of all the N_D possible designs, respectively.

To improve the computational efficiency, SS can be combined with the expanded RBD. SS divides sample space Ω of uncertain parameters (including $\mathbf{d}$ and $\mathbf{x}$ for an expanded RBD) into $m+1$ individual subsets $\{\Omega_i, i = 0, 1, 2,..., m\}$ by the intermediate threshold values $\{y_i, i = 1, 2,..., m\}$ of the driving variable Y. In addition, the conditional probability $P(D^{(k)}|F)$ is given by

$$P\left(D^{(k)} \mid F\right)=\sum_{i=0}^{m} P\left(D^{(k)} \mid F \cap \Omega_i\right) P\left(\Omega_i \mid F\right) \tag{6.51}$$

where $P(D^{(k)}|F\cap\Omega_i)$ is the conditional probability of $D^{(k)}$ given sampling in Ω_i and the occurrence of failure, and it is expressed as the ratio of the number, $n_{f,i}^{(k)}$, of failure samples for $D^{(k)}$ in Ω_i over the total failure sample number $n_{f,i}$ in Ω_i, i.e., $P(D^{(k)}|F\cap\Omega_i) = n_{f,i}^{(k)}/n_{f,i}$. Subsequently, $P(F|D^{(k)})$ is obtained in accordance with Eq. (6.50). Compared with direct MCS, SS efficiently generates a large number of failure samples to calculate $P(F)$ and $P(D^{(k)}|F)$ to improve the accuracy and resolution of $P(F|D^{(k)})$ for RBD (Wang & Cao, 2013).

Alternatively, GSS can efficiently estimate the failure probabilities of various limit states of all possible designs by a single simulation run without the Bayesian analysis. In the context of GSS, each possible design $D^{(k)}$ in the design space can be viewed as an equivalent failure event for RBD. The major difference between GSS and SS lies in determining intermediate events and selecting "seed" samples for conditional sampling. GSS adopts a design response vector $\mathbf{Y}$ to define unified intermediate events instead of a driving variable Y adopted in SS, and it is written as (Yang et al., 2021)

$$\mathbf{Y}=\left[\mathbf{Y}^{(1)},\mathbf{Y}^{(2)},\ldots,\mathbf{Y}^{(N_D)}\right] \tag{6.52}$$

where $\mathbf{Y}^{(k)}=\left[\mathrm{Y}_1^{(k)},\mathrm{Y}_2^{(k)},\ldots,\mathrm{Y}_{NM}^{(k)}\right]$, $k = 1, 2,..., N_D$, is a vector composed of the N_M responses of the kth possible design corresponding to the N_M limit states, and its jth component $\mathrm{Y}_j^{(k)}, j = 1, 2,..., N_M$ represents the response (e.g., the reciprocal of the safety factor, $1/FS_j^{(k)}$) of the jth limit state of the kth possible design.

Using GSS, F (i.e., the union of failure events of all possible designs) is progressively explored by simulating a number of nested intermediate events F_i (i.e., the union of intermediate failure events $F_j^{(k)}$ of the N_D possible designs). As GSS proceeds, the failure domains of failure events $F_j^{(k)}$ with relatively large $P\left(F_j^{(k)}\right)$ values are arrived at first. After $F_j^{(k)}$ is reached, its corresponding intermediate failure event can then be dropped from F_i in the following simulation levels. This reduces the number of limit state function calls (or, equivalently, the number of components in $\mathbf{Y}$) during the following simulation levels as i increases, thus providing computational efficiency.

6.6.2.2 Approximation of Design Point

Despite that the MCS-based full probabilistic design does not require the design point for the purpose of decision-making, the design point of each component of the failure modes can be approximately determined from the failure samples according to its probabilistic interpretation, i.e., the most probable failure point. In general, this can be accomplished using the expanding ellipsoid perspective of the first-order reliability method in the space of original random variables, i.e., the x-space (Low & Tang, 2007; Gao et al., 2019). Alternatively, if random variables follow multivariate normal distribution, the design point can simply be taken as the failure sample with the maximum value of the joint probability density function.

Note that the accuracy of the design point determined from the failure samples relies on the number of failure samples determined by MCS, and it generally improves as the number of failure samples increases. In comparison with direct MCS, an advanced simulation method such as SS and GSS can efficiently generate a large number of failure samples with relative ease. More importantly, the fact that SS and GSS explore the sample space from safe to failure domains in a progressive manner helps generate failure samples close to the limit state surface, thus improving the accuracy of the estimated design point information.

6.7 GEOTECHNICAL RELIABILITY ANALYSIS ACCOUNTING FOR SOIL SPATIAL VARIABILITY

In this section, a simplified geotechnical reliability analysis dealing with soil spatial variability is illustrated. Figure 6.11 shows an undrained soil slope (Cho, 2010). The spatially variable undrained shear strength c_u is modeled as a random field. The statistical properties of the soil parameters are summarized in Table 6.1.

Based on the mean value of c_u, the safety factor can be obtained by the FEM strength reduction method using the commercial software Abaqus. The FS is 1.357 and the reliability index is 0.867 using the FORM with the HLRF-*x* algorithm when c_u is simply taken as a random variable. Since the soil properties are not always the same at different spatial locations and there is a certain correlation between different locations, the random variable analysis cannot reflect the realistic spatial variability of the soil. Given the spatial variability characteristics, the random field of c_u can be represented using the well-known Karhunen–Loeve (KL) expansion technique that is usually truncated with M limited number of random variables. For illustration, Figure 6.12 shows a typical KL random field realization of the example slope, where the random field of c_u is characterized with an exponential square autocorrelation function of horizontal and vertical autocorrelation distances, respectively, to be $l_h = 20$ m and $l_v = 2$ m.

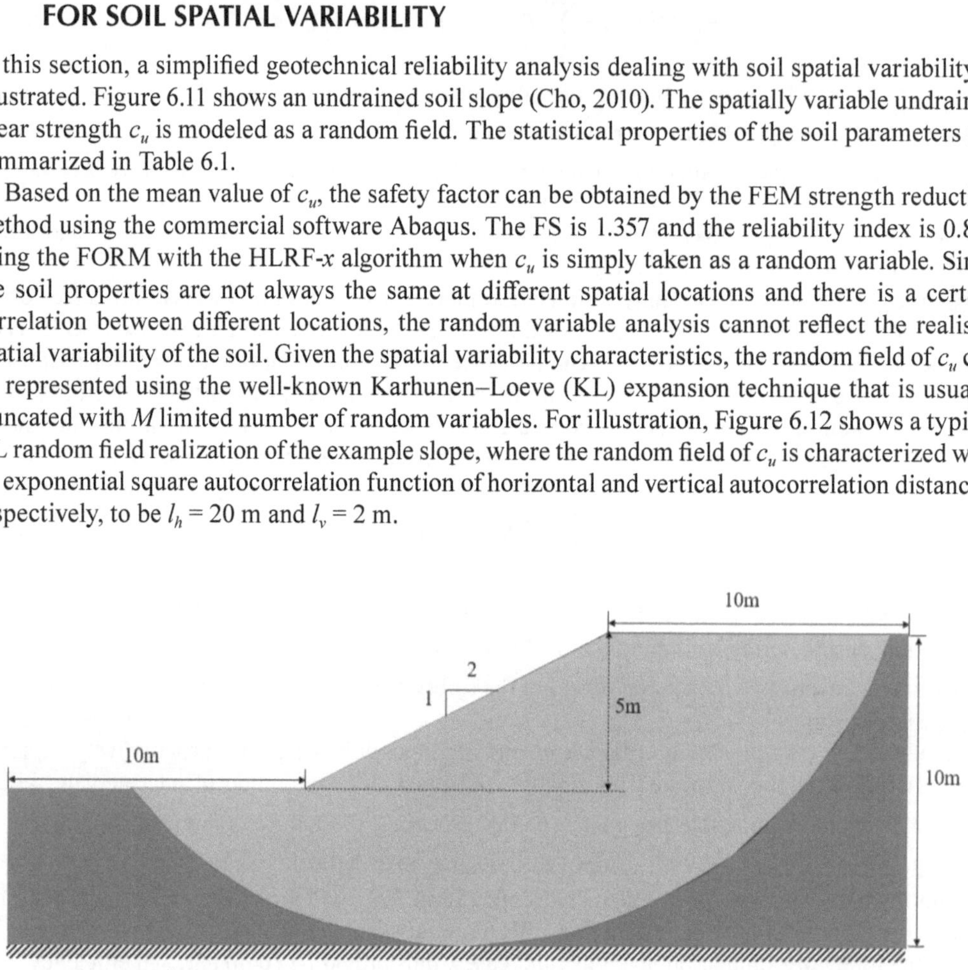

FIGURE 6.11 Cross section of an undrained soil slope.

TABLE 6.1
Statistical Properties of Soil Parameters

Soil properties	Distribution	Mean	COV
Unit weight γ_{sat} (kN/m^3)	–	20	–
Undrained shear strength c_u (kPa)	Lognormal	23	0.3

FIGURE 6.12 Once typical realization of random fields of the example slope.

Combining the KL expansion for random field modeling and FORM for practical reliability analysis, a new framework of KL-FORM works for this realistic geotechnical reliability analysis. In this example, the KL truncation term $M = 12$ suffices for the random field representation with accuracy in terms of the ratio of energy <0.5. Based on HLRF-x iterative computations, the FORM reliability index is 1.026.

Since the autocorrelation distance is a very important parameter of the random field, the influence of the autocorrelation length on the reliability of the example slope is shown in Figure 6.13. It can be found that the slope reliability is more sensitive to the vertical autocorrelation length compared to the horizontal autocorrelation length. With the increase in the vertical autocorrelation length, the reliability index decreases significantly, reducing to a random variable model.

6.8 NOTE ON GEOTECHNICAL RELIABILITY ANALYSIS APPLICATIONS VIA STAND-ALONE NUMERICAL CODES

It is obvious that for many geotechnical problems the performance function evaluation is not realistic and various stand-alone commercial software must be resorted to. For general commercial software that are only with deterministic simulation features, practical geotechnical reliability analysis can be fulfilled using the aforementioned techniques for dealing with implicit performance functions, e.g., the response surface method, the surrogate model–based MCS, and the direct FORM coupling method with HLRF recursive algorithms. For example, Low(Low, 2021) demonstrated that the ubiquitous MC-Excel is a promising platform for implementing various geotechnical reliability analyses when bridging with other deterministic geotechnical software. Some specific geotechnical software have also developed embedded probabilistic simulation features. For example, the Bentley's subproduct *PLAXIS/LE* (earlier known as SOILVISION/*SVSLOPE*) (https://bentley.com/en/products/product-line/geotechnical-engineering-software/plaxis-le) enables the probability of failure analysis using MCS or alternate point estimate methods (APEM). It is worth noting that APEM are based on first-order

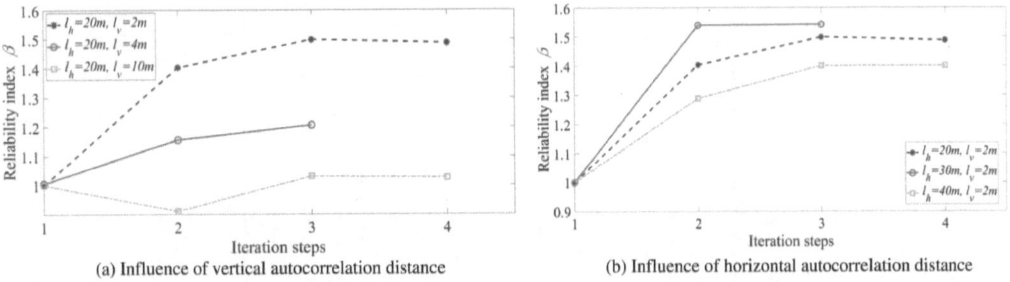

FIGURE 6.13 The influence of autocorrelation distances on slope reliability.

TABLE 6.2
Methods, Implementations, and Applications of MCS-Based Geotechnical Reliability Analysis

Methods		Software implementations	
Simulation algorithm	Response surface	Probability and statistics	Geotechnical numerical simulations
MCS SS LHS IS LS WUS	CPRS PCE Kriging NN SVM	MATLAB Excel Python UQLAB COSSAN	Abaqus FLAC Plaxis GeoStudio Ansys RS2 Slide2&3
MCS: slopes, retaining structures, foundations, tunnels **SS**: slopes, retaining structures, foundations **LHS**: slopes, retaining structures, foundations **IS**: slopes, retaining structures, foundations **LS**: foundations **WUS**: slopes, foundations	**CPRS**: slopes, tunnels **PCE**: slopes, retaining structures, foundations, tunnels **Kriging**: slopes, foundations **NN**: slopes, retaining structures, foundations **SVM**: slopes, retaining structures, foundations, tunnels	**MATLAB**: slopes, retaining structures, foundations, tunnels **Excel**: slopes, retaining structures, foundations, tunnels **Python**: slopes **UQLAB**: slopes, foundations, tunnels **COSSAN**: foundations	**ABAQUS**: slopes, foundations, tunnels **FLAC**: slopes, retaining structures, foundations, tunnels **PLAXIS**: slopes, retaining structures, foundations **GeoStudio**: slopes **ANSYS**: slopes, tunnels **RS2**: slopes, retaining structures, foundations, tunnels **Slide2 & Slide3**: 2D and 3D slopes

Source: Adapted from Phoon et al. (2022).

Notes:

(a) MCS, routine Monte Carlo simulation; LHS, Latin hypercube sampling; LS, line sampling; IS, importance sampling; SS, subset simulation; WUS, weighted uniform simulation.

(b) CPRS, classical polynomial response surfaces; PCE, polynomial chaos expansions, SVM, support vector machines; NN, neural networks.

(c) UQLAB (www.uqlab.com/); COSSAN (https://cossan.co.uk/); Excel (www.microsoft.com/); MATLAB (https://ww2.mathworks.cn/); Python (www.python.org/).

(d) ABAQUS (www.abaqus.com/); ANSYS (www.ansys.com/); FLAC (www.itascacg.com/); Geostudio (www.geoslope.com/); PLAXIS (www.bentley.com/); Rocscience (www.Rocscience.com).

second-moment (FOSM) principles and enable the calculation of probability of failure with a greatly reduced number of model runs compared to Monte Carlo. Other routine MCS-embedded geotechnical software include Geostudio's *Geoslope/W* (https://geoslope.com/products/slope-w), Rocscience's *Slide2* (https://www.rocscience.com /software/slide2), and Optumce's *OPTUM*^{G2} (https://optumce.com/products/optumg2/).

The geotechnical reliability analysis by surrogate model–based MCS in combination with various stand-alone software implementations has garnered increasing attention in recent years. See Table 6.2 for more details on typical geotechnical reliability applications of the abovementioned software facilitating risk assessment for large realistic geotechnical problems.

REFERENCES

Aoues, Y. & Chateauneuf, A. 2010. Benchmark study of numerical methods for reliability-based design optimization. *Structural and Multidisciplinary Optimization*, 41(2), 277–294.

Au, S.-K. & Beck, J. L. 2001. Estimation of small failure probabilities in high dimensions by subset simulation. *Probabilistic Engineering Mechanics*, 16(4), 263–277.

Au, S.-K. & Wang, Y. 2014. *Engineering risk assessment with subset simulation*. John Wiley & Sons.

Baecher, G. B. & Christian, J. T. 2005. *Reliability and statistics in geotechnical engineering*. John Wiley & Sons.

Bjerager, P. 1990. On computation methods for structural reliability analysis. *Structural Safety*, 9(2), 79–96.

Cao, Z.-J., Peng, X., Li, D.-Q. & Tang, X.-S. 2019. Full probabilistic geotechnical design under various design scenarios using direct Monte Carlo simulation and sample reweighting. *Engineering Geology*, 248, 207–219.

Casagrande, A. 1965. Role of the calculated risk in earthwork and foundation engineering. *Journal of the Soil Mechanics and Foundations Division*, 91(4), 1–40.

Chan, C. L. & Low, B. K. 2011. Practical second-order reliability analysis applied to foundation engineering. *International Journal for Numerical and Analytical Methods in Geomechanics*.

Chan, C. L. & Low, B. K. 2012. Probabilistic analysis of laterally loaded piles using response surface and neural network approaches. *Computers and Geotechnics*, 43, 101–110.

Ching, J. & Phoon, K.-K. 2012. Value of geotechnical site investigation in reliability-based design. *Advances in Structural Engineering*, 15(11), 1935–1945.

Ching, J., Phoon, K.-K. & Hu, Y.-G. 2009. Efficient evaluation of reliability for slopes with circular slip surfaces using importance sampling. *Journal of Geotechnical and Geoenvironmental Engineering*, 135(6), 768–777.

Cho, S. E. 2009. Probabilistic stability analyses of slopes using the ANN-based response surface. *Computers and Geotechnics*, 36(5), 787–797.

Cho, S. E. 2010. Probabilistic assessment of slope stability that considers the spatial variability of soil properties. *Journal of Geotechnical and Geoenvironmental Engineering*, 136(7), 975–984.

Coates, R. C., Coutie, M. & Kong, F. K. 1994. *Structural analysis*, 3rd ed. Chapman and Hall.

Cornell, C. A. 1969. A probability-based structural code. *Journal of the American Concrete Institute*, 66, 974–985.

Depina, I., Le, T. M. H., Fenton, G. & Eiksund, G. 2016. Reliability analysis with metamodel line sampling. *Structural Safety*, 60, 1–15.

Ditlevsen, O. 1979b. Narrow reliability bounds for structural systems. *Journal of Structural Mechanics*, 7(4), 453–472.

Fonseca, J. R., Friswell, M. I. & Lees, A. W. 2007. Efficient robust design via Monte Carlo sample reweighting. *International Journal for Numerical Methods in Engineering*, 69(11), 2279–2301.

Gao, G.-H., Li, D.-Q., Cao, Z.-J., Wang, Y. & Zhang, L. 2019. Full probabilistic design of earth retaining structures using generalized subset simulation. *Computers and Geotechnics*, 112, 159–172.

Ghohani Arab, H., Rashki, M., Rostamian, M., Ghavidel, A., Shahraki, H. & Keshtegar, B. 2019. Refined first-order reliability method using cross-entropy optimization method. *Engineering with Computers*, 35(4), 1507–1519.

Haldar, A. & Mahadevan, S. 2000. *Probability, reliability and statistical methods in engineering design*. Wiley.

Hasofer, A. M. & Lind, N. C. 1974. Exact and invariant second moment code format. *Journal of the Engineering Mechanics, ASCE*, 100(1), 111–121.

Huang, J. & Griffiths, D. V. 2011. Observations on FORM in a simple geomechanics example. *Structural Safety*, 33(1), 115–119.

Ibrahim, Y. 1991. Observations on applications of importance sampling in structural reliability analysis. *Structural Safety*, 9(4), 269–281.

Ji, J. & Kodikara, J. K. 2015. Efficient reliability method for implicit limit state surface with correlated non-Gaussian variables. *International Journal for Numerical and Analytical Methods in Geomechanics*, 39(17), 1898–1911.

Ji, J. & Low, B. K. 2012. Stratified response surfaces for system probabilistic evaluation of slopes. *Journal of Geotechnical and Geoenvironmental Engineering, ASCE*, 138(11), 1398–1406.

Ji, J. & Wang, L.-P. 2022. Efficient geotechnical reliability analysis using weighted uniform simulation method involving correlated nonnormal random variables. *Journal of Engineering Mechanics*, 148(6), 06022001.

Ji, J., Zhang, C., Gao, Y. & Kodikara, J. 2018. Effect of 2D spatial variability on slope reliability: A simplified FORM analysis. *Geoscience Frontiers*, 9(6), 1631–1638.

Ji, J., Zhang, C., Gao, Y. & Kodikara, J. 2019. Reliability-based design for geotechnical engineering: An inverse FORM approach for practice. *Computers and Geotechnics*, 111, 22–29.

Ji, J., Zhang, C., Gui, Y., Lü, Q. & Kodikara, J. 2017. New observations on the application of LS-SVM in slope system reliability analysis. *Journal of Computing in Civil Engineering*, 31(2), 06016002.

Kang, F., Han, S., Salgado, R. & Li, J. 2015. System probabilistic stability analysis of soil slopes using Gaussian process regression with Latin hypercube sampling. *Computers and Geotechnics*, 63, 13–25.

Keshtegar, B. & Chakraborty, S. 2018. A hybrid self-adaptive conjugate first order reliability method for robust structural reliability analysis. *Applied Mathematical Modelling*, 53, 319–332.

Keshtegar, B. & Meng, Z. 2017. A hybrid relaxed first-order reliability method for efficient structural reliability analysis. *Structural Safety*, 66, 84–93.

Koutsourelakis, P.-S., Pradlwarter, H. J. & Schueller, G. I. 2004. Reliability of structures in high dimensions, part I: Algorithms and applications. *Probabilistic Engineering Mechanics*, 19(4), 409–417.

Koyluoglu, H. U. & Nielsen, S. R. K. 1994. New approximations for SORM integrals. *Structural Safety*, 13(4), 235–246.

Li, D., Chen, Y., Lu, W. & Zhou, C. 2011. Stochastic response surface method for reliability analysis of rock slopes involving correlated non-normal variables. *Computers and Geotechnics*, 38(1), 58–68.

Li, D.-Q., Jiang, S.-H., Cao, Z.-J., Zhou, W., Zhou, C.-B. & Zhang, L.-M. 2015a. A multiple response-surface method for slope reliability analysis considering spatial variability of soil properties. *Engineering Geology*, 187, 60–72.

Li, H. & Foschi, R. O. 1998. An inverse reliability method and its application. *Structural Safety*, 20(3), 257–270.

Li, H.-S., Ma, Y.-Z. & Cao, Z. 2015b. A generalized Subset Simulation approach for estimating small failure probabilities of multiple stochastic responses. *Computers and Structures*, 153, 239–251.

Liao, W. & Ji, J. 2021. Time-dependent reliability analysis of rainfall-induced shallow landslides considering spatial variability of soil permeability. *Computers and Geotechnics*, 129, 103903.

Liu, P.-L. & Der Kiureghian, A. 1991. Optimization algorithms for structural reliability. *Structural Safety*, 9(3), 161–177.

Liu, X., Cao, Z.-J., Li, D.-Q. & Wang, Y. 2022. Adaptive Monte Carlo simulation method and its applications to reliability analysis of series systems with a large number of components. *ASCE-ASME Journal of Risk and Uncertainty in Engineering Systems, Part A: Civil Engineering*, 8(1), 04021075.

Liu, X., Li, D.-Q., Cao, Z.-J. & Wang, Y. 2020. Adaptive Monte Carlo simulation method for system reliability analysis of slope stability based on limit equilibrium methods. *Engineering Geology*, 264, 105384.

Low, B. K. 2017. Efficient FORM procedure and geotechnical reliability-based design. *Geotechnical Safety and Reliability*.

Low, B. K. 2021. *Reliability-based design in soil and rock engineering: Enhancing partial factor design approaches*. CRC Press.

Low, B. K. & Phoon, K.-K. 2015. Reliability-based design and its complementary role to Eurocode 7 design approach. *Computers and Geotechnics*, 65, 30–44.

Low, B. K. & Tang, W. H. 2004. Reliability analysis using object-oriented constrained optimization. *Structural Safety*, 26(1), 69–89.

Low, B. K. & Tang, W. H. 2007. Efficient spreadsheet algorithm for first-order reliability method. *Journal of Engineering Mechanics*, 133(12), 1378–1387.

Low, B. K., Zhang, J. & Tang, W. H. 2011. Efficient system reliability analysis illustrated for a retaining wall and a soil slope. *Computers and Geotechnics*, 38(2), 196–204.

Lü, Q., Xiao, Z.-P., Ji, J. & Zheng, J. 2017. Reliability based design optimization for a rock tunnel support system with multiple failure modes using response surface method. *Tunnelling and Underground Space Technology*, 70, 1–10.

Mckay, M., Beckham, R. & Conover, W. 1979. A comparison of three methods for selecting values of input variables in the analysis of output from a computer code, 1979. *Technometrics*, 21, 21.

Olsson, A., Sandberg, G. & Dahlblom, O. 2003. On Latin hypercube sampling for structural reliability analysis. *Structural Safety*, 25(1), 47–68.

Periçaro, G. A., Santos, S. R., Ribeiro, A. A. & Matioli, L. C. 2015. HLRF–BFGS optimization algorithm for structural reliability. *Applied Mathematical Modelling*, 39(7), 2025–2035.

Phoon, K.-K., Cao, Z.-J., Ji, J., Leung, Y. F., Najjar, S., Shuku, T., Tang, C., Yin, Z.-Y., Ikumasa, Y. & Ching, J. 2022. Geotechnical uncertainty, modeling, and decision making. *Soils and Foundations*, 62(5), 101189.

Phoon, K.-K., Kulhawy Fred, H. & Grigoriu Mircea, D. 2003. Development of a reliability-based design framework for transmission line structure foundations. *Journal of Geotechnical and Geoenvironmental Engineering*, 129(9), 798–806.

Pradlwarter, H., Schueller, G. I., Koutsourelakis, P.-S. & Charmpis, D. C. 2007. Application of line sampling simulation method to reliability benchmark problems. *Structural Safety*, 29(3), 208–221.

Rackwitz, R. 2001. Reliability analysis: A review and some perspectives. *Structural Safety*, 23(4), 365–395.

Rackwitz, R. & Fiessler, B. 1978. Structural reliability under combined random load sequences. *Computers and Structures*, 9(5), 484–494.

Ramesh, R. B., Mirza, O. & Kang, W.-H. 2017. HLRF-BFGS-based algorithm for inverse reliability analysis. *Mathematical Problems in Engineering*, 2017, 4317670, 1–15.

Rashki, M., Miri, M. & Azhdary Moghaddam, M. 2012. A new efficient simulation method to approximate the probability of failure and most probable point. *Structural Safety*, 39, 22–29.

Shinozuka, M. 1983. Basic analysis of structural safety. *Journal of Structural Engineering, ASCE*, 109(3), 721–740.

Stein, M. 1987. Large sample properties of simulations using Latin hypercube sampling. *Technometrics*, 29(2), 143–151.

Wang, Y. 2011. Reliability-based design of spread foundations by Monte Carlo simulations. *Géotechnique*, 61(8), 677–685.

Wang, Y. 2012. Uncertain parameter sensitivity in Monte Carlo simulation by sample reassembling. *Computers and Geotechnics*, 46, 39–47.

Wang, Y., Au, S.-K. & Kulhawy, F. H. 2011. Expanded reliability-based design approach for drilled shafts. *Journal of Geotechnical and Geoenvironmental Engineering*, 137(2), 140–149.

Wang, Y. & Cao, Z. 2013. Expanded reliability-based design of piles in spatially variable soil using efficient Monte Carlo simulations. *Soils and Foundations*, 53(6), 820–834.

Wang, Y. & Cao, Z. 2014. Practical reliability analysis and design by Monte Carlo Simulation in spreadsheet. *Risk and Reliability in Geotechnical Engineering*, pp. 301–335.

Wang, Y., Cao, Z. & Au, S.-K. 2010. Efficient Monte Carlo simulation of parameter sensitivity in probabilistic slope stability analysis. *Computers and Geotechnics*, 37(7–8), 1015–1022.

Wang, Z., Broccardo, M. & Der Kiureghian, A. 2016. An algorithm for finding a sequence of design points in reliability analysis. *Structural Safety*, 58, 52–59.

Wong, F. S. 1985. Slope reliability and response surface method. *Journal of Geotechnical Engineering - ASCE*, 111(1), 32–53.

Yang, Y.-J., Li, D.-Q., Cao, Z.-J., Gao, G.-H. & Phoon, K.-K. 2021. Geotechnical reliability-based design using generalized subset simulation with a design response vector. *Computers and Geotechnics*, 139, 104392.

Zeng, P., Jimenez, R. & Li, T. 2016. An efficient quasi-Newton approximation-based SORM to estimate the reliability of geotechnical problems. *Computers and Geotechnics*, 76, 33–42.

Zeng, P., Li, T., Jimenez, R., Feng, X. & Chen, Y. 2018. Extension of quasi-Newton approximation-based SORM for series system reliability analysis of geotechnical problems. *Engineering with Computers*, 34(2), 215–224.

Zhang, J., Huang, H. & Phoon, K. 2013. Application of the Kriging-based response surface method to the system reliability of soil slopes. *Journal of Geotechnical and Geoenvironmental Engineering*, 139(4), 651–655.

Zhang, J., Zhang, L. & Tang, W. H. 2009. Bayesian framework for characterizing geotechnical model uncertainty. *Journal of Geotechnical and Geoenvironmental Engineering*, 135(7), 932–940.

Zhang, Y. & Kiureghian, A. D. 1995. Two improved algorithms for reliability analysis. In *Reliability and optimization of structural systems*. Proceedings of the sixth IFIP WG7. 5 working conference on reliability and optimization of structural systems, edited by R. Rackwitz, G. Augusti and A. Borri, pp. 297–304. Springer.

Zhang, Z. & Ji, J. 2022. Geotechnical RBDO: Coupling the inverse reliability algorithm with multi-objective reliability-based design optimization of geotechnical systems. *Computers and Geotechnics*, 152, 105005.

Zhao, H.-B. 2008. Slope reliability analysis using a support vector machine. *Computers and Geotechnics*, 35(3), 459–467.

Zhao, H., Ru, Z., Chang, X., Yin, S. & Li, S. 2014. Reliability analysis of tunnel using least square support vector machine. *Tunnelling and Underground Space Technology*, 41, 14–23.

Zhao, Y.-G. & Ono, T. 1999. A general procedure for first/second-order reliability method (FORM/SORM). *Structural Safety*, 21(2), 95–112.

Zhu, S.-P., Keshtegar, B., Chakraborty, S. & Trung, N.-T. 2020. Novel probabilistic model for searching most probable point in structural reliability analysis. *Computer Methods in Applied Mechanics and Engineering*, 366, 113027.

7 Reliability Analysis with Reduced Order Model

Yu Otake and Taiga Saito

ABSTRACT

This chapter delves into the utilization of reduced order models (ROMs) for assessing system performance in geotechnical engineering, specifically focusing on robustness and resilience. In recent years there has been a growing interest in these topics, and surrogate modeling is considered a promising tool for enhancing reliability analysis by capturing spatiotemporal information embedded in computationally demanding geotechnical engineering mechanics calculations. This chapter examines two fundamental ROM methods, namely proper orthogonal decomposition (POD) and dynamic mode decomposition (DMD), to extract eigenmode functions that represent spatiotemporal dynamics information. These ROM methods facilitate low-cost simulations by constructing a reduced model that accurately reproduces the original data using a small number of spatial mode functions. This chapter presents the theoretical background and formulation of POD and DMD and provides a fundamental example problem to introduce the specific features of ROMs. Moreover, the study employs ROM analysis with POD as the key technology to evaluate the seismic behavior of a simple embankment on a liquefiable sand layer. The results of this analysis demonstrate that the proposed framework can accurately capture the dynamic behavior of the embankment and provide insight into the factors that affect its stability. Furthermore, the study applies the DMD method in combination with feedback control theory (DMD-C) to predict wall displacement in large-scale earth-retaining excavation works. The results show that the proposed approach can accurately predict wall displacement and provide a cost-effective alternative to traditional numerical simulations. In summary, this chapter highlights the potential of ROM analysis to contribute to reliability analysis and evaluate system performance in geotechnical engineering. ROMs offer a significant advantage in reducing the computational cost of numerical simulations while preserving the spatiotemporal information contained in the data. Therefore, ROMs have the potential to be a valuable tool for engineers and researchers in the field of geotechnical engineering.

7.1 INTRODUCTION

Geotechnical engineers have recognized the importance of new performance concepts, such as resilience and robustness, in structural design (Bruneau et al., 2003; ISO 2394:2015; Argyroudis et al., 2019; Otake and Honjo, 2022). However, addressing issues related to system performance solely through reliability analysis in geotechnical engineering can be challenging since these concepts apply to a group of structures rather than to a single structure. To contribute to these system performance assessments, it is necessary to incorporate the spatiotemporal features of geotechnical structures into reliability analysis. This approach will enable us to provide information related to recovery time and failure modes (spatiotemporal features) for the target geotechnical structures to the management side of the system, along with any uncertainty.

Therefore, we are interested in the reduced order model (ROM), which was developed in fluid analysis, and its potential application in geotechnical engineering. Analytical methods, such as proper orthogonal decomposition (POD) and dynamic mode decomposition (DMD), extract

DOI: 10.1201/9781003333586-9

eigenmode functions representing the spatiotemporal information of the dynamics under analysis. Low-cost simulations can be performed by extracting only the dominant components from the obtained modes and creating a ROM that reconstructs the original data with a small number of spatial mode functions (i.e., eigenmodes).

The dynamics of interest are fundamentally governed by certain complex partial differential equations in space and time. Reduced order modeling is a technique that involves extracting spatial mode functions from these complex dynamics and transforming physical laws into ordinary differential equations for the expansion coefficients of the mode functions. In other words, ROM is a surrogate model that retains information regarding time and space. Therefore, ROM is expected to be helpful in the rational control of rare events, such as conducting performance evaluations using numerical analyses with high computational costs, an efficient search of rare events, and real-time control during disasters.

This chapter discusses the most promising applications of reliability-based design in geotechnical engineering, including proper orthogonal decomposition (Holmes et al., 1996; Chatterjee, 2000; Berkooz et al., 1993; Kaiser et al., 2014; Fukutani et al., 2021) and dynamic mode decomposition (Schmid, 2010, 2011; Rowley et al., 2009; Kutz et al., 2015; Kaneko et al., 2019; Arai et al., 2021; Shioi et al., 2023). Specifically, Sections 7.1 and 7.2 outline the theoretical background and formulation of POD and DMD. In Section 7.3, a simple example problem is presented using the results of an analysis of a one-dimensional (1D) consolidation equation (heat conduction equation) to introduce the specific features of ROM. Furthermore, Section 7.4 presents a case study that evaluates the seismic behavior of a simple embankment on a liquefiable sand layer (Otake et al., 2022). This case study includes the overall framework of the reliability analysis method using ROM analysis with POD as the key technology. Finally, Section 7.5 applies the proposed framework to predict wall displacement in large-scale earth-retaining excavation works (Saito et al., 2022). A reliability analysis is presented that combines DMD with control (DMD-C) (Proctor et al., 2014), which adds a control term to the DMD presented in Section 7.3 with elastic beam theory on an elastic foundation. This chapter also discusses the possibility of integrating the idea of DMD-C with existing knowledge in civil and geotechnical engineering.

7.2 OUTLINE OF ROM

7.2.1 POD

POD (Holmes et al., 1996; Chatterjee, 2000; Berkooz et al., 2021) is an analytical method used to extract eigenmode functions that represent the spatial characteristics of multidimensional time-series data. This book focuses on the application of ROM in reliability analysis and assumes that the state vector is obtained from a numerical analysis.

A vector of state quantities at an arbitrary time t_k is represented as $\underline{z}(t_k) \in \mathbb{R}^n$. The state vector is a collection of the spatial distributions of the physical quantities to be analyzed, also referred to as snapshots. As the state vectors are observed discretely (i.e., digital output as some numerical analysis results), they are treated as discrete quantities. However, for future prediction, they should be interpreted as continuous quantities. In this book, the state vector is defined by distinguishing between the continuous $\underline{z}(t_k)$ and discrete $\underline{z}_k$ quantities.

$$\underline{z}(t_k) = \underline{z}_k \tag{7.1}$$

where the vectors $\underline{z}(t_k)$ and $\underline{z}_k$ represent physical quantities such as displacements and stresses, respectively, at n grid or integration points at time $t_k \left(= \Delta t \cdot k\right)$, and they are arranged vertically with k being a positive integer. To apply the POD method, a data matrix $\boldsymbol{Z} \in \mathbb{R}^{n \times m}$ is initially defined, where the state vectors are arranged vertically by time, as follows:

$$\boldsymbol{Z} = \begin{bmatrix} | & | & & | \\ \underline{z}_0 & \underline{z}_1 \cdots & & \underline{z}_{m-1} \\ | & | & & | \end{bmatrix} \in \mathbb{R}^{n \times m} \tag{7.2}$$

When constructing the data matrix $\boldsymbol{Z}$, it is common to subtract the time-averaged value of $\underline{z}_k$, denoted by $\frac{1}{m}\Sigma_m(\underline{z}_k)$, in advance. Although not strictly required, this step is desirable because it clarifies the statistical interpretation of POD, which will be discussed later. The resulting data matrix is then transformed into a square matrix using the following operations: ($\boldsymbol{C}_z \in \mathbb{R}^{n \times n}, \boldsymbol{D}_z \in \mathbb{R}^{m \times m}$). Subsequently, the eigenvalues and eigenvectors of this matrix are computed, thereby enabling the extraction of the spatial and temporal features of the target dynamics as a set of orthogonal mode functions.

$$\boldsymbol{C}_z = \boldsymbol{Z}\boldsymbol{Z}^T = \boldsymbol{U}\Lambda_{cz}\boldsymbol{U}^T \tag{7.3}$$

$$\boldsymbol{D}_z = \boldsymbol{Z}^T\boldsymbol{Z} = \boldsymbol{V}\Lambda_{dz}\boldsymbol{V}^T \tag{7.4}$$

where $\boldsymbol{U} \in \mathbb{R}^{n \times n}$ is the eigenvector matrix representing spatial features, known as the POD mode functions, while $\boldsymbol{V} \in \mathbb{R}^{m \times m}$ is the eigenvectors representing the temporal evolution features. Both $\Lambda_{cz} \in \mathbb{R}^{n \times n}$ and $\Lambda_{dz} \in \mathbb{R}^{m \times m}$ are the diagonal matrices of the eigenvalues; the non-zero eigenvalues of both matrices are coincident. The basic idea behind POD is to extract these spatial and temporal evolution features to facilitate low-dimensional simulations. In general, for computational efficiency, the features are typically extracted using the singular value decomposition (SVD) of the data matrix Z, given as follows:

$$\begin{aligned} Z = U\Sigma V^T &= \begin{bmatrix} U_r U_{rem} \end{bmatrix} \begin{bmatrix} \Sigma_r & 0 \\ 0 & \Sigma_{rem} \end{bmatrix} \begin{bmatrix} V_r^T \\ V_{rem}^T \end{bmatrix} \\ &\approx U_r \Sigma_r V_r^T = U_r \Gamma_r = Z_r \end{aligned} \tag{7.5}$$

where $\boldsymbol{U} \in \mathbb{R}^{n \times n}$ is the matrix of columnar left singular vectors and $\boldsymbol{V} \in \mathbb{R}^{m \times m}$ is the matrix of columnar right singular vectors, which matches the orthogonal mode functions of the previous square matrix. In addition, $\Sigma \in \mathbb{R}^{n \times m}$ is a matrix of singular values arranged in diagonal components that coincides with the square root of the eigenvalues of the previous square matrix (Brunton and Kutz, 2022). Based on these equations, the components of each matrix are described as follows:

$$\boldsymbol{U} = \begin{bmatrix} | & | & & | \\ \underline{u}_1 & \underline{u}_2 \cdots & & \underline{u}_n \\ | & | & & | \end{bmatrix} \tag{7.6}$$

$$\Sigma = \begin{bmatrix} \sigma_1 & & & 0 \\ & \sigma_2 & & \\ & & \ddots & \\ 0 & & & \sigma_n \end{bmatrix} = \Lambda_{cz}^{1/2} \tag{7.7}$$

$$\boldsymbol{V} = \begin{bmatrix} | & | & & | \\ \underline{v}_1 & v_2 \cdots & & \underline{v}_m \\ | & | & & | \end{bmatrix} \tag{7.8}$$

where $\underline{u}_j \in \mathbb{R}^n$ and $\underline{v}_j \in \mathbb{R}^m$ denote the jth eigenvectors, while σ_j is the jth singular value. The ROM is constructed by truncating the number of POD mode functions used in the simulation to "r". Subscript r in each matrix refers to the submatrix used in the simulation, whereas subscript rem refers to the truncated submatrix. Considering the contribution rate (Γ) of the POD mode function, the matrix product $\Gamma = \Sigma V \in \mathbb{R}^{n\times m}$ yields the expansion coefficients of each POD mode function. The data matrix Z is approximately reconstructed by the truncated POD mode function $\boldsymbol{U}_r \in \mathbb{R}^{n\times r}$ and the expansion coefficient matrix $\Gamma_r \in \mathbb{R}^{n\times r}$. Singular value decomposition is mathematically equivalent to the principle component analysis (PCA) (Pearson, 1901) and the Karhunen–Loeve (KL) expansion (KLT) (Karhunen, 1947; Stark and Woods, 1986). The reduced order data matrix $\boldsymbol{Z}_r = \boldsymbol{U}_r \Gamma_r$ can be expressed by Eq. (7.9), which can be interpreted as finding the spatial feature $\boldsymbol{U}_r$ and the time evolution feature $\Gamma_r \in \mathbb{R}^{n\times r}$ simultaneously, with the minimum least squares error.

$$\arg\min_{\tilde{Z}} \left\| \boldsymbol{Z} - \boldsymbol{Z}_r \right\|_F = \boldsymbol{U}_r \Gamma_r \quad s.t. \operatorname{rank}\left(\boldsymbol{Z}_r\right) = r \tag{7.9}$$

where $\|\cdot\|_F$ is the Frobenius norm. Based on the above preparations, the vector of the physical quantities of interest $\underline{z}(\boldsymbol{x}, t_k)$ for all grid points (integration points) at a certain time t_k can be approximated by Eq. (7.10):

$$\underline{z}(\boldsymbol{x}, t_k) = \overline{\underline{z}}(\boldsymbol{x}, t_k) + \sum_{j=1}^{r} \underline{u}_j(\boldsymbol{x}) \gamma_j(t_k) \tag{7.10}$$

where $\overline{\underline{z}}(\boldsymbol{x}, t_k)$ is the time average of the physical quantity of interest calculated prior to POD implementation; $\underline{u}_j(\boldsymbol{x})$ is the jth POD mode function; and $\gamma_j(t_k)$ is the expansion coefficient representing the magnitude of the mode at a certain time t_k and is the (j, t_k) component of Γ_r.

7.2.2 DMD

In POD, as explained previously, time and space correlations are computed concurrently, resulting in time and space modes that are often poorly separated, and multiple mode combinations may be necessary to capture a single system frequency. Moreover, understanding the time evolution features from POD is often challenging since they are obtained solely as a series of expansion coefficients. On the other hand, DMD (Schmid, 2010, 2011; Rowley et al., 2009; Kutz et al., 2015; Kaneko et al., 2019; Arai et al., 2021) is a mode decomposition method that focuses on the temporal evolution of time-series data. The eigenmodes, which represent the spatial characteristics, and the associated frequency features of the temporal evolution are approximated by an exponential function in DMD. Consequently, DMD has the advantage of providing information about mode stability, including divergence, damping, and oscillation over time. The fundamental theory of DMD is concisely and comprehensively explained in the references (Procter et al., 2014; Brunton and Kutz, 2022; TTRS). In the following, we provide a summary of the basic theory of DMD as presented in these references.

DMD starts with the construction of data matrices $\boldsymbol{Z}' \in \mathbb{R}^{n\times(m-1)}$ and $\boldsymbol{Z}'' \in \mathbb{R}^{n\times(m-1)}$ by shifting the same physical quantity as POD by one time step, as shown in Eqs (7.11) and (7.12).

$$\boldsymbol{Z}' = \begin{bmatrix} | & | & & | \\ \underline{z}_0 & \underline{z}_1 & \cdots & \underline{z}_{m-2} \\ | & | & & | \end{bmatrix} \in \mathbb{R}^{n\times(m-1)} \tag{7.11}$$

$$Z'' = \begin{bmatrix} | & | & & | \\ \underline{z}_1 & \underline{z}_2 & \cdots & \underline{z}_{m-1} \\ | & | & & | \end{bmatrix} \in \mathbb{R}^{n \times (m-1)} \tag{7.12}$$

The above equation represents a matrix of $(m - 1)$ vectors that are shifted by one time step from the original matrix. The correlation (covariance) structure of Z' and Z'' is utilized to determine the time evolution.

$$Z'' \approx AZ'; Z''(Z')^T \approx A \tag{7.13}$$

However, the time evolution of the dynamics of interest generally follows a non-linear function with non-stationarity, while DMD approximates the time evolution of the dynamics of interest to a linear matrix (linear map) with stationarity. Thus, the simple DMD approach may not always be applicable to all problems, and it may not be feasible to derive an appropriate reduced model based on the characteristics of the physics under consideration. To address this challenge, Koopman operator theory (Koopman, 1931) provides a generalized approach. Koopman operators are linear operators that retain information about the original non-linear dynamical system, allowing the analysis of non-linear dynamical systems based on the spectral properties of the linear operators. Although research on this subject has been active in recent years (Rowley et al., 2009; Mezić, 2013), generalized methods have not yet been established, and this problem is beyond the scope of this book.

Next, we describe the characteristics of DMD using a continuous dynamical system. Generally, the time evolution of the n-dimensional state vector at time t, denoted by $z(t) \in \mathbb{R}^n$, is governed by a non-linear and non-autonomous function $f(\cdot)$,

$$\frac{d\underline{z}}{dt} = f(\underline{z}, t; \underline{\beta}) \tag{7.14}$$

where $\underline{\beta}$ is the parameter vector of the dynamical system. However, since the equation cannot be solved analytically, DMD approximates it by linearizing it in a suitable manner and transforming it into an autonomous system represented by

$$\frac{d\underline{z}}{dt} \approx A\underline{z} \tag{7.15}$$

where A is the time transition matrix of the state vector. Since this equation is a first-order linear differential equation, it can be solved analytically as follows:

$$\underline{z}(t) = \underline{z}(0)\exp(At) \tag{7.16}$$

DMD constructs a reduced order model by approximately computing the eigenvalues and eigenmode functions of A from the data. Since the time evolution structure is determined from the data, it provides a natural approach to predict the time evolution of the target dynamics. In addition, by analyzing the eigenvalues of A, the stability of the system can be assessed, as is generally the case in the dynamics of continuous systems.

The matrix can also be expanded to a discrete dynamical system, given by

$$\underline{z}_{k+1} = A_d \underline{z}_k = A_d{}^k \underline{z}_0 \tag{7.17}$$

where $\underline{z}_k \in \mathbb{R}^n$ is the state vector at time k and A_d is the time transition matrix of the discrete system. By computing the powers of A_d, we can obtain the state vector $\underline{z}_k$ at any time. Then, to distinguish between the continuous and discrete systems, we describe their respective time evolution matrices as (A_c, A_d), respectively.

$$\begin{aligned} \underline{z}\left((k+1)\Delta t\right) &= A_d \underline{z}\left(k\Delta t\right) \\ &= \exp\left(A_c \Delta t\right) \underline{z}\left(k\Delta t\right) \end{aligned} \tag{7.18}$$

Furthermore, considering that the discrete time is integer-valued, we can derive

$$\begin{aligned} A_d &= \exp\left(A_c \Delta t\right) \\ &= U_c \exp(\Lambda_c \Delta t) U_c^{-1} \end{aligned} \tag{7.19}$$

where $U_c \in \mathbb{C}^{n \times n}$ is the matrix aligning the eigenvectors of A_c; and $\Lambda_c \in \mathbb{C}^{n \times n}$ is the corresponding eigenvalue diagonal matrix. Both sides of Eq. (7.19) can be diagonalized by the eigenvectors U_c, yielding

$$\begin{aligned} \Lambda_d &= U_c^{-1}\left(U_c \exp(\Lambda_c \Delta t) U_c^{-1}\right) U_c \\ &= \exp(\Lambda_c \Delta t) \end{aligned} \tag{7.20}$$

where $\Lambda_d \in \mathbb{C}^{n \times n}$ is a diagonal matrix of the eigenvalues of the discrete system aligned in diagonal terms. Therefore, the eigenvalues of the continuous system can be converted from the eigenvalues of the discrete system using

$$\lambda_{c,j} = \frac{\ln \lambda_{d,j}}{\Delta t} = \omega_j \tag{7.21}$$

where $\lambda_{c,j}$ is the eigenvalue of the jth continuous system and $\lambda_{d,j}$ is the eigenvalue of the corresponding discrete system. Based on the properties of the eigenvalues and eigenvectors of the discrete system, the stability of the system under consideration can be determined based on the dynamics of the continuous system. The eigenvalues of the continuous system are denoted by ω_j.

Based on the preceding preparation, the process of identifying the time evolution matrix A from the data matrix is described. As the data matrix Z may not necessarily be non-singular, determining A satisfies the objective function in Eq. (7.22).

$$A \triangleq \underset{A}{\arg\min} \parallel Z'' - AZ' \parallel_F \tag{7.22}$$

Therefore, based on the SVD of Z, A is determined by the following process:

$$\begin{aligned} Z' &= U\Sigma V^* \\ A &\approx Z'' V_r \Sigma_r^{-1} U_r^* \end{aligned} \tag{7.23}$$

where * is the complex conjugate transpose and subscript r denotes the number of ranks reduced during the SVD.

The next step involves finding the eigenvalues and eigenvectors of the matrix A. However, solving the eigenvalue problem in its current form is not recommended due to the large size of matrix A in practical calculations. Therefore, projecting the matrix $\tilde{A} \in \mathbb{R}^{r \times r}$ onto POD allows obtaining U_r, and solving the eigenvalue problem for this matrix reduces the computational cost.

$$\tilde{A} = U_r^* A U_r = U_r^* Z'' V_r \Sigma_r^{-1} \tag{7.24}$$

$$\tilde{A} W = W \Lambda \tag{7.25}$$

The eigenvalue matrix W is identical to the original matrix $\tilde{A}$ due to transformation similarities. Equation (7.26) describes the DMD mode obtained based on the aforementioned process:

$$\Phi_r = Z''V_r\Sigma_r^{-1}W \tag{7.26}$$

This method of DMD mode identification is called Exact DMD (Tu et al., 2014). In contrast, the original DMD employed the method obtained from Projected DMD ($\Phi = U_rW$) (Schmid, 2010; Schmid et al., 2011). However, in recent years, the above method has commonly been applied as Exact DMD is considered to capture the time evolution characteristics of the data more rigorously. Exact DMD is calculated using both Z' and Z'', whereas Projected DMD uses only Z'.

The prediction model for discrete dynamical systems is described by

$$\underline{z}_k = \Phi_r\Lambda^k\Phi\underline{z}_0 \tag{7.27}$$

Based on the correspondence between the discrete and continuous systems, $\Omega = \ln(\Lambda)/\Delta t$, Eq. (7.27) can be converted into a prediction model for the state quantity $\underline{z}(t)$ of a continuous dynamical system at any time.

$$\begin{aligned}\underline{z}(t) &= \Phi_r\exp(\Omega t)\Phi^{\dagger}\underline{z}_0 \\ &= \Phi_r\exp(\Omega t)\underline{b}\end{aligned} \tag{7.28}$$

where $\underline{b} = \Phi^{\dagger}\underline{z}_0$ is defined as the initial vector of expansion coefficients for the DMD mode. Finally, like POD, the mode expansion of the DMD mode is given by

$$\begin{aligned}\underline{z}(t) &= \sum_{j=0}^{r}\underline{\phi}_j(x)b_j(t) \\ &= \sum_{j=0}^{r}\underline{\phi}_j(x)b_j(0)\exp(w_jt)\end{aligned} \tag{7.29}$$

Unlike POD, which directly models snapshots of a system, DMD models the system's time evolution. In principle, if a model is built using data for one period of the target dynamics, it can predict the behavior for any number of subsequent periods. Furthermore, DMD can be applied to extrapolation and interpolation problems, making it a more advantageous predictive model than POD. However, the DMD formulation requires that the geophysical system under consideration be stationary and linear, while the non-linearity of geomaterials poses a major challenge in predicting geotechnical behavior, as previously discussed. Additionally, an alternative method for calculating the contribution rate of each mode involves taking the diagonal elements of matrix $\boldsymbol{D}$ as the corresponding DMD mode's contribution. This method is similar to that used for POD and is described by

$$Z \approx \Phi_r D V_{and} \tag{7.30}$$

$$\Phi_r = \begin{bmatrix} | & | & & | \\ \underline{\phi}_1 & \underline{\phi}_2 & \cdots & \underline{\phi}_r \\ | & | & & | \end{bmatrix} \in \mathbb{C}^{n\times r} \tag{7.31}$$

$$D = \begin{bmatrix} d_1 & & & 0 \\ & d_2 & & \\ & & \ddots & \\ 0 & & & d_r \end{bmatrix} \in \mathbb{C}^{r\times r} \tag{7.32}$$

$$V_{and} = \begin{bmatrix} \lambda_1^{(0)} & \lambda_1^{(1)} & \cdots & \lambda_1^{(m-1)} \\ \lambda_2^{(0)} & \lambda_2^{(1)} & \cdots & \lambda_2^{(m-1)} \\ \vdots & \vdots & \ddots & \vdots \\ \lambda_r^{(0)} & \lambda_r^{(1)} & \cdots & \lambda_r^{(m-1)} \end{bmatrix} \in \mathbb{C}^{r\times(m-1)} \tag{7.33}$$

7.3 CHARACTERIZATION OF ROM BASED ON THE ANALYSIS OF LINEAR AND HOMOGENEOUS DIFFERENTIAL EQUATIONS

7.3.1 Problem Setting

The clay layer was assumed to be horizontally stratified, as illustrated in Figure 7.1, with sand layers stratified above and below the clay layer. It was assumed that the clay layer was drained on both sides. An equally distributed static load of p_0 (kN/m^2) was applied to the surface of the top sand layer, resulting in the excess pore water pressure reaching p_0 throughout the entire clay layer before dissipating. The spatial distribution of excess pore water pressure followed Terzaghi's consolidation equation (Terzaghi, 1925), which represents a one-dimensional heat conduction equation and a linear homogeneous equation in the depth direction. We applied the POD and DMD techniques to the time-series variation of the depth profile of excess pore water pressure as the state vector.

Terzaghi's consolidation equation is expressed as follows:

$$\frac{\delta u(x,t)}{\delta t} = c_v \frac{\partial^2 u(x,t)}{\partial t^2} \tag{7.34}$$

where $u(x,t)$ is the excess pore water pressure u (kN/m^2) at depth x (m) and time t (s); and c_v is the consolidation coefficient (m^2/day), set to 0.025. We assumed double-sided drainage ($H' = 2H = 10$ m), where H' is the thickness of the clay layer, and H is the drainage distance of the clay layer. We assumed that the excess pore water pressure in the entire clay layer at the beginning of the analysis was equal to the top load $p_0 = 1$ kN/m^2.

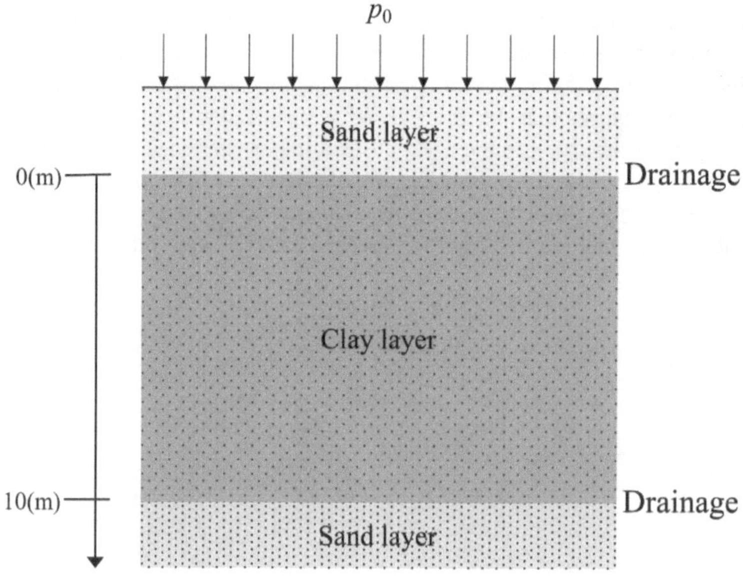

FIGURE 7.1 Analytical model applying the 1D consolidation equation.

$$u(0,t) = u(2H,t) = 0 \tag{7.35}$$

$$u(x,0) = p_0 = 1\left(\text{kN}/\text{m}^2\right) \tag{7.36}$$

We obtained the analytical solution of the consolidation equation for the above boundary and initial conditions based on the Fourier series expansion.

$$\begin{aligned} u(x,t) &= \sum_{i=1}^{\infty} \frac{4p_0}{(2i-1)\pi} \cdot \sin\frac{(2i-1)\pi x}{2H} \cdot \exp\left[\left\{-\frac{(2i-1)\pi}{2H}\right\}^2 c_v t\right] \\ &= \sum_{i=1}^{\infty} \psi_i(x)\Upsilon_i(t) \end{aligned} \tag{7.37}$$

Equation (7.37) can be expanded as

$$\begin{aligned} u(x,t) &= \frac{4p_0}{\pi}\sin\frac{\pi x}{2H}\exp\left[\left(-\frac{\pi}{2H}\right)^2 c_v t\right] \\ &+ \frac{4p_0}{3\pi}\sin\frac{3\pi x}{2H}\exp\left[\left(-\frac{3\pi}{2H}\right)^2 c_v t\right] \\ &+ \frac{4p_0}{5\pi}\sin\frac{5\pi x}{2H}\exp\left[\left(-\frac{5\pi}{2H}\right)^2 c_v t\right] \\ &+ \cdots \end{aligned} \tag{7.38}$$

We divided the analytical model into elements at 0.1 m intervals in the depth direction, and we considered the spatial distribution of excess pore water pressure for $n = 101$. The excess pore water pressure was solved in 51 steps ($m = 51$) with $\Delta t = 16$ days and stored in the state quantity matrix $z_k \in \mathbb{R}^n$.

$$z_k = z(x, t_k) = u(x, t_k) \tag{7.39}$$

$$Z = [z_1 \; z_2 \cdots z_m] \tag{7.40}$$

Next, we applied POD and DMD to this state quantity data matrix and evaluated the accuracy of the reconstruction of the analytical solution.

7.3.2 Result and Consideration

In this section, we present Figure 7.2, which shows the analytical solution of the consolidation equation. As mentioned earlier, the consolidation equation is a homogeneous linear differential equation, and its solution is described as a superposition of Fourier series expansions $\psi_i(x)$.

$$\psi_i(x) = \frac{4p_0}{(2i-1)\pi} \cdot \sin\frac{(2i-1)\pi x}{2H} \tag{7.41}$$

Figure 7.2b also shows the time evolution $\Upsilon_i(t)$ (exponential function) of each Fourier series.

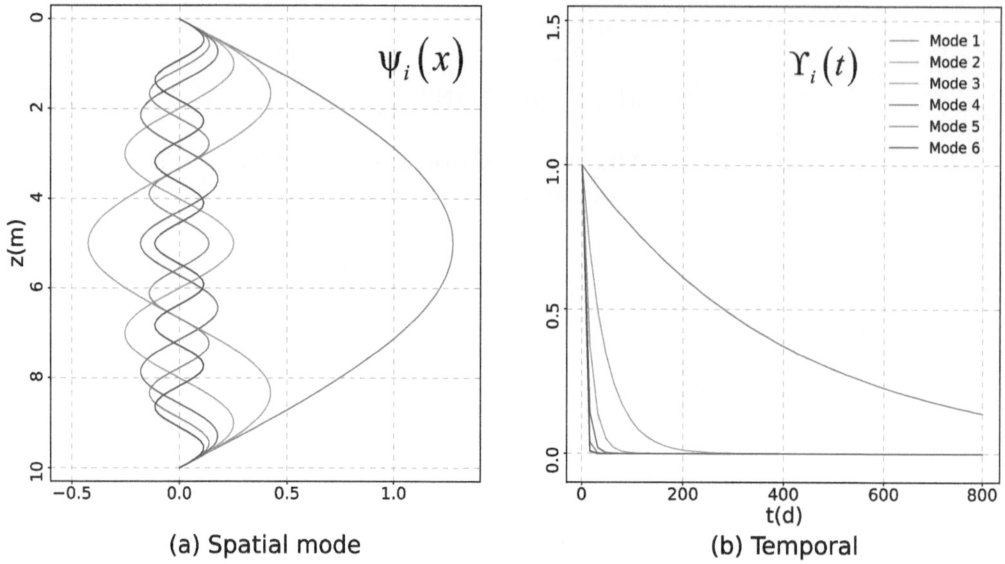

FIGURE 7.2 Analytical solution of 1D consolidation equation.

$$\Upsilon_i(t) = \exp\left[\left\{-\frac{(2i-1)\pi}{2H}\right\}^2 c_v t\right] \tag{7.42}$$

The analytical solution of the consolidation equation is known to be characterized by the fact that $i = 1$ (Mode 1) is dominant and there is rapid damping of the other spatial features. Figure 7.3 shows the results of mode decomposition by POD. Figure 7.3a and b show $\sigma_i \cdot u_i(x)$ and v_i, respectively. Similarly, Figure 7.4 shows the results of mode decomposition by DMD, depicting $b_i\Phi_i(x)$ and $\exp(w_i t)$ in Figure 7.4a and b, respectively. Both POD and DMD are dominated by $i = 1$ (Mode 1), which is a characteristic of the consolidation equation solution.

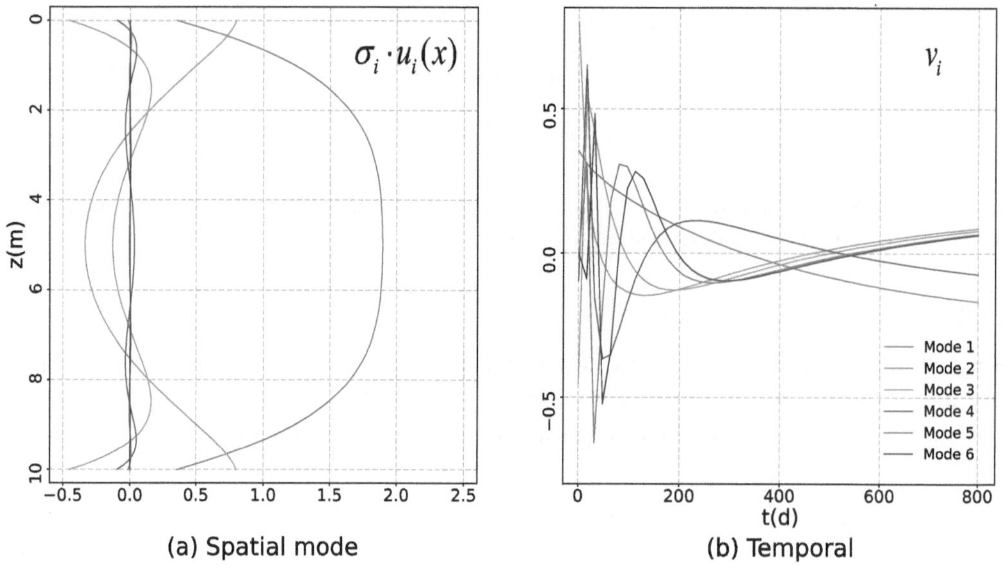

FIGURE 7.3 Data-driven solution by POD mode decomposition.

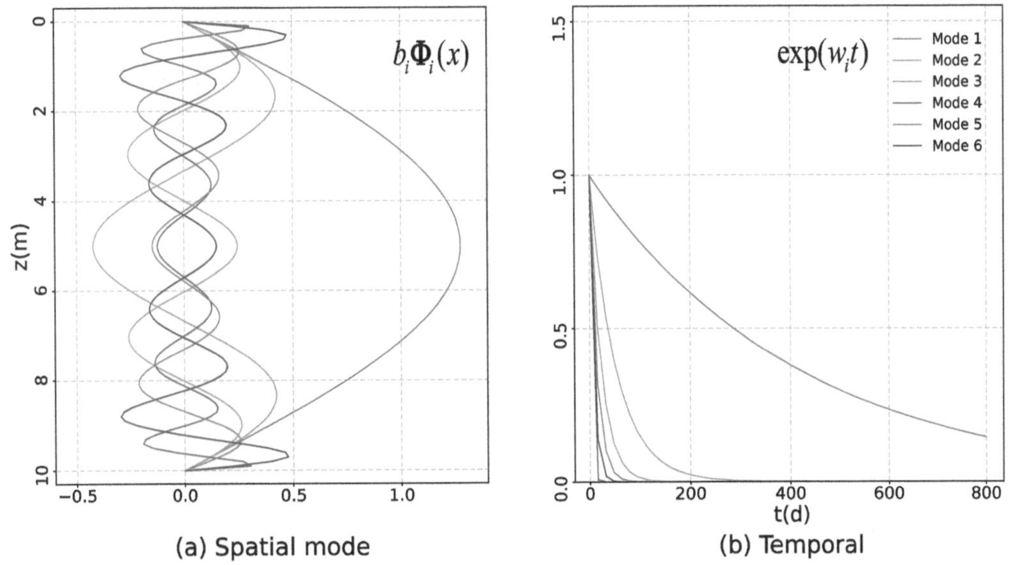

FIGURE 7.4 Data-driven solution by DMD mode decomposition.

For DMD, we observed that both the spatial distribution and time evolution characteristics are in good agreement with the analytical solution. As mentioned earlier, POD extracts the spatial features of snapshots through singular value decomposition and reconstructs them through the time evolution of their component coefficients. Therefore, the analytical solution is difficult to recover because the time evolution structure is not functionalized.

However, the dynamic mode decomposition method was initially developed based on the assumption of linear homogeneous differential equations representing time-invariant systems, which allows for an analytical solution. However, this assumption of stationarity and linearity may not hold for the physics of interest. Proper orthogonal decomposition is a more general method compared to DMD and is considered advantageous in terms of computational simplicity. While DMD is considered an equation-discovering model for differential equations, assuming both extrapolation and interpolation, it models the time evolution characteristics. In both POD and DMD simulations, the number of mode functions, denoted by r, is generally significantly reduced compared to the n and m dimensions. Both methods can be considered approximations of ordinary differential equations whose time evolution embeds the spatial information of time- and space-dependent partial differential equations governing the physics of interest in the form of a small number of spatial mode functions.

7.4 APPLICATION IN THE PREDICTION OF EMBANKMENT SETTLEMENT ON LIQUEFIED GROUND

7.4.1 Outline of the Proposed Framework

We strive to develop reliability analyses and risk assessments that retain the spatiotemporal information of the target structures. In the future, we aim to develop a new design framework that mathematically controls the failure modes of the target structures. The basis of this framework should follow the two-stage seismic design approach developed in the field of structural engineering in Japan that uses methods such as the horizontal load-bearing capacity method applied to earthquakes. This design method limits the plasticized areas of a structure and guides the failure mode of the structural system toward one that is easily repairable. The concept includes the construction of a robust structural system for rare events for which design conditions are difficult to ascertain. Since the failure modes of geotechnical structures, particularly soil structures, such as shear and bending failures, are difficult to categorize, the introduction of sophisticated numerical analyses (i.e., dynamic effective stress analysis) to directly recognize the failure mode is essential. However, these analyses are computationally intensive in the design calculations.

Otake et al. (2022) attempted to create an efficient framework for reliability analyses while preserving the spatiotemporal information contained in the numerical analysis using ROM analysis. However, evaluating the influence of other important factors such as the uncertainty of external forces, the spatial distribution of soil profiles, and model errors in numerical analyses is a future endeavor and beyond the scope of this book. Figure 7.5 shows the proposed reliability analysis framework. This section provides an outline of the framework and clarifies issues related to it. The proposed analysis framework is divided into two main parts: "initial data processing" and "analysis". This framework provides a simplified reliability analysis framework with spatiotemporal features that can evaluate failure modes and follows traditional simplified reliability analysis concepts, namely the first-order second-moment method (FOSM; Cornell, 1969) and the first-order reliability method (FORM; Hasfer, 1974), which are based on a linear approximation of the mean values and design point through ROM analysis.

1. **Initial data processing**

Our aim is to collect input parameter sets for the constitutive laws used in numerical analyses from reliable experts that encompass a diverse range of conditions. Each input parameter set is modeled as a multivariate normal distribution, and its eigenvalues and eigenvectors are calculated using principal component analysis (PCA) of the covariance matrix. We reduce the information to lower-dimensional data by considering reconstruction using only the main eigenvectors. A detailed description of the initial data processing is not included in this book due to space limitations, but interested readers can refer to Otake et al. (2022).

2. **Analysis**

We use the multivariate normal distribution obtained during the initial data processing as a prior distribution for performing Bayesian analysis based on partial information and constraint conditions obtained on-site to produce a posterior distribution (multivariate normal distribution) of the input parameters that reflect the on-site features. We perform design experiments (DEs) using this posterior distribution and obtain numerical analysis time-series data (i.e., a collection of snapshots from multiple cases based on the time point) of various cases (i.e., numerical experiments). We calculate the eigenvalues and eigenvectors by using proper orthogonal decomposition of the numerical analysis time-series data covariance matrix.

Although PCA and POD are consistent in theory, we distinguish between them in this book. We use PCA for the input parameter set and POD for the numerical analysis of the time-series data. We observe and consider the features that appear in the main modes and component coefficients, similar to the engineering features of the input parameters. Furthermore, we design a reduced order model that considers the spatiotemporal information by conducting a simple linear regression analysis (i.e., linear combination) of the reduced number of dimensions of the input parameters and the spatial distribution of the analysis result (i.e., response value). We can analyze uncertainty propagation using the expected value characteristics by expressing them using a linear combination. We perform a reliability analysis (uncertainty propagation analysis of the input parameter setting uncertainty) using the expected value characteristics by dimensionally reducing the input parameter set and analysis results and by utilizing their linear combination.

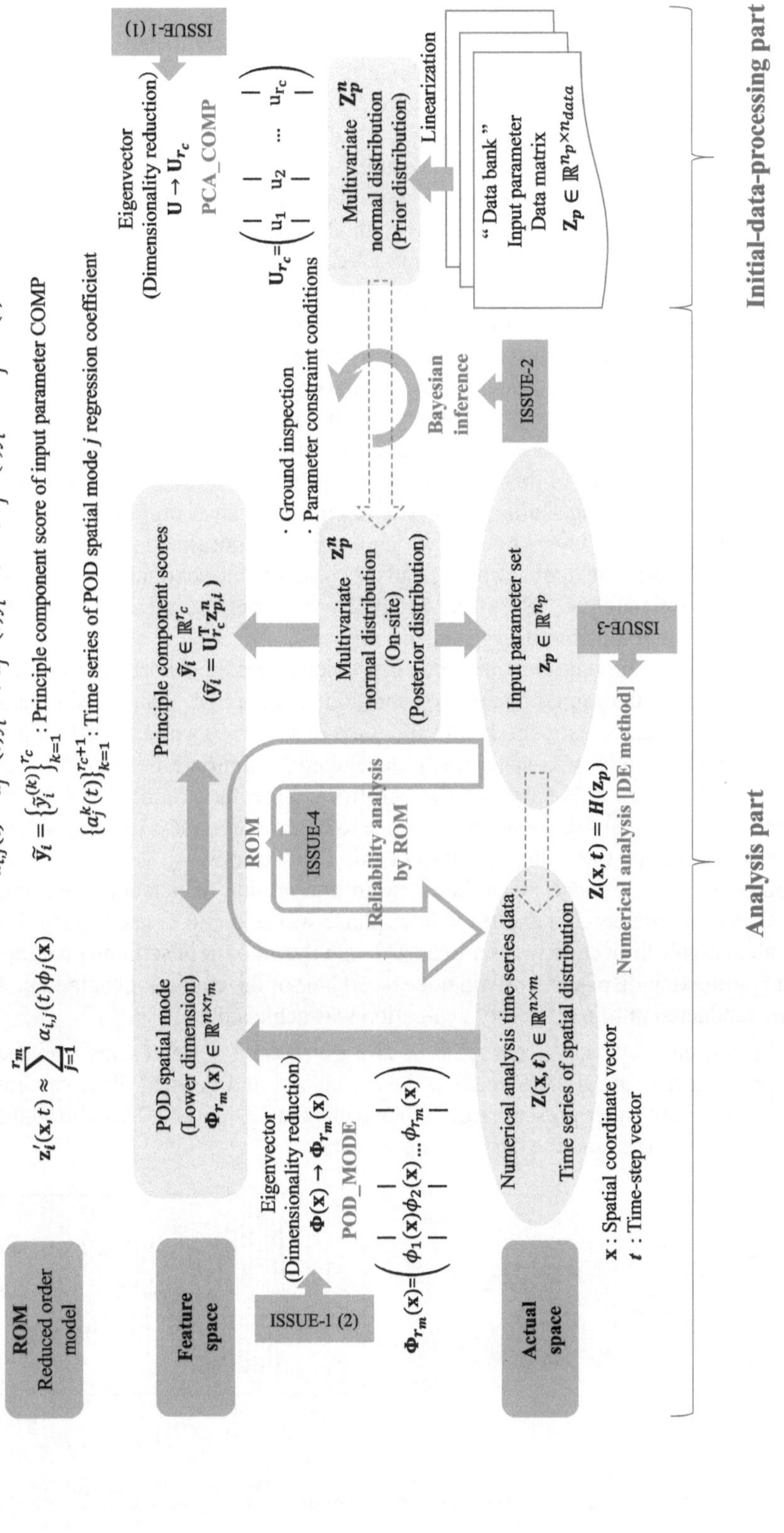

FIGURE 7.5 Proposed framework of geotechnical reliability analysis with ROM. (Source: Otake et al., 2022)

7.4.2 Target Structure and Basic Analysis Method

Figure 7.6a shows a finite element model (FEM) of a simple soil section consisting of 464 nodes and 424 elements. This model was developed for dynamic centrifugal model tests (50 G) at the Public Works Research Institute (PWRI), Japan, and employed the effective stress dynamic FEM (Liquefaction analysis method LIQCA development group, 2007; Oka et al., 1994, 1999). This numerical analysis requires physical input parameters that can be measured and fitting parameters that are difficult to set objectively. However, it can realistically evaluate the seismic behavior (spatial distribution of excess pore water pressure, acceleration, velocity, and displacement) of soil structures such as riverbank embankments on a liquefiable sand layer, if the parameters are appropriately set (Oka et al., 1994, 1999; Otsushi et al., 2010a, 2010b; Otake and Honjo, 2012). The analysis time is relatively long, and surrogate model calculations are essential for the reliability analysis. However, 14–15 input parameter sets are required, and connection to reliability analysis is difficult because these sets have a spatial distribution, which is not addressed in this document. The spatial variability and statistical estimation error of the geotechnical parameter are important uncertainties, and it is challenging to integrate a reduced order model with this problem. Thus, this issue falls beyond the scope of this book. Future research will focus on this point. The geotechnical parameters in this book are fixed for each layer, and the simulation of information compression of the numerical results is the focus of this book. The dimensions of Figure 7.6a were converted to the scale of an actual structure using scaling laws. The topmost soil layer represents an embankment, followed by a liquefiable sand layer (both of which comprise Edosaki sand set to approximately D_r=50% immediately below the embankment) and an unliquefiable sand layer (consolidated No. 7 silica sand set to approximately D_r=90%). Additionally, the experimental soil vessel's boundaries were set as rigid walls that did not undergo shear deformation.

To match the experimental requirements, the FEM's sides were fixed horizontally and free vertically, and a bottom-fixed undrained boundary condition was applied. Plane strain elements were utilized, and the cyclic elastoplastic model of sand (Oka et al., 1999) was adopted for all elements. The cyclic elastoplastic model was specifically developed to simulate liquefaction using a generalized non-associated flow rule based on the non-linear kinematic hardening rule. This model considers over-consolidation of the boundary surface to control dilatancy, the plastic strain dependency of shear stiffness, and the fading memory of initial anisotropy.

Figure 7.6b presents the time series of the input ground motion, which consisted of a principal shock of 1.5 G (approximately 5 s) and three later-phase waves of 1.0 G used as rigid foundation inputs. The calculation's limit state was set to ensure that the amount of settlement at the top of the embankment (y-direction displacement) did not exceed half of the embankment height. A series of analyses were conducted until a reliability evaluation was achieved.

The state vector, denoted as $z_k = z(x, t_k) \in \mathbb{R}^n$ at a given time t_k, represents a snapshot of the FEM. It comprises three partial state vectors: the horizontal and vertical displacements of each node, and the excess pore water pressure ratio of each element. Equation (7.43) shows the relationship between the state vector and the partial state vectors:

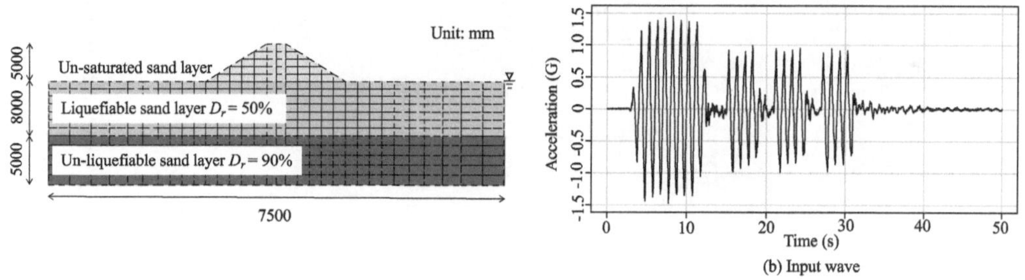

FIGURE 7.6 Target structure diagram of the (a) FEM and (b) input wave. (Source: Otake et al., 2022)

$$z_k = z(x,t_k) = \left[z_{d_x}^T(x_n,t_k) \quad z_{d_y}^T(x_n,t_k) \quad z_{epwp}^T(x_e,t_k) \right]^T \tag{7.43}$$

where $z_{d_x}(x_n,t_k)$, $z_{d_y}(x_n,t_k)$, and $z_{epwp}(x_e,t_k)$ are the spatial coordinate vectors of the FEM nodes, FEM elements, and the node and element that correspond to the partial state vector, respectively. The total dimension number of the partial state vector, denoted as n, is equal to the number of nodes n_n multiplied by two, plus the number of elements n_e. The analysis results are obtained from time t_1 to t_m and output at an interval of Δt. At each time point, the state vector $z(x,t_1)$, $z(x,t_2)$, ..., $z(x,t_m)$ is entered into the data matrix $Z \in \mathbb{R}^{n \times m}$ in the order of the time series shown in the following equation:

$$Z = \left[z(x,t_1) \quad z(x,t_2) \quad \cdots \quad z(x,t_m) \right] \tag{7.44}$$

7.4.3 Results and Considerations

1. **Mode decomposition of numerical analysis results**

The contribution of POD spatial mode 1 was extremely high, providing 98% of the information, followed by POD spatial mode 3. Figure 7.7 shows the POD spatial modes 1–5 in ascending order of eigenvalues. The deformation modes (i.e., horizontal and vertical displacements) are represented as nodal displacements, and the excess pore water pressure ratio modes are shown as the shade of coloring of the elements. Figure 7.8 shows the relationship between the time-series component coefficient of POD spatial modes 1–3 and the input waveform.

As shown in Figure 7.7a, the embankment in POD spatial mode 1 exhibits settling tendencies and high excess pore water pressure tendencies in all parts except those immediately below the embankment body where the confining pressure is high. Figure 7.8 indicates that the time-series component coefficient controls the embankment behavior after 12 s, which is when the first wave ends (i.e., when most of the foundation ground has liquefied). This can be interpreted as a fundamental settlement mode that governs the target structure and accompanies the embankment foundation ground liquefaction, and is henceforth referred to as the "Settlement-MODE".

As seen in Figure 7.7b, POD spatial mode 2 exhibits settling tendencies like those of POD spatial mode 1; however, the excess pore water pressure ratio behavior differs. The excess pore water pressure ratio is particularly high in the local areas at the foot of the embankment body slope. The time-series component coefficient in Figure 7.8 indicates significant contributions to the embankment behavior during the 12 s from the start of the vibrations to the end of the first wave when most of the foundation ground had liquefied. However, its subsequent contributions were small. This is referred to as the "Undrained shear-MODE", where shearing is repeated under undrained conditions.

As seen in Figure 7.7a–c, POD spatial mode 3 differed from POD spatial modes 1 and 2, and mode 3 did not contribute to the residual displacement of the embankment body. The lower foot of the slope and the fixed-side boundary areas had high excess pore water pressure ratios, and this is referred to as the "Vibration-MODE".

As shown in Figure 7.7, the physical meanings of POD spatial modes 4 and 5 are challenging to interpret from an engineering perspective. Therefore, the alternative model constructed using POD spatial modes 1–3 and the number of dimensions (i.e., horizontal displacement n_n, vertical displacement n_n, and excess pore water pressure ratio n_e dimensions) significantly decreased from $2n_n + n_e = 1352$ to 3.

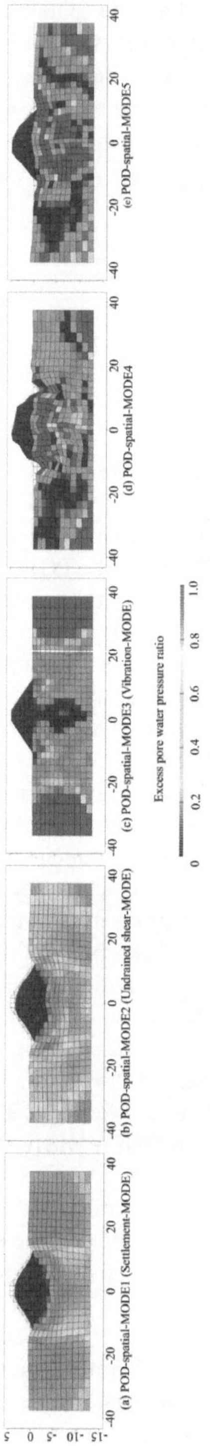

FIGURE 7.7 Principal POD mode functions. (Source: Otake et al., 2022)

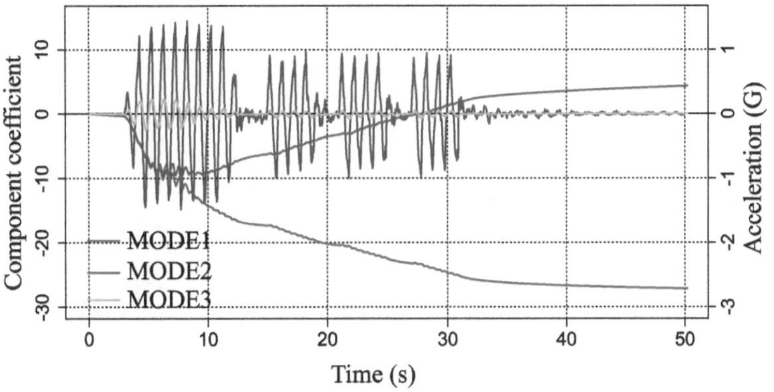

FIGURE 7.8 POD mode temporal component coefficient time series. (Source: Otake et al., 2022)

2. **Reliability analysis using POD ROM**

Figure 7.9 illustrates the spatial distribution of deformation plots and excess pore water pressure ratios (represented by a color gradient in each element) at time points of 5, 7.5, 10, 20, 30, and 40 s. The ROM simulation results, obtained under the same conditions as those shown in Figure 7.9a, are depicted in Figure 7.9b. Both the FEM analysis and ROM simulations are qualitatively consistent with each other at each time point. The results of applying a time-series correction factor to each POD spatial mode are shown in Figure 7.9c and d. Figure 7.9b was obtained by combining these three POD spatial modes.

Figure 7.10a focuses on the deformation of the top of the embankment body and the excess pore water pressure ratio immediately below the embankment body for the average value case of the prior distribution (DE-1). This figure compares the time-series behavior of the FEM analysis and ROM simulation. From left to right, Figure 7.10a shows the positions in the embankment where the structural displacements and time series are compared, the time series of the foundation input waveform, the time series of the excess pore water pressure ratio immediately below the embankment, the time series of the horizontal displacement (x-displacement) of the top end of the embankment body, and the time series of the vertical displacement (y-displacement) of the top of the embankment body. The FEM analysis and ROM simulation results are represented by solid and dotted lines, respectively. The results confirm that the ROM accurately simulates the FEM analysis results at each time point. Figure 7.10b shows the same comparisons for DE-2. Minor deviations are observed between the FEM analysis and ROM simulation results in the time series of the excess pore water pressure ratio. Overall, the ROM simulations accurately characterize the features of the time-series behavior.

Figure 7.11a presents the results of the ROM simulation for the average value of the prior distribution (DE-1) and the uncertainty propagation analysis. The uncertainty analysis results are displayed as a range of $\pm 1\sigma$ (hatched areas) of the ROM analyses (grey hatching). Figure 7.11a illustrates, from left to right, the embankment structure, the time series of the excess pore water pressure ratio just below the embankment, the time series of the horizontal displacement of the top of the embankment (x-direction), the time series of the vertical displacement of the top of the embankment (y-direction), and the failure probability with one-half of the embankment height as the settlement limit. The failure probability was determined at an arbitrary time point by setting the displacement limit values as random variables and by an analytical calculation using the expected value characteristics (as employed in FOSM and FORM). The failure probability increased dramatically between the second and third later-phase waves and reached approximately 10^{-1} at the end of the analysis (50 s). Similar to Figure 7.11a, Figure 7.11b shows the results of the ROM and the uncertainty propagation analyses corresponding to the posterior distribution based on a portion

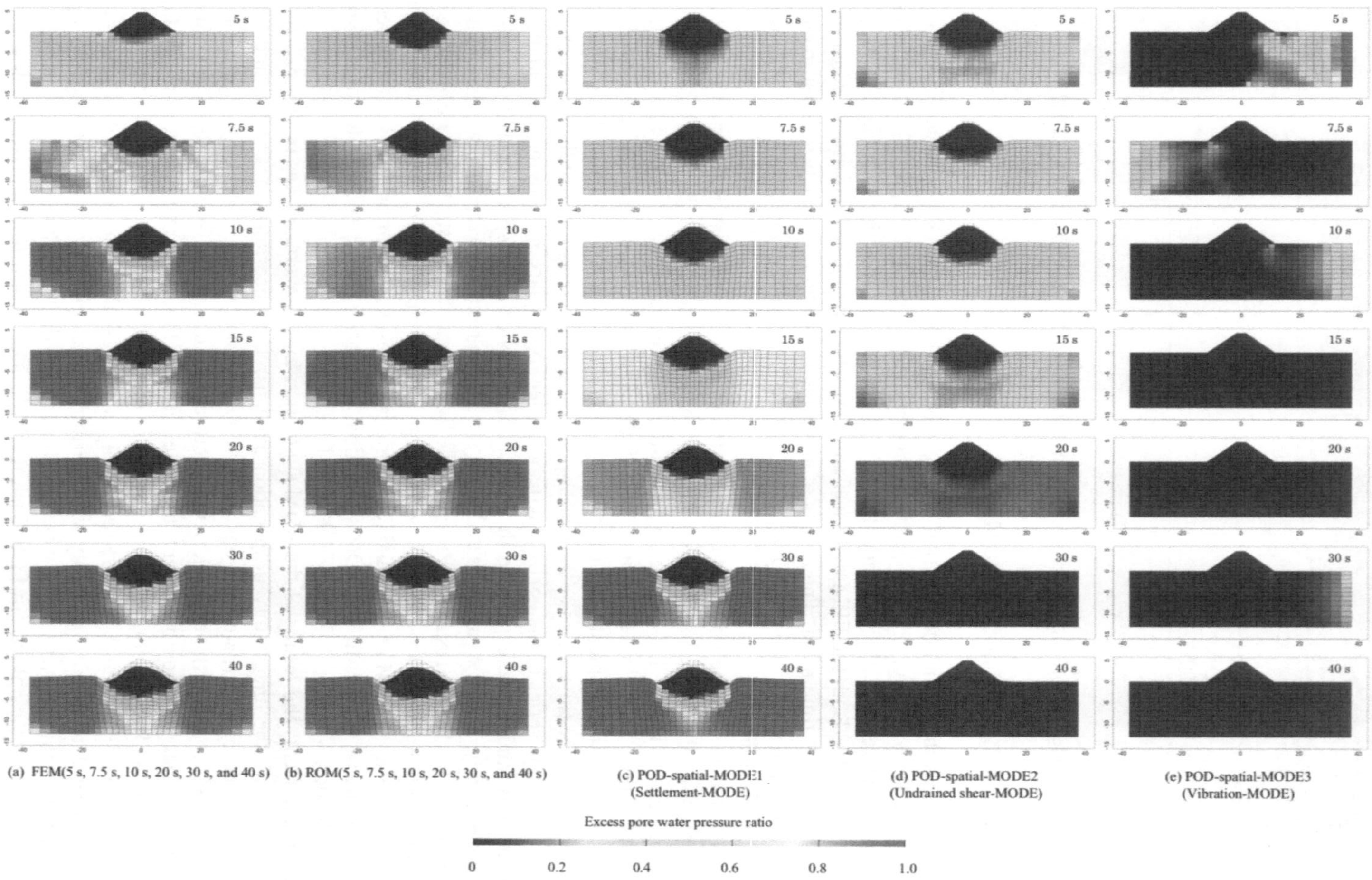

FIGURE 7.9 FEM and ROM comparisons for each POD mode (5, 7.5, 10, 20, 30, and 40 s). (a) FEM analysis, (b) ROM, (c) MODE 1, (d) MODE 2, and (e) MODE 3. (Source: Otake et al., 2022)

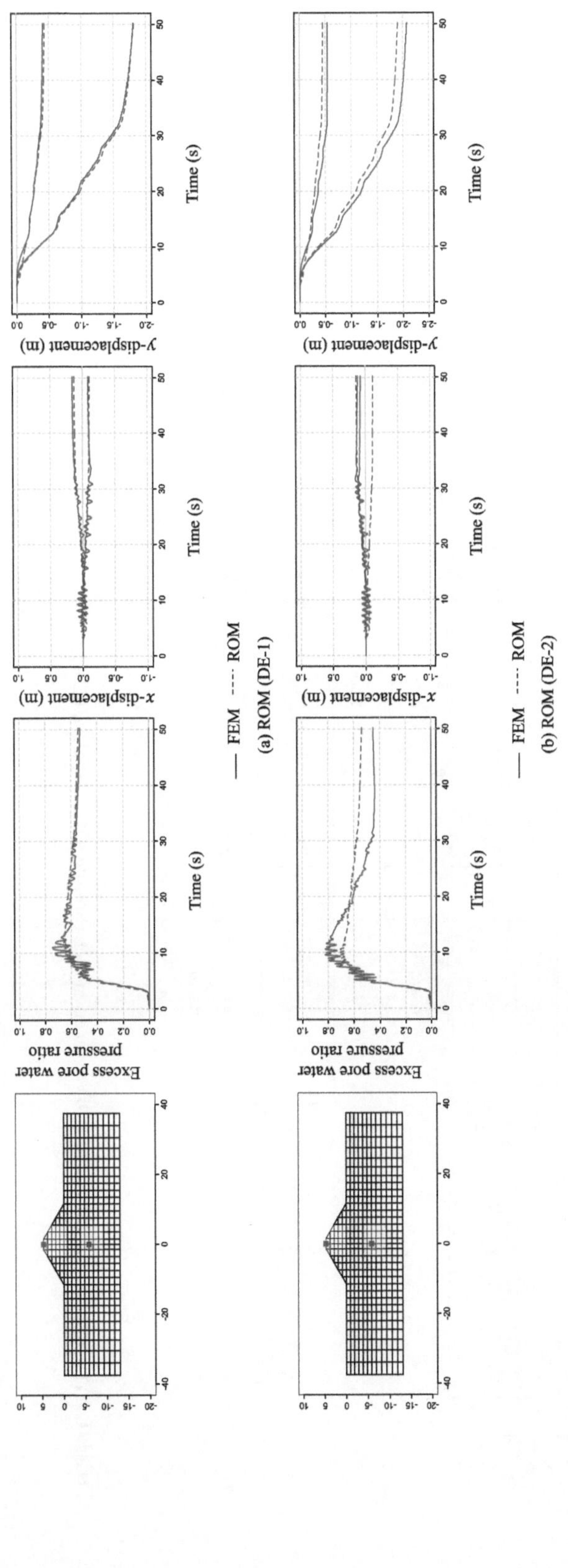

FIGURE 7.10 Reconstruction performance of ROM results. (a) DE-1 and (b) DE-2. (Source: Otake et al., 2022)

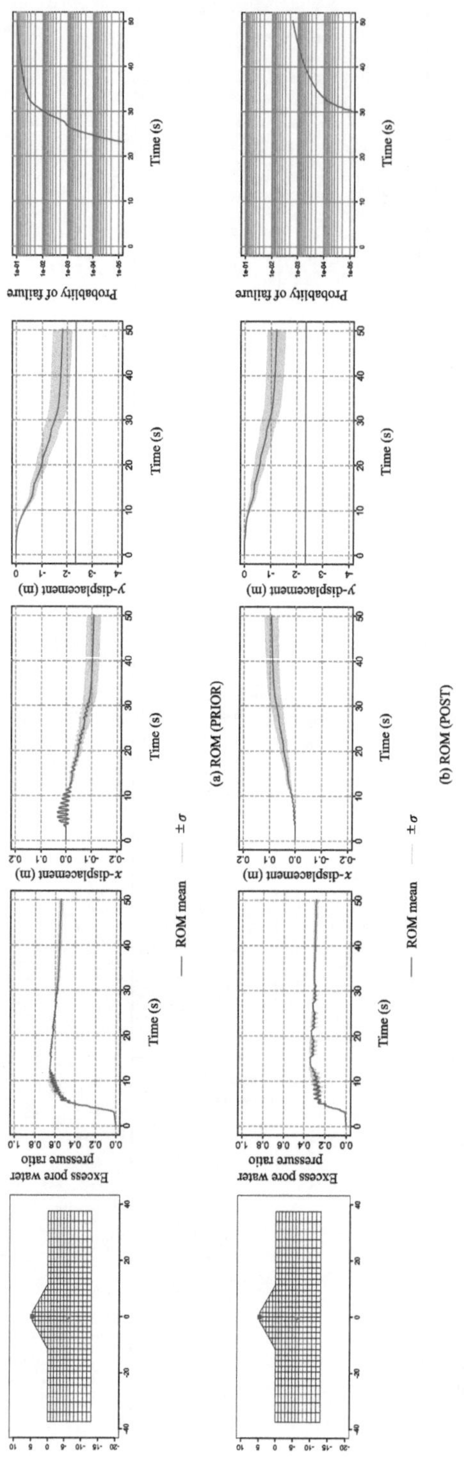

FIGURE 7.11 Reliability analysis of POD reduced model. (Source: Otake et al., 2022)

of the information in DE-1 (e_0, G_m, $\sigma m'$, M_f^*). The magnitude and trend of the failure probability time series vary considerably, depending on the input parameters, leading to different outcomes for the two cases.

Figure 7.12 illustrates the sensitivity coefficient time series for COMP1, COMP2, and COMP3 at the top of the embankment. COMP1 is interpreted as the "comprehensive index of rigidity", making a significant contribution to the overall analysis time. COMP2 is interpreted as the "comprehensive index of dilatancy", occurring after the main shaking, and its contribution gradually increases after the soil foundation liquefies. COMP3 is interpreted as the "comprehensive index of stickiness after reaching the phase transformation line" and tends to exhibit a significant contribution during the main dynamics and gradually decreases as the contribution of COMP2 increases. Thus, the sensitivity coefficient time series are an indicator of the physical properties of each COMP.

Figure 7.13 shows the spatial distribution time series for the variance and sensitivity coefficients at 5, 7.5, 10, 20, 30, and 40 s. Figure 7.13 presents, from left to right, the spatial distribution of the horizontal displacement (x-displacement), vertical displacement (y-displacement), excess pore water pressure ratio, and sensitivity coefficient of C_1^2: COMP1, C_2^2: COMP2, and C_3^2: COMP3. The variance of the calculated engineering indices (i.e., displacement and stress) and the sensitivity coefficients corresponding to each COMP were calculated as spatiotemporal information. Thus, the spatial characteristics of each calculated parameter can be observed at an arbitrary time, as shown in Figure 7.13. Based on this information, the difficulty in predicting the physical behavior of a facility from an engineering perspective can be analyzed. Furthermore, this provides critical information for decision-making while exploring locations for additional soil investigations or monitoring.

7.5 APPLICATION IN THE INFORMATION-ORIENTED CONSTRUCTION CONTROL OF RETAINING WALLS

7.5.1 Introduction

Otake et al. (2019) and Saito et al. (2022) investigated the applicability of DMD-C to the information-oriented construction control of retaining walls. Information-oriented construction control aims to ensure the safe management of excavation through real-time measurements from displacement measurement centers installed in earth-retaining walls. It is desirable to quickly calculate the optimal decision-making process based on the deformation distribution and supporting system conditions of the retaining wall. As a fundamental study, the possibility of surrogate modeling the

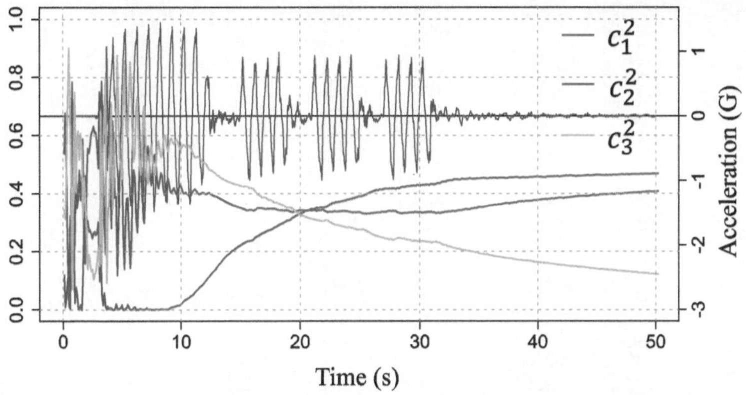

FIGURE 7.12 Sensitivity coefficient time series at the top of the embankment (C_1^2, C_2^2, and C_3^2). COMP1: comprehensive index of rigidity; COMP2: comprehensive index of dilatancy; and COMP3: comprehensive index of persistence after reaching the phase transformation line. (Source: Otake et al., 2022)

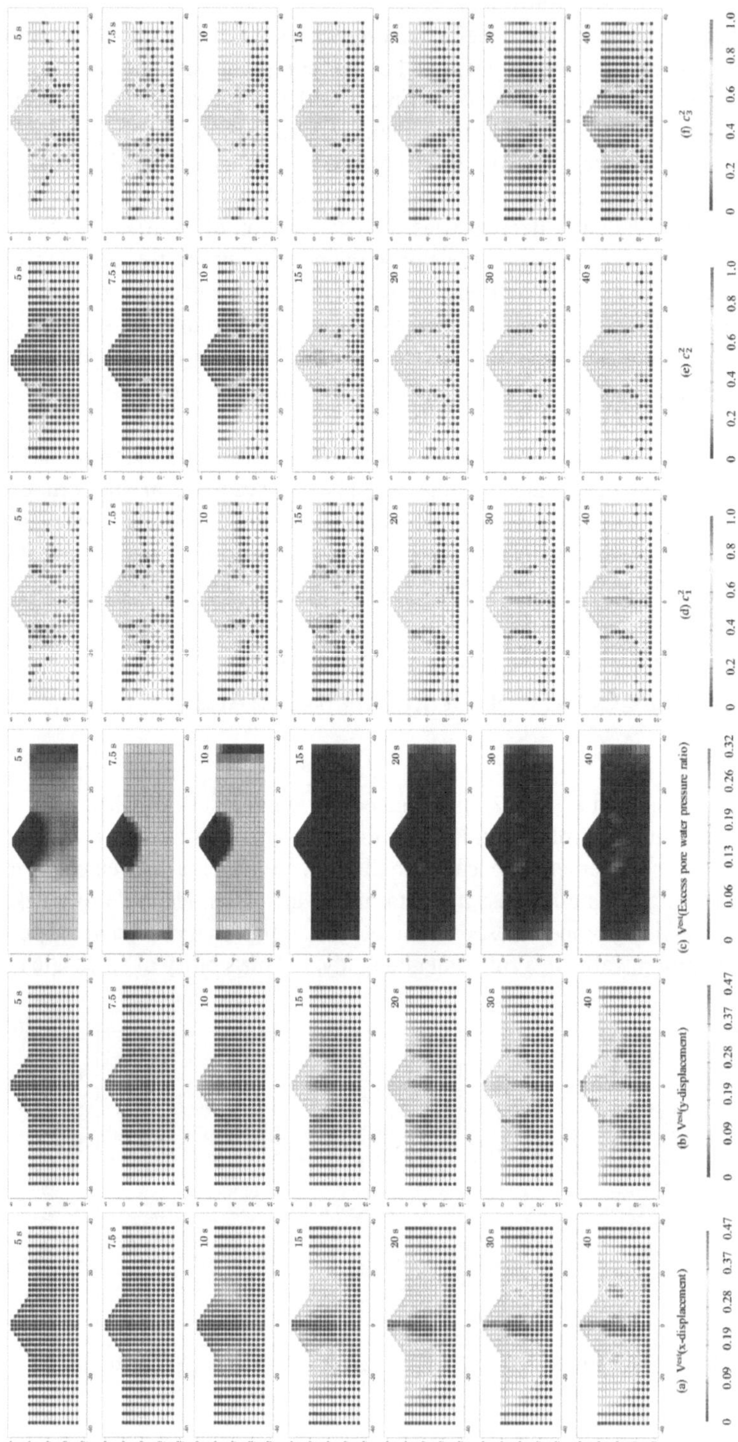

FIGURE 7.13 Spatial distributions for the variance and sensitivity coefficients. (Source: Otake et al., 2022)

behavior of an earth-retaining wall with installed supports is discussed. To consider arbitrary supporting effects, DMD-C, which is described as DMD with an additional control term, is utilized. This section presents the formulation of DMD-C and discusses the potential of combining data- and model-driven modeling with the theory of beams on elastic foundations.

7.5.2 Target Structure

Figure 7.14a depicts a schematic of an earth-retaining wall, while Table 7.1 lists its basic specifications. A hypothetical excavation was carried out with a wall length of $L = 20$ m and an excavation depth of $h = 10$ m. The displacement behavior of the wall was evaluated under increasing overburden load q, from 0 to 30 kN/m^2 at regular intervals of $\Delta q = 1$ kN/m^2 on completion of excavation. The displacement calculation method used in this study was based on the elastoplastic method prescribed in the Japanese design standards (JRA, 2017). A conceptual diagram of the calculation model is presented in Figure 7.14b. In this model, the retaining wall was represented as a beam, while the ground was modeled as a spring. The effective lateral pressure, obtained by subtracting the earth pressure at rest from the active earth pressure, was applied to the back of the wall. The ground reaction force on the excavation side (resistance side) was modeled as a fully elastoplastic spring with shear strength as the upper limit. Deformation coefficients were spatially varied, as depicted in Figure 7.15. Subsequently, a linear system surrogate model was established for the four spatial distributions generated by the stochastic process, and the effect of variation on the linear system surrogate model is discussed in detail.

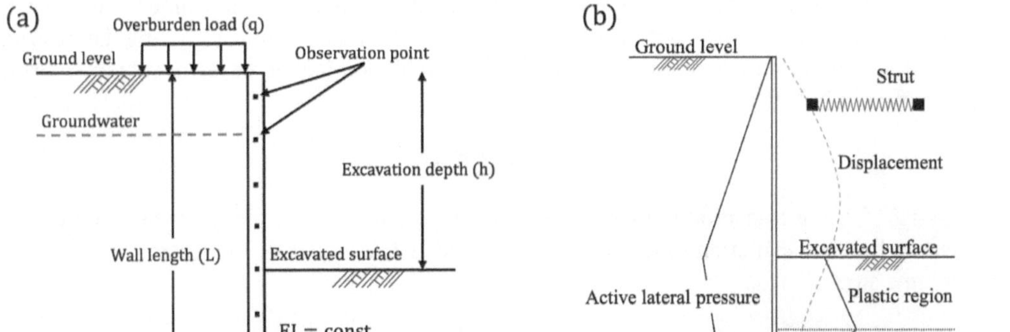

FIGURE 7.14 (a) Schematic of target excavation section and (b) conceptual diagram of elastoplastic method.

TABLE 7.1
Structural and Geotechnical Parameters

Parameter	Value	Unit
Wall length	20	m
Observation points	41	
Groundwater level	–1	m
Young's modulus	2.00×10^8	kN/m^2
Second moment of area	6.89×10^{-4}	m^2
Excavation depth	10	m

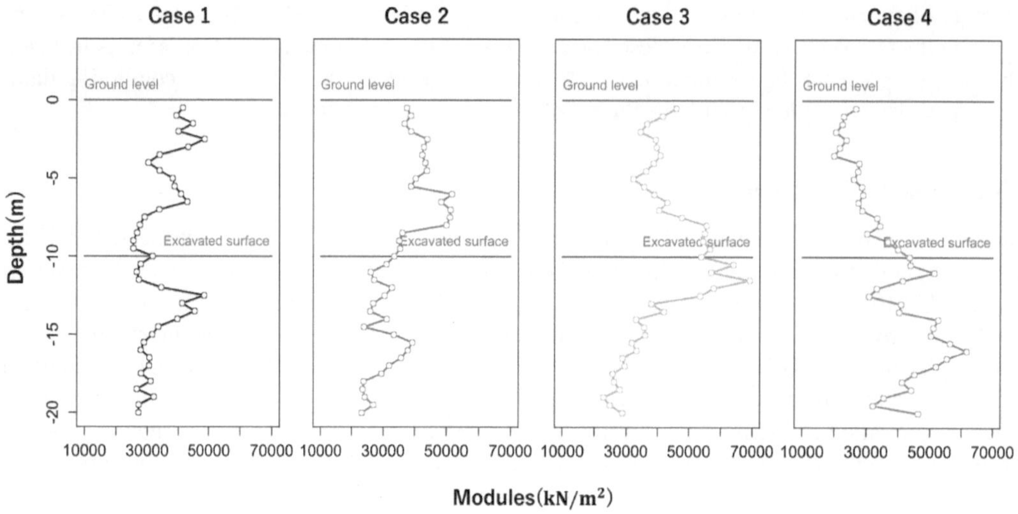

FIGURE 7.15 Four geotechnical scenarios (spatial distribution of deformation coefficients).

1. **DMD-C**

DMD-C characterizes the relationship between the state z_k of the system at an arbitrary time, the state z_{k+1} of the system at a future time, and the control value ξ_k at an arbitrary time by identifying the time evolution characteristics of the underlying observation data, while excluding the external control and its time evolution characteristics. The time evolution law of a system subject to exogenous forcing can be given as

$$Z'' = AZ' + B\Xi \tag{7.45}$$

where $\boldsymbol{A} \in \mathbb{R}^{n \times n}$ is the transition matrix of the state vector; $\boldsymbol{B} \in \mathbb{R}^{n \times l}$ is the time evolution matrix representing the time evolution of the forcing term; and Ξ is a matrix of the forcing term vectors for each time step.

$$\Xi = \begin{pmatrix} | & | & & | \\ \xi_1 & \xi_2 & \cdots & \xi_{m-1} \\ | & | & & | \end{pmatrix} \in \mathbb{R}^{l \times (m-1)} \tag{7.46}$$

The number of dimensions of the control term vector is denoted by l, and from Eqs (7.45) and (7.46), Eq. (7.47) is derived:

$$Z'' = AZ' + B\Xi = \begin{bmatrix} A\, B \end{bmatrix} \begin{bmatrix} Z' \\ \Xi \end{bmatrix} = G\Omega \tag{7.47}$$

where $G = \begin{bmatrix} A\, B \end{bmatrix}$ and $\Omega = \begin{bmatrix} Z'\Xi \end{bmatrix}^T$ are the operator matrix and feedthrough matrix, respectively. From Eq. (4.47), the operator matrix can be derived as

$$G = Z''\Omega^{-1} \tag{7.48}$$

However, since the matrix Ω is composed of observed data, it is generally a singular matrix and its regularity is not guaranteed. Therefore, to find a matrix such that Eq. (7.48) holds approximately, the following minimization problem is solved:

$$G \triangleq \arg\min_{A,B} \left\| Z'' - G\Omega \right\|_F \tag{7.49}$$

where $\|\cdot\|_F$ is the Frobenius norm. The matrix satisfying Eq. (7.49) can be computed based on singular value decomposition. SVD is applied to the matrix Ω to obtain an approximate matrix $\tilde{G}$ of Ω:

$$\Omega \approx \tilde{U}\tilde{\Sigma}\tilde{V}^* \tag{7.50}$$

$$\tilde{G} = Z''\tilde{V}\tilde{\Sigma}^{-1}\tilde{U}^* \tag{7.51}$$

where * denotes the complex conjugate transpose. Thereafter, to derive the approximate matrices ***A*** and ***B*** of the operator matrix, the left singular matrix $\tilde{U}$ in Eq. (7.50) is divided into two elements as follows:

$$\tilde{U}^* = \left[\tilde{U}_1^* \tilde{U}_2^*\right] \tag{7.52}$$

where $\tilde{U}_1^* \in \mathbb{R}^{r \times n}, \tilde{U}_2^* \in \mathbb{R}^{r \times l}$ are the matrices containing the leading and trailing columns of $\tilde{U}$, respectively; and r is the number of dimensions after dimension reduction in the singular value decomposition. Using the results of Eqs (7.50) and (7.52), we compute the approximate matrices $\bar{A}$ and $\bar{B}$ of the operator matrix by Eqs (7.53) and (7.54):

$$\bar{A} = Z''\tilde{V}\tilde{\Sigma}^{-1}\tilde{U}_1^* \tag{7.53}$$

$$\bar{B} = Z''\tilde{V}\tilde{\Sigma}^{-1}\tilde{U}_2^* \tag{7.54}$$

Using Eqs (7.53) and (7.54) to reconstruct the state quantities, Eq. (7.55) is derived:

$$Z'' \approx \bar{A}Z' + \bar{B}\Xi \tag{7.55}$$

The operator matrix can be identified by the above process.

2. **Modeling deformation time evolution of a wall in a cantilever state**

For a specific ground scenario i, the time evolution of the distribution of wall displacement, $\underline{u}_i$ (the state vector), can be approximated by a linear system using Eq. (7.56), which also stores the displacement and rotation angles of each node:

$$\frac{d}{dt}\underline{u}_i = A_i\underline{u}_i + B\underline{f} \tag{7.56}$$

In Eq. (7.56), $A_i \in \mathbb{R}^{2n \times 2n}$ represents the time evolution matrix of the wall displacement vector $\underline{u}_i \in \mathbb{R}^{2n}$, and $\underline{f} \in \mathbb{R}^{2n}$ is the forced load vector used to control wall deformation. Furthermore, $\underline{f} \in \mathbb{R}^{2n}$ is a matrix of the column-wise array of retaining-wall deformation vectors for a unit forcing load from an arbitrary point, known as the forcing load operator. This operator is assumed to be determined solely from the wall specifications.

$$B = K_w\left(E_w, I_w, L_w\right)^{-1} \tag{7.57}$$

In Eq. (7.57), $K_w\left(E_w, I_w, L_w\right)^{-1}$ is the stiffness matrix when the wall is modeled as an elastic beam. This can be calculated from the stiffness $\left(E_w, I_w\right)$ and length $\left(L_w\right)$ of the wall, which are fundamental parameters in the design of earth-retaining walls and are given as preconditions at the beginning of the design process. Therefore, the forcing load operator is treated as a known parameter. In other words, the construction of a linear system surrogate model is impeded by the determination of the time evolution matrix, A_i, of the wall displacement vector; A_i is assumed to depend on the displacement level of the wall because it may be affected by the non-linearity of the ground. Additionally, because it depends on the strut placement plan (placement position and time), which can have countless combinations, it is difficult to determine a general-purpose matrix.

Our focus is on the time evolution matrix $A_i \in \mathbb{R}^{2n\times 2n}$ of the displacement vector $\underline{u}_{0,i}$ of the wall in the cantilever state (no struts in place). The temporal variation of the wall displacement distribution obtained through the elastoplastic method analysis is determined using the DMD process, as shown in Eqs (7.58) and (7.59).

$$\frac{d}{dt}\underline{u}_{0,i} \approx A_i \underline{u}_{0,i} = \Phi_{r,i}\Lambda_{r,i}\Phi_{r,i}^{\dagger}\underline{u}_{0,i} \tag{7.58}$$

$$A_i \triangleq \underset{A_i}{\arg\min}\left\| U_i' - A_i U_i \right\|_F \tag{7.59}$$

In Eq. (7.58), $\Phi_{r,i} \in \mathbb{C}^{2n\times r}$ is a time-independent DMD spatial mode function, and $\Lambda_{r,i} \in \mathbb{C}^{r\times r}$ is a matrix with eigenvalues arranged in diagonal terms. The DMD spatial mode function is approximated in a dimensionally compressed manner by extracting the r largest eigenvalues. The data matrices $U_i \in \mathbb{R}^{2n\times(m-1)}$ and $U_i' \in \mathbb{R}^{2n\times(m-1)}$ are defined as columnar matrices of wall displacement vectors obtained using the elastoplastic method.

$$U_i = \begin{pmatrix} | & | & & | \\ \underline{u}_{0,i}(t_0) & \underline{u}_{0,i}(t_1) & \cdots & \underline{u}_{0,i}(t_{m-1}) \\ | & | & & | \end{pmatrix} \tag{7.60}$$

$$U_i' = \begin{pmatrix} | & | & & | \\ \underline{u}_{0,i}(t_1) & \underline{u}_{0,i}(t_2) & \cdots & \underline{u}_{0,i}(t_m) \\ | & | & & | \end{pmatrix} \tag{7.61}$$

Equations (7.60) and (7.61) define $\underline{u}_0(t_k)$ as the displacement vector of the wall at time t. Since we focus only on the deformation behavior in the cantilever state, r is assumed to be extremely small. Based on the above equations, the wall displacement vector $\underline{u}_0(t_k)$ at a certain time k in the self-supporting state can be calculated using Eq. (7.62) with the initial wall displacement vector $\underline{u}_0(t_0)$:

$$\underline{u}_{0,i}(t_k) = \Phi_{r,i}\Lambda_{r,i}^{k}\Phi_{r,i}^{\dagger}\underline{u}_{0,i}(t_0) \tag{7.62}$$

3. **Modeling of the deformation time evolution of an earth-retaining wall in a cut-beam configuration**

In the elastoplastic method, the wall is modeled as a beam while the ground is modeled as a spring. The subgrade reaction force on the excavation side (resistance side) is represented as a fully elastoplastic model with the shear strength serving as the upper limit. However, the non-linearity of the ground is approximated using an equivalent linear model, expressed by Eqs (7.63) and (7.64):

$$P_{sa} = K_{0,i}\underline{u}_{0,i} \tag{7.63}$$

$$K_{0,i} = K_w + K_{c,i}^{EL} \tag{7.64}$$

where $P_{sa} \in \mathbb{R}^{2n}$ represents the external force vector (effective main active lateral pressure vector); $K_{0,i} \in \mathbb{R}^{2n\times 2n}$ represents the stiffness matrix, expressed as a linear sum of the stiffness matrix of the earth-retaining wall, $K_w \in \mathbb{R}^{2n\times 2n}$, and the distributed van matrix of the ground, $K_{c,i}^{EL}$; superscript EL refers to equivalent linear; and $K_{c,i}^{EL}$ represents the distribution spring matrix of the equivalent linearized ground, assuming that the ground is linear and possesses no spatial variability. The $2n \times 2n$ matrix contains only one unknown quantity as it assumes linearity and no spatial variability. Using a particle filter, $K_{c,i}^{EL}$ is determined with Eq. (7.65) as the objective function:

$$K_{c,i}^{EL} \triangleq \underset{K_{c,i}^{EL}}{\arg\min} \left\| U_{NL,i} - U_{EL,i} \right\|_F \tag{7.65}$$

where $U_{NL,i} \in \mathbb{R}^{2n\times m}$ is the data matrix of the wall displacements based on the elastoplastic method; and $U_{EL,i} \in \mathbb{R}^{2n\times m}$ is the data matrix of the wall displacements based on a linear analysis using $K_{c,i}^{EL}$. The assumption of ground linearity with no spatial variability is akin to that utilized in the displacement method for pile foundation design. The reason for adopting this assumption is that the predicted retaining wall behavior is typically based on low displacement levels, and the wall's displacement is deemed dependent on the soil's properties within a few meters of the bottom of the excavation, as evidenced by beam theory on an elastic foundation. Similarly, the displacement vector for a cut beam is represented by Eqs (7.66) and (7.67):

$$P_{sa} = K_i \underline{u_0} \tag{7.66}$$

$$K_i = K_w + K_{c,i}^{EL} + K_{st} \tag{7.67}$$

where $K_{st} \in \mathbb{R}^{2n\times 2n}$ represents the stiffness matrix of the cut beam, and the stiffness of the strut $(E_{st}A_{st})$ is located at the strut's position (diagonal term) and zero elsewhere. As P_{sa} does not depend on the presence or absence of struts, the displacement vector when struts are in place can be determined using Eq. (7.68):

$$\underline{u_i} = K_i^{-1} K_{0,i} \underline{u_{0,i}} = T_{k,i} \underline{u_{0,i}} \tag{7.68}$$

The linear system surrogate model can be expressed as follows:

$$\frac{d}{dt}\underline{u_i} = A_i \underline{u_i} + B\underline{f} = A_i T_{k,i} \underline{u_{0,i}} + B\underline{f} \tag{7.69}$$

Since the proposed model is based on DMD learning for the cantilever state, it is expected to exhibit significant dimensionality compression. The wall displacements under various conditions can be calculated by merely transforming the wall displacements in the cantilever state using a known stiffness matrix.

4. **Extending the model with uncertainty**

Assuming that the spatial variability of soil is modeled using stochastic processes, we generated multiple soil scenarios. For each scenario, we conducted an elastoplastic analysis of the cantilever state and approximated the dynamic mode decomposition using Eq. (7.58). The results of the analysis of the four soil scenarios described below indicate that $\Phi_{r,i}$ is insensitive to the spatial distribution of soil variability and is roughly similar, as described later.

Therefore, we have employed a certain reference spatial mode function $\Phi_{r,ref}$ for the DMD spatial mode function and aggregated the changes in wall behavior for different ground scenarios into eigenvalues, as shown in Eq. (7.70):

$$\begin{aligned} \underline{u}_{0,i}(t_k) &= \Phi_{r,i}\Lambda_{r,i}^{k}\Phi_{r,i}^{\dagger}\underline{u}_{0,i}(t_0) \\ &\approx \Phi_{r,ref}\Lambda_{r,i}^{'k}\Phi_{r,ref}^{\dagger}\underline{u}_{0,i}(t_0) \\ &= \Phi_{r,ref}\underline{x}_{k,i} \end{aligned} \tag{7.70}$$

$$\underline{x}_{k,i} = \Lambda_{r,i}^{'k}\Phi_{r,ref}^{\dagger}\underline{u}_{0,i}(t_0) \tag{7.71}$$

where $\Lambda_{r,i}^{'}$ is converted using a particle filter with the following objective function:

$$\Lambda_{r,i}^{'} \triangleq \underset{\Lambda_{r,i}^{'}}{\arg\min} \left\| U_i^{'} - \Phi_{r,ref}\Lambda_{r,i}^{'}\Phi_{r,ref}^{\dagger}U_i \right\|_F \tag{7.72}$$

Based on the above equations, a simple stochastic model was constructed by incorporating the effect of soil spatial variability into the eigenvalue matrix $\Lambda_{r,i}^{'}$. To treat $\underline{u}_0(t_k)$ and $\underline{x}_k$ as random variable vectors, they were rewritten as $\underline{\dot{u}}_0(t_k)$ and $\underline{\dot{x}}_k$. Since the statistics of these random variables cannot be determined in advance, soil spatial variability scenarios were generated based on stochastic process theory, which simulates the soil characteristics at the target site. The statistics were calculated based on the numerical analysis results while considering the soil spatial variability.

Although the value of $\Phi_{r,ref}$ is arbitrary, it is assumed to be the expected value of the soil scenario. If the expected value $E\left[\underline{\dot{x}}_k\right]$ and the covariance matrix $Cor\left[\underline{\dot{x}}_k\right]$ of $\underline{\dot{x}}_k$ can be obtained based on the numerical analysis, as shown in Eqs (7.73) and (7.74), the expected value and covariance matrix of $\underline{\dot{u}}_{0,i}(t_k)$ can be calculated using Eqs (7.75) and (7.76):

$$E\left[\underline{\dot{x}}_k\right] = E\left[\mathrm{Re}\left[\underline{\dot{x}}_k\right]\right] + E\left[\mathrm{Im}\left[\underline{\dot{x}}_k\right]\right]j \tag{7.73}$$

where j is the imaginary unit.

$$Cor\left[\underline{\dot{x}}_k\right] = E\left[(\underline{\dot{x}}_k - E\left[\underline{\dot{x}}_k\right])(\underline{\dot{x}}_k - E\left[\underline{\dot{x}}_k\right])^H\right] \tag{7.74}$$

$$E\left[\underline{\dot{u}}_0(t_k)\right] = \Phi_{r,ref}E\left[\underline{\dot{x}}_k\right] \tag{7.75}$$

$$Cor\left[\underline{\dot{u}}_0(t_k)\right] = \Phi_{r,ref}Cor\left[\underline{\dot{x}}_k\right]\Phi_{r,ref}^H \tag{7.76}$$

where superscript H denotes the conjugate complex. The time evolution of the strut arrangement is approximated using Eq. (7.73).

$$E\left[\underline{\dot{u}}(t_k)\right] \approx T_{k,ref}E\left[\underline{\dot{u}}_0(t_k)\right] + B\underline{f} \tag{7.77}$$

$$Cor\left[\underline{\dot{u}}(t_k)\right] \approx T_{k,ref}Cor\left[\underline{\dot{u}}_0(t_k)\right]T_{k,ref}^T \tag{7.78}$$

where $T_{k,ref}$ is the transformation matrix based on $K_{c,ref}^{EL}$ and is identified from the ground scenario of the reference case; and $K_{c,ref}^{EL}$ is associated with the displacement level at the bottom of the excavation during the learning process. The convergence $K_{c,ref}^{EL}$ is obtained by a simple calculation used in the equivalent linear ground response analysis SHAKE (Schnable et al., 1972). The above equation is an approximate solution because it does not reflect the variation in $T_{k,i}$ for each ground scenario. This assumption is made because $T_{k,i}$ is assumed to reflect the effect of K_{st} more strongly than $K_{c,i}^{EL}$ and as a convenience assumption to limit the particle filter to only one reference case. The accuracy of this approximation needs to be examined in future studies.

Therefore, if the mean vector and covariance matrix of the eigenvalues can be determined at arbitrary steps in the cantilever state during the learning process, the uncertainty in the displacement distribution under various design conditions can be computed immediately.

7.5.3 Results and Considerations

Figure 7.16 illustrates the results of calculating the DMD spatial mode functions and time evolution (eigenvalues) from the elastoplastic analysis, assuming the cantilever state for the four ground scenarios. The number of significant modes was determined to be two, since the cantilever state served as the training target, and the unknowns (parameters) of the predictive model were the eigenvalues of the two modes. The number of mode functions was determined qualitatively by examining the magnitude of singular values. These results demonstrate that the effect of soil spatial variability is insensitive to the spatial mode function, and there is no significant difference in the spatial mode function. Figure 7.17 depicts the superimposition of the elastoplastic method and proxy calculation values for the cut-beam configuration. The proposed method nearly perfectly replicated the elastoplastic method results, thereby confirming the efficacy of the proposed method. Figure 7.18 shows the range of expected values and standard deviation of the predictions based on Eqs (7.73) and (7.74) under the same conditions as those in Figure 7.16. The validity of the proposed method is confirmed by its ability to encompass the responses of four cases of ground scenarios.

In this book, we establish a linear system–type surrogate model to efficiently predict the displacement behavior of an earth-retaining wall when the surface load is increased after excavation is completed. The proposed method is developed to construct an autonomous control model for large-scale earth-retaining walls. The linear system–type surrogate model integrates DMD and beam theory on an elastic foundation. We analyzed four simple soil heterogeneity scenarios, and their efficacy was verified numerically. As the linear system model has a high affinity with optimization and control theories, we believe it will be an essential elemental technique for constructing autonomous control models. In the future, we plan to validate the model using actual observation records and extend the model to predict the displacement behavior of earth-retaining walls during excavation. Additionally, we intend to develop a robust design method for the initial design of struts and extend the model to a real-time control model during construction.

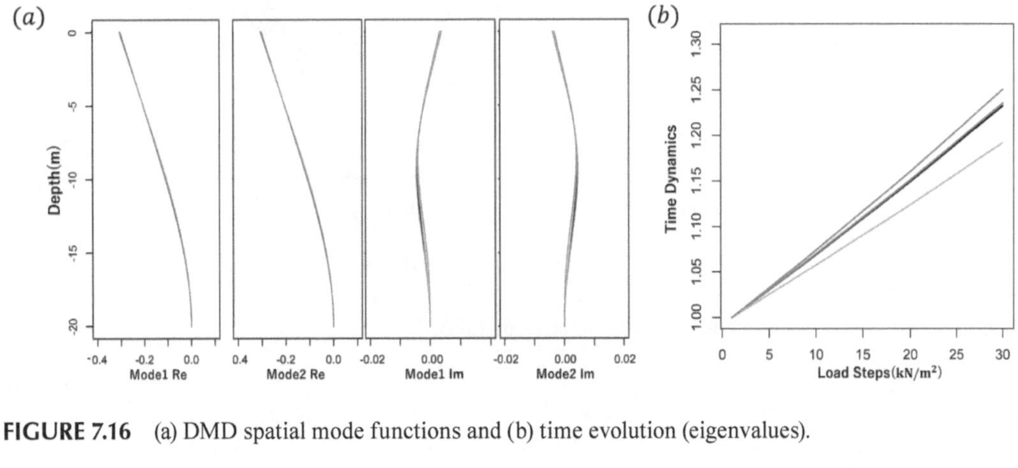

FIGURE 7.16 (a) DMD spatial mode functions and (b) time evolution (eigenvalues).

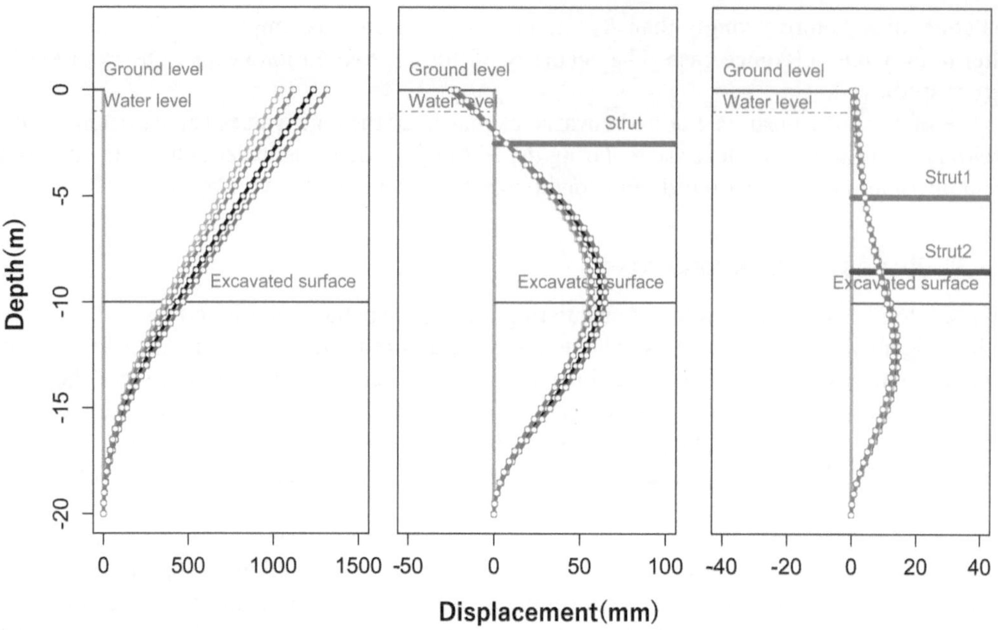

FIGURE 7.17 Comparison of elastoplastic method and proxy calculation values.

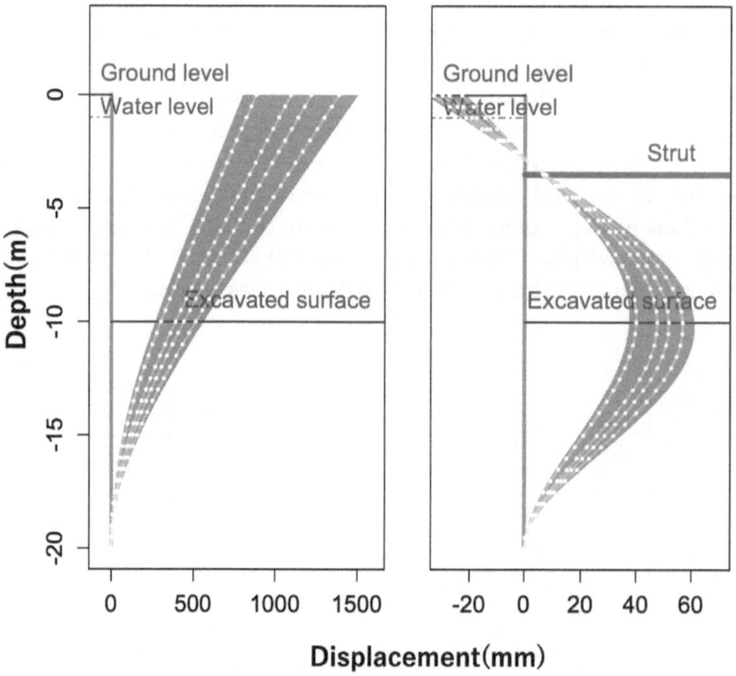

FIGURE 7.18 Range of one standard deviation of predictions.

7.6 SUMMARY

The application of reduced order models in geotechnical engineering for reliability design has been limited, and thus motivated the writing of this chapter. The chapter primarily focuses on the fundamental theories of proper orthogonal decomposition and dynamic mode decomposition for ROMs and their applicability in geotechnical engineering through various example problems.

The authors acknowledge that an ideal method is required to linearly approximate the mechanical behavior of soils exhibiting strong material non-linearity for the effective application of ROMs in reliability analysis. Therefore, future research should address this need to improve the objectivity of surrogate modeling using ROMs and enhance the reliability of extrapolation estimates. This issue cannot be resolved with conventional transformations, and it may be necessary to consider non-linear mapping using machine learning methods. From this perspective, the authors highlight ongoing research on DMD in connection with the Koopman Operator (e.g., Brunton et al., 2021), which has the potential to map to an ideal stationary and linear space. This approach might be able to interpret the rational feature space in "digital twin" in geotechnical design. The application of this research in geotechnical engineering would be an essential development that could provide a powerful link with traditional reliability analysis and design methods based on linear systems. Moreover, it is interpreted as having valuable properties, such as a high affinity with risk analysis and optimization problems in planning.

In summary, this chapter presents the potential of ROMs, particularly POD and DMD, in geotechnical engineering reliability design and emphasizes the need for further research in developing an ideal method to linearly approximate the mechanical behavior of soils exhibiting strong material non-linearity. The authors' recognition of this limitation in ROMs provides insights and direction for future research in this field. The potential application of POD and DMD in geotechnical engineering represents an exciting prospect that could advance the field in the coming years.

REFERENCES

Arai, Y., Muramatsu, S., Yasuda, H., Hayasaka, K. and Otake, Y. 2021. Sparse-coded dynamic mode decomposition on graph for prediction of river water level distribution. In *Proceedings of the IEEE International Conference on Acoustics, Speech and Signal Processing*, 3225–3229. https://ieeexplore.ieee.org/document/9414533

Argyroudis, S. A., Mitoulis, S., Winter, M. G. and Kaynia, A. M. 2019. Fragility of transport assets exposed to multiple hazards: State-of-the-art review toward infrastructural resilience. *Reliability Engineering and System Safety*, 191, 106567.

Berkooz, G., Holmes, P. and Lumley, J. L. 1993. The proper orthogonal decomposition in the analysis of turbulent flows. *Annual Review of Fluid Mechanics*, 25(1), 539–575.

Bruneau, M., Chang, S. E., Eguchi, R. T., et al. 2003. A framework to quantitatively assess and enhance the seismic resilience of communities. *Earthquake Spectra*, 19(4), 733–752.

Brunton, S., Budišić, M., Eurika, K. and Kutz, J. 2021. Modern Koopman theory for dynamical systems. arXiv, 2102.12086.

Brunton, S. and Kutz, J. 2022. *Data-Driven Science and Engineering: Machine Learning, Dynamical Systems, and Control* (2nd ed.). Cambridge: Cambridge University Press.

Chatterjee, A. 2000. An introduction to the proper orthogonal decomposition. *Current Science*, 78(7), 808–817.

Cornell, C. A. 1969. A probability-based structural code. *Proceedings of the ACI Journal*, 66(12), 974–985.

Fukutani, Y., Moriguchi, S., Terada, K. and Otake, Y. 2021. Time-dependent probabilistic tsunami inundation assessment using mode decomposition to assess uncertainty for an earthquake scenario. *Journal of Geophysical Research: Oceans*, 126(7), 1-29.

Ghanem, R. and Spanos, P. 1991. *Stochastic Finite Elements: A Spectral Approach*. New York: Springer-Verlag.

Hasofer, A. M. and Lind, N. C. 1974. Exact and invariant second moment code format. *Journal of the Engineering Mechanics Division, ASCE*, 100(1), 111–121.

Holmes, P., Lumley, J. and Berkooz, G. 1996. Proper orthogonal decomposition. In *Turbulence, Coherent Structures, Dynamical Systems and Symmetry.* 2nd ed., 86–128, Cambridge: Cambridge University Press.

ISO 2394:2015. 2015. General principles on reliability for structures.

Japan Road Association (JRA). 2017. Specifications for highway bridges and commentary, lower structure edition.

Kaiser, E., Noack, B., Cordier, L., Spohn, A., Segond, M., Abel, M., et al. 2014. Cluster-based reduced-order modelling of a mixing layer. *Journal of Fluid Mechanics*, 754, 365–414.

Kaneko, Y., Muramatsu, S., Yasuda, H., Hayasaka, K., Otake, Y., Ono, S. and Yukawa, M. 2019. Convolutional-sparse-coded dynamic mode decomposition and its application to river state estimation, *ICASSP.* In *Proceedings of the IEEE International Conference on Acoustics, Speech and Signal Processing*, 8683848, 1872–1876.

Karhunen, K. 1947. Über lineare Methoden in der Wahrscheinlichkeitsrechnung. *Annales Academiae Scientiarum Fennicae Mathematica*, 37, 1–79.

Koopman, B. O. 1931. Hamiltonian systems and transformations in Hilbert space. *Proceedings of the National Academy of Sciences of the United States of America*, 17(5), 315–318.

Kutz, J., Fu, X. and Brunton, S. 2015. Multi-resolution dynamic mode decomposition. arXiv, 1506.00564.

Liquefaction analysis method LIQCA development group, 2007. LIQCA2D07, 2007 Published Edition.

Mezić, I. 2013. Analysis of fluid flows via spectral properties of the Koopman operator. *Annual Review of Fluid Mechanics*, 45(1), 357–378.

Oka, F., Yashima, A., Shibata, T., Kato, M. and Uzuoka, R. 1994. FEM-FDM coupled liquefaction analysis of porous soil using an Elasto-plastic Model. *Applied Scientific Research*, 52(3), 209–245.

Oka, F., Yashima, A., Tateishi, A., Taguchi, Y. and Yamashita, A. 1999. A cyclic Elasto-plastic constitutive model for sand considering a plastic-strain dependence of the shear modulus. *Géotechnique*, 49(5), 661–680.

Otake, Y. and Honjo, Y. 2012. Reliability based design on long irrigation channel considering the soil investigation locations. *Geotechnical Special Publication*, 225, 2836–2845.

Otake, Y. and Honjo, Y. 2022. Challenges in geotechnical design revealed by reliability assessment: Review and future perspectives. *Soils and Foundations*, 62(3), 101129.

Otake, Y., Kodama, S. and Watanabe, S. 2019. Improvement in the information-oriented construction of temporary soil-retaining walls using sparse modeling. *Underground Space*, 4(3), 210–224.

Otake, Y., Shigeno, K., Higo, Y. and Muramatsu, S. 2022. Practical dynamic reliability analysis with spatiotemporal features in geotechnical engineering. *Georisk: Assessment and Management of Risk for Engineered Systems and Geohazards*, 16(4), 662–677.

Otsushi, K., Kato, T., Hara, T., Yashima, A., Otake, Y., Sakanashi, K. and Honda, A. 2010a. Study on a liquefaction countermeasure for flume structure by sheet-pile with drain. *Geotechnical Society of Singapore - International Symposium on Ground Improvement Technologies and Case Histories*, ISGI'09, 437–443.

Otsushi, K., Kato, T., Hara, T., Yashima, A., Otake, Y., Sakanashi, K. and Honda, A. 2010b. Analytical study on mitigation of liquefaction-related damage to flume channel using sheet-pile with drain. *Geotechnical Special Publication*, 199, 3062–3071.

Pearson, K. F. R. S. 1901. LIII. On lines and planes of closest fit to systems of points in space. *The London, Edinburgh, and Dublin Philosophical Magazine and Journal of Science*, 2(11), 559–572.

Proctor, J., Brunton, S. and Kutz, J. 2014. Dynamic mode decomposition with control. arXiv, 1409.6358.

Rowley, C. W., Mezic, I., Bagheri, S., Schlatter, P. and Henningson, D. S. 2009. Spectral analysis of nonlinear flows. *Journal of Fluid Mechanics*, 641, 115–127.

Saito, T., Kodama, S. and Otake, Y. 2022. Linear-system-type surrogate model for large-scale earth-retaining work based on dynamic mode decomposition. In *Proceedings of the 8th International Symposium for Geotechnical Safety & Risk*, 17-008.

Schmid, P. J. 2010. Dynamic mode decomposition of numerical and experimental data. *Journal of Fluid Mechanics*, 656, 5–28.

Schmid, P. J., Li, L., Juniper, M. P. and Pust, O. 2011. Applications of the dynamic mode decomposition. *Theoretical and Computational Fluid Dynamics*, 25(1–4), 249–259.

Schnabel, P., Lysmer, J. and Seed, H. 1972. Shake - A computer program for earthquake analysis of horizontally layered sites. Earthquake Engineering Research Center, University of California, Berkeley, Report, EERC 72-12, 257–265.

Shioi, A., Otake, Y., Yoshida, I., Muramatsu, S. and Ohno, S. 2023. Data-driven approximation of geotechnical dynamics to an equivalent single-degree-of-freedom vibration system based on dynamic mode decomposition. *Georisk: Assessment and Management of Risk for Engineered Systems and Geohazards*, 17(1), 77–97.

Stark, H. and Woods, J. W. 1986. *Probability, Random Processes, and Estimation Theory for Engineers.* Prentice-Hall New Jersey.

Terzaghi, K. 1925. Erdbaumechanik auf boden physikalischer Grundlage. *Franz Deuticke*, 140–146.

TTRS, Totsutotsutoshiteroutosezu. https://iqujack-lequina.hatenablog.com/entry/2018/05/20/%E5%8B%95%E7%9A%84%E3%83%A2%E3%83%BC%E3%83%89%E5%88%86%E8%A7%A3%E3%81%AB%E9%96%A2%E3%81%99%E3%82%8B%E8%A6%9A%E6%9B%B8/ 24 Feb 2023, in Japanese.

Tu, J. H., Rowley, C. W., Luchtenburg, D. M., Brunton, S. and Kutz, J. 2014. On dynamic mode decomposition: Theory and applications. *American Institute of Mathematical Sciences (AIMS), Journal of Computational Dynamics*, 1(2), 391–421.

8 Stochastic Finite Element Methods for Slope Stability Analysis and Risk Assessment

Shui-Hua Jiang, Te Xiao and Dian-Qing Li

ABSTRACT

Slope failures or landslides are one type of major geo-hazards worldwide. Quantitative slope reliability and risk assessment have become effective methods for mitigating landslide hazards. A key task of quantitative landslide risk assessment is to evaluate the probability and consequence of slope failure, which relies on understanding geological conditions, slope stability modeling, uncertainty quantification and propagation, etc. It can be rationally accomplished under a probabilistic framework through stochastic finite element methods. This chapter introduces recent developments in stochastic finite element methods for slope reliability and risk assessment, implementation procedures, and computational issues of stochastic finite element methods. Three slope examples are investigated to illustrate the effectiveness of the proposed methods. The effects of the spatial variability of geomaterials on the failure mode and the probability and risk of slope failure are systematically explored. The presented methods are expected to provide versatile and promising tools for the reliability and risk assessment of complex slopes.

8.1 INTRODUCTION

It is widely recognized that various sources of uncertainties exist in geotechnical engineering (e.g., Phoon, 2020; Chwała et al., 2022), such as the inherent spatial variability of geomaterials, soil stratigraphic uncertainty, transformation uncertainty, and model uncertainty. The inherent spatial variability of geomaterials is identified as the most dominant source of uncertainties due to natural geological, environmental, physical, and chemical processes (e.g., Li and Lumb, 1987; Vanmarcke, 1977; Phoon and Kulhawy, 1999a, 1999b), which has received considerable attention in slope reliability and risk analysis (Jiang et al., 2022). However, it is not a trivial task to conduct a reliability and risk analysis of complex slopes when the inherent spatial variability of geomaterials is considered. This is because the spatial variability of geomaterials needs to be numerically represented in terms of a large number of random variables through random field discretization (Vanmarcke, 2010). Moreover, a rigorous deterministic slope stability analysis using the finite element method (FEM) or the finite difference method (FDM) is usually computationally expensive. How to effectively evaluate the reliability and risk of slopes accounting for multi-source uncertainties (including the inherent spatial variability of geomaterials) remains an open question.

Direct Monte Carlo simulation (MCS) is the most robust and versatile method in slope reliability analysis accounting for soil spatial variability, but it suffers from a lack of efficiency due to the fact that a large number of realizations are needed to accurately estimate the probability of slope failure. To alleviate the computational burden, approximation methods such as the first-order second-moment (FOSM) method and the first-order reliability method (FORM) have been applied (e.g., El-Ramly et al., 2002; Ji et al., 2012; Low, 2015; Xiao et al., 2017a). However, these approximation methods are inaccurate for slope reliability problems involving compound failure modes

 DOI: 10.1201/9781003333586-10

and may suffer from the curse of dimensionality when thousands of random variables are used to model the soil spatial variability. The stochastic finite element method (SFEM) (e.g., Sudret and Der Kiureghian, 2000; Ghanem and Spanos, 2003) is particularly developed for reliability analysis with spatial variability. It integrates finite element analysis and random field theory, mostly in the framework of MCS, and has wide applications in slope reliability analysis (e.g., Griffiths and Fenton, 2004; Huang et al., 2010a; Jiang et al., 2014). Within the SFEM, realizations of random variables or random fields are first generated based on the statistics of geotechnical parameters, and are then mapped onto the finite element mesh, with each element having different values. Finally, the most critical slip surface in each slope realization which may not be a circular shape is sought out through finite element analysis and the corresponding factor of safety is calculated. The main challenge of the SFEM is the high computational cost because a substantial number of finite element model evaluations are needed to obtain the slope reliability or risk analysis results with sufficient accuracy, particularly when the probability of slope failure is small, the slope size is large, and the geological condition is highly complicated. In addition, the conventional SFEM may require modifications of existing FEM/FDM codes to conduct the random field modeling of materials and reliability assessment. It is relatively difficult to directly apply the SFEM in engineering practices as most engineers do not have access to the source codes of commercial software packages.

To this end, this chapter introduces two types of SFEM that have recently been developed, namely the non-intrusive stochastic finite element method (NISFEM) and the collaborative stochastic finite element method (CSFEM), for reliability and risk assessment of complex slopes. In both SFEMs, the deterministic finite element analysis of slope stability is decoupled from the reliability and risk analysis, and users are not required to modify existing deterministic finite element codes, which can be used as "black boxes". This configuration effectively removes the hurdle of reliability or risk computational algorithms and allows geotechnical practitioners to focus on the deterministic slope stability analysis with which they are familiar, facilitating a wider application of the SFEM.

8.2 NON-INTRUSIVE STOCHASTIC FINITE ELEMENT METHOD

The reliability and/or risk problems of complex slopes often involve high-dimensional variables, small probabilities, non-linear limit state functions, and multiple failure modes due to soil spatial variability. One has to resort to numerical methods to evaluate the output responses of slopes such as the factor of safety, pore water pressure, and/or displacement. These output responses of slopes cannot be explicitly expressed as the functions of uncertain input parameters. Direct MCS has gained wide popularity in slope reliability analysis by virtue of its simplicity and flexibility, but it necessitates a substantial number of finite element model evaluations to obtain the estimate with satisfactory accuracy. In this regard, the NISFEM provides an efficient strategy for addressing the computational inefficiency of MCS.

The basic idea of the NISFEM is to significantly reduce the evaluations of a finite element model by constructing an effective surrogate model (or a response surface model [RSM]) for the output responses of slopes using the quadratic polynomial, Hermite polynomial chaos expansion (HPCE), back-propagation neural network (BPNN), or other techniques. For slope stability problems, the surrogate model is constructed to fit the implicit function between the factor of safety (*FS*) of a slope and uncertain input parameters. After that, MCS can be used to estimate the probability and risk of slope failure. In the execution of MCS, the surrogate model is adopted to evaluate the *FS* for each random sample explicitly and efficiently, and the conventional deterministic slope analysis is avoided. In this way, the computational cost used for each random sample is substantially reduced.

8.2.1 Deterministic Slope Stability Analysis

The finite element method/finite difference method is a popular technique for deterministic slope stability analysis because it can account for the stress–strain behavior of geomaterials, seek out the weakest failure path, and calculate the corresponding *FS* using the shear strength reduction method

(SSRM) (e.g., Zienkiewicz et al., 1975; Griffiths and Lane, 1999). This method can address a major limitation of the limit equilibrium method (LEM), i.e., the need of assuming the location and shape of the sliding surface a priori. Commercial finite element/finite difference software packages, such as ABAQUS, PLAXIS, and FLAC3D, have implemented the SSRM for slope stability analysis.

The SSRM calculates the *FS* of a slope by progressively reducing (or increasing) the shear strengths of the geomaterials until a state of limit equilibrium in the system is reached. The bisection method can be used to search for the critical shear strength reduction factor, in which the upper and lower limits of the reduction factor are defined first and the average reduction factor is then used in the subsequent calculations to search for a new reduction factor. Mathematically, the SSRM can be written as (Griffiths and Lane, 1999)

$$\begin{cases} \tan\varphi_f = \dfrac{\tan\varphi}{K} \\ c_f = \dfrac{c}{K} \end{cases} \tag{8.1}$$

where K is the reduction factor; φ and c are the friction angle and cohesion of soils, respectively; and φ_f and c_f are the reduced friction angle and reduced cohesion, respectively. The reduction factor K to disturb equilibrium is taken as the factor of safety. In the literature, there exist several criteria to determine the slope failure or the termination of the shear strength reduction, such as the shear/total energy dissipation, the continuity of plastic regions and displacement velocity fields, and the non-convergence in the numerical calculations. The critical slip surface with the minimum *FS* determined using the SSRM varies spatially when the spatial variability of soil properties is considered (e.g., Huang et al., 2010a; Jiang et al., 2015), which can induce numerous failure modes in a slope.

8.2.2 Construction of Surrogate Model

The slope reliability and risk assessments usually require computing the values of *FS* repeatedly for many sets of inputs, which is time-consuming for a complex slope. To reduce the computational cost, the quadratic polynomial, HPCE, BPNN, and other techniques are often adopted to construct an explicit function (surrogate model) between the *FS* and input parameters (e.g., Xu and Low, 2006; Zhang et al., 2011; Jiang et al., 2014, 2015; Wang and Goh, 2021; Wang, 2022). They are briefly introduced as follows.

8.2.2.1 Quadratic Polynomial

A quadratic polynomial without cross terms is widely adopted to establish the surrogate model between the *FS* of a slope and input random variables due to its simplicity. For instance, the quadratic polynomial–based surrogate model of *FS* for the *j*th failure mode can be expressed as follows (e.g., Xu and Low, 2006; Zhang et al., 2011; Jiang and Huang, 2016):

$$FS_j(\boldsymbol{x}) = a_{1,j} + \sum_{i=1}^{n} b_{i,j}x_i + \sum_{i=1}^{n} c_{i,j}x_i^2 \tag{8.2}$$

where $FS_j(\boldsymbol{x})$, $j = 1, 2, \ldots, N_r$, is the *FS* for the *j*th failure mode; N_r is the number of considered failure modes; $\boldsymbol{x} = (x_1, x_2, \cdots, x_n)^{\mathrm{T}}$ is a vector of input random variables in the physical space corresponding to those used to discretize the random fields; n is the number of input random variables; and $\boldsymbol{A}_j = (a_{1,j}, b_{1,j}, \cdots, b_{n,j}, c_{1,j}, \cdots, c_{n,j})^{\mathrm{T}}$ is a vector of unknown coefficients with a size of $N_c = 2n + 1$.

To determine the unknown coefficients in Eq. (8.2), a sampling design method using $(2n + 1)$ combinations proposed by Bucher and Bourgund (1990) is widely employed. The values of *FS* for the *j*th failure mode, j=1, 2, …, N_r, are first evaluated at $N_c = 2n + 1$ samples as follows: $\{\mu_{X_1}, \mu_{X_2}, \ldots, \mu_{X_n}\}$, $\{\mu_{X_1} \pm k\sigma_{X_1}, \mu_{X_2}, \ldots, \mu_{X_n}\}$, …, $\{\mu_{X_1}, \mu_{X_2}, \ldots, \mu_{X_i} \pm k\sigma_{X_i}, \ldots, \mu_{X_n}\}$, …, and $\{\mu_{X_1}, \mu_{X_2}, \ldots, \mu_{X_n} \pm k\sigma_{X_n}\}$, where μ_{X_i} and σ_{X_i} are the mean and standard deviation of the *i*th variable, respectively, and *k* is a sampling factor that is normally set to be 2. In this way, a system of N_c linear algebraic equations can be established, i.e., $\boldsymbol{FS}_j = \boldsymbol{HA}_j$, where $\boldsymbol{FS}_j$ is an $N_c \times 1$ vector of all results of *FS* and $\boldsymbol{H}$ is an $N_c \times N_c$ matrix of the selected samples. Then, the unknown coefficients can obtained by $\boldsymbol{A}_j = \boldsymbol{H}^{-1}\boldsymbol{FS}_j$. Applying a similar method, multiple quadratic polynomial–based surrogate models can be obtained for different failure modes and collectively taken as an explicit function between the *FS* of a slope and the input random variables. Note that the obtained surrogate models using the quadratic polynomial do not involve the realizations of random variables or random fields, thus they do not rely on the statistics (e.g., mean, coefficient of variation [COV], marginal distribution, and scale of fluctuation) of soil properties. Based on these, the limit state function for slope reliability analysis can be derived as follows:

$$g(\boldsymbol{x}) = \min_{j=1,2,\ldots,N_r} FS_j(\boldsymbol{x}) - 1.0 \tag{8.3}$$

where $\min_{j=1,2,\ldots,N_r} FS_j(\boldsymbol{x})$ is the minimum *FS* value of the surrogate models for all failure modes at a given realization of random variables or random fields in the physical space. The values of $g(\boldsymbol{x})$ can be readily computed by substituting various random realizations into Eq. (8.3) without performing deterministic slope stability analyses again.

8.2.2.2 Hermite Polynomial Chaos Expansion

The HPCE is similar to the quadratic polynomial but has more complicated polynomial forms. Therefore, it can better represent highly non-linear performance functions. The HPCE is often associated with the Karhunen–Loève expansion (KLE), a series expansion method for random field generation. Using the HPCE, the *FS* for a given slope failure mode can be calculated as (e.g., Jiang et al., 2014, 2015)

$$FS_j(\xi) = a_0\Gamma_0 + \sum_{i_1=1}^{N} a_{i_1}\Gamma_1(\xi_{i_1}) + \sum_{i_1=1}^{N}\sum_{i_2=1}^{i_1} a_{i_1,i_2}\Gamma_2(\xi_{i_1},\xi_{i_2}) + \sum_{i_1=1}^{N}\sum_{i_2=1}^{i_1}\sum_{i_3=1}^{i_2} a_{i_1,i_2,i_3}\Gamma_3(\xi_{i_1},\xi_{i_2},\xi_{i_3}) + \ldots \tag{8.4}$$

where $j = 1, 2, \ldots, N_r$; $n = M \times N_F$ is the number of random variables in the standard normal space, in which *M* is the number of truncated KLE terms and N_F is the number of random fields; $a_0, a_{i_1}, a_{i_1,i_2}, a_{i_1,i_2,i_3}, \ldots$ are the unknown coefficients; $\Gamma_{j_p}(\cdot)$, $j_p = 1, 2, 3, \ldots$ are Hermit polynomials with j_p degrees of freedom; and $\xi = (\xi_1, \xi_2, \ldots, \xi_n)^{\mathrm{T}}$ are a set of independent standard normal random variables used to discretize the random fields using KLE. From a physical point of view, Eq. (8.4) is a surrogate of the slope stability analysis that involves uncertain input parameters.

For the n_{PCE}th order HPCE, there are a total of $N_c = (n + n_{PCE})!/(n! \times n_{PCE}!)$ unknown coefficients (i.e., $a_0, a_{i_1}, a_{i_1,i_2}, a_{i_1,i_2,i_3}, \ldots$) in Eq. (8.4), which are also required to be determined for constructing the surrogate models. Usually, N_p (e.g., $N_p \approx [2 \sim 3]N_c$) realizations of random variables or random fields of geotechnical parameters are generated using the KLE, and their corresponding N_p values of *FS* are calculated using SSRM to determine the unknown coefficients for each failure mode. Similar to quadratic polynomial, same samples can be used repeatedly to construct surrogate models for all N_r failure modes. Finally, the N_r surrogate models are collectively used as a surrogate of the deterministic slope stability analysis to explicitly and efficiently evaluate the *FS* for each random sample. Similarly, the limit state function for slope reliability analysis can be derived as follows:

$$g(\xi) = \min_{j=1,2,\ldots,N_r} FS_j(\xi) - 1.0 \tag{8.5}$$

In addition to different surrogate models used, sample space is another major difference between Eqs (8.3) and (8.5), namely $\boldsymbol{x}$ space and $\boldsymbol{\xi}$ space, respectively.

8.2.2.3 Back-Propagation Neural Network

A neural network whose architecture naturally mimics the information processing and optimization strategies of human brains is another feasible model to replace the time-consuming numerical calculation of slope stability. A neural network typically consists of a few interconnected neurons that form an input layer, hidden layers, and an output layer (Anderson, 1995). Through weighted connections, the input information stored in the neurons of the input layer is redistributed in the hidden neurons and then transformed into the response of the neural network in terms of the neurons of the output layer. The BPNN is commonly adopted in slope reliability analysis (e.g., Chen et al., 2020; He et al., 2020). The dataset used in the BPNN is divided into the training dataset for establishing the trained model, the validation dataset for determining the model architecture, and the testing dataset for assessing the model generalizability to construct the correct input–output mapping.

As shown in Figure 8.1, a three-layer BPNN topology contains five neurons in the input layer, nine neurons in the hidden layer, and one neuron in the output layer. The number of neurons in the input layer is equal to the number of discretized independent standard normal random variables. Accordingly, the number of neurons in the output layer is equal to the number of output responses of the slope. Although the mapping relationship between the output and the input is difficult to depict using an analytical function, it can easily be expressed using a BPNN. For the three-layer BPNN structure shown in Figure 8.1, the basic neural network function can be expressed as

$$FS(\xi) = \lambda_{23}\left[b_0 + \sum_{k=1}^{N_{hid}} w_k \lambda_{12}\left(b_k + \sum_{i=1}^{N_{in}} w_{ik}\xi_i\right)\right] \tag{8.6}$$

where N_{in} and N_{hid} are the numbers of neurons in the input layer and hidden layer, respectively; w_k and b_0 are the weight coefficient and threshold (bias) of the connection between the neuron H in the hidden layer and the neuron FS in the output layer, respectively; w_{ik} and b_k are the weight coefficient and threshold (bias) of the connection between the neuron ξ_i in the input layer and the neuron H in the hidden layer, respectively; λ_{12} is the transfer function between the input layer and the hidden layer; and λ_{23} is the transfer function between the hidden layer and the output layer.

Similarly, the BPNN needs to be trained using the samples obtained from the evaluations of a numerical model before it is used as a predictive model. The Latin hypercube sampling (LHS) technique is an efficient and suitable technique to generate the training samples, which can significantly

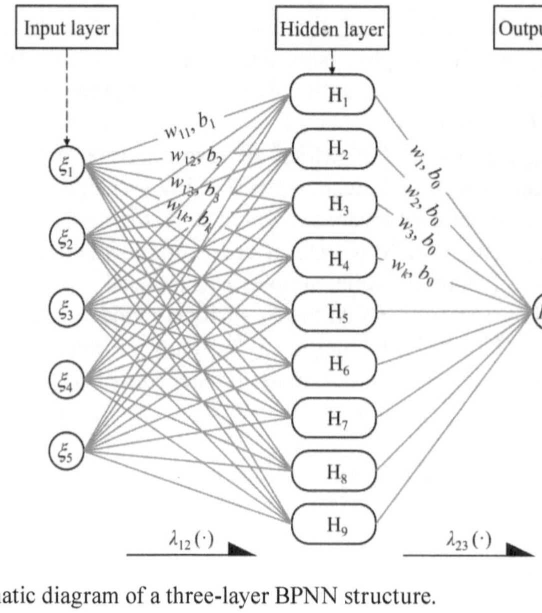

FIGURE 8.1 Schematic diagram of a three-layer BPNN structure.

reduce the number of trials. In the training of the BPNN-based model, the input and output datasets are generally normalized in the range of [–1, 1]. No accepted rule exists for precisely determining the number of neurons in the hidden layer, and a trial-and-error procedure is often adopted. Training of the BPNN model typically starts with the minimum number of neurons in the hidden layer. The hidden size is then augmented in increments of one neuron per training. Finally, an optimal number of neurons in the hidden layer is determined using the validation dataset according to the mean squared error obtained from each training. The first-order gradient descent–based back-propagation algorithm is then used to estimate the weight coefficients and thresholds (biases) in Eq. (8.6). After that, the limit state function for slope reliability analysis can also be derived as

$$g(\xi) = FS(\xi) - 1.0 \tag{8.7}$$

In this way, the values of $g(\xi)$ in slope reliability analysis can be computed by substituting the independent standard normal random samples into Eq. (8.7) without performing deterministic slope stability analyses.

8.2.3 Estimation of Probability of Slope Failure

The stability performance of a slope is commonly defined by the limit state function g, as mentioned earlier. The slope failure occurs when $g < 0$. The probability of slope failure P_f can be estimated as follows by computing the volume of $f(\boldsymbol{x})$ within the failure domain defined by the limit state function (e.g., Ang and Tang, 2007):

$$P_f = P\left[g(\boldsymbol{x}) < 0\right] = \int \cdots \iint_{g(\boldsymbol{x})<0} f(\boldsymbol{x})d\boldsymbol{x} \tag{8.8}$$

where $f(\boldsymbol{x})$ is the joint probability density function of $\boldsymbol{x}$. It is practically impossible to evaluate the n-fold integral in Eq. (8.8) because complete probabilistic information on the geotechnical parameters is often unavailable. Thus, simulation methods are commonly used to evaluate this integral. Having constructed the surrogate model between the *FS* of a slope and uncertain input parameters using the method outlined in Section 8.2.2, MCS can be adopted to efficiently estimate the probability of slope failure as follows:

$$P_f = \frac{1}{N_{sim}} \sum_{i=1}^{N_{sim}} I\left[FS(\boldsymbol{x}_i) < 1.0\right] \tag{8.9}$$

where N_{sim} is the total number of samples and $I(.)$ is the indicator function. For a given random sample, the indicator function is taken as 1 when $FS < 1.0$, otherwise it is equal to zero; $\boldsymbol{x}_i$ is the ith realization of random variables or random fields. For example, for each random sample during MCS, the surrogate model is employed to evaluate the *FS* explicitly and efficiently. As such, conventional deterministic slope analyses are not invoked again, and the computational cost used for each random sample is substantially reduced. In addition to MCS, the LHS, subset simulation (SS), importance sampling (IS) and other approximation methods, such as FOSM and FORM, can also be employed to estimate the probability of slope failure based on the trained surrogate model.

8.2.4 Quantitative Risk Assessment of Slope Failure

The risk assessment of slope failure involves the estimation of slope failure probability as well as the corresponding failure consequence. The overall risk of slope failure along multiple failure modes can be evaluated as (Zhang and Huang, 2016)

$$R = \int \cdots \iint_{g(\boldsymbol{x})<0} C_m(\boldsymbol{x}) f(\boldsymbol{x})d\boldsymbol{x} \tag{8.10}$$

where $C_m(\boldsymbol{x})$ denotes the consequence induced by slope failure. Similarly, it is practically impossible to evaluate the n-fold integrals in Eq. (8.10) since complete probabilistic information on the geotechnical parameters is often unavailable. To this end, MCS is used to approximate the R in Eq. (8.10) as follows (Jiang et al., 2017):

$$R \approx \frac{1}{N_{sim}} \sum_{i=1}^{N_{sim}} C_m(\boldsymbol{x}_i) I\left[FS(\boldsymbol{x}_i) < 1.0\right] = \frac{1}{N_{sim}} \sum_{j=1}^{N_r} C_m^j n_f^j \tag{8.11}$$

where C_m^j is the consequence corresponding to the jth failure mode, $j = 1, 2, \ldots, N_r$; n_f^j is the number of failure samples associated with the jth failure mode. Additionally, the contribution of each failure mode to the R can be quantified as

$$w_R^j = \frac{C_m^j n_f^j}{N_{sim} R} \tag{8.12}$$

To measure the accuracy in the estimated R, the coefficient of variation of R, COV_R, is proposed and defined as (e.g., Zhang and Huang, 2016; Jiang et al., 2017)

$$\mathrm{COV}_R = \sqrt{\frac{N_{sim} - n_f + \sum_{j=1}^{N_r} n_f^j \left(\frac{C_m^j}{R} - 1 \right)^2}{N_{sim}\left(N_{sim} - 1\right)}} \tag{8.13}$$

where n_f is the total number of failure samples for the slope, $n_f = \sum_{j=1}^{N_r} n_f^j$. To yield a good estimate of R, N_{sim} should be large enough such that the COV_R is below a commonly agreed value (e.g., 10%).

Note that the individual consequence needs to be assessed for each failure mode in landslide risk assessment using Eq. (8.11) (Huang et al., 2013). Based on the probabilities of failure underlying the N_r failure modes and the correlation coefficients among them, the bimodal bounds of P_f ignoring and considering the correlations among the slope failure modes can be obtained. By incorporating the consequences associated with different failure modes, the bimodal bounds of R ignoring and considering the correlations among the slope failure modes can be derived, respectively, as follows:

$$\max\left(P_{fi} C_m^i\right) \le R \le \sum_{i=1}^{N_r} P_{fi} C_m^i \tag{8.14}$$

$$P_{f1} C_m^1 + \sum_{i=2}^{N_r} \max\left(P_{fi} C_m^i - \sum_{j=1}^{i-1} P_{fij} C_m^{ij}, 0 \right) \le R \le \sum_{i=1}^{N_r} P_{fi} C_m^i - \sum_{i=2}^{N_r} \max_{j<i}\left(P_{fij} C_m^{ij}\right) \tag{8.15}$$

where P_{fi} is the probability of slope failure corresponding to the ith failure mode; P_{fij} is the joint probability of the ith and jth failure modes and its estimation can be referred to Ditlevsen (1979); C_m^{ij} is the consequence corresponding to slope failing along the ith and jth failure modes simultaneously, which is approximately taken as $\left(C_m^i + C_m^j\right)/2$. The volume V (or area in the two-dimensional [2D] case) of the sliding mass is often taken as an equivalent index to quantify the consequence of slope failure. To ensure a more complete and realistic assessment of the consequence of slope failure, the progressive failure of a slope should be modeled through large deformation analyses so that the true consequence of slope failure, including the runout distance, retrogression distance, and sliding volume of landslide, can be assessed.

Based on the above, Figure 8.2 presents the implementation flowchart of the NISFEM for slope stability and risk assessment. It should be highlighted that the implementation procedure has been

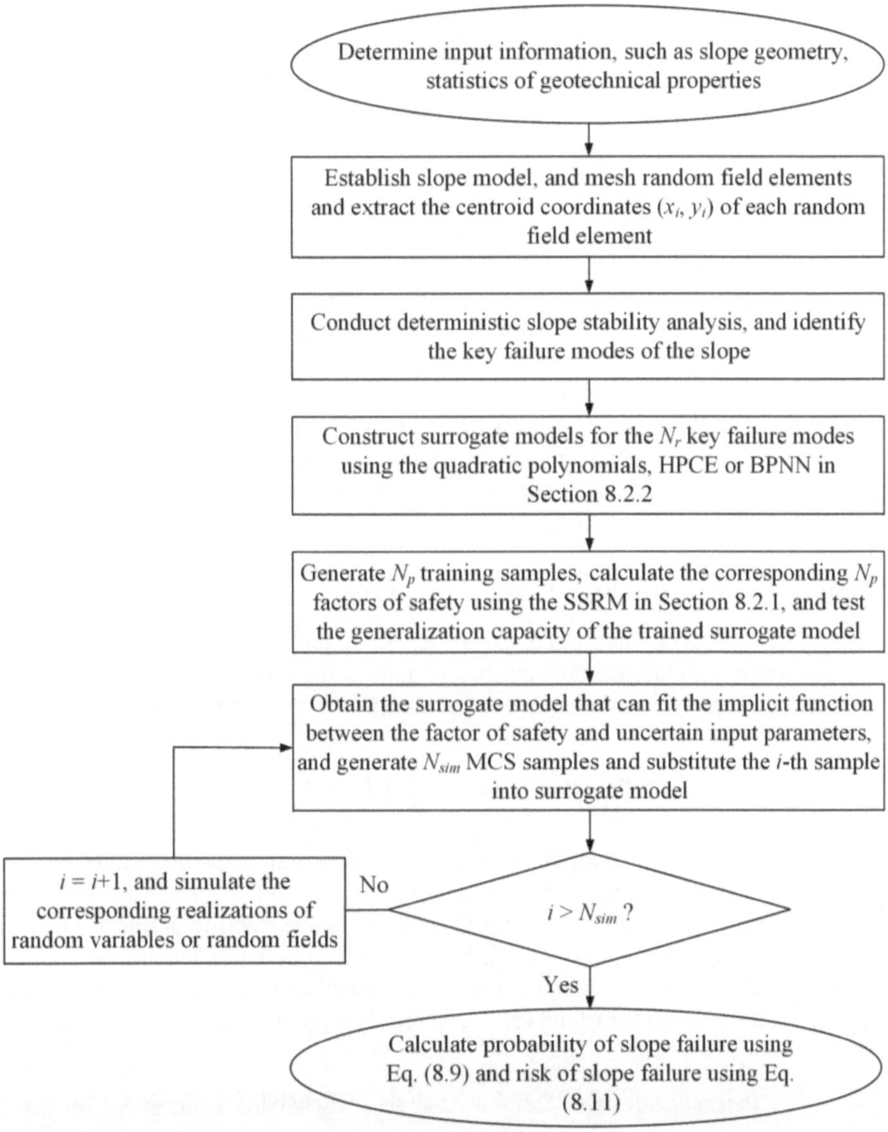

FIGURE 8.2 Implementation flowchart of NISFEM. (Source: Adapted from Jiang et al., 2017)

programmed as a user-friendly software package NIGPA (Non-Intrusive Geotechnical Probabilistic Analysis) with the aid of commercial software packages for deterministic slope stability analysis, such as ABAQUS, PLAXIS, FLAC3D, and SLOPE/W. By this means, the deterministic slope stability analysis is intelligently decoupled from the reliability and risk analyses (including the construction of surrogate models and MCS for uncertainty propagation), so that the reliability and risk assessments of slope stability can be executed as an extension of the deterministic slope stability analysis in a non-intrusive manner.

8.3 COLLABORATIVE STOCHASTIC FINITE ELEMENT METHOD

The CSFEM is another variant of the SFEM that improves computational efficiency by jointly utilizing two deterministic analysis models (Li et al., 2016a; Xiao et al., 2016, 2017b): (i) a relatively simple model with high computational efficiency, and (ii) a complex model with accurate computational responses. Taking slope stability analysis as an example, engineers are concerned with the probability P_f and risk R of slope failure that the FS is smaller than a given threshold fs (i.e., fs = 1.0), which is usually a rare event for a well-designed slope. Many deterministic models have been developed for slope stability analysis, among which the simple and complex models could be a LEM model versus a FEM model (Li et al., 2016a), a surrogate model versus a FEM model (Zhou et al., 2021), a coarsely meshed FEM model versus a finely meshed FEM model (Xiao et al., 2016), or a 2D FEM model versus a three-dimensional (3D) FEM model (Xiao et al., 2020). The simple model and complex model should be correlated to a reasonable degree as they share the same slope geometry, boundary condition, underground stratification, and soil properties. Therefore, the simple model–based SFEM could provide valuable guidance for the complex model–based SFEM. Inspired by this, the CSFEM consists of two major steps as shown in Figure 8.3: (1) preliminary analysis using a simple model and SS (Au and Beck, 2001) to efficiently approximate slope reliability and risk; and (2) target analysis using a complex model and the response conditioning method (Au, 2007) to achieve efficient and consistent reliability and risk assessment. To facilitate understanding, subscripts p and t in the following sections denote the estimates obtained from the preliminary and target analyses of the CSFEM, respectively.

8.3.1 Preliminary Analysis Using Subset Simulation

The SS (Au and Beck, 2001) stems from the idea that a small probability of failure can be expressed as a product of larger conditional probabilities of failure for some intermediate failure events, thereby converting a rare event into a sequence of more frequent events as follows:

$$P_f = P\left(F_m\right) = P\left(F_1\right)\prod_{k=2}^{m} P\left(F_k \middle| F_{k-1}\right) \tag{8.16}$$

where $F_k = \{g(\boldsymbol{x}) < g_k, k = 1, 2, \ldots, m\}$ are a set of intermediate failure events defined by a decreasing sequence of intermediate threshold values $g_1 > g_2 > \ldots > g_m = 0$, respectively; $g(\boldsymbol{x})$ is the limit state function and it can be taken as $g(\boldsymbol{x}) = FS(\boldsymbol{x}) - 1.0$ for a slope stability problem; $P(F_1) = P[g(\boldsymbol{x}) < g_1]$ and $P(F_k|F_{k-1}) = P[g(\boldsymbol{x}) < g_k|g(\boldsymbol{x}) < g_{k-1}]$, $k = 2, 3, \ldots, m$. In the implementation of the SS, the sample space is divided into $m + 1$ mutually exclusive and collectively exhaustive subsets Ω_k, $k = 0, 1, \ldots, m$, by the m intermediate threshold values, where $\Omega_0 = \{g(\boldsymbol{x}) \geq g_1\}$, $\Omega_k = \{g_{k+1} \leq g(\boldsymbol{x}) < g_k\}$ for

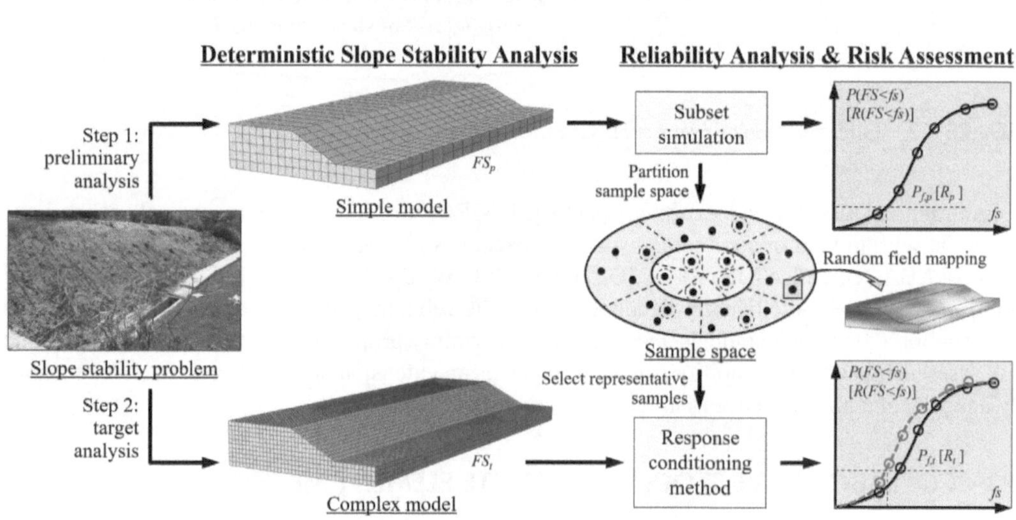

FIGURE 8.3 CSFEM for a slope stability problem. (Source: Adapted from Xiao et al., 2017b)

$k = 1, 2, \ldots, m-1$, and $\Omega_m = \{g(\boldsymbol{x}) < g_m\}$. To implement the SS, $\{g_1, g_2, \ldots, g_{m-1}\}$ can be determined adaptively so that the estimates of $P(F_1)$ and $P(F_k|F_{k-1})$, $k = 2, 3, \ldots, m-1$, always correspond to a common specified value of a conditional probability p_0 and $P(F_m|F_{m-1})$ can be estimated by the rate of failure samples ($g_m = 0$) in the last subset.

The SS generates a total of $mN(1 - p_0) + Np_0$ random samples, in which N is the number of samples at per subset level. These samples falling in different subsets have different probability weights w_k, which can be quantified by the occurrence probability of each subset, $P(\Omega_k)$ (i.e., $P(\Omega_k) = p_0^k(1 - p_0)$ for $k = 0, 1, \ldots, m-1$, and p_0^m for $k = m$), to the sample size in each subset, N_k (i.e., $N_k = (1 - p_0)N$ for $k = 0, 1, \ldots, m-1$, and $N_k = Np_0$ for $k = m$), namely $w_k = P(\Omega_k)/N_k$. This is different from the conventional MCS, in which the samples have the same weight of $1/N_t$. According to the total probability theorem, the preliminary probability of slope failure $P_{f,p}$ (i.e., Eq. [8.16]), based on the simple deterministic model can be calculated as

$$P_{f,p} = \sum_{k=0}^{m} P(F|\Omega_k)P(\Omega_k) = \sum_{k=0}^{m}\left[\frac{\sum_{j=1}^{N_k} I_{p,k}^{(j)}}{N_k}\right]P(\Omega_k) = \sum_{k=0}^{m}\sum_{j=1}^{N_k} I_{p,k}^{(j)} w_{p,k}^{(j)} \tag{8.17}$$

where $I_{p,k}^{(j)} = I_p(\boldsymbol{x}_k^{(j)})$ is the indicator of failure of sample $\boldsymbol{x}_k^{(j)}$ using the simple model (i.e., $I_{p,k}^{(j)} = 1$, if $g(\boldsymbol{x}_k^{(j)}) < 0$; otherwise $I_{p,k}^{(j)} = 0$); $\boldsymbol{x}_k^{(j)}$ is the jth sample falling in Ω_k; and $w_{p,k}^{(j)} = w_k = P(\Omega_k)/N_k$ is the sample weight of $\boldsymbol{x}_k^{(j)}$ in the preliminary analysis. Equation (8.17) is like the importance sampling but with a higher applicability to the high-dimensional problems. Similarly, the preliminary risk of slope failure R_p can be estimated using the same samples as

$$R_p = \sum_{k=0}^{m}\sum_{j=1}^{N_k} C_{p,k}^{(j)} w_{p,k}^{(j)} \tag{8.18}$$

where $C_{p,k}^{(j)}$ is the failure consequence corresponding to the jth sample in Ω_k based on the simple model. The consequence of slope failure is evaluated only when the *FS* calculated by a simple model, FS_p, is less than 1.0; otherwise $C_{p,k}^{(j)} = 0$. It can be proved that Eq. (8.18) equals the conventional definition of R, namely $R = P_f \times \bar{C}$ (Li et al., 2016b), where $\bar{C}$ is the average consequence of slope failure.

Although the preliminary analysis based on a simple model only provides approximate estimates of P_f and R, it can be executed with acceptable computational effort in engineering practice and provides valuable information and insights (e.g., Ω_k, $k = 0, 1, \ldots, m$, and random samples in these subsets) for understanding the slope stability performance. Such information can be incorporated into a more realistic complex model–based target analysis to obtain refined and consistent reliability and risk estimates.

8.3.2 Target Analysis Using Response Conditioning Method

For a complex model, the total sample size [i.e., $mN(1 - p_0) + Np_0$] in the SS is still not a small number. Considering that samples in the close neighborhood may have similar performance, it is reasonable to select some samples as the representative samples in the small sample space, as shown in Figure 8.4, which is referred to as the sub-binning strategy in the response conditioning method (Au, 2007). Specifically, Ω_k is further divided into N_s ($N_s << N$) equal sub-bins Ω_{kj}, $j = 1, 2, \ldots, N_s$, according to the FS_p values. One of the N_k/N_s samples in each Ω_{kj} is then randomly selected as a representative sample and is used as the input of the complex model to recalculate the safety margin of slope stability. By this means, the target probability of slope failure $P_{f,t}$ and the target risk of slope failure R_t based on the complex model are updated as

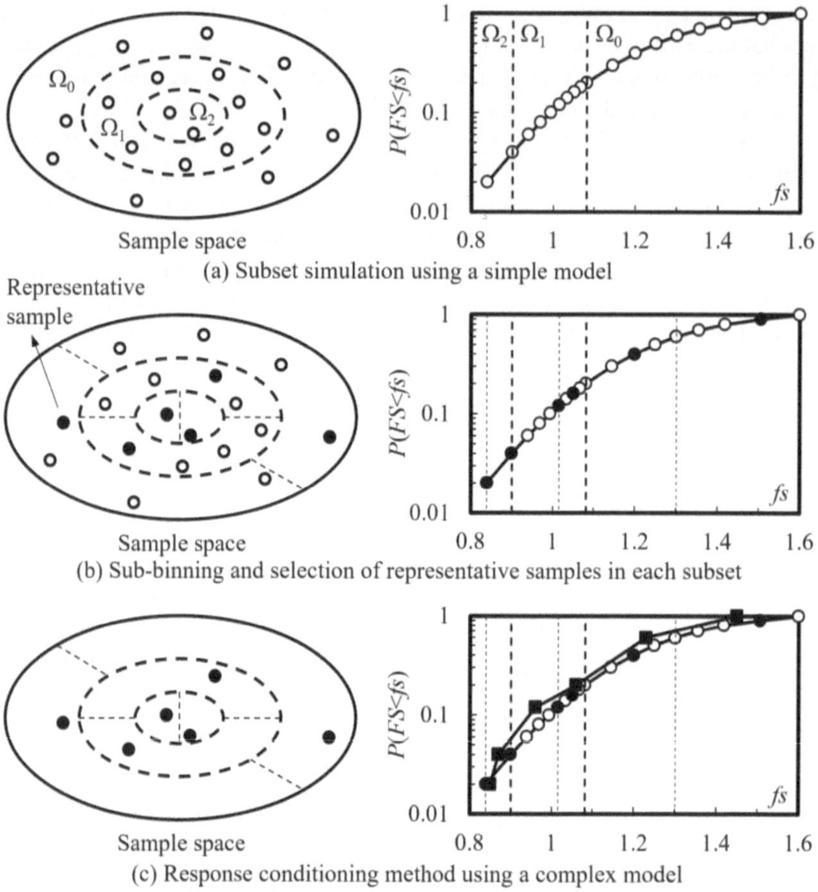

FIGURE 8.4 Schematic diagram of a subset simulation and response conditioning method. (Source: Adapted from Xiao et al., 2016)

$$P_{f,t} = \sum_{k=0}^{m} \sum_{j=1}^{N_s} I_{t,k}^{(j)} w_{t,k}^{(j)} \tag{8.19}$$

$$R_t = \sum_{k=0}^{m} \sum_{j=1}^{N_s} C_{t,k}^{(j)} w_{t,k}^{(j)} \tag{8.20}$$

where $w_{t,k}{}^{(j)} = P(\Omega_k)/N_s$ is the updated sample weight of $\boldsymbol{x}_k{}^{(j)}$ in the target analysis; and $I_{t,k}{}^{(j)}$ and $C_{t,k}{}^{(j)}$ are the indicator function and consequence of slope failure corresponding to the representative sample in Ω_{kj} based on the complex model, respectively.

Note that Eqs (8.19) and (8.20) are respective analogues of Eqs (8.17) and (8.18). Using the response conditioning method, only $(m + 1)N_s$ analyses of the complex model are required for estimating $P_{f,t}$ and R_t. This number is much less than that [i.e., $mN(1 - p_0) + Np_0$] required for directly performing the SS based on the complex model. The computational cost is substantially reduced by incorporating the information generated using the SS and the simple model in the preliminary analysis. More importantly, it can be shown that the target estimates using Eqs (8.19) and (8.20) are asymptotically unbiased (Au, 2007) compared with the conventional SFEM based on the same complex model.

8.3.3 Computational Issues in CSFEM

The CSFEM integrates the advantages of simple model–based preliminary analysis (i.e., computationally more efficient) and complex model–based target analysis (i.e., theoretically more accurate) through the sample space of random simulation. In this regard, the same uncertainty modeling strategy should be adopted in the preliminary and target analyses; otherwise a connection between the two uncertainty modeling strategies should be established. For example, if the spatial variability of soil properties is considered, the same random field modeling techniques should be adopted, regardless of the location-based methods (e.g., covariance matrix decomposition method) or the series expansion methods (e.g., KLE) (e.g., Li et al., 2019; Zhang et al., 2021); if 2D and 3D analyses are taken as the simple and complex models, respectively, the spatial variability should be modeled first in the 3D space and then exported as a 2D profile for the 2D analysis so that the two models can be correlated.

The computational effort of the CSFEM consists of two parts. The first part involves the evaluation of the $mN(1 - p_0) + Np_0$ analyses using the simple model in the preliminary analysis, and the second part involves the evaluation of the $(m + 1)N_s$ analyses using the complex model in the target analysis. Let d denote the ratio of the computational effort using the simple model over that using the complex model. The total computational effort of the CSFEM can be expressed in terms of the equivalent number N_T of the complex model as follows:

$$N_T = (m+1)N_s + d\left[mN(1-p_0) + Np_0\right] \tag{8.21}$$

The value of d depends on the models adopted in the calculation. When the value of d is relatively small, which means that the simple model is much more efficient than the complex model, the computational effort of the CSFEM mainly comes from that used for the $(m + 1)N_s$ complex analyses in the target analysis, which relies on N_s. Typically, N_s is small compared with N. To further improve efficiency, parallel computing strategies are introduced into the CSFEM for both deterministic analysis and uncertainty propagation (i.e., SS and response conditioning method). Although the total computational efforts of parallel computing and serial computing are equal in terms of sample size, parallel computing can reduce computational time because multiple computational tasks can be executed simultaneously. Samples from different Markov chains (i.e., Np_0) can be parallelized for the SS, and all selected samples [i.e., $(m + 1)N_s$] can be parallelized for the response conditioning method because they have been determined before the target analysis.

The computational error of the CSFEM has three parts: (1) errors in the sample space partition based on the simple model in the preliminary analysis; (2) errors in the choice of representative samples to characterize the entire sub-space in the target analysis; and (3) errors in the correlation between the simple and complex models. The first errors can be minimized by improving the accuracy of SS (i.e., with a larger N value), whose variability can be estimated approximately (Au and Beck, 2001). The second errors can be minimized by increasing the sample space partition of the response conditioning method (i.e., with a larger N_s value), as discussed in Li et al. (2016a). The third errors reduce as the correlation between the simple model and complex model increases. Although it is difficult to establish the relationship between the model correlation and estimation variability, the correlation can be taken as an indicator to select a proper simple model in the CSFEM.

Note that there are $(m + 1)N_s$ samples evaluated by both the simple and complex models. Let FS_S and FS_C be the factors of safety evaluated by the simple and complex models, respectively. The correlation coefficient ρ between the two models can be calculated as

$$\rho = \frac{\mathrm{E}(FS_S \times FS_C) - \mathrm{E}(FS_S)\mathrm{E}(FS_C)}{\sqrt{\mathrm{D}(FS_S)}\sqrt{\mathrm{D}(FS_C)}} \tag{8.22}$$

where E(·) and D(·) are sample expected value and sample variance, respectively. As the $(m + 1)N_s$ samples are distributed in different sample spaces, Ω_k, $k = 0, 1, \ldots, m$, with different probability weights, a weighted summation should be adopted as follows:

$$\mathrm{E}(FS_S \times FS_C) = \sum_{k=0}^{m} \mathrm{E}(FS_S \times FS_C)_k P(\Omega_k) \tag{8.23}$$

$$\mathrm{E}(FS_M) = \sum_{k=0}^{m} \mathrm{E}(FS_M)_k P(\Omega_k) \tag{8.24}$$

$$\mathrm{D}(FS_M) = \sum_{k=0}^{m} \mathrm{E}(FS_M^2)_k P(\Omega_k) - \mathrm{E}(FS_M)^2 \tag{8.25}$$

where $\mathrm{E}(\cdot)_k$ is the expected value of the responses of the N_s samples selected in Ω_k; and FS_M represents FS_S or FS_C. After substituting Eqs (8.23)–(8.25) into Eq. (8.22), the correlation between the simple and complex models can be obtained. If the correlation coefficient is relatively small (e.g., <0.6), which may be caused by unreasonable simplifications in the simple model, the estimated $P_{f,t}$ and R_t using the CSFEM should be used with caution as the estimation variation would be high. Some adaptive techniques (Zhou et al., 2021) should be adopted to update the simple model to achieve a higher correlation and therefore a smaller variation.

8.4 EXAMPLE #1: A TWO-LAYERED COHESIVE SLOPE

This section applies the two proposed SFEMs to analyze the reliability of a two-layered cohesive slope, which has been studied by Huang et al. (2013), Li et al. (2016a, 2016b), Jiang and Huang (2016), Zhang and Huang (2016), and Jiang et al. (2017). The slope model and source codes of slope reliability and risk analysis for Example #1 are available as soft copy at: www.routledge.com/Uncertainty-Modeling-and-Decision-Making-in-Geotechnics/Phoon-Shuku-Ching/p/book/9781032367491. As shown in Figure 8.5, the cohesive slope has a height of 24 m and a slope angle of 36.9°. The cohesive soil layer extends to 28 m below the top of the slope and has a unit weight of 19 kN/m³. The undrained shear strengths of the two layers, S_{u_1} and S_{u_2}, are modeled as lognormal random fields. In the upper soil layer, the random field S_{u_1} has a mean value of 80 kPa and a COV of 0.3. In the lower soil layer, the random field S_{u_2} has a mean value of 120 kPa and a COV of 0.3.

To enable the finite difference analysis of slope stability, Young's modulus and Poisson's ratio of the two soil layers are assumed to be 100 MPa and 0.3, respectively (Griffiths and Lane, 1999). Using the information pertaining to the slope geometry and soil properties, a finite difference model of the slope is created using FLAC3D in this example. As shown in Figure 8.6, the upper

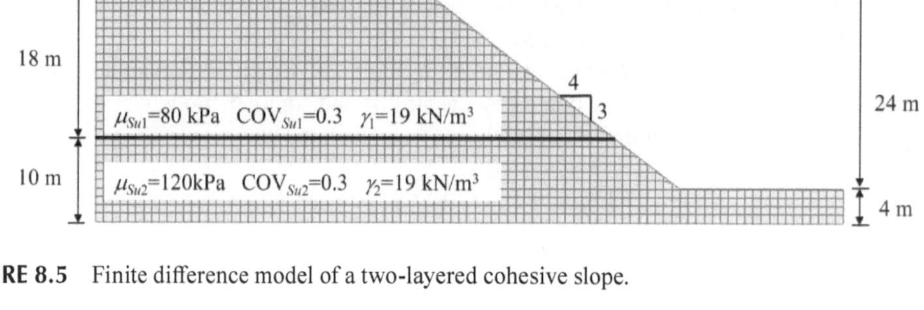

FIGURE 8.5 Finite difference model of a two-layered cohesive slope.

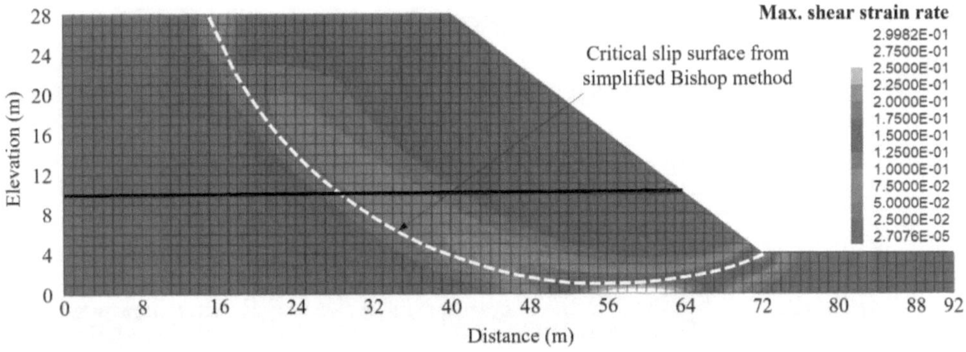

FIGURE 8.6 Deterministic slope stability analysis result.

layer and lower layer are discretized into 954 and 782 eight-node quadrilateral elements with a side length of 1.0 m, respectively, which are degenerated into three-node triangular elements along the slope. The boundary conditions are rollers on both lateral boundaries and full fixity at the base. Based on the mean values of S_{u_1} and S_{u_2}, the finite difference analysis of slope stability is performed using an elastic-perfectly plastic constitutive model with a Mohr–Coulomb failure criterion. The nominal value of *FS* calculated using the SSRM is equal to 1.418, which is close to the value (i.e., 1.443) reported in Li et al. (2016a, 2016b) calculated using the FEM in ABAQUS and the value (i.e., 1.448) reported in Jiang and Huang (2016) using the LEM (i.e., simplified Bishop method). As shown in Figure 8.6, the maximum shear strain contour obtained from the FDM matches with the critical slip surface obtained from the simplified Bishop method. The failure consequence (i.e., sliding mass volume in this example) corresponding to the critical slip surface is about 680.6 m^2.

To account for the inherent spatial variability of S_{u_1} and S_{u_2}, a 2D single exponential autocorrelation function with horizontal and vertical scales of fluctuation of $\delta_h = 24$ m and $\delta_v = 2.4$ m is used. The random field meshes in two soil layers are consistent with the corresponding finite element meshes, as shown in Figure 8.5. Based on the KLE and the required maximum error in the random field discretization (i.e., expected energy ratio $\geq 85\%$), $M = 115$ and 270 KLE terms are needed for the discretization of the random fields of S_{u_1} and S_{u_2}, respectively. Figures 8.7a and b show two typical realizations of the random fields of S_{u_1} and S_{u_2}.The dark and light shaded regions indicate areas of high and small undrained shear strengths, respectively. The corresponding results of slope stability analysis obtained from the FDM, including the values of *FS* (i.e., 1.222 and 1.002), the critical slip surfaces identified by the K-means clustering method (e.g., Huang et al., 2013) (see the dash line), and the failure consequences of about 689.5 m^2 and 902.8 m^2, are also plotted in Figures 8.7a and b, respectively.

8.4.1 Reliability Analysis Results

The BPNN is used as the surrogate model between the *FS* of the slope and the 385 (i.e., 115+270) independent standard normal random variables when the NISFEM is first adopted to evaluate the reliability of the slope. The numbers of neurons in the input layer and the output layer are 385 and 1, respectively. Based on the tradeoff between computational accuracy and efficiency, 2700 sets of independent standard normal random samples are generated using the LHS technique and the corresponding values of *FS* are calculated using the SSRM in FLAC3D. They are selected as the training samples for establishing the BPNN model. The first-order gradient descent–based back-propagation algorithm is then used to estimate the weight coefficients and thresholds (biases) in Eq. (8.6). The Levenberg–Marquardt-type function is selected as the activation function. Based on these configurations, the number of neurons in the hidden layer is determined as 22 after a

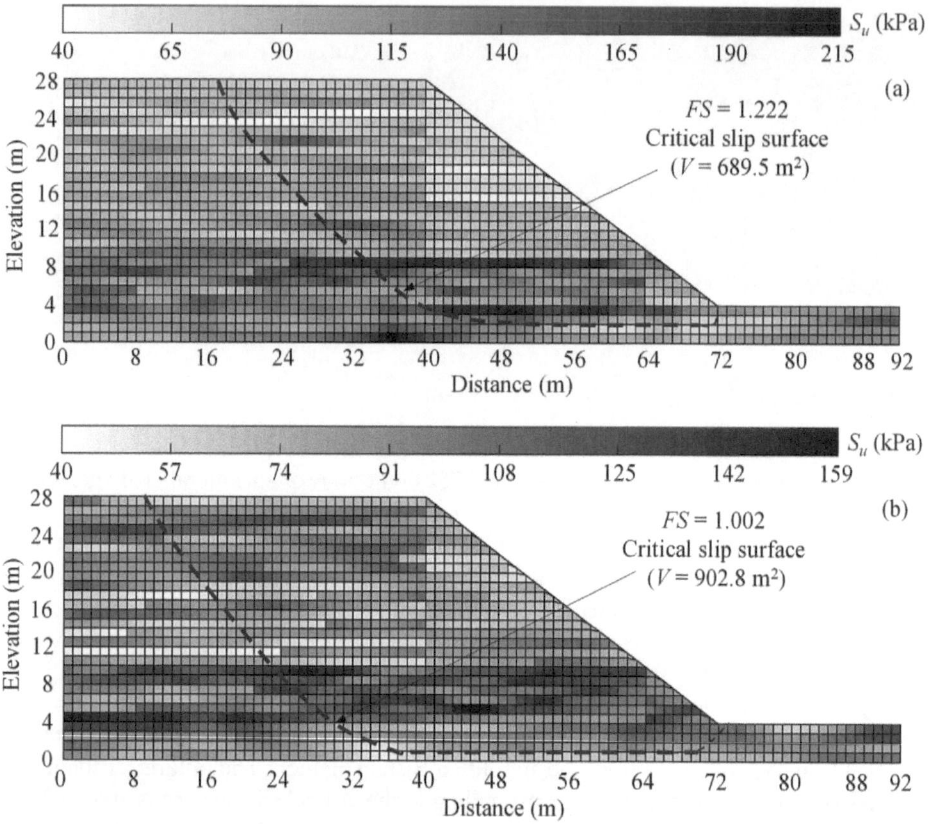

FIGURE 8.7 Two random realizations of undrained shear strength and slope stability results.

trial-and-error test is conducted. In this case, the obtained mean squared error using the validation dataset remains almost unchanged as the number of neurons in the hidden layer increases.

To verify the generalization capacity of the BPNN-based surrogate model, the *FS* of the slope obtained from the BPNN-based surrogate model is compared with that determined from the original deterministic analysis (e.g., the FDM) of slope stability. Figure 8.8 compares the values of *FS* obtained from the BPNN-based surrogate model versus those determined from the FDM using the testing dataset including 150 LHS samples, which were not involved in the training of the BPNN-based surrogate model. As observed from Figure 8.8, the values of *FS* obtained from the two approaches are in good agreement. The data clusters distribute tightly along the 1:1 line with a coefficient of determination R^2 of 0.928. It indicates that the BPNN-based surrogate model can well approximate the slope stability model in the entire sampling space.

The probability of slope failure is then calculated using the LHS with 100,000 samples predicted using the trained BPNN-based surrogate model, and the value is 1.03×10^{-4}. As shown in Table 8.1, the probability of slope failure calculated from the NISFEM is well consistent with the values (i.e., 1.83×10^{-4}, 1.71×10^{-4}, and 1.79×10^{-4}) calculated from the MCS ($N_{sim} = 60{,}000$), the SS ($N = 500$ and $p_0 = 0.1$), and the CSFEM (LEM in preliminary analysis with $N = 5000$, $p_0 = 0.1$ and FEM in target analysis with $N_s = 50$) in conjunction with the FEM, respectively. The probability of slope failure obtained from the NISFEM is slightly smaller than values (i.e., 4.86×10^{-4} and 5.13×10^{-4}) calculated from the SS and MCS in conjunction with the LEM as reported in Li et al. (2016a) and Jiang et al. (2017), respectively. This is not surprising since the adopted FEM/FDM in the NISFEM can automatically locate the most dangerous slip surface which may not be considered in the LEM.

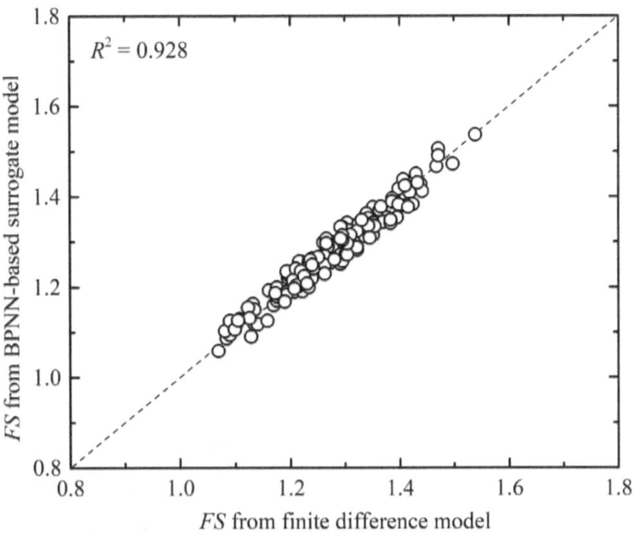

FIGURE 8.8 Validation of BPNN-based surrogate model.

TABLE 8.1
Comparison of Slope Reliability Analysis Results in Example #1

Reliability method	Deterministic method	N_{sim}	P_f	Source
MCS	FEM	60,000	1.83×10^{-4}	Li et al. (2016a)
SS	FEM	1850	1.71×10^{-4}	Li et al. (2016a)
SS	LEM	1850	4.86×10^{-4}	Li et al. (2016a)
MCS	LEM	300,000	5.13×10^{-4}	Jiang et al. (2017)
NISFEM	FDM	2700	1.03×10^{-4}	This chapter
CSFEM	LEM+FEM	304	1.79×10^{-4}	This chapter

Note: MCS: Monte Carlo simulation; SS: subset simulation; NISFEM: non-intrusive stochastic finite element method; CSFEM: collaborative stochastic finite element method; LEM: limit equilibrium method; FEM: finite element method; FDM: finite difference method.

8.4.2 Risk Analysis Results

In this section, the NISFEM with N_{sim} = 300,000 MCS samples is further employed to evaluate the risk R of slope failure. By adopting the approach developed by Jiang et al. (2017) for the identification of key failure modes of a slope, a total of $N_r = 74$ key failure modes are identified. The R estimated using Eq. (8.11) based on these $N_r = 74$ key failure modes is 0.334 m^2. The fluctuation on the R falls within a tolerable range and the estimated $COV_R = 8.2\%$ is below 10%, which indicates satisfactory accuracy.

To explore the effects of horizontal and vertical spatial variability on the risk of slope failure, Figures 8.9a and b show the values of R estimated from the NISFEM as a function of Θ_h and Θ_v, respectively. Here, the dimensionless spatial correlation lengths are used, which are defined as $\Theta_h = \delta_h / H$ and $\Theta_v = \delta_v / H$, respectively, to diminish the impact of slope height. In Figure 8.9a, the results are obtained at Θ_h varying from 0.5 to 2.5 and Θ_v = 0.1. In Figure 8.9b, the results are obtained at Θ_h = 1.0 and Θ_v varying from 0.05 to 0.25. The corresponding δ_h and δ_v are in the ranges of [12 m, 60 m] and [1.2 m, 6.0 m], respectively, which are in agreement with the

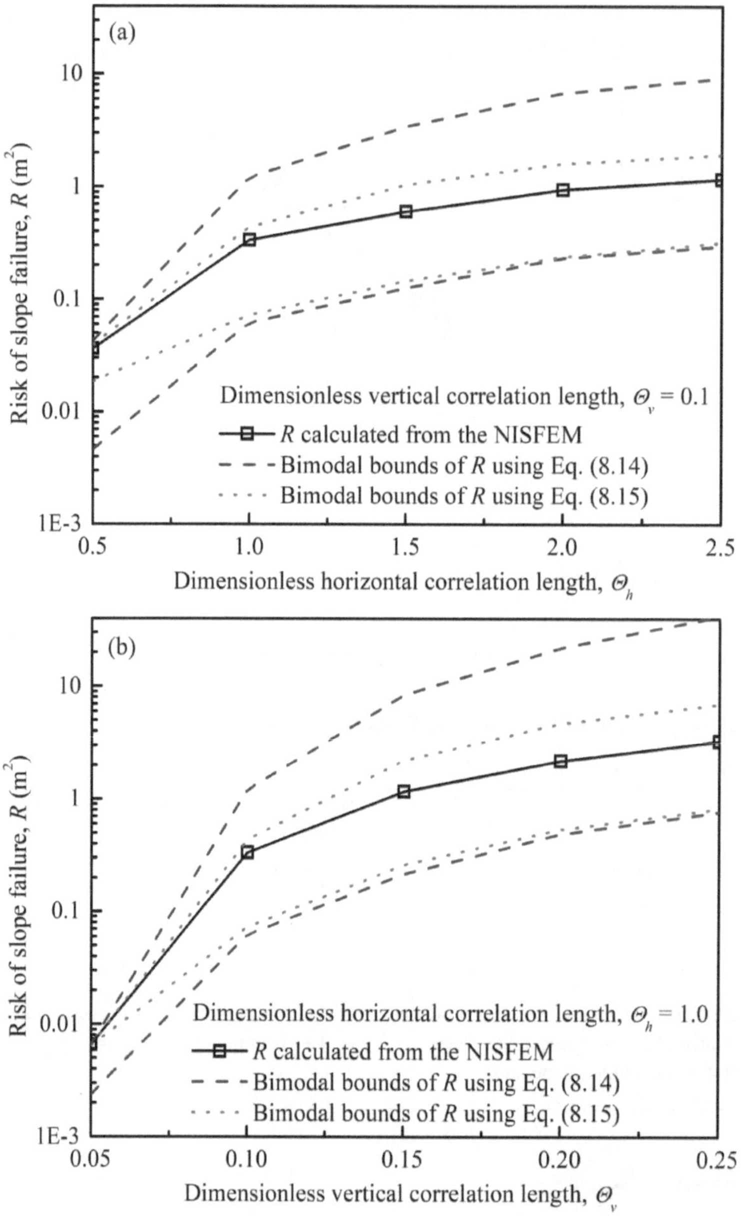

FIGURE 8.9 Effects of spatial correlation lengths on slope failure risk: (a) horizontal correlation; (b) vertical correlation.

typical ranges of the scales of fluctuation reported in the geotechnical literature (e.g., Phoon and Kulhawy, 1999a). The bimodal bounds of R calculated for the various values of Θ_h and Θ_v using Eqs (8.14) and (8.15) are also plotted in Figure 8.9 for comparison. The values of R obtained from the NISFEM match with the bimodal bounds of R. The results indicate that the influence of the correlations among the slope failure modes on the risk analysis can be well accounted for by the NISFEM. The obtained bimodal bounds of R become wider as the Θ_h or Θ_v increases, which is attributed to the stronger correlations among the failure modes for more spatially smoothly transitioned soil properties. The bimodal bounds of R become much wider if the correlations are ignored or not well addressed, and hence a significant biased estimate of R could be induced (see the dash lines in Figure 8.9). Additionally, the vertical spatial variability affects the R more profoundly than the horizontal spatial variability in such a 2D problem, which is in accordance with the influence of the spatial variability on the probability of slope failure (e.g., Ji et al., 2012; Jiang et al., 2015).

Figure 8.10 further compares the values of R estimated from the NISFEM and the CSFEM as Θ_v varies from 0.1 to 1.0 for a given $\Theta_h = 1.0$. The values of R estimated from the NISFEM are marginally larger than those obtained from the CSFEM, as the deterministic *FS* of the FDM in the NISFEM (i.e., 1.418) is slightly lower than that of the FEM in the CSFEM (i.e., 1.443). Such results indicate that the proposed two SFEMs can effectively evaluate the risk of slope failure considering the 2D spatial variation of soil properties. In addition, both can properly incorporate the influence of the spatial variability of soil properties on the distribution of slope key failure modes. For a given $\Theta_h = 1.0$, the N_r increases from 3 to 283 as Θ_v varies from 0.05 to 0.5, and then slightly decreases to 270 at $\Theta_v = 1.0$.

Moreover, the contributions of each key failure mode to the probability and risk of slope failure can be quantified using the NISFEM. As stated previously, the number of failure samples associated with a failure mode can reflect the contribution of the failure mode to the probability of slope failure. Figure 8.11a depicts the histogram of the number of failure samples associated with the 74 key failure modes and the slope system failure. Figure 8.11b presents the histogram of the weights of the 74 key failure modes on the risk of slope failure. As observed from Figure 8.11, the contributions of the same failure mode to the probability and risk of slope failure are generally consistent, but the contributions of different key failure modes to the risk of slope failure differ significantly. The key failure modes associated with deep sliding typically contribute more to the risk of slope failure than those associated with shallow sliding. The first key failure mode contributes the least to the risk of slope failure (e.g., $w_R^1 = 0.25\%$), whereas the weight of the 46th key failure mode on the risk of slope failure is the largest (e.g., $w_R^{46} = 7.56\%$) (see Figure 8.11b). It is deduced that the location of the 46th key failure mode can provide an important implication for working out mitigation measures for landslide risk. It is also worthwhile highlighting that the failure mode identified from the deterministic analysis is not the failure mode that contributes the greatest to the risk of slope failure in this example.

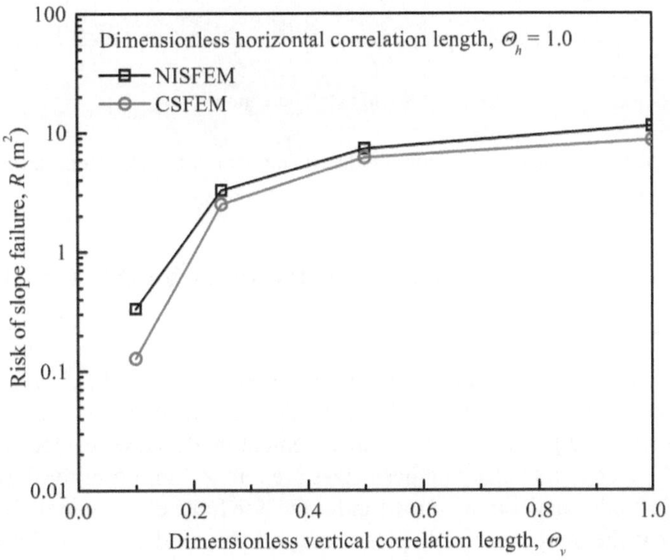

FIGURE 8.10 Comparison of NISFEM and CSFEM.

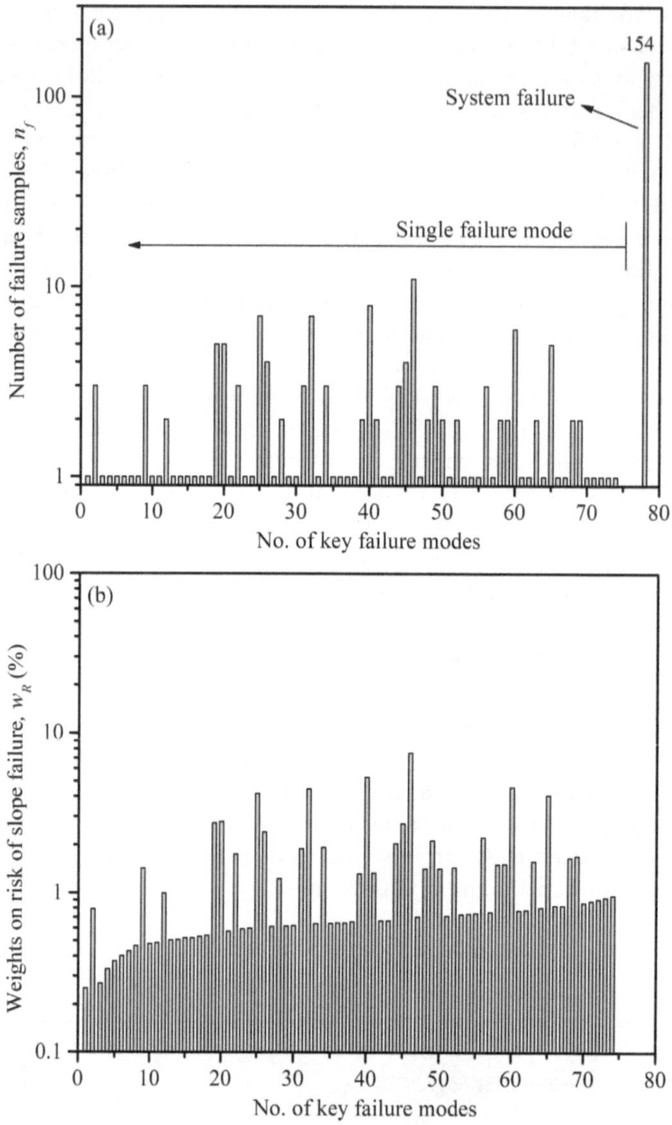

FIGURE 8.11 Frequency and weight of failure modes: (a) number of failure samples; (b) weights on risk of slope failure.

8.5 EXAMPLE #2: A 3D LONG SLOPE IN SPATIALLY VARYING SOILS

8.5.1 Benchmark Analysis

This section applies the CSFEM to evaluate the failure probability and risk of a 3D long slope in spatially varying soils. As shown in Figure 8.12, the slope has a height of 6 m, a slope angle (α) of about 26.6°, and a length (B) of 100 m. Two finite element models are developed in ABAQUS, as shown in Figure 8.13. The finite element mesh sizes are 2 m × 2 m × 5 m for the coarse finite element model (simple model) and 1 m × 1 m × 1 m for the fine finite element model (complex model). For the soil property, the elastic-perfectly plastic constitutive model with Mohr–Coulomb failure criterion is used in both finite element analyses. The undrained shear strength S_u is considered to be lognormally distributed with a mean value of 30 kPa and a COV of 0.3. The spatial variability

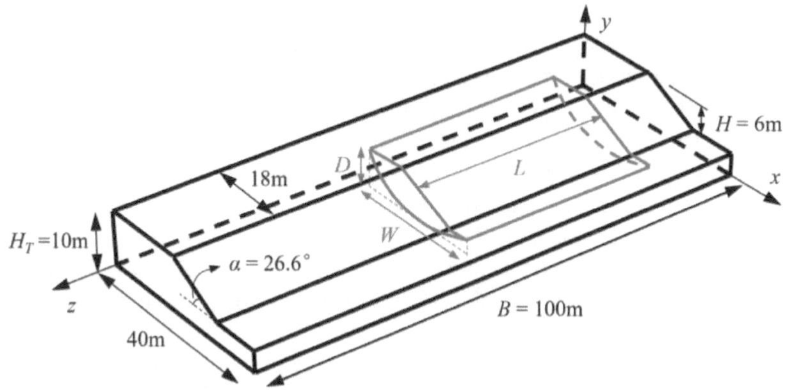

FIGURE 8.12 3D undrained slope example.

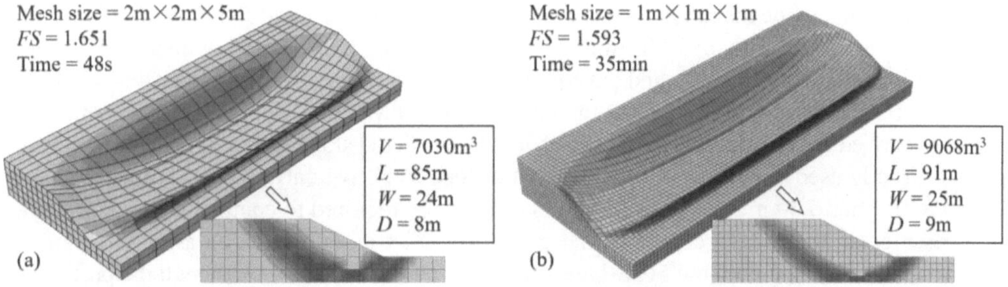

FIGURE 8.13 Deterministic slope stability results: (a) simple model; (b) complex model.

of S_u is modeled using a squared exponential autocorrelation function with horizontal and vertical autocorrelation distances of l_h = 20 m and l_v = 2 m, respectively. The unit weight, Young's modulus, and Poisson's ratio of soil are 20 kN/m³, 100 MPa, and 0.3, respectively.

Figure 8.13 shows the results of the deterministic slope stability analysis based on the mean value of S_u, which take 48 s and 35 min, respectively, on a desktop with 8 GB RAM and one Intel Core i3 CPU clocked at 3.3 GHz. The values of *FS* calculated by the coarse and fine finite element models using the SSRM in ABAQUS are 1.651 and 1.593, respectively, and the corresponding sliding mass volumes (*V*) are 7030 m³ and 9068 m³, respectively. The failure modes (i.e., critical slip surfaces) identified by the two models are similar and nearly cylindrical. Their sliding mass lengths are almost the same as the slope length. These results appear to be similar to those of the 2D analysis, namely, sliding along the entire slope length from the 3D perspective. This is because the slope is relatively long and the soil is homogeneous without considering the spatial variability, which basically satisfies the assumptions adopted in the 2D analysis. To better understand the slope failure mode, several shape characteristics of the slip surface are also evaluated, such as the sliding mass length (*L*), the sliding mass width (*W*), and the sliding mass depth (*D*), as defined in Figure 8.12. All these values are referred to as the maximum value along one particular direction. If there is more than one sliding mass along the Z direction, the *V* and *L* are taken as the summation of all sliding masses. The coarse finite element model slightly overestimates the *FS* and underestimates the scale of sliding mass in all directions. This may lead to unconservative estimates of P_f and *R* in the probabilistic slope stability analysis.

The failure mechanism of such a 3D spatially varying slope is significantly different from that of a 3D homogeneous slope. A typical random field realization of the slope is shown in Figure 8.14

using the expansion optimal linear estimation approach (Sudret and Der Kiureghian, 2000). The corresponding *FS* of the 3D slope stability analysis calculated by the fine finite element model is 0.741, which implies that the slope fails. Its slip surface is nearly spherical with a small sliding mass length (i.e., 24 m) located from 19.5 m to 43.5 m in the Z direction. The 3D heterogeneous slope considering the soil spatial variability models predicts real slope failure events that are more realistic than the 3D homogeneous slope in terms of the shape, location, and length of the slip surface. Several cross sections are extracted from the 3D realization to perform 2D finite element analyses. As shown in Figure 8.14, the values of 2D *FS* and slip surfaces vary along the Z direction of the slope. Although the 2D analysis could be more conservative than the 3D analysis based on the cross section with the minimal 2D *FS*, the location of the 3D critical slip surface would remain unknown if the 3D analysis is not performed.

The CSFEM is applied to estimate the probability and risk of slope failure, with $m = 4$, $N = 500$, and $p_0 = 0.1$ in the preliminary analysis using the coarse finite element model and $N_s = 25$ in the target analysis using the fine finite element model. The sliding mass volume is taken as an equivalent index to quantify the slope failure consequence in this study. The preliminary analysis yields $P_{f,p} = 8.84 \times 10^{-4}$ and $R_p = 1.77$ m^3 using 1850 coarse finite element analyses and requires about 7 h using parallel computing, while the target analysis updates $P_{f,t} = 2.80 \times 10^{-3}$ and $R_t = 7.09$ m^3 with 125 fine finite element analyses in about 27 h using parallel computing. In total, approximately 34 h (or 1.4 days) are required by the CSFEM, and the equivalent number of the complex model analyzed is 162 (where d is 1/50 on average). To validate the results, a direct MCS-based NISFEM run with 10,000 samples is carried out to calculate the P_f and R of the considered slope, where the fine finite element model is directly used to perform the deterministic slope stability analysis. The estimated P_f and R are 3.20×10^{-3} and 7.00 m^3, respectively, and 89.9 days are required to complete the reliability and risk assessments. Figure 8.15 compares the cumulative distribution function and the cumulative risk function of *FS* using both the MCS-based NISFEM and the CSFEM. Although the preliminary analysis of the CSFEM underestimates the probability and risk of slope failure, the target analysis of the CSFEM corrects the results to be consistent with those of the conventional MCS-based NISFEM for all *fs* values. These results validate that the CSFEM can produce unbiased estimates

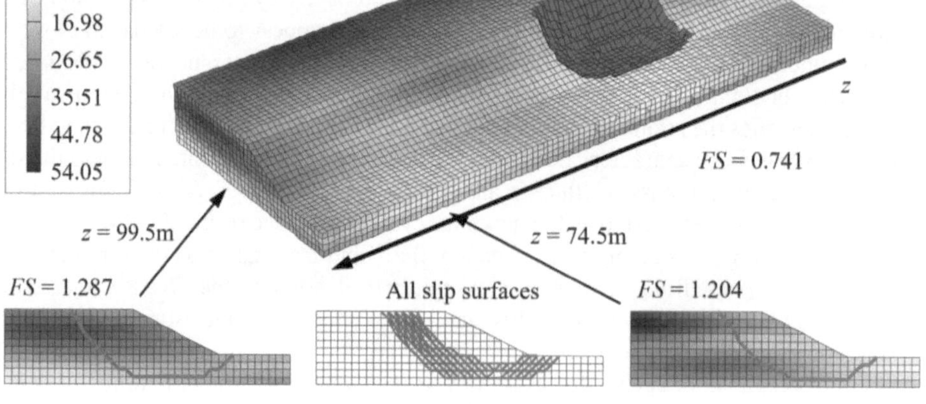

FIGURE 8.14 Results of the 2D and 3D analyses for a typical random field realization.

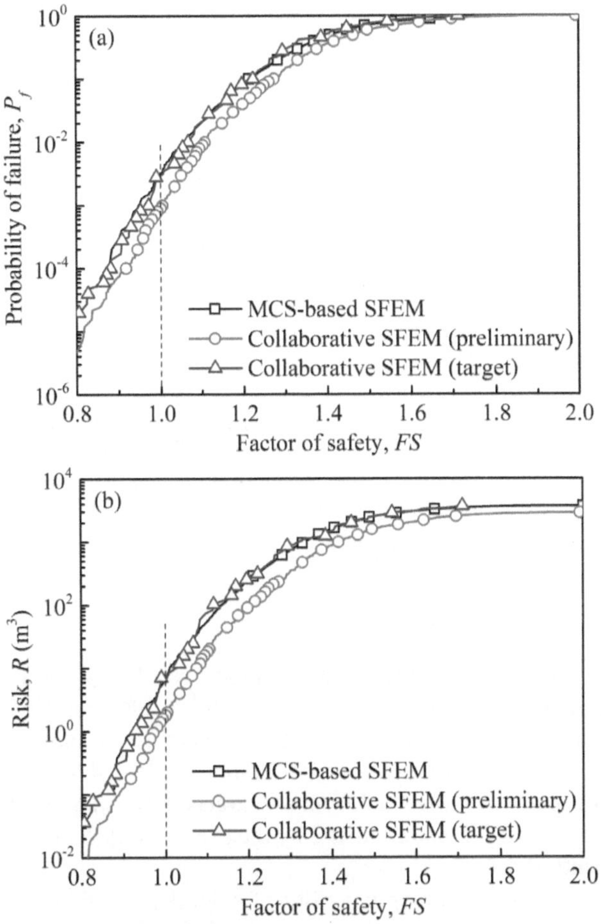

FIGURE 8.15 Comparison of MCS-based SFEM and CSFEM: (a) cumulative distribution function; (b) cumulative risk function.

of probability and risk of slope failure compared with the MCS-based NISFEM with significantly reduced computational cost by over 60 times in this example.

Figure 8.16 compares the finite element responses (i.e., *FS*, *V*, *L*, *W*, and *D*) of the selected 125 representative samples calculated by both the coarse and fine finite element models, and illustrates the 1:1 lines and the respective linear regression lines for reference. To enable a fair comparison, all the responses are normalized by the deterministic results of the fine finite element model, that is, $FS = 1.593$, $V = 9068$ m^3, $L = 91$ m, $W = 25$ m, and $D = 9$ m. Although the linear regression lines do not overlap with the 1:1 line, these finite element responses are well correlated. The *FS* has the highest correlation coefficient of 0.99, followed by *V* (0.96), *L* (0.93), *W* (0.82), and *D* (0.75). The high correlations indicate that the coarse finite element model used in the preliminary analysis is appropriate and can well reflect the main features, particularly the *FS*, of the fine finite element model. In addition, similar to the deterministic slope stability analysis, using the coarse finite element model generally leads to overestimations of *FS* and *W* and underestimations of *V*, *L*, and *D*, which subsequently results in underestimations of P_f and *R*. Such differences become more obvious as the responses increase, which further justifies the need for a corrected reliability analysis based on the results of a preliminary analysis of the CSFEM and the acknowledgment of the difference between the deterministic analyses of simple and complex models (Xiao et al., 2020).

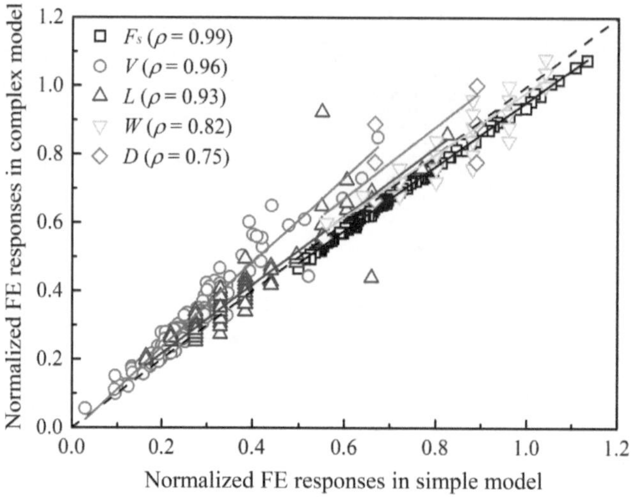

FIGURE 8.16 Correlation between a simple model and a complex model.

8.5.2 Effect of 3D Spatial Variability

The effects of spatial variability on 3D slope failure mode, reliability, and risk are explored with the aid of the CSFEM. In addition to the nominal case with $l_h = 20$ m and $l_v = 2$ m, eight cases with different autocorrelation distances are also considered, including (i) four cases with $l_h = [10, 40, 80, 120]$ m and $l_v = 2$ m, and (ii) four cases with $l_h = 20$ m and $l_v = [1, 4, 8, 12]$ m. For simplicity, all results presented in this section are obtained from the target analysis in the CSFEM with $fs = 1$. To make a fair comparison, l_h and l_v are normalized by the slope length B and the nominal height H_T (see Figure 8.12), respectively.

Figure 8.17a shows the variation of probability of slope failure as a function of normalized autocorrelation distances. When the normalized autocorrelation distance increases from 0.1 to 1.2, P_f increases by several orders of magnitude, and the influence weakens when l_h exceeds half of the slope length or l_v exceeds the slope height. In addition, the vertical spatial variability has a greater impact on P_f than the horizontal spatial variability. By performing statistical analysis on the *FS*, it can be found that, for such a small probability problem, the increase in P_f is mainly attributed to the increase in the variance of *FS* (see Figure 8.17b). As a matter of fact, the increasing variation of *FS* is further related to the spatial average phenomenon in spatially variable soils (Vanmarcke, 1977), which suggests that the variance for soil property along a particular slip surface reduces as the autocorrelation distance decreases.

Regarding the failure mechanism of the slope, the variation of average volume, length, width, and depth of sliding mass are shown in Figures 8.17c and d. Apparently, the horizontal and vertical spatial variabilities have conflicting influences on the average sliding mass volume and length in this example. Both responses increase as the normalized horizontal autocorrelation distance increases, and slightly decrease as the normalized vertical autocorrelation distance increases. The variation of the average sliding mass volume is relatively smaller than that of the average sliding length. Note that the average sliding mass volume is equivalent to the average failure consequence $\bar{C}$ in this study. Therefore, R (i.e., $P_f \times \bar{C}$) is more sensitive to P_f than $\bar{C}$. Compared with the 2D slope risk assessment (Li et al., 2016b), the effects of vertical spatial variability on P_f, $\bar{C}$, and R of a 3D slope are consistent with the observations in the 2D analysis. This implies that the 2D analysis can properly account for the vertical spatial variability.

In addition, since the average sliding mass width and depth remain almost unchanged as the normalized autocorrelation distance varies in this example (see Figure 8.17d), the variation of the

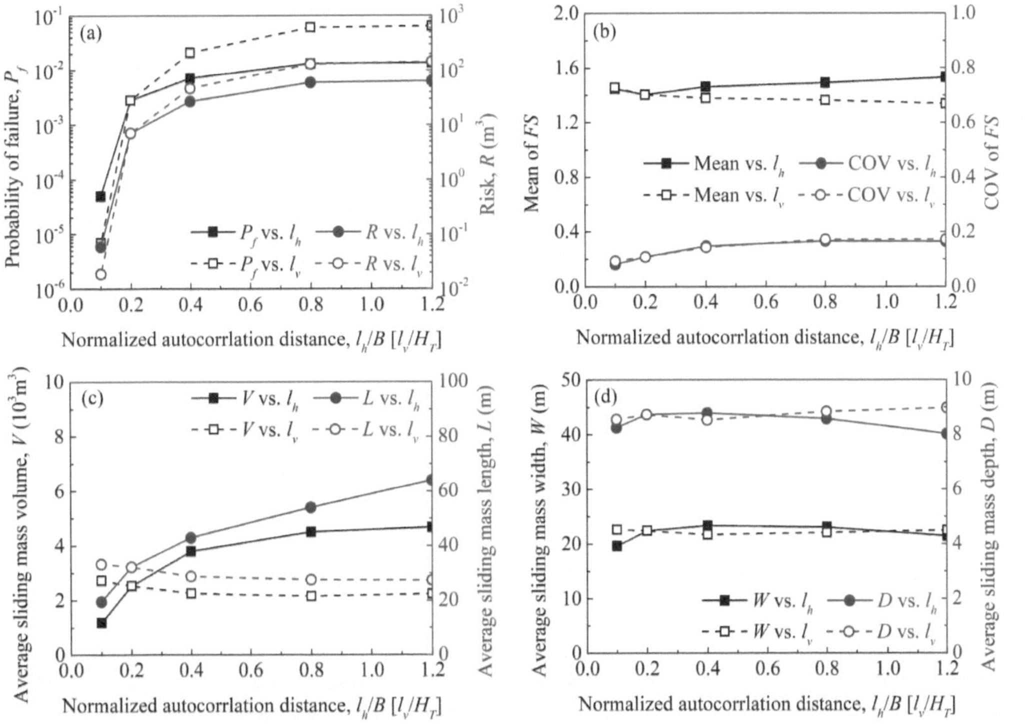

FIGURE 8.17 Effects of 3D spatial variability on slope stability: (a) failure probability and risk; (b) mean and COV of safety factor; (c) average sliding mass volume and length; (d) average sliding mass width and depth.

sliding mass volume is dominated by the variation of the sliding mass length. On the one hand, this indicates that the horizontal spatial variability in the Z direction, instead of that in the X direction, affects the failure mechanism and average failure consequence of the 3D slope. Since the 2D analysis only accounts for the spatial variability in the X and Y directions, rather than along the Z direction, the effect of horizontal spatial variability is commonly undervalued in the previous literature based on the 2D analysis. Such an effect can be properly incorporated into the 3D slope risk assessment with 3D spatial variability modeling. On the other hand, it also indicates that the horizontal spatial variability has a greater impact on the slope failure mechanism than the vertical spatial variability. At least in this example, the vertical spatial variability has limited influence on all the shape characteristics (i.e., volume, length, width, and depth) of sliding mass. The location of the sliding mass is dominated by the horizontal spatial variability as well.

Based on a sensitivity study, it is also interesting to note that the shape of the slip surface differs as the spatial variability varies, particularly as the horizontal spatial variability varies. For the 3D slope stability analysis using the LEM, an ellipsoidal slip surface (spherical in particular) or a cylindrical slip surface is widely assumed. They appear to be reasonable under some given conditions. Figure 8.18 demonstrates two random field realizations and the corresponding critical slip surfaces. The slip surface is nearly spherical when the spatial variability is significant (i.e., small autocorrelation distance) (see Figure 8.18a), and it turns to be nearly cylindrical when the spatial variability becomes weak (see Figure 8.18b). Therefore, it is difficult to determine the shape of the slip surface in advance, especially when the spatial variability of soil properties is considered. Inappropriate assumptions of the shape of the slip surface may fail to locate the most dangerous failure mode and overestimate the *FS* of the slope. Due to the 3D finite element analysis, there is no need to assume the shape and location of the slip surface prior to the analysis of the CSFEM.

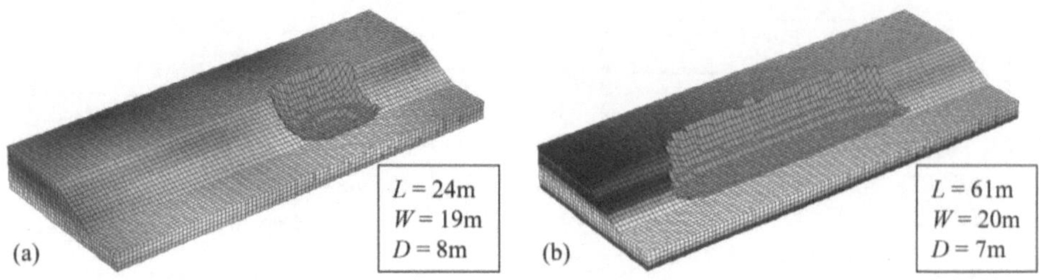

FIGURE 8.18 Effects of 3D spatial variability on slope failure modes: (a) nearly spherical (l_h = 20 m, l_v = 2 m); (b) nearly cylindrical (l_h = 80 m, l_v = 2 m).

To sum up, both the horizontal and vertical spatial variabilities have significant, but different, impacts on the reliability, risk, and failure mechanisms of a 3D slope. The slope reliability and risk are more dominated by the vertical spatial variability, while the slope failure mechanism, such as the volume, length, and location of sliding mass, is more affected by the horizontal spatial variability.

8.6 EXAMPLE #3: A 3D LARGE-SCALE ENGINEERED SLOPE

8.6.1 Slope Overview

This section applies the NISFEM to evaluate the reliability of the left abutment slope, a 3D large-scale engineered slope, at Jinping I Hydropower Station, which is located in a deep valley of the Yalong River in the southwest of China. This slope is one of the engineered rock slopes with the largest excavation scale in rock engineering, which contains the following three main features (CHIDI, 2003; Xu et al., 2009, 2011; Qi et al., 2010, 2012; Song et al., 2010; Zhou et al., 2011): (1) A large-scale model range with the relative height difference of the natural slope reaching 1000–1700 m, the dip angle varying from 55° to 70° near elevation 1850 m, the total excavation height being approximately 530 m (elevations from 2110 m to 1580 m), the maximum excavation width being up to 350 m, and the total excavation volume of materials reaching 5.50 million m^3. (2) Very complicated geological structures in this area, including a V-shaped deep valley, high geo-stresses, strong unloading-induced stress releases, complex engineering and technical conditions, inter-layer extrusion zones, and deep fractures. Furthermore, a large number of deep fractures and faults, such as faults $f_{42\text{-}9}$, f_5, f_8, and lamprophyre dike *X*, could form huge latent unstable blocks that impact the slope stability. (3) The use of a complex physical-mechanical model introduces complications to deriving the numerical solution of *FS*. The slope after full excavation and support is shown in Figure 8.19.

These special geological features are rare in other hydropower projects around the world, which make it very challenging for the slope stability analysis. Substantial 2D and 3D stability analyses of the slope were performed using the LEM (Song et al., 2010), FEM (Xu et al., 2011), and block theory (Huang et al., 2010b; Sun et al., 2015). However, few attempts have been made to investigate the 3D reliability of this slope taking into account the uncertainties of the shear strength parameters of rock masses and discontinuities. As reported in the literature (e.g., Xu et al., 2009; Huang et al., 2010b; Song et al., 2010; Zhou et al., 2011; Qi et al., 2012), the overall stability of this slope is mainly governed by a deformed and cracked rock mass, as concluded from the spatial distribution and intersections of the weak structural features in the left abutment slope. Thus, the wedge failure mode of the engineered slope below elevation 2110 m is selected as the case study for the 3D slope reliability analysis. The toppling failure mode of the natural slope above elevation 2110 m is not considered in the present study.

It is very difficult to accurately determine the geomechanical parameters of a rock mass since it may exhibit extremely complex characteristics due to long-term geological processes. In

FIGURE 8.19 The Jinping I left abutment slope after full excavation and support.

engineering practice, the geomechanical parameters of a rock mass used in the numerical analysis of slope stability are usually derived from in situ tests and laboratory tests supplemented with engineering judgment from similar projects (e.g., Miranda et al., 2009; Xu et al., 2011). According to the geological investigation conducted by CHIDI (2003), the geomechanical parameters of rock masses and discontinuities are summarized in Table 8.2, respectively. These parameters will be used for the following deterministic and probabilistic slope stability analyses.

8.6.2 Deterministic Stability Analysis of Rock Slope

The stress field within the slope during the construction period is simulated using a finite difference model constructed in FLAC3D, as shown in Figure 8.20, in which the X direction is along the river, the Y direction is perpendicular to the river, and the Z direction trends upwards. The finite difference grid consists of 75,647 grid points. The rock masses and discontinuities are modeled by a total of 46,875 eight-node hexahedral zones, 773 six-node wedge zones, and 135,754 four-node tetrahedron zones. Following Zienkiewicz et al. (1975) and Qi et al. (2012), a Mohr–Coulomb failure criterion is adopted to represent the stress–strain behavior of the rock masses and discontinuities. Movements are allowed vertically on the four lateral boundaries in the X and Y directions, while restraint is imposed in the three directions on the base boundary. The faults $f_{42\text{-}9}$, f_5, f_8, f_2, deep fractures $SL_{44\text{-}1}$–$SL_{44\text{-}9}$, and lamprophyre dike *X* are modeled as special zones with assigned shear strength rather than as interfaces between the rock blocks for simplicity. In addition, the topography, stratigraphic boundaries, unloading or weathering boundaries, and the excavation face are also modeled explicitly.

TABLE 8.2
Geomechanical Parameters for Rock Masses and Discontinuities

	No.	Name	γ (kN/m³)	*E* (GPa)	*v*	*c* (MPa)	φ (°)
Rock mass	1	Sandy slate in toppling deformation zone	27	1.0	0.30	0.4	30.96
	2	Strongly unloading sandy slate	27	2.0	0.28	0.6	34.99
	3	Weakly unloading sandy slate	27	3.0	0.35	0.9	45.57
	4	Thick-bedded sandy slate	27	9.0	0.30	1.5	46.94
	5	Strongly unloading marble	27	2.0	0.30	0.6	34.99
	6	Weakly unloading marble	27	4.0	0.27	0.9	45.57
	7	Second block marble	27	11.0	0.25	1.5	46.94
	8	Fresh sandy slate or marble	27	21.0	0.3	2.0	53.47
	9	Reinforced concrete	27	21.0	0.167	1.0	45.0
Structural plane	1	Fault $f_{42\text{-}9}$	27	0.375	0.35	0.02	16.71
	2	Lamprophyre dike *X* (above elevation of 1680 m)	27	1.00	0.20	0.02	16.71
	3	Lamprophyre dike *X* (below elevation of 1680 m)	27	3.00	0.30	0.45	27.02
	4	Deep fracture $SL_{44\text{-}1}$	27	1.25	0.35	0.10	24.23
	5	Fault f_2, f_5, f_8	27	0.45	0.28	0.02	16.71

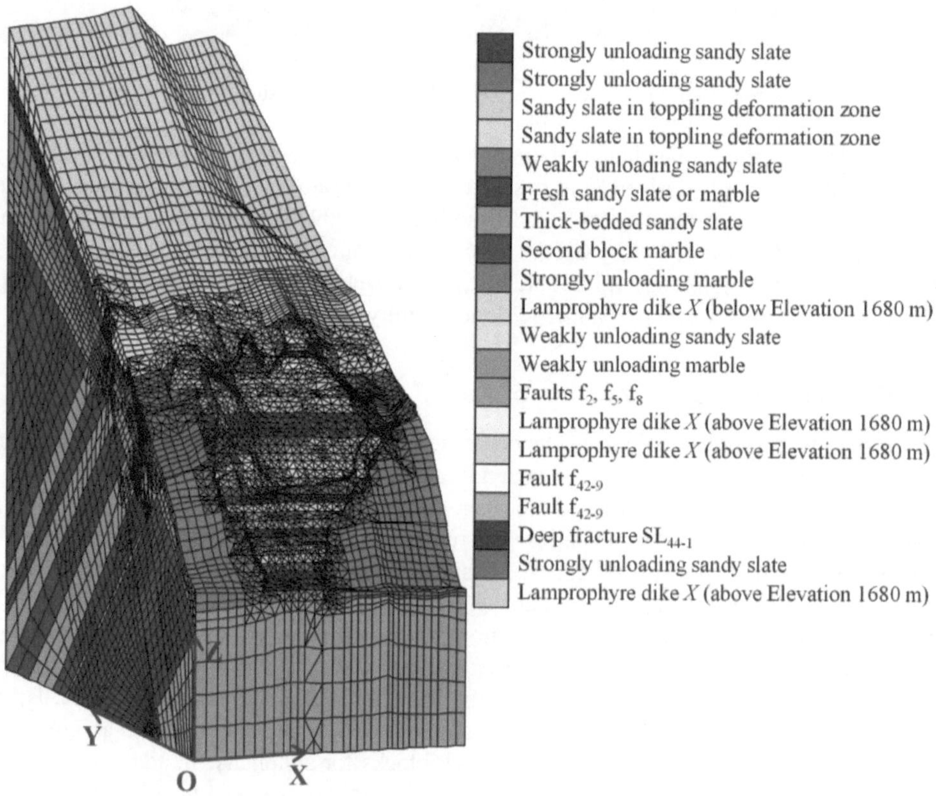

FIGURE 8.20 3D finite difference model of the Jinping I left abutment slope.

The excavation process of the left abutment slope lasted for more than three years, starting in September 2005, with the dam crest (at elevation 1885 m) reached in June 2007 and the concrete cushion foundation platform (at elevation 1730 m) completed in August 2008. The slope excavation was completed when the dam foundation pit (at elevation 1580 m) was excavated in August 2009. The slope excavation process is simulated using FLAC3D, which includes 25 steps. Step 0 represents the stability of the natural slope. The slope is then cut to elevations 1960 m, 1885 m, 1855 m, 1810 m, and 1780 m corresponding to Steps 3, 5, 7, 10, and 12, respectively. Finally, the slope is excavated to elevation 1580 m in Step 25. The excavated materials at the various stages of the excavation are represented by the null model available in FLAC3D.

To control the deformation and ensure the stability of the crushed rock masses within the shallow slope, a large number of prestressed cables with design pullout loads of 2000 kN and 3000 kN and lengths of 40 m, 60 m, and 80 m are installed. For all the prestressed cables, the length of the anchored part is 12 m, the inclination is 8°, and the horizontal and vertical spacings are 4 m × 4 m or 6 m × 6 m. A total of 2815 and 930 bunches of prestressed cables are simulated above and below elevation 1885 m, respectively. The basic mechanical and geometric parameters of the prestressed cables are listed in Table 8.3. A layout of the prestressed cables with the design pullout load of 3000 kN is shown in Figure 8.21. The cable elements in FLAC3D are utilized to simulate the stabilization effects of the prestressed cables.

To further improve the slope stability, three shear-resistant concrete plugs (excavated tunnels backfilled with reinforced concrete) with a size of 9 m (width) × 10 m (height) were constructed at elevations 1883 m, 1860 m, and 1834 m, respectively, to cut through the crush zones of the fault $f_{42\text{-}9}$. The corresponding lengths are 110 m, 90 m, and 78 m, respectively. Figure 8.22 shows the locations of the three shear-resistant concrete plugs and the interrelation with the boundaries of the deformed and cracked rock masses which include the downstream fault $f_{42\text{-}9}$ as the main slip surface, the deep fracture $S_{L44\text{-}1}$ as the upstream cut surface, the lamprophyre dike *X* forming the rear boundary, and the fault f_5 forming the excavation face. The stabilization and shear-resistant effects of the three shear-resistant concrete plugs are modeled by the entity elements assigned with the shear strength of reinforced concrete, as shown in Table 8.2.

As mentioned in Section 8.2.1, the FDM not only accounts for the stress–strain behavior of rock masses and discontinuities, but also simulates the progressive failure process of a slope without the assumption of the shape and location of the failure surface in advance. The SSRM in FLAC3D is used to evaluate the global *FS* of the rock slope during construction. The convergence criterion for the strength reduction analysis is the nodal unbalanced force: the sum of forces acting on a node from its neighboring elements. The computation is considered to reach convergence when the ratio of the nodal unbalanced force is less than 10^{-5} at a specified calculation step (e.g., 5500). As mentioned earlier, the reinforced system of the slope is mainly composed of the prestressed cables and three shear-resistant concrete plugs at elevations 1883 m, 1860 m, and 1834 m, respectively, as shown in Figure 8.23. To account for the effect of the reinforced system on the deformation and stability of the slope during construction, four reinforcement cases during slope excavation are considered as follows: (1) combined supports with the prestressed cables and three shear-resistant concrete plugs; (2) the prestressed cables fail but the three shear-resistant concrete plugs perform well; (3) the three shear-resistant concrete plugs fail but the prestressed cables perform well; and (4) no reinforcements.

Figure 8.24a compares the values of *FS* of the slope under the four reinforcement cases. The seven points in each curve denote the natural slope and the slope excavated to elevations 1960 m, 1885 m, 1855 m, 1810 m, 1780 m, and 1580 m, respectively. For the case of the unreinforced excavated slope in Figure 8.24a, the *FS* remains almost the same at about 1.5 before the slope is excavated to elevation 1885 m in June 2007. The reason is that the marble outside faults f_5 and f_8 could support the deformed and cracked rock masses like a "rock wall", and the excavation process plays a role in reducing the surcharge loads, thereby improving the slope stability. When the slope is excavated to elevation 1810 m in March 2008, the *FS* decreases to 1.264 due to the gradual

TABLE 8.3
Basic Parameters of Prestress Cables

Cable type	Pullout load (kN)	Young's modulus of strand (GPa)	Strand yield strength (MPa)	Number of strands	Nominal cross-sectional area (mm^2)	Hole diameter (mm)	Cable diameter (mm)	Anchored part length (m)
DKDF-2000	2000	180	1860	12 NOS ϕ15.24 mm	2188.92	ϕ140	ϕ90	12
DKDF-3000	3000	180	1860	19 NOS ϕ15.24 mm	3465.79	ϕ165	ϕ110	12

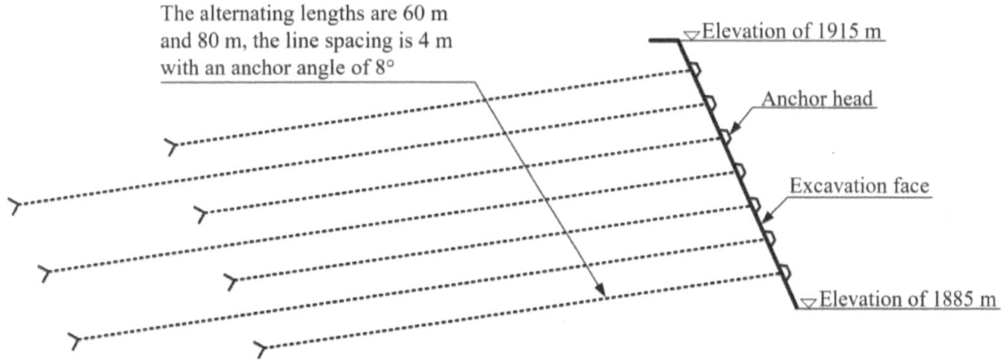

FIGURE 8.21 Layout of prestressed cables with the design pullout load of 3000 kN.

exposure of the boundaries of the deformed and cracked rock masses as the excavation proceeds. When the slope is excavated to elevation 1780 m in May 2008, fault $f_{42\text{-}9}$ is fully exposed at the excavation face. Consequently, the *FS* dramatically decreases to 1.186, which is below the minimal *FS* of 1.25–1.3 required for the slope to function adequately under the permanent condition (Song et al., 2010). These results indicate that slope stability cannot be guaranteed in this case. The safety performance of the slope will be adversely impacted if other factors such as excavation disturbances and unloading, rainfall, and earthquakes are also considered. Therefore, reinforcement measures are urgently needed to ensure the slope safety.

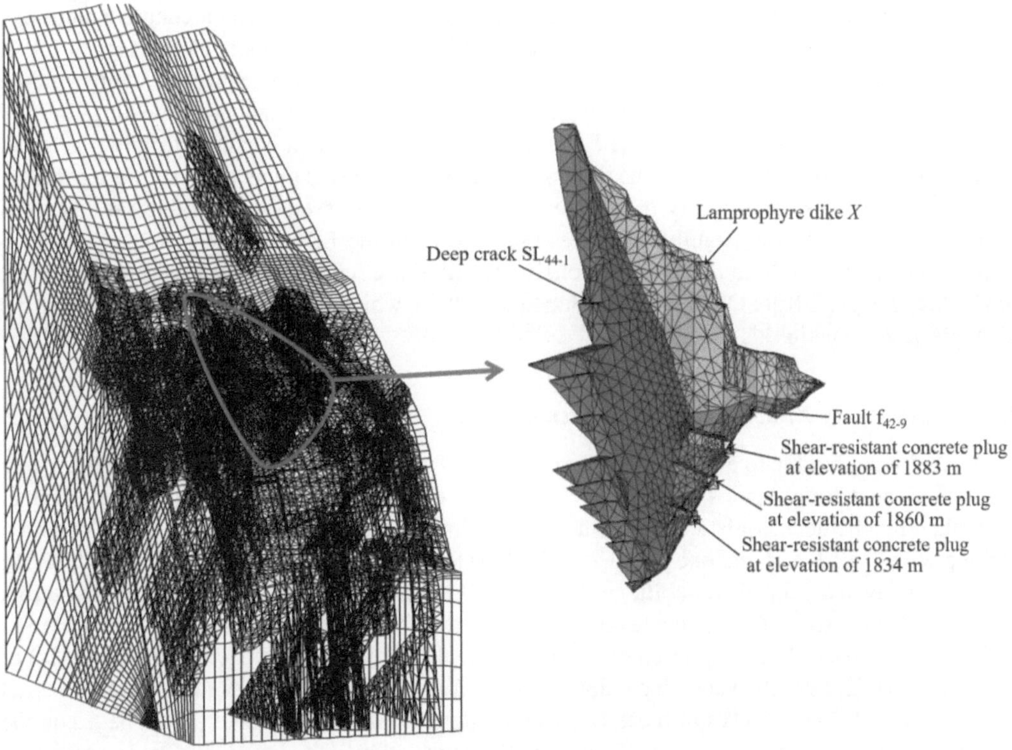

FIGURE 8.22 Layout of three shear-resistant concrete plugs in the deformed and cracked rock masses. (Source: After Zhou et al., 2011)

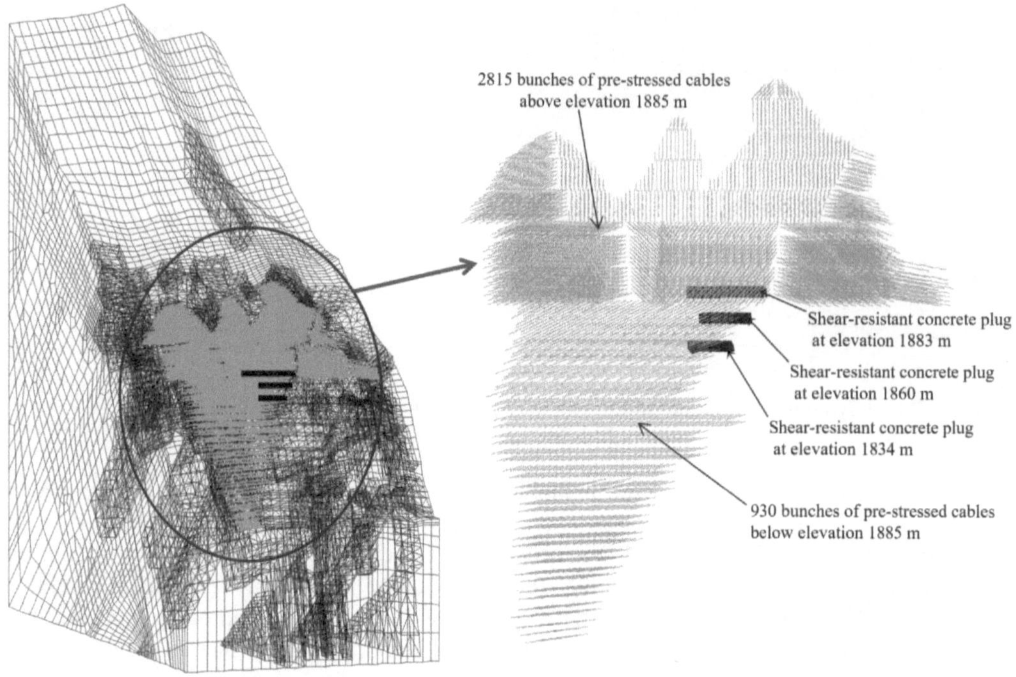

FIGURE 8.23 Layout of the reinforced system.

The combined support using the prestressed cables and three shear-resistant concrete plugs can be used to reinforce the rock slope. In this case, the *FS* of the excavated slope increases to 1.498, which indicates that the combined stabilization measures can effectively ensure the stability of the Jinping I left abutment slope. Alternatively, if only the three shear-resistant concrete plugs or only the prestressed cables perform well, the *FS* of the excavated slope will decrease to 1.342 or 1.306, respectively. It can be observed that the three shear-resistant concrete plugs have a more significant effect on improving slope stability in comparison with the prestressed cables. Table 8.4 compares the results of the 3D slope stability analyses obtained from the FDM and block theory. The *FS* calculated using the FDM in this study is consistent with those reported in the literature (Xu et al., 2009; Huang et al., 2010b; Qi et al., 2012; Sun et al., 2015), which demonstrates the validity of the 3D numerical model in this study.

8.6.3 Reliability Analysis Results of Rock Slope

It is virtually impossible to evaluate the 3D reliability of such a large-scale engineered slope using MCS since the deterministic slope stability analysis is quite time-consuming. The computational time taken by one deterministic stability analysis of this reinforced rock slope is about 30 h on a desktop with 8 GB RAM and one Intel Core i3 CPU clocked at 3.3 GHz. To this end, the NISFEM is adopted to evaluate the 3D reliability of the slope. The quadratic polynomial without cross terms is employed to construct the surrogate model between the *FS* of the slope and the random variables in the physical space. The shear strength parameters (cohesions c_1–c_6 and friction angles φ_1–φ_6) of rock masses (IV2, IV1, III2) and discontinuities ($f_{42\text{-}9}$, X, $SL_{44\text{-}1}$) are treated as random variables, as shown in Table 8.5. The COVs of these 12 shear strength parameters are determined based on the values reported in the literature (e.g., Phoon and Kulhawy, 1999a; Tang et al., 2012). The statistics of the basic random variables are summarized in Table 8.5. Following Tang et al. (2012), lognormal distributions are used to model the geomechanical properties to avoid negative values. To estimate

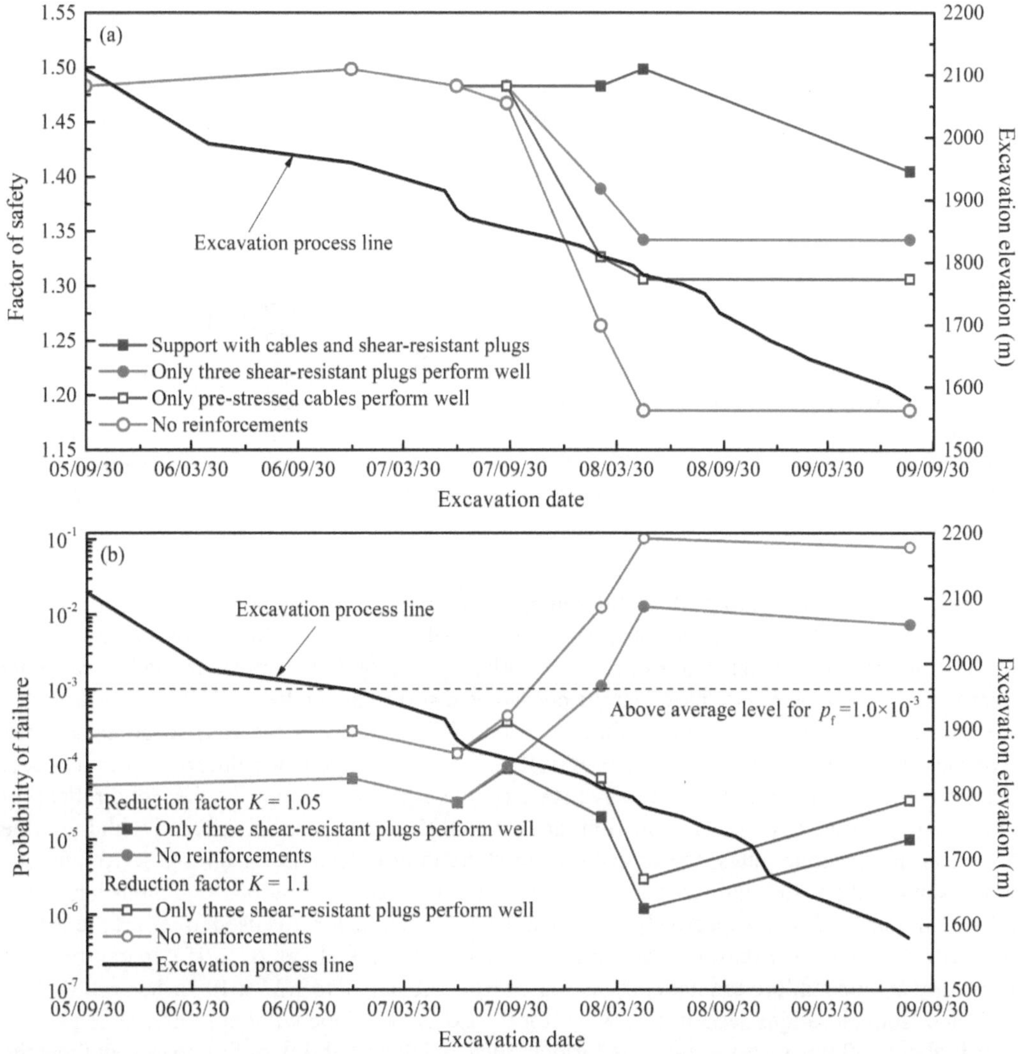

FIGURE 8.24 Variation of slope stability during construction: (a) factor of safety; (b) probability of failure.

TABLE 8.4
Comparison of Deterministic 3D Slope Stability Analyses in Example #3

	Factor of safety			
Method	Nature slope	No reinforcements after full excavation	Support after full excavation	Source
FDM	1.483	1.186	1.405	This chapter
FDM	1.277	1.152	1.385	Xu et al. (2009)
FDM	1.45	1.05	1.35	Qi et al. (2012)
Block theory	1.413	–	–	Huang et al. (2010b)
Block theory	1.56	1.104	1.43	Sun et al. (2015)

TABLE 8.5
Statistics of Shear Strength Parameters

Material	Variable	Mean	COV	Distribution
Fault $f_{42\text{-}9}$	c_1 (MPa)	0.02	0.3	Lognormal
	φ_1 (°)	16.7	0.2	Lognormal
Lamprophyre dike X (above elevation 1680 m)	c_2 (MPa)	0.02	0.25	Lognormal
	φ_2 (°)	16.7	0.15	Lognormal
Deep fracture $SL_{44\text{-}1}$	c_3 (MPa)	0.1	0.22	Lognormal
	φ_3 (°)	24.228	0.14	Lognormal
Class IV2 rock mass (sandy slate in toppling deformation zone)	c_4 (MPa)	0.4	0.2	Lognormal
	φ_4 (°)	30.96	0.12	Lognormal
Class IV1 rock mass (strongly unloading sandy slate and marble)	c_5(MPa)	0.6	0.18	Lognormal
	φ_5 (°)	35	0.10	Lognormal
Class III2 rock mass (weakly unloading sandy slate and marble)	c_6 (MPa)	0.9	0.15	Lognormal
	φ_6 (°)	45.57	0.08	Lognormal

the unknown coefficients in Eq. (8.2), the sampling factor $k = 2$ is used according to Zhang et al. (2011). For each reinforcement case, only a total of 25 evaluations of the slope stability model using the FDM are required. After that, direct MCS with N_{sim} = 1,000,000 samples is used to calculate the probability of slope failure based on the constructed surrogate model.

Figure 8.24b compares the probabilities of slope failure between the case wherein only the three shear-resistant concrete plugs perform well and the case wherein no reinforcements are used. It can be observed that the three shear-resistant concrete plugs can improve the slope reliability substantially. For the case with a reduction factor $K = 1.05$ and the slope reinforced with the three shear-resistant concrete plugs, the probability of slope failure decreases from 5.3×10^{-5} for the natural slope to 3.1×10^{-5} and 1.2×10^{-6} for the slope excavated to elevations 1885 m in June 2007 and 1780 m in May 2008, respectively. Note that the reduction factor K is used to reflect the disturbance effect induced by excavation blasting on the slope stability (Li et al., 2015). In contrast, the probability of slope failure dramatically increases from 3.1×10^{-5} to 1.27×10^{-2} when the slope is excavated from elevations 1885 m to 1780 m for the excavated slope without reinforcements. The latter is about 400 times greater than the former. Such a high probability of failure cannot meet the slope safety requirement of "above average performance level" (i.e., $P_f = 1.0 \times 10^{-3}$) as suggested by the US Army Corps of Engineers (1997). In addition, the probability of slope failure significantly increases as the reduction factor K increases from 1.05 to 1.1, which indicates that the disturbance induced by rock bursts is an important adverse factor affecting the slope reliability. Therefore, effective measures should be taken to minimize the disturbance effects during excavation blasting. It can also be observed from Figure 8.24b that the probability of slope failure will significantly increase if no reinforcements are used during the slope excavation.

8.7 PRACTICAL STOCHASTIC FINITE ELEMENT ANALYSIS SOFTWARE

To facilitate practical applications of SFEMs, several software packages have been developed, such as COSSAN (Schuëller and Pradlwarter, 2006), UQLab (Marelli and Sudret, 2014), RFEM (Griffiths and Fenton, 2004), and NIGPA (Li et al., 2016c), among which the NIGPA is particularly designed for geotechnical engineering.

Developed by Wuhan University, the NIGPA (Non-Intrusive Geotechnical Probabilistic Analysis) is a versatile and user-friendly MATLAB toolbox that aims to evaluate the reliability of geotechnical structures in a non-intrusive manner. It is deliberately framed for the practical use of engineers who are well-trained in deterministic geotechnical analyses but have limited training in probability theory and statistics. With the NIGPA, engineers only need to develop a deterministic analysis model that they are familiar with using commonly used commercial software, and to input the estimated statistics and probability distributions of soil properties. After that, the NIGPA will simulate soil properties stochastically based on uncertainty modeling, execute the deterministic model repeatedly through uncertainty propagation, and return the reliability and risk analysis results as outputs. The procedures are implemented in a non-intrusive manner, as shown schematically in Figure 8.25. Although a thorough understanding of probability theory, statistics, and reliability algorithms is always advantageous, it is not a prerequisite for engineers to use the NIGPA.

The NIGPA consists of four major modules as shown in Figure 8.26, including deterministic modeling, uncertainty modeling, uncertainty propagation, and result visualization. These four modules correspond exactly to the four analysis procedures of NIGPA. Among them, the first three modules are the key modules of NIGPA. For geotechnical stability analyses, the LEM and FEM/FDM are commonly used numerical approaches, which have been successfully applied in the probabilistic analysis of various geotechnical structures. In the NIGPA, the deterministic analysis is carried out using third-party commercial software in a non-intrusive manner. The commercial software SLOPE/W and ABAQUS/FLAC3D are used for the LEM and FEM/FDM, respectively. The commercial software should be successfully installed before using the NIGPA. Additional optional commercial software will be supported in the future version of the NIGPA. In addition, the surrogate models, including the quadratic polynomial, HPCE, and BPNN, can also be used in conjunction with the LEM or FEM/FDM in the NIGPA to perform deterministic analyses with reduced computational time.

For uncertainty modeling, the inherent variability of soil properties is one of the major uncertainties that significantly affect geotechnical reliability and risk. The NIGPA provides two types of probabilistic models for uncertain soil parameters, namely the random variable model (RVM) and the random field model (RFM). The former ignores the spatial variability of soil properties, while the latter explicitly models it using the random field theory. Seven commonly used probabilistic analysis approaches are provided in the NIGPA for uncertainty propagation and reliability analysis, including MCS, LHS, SS, IS, RSM, FOSM, and FORM. Among these, MCS, LHS, and SS are available for both the RVM and RFM due to their flexibility in high-dimensional reliability problems. The remaining approaches are recommended to be used only when the

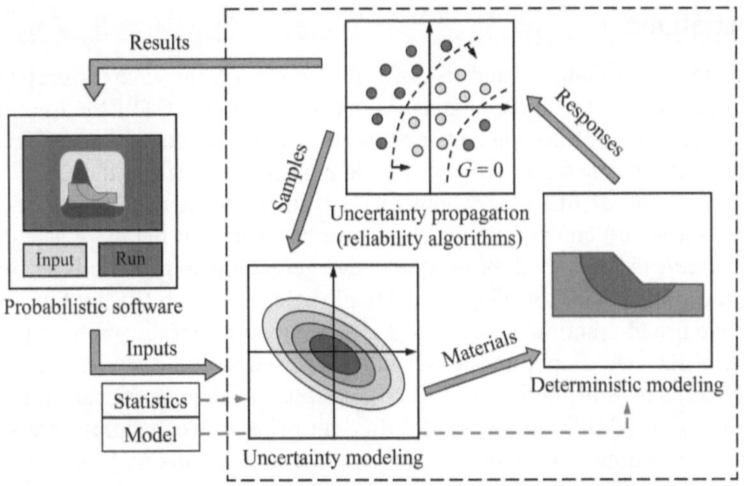

FIGURE 8.25 Schematic diagram of non-intrusive manner for probabilistic analysis.

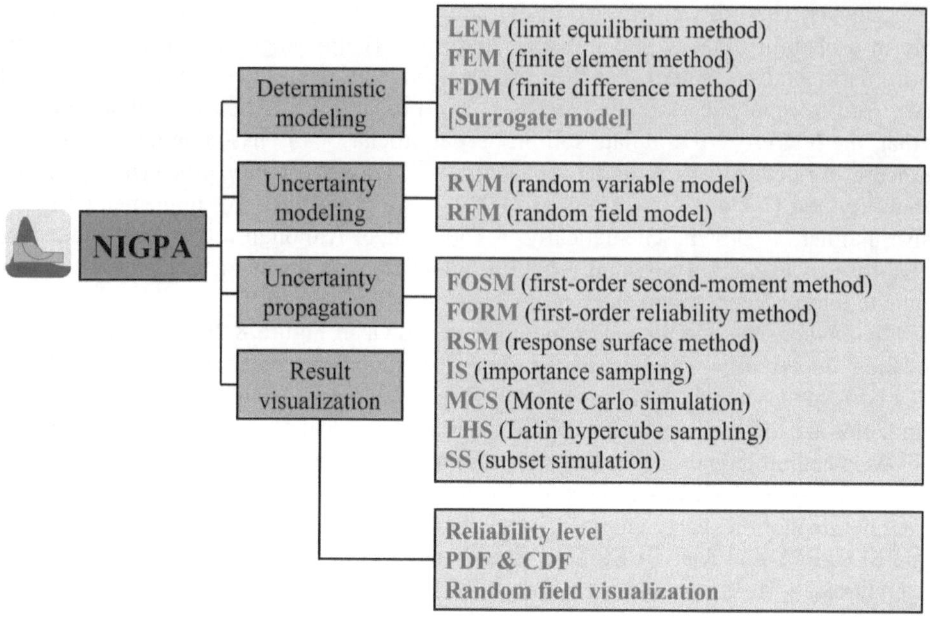

FIGURE 8.26 Framework of NIGPA.

RVM is adopted in the NIGPA. Details of these reliability algorithms can be referred to Zhang et al. (2021).

The NIGPA has four major interfaces as shown in Figure 8.27, corresponding to the four above-mentioned modules. Real-time feedback will be displayed in the message area throughout the execution of the analysis.

Interested readers are referred to Li et al. (2016c) for more details of the NIGPA. The source codes of NIGPA (version 1.1) are available as soft copy at: www.routledge.com/Uncertainty-Modeling-and-Decision-Making-in-Geotechnics/Phoon-Shuku-Ching/p/book/9781032367491. The NIGPA provides a good example for beginners and practitioners to learn SFEM. Hopefully, the NIGPA can contribute to promoting the application of probabilistic analysis in geotechnical practice.

8.8 CONCLUSIONS

To effectively assess the reliability and risk of complex slopes, this chapter proposes two types of stochastic finite element methods, namely the NISFEM and the CSFEM. To reduce the time-consuming evaluations of finite element models, the NISFEM constructs effective surrogate models for slope output responses from finite element models, while the CSFEM utilizes an efficient simple model to facilitate the reliability analysis based on a complex model. The implementation procedures and computational issues are presented in detail. A user-friendly software package NIGPA is developed for non-intrusive geotechnical probabilistic analysis, which takes widely used commercial software for deterministic slope stability analysis as a black box and thus can have broad practical applications. Three slope examples, including a two-layered cohesive slope, a 3D long slope, and a 3D large-scale engineered slope, are investigated to illustrate the effectiveness of the two SFEMs. The effects of the spatial variability of geotechnical parameters on the failure mode, probability, and risk of slope failure are systematically explored. The presented methods can provide versatile and promising tools for reliability and risk assessment of complex slopes.

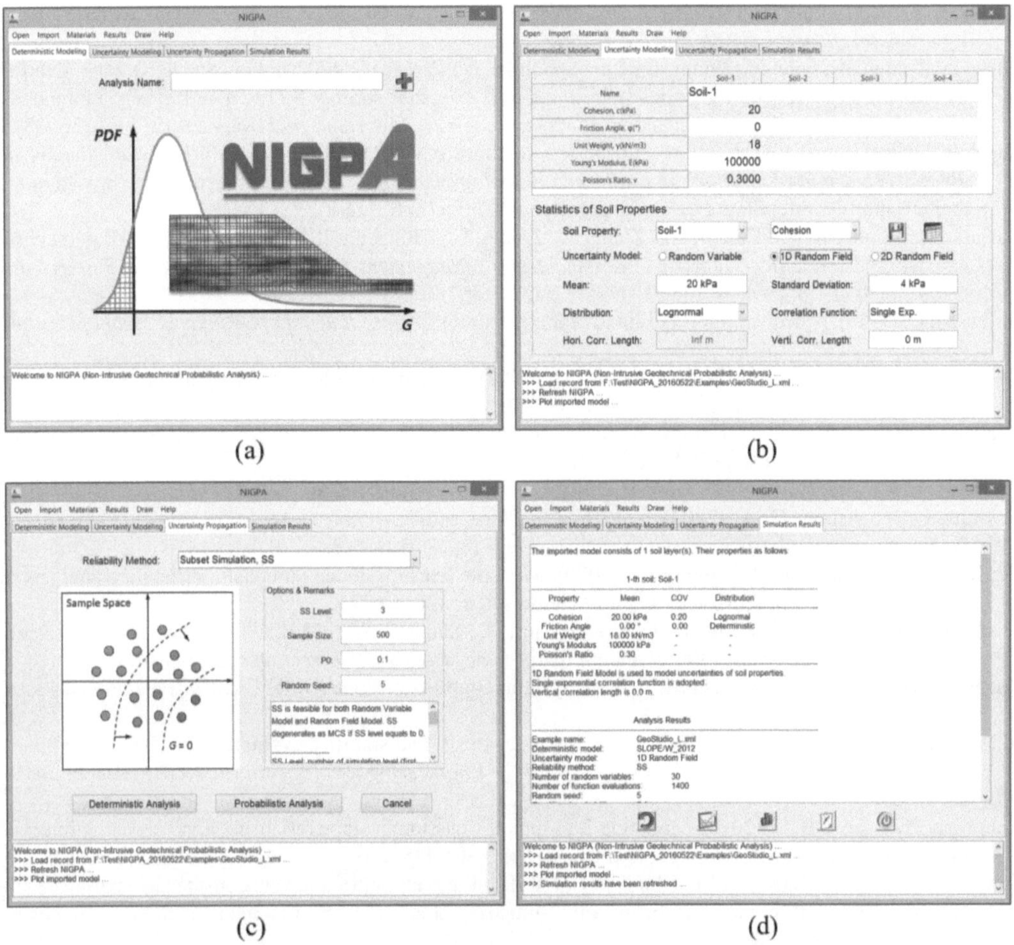

FIGURE 8.27 Major interfaces of NIGPA: (a) deterministic modeling; (b) uncertainty modeling; (c) uncertainty propagation; (d) result visualization.

ACKNOWLEDGMENTS

This work was supported by the National Natural Science Foundation of China (Grant Nos. 52222905, 52179103, 42272326 and 41972280) and the Jiangxi Provincial Natural Science Foundation (Grant Nos. 20224ACB204019 and 20232ACB204031). The financial supports are gratefully acknowledged. Valuable comments on this work from Dr Ze-Zhou Wang in the National University of Singapore are also acknowledged.

REFERENCES

Anderson, J. A. (1995). *An Introduction to Neural Networks*. MIT Press.

Ang, H. S., Tang, W. H. (2007). *Probability Concepts in Engineering: Emphasis on Applications to Civil and Environmental Engineering*. 2nd edition. New York: John Wiley and Sons.

Au, S. K., Beck, J. L. (2001). Estimation of small failure probabilities in high dimensions by subset simulation. *Probabilistic Engineering Mechanics*, 16(4), 263–277.

Au, S. K. (2007). Augmenting approximate solutions for consistent reliability analysis. *Probabilistic Engineering Mechanics*, 22(1), 77–87.

Bucher, C. G., Bourgund, U. A. (1990). A Fast and efficient response surface approach for structural reliability problems. *Structural Safety*, 7(1), 57–66.

Chen, L., Zhang, W., Gao, X., Wang, L., Li, Z., Böhlke, T., Perego, U. (2020). Design charts for reliability assessment of rock bedding slopes stability against bi-planar sliding: SRLEM and BPNN approaches. *Georisk: Assessment and Management of Risk for Engineered Systems and Geohazards*, 16(2), 360–375.

Chengdu Hydroelectric Investigation and Design Institute (CHIDI). (2003). *Feasibility Study Report on Jinping I Hydropower Station (3): Engineering Geology*. Chengdu: Chengdu Hydroelectric Investigation and Design Institute, State Power Corporation of China (in Chinese).

Chwała, M., Phoon, K. K., Uzielli, M., Zhang, J., Zhang, L., Ching, J. (2022). Time capsule for geotechnical risk and reliability. *Georisk: Assessment and Management of Risk for Engineered Systems and Geohazards*, 1–28. DOI: 10.1080/17499518.2022.2136717.

Ditlevsen, O. (1979). Narrow reliability bounds for structural systems. *Journal of Structural Mechanics*, 7(4), 453–472.

El-Ramly, H., Morgenstern, N. R., Cruden, D. M. (2002). Probabilistic slope stability analysis for practice. *Canadian Geotechnical Journal*, 39(3), 665–683.

Ghanem, R. G., Spanos, P. D. (2003). *Stochastic Finite Element: A Spectral Approach*. Revised Version. Mineola, New York: Dover Publication, Inc.

Griffiths, D. V., Lane, P. A. (1999). Slope stability analysis by finite elements. *Géotechnique*, 49(3), 387–403.

Griffiths, D. V., Fenton, G. A. (2004). Probabilistic slope stability analysis by finite elements. *Journal of Geotechnical and Geoenvironmental Engineering*, 130(5), 507–518.

He, X., Xu, H., Sabetamal, H., Sheng, D. (2020). Machine learning aided stochastic reliability analysis of spatially variable slopes. *Computers and Geotechnics*, 126, 103711.

Huang, J., Lyamin, A. V., Griffiths, D. V., Krabbenhoft, K., Sloan, S. W. (2013). Quantitative risk assessment of landslide by limit analysis and random fields. *Computers and Geotechnics*, 53, 60–67.

Huang, J., Griffiths, D. V., Fenton, G. A. (2010a). System reliability of slopes by RFEM. *Soils and Foundations*, 50(3), 343–353.

Huang, R., Lin, F., Yan, M. (2010b). Deformation mechanism and stability evaluation for the left abutment slope of Jinping I hydropower station. *Bulletin of Engineering Geology and the Environment*, 69(3), 365–372.

Ji, J., Liao, H. J., Low, B. K. (2012). Modeling 2-D spatial variation in slope reliability analysis using interpolated autocorrelations. *Computers and Geotechnics*, 40, 135–146.

Jiang, S. H., Li, D. Q., Zhang, L. M., Zhou, C. B. (2014). Slope reliability analysis considering spatially variable shear strength parameters using a non-intrusive stochastic finite element method. *Engineering Geology*, 168, 120–128.

Jiang, S. H., Li, D. Q., Cao, Z. J., Zhou, C. B., Phoon, K. K. (2015). Efficient system reliability analysis of slope stability in spatially variable soils using Monte Carlo simulation. *Journal of Geotechnical and Geoenvironmental Engineering*, 141(2), 04014096.

Jiang, S. H., Huang, J. (2016). Efficient slope reliability analysis at low-probability levels in spatially variable soils. *Computers and Geotechnics*, 75, 18–27.

Jiang, S. H., Huang, J., Yao, C., Yang, J. (2017). Quantitative risk assessment of slope failure in 2-D spatially variable soils by limit equilibrium method. *Applied Mathematical Modelling*, 47, 710–725.

Jiang, S. H., Huang, J. S., Griffiths, D. V., Deng, Z. P. (2022). Advances in reliability and risk analyses of slopes in spatially variable soils, a state-of-the-art review. *Computers and Geotechnics*, 141, 104498.

Li, D. Q., Jiang, S. H., Cao, Z. J., Zhou, C. B., Li, X. Y., Zhang, L. M. (2015). Efficient 3-D reliability analysis of the 530 m high abutment slope at Jinping I Hydropower Station during construction. *Engineering Geology*, 195, 269–281.

Li, D. Q., Xiao, T., Cao, Z. J., Zhou, C. B., Zhang, L. M. (2016a). Efficient and consistent reliability analysis of soil slope stability using both limit equilibrium analysis and finite element analysis. *Applied Mathematical Modelling*, 40(9–10), 5216–5229.

Li, D. Q., Xiao, T., Cao, Z. J., Phoon, K. K., Zhou, C. B. (2016b). Enhancement of random finite element method in reliability analysis and risk assessment of soil slopes using Subset Simulation. *Landslides*, 13(2), 293–303.

Li, D. Q., Xiao, T., Liu, X., Cao, Z. J. (2016c). Introduction to NIGPA (non-intrusive geotechnical probabilistic analysis) (version 1.1). Wuhan University, Wuhan. DOI: 10.13140/RG.2.2.13827.12327.

Li, D. Q., Xiao, T., Zhang, L. M., Cao, Z. J. (2019). Stepwise covariance matrix decomposition for efficient simulation of multivariate large-scale three-dimensional random fields. *Applied Mathematical Modelling*, 68, 169–181.

Li, K. S., Lumb, P. (1987). Probabilistic design of slopes. *Canadian Geotechnical Journal*, 24(4), 520–535.

Low, B. K. (2015). Reliability-based design, practical procedures, geotechnical examples, and insights. In Phoon, K. K., Ching, J. (Eds.), *Risk and Reliability in Geotechnical Engineering*. Boca Raton, FL: CRC Press, pp. 385–424.

Marelli, S., Sudret, B. (2014). UQLab: A framework for uncertainty quantification in MATLAB. In Beer, M., Au, S. K., Hall, J. W. (Eds.), *Vulnerability, Uncertainty, and Risk: Quantification, Mitigation, and Management*pp. 2554–2563, Reston, VA: ASCE.

Miranda, T., Correia, A. G., Sousa, L. R. (2009). Bayesian methodology for updating geomechanical parameters and uncertainty quantification. *International Journal of Rock Mechanics and Mining Sciences*, 46(7), 1144–1153.

Phoon, K. K., Kulhawy, F. H. (1999a). Characterization of geotechnical variability. *Canadian Geotechnical Journal*, 36(4), 612–624.

Phoon, K. K., Kulhawy, F. H. (1999b). Evaluation of geotechnical property variability. *Canadian Geotechnical Journal*, 36(4), 625–639.

Phoon, K. K. (2020). The story of statistics in geotechnical engineering. *Georisk: Assessment and Management of Risk for Engineered Systems and Geohazards*, 14(1), 3–25.

Qi, S., Wu, F., Zhou, Y., Song, Y., Gong, M. (2010). Influence of deep seated discontinuities on the left slope of Jinping I hydropower station and its stability analysis. *Bulletin of Engineering Geology and the Environment*, 69(3), 333–342.

Qi, Z. F., Jiang, Q. H., Tang, Z. D., Zhou, C. B. (2012). Stability analysis of abutment slope at left bank of Jinping I hydropower project during construction. *Rock and Soil Mechanics*, 33(2), 531–538. (in Chinese).

Schuëller, G. I., Pradlwarter, H. J. (2006). Computational stochastic structural analysis (COSSAN) – A software tool. *Structural Safety*, 28(1–2), 68–82.

Song, S. W., Xiang, B. Y., Yang, J. X., Feng, X. M. (2010). Stability analysis and reinforcement design of high and steep slopes with complex geology in abutment of Jinping I hydropower station. *Chinese Journal of Rock Mechanics and Engineering*, 29(3), 442–458. (in Chinese).

Sudret, B., Der Kiureghian, A. (2000). Stochastic finite element methods and reliability: A state-of-the-art report. Technical Report No. UCB/SEMM-2000/08, University of California, Berkeley.

Sun, G., Zheng, H., Huang, Y. (2015). Stability analysis of statically indeterminate blocks in key block theory and application to rock slope in Jinping-I hydropower station. *Engineering Geology*, 186, 57–67.

Tang, X. S., Li, D. Q., Chen, Y. F., Zhou, C. B., Zhang, L. M. (2012). Improved knowledge-based clustered partitioning approach and its application to slope reliability analysis. *Computers and Geotechnics*, 45, 34–43.

U.S. Army Corps of Engineers. (1997). Introduction to probability and reliability methods for use in geotechnical engineering. Department of the Army, Washington, DC, Technical Letter No. 1110-2-547.

Vanmarcke, E. H. (1977). Probabilistic modeling of soil profiles. *Journal of the Geotechnical Engineering Division*, 3(11), 1227–1246.

Vanmarcke, E. H. (2010). *Random Fields: Analysis and Synthesis*. Revised and Expanded New Edition. Singapore: World Scientific Publishing Co., Pte. Ltd.

Wang, Z. Z., Goh, S. H. (2021). Novel approach to efficient slope reliability analysis in spatially variable soils. *Engineering Geology*, 281, 105989.

Wang, Z. Z. (2022). Deep learning for geotechnical reliability analysis with multiple uncertainties. *Journal of Geotechnical and Geoenvironmental Engineering*, 148(4), 06022001.

Xiao, T., Li, D. Q., Cao, Z. J., Au, S. K., Phoon, K. K. (2016). Three-dimensional slope reliability and risk assessment using auxiliary random finite element method. *Computers and Geotechnics*, 79, 146–158.

Xiao, T., Li, D. Q., Cao, Z. J., Tang, X. S. (2017a). Full probabilistic design of slopes in spatially variable soils using simplified reliability analysis method. *Georisk: Assessment and Management of Risk for Engineered Systems and Geohazards*, 11(1), 146–159.

Xiao, T., Li, D. Q., Cao, Z. J., Au, S. K., Tang, X. S. (2017b). Auxiliary random finite element method for risk assessment of 3-D slope. In *Georisk 2017: Reliability-Based Design and Code Developments (GSP 283), Denver, United States*, June 4–7, 2017, pp. 120–129.

Xiao, T., Zhang, L. M., Li, D. Q. (2020). Consistent geotechnical reliability analysis with a simple deterministic model. In *Proceedings of 7th Asian-Pacific Symposium on Structural Reliability and Its Applications (APSSRA2020), Tokyo, Japan*, Oct 5–7, 2020, pp. 448–452.

Xu, B., Low, B. K. (2006). Probabilistic stability analyses of embankments based on finite-element method. *Journal of Geotechnical and Geoenvironmental Engineering*, 132(11), 1444–1454.

Xu, N. W., Tang, C. A., Li, L. C., Zhou, Z., Sha, C., Liang, Z. Z., Yang, J. Y. (2011). Microseismic monitoring and stability analysis of the left bank slope in Jinping first stage hydropower station in southwestern China. *International Journal of Rock Mechanics and Mining Sciences*, 48(6), 950–963.

Xu, Q., Zhang, D. X., Zheng, G. (2009). Failure mode and stability analysis of left bank abutment high slope at Jinping I hydropower station. *Chinese Journal of Rock Mechanics and Engineering*, 28(6), 1183–1192. (in Chinese).

Zhang, J., Zhang, L. M., Tang, W. H. (2011). New methods for system reliability analysis of soil slopes. *Canadian Geotechnical Journal*, 48(7), 1138–1148.

Zhang, J., Huang, H. W. (2016). Risk assessment of slope failure considering multiple slip surfaces. *Computers and Geotechnics*, 74, 188–195.

Zhang, J., Xiao, T., Ji, J., Zeng, P., Cao, Z. J. (2021). *Geotechnical Reliability Analysis: Theories, Methods, and Algorithms*. Shanghai: Tongji University Press.

Zhou, C. B., Chen, Y. F., Jiang, Q. H., Lu, W. B. (2011). A generalized multi-field coupling approach and its application to stability and deformation control of a high slope. *Journal of Rock Mechanics and Geotechnical Engineering*, 3(3), 193–206.

Zhou, Z., Li, D. Q., Xiao, T., Cao, Z. J., Du, W. (2021). Response surface guided adaptive slope reliability analysis in spatially varying soils. *Computers and Geotechnics*, 132, 103966.

Zienkiewicz, O. C., Humpheson, C., Lewis, R. W. (1975). Associated and non-associated visco-plasticity and plasticity in soil mechanics. *Géotechnique*, 25(4), 671–689.

9 Reliability-Based Design with Numerical Models

Shadi Najjar and Imad El-Chiti

ABSTRACT

Geotechnical engineering design is affected by varying sources of uncertainties that dictate the level of risk associated with the design. The main goal of reliability-based design (RBD) is to facilitate and improve decision-making in the presence of uncertainty. Despite the fact that several design codes have adopted RBD principles in geotechnical practice, it remains challenging for geotechnical engineers in general, and numerical engineers in particular, to embrace RBD in daily design practice. This chapter aims at breaking the barrier between the practicing numerical engineer and reliability-based design to encourage practitioners to use RBD tools to quantify risk and inform decision-making. To this end, practical RBD methodologies that could be used in tandem with numerical tools are showcased and a practical RBD design example that illustrates the advantages and limitations of each method is presented.

9.1 INTRODUCTION

Reliability-based design (RBD) is a rational methodology that allows practitioners to design geotechnical components and systems to achieve a target probability of failure. Similar to the conventional deterministic design method, designers can change typical design parameters (e.g., the length of a pile, the width of a footing, or the spacing of soil anchors) to guarantee the "safety" of the design. The only difference is that safety is ensured through the probability of failure that incorporates the different sources of uncertainty that affect the performance of the geotechnical design. As such, reliability-based design is considered the inverse problem of reliability analysis. The latter computes the probability of failure or reliability index for a given design. The former searches for a design to produce a target reliability index.

The main goal of RBD is to facilitate and improve decision-making. While the attention in RBD has generally been focused on the format and calibration of design-checking equations, the basic premise of a reliability-based design is that the target of the design should achieve an acceptable level of reliability. With this premise in mind, numerical designers can utilize a range of possible alternatives in order to achieve the desired reliability. The objective of this chapter is to present practical methods and information that can be used to this end. An illustrative example is also included to show how RBD can be conducted by the practicing engineer using conventional numerical tools.

This chapter begins with a discussion of the concept of reliability-based design and target probabilities of failure. Next, methods that are available for the numerical geotechnical engineer to conduct simplified RBD analyses are considered and their limitations are presented. Finally, practical RBD methodologies that could be used in tandem with numerical tools are showcased and a practical RBD design example that illustrates the advantages and limitations of each method is presented. The objective of the chapter is to break the barrier between the practicing numerical engineer and reliability-based design and to encourage practitioners to use RBD tools to quantify risk and inform decision-making.

DOI: 10.1201/9781003333586-11

9.2 TARGET PROBABILITY OF FAILURE AND RELIABILITY INDEX

The target of reliability-based design is to produce designs with an adequate and uniform probability of failure. In essence, RBD utilizes the concept of a target probability of failure in the same way as traditional geotechnical design utilizes the concept of the factor of safety (FOS) to control risk. As a result, the selection of a rational target probability of failure is critical in RBD.

Target probabilities of failure and reliability indices are expected to yield an acceptable level of risk. Since risk is a reflection of the product of the probability of failure and the consequence of failure, it is natural to expect that target reliability indices should be higher for problems that consider the ultimate limit state (catastrophic failure with high consequences) and lower for problems that consider the serviceability limit state (excessive distortions with low consequences). In Clause 8.4 of the ISO 2394:2015 code, it is mentioned that the target probability of failure should be governed by the consequence and nature of the failure, economic losses, social inconvenience, effects on the environment, sustainable use of natural resources, and the amount of expense and effort required to reduce the probability of failure.

EN1990:2002 recommends three target reliability levels for the ultimate limit state with target reliability indices of 3.3, 3.8, and 4.3 for a 50-year reference period, for 3 consequence classes that use the loss of human life in addition to economic, social, or environmental consequences as a basis for evaluating risk (Table 9.1). The corresponding target reliability index for the serviceability limit state is as low as 1.5 for a 50-year reference period.

ISO 2394:2015 recommends a β of 1.5, 2.3, and 3.1 for the serviceability limit state with the design life taken as the reference period (see failure consequence "some" in Table 9.2). For the ultimate limit state, β values of 3.1, 3.8, and 4.3 (see failure consequence "great" in Table 9.2) are recommended. ISO Annex D provides guidance for determining the target reliability index for structural design. The recommended target reliability index varies depending on the consequence of failure and the level of uncertainty. For most cases, a target reliability index of 3.8 is commonly used, but for more critical structures with severe failure consequences, a higher target β is recommended.

The target reliability index for the ultimate limit state design of bridges and other structures based on the Load and Resistance Factor Design (LRFD) version of the American Association of Highway and Transportation Officials (AASHTO) Bridge Design Specifications is 3.5 for

TABLE 9.1
Reliability Classification and Recommended Target Reliability Indices in EN 1990

		Reliability Index, β for Reference Period	
Reliability Class	**Failure Consequences**	**β (1 year)**	**β (50 years)**
RC-3 (High)	High	5.2	4.3
RC-2 (Normal)	Medium	4.7	3.8
RC-1 (Low)	Low	4.2	3.3

TABLE 9.2
Life-Time Target Reliability Indices in Accordance with ISO 2394:2005

	Life-Time Reliability Index, β for Failure Consequences			
Relative Costs of Safety Measures	**Small**	**Some**	**Moderate**	**Great**
High	0	1.5	2.3	3.1
Moderate	1.3	2.3	3.1	3.8
Low	2.3	3.1	3.8	4.3

superstructures with a 75-year design life. The calibration of geotechnical systems was conducted by the Transportation Research Board (TRB 2004; TRB 2010) which introduced the concept of redundancy for deep foundations and adopted a life-time target reliability between 2.0 and 2.5 for redundant deep foundations and between 3.0 and 3.5 for non-redundant foundation elements. In commentary C10.5.5.2.1, AASHTO 2020 states that

> resistance factors for bridges and other structure designs have been derived to achieve a reliability index, β, of 3.5, an approximate probability of failure, P_f, of 1 in 5,000. However, past geotechnical design practice has resulted in an effective reliability index, β, of 3.0, or an approximate probability of a failure of 1 in 1,000, for foundations in general, and for highly redundant systems, such as pile groups, an approximate reliability index, β, of 2.3, an approximate probability of failure of 1 in 100.

The justification for the lower β value is that if one load-carrying element in these systems fails, the load can be shed to other elements in the system.

Continuous efforts are ongoing by the makers of codes to rationalize and refine the target reliability indices. In 2019, ISO published an International Standard (ISO 22111:2019) "Bases for design of structures – General requirements" that includes a compilation of β values for a 1-year reference period. An adapted version of the compiled target reliability indices is presented in Table 9.3 to show the methodologies being envisaged by different codes in defining the target reliability level of structures.

9.3 SIMPLIFIED RBD IN DESIGN CODES

RBD in Structural Design Codes

The theoretical basis of the reliability-based design approach has long been established in the structural design community (Ang and Cornell 1974; Ellingwood and Ang 1974; Ravindra et al. 1974; Ravindra and Galambos 1978). In these early works, the probability of failure is used as a relative measure of the safety of a design. However, to make the application of the method simple to practitioners, the equation that governs the design and ensures the achievement of the target probability of failure is written in the LRFD format:

$$\varnothing R_n > \sum_{i=1}^{n} \gamma_i S_{ni} \tag{9.1}$$

Where R_n is the nominal resistance, S_{ni} are the load components (live load, dead load, and wind load), and Φ and γ_i are resistance and load factors, respectively, that ensure that the target probability of failure is achieved. Ellingwood (2000) states that "LRFD represents the first attempt in the United States to implement rational probabilistic thinking in the context of a modern limit states structural code".

In LRFD, load factors are generally greater than 1.0 and resistance factors are generally less than 1.0, to serve the purpose of accounting for the uncertainties inherent in the determination of the strength and load effects due to variation in the loads, material properties, accuracy of the modeling and analysis, etc. The loads and the load factors (γ_i) in the LRFD design criterion are generally set by design codes and the task of materials specification groups is to develop resistance factors, which are consistent with the target reliabilities (Figure 9.1). The procedure by which load γ_i and resistance factors Φ were established in the early LRFD community involved (1) collection and analysis of statistical data on loads and strengths, (2) reliability assessment of structural members for which it was believed that the current design was acceptable, (3) selection of reliability targets, and (4) setting load criteria to meet those targets (Ellingwood 1982).

RBD in Geotechnical LRFD: Advantages and Limitations

The application of RBD in geotechnical engineering is traced back to the 1990s through the works of Barker et al. (1991) and Phoon et al. (1995). In its development stage, geotechnical RBD adopted

TABLE 9.3
Compilation of Target Reliability Indices in Different Codes (Adapted from ISO 22111:2019)

ISO 2394	Class 1	Class 2	Class 3	Class 4	Class 5
Consequences	Insignificant	No Societal Impact	Societal Significance	Several Losses	Catastrophic
Buildings	Low Rise	Smaller Buildings	Most Residential	High rises	National Significance
Fatalities	–	< 5	< 50	< 500	> 500
Target $\beta_{1,e}$ based on economic optimization	–	3.7	4.2	4.4	–
Target $\beta_{1,l}$ based on life safety (Life Quality Index)	–	3.1	3.7	4.2	–
EN 1999	Class	RC1	RC2	RC3	–
Life loss / Economic	–	Low / Small	Medium / Considerable	High / Very Great	–
Target β_1 Terms	–	4.2	4.7	5.2	–
Consequence Class (Robustness)	Type of Building	Single House; Agricultural	Low: less 5 Storeys High: 5 to 15 Storeys	Public in Significant Numbers	–
ASCE-7	Risk Category	RC I	RC II	RC III	RC IV
Buildings	–	Low Risk to Life	All Structures Excluding RCI, III, IV	Pose substantial life risk; hazardous materials	Essential facilities; posing substantial hazard-
–	Lives placed at risk	< 3	3–300	300–5000	5000–100000
–	Failure not sudden or progressive	3.7	4.0	4.2	4.4
–	Sudden or progressive	4.0	4.4	4.6	4.8
–	Sudden and progressive	4.4	4.8	5.0	5.2

Note: reliability index values cited in ASCE SEI 7 are based upon a 50 year reference period. Tabulated values are converted here to 1 year reference period based on the approximate relationship $\Phi(\beta_{50}) \approx [\Phi(\beta_1)]^{50}$

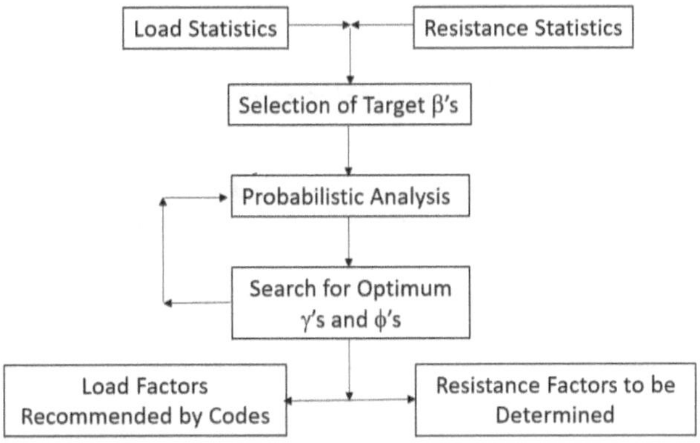

FIGURE 9.1 Calibrating load and resistance factors (simplified from Galambos and Rivandra 1981).

structural LRFD concepts (Paikowsky 2004) through simple closed-form solutions that express the resistance factor, φ, for geotechnical design (Barker et al. 1991):

$$\varnothing = \frac{\lambda_R\left(\Sigma\gamma_i Q_i\right)\sqrt{\dfrac{1+COV_Q^2}{1+COV_R^2}}}{Q exp\left\{\beta_T\sqrt{ln\left[\left(1+COV_R^2\right)\left(1+COV_Q^2\right)\right]}\right\}} \tag{9.2}$$

Where λ_R is the bias for resistance (ratio of actual resistance to predicted resistance), COV_Q and COV_R are the coefficients of variation of the load and the resistance, and β_T is the target reliability index. For the case where the load is broken down into a dead and live load as is the convention in many geotechnical design applications, Eq. (9.2) becomes:

$$\varnothing = \frac{\lambda_R\left(\dfrac{\gamma_D Q_D}{Q_L}+\gamma_L\right)\sqrt{\dfrac{1+COV_{QD}^2+COV_{QL}^2}{1+COV_R^2}}}{\left(\dfrac{\lambda_{QD}Q_D}{Q_L}+\lambda_{QL}\right)\exp\left\{\beta_T\sqrt{ln\left[\left(1+COV_R^2\right)\left(1+COV_{QD}^2+COV_{QL}^2\right)\right]}\right\}} \tag{9.3}$$

Where λ_{QD} and λ_{QL} are the bias of the dead and live load, respectively and COV_{QD} and COV_{QL} are the coefficients of variation of the dead load and live load, respectively. The bias factor for the resistance (λ_R) is typically assumed to be a random variable, whose statistics are obtained from load test databases (e.g., Phoon and Tang 2019).

One of the first design codes that adopted the LRFD approach for geotechnical design is the AASHTO LRFD Bridge Design Specification. In its latest 9th Edition (AASHTO 2020), the code recommends resistance factors for the ultimate limit state for (1) shallow foundations (Table 10.5.5.2.2-1), (2) driven piles (Table 10.5.5.2.3-1), (3) drilled shafts (Table 10.5.5.2.4-1), (4) micropiles (Table 10.5.5.2.5-1), (5) retaining walls (Table 11.5.7-1), and (6) overall stability of slopes (article 11.6.3.7). The load factors associated with the strength limits state are presented in the provisions of Article 3.4.1.

A compilation of some resistance factors that are published in AASHTO 2020 is presented in Table 9.4. In general, resistance factors for bridges and other structures have been derived to achieve a target reliability index, β, of 3.5 which corresponds to an approximate probability of failure of 1 in 5,000. However, section 10.5.5.2.1 in AASHTO 2020 states clearly that "not all of the resistance factors provided in this Article have been derived using statistical data from which a specific β value can be estimated since such data were not always available. In those cases where data were not available, resistance factors were estimated through calibration by fitting to past

TABLE 9.4
Resistance Factors as Recommended in AASHTO 2020 (Adapted from Phoon et al. 2022)

Method/Soil/Condition			Resistance Factor
Shallow Foundations	Bearing Resistance	Theoretical Bearing Resistance	0.5
		Theoretical method (Munfakh et al., 2001), in sand using CPT	0.5
		Theoretical method (Munfakh et al., 2001), in sand using SPT	0.45
		Semi-empirical methods (Meyerhof, 1957), all soils	0.45
		Footings on rock	0.45
		Plate Load Test	0.55
	Sliding	Pre-Cast Concrete placed on sand	0.9
		Cast-in-Place Concrete on sand	0.8
		Cast-in-Place or pre-cast Concrete on Clay	0.85
		Soil on soil	0.9
		Passive earth pressure component of sliding resistance	0.5
Driven Piles	Nominal Bearing Resistance of Single Pile—Static Analysis Methods	Side Resistance and End Bearing: Clay and Mixed Soils	–
		α-method (Tomlinson, 1987; Skempton, 1951)	0.35
		β-method (Esrig & Kirby, 1979; Skempton, 1951)	0.25
		λ-method (Vijayvergiya & Focht, 1972)	0.40
		–	–
		Side Resistance and End Bearing: Sand	–
		Nordlund/Thurman Method (Hannigan et al., 2005)	0.45
		SPT-method (Meyerhof)	0.30
		CPT-method (Schmertmann)	0.50
		End bearing in rock (Canadian Geotech. Society, 1985)	0.45
	Uplift Resistance of Single Piles	Nordlund Method	0.35
		α-method	0.25
		β-method	0.20
		λ-method	0.30
		SPT-method	0.25
		CPT-method	0.40
		Static load test	0.60
		Dynamic test with Signal Matching	0.50
Drilled Shafts (single compressive)	Side resistance clay	α-method (Brown et al., 2010)	0.45
	Tip resistance clay	Total Stress (Brown et al., 2010)	0.4
	Side resistance sand	σ-method (Brown et al., 2010)	0.55
	Tip resistance sand	Brown et al. (2010)	0.5
	Side resistance IGM	Brown et al. (2010)	0.6
	Tip resistance IGM	Brown et al. (2010)	0.55
	Side resistance rock	Kulhawy et al. (2005) Brown et al. (2010)	0.55
	Side resistance rock	Carter and Kulhawy (1988)	0.5
	Tip resistance rock	Canadian Geotechnical Society	0.5
Drilled Shafts (uplift)	Clay	α-method (Brown et al., 2010)	0.35
	Sand	β-method (Brown et al., 2010)	0.45
	Rock	Kulhawy et al. (2005), Brown et al. (2010)	0.4
Drilled Shaft Static Test	All soils	Compression	0.7
	All soils	Uplift	0.6

allowable stress design safety factors, e.g., the AASHTO Standard Specifications for Highway Bridges (2002)."

Despite the advancement made in simplified LRFD codes in adopting a geotechnical reliability-based design approach in design, it is clear that traditional LRFD (as implemented in the AASHTO code for example) that prescribes a single value to each resistance factor does not meet the needs of geotechnical RBD. Starting from the mid-1990s, studies have increasingly recognized the need to accommodate the distinctive features of geotechnical practice, such as the wide range of site conditions and the associated range of testing methods and transformation models that evolved to handle diverse site conditions (Phoon 2016). Two important factors that heavily influence the LRFD calibration in its geotechnical version are:

1. The inability to maintain a relatively uniform level of reliability over the wide range of COVs that are typically encountered in soil and rock properties.
2. Lumping all sources of uncertainty in the geotechnical resistance in one bias factor that affects the calibration process. The statistics of the bias factor are generally highly sensitive to geometry, soil properties, and the model used to predict the resistance. In this case, it is debatable whether the bias factor could lead to representative resistance factors across a wide range of design conditions.

The Canadian Highway Bridge Design Code (Edition 2019) realized the importance of maintaining a uniform level of reliability by acknowledging the different degrees of understanding of the site and modeling considerations. This was reflected in the code by adopting resistance factors (Table 9.5) that depend on the "degree of understanding", defined to be one of three categories (low, typical, or high). Fenton et al. (2016) and Phoon (2017) state that site understanding is reflected in the COV of the geotechnical properties and refers to how well the ground is known and how confident the designer is in the model used to predict the geotechnical resistance. Table 9.6 shows an example of how the soil property variability could be divided into a three-tier system for reliability calculations (Phoon and Kulhawy 2008).

Two creative approaches were proposed by Phoon et al. (2003) and Ching and Phoon (2011) to resolve the inability of simplified RBD to maintain a uniform level of reliability over a wide range of soil and design parameters. The first approach (Phoon et al. 2003) is referred to as the Multiple Load and Resistance Factor Design (MRFD) method. The method allows the utilization of multiple resistance factors for the different components of geotechnical resistance, segmenting the range of influential design parameters, and conducting reliability-based calibration at several calibration points in the domain to ensure that the calibrated resistance factors for each segment provide a uniform target reliability level across the whole domain. The second approach (Ching and Phoon 2011) is referred to as the Quantile Value Method (QVM). The method allows for the maintaining of a relatively uniform level of reliability over a wide range of COVs without changing the resistance factor. This is possible through the use of novel quantile-based characteristic design values that are selected to reflect the "degree of understanding" in the soil properties while maintaining a target level of reliability in the proposed design.

The MRFD is a natural extension to LRFD and aims at achieving greater uniformity in the reliability across a large domain of input design parameters (Phoon et al. 2003). The use of multiple resistance factors in MRFD conforms to the rationale for multiple load factors in LRFD. Table 9.7 shows an example of how three resistance factors (Ψ_{su}, Ψ_{tu}, and Ψ_{w}) could be applied to the undrained uplift side resistance, uplift tip resistance, and weight of a shallow foundation in the MRFD. For this example, the use of multiple resistance factors results in a relatively narrow range of target reliability indices (3.0 and 3.3) for COVs ranging from 30% to 90% for the operative horizontal stress coefficient, K. In comparison, the range of reliability indices for the traditional LRFD is between 2.6 and 3.7 (Phoon et al. 2003). It is worth noting that the MRFD method allows for increasing the degree of uniformity in the reliability level by reducing the sizes of the

TABLE 9.5
Geotechnical Resistance Factors in the 2019 Canadian Highway Bridge Code, CSA 2019 (Adapted from Phoon et al. 2022)

Application	Limit State	Test Method/ Model	Degree of Understanding		
			Low	Typical	High
Shallow foundations	Bearing, φ_{gu}	Analysis	0.45	0.50	0.60
		Scale model test	0.50	0.55	0.65
	Sliding, φ_{gu}				
	Frictional	Analysis	0.70	0.80	0.90
		Scale model test	0.75	0.85	0.95
	Sliding, φ_{gu}				
	Cohesive	Analysis	0.55	0.60	0.65
		Scale model test	0.60	0.65	0.70
	Passive resistance, φ_{gu}	Analysis	0.40	0.50	0.55
	Settlement or lateral movement, φ_{gs}	Analysis	0.70	0.80	0.90
		Scale model test	0.80	0.90	1.00
Deep foundations	Compression, φ_{gu}	Static analysis	0.35	0.40	0.45
		Static test	0.50	0.60	0.70
		Dynamic analysis	0.35	0.40	0.45
		Dynamic test	0.45	0.50	0.55
	Tension*, φ_{gu}	Static analysis	0.20	0.30	0.40
		Static test	0.40	0.50	0.60
	Lateral, φ_{gu}	Static analysis	0.45	0.50	0.55
		Static test	0.45	0.50	0.55
	Settlement or lateral deflection, φ_{gs}	Static analysis	0.70	0.80	0.90
		Static test	0.80	0.90	1.00
Ground Anchors	Pull-out, φ_{gu}	Analysis	0.35	0.40	0.50
		Test	0.55	0.60	0.65
Internal MSE reinforcement	Rupture, φ_{gu}	Steel strip	0.65	0.70	0.75
		Steel grid	0.55	0.60	0.65
		Geosynthetic	0.80	0.85	0.90
	Pull-out, φ_{gu}	Analysis	0.80	0.85	0.90
Retaining systems	Bearing, φ_{gu}	Analysis	0.45	0.50	0.60
	Overturning†, φ_{gu}	Analysis	0.45	0.50	0.55
	Base sliding, φ_{gu}	Analysis	0.70	0.80	0.90
	Facing sliding, φ_{gu}	Test	0.75	0.85	0.95
	Connections, φ_{gu}	Test	0.65	0.70	0.75
	Settlement, φ_{gs}	Analysis	0.70	0.80	0.90
	Deflection/tilt, φ_{gs}	Analysis	0.70	0.80	0.90

(Continued)

TABLE 9.5 (CONTINUED)
Geotechnical Resistance Factors in the 2019 Canadian Highway Bridge Code, CSA 2019 (Adapted from Phoon et al. 2022)

Application	Limit State	Test Method/ Model	Degree of Understanding		
			Low	Typical	High
Embankments (fill)	Bearing, φ_{gu}	Analysis	0.45	0.50	0.60
	Sliding, φ_{gu}	Analysis	0.70	0.80	0.90
	Global stability - temporary, φ_{gu}	Analysis	0.70	0.75	0.80
	Global stability - permanent, φ_{gu}	Analysis	0.60	0.65	0.70
	Settlement, φ_{gs}	Analysis	0.70	0.80	0.90
		Test	0.80	0.90	1.00

* Where maximum frost penetration depth is used, a resistance factor of 1.0 shall be used for tensile resistance to uplift.
† Does not apply to MSE walls

TABLE 9.6
Three-Tier Classification of Soil Property Variability for Reliability Calibration (Adapted from Phoon and Kulhawy 2008)

Geotechnical Parameter	Property Variability	COV (%)
Undrained Shear Strength	Low	10–30
	Medium	30–50
	High	50–70
Effective Stress Friction Angle	Low	5–10
	Medium	10–15
	High	15–20
Horizontal Stress Coefficient	Low	30–50
	Medium	50–70
	High	70–90

calibration domains. If the size is excessively reduced, this may result in an RBD format that is cumbersome and impractical for the design engineer. A practical MRFD should aim at balancing between achieving uniformity in the target reliability index and keeping the format simple enough to be used in practice.

In the QVM method (Ching and Phoon 2011), parameters affecting the problem are grouped as stabilizing random variables (e.g., soil strength) and destabilizing random variables (e.g., load), and the main premise is to reduce stabilizing variables to their η quantile and increase the destabilizing variables to their (1-η quantile) to define their design values. The power of the "quantile" approach is that the design values will automatically reflect the variation in the COV of the design variables in the simplified RBD problem, eliminating the non-uniformity in the resulting reliability index.

Ching et al. (2015) introduced the concept of an effective random dimension (ERD) to extend the reach of the QVM method to a layered soil system. Layered soil systems have been shown to lead to challenges in calibrating resistance factors in the simplified RBD approach. The use of a constant η fails to produce uniform reliability between a multi-layered profile and a single soil

TABLE 9.7
Uplift Resistance Factors for Multiple Resistance Factor Design for Undrained Loading Conditions (Adapted from Phoon et al. 2003)

Depth/Width	Width (m)	Coefficient of Variation of K (%)	Ψ_{su}	Ψ_{tu}	Ψ_{w}
1–2	1–2	30–50	0.35	0.26	0.64
		50–70	0.28	0.27	0.65
		70–90	0.22	0.27	0.65
–	2–3	30–50	0.34	0.24	0.59
		50–70	0.27	0.24	0.6
		70–90	0.21	0.25	0.6
2–3	1–2	30–50	0.33	0.24	0.61
		50–70	0.25	0.24	0.63
		70–90	0.2	0.23	0.65
–	2–3	30–50	0.32	0.21	0.57
		50–70	0.24	0.23	0.58
		70–90	0.19	0.24	0.59

profile. This was traced by Ching et al. (2015) to "redundancy" in the multi-layered system, where underestimation of design values of soil properties in one of the layers could be counterbalanced by an overestimation in the design values of the second layer, which ultimately reduces the probability of failure of a deep foundation that is supported in the two layers compared to the single layer case.

This issue of "variable redundancy" can be present in different types of geotechnical problems, but can be minimized by determining the design quantile η as a function of the ERD (number of independent layers, n_L for the example multi-layered pile problem) and the target reliability index β_T such that $\eta=\Phi^{-1}(-\beta_T/(n_L)^{0.5})$, where Φ^{-1} is the inverse of the standard normal function. For general problems, the ERD links the quantile used to define the characteristic values of the input parameters (η) with the reliability index (β) of the design produced by the QVM method. The QVM method holds a lot of promise in producing simplified RBD design formats that are robust and amenable to further simplification to be directly utilized by the practicing engineer (Yang and Ching 2020).

Applicability of RBD as Implemented in Design Codes to Numerical Modeling

In geotechnical practice, the implementation of RBD through "design codes" remains to be the most practical and convenient approach for geotechnical designers to embrace the advantages of reliability theory. This is mainly because geotechnical designs that are based on RBD codes do not require knowledge of reliability theory.

Since most non-trivial design problems in geotechnical engineering are being increasingly modeled in practice using numerical tools and software, there is a need for simple methodologies that would allow numerical designers to adopt RBD in design problems in which the performance is quantified using numerical tools. The obvious path is for the numerical designer to adopt RBD design codes as a venue for design. The main limitation in this path is that the majority of design codes have been calibrated specifically for traditional empirical performance prediction models.

For illustration, the sample resistance factors that are recommended in AASHTO (Table 9.4) are specifically calibrated and recommended for certain prediction models. These prediction models portray model biases and uncertainties that will probably differ from predictions obtained using robust numerical tools or software packages that may be more capable of (1) reflecting the actual subsurface soil profile, (2) utilizing advanced constitutive models, and (3) modeling the complex geometries and variations in the applied loads/stresses. The literature does not reflect any attempt in RBD design codes to calibrate and present resistance factors or partial factors that are specifically tailored for geotechnical performance prediction using numerical tools. As such, the adoption of current RBD design codes (with current resistance and partial factors) as a basis for design using numerical tools is a research problem that has yet to be tackled.

Since the basic premise of a reliability-based design is that the design should achieve an acceptable level of reliability, there is a need to equip numerical designers with a range of possible tools and alternatives to allow them to achieve the desired reliability in their design. Section 9.4 will portray practical RBD methodologies and procedures that could be used in tandem with numerical tools to encourage practitioners to use RBD principles to quantify risk and inform decision-making.

9.4 RELIABILITY-BASED DESIGN USING NUMERICAL MODELS

Problem-Specific RBD

The challenge of producing designs that exhibit a uniform reliability level may be addressed by adopting RBD methods that explicitly aim at quantifying the probability of failure or reliability index for a given design problem. In these methods, the designer models all sources of uncertainty in the input parameters affecting the load and resistance and conducts a reliability analysis to quantify the probability of failure of a given design. If the design does not produce the target reliability level, the designer repeats the reliability analysis with different parameters until the target reliability is achieved.

The main challenge in adopting a problem-specific reliability-based design approach is the computational cost. To achieve a final design, the designer must solve a number of forward reliability analysis problems to quantify the probability of failure for a number of design configurations (typically by changing the dimensions of structural elements) until the target reliability level is met. The computational cost in a problem-specific RBD approach is highly dependent on the level of complexity of the model or performance function used to predict the limit state that will govern the design. This is particularly true when numerical analyses in the form of finite element, finite difference, or discrete element methods are utilized to predict the limit state. In these problems, the implicit computational cost that is incurred to evaluate the output for a single deterministic analysis is amplified by the need to call the numerical model multiple times to (1) conduct a reliability analysis to assess the probability of failure for a single design configuration (generally using Monte Carlo Simulations) and (2) repeat the reliability analysis for a number of design configurations to achieve the target reliability level.

Adopting problem-specific RBD design approaches in lieu of simplified LRFD or MRFD methods is advantageous if the computational cost can be reduced while maintaining accuracy in the predictions of the probability of failure. The main drawback of full RBD methods is that, unlike simplified RBD methods, they are computationally demanding, theoretically involved, and require adequate knowledge of reliability theory. These drawbacks are particularly present when the performance of geotechnical systems is evaluated using numerical tools. The next section presents a detailed depiction of the different RBD approaches that could be adopted by the numerical engineer in geotechnical design practice. The aim is to showcase the advantages and limitations of RBD methods in tackling problems that require numerical modeling in geotechnical design practice. While the focus is on problems that require numerical models, the methods and tools described are also applicable to any other performance function.

RBD Using Monte Carlo Simulation

Monte Carlo Simulation (MCS) methods have been proven to be effective in quantifying the reliability of geotechnical components and systems. In these methods, a large number of realizations

(N) are simulated from the joint probability distribution of input parameters (X). Each realization of parameters is then used as input to the performance function to evaluate the desired output response (Y), leading to N values for Y. These values of Y are then used to evaluate the probability of failure or unsatisfactory performance by counting the number of realizations in which the output Y exceeds a specified limit state. For example, the output response could be the FOS of a slope or the lateral displacement of a shoring system (δ). For the case of the slope, the probability of failure could be defined as the number of simulations in which the FOS is less than unity, divided by the total number of simulated cases. For the case of the shoring system, the probability of unsatisfactory performance could be defined as the number of simulations in which δ was larger than a prescribed allowable horizontal displacement, divided by the total number of simulations.

Simulating N realizations from a joint probability density function of input random variables X can be implemented in most programming languages using dedicated functions. Historically, the main challenge in implementing Monte Carlo simulations within numerical tools or software has been in (1) automating the process of assigning input parameters to the numerical model and (2) automating the retrieval of the output. Automation is needed to cater to the large number of calls to the numerical model in an MCS framework. It is clearly not feasible that a user manually inputs thousands of parameters into the numerical package to evaluate the probability of failure.

In recent years, many of the main software packages (e.g., FLAC, PLAXIS, SLIDE, SLOPE-W, among others) that geotechnical engineers rely on to analyze and design geotechnical systems have realized the importance of facilitating the implementation of RBD within their numerical frameworks. For illustration, FLAC (ITASCA Consulting Group 2022) contains a powerful built-in programming language, FISH (short for FLAC-ish), that enables the user to define new variables and functions and offers a unique capability to users to tailor analyses to suit their specific needs. The simulations of input random variables or random fields are conducted in any programming language (Excel, Matlab, or R-language) and FISH functions inside FLAC are used to map the input values to layers (for case of random variables) or elements (for case of random fields). The random input parameters are stored in separate text files and exported to FLAC after drawing the geometry of the model. The input properties are mapped to the corresponding elements and the model is run in FLAC to calculate the different required outputs for the target number of simulations.

PLAXIS (Bentley Systems 2021) also allows for performing probabilistic analyses within a finite element framework. This requires the user to write specific codes in PYTHON language to communicate with the PLAXIS engine to allow for feeding input parameters and retrieving output parameters. Probabilistic analyses can also be performed using Slide3 (ROCSCIENCE INC. 2021) and SLOPE/W (Geo-Slope 2022) for various applications of slope stability. In both software packages, the probabilistic framework includes an in-house Monte Carlo Simulation engine that can allow for simulating random soil properties, pore-water pressure conditions, surcharge and point loads, and reinforcement parameters. The output is an evaluation of the probability of failure of slopes, which would allow for the RBD of the slope using a trial and error procedure in which the slope geometry or any other design parameter is varied until the target reliability index is achieved.

It should be noted that the majority of the computational time for the reliability analysis of real-world problems with numerical tools is related to the evaluations of the limit-state functions. This is due to the fact that for such problems, a considerable computational time is expected for the single finite element or finite difference calculation. Therefore, there is a clear need for a compromise between accuracy and efficiency in the reliability analysis algorithm, whereby achieving efficiency mandates limiting the number of evaluations of the limit-state functions. The methods described in the following sections aim at improving computational efficiency in RBD while attempting to limit the deterioration in accuracy as a result of limiting the calls to the performance function.

RBD using Response Surface Methods

In the previous section, it is shown that a large number of numerical evaluations of the limit-state function (i.e. calls to the numerical solver) may be required for an accurate estimation of the probability of failure using MCS. Response surface methods are based on approximating the limit-state function using a simple mathematical model. Then the reliability analysis can be performed using an analytical expression instead of the true limit-state function. This approach may reduce the computational effort considerably, provided that a sufficiently accurate approximation of the limit-state function is built with a limited number of calls to the numerical solver.

Let g(X) be the approximation of the true limit-state function in the basic random variable space X. Typically g(X) is of quadratic polynomial form that could be implemented with or without mixed terms (Faravelli 1998; Papaioannou 2012). The quadratic response surface without mixed terms is presented in Eq. (9.4):

$$g(X) = c_o + \sum_{i=1}^{n} c_i x_i + \sum_{i=1}^{n} c_{ii} x_i^2 \tag{9.4}$$

where the (1 + 2n) coefficients c are to be determined. Eq. (9.4) can be enriched by adding the mixed terms and therefore accounting for possible interaction between the random variables such that:

$$g(X) = c_o + \sum_{i=1}^{n} c_i x_i + \sum_{i=1}^{n} c_{ii} x_i^2 + \sum_{i=1}^{n} \sum_{j=i+1}^{n} c_{ij} x_i x_j \tag{9.5}$$

where in this case, the total number of unknown coefficients is [1 + 2n + n(n – 1)/2]. The unknown coefficients "c" are determined by the least squares method. First, a set of experimental design points are chosen, for which the exact value of the limit-state function is computed. The coefficients c are found by requiring the sum of squares of the differences between the value of the function and the computed actual value at the K experimental points to be minimum.

Several experimental designs have been proposed for the selection of the experimental points and their number (see Figure 9.2 for the case of n = 3). Generally, experimental points are chosen to be the mean of the design variables in addition to a number of grid points that are centered around the mean. In its simplest form (Figure 9.2a), the experimental grid can consist of the mean value and two additional points x_i such that:

$$x_i = \mu_x \pm k\sigma_{xi} \tag{9.6}$$

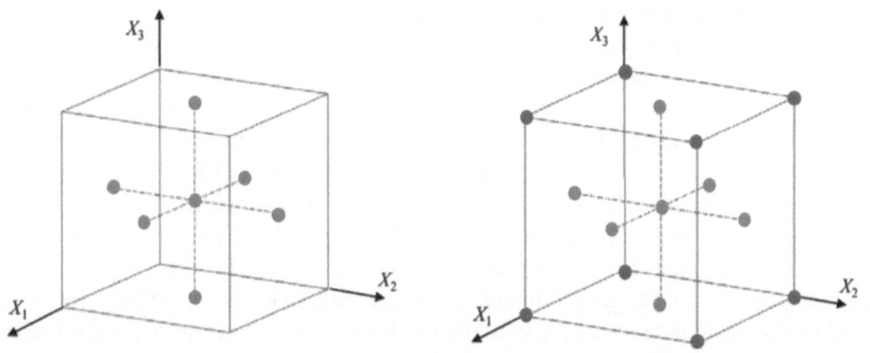

FIGURE 9.2 Experimental design points for case of n = 3 variables with (a) simple axial design, and (b) central composite design (adapted from Papaioannou 2012)

This approach leads to K = 1 + 2n points, which coincides with the number of unknowns *c* for the model of Eq. (9.4). The quality of the approximation is shown to be strongly dependent on the choice of *k*. For most geotechnical reliability problems, k is chosen as 2 or 3.

For the quadratic response surface with mixed terms, see Eq. (9.5), the least expensive choice for the experimental points is that presented in Figure 9.2b which shows a central composite design leading to $K = 1 + 2n + 2^n$. For illustration, it could be concluded that for the case of three input random variables, the number of experimental points needed to calibrate the response surface models in Eqs. (9.4) and (9.5) are K = 8 and 15 points, respectively.

Li et al. (2016) extended the application of the response surface method to problems involving random soil parameters that are described by random fields. In this method, the response parameter G(*X*) is estimated from the quadratic polynomial response surface with $X = (x_1,\cdots, x_i,\cdots, x_n)$ being the vector of input random variables or random field elements. For reliability-based design problems involving multiple failure modes (e.g., slopes with multiple soil layers), Li et al. (2016) recommended the use of multiple response surface methods. In these methods, multiple quadratic response surfaces are determined for all the potential slip surfaces to approximate the relationship between the response parameter (*Y*) and the random variables (or random fields). Each response surface corresponds to the original performance function of one potential slip surface. After determining the parameters of the multiple quadratic response surfaces using regression, Monte Carlo simulations are used to generate N realizations of the random variables or random fields (*X*), and G(*X*) is evaluated for each response surface to determine the output *Y* and to determine the probability of unsatisfactory performance or failure. For example, if the response surface is aimed at predicting the FOS of a slope, the probability of failure of the slope is evaluated from the Monte Carlo simulations by counting the number of realizations where the FOS from the response surfaces (minimum FOS among all potential failure surfaces) is less than 1.0 and dividing it by the total number of simulations.

Once the response surface is determined for a given performance function based on the numerical model, it could be used to quantify the probability of failure using two different approaches. The first approach makes use of the First-Order Reliability Method (FORM) to determine the reliability index using the explicit performance function. A detailed depiction of the formulation behind FORM and its application in evaluating the reliability index can be found in Chapter 6 and will not be repeated here. If FORM (Low 2021) is to be used as a basis for RBD using response surfaces, the derivatives/gradients of the performance function can be determined from the quadratic polynomial and used to find the design point and reliability index as per the methods discussed in Chapter 6. Alternatively, the response surface could be used as a basis for conducting Monte Carlo simulations to quantify the reliability index. MCS using the quadratic response surface is computationally less expensive compared to MCS using the original numerical model.

Irrespective of the method used to calculate the reliability index, RBD using response surfaces will still require a systematic trial and error procedure in which the RBD analysis is repeated for multiple design parameters to yield a final design that meets a target reliability index. These design parameters are generally geometric (size of a footing, diameter or length of a pile, height of an embankment, thickness of a raft, etc.). A systematic approach for converging on a final design that satisfies the target reliability index may involve establishing a realistic range for the design parameter and dissecting the range into practical increments. The reliability analysis could be repeated using the response surface for all the possible options within the range to establish a relationship between the design parameter and the reliability index. The relationship could then be used to find the optimal design parameter that achieves the desired reliability level.

The above approach for design is often referred to as full RBD. In its simplest form, full RBD can be used to back-calculate deterministic "required factors of safety" for a problem-specific target reliability index that is appropriate to the importance of the structure under consideration. This approach was used by Najjar and Gilbert (2009), Bou Diab et al. (2018), Kahiel et al. (2017), Najjar et al. (2017), and Najjar et al. (2020) among others for driven piles, footings on fiber-reinforced clay, footings on aggregate piers, design of pile load test programs, and undrained slopes.

Two separate extensions/developments to the full RBD method were introduced in the last decade by Wang and Cao (2013) and Juang et al. (2013) through the "expanded" RBD method and the "robust" RBD method, respectively. In the expanded full RBD approach, basic design parameters, such as diameter (D) and length (L) of a pile, are formulated artificially as discrete uniform random variables and the design process becomes one in which failure probabilities are developed for various combinations of diameter and length conditional on achieving the target probability of failure using MCS or Response Surfaces. Feasible designs that satisfy the target reliability levels are defined, and the design with the minimum construction cost is selected as the final design. Cao et al. (2019) provided a summary of MCS-based full RBD methods and their implementation in spreadsheets, and values of MCS samples for geotechnical RBD are discussed. On the other hand, the essence of the Robust RBD as presented by Juang et al. (2013) is to minimize the variation of the probability of failure caused by the uncertainty in the estimated sample statistics of soil parameters by adjusting the design parameters of the problem under consideration. For the case of a drilled shaft example, the diameter or length is selected so as to increase the robustness of the design. Robustness measures may include the standard deviation of the probability of failure, which together with the cost of the foundation, could be considered as design objectives. The best design is selected based on a tradeoff relationship between cost and robustness.

RBD Using Point Estimate Methods

The Point Estimate Method (PEM) was introduced by Rosenblueth (1975) to estimate the moments of a random variable (Y) that is a function of a number of input random variables (*X*). The use of the PEM method in geotechnical engineering is discussed by Baecher and Christian (2005), Harr (1987), and Wolff (1996). The method requires 2^N calculations (M) of the output parameter *Y*, where *N* is the number of input random variables. The response *Y* is calculated using combinations of input variables, where each is a standard deviation unit (ξ) above or below the mean.

As a first step in the application of PEM, the output parameter *Y* is calculated for all combinations of input parameters taken at one standard deviation unit (ξ^-) below their mean or one standard deviation unit (ξ^+) above their mean. For illustration, in the case of two input variables, four combinations of input parameters will lead to four calculated values (M = 4) of the output parameter *Y*. Each calculated value (Y_i) is assigned a weighting probability (P_i). For the case of non-correlated input variables, the weighting probability is constant and equal to $P_i = 1/M$.

The mean (μy) and standard deviation (σy) of the output parameter *Y* can then be calculated from the different combinations such that:

$$\mu_y = \sum_{i=1}^{2^n} P_i Y_i \tag{9.7}$$

$$\sigma_y = \sqrt{\sum_{i=1}^{2^n} P_i \left(Y_i - \mu_y\right)^2} \tag{9.8}$$

With both the mean and standard deviation of the output performance parameter known, the probability of unsatisfactory performance or failure can be calculated by assuming either a normal or lognormal distribution to the output parameter *Y*.

If the input variables are correlated, the weights associated with the variables will need to be adjusted to account for the correlation. To obtain a clear definition of the weights, Rosenblueth used a set of + and − as subscripts. The sign is positive when the evaluation point is considered above the mean value and negative when it is below the mean value. The general equation that allows for the calculation of the weights is:

$$P_{(s1,s2,..sn)} = \frac{1}{(2)^n}\left[1+\sum_{i=1}^{n-1}\sum_{j=i+1}^{n} s_i s_j \rho_{ij}\right] \tag{9.9}$$

where s_i is positive when the value of the ith variable is above the mean and negative when the value is below the mean, and *i* is an appropriate combination of + and − signs indicating the location of X_i . For uncorrelated variables, Eq. (9.9) reduces to $P_i = 1/(2)^n$.

For illustration, if two variables are correlated with the correlation coefficient ρ, the weights will be $p_i = (1 + \rho)/4$ and $(1-\rho)/4$. For three correlated variables, where ρ_{12} is the correlation coefficient between X_1 and X_2 and so on, the weights are defined as:

$$P_{+++} = P_{---} = \frac{1}{8}\left(1+\rho_{12}+\rho_{23}+\rho_{31}\right) \tag{9.10}$$

$$P_{++-} = P_{--+} = \frac{1}{8}\left(1+\rho_{12}-\rho_{23}-\rho_{31}\right) \tag{9.11}$$

$$P_{+-+} = P_{-+-} = \frac{1}{8}\left(1-\rho_{12}-\rho_{23}+\rho_{31}\right) \tag{9.12}$$

$$P_{+--} = P_{-++} = \frac{1}{8}\left(1-\rho_{12}+\rho_{23}-\rho_{31}\right) \tag{9.13}$$

For each input variable, the evaluation locations are determined with the use of standard deviation units (ξ), with ξ being a function of the skewness of the distribution of the input variable. For a normal distribution, the skewness is zero so the standard deviation unit is 1 ($\xi = 1$). For a random variable with a non-zero skewness (ν_{xi}), the standard deviation units are given by:

$$\xi_{(x_{i+})} = \frac{\nu_{xi}}{2}+\sqrt{1+\left(\frac{\nu_{xi}}{2}\right)^2} \text{ and } \xi_{(x_{i-})} = \xi_{(x_{i+})} - \nu_{xi} \tag{9.14}$$

The locations of the evaluation points are then calculated with:

$$x_{i+} = \mu_{xi} + \xi_{(x_{i+})}\sigma_{xi} \text{ and } x_{i-} = \mu_{xi} - \xi_{(x_{i-})}\sigma_{xi} \tag{9.15}$$

When the skewness is zero (e.g. normal distribution), ξ is equal to 1 and Eq. (9.15) indicates that x_{i+} x_{i+} and x_{i-} are taken as one standard deviation away from the mean value. For a lognormal distribution where the skewness is non-zero (the skewness ν_{xi} may be calculated as a function of the COV as per Eq. (9.16) by Benjamin and Cornell 2013), ξ will need to be calculated using Eq. (9.14) and used to calculate the evaluation points as per Eq. (9.15).

$$\nu_{xi} = 3COV_{xi} + \left(COV_{xi}\right)^3 \tag{9.16}$$

After all input variables are defined and the calculation has been performed, the output statistics can be computed. Typical output statistics of PEM are the mean (μ_Y), standard deviation (σ_Y), and possibly the skewness (ν_Y) such that:

$$\mu_y = \sum_{i=1}^{2^n} P_i y_i \quad \sigma_y = \sqrt{\sum_{i=1}^{2^n} P_i\left(y_i - \mu_y\right)^2} \quad v_y = \frac{1}{\sigma_y^{\,3}} \sum_{i=1}^{2^n} P_i\left(y_i - \mu_y\right)^3 \tag{9.18}$$

Christian and Baecher (1999) suggested that PEM should not be used for moments higher than the second moment.

RBD Using the First-Order Second Moment Method (FOSM)

The "first-order second moment" (FOSM) method is a simple reliability method in which only the first two "moments" (the mean and the standard deviation) of input and output variables are considered in the analysis. The application of the FOSM method in geotechnical engineering has been described by Wolff (1994), U.S. Army Corps of Engineers (1997), Duncan (2000), Baecher and Christian (2005), and Duncan and Sleep (2015). When using FOSM, 2N + 1 calculations of the output variable (e.g., FOS or settlement) are required, where N is the number of input random variables affecting the problem. In the FOSM method, Taylor Series are used to compute the first two moments (mean and COV) of the output response. Then, a theoretical probability distribution (usually normal or lognormal) is assumed to model the uncertainty in the response and compute the probability of failure.

FOSM requires only knowledge about the mean and standard deviations of the input variables. The mean of the output parameter (μ_Y) is calculated by evaluating the performance function with the mean values of all input variables. The standard deviation of the output parameter (σ_Y) is calculated using Eq. (9.19) such that:

$$\sigma_y = \sqrt{\sum_{i=1}^{n}\left(\frac{\Delta Y_i}{2}\right)^2} \quad \text{where } \Delta Y_i = Y_i^+ - Y_i^- \tag{9.19}$$

Y_i^+ is the output parameter calculated with the value of the ith input parameter increased by one standard deviation from its mean value and Y_i^- is the output parameter calculated with the value of the ith input parameter decreased by one standard deviation from its mean value. In calculating Y_i^+ and Y_i^-, all other input parameters are kept at their mean values.

With both the mean and standard deviation of the output performance parameter known, the probability of unsatisfactory performance or failure can be calculated by assuming either a normal or lognormal distribution to the output parameter *Y*. For example, if the output parameter is the FOS, the probability of failure can be calculated as the probability that FOS is less than 1, using the cumulative distribution function of the assumed distribution. If the output parameter is settlement, the probability of unsatisfactory performance could be calculated as the probability that settlement is greater than $\delta_{allowable}$, where $\delta_{allowable}$ is the tolerable settlement for the structure under consideration.

Duncan and Sleep (2015) compared the results of FOSM to the results from Monte Carlo simulations for a problem involving a retaining wall subjected to three limit states (sliding on sand, sliding on clay, and bearing capacity). Their results indicated that FOSM results did not agree well with the Monte Carlo results, and in the case of the bearing capacity failure mode, they differed by nearly an order of magnitude. They concluded that while the FOSM method offers a relatively easy way to compute probabilities of failure, the results can differ significantly from the results of the Monte Carlo method, which is considered to be a more accurate method. In addition, the Taylor Series method requires that the form of the safety factor distribution be assumed, and there is no logical way to determine the best form of this distribution. Duncan and Sleep (2015) concluded that the strongest point in favor of using the FOSM method is its simplicity, but probabilities of failure computed this way should be viewed as rough estimates that may be higher or lower than the real values.

9.5 RBD EXAMPLE USING A NUMERICAL TOOL

Problem Definition

A practical reliability–based design example for a strip footing supported on a two–layered clay system is presented to illustrate the methodology in which the simplified methods presented in this chapter could be used in practical design scenarios in which numerical tools are used to predict performance. The footing in this problem is supported on a stiff clay that is supported on a natural soft clay layer. The thickness of the stiff clay layer is assumed to be 1.5 m, while that of the soft clay is assumed to be infinite. The nominal undrained shear strength values that are needed to predict the bearing capacity of the foundation are presented in Figure 9.3. Also shown in the figure are the nominal values of the modulus of elasticity that is needed to predict the settlement of the footing. The nominal design load on the strip footing is estimated to be 147 kN per linear meter. The unit weights of the two clay layers are set to be equal to 19 and 16 KN/m³ for the stiff and soft layers, respectively.

Deterministic Design Using PLAXIS 2D

The finite element package PLAXIS 2D was used to model the response of the footing and determine the design width B of the foundation. An initial design trial with a footing width of $B =$ 1.4 m indicates a nominal ultimate force of about 520 kN/m. For a nominal strip footing design load of 147 kN per linear meter, the deterministic design FOS for a 1.4 m wide footing is greater than 3.0 (~ 3.5), which is in line with typical factors of safety recommended for shallow foundations. The settlement response under a nominal force of 147 kN/m as determined from the numerical model indicates a footing settlement of about 3.5 cm. Assuming a settlement criterion of 2.5cm as is the convention in the deterministic design of shallow footings, the design of the footing is shown to be inadequate from a serviceability limit state perspective.

A number of additional deterministic calls to the numerical model with footing widths that are larger than 1.4 m indicates that a footing width of $B = 3.0$ m is needed to achieve a footing settlement of 2.5 cm. For this larger footing width, the ultimate bearing capacity increases to 625 kN/m and the FOS against bearing capacity failure increases to around 4.2, indicating that the design of this strip footing is governed by serviceability criteria.

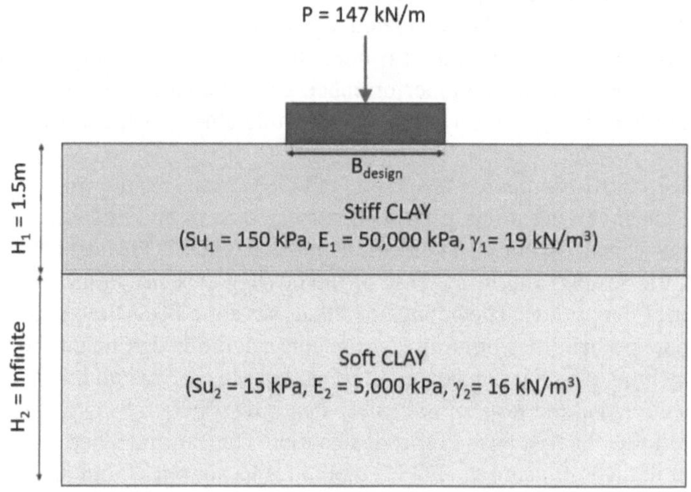

FIGURE 9.3 Illustrative practical RBD design example.

Reliability-Based Design of Strip Footing Using MCS

In reality, the footing load P, the clay undrained shear strength Su_1 and Su_2, and the clay modulus of elasticity E_1 *and* E_2 are parameters that portray uncertainty. As such, a realistic representation of the risk associated with the proposed deterministic footing design requires an evaluation of the probability of failure, particularly for the serviceability limit state which seems to control the design for the problem under consideration. Based on RBD principles, the footing should be designed to ensure a target reliability index against ultimate and serviceability limit states.

Table 9.8 shows the statistical characterization of the parameters affecting the serviceability limit state design of the footing (load and moduli of the two layers). For simplicity, all parameters were assumed to follow a lognormal distribution. The mean of each parameter (including the design load) was estimated by applying a bias factor (ratio of mean to nominal) to the nominal values used in the deterministic analysis. Bias factors for loads are generally less than 1.0 while those for shear strength and moduli are generally greater than 1.0 to reflect the conservatism that is inherent in the designer's selection of design parameters. For soil parameters, nominal or characteristic values are generally chosen to reflect a cautious estimate of the mean or a given percentile/fractile that reflects the uncertainty in the design parameter (e.g., 5th percentile). This will lead to bias factors that are greater than 1.0 for soil parameters reflecting strength and stiffness. For the problem under consideration, the bias factor was assumed to be 1.18 for all soil parameters. The coefficient of variation assumed for each input parameter is also shown in Table 9.8. Since the RBD problem that is considered adopts random variables to represent the uncertainty in the soil parameters, the adopted COVs could be assumed to reflect the uncertainty in the spatial average of the soil property. Applying the bias factors to the deterministic nominal values for P, E_1, and E_2 yields the statistics shown in Table 9.8.

Within a probabilistic framework in which input parameters are considered as random variables, the settlement of the footing will also be a random variable. The probability of exceeding the serviceability limit state of the footing can be calculated as:

$$P_f = Prob\left(\delta > \delta_{all}\right) = \Phi\left(-\beta_{SLS}\right) \tag{9.20}$$

Where P_f is the probability of failure for the serviceability limit state (SLS), δ is the footing vertical settlement, and δ_{all} is the tolerable footing settlement that, if exceeded, will lead to serviceability problems in the structure. Although a settlement of 2.5 cm was selected as a basis for the deterministic design of the footing (as is the convention in practice), the actual tolerable settlement that dictates the exceedance of an SLS varies in a wide range (Zhang and Ng 2005). For shallow foundations, Zhang and Ng (2005) report that different structures may have different limits of allowable displacement. The limiting tolerable displacements are affected by the type and size of the structure, the properties of the structural materials, and the properties of subsurface soils. Based on an assessment of 300 buildings, Zhang and Ng (2005) concluded that the limiting tolerable movement for shallow foundations is a random variable with a mean of 12.9 cm and a coefficient of variation of 0.56. They recommend several methods for determining a single representative tolerable movement for design. One of the methods is based on adopting a tolerable settlement that

TABLE 9.8
Statistical Characteristics of the Input Random Variables for the Footing Design

Soil Properties	Distribution	Bias	Mean	COV
Strip Footing Load, P (kN/m)	Lognormal	0.9	132	0.15
Modulus of Elasticity of Stiff Clay, E_1 (kN/m^2)	Lognormal	1.18	59,000	0.2
Modulus of Elasticity of Soft Clay, E_2 (kN/m^2)	Lognormal	1.18	5,900	0.3

is calculated as a percentile. A 10th percentile tolerable movement (δ_{all} = 3.5cm) is adopted in this chapter as a basis for computing the SLS probability of failure and reliability index using Eq. (9.20).

Fifty thousand Monte Carlo simulations were used to evaluate the probability of failure for the deterministic design (B = 3.0m). The simulation of input variables was conducted in an Excel spreadsheet, saved into text files, and fed into PLAXIS 2D through a PYTHON script. Since the problem is governed by the SLS, and to reduce the computational cost, only E_1, E_2, and P were assumed to be random variables, while Su_1 and Su_2 were taken as deterministic. The fifty thousand simulations were divided into 10 sets of 5000 simulations each. The probability of failure was established from Eq. (9.20) by counting the number of simulations in which δ was greater than δ_{all} (3.5 cm) and dividing by 5000. The reliability indices that were obtained for each set were then averaged and reported as the solution for β_{SLS}.

Results of the reliability analysis conducted for the deterministic design with B = 3.0m indicated a probability of failure of 1.65% with an associated β_{SLS} of 2.13. The basis for RBD is to achieve a design with a given target reliability index. Adopting the recommendations of ISO 2394:1998 with the failure consequences assumed to be "some" for the serviceability limit state, the target SLS reliability index could be taken as 1.5 (if the relative costs of safety measures are high) or 2.3 (if the relative costs of safety measure are moderate). Both targets will be used in the RBD design of the strip footing in this illustrative example to showcase the impact of the target reliability index on design.

To achieve a design that satisfies the target β_{SLS}, the reliability analysis was repeated using MCS for footing design widths of 1.4 m, 2.0 m, 2.4 m, 2.6 m, 2.8 m, 3.2 m, and 3.6 m. The resulting β_{SLS} and probability of failure p_f are plotted versus the footing width in Figure 9.4. Results of the RBD using MCS indicate that footing widths of 1.9 m and 3.5 m are needed to achieve target β_{SLS} values of 1.5 and 2.3, respectively. The computational cost associated with the RBD for this strip footing using MCS can be computed as 8 x 5000 x 10 = 400,000 calls to the performance function in PLAXIS 2D. With each call taking around 5 seconds (given that the finite element model is relatively simple for the problem under consideration), the total computational time needed to design the footing with RBD using Monte Carlo Simulations is around 23 days.

These results clearly indicate that the utilization of direct MCS in RBD using numerical tools could be prohibitive from a computational cost perspective. Several adaptations of the MCS method have been established to limit the computation cost. These include Latin Hypercube sampling, line sampling, importance sampling, and subset simulation, among others (see Chapter 6 for the formulation of each method). Utilizing these methods in daily practice remains a mathematical challenge.

Reliability-Based Design of Strip Footing Using Response Surfaces

The quadratic response surfaces of Eqs. (9.4) (without mixed terms) and (9.5) (with mixed terms) were used to estimate β_{SLS} at a lower computational cost. Assuming three random input parameters (P, E_1, and E_2), the response surfaces can be expressed as:

$$\delta = g(X) = c_o + c_1P + c_2E_1 + c_3E_2 + c_4P^2 + c_5E_1^2 + c_6E_2^2 \tag{9.21}$$

$$\delta = g(X) = c_o + c_1P + c_2E_1 + c_3E_2 + c_4P^2 + c_5E_1^2 + c_6E_2^2 + c_{12}PE_1 + c_{13}PE_2 + c_{23}E_1E_2 \tag{9.22}$$

Where P in Eqs. (9.21) and (9.22) is the applied footing load. Seven experimental points were used to calibrate the simple quadratic response surface function of Eq. (9.21) and fifteen points to calibrate the response surface with mixed terms, see Eq. (9.22). The footing settlement was calculated using PLAXIS 2D for each of the experimental points and used to calibrate the coefficients of the

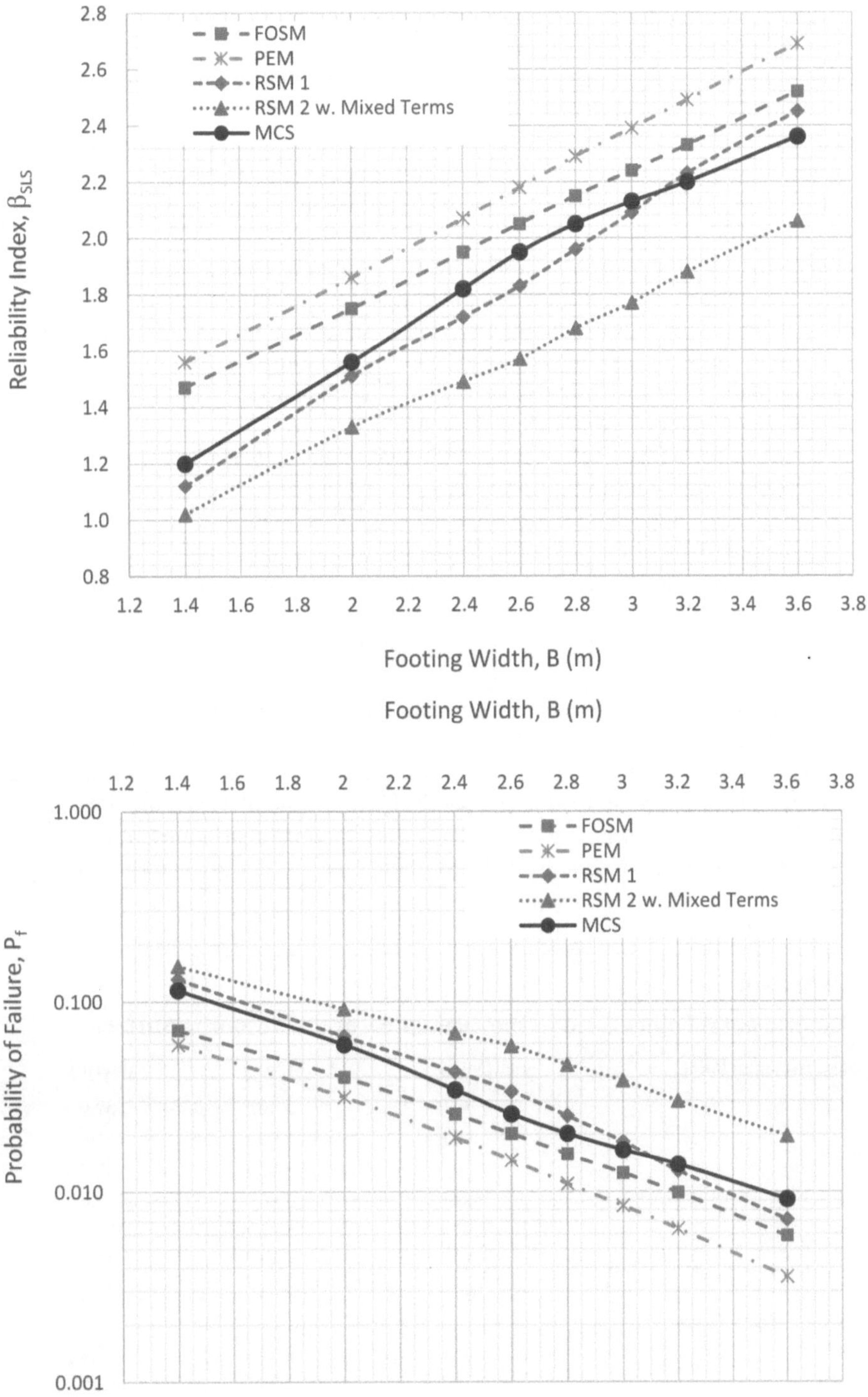

FIGURE 9.4 Variation of p_f and β_{SLS} with footing width for different RBD methods.

response surface. The experimental points and the resulting coefficients of the response surfaces are presented in Tables 9.9 and 9.10, respectively.

The reliability indices (β_{SLS}) corresponding to each potential footing design width were calculated using Monte Carlo Simulations, with the response surfaces in Table 9.10 used in lieu of PLAXIS 2D to predict the footings settlement. For each footing width considered, 10 sets of 5000 MCS were utilized with the response surfaces to estimate the average β_{SLS} as a function of footing width. The resulting reliability indices and probabilities of failure are presented in Figure 9.4.

For all footing widths considered, results indicated that β_{SLS} as determined from the simple quadratic response surface were relatively close to the β_{SLS} computed with direct MCS using PLAXIS. Interestingly, the computed β_{SLS} for the response surface with mixed terms underestimated the β_{SLS} computed from the direct MCS using PLAXIS for all footing widths considered. The response surface with mixed terms overestimated the probability of failure by 50% to 100%. This result is counterintuitive since the expectation is that a more mathematically rigorous response surface should perform better in terms of model prediction.

The difference in the RBD results between the two response surfaces translates into a difference in the design footing width required to achieve the target reliability indices of 1.5 and 2.3. The results in Figure 9.4 indicate that footing widths of 2.0 m (compared to 1.9 m for direct MCS) and 3.4 m (compared to 3.5 m for direct MCS) are needed to achieve target β_{SLS} values of 1.5 and 2.3, respectively for the simple quadratic response surface (no mixed terms). For the response surface with mixed terms, the corresponding design footing widths are 2.4 m (compared to 1.9 m for direct MCS) and 4.0 m (compared to 3.5 m for direct MCS) for target β_{SLS} values of 1.5 and 2.3, respectively.

It should be noted that the better accuracy of the reliability predictions of the simple quadratic model should not be generalized to all RBD problems. A comparison between the settlement predictions of the simple quadratic RSM and that with mixed terms is presented in Figure 9.5 for footings with widths of 2.0 m and 3.0 m, respectively. In a general sense, the data indicates that the RSM model with mixed terms yields settlement predictions that are more coherent and less scattered compared to the simple quadratic RSM, indicating a superior performance in predicting

TABLE 9.9
Experimental Points Used in the Evaluation of the Response Surfaces

Experimental Point	E_1 (MPa)	E_2 (MPa)	P (kN/m)
1	59.0	5.90	132.0
2	59.0	5.90	171.6
3	59.0	5.90	92.4
4	82.6	5.90	132.0
5	35.4	5.90	132.0
6	59.0	9.44	132.0
7	59.0	2.36	132.0
8 to 15 Mixed Terms			
8	82.6	9.44	92.4
9	35.4	2.36	92.4
10	35.4	9.44	92.4
11	82.6	2.36	92.4
12	82.6	9.44	171.6
13	35.4	2.36	171.6
14	35.4	9.44	171.6
15	82.6	2.36	171.6

TABLE 9.10
Coefficients of Response Surfaces (E_1 and E_2 in MPa, P in kN/m)

Response Surface	B = 1.4 m	B = 2.0 m	B = 2.4 m	B = 2.6 m	B = 2.8 m	B = 3.0 m	B = 3.2 m	B = 3.6 m
Simple								
c_o	5.59	5.24	4.99	4.87	4.74	4.63	4.50	4.27
c_1	0.02	0.01	0.01	0.01	0.01	0.01	0.01	– 0.01
c_2	– 0.03	– 0.03	– 0.02	– 0.02	– 0.02	– 0.02	– 0.02	– 0.02
c_3	– 1.08	– 1.03	– 0.99	– 0.97	– 0.95	– 0.93	– 0.91	-0.88
c_4	0.00001	0.00001	0.00001	0.00001	0.00001	0.00001	0.00001	0.00001
c_5	0.00014	0.00012	0.00011	0.00010	0.00010	0.00009	0.00008	0.00007
c_6	0.06	0.06	0.05	0.05	0.05	0.05	0.05	0.05
Mixed Terms								
c_o	2.59	2.40	2.30	2.25	2.20	2.15	2.10	2.00
c_1	0.05	0.04	0.04	0.04	0.04	0.04	0.04	0.04
c_2	– 0.03	– 0.03	– 0.02	– 0.02	– 0.02	– 0.02	– 0.02	– 0.02
c_3	– 0.82	– 0.78	– 0.75	– 0.73	– 0.72	– 0.70	– 0.69	– 0.66
c_4	0.00003	0.00002	0.00002	0.00002	0.00002	0.00002	0.00002	0.00002
c_5	0.00018	0.00014	0.00013	0.00012	0.00011	0.00011	0.00010	0.00009
c_6	0.06	0.06	0.05	0.05	0.05	0.05	0.05	0.05
c_{12}	0.00016	– 0.00014	– 0.00013	– 0.00012	– 0.00012	– 0.00011	– 0.00011	– 0.00010
c_{13}	– 0.00333	– 0.00310	– 0.00296	– 0.00288	– 0.00280	– 0.00274	– 0.00267	– 0.00254
c_{23}	0.00243	0.00225	0.00211	0.00204	0.00197	0.00190	0.00184	0.00171

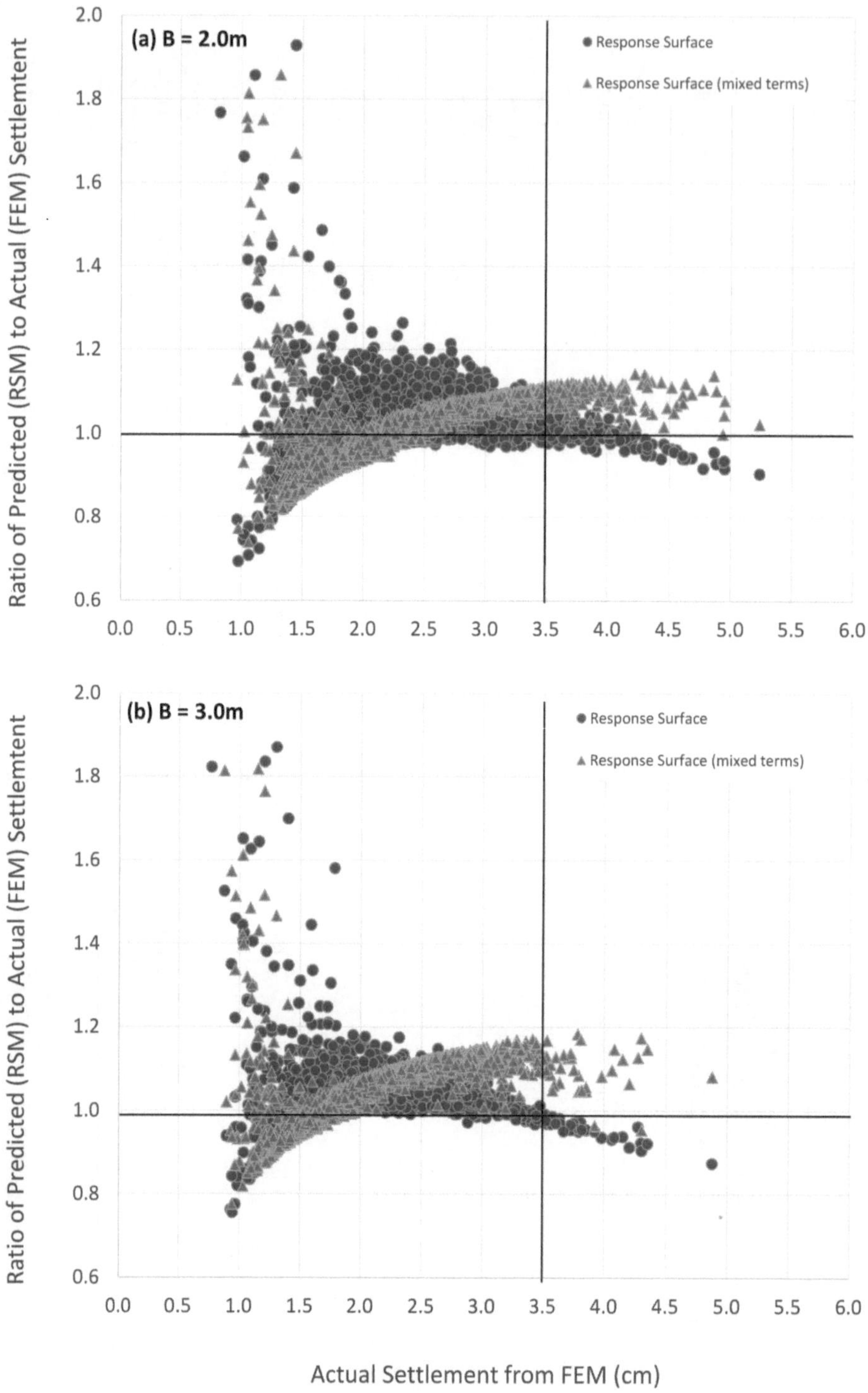

FIGURE 9.5 Comparison between settlement predictions from RSM and MCS.

settlement. However, for the specific range of settlement that affects the probability of failure (footing settlements that are greater than 3.5 cm), the quadratic surface with mixed terms consistently overestimates the settlement, leading to a higher probability of failure. The simple quadratic surface on the other hand accurately estimates settlement in this important range, leading to more reliable probabilities of failure for the particular problem under consideration.

The computational cost associated with RBD using response surface methods is significantly smaller than that required for RBD with direct MCS in PLAXIS 2D. For each footing width and response surface type, 15 calls to PLAXIS 2D were needed to establish the settlement values of the experimental points (around 1.5 minutes of computational time). Once these settlement values were determined, the response surfaces were calibrated (5 seconds using the Solver function in Excel). This was followed by running 10 sets of 5000 MC simulations in Excel using the response surface as the performance function (another 5 seconds in Excel) to determine the average reliability index. As such, the total computation cost for running RBD using response surfaces for the problem under consideration is less than 15 minutes for the eight different footing widths considered.

Reliability-Based Design of Strip Footing Using Point Estimate Method (PEM)

The first step in applying the PEM method is determining the evaluation points (x_{i+} and x_{i+}) for each random variable considered. Since x_{i+} and x_{i+} are a function of the standard deviation units (ξ) which are in turn a function of the skewness of the distribution of the input variable, the skewness (ν_{xi}) needs to be determined. Since all parameters follow a lognormal distribution, ν_{xi} is calculated for P, E_1, and E_2 as a function of the COV using Eq. (9.16) leading to $\nu_P = 0.45$, $\nu_{E1} = 0.61$, and $\nu_{E2} = 0.93$. Based on these values for skewness, the standard deviation units ξ^+ and $\xi^-$are calculated from Eq. (9.14) with ξ^+ values of 1.25, 1.35, and 1.57 and ξ^- values of 0.8, 0.74, and 0.64 for input parameters P, E_1, and E_2, respectively. The eight experimental points needed to calculate the mean and the variance of the footing settlement using the PEM method can now be calculated using Eq. (9.15). The points and the settlements computed using PLAXIS are presented in Table 9.11.

The settlements computed in PLAXIS for the different experimental points are used to calculate the mean and standard deviation of the settlement, δ using Eq. (9.18). This requires the computation of the weighting factor P. Since the input parameters are assumed to be statistically independent in this problem, the weighting factors can be calculated as 1/8 from Eq. (9.9) which reduces to $P_i = 1/(2)^n$, with n being equal to 3.

The resulting means and standard deviations of δ are presented in Table 9.11. The computation of β_{SLS} requires an assumption regarding the probability distribution of the footing displacement, δ. Since MCS was conducted as part of this study, the results of the simulations could be used to inform the probability distribution of the resulting settlement. Figure 9.6 shows the distribution of settlement from MCS for the case with B = 3.0 m. It is clear that the distribution of the footing settlement follows a theoretical lognormal distribution. As a result, the settlement in the PEM method will be assumed to follow a lognormal distribution in calculating the probability of failure and reliability index based on Eq. (9.20). Note that if the results from Monte Carlo simulations were not available, the assumption of a lognormal distribution for the footing settlement may not be verified.

The reliability indices calculated using PEM for all the footing widths are presented in Table 9.11 and Figure 9.4. For all footing widths considered, results indicated that β_{SLS} as determined from the PEM was consistently higher than β_{SLS} computed with direct MCS using PLAXIS, with β_{SLS} using PEM exceeding the β_{SLS} from direct MCS by a value of 0.23 to 0.36. For illustration, the probabilities of exceeding the SLS as determined by the PEM are 40% to 60% smaller than the probabilities of failure calculated by a direct MCS method.

The impact of the difference in the resulting β_{SLS} on the RBD indicates that footing widths of 1.3 m (compared to 1.9 m for direct MCS) and 2.8 m (compared to 3.5 m for direct MCS) are needed to achieve target β_{SLS} values of 1.5 and 2.3, respectively. The results of the PEM could be considered to be unconservative since they yield design widths that are smaller than those required to achieve

TABLE 9.11
Experimental Points and Reliability Results of the PEM (B = 1.4, 2.0, 2.6, 3.2 and 3.6)

Point	E_1 (MPa)	E_2 (MPa)	P(kN/m)	δ (Plaxis)	P_i	Mean δ	Stdev δ	β_{SLS} (Lognormal)
Footing Width = 1.4 m								
1	74.9	8.67	116.2	1.49	0.13	2.38	0.65	1.56
2	50.3	4.77	116.2	2.55	0.13			
3	50.3	8.67	116.2	1.67	0.13			
4	74.9	4.77	116.2	2.26	0.13			
5	74.9	8.67	156.8	2.08	0.13			
6	50.3	4.77	156.8	3.55	0.13			
7	50.3	8.67	156.8	2.34	0.13			
8	74.9	4.77	156.8	3.15	0.13			
Footing Width = 2.0 m								
1	74.9	8.67	116.2	1.36	0.13	2.19	0.61	1.86
2	50.3	4.77	116.2	2.34	0.13			
3	50.3	8.67	116.2	1.52	0.13			
4	74.9	4.77	116.2	2.10	0.13			
5	74.9	8.67	156.8	1.90	0.13			
6	50.3	4.77	156.8	3.26	0.13			
7	50.3	8.67	156.8	2.12	0.13			
8	74.9	4.77	156.8	2.91	0.13			
Footing Width = 2.6 m								
1	74.9	8.67	116.2	1.25	0.13	2.00	0.56	2.18
2	50.3	4.77	116.2	2.15	0.13			
3	50.3	8.67	116.2	1.38	0.13			
4	74.9	4.77	116.2	1.94	0.13			
5	74.9	8.67	156.8	1.73	0.13			
6	50.3	4.77	156.8	2.98	0.13			
7	50.3	8.67	156.8	1.92	0.13			
8	74.9	4.77	156.8	2.67	0.13			
Footing Width = 3.2 m								
1	74.9	8.67	116.2	1.14	0.13	1.84	0.51	2.49
2	50.3	4.77	116.2	1.98	0.13			
3	50.3	8.67	116.2	1.25	0.13			
4	74.9	4.77	116.2	1.79	0.13			
5	74.9	8.67	156.8	1.58	0.13			
6	50.3	4.77	156.8	2.73	0.13			
7	50.3	8.67	156.8	1.74	0.13			
8	74.9	4.77	156.8	2.46	0.13			
Footing Width = 3.6 m								
1	74.9	8.67	116.2	1.08	0.13	1.73	0.49	2.69
2	50.3	4.77	116.2	1.87	0.13			
3	50.3	8.67	116.2	1.18	0.13			
4	74.9	4.77	116.2	1.70	0.13			
5	74.9	8.67	156.8	1.49	0.13			
6	50.3	4.77	156.8	2.58	0.13			
7	50.3	8.67	156.8	1.64	0.13			
8	74.9	4.77	156.8	2.33	0.13			

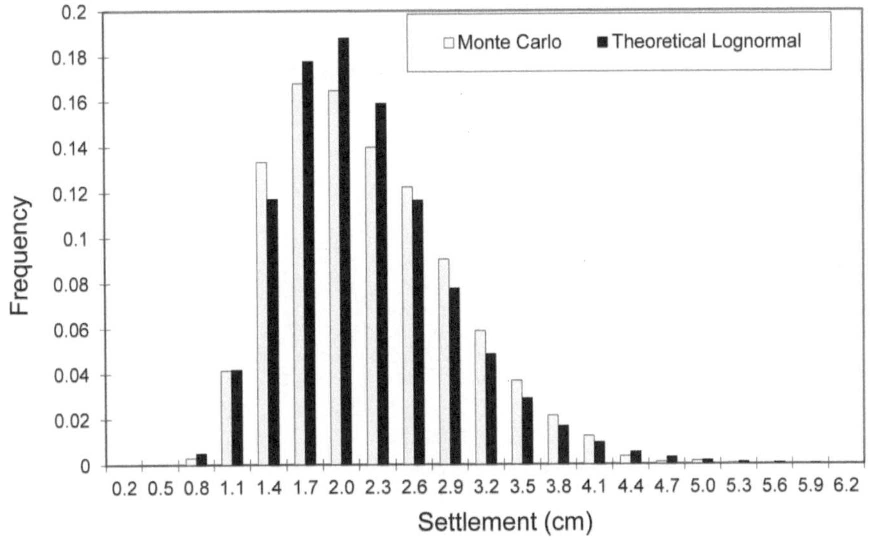

FIGURE 9.6 Comparison between settlements from MCS and theoretical lognormal.

the target reliability index. The advantage of using PEM is the small computational cost which was significantly smaller than that required for RBD with direct MCS in PLAXIS 2D. For each footing width, only eight calls to PLAXIS 2D were needed to establish the settlement values of the experimental points (around one minute of computational time). Once these settlement values were determined, the reliability index was instantaneously computed using Eq. (9.20), assuming that the footing settlement is lognormally distributed. As such, the total computational cost for running RBD using PEM for the problem under consideration is in the order of eight minutes for the eight different footing widths considered.

Reliability-Based Design of Strip Footing Using FOSM

For the footing design problem under consideration, the application of FOSM requires the determination of seven evaluation points and knowledge about the mean and standard deviation of the input variables. The first evaluation point is the mean of all random variables. Six additional evaluation points are taken as the mean of every input variable plus or minus its standard deviation (see Table 9.12). The settlement at each evaluation point is predicted using a deterministic call to PLAXIS 2D. The mean of the footing settlement ($\mu\delta$) is calculated with an additional call to PLAXIS using the mean of every input variable. The standard deviation of the footing settlement ($\sigma\delta$) is calculated using Eq. (9.19).

The resulting means and standard deviations of δ as obtained from FOSM are also included in Table 9.12. As in the case of PEM, the computation of β_{SLS} in FOSM requires an assumption regarding the probability distribution of the footing displacement, δ. As a result, the settlement will be assumed to follow a lognormal distribution in calculating the probability of failure and reliability index based on Eq. (9.20).

The reliability indices calculate using FOSM for all the footing widths are presented in Table 9.12 and Figure 9.4. As in the case of the PEM method, β_{SLS} as determined from FOSM was consistently higher than β_{SLS} computed with direct MCS using PLAXIS. The β_{SLS} predicted from FOSM exceeded the β_{SLS} from direct MCS by a value of 0.1 to 0.29. For illustration, the probabilities of exceeding the SLS as determined by the FOSM are 21% to 38% smaller than the probabilities of failure calculated by a direct MCS method.

TABLE 9.12
Experimental Points and Reliability Results of the FOSM (B =1.4, 2.0, 2.6, 3.2 and 3.6)

Point	E_1 (MPa)	E_2 (MPa)	P (kN/m)	δ (Plaxis)	Mean δ	Stdev δ	β_{SLS} (Lognormal)
Footing Width = 1.4 m							
1	59.0	5.90	132.0	2.41	2.41	0.692	1.47
2	59.0	5.90	151.8	2.82			
3	59.0	5.90	112.2	2.01			
4	70.8	5.90	132.0	2.28			
5	47.2	5.90	132.0	2.57			
6	59.0	7.67	132.0	2.00			
7	59.0	4.13	132.0	3.09			
Footing Width = 2.0 m							
1	59.0	5.90	132.0	2.21	2.21	0.644	1.75
2	59.0	5.90	151.8	2.58			
3	59.0	5.90	112.2	1.85			
4	70.8	5.90	132.0	2.10			
5	47.2	5.90	132.0	2.35			
6	59.0	7.67	132.0	1.83			
7	59.0	4.13	132.0	2.86			
Footing Width = 2.6 m							
1	59.0	5.90	132.0	2.03	2.03	0.594	2.05
2	59.0	5.90	151.8	2.36			
3	59.0	5.90	112.2	1.70			
4	70.8	5.90	132.0	1.93			
5	47.2	5.90	132.0	2.15			
6	59.0	7.67	132.0	1.67			
7	59.0	4.13	132.0	2.63			
Footing Width = 3.2 m							
1	59.0	5.90	132.0	1.86	1.86	0.549	2.33
2	59.0	5.90	151.8	2.16			
3	59.0	5.90	112.2	1.56			
4	70.8	5.90	132.0	1.77			
5	47.2	5.90	132.0	1.96			
6	59.0	7.67	132.0	1.52			
7	59.0	4.13	132.0	2.42			
Footing Width = 3.6 m							
1	59.0	5.90	132.0	1.75	1.75	0.521	2.52
2	59.0	5.90	151.8	2.04			
3	59.0	5.90	112.2	1.48			
4	70.8	5.90	132.0	1.68			
5	47.2	5.90	132.0	1.85			
6	59.0	7.67	132.0	1.44			
7	59.0	4.13	132.0	2.30			

The impact of the difference in the resulting β_{SLS} on the RBD indicates that footing widths of 1.5 m (compared to 1.9 m for direct MCS) and 3.2 m (compared to 3.5 m for direct MCS) are needed to achieve target β_{SLS} values of 1.5 and 2.3, respectively. The results of the PEM could be considered to be unconservative since they yield design widths that are smaller than those required to achieve the target reliability index, with the advantage being the relatively small computational cost for running RBD (eight minutes for the eight different footing widths considered).

9.6 CONCLUSION

With the advancement of technology and computational power, geotechnical engineers are increasingly relying on numerical software for modeling complex geotechnical problems. Current design codes are generally tailored for design solutions that are based on commonly used empirical models. The design-checking equations in simplified RBD codes may not reflect the true reliability level that is inherent in designs that are based on numerical performance predictions.

This chapter summarized available RBD methods that can be used by the numerical engineer to design geotechnical components and systems that achieve a target reliability level. The formulation of the available methods is presented and their use is illustrated through a practical design example that is solved using finite elements.

Results indicate that the robust direct Monte Carlo simulation method remains to be the most accurate method in RBD. The practical design example which involved a strip footing on a two-layered soil system clearly showed that the main limitation of the MCS method is computational cost, which may be prohibitive even for a simple design problem that targets the serviceability limit state (relatively small target reliability indices).

This indicates that for RBD to be implemented in practice with numerical models, simplified tools should be available to cut down on the computational cost by limiting the number of calls to the numerical performance function. In the design example, the response surface method was shown to be an efficient RBD method that provides a compromise between efficiency and accuracy. Interestingly, the simple quadratic response surface outperformed the more complex response surface that utilized cross terms. The simple response surface provided RBD designs that were relatively similar to those obtained through direct MCS. On the other hand, the simpler point estimate method (PEM) and First-Order Second Moment (FOSM) method resulted in slightly unconservative designs for the problem under consideration.

The goal of this chapter was to break the boundary between the practicing numerical geotechnical engineer and reliability-based design. RBD principles should infiltrate into day-to-day geotechnical practice and should not be confined to simplified design codes. This chapter clearly illustrates that simple mathematical tools are available to support the numerical engineer in decision-making in the presence of uncertainty. While simplified RBD methods may compromise robustness to reduce the computational cost, they can still be used to portray the impact of uncertainty on risk.

REFERENCES

AASHTO. 2002. *Standard Specifications for Highway Bridges*, 17th ed., Washington, DC.

AASHTO. 2020. *LRFD Bridge Design Specifications*, 9th ed., Washington, DC.

Abou Diab, A., Najjar, S. S. & Sadek, S. 2018. Reliability-based design of spread footings on fibre-reinforced clay. *Georisk: Assessment and Management of Risk for Engineered Systems and Geohazards*, 12(2), 135–151.

Ang, A. H. S. & Conell, C. A. 1974. Reliability bases of structural safety and design. *Journal of Structural Division, ASCE*, 100(9), 1755–1769.

ASCE SEI 7. 2022. *Minimum Design Loads for Buildings and Other Structures*. American Society of Civil Engineers, Structural Engineering Institute, 7.

Baecher, G. B. & Christian, J. T. 2005. *Reliability and Statistics in Geotechnical Engineering*. John Wiley & Sons, Chichester, England.

Barker, R. M., Duncan, J. M., Rojiani, K. B., Ooi, P. S. K., Tan, C. K. & Kim, S. G. 1991. *Manuals for the Design of Bridge Foundations. Transportation Research Board, NCHRP-343*, Washington, DC: National Research Council.

Benjamin, J. R. & Cornell, C. A. 2013. *Probability, Statistics, and Decision for Civil Engineers*.

Bentley Systems. 2021. PLAXIS 2D (Version 2021) [Software]. Bentley Systems, Inc.

Brown, D. A., Turner, J. P. & Castelli, R. J. 2010. *Drilled Shafts: Construction Procedures and LRFD Design Methods—Geotechnical Engineering Circular No. 10*, FHWA NHI-10–016. Washington, DC: Federal Highway Administration, U.S. Department of Transportation.

Cao, Z. J., Gao, G. H., Li, D. Q. & Wang, Y. 2019. Values of Monte Carlo samples for geotechnical reliability-based design. In *Proceedings, 7th International Symposium on Geotechnical Safety and Risk* (ISGSR 2019), Taipei, Taiwan, 86–95.

Carter, J. P. & Kulhawy, F. H. 1988. *Analysis and Design of Foundations Socketed into Rock, Report No. EL-5918*. Empire State Electric Engineering Research Corporation and Electric Power Research Institute, New York, 158.

Ching, J. & Phoon, K. K. 2011. A quantile-based approach for calibrating reliability-based partial factors. *Structural Safety*, 33(4–5), 275–285.

Ching, J., Phoon, K. K. & Yang, J. J. 2015. Role of redundancy in simplified geotechnical reliability-based design – A quantile value method perspective. *Structural Safety*, 55, 37–48.

Christian, J. T. & Baecher, G. B. 1999. Point-estimate method as numerical quadrature. *Journal of Geotechnical and Geoenvironmental Engineering – ASCE*, 125(9), 779–786.

CSA S6:19. 2019. *Canadian Highway Bridge Design Code*. Canadian Standards Association.

Duncan, J. M. 2000. Factors of safety and reliability in geotechnical engineering. *Journal of Geotechnical Engineering, ASCE*, 126(4), 307–316.

Duncan, J. M. & Sleep, M. D. 2015. Evaluating reliability in geotechnical engineering. In *Risk and Reliability in Geotechnical Engineering*, 131–178.

Ellingwood, B. R. 1982. Safety checking formats for limit states design. *Journal of Structural Division, ASCE*, 108(7), 1481–1493.

Ellingwood, B. R. 2000. LRFD: Implementing structural reliability in professional practice. *Engineering Structures*, 22(2), 106–115.

Ellingwood, B. R. & Ang, A. H. S. 1974. Risk-based evaluation of design criteria. *Journal of Structural Division, ASCE*, 100(9), 1771–1778.

EN 1990:2002. *Eurocode – Basis of Structural Design*.

Esrig, M. E. & Kirby, R. C. 1979. Advances in general effective stress method for the prediction of axial capacity for driven piles in clay. In *Proceedings, 11th Annual Offshore Technology Conference*, Houston, TX, 437–449.

Faravelli, L. 1989. Response surface approach for reliability analysis. *Journal of Engineering Mechanics, ASCE*, 115(12), 2763–2781.

Fenton, G. A., Naghibi, F., Dundas, D., Bathurst, R. J. & Griffiths, D. V. 2016. Reliability-based geotechnical design in 2014 Canadian Highway Bridge Design Code. *Canadian Geotechnical Journal*, 53(2), 236–251.

Galambos, T. V. & Ravindra, M. K. 1981. Load and resistance factor design. *Engineering Journal, AISC*, 18(3), 78–84.

Geo-Slope International Ltd. 2022. SLOPE/W (Version 2022) [Software]. Calgary, Alberta, Canada.

Hannigan, P. J., Goble, G. G., Thendean, G., Likins, G. E. & Rausche, F. 2006. *Design and Construction of Driven Pile Foundations*. FHWA-NHI-05-042 and NHI-05-043, Federal Highway Administration, U.S. Department of Transportation, Washington, DC, Vols. I and II.

Harr, M. E. 1987. *Reliability-Based Design in Civil Engineering*. New York: McGraw-Hill.

ISO. 2015. ISO 2394:2015. *General Principles on Reliability for Structures*. Geneva: International Organization for Standardization.

ISO. 2015. ISO 690:2010. *Information and Documentation – Guidelines for Bibliographic References and Citations to Information Resources. Annex D (Informative)*. Geneva: International Organization for Standardization.

ISO 22111:2019. *Bases for Design of Structures – General Requirements*. Geneva: International Organization for Standardization.

ITASCA Consulting Group Inc. 2022. FLAC (Version 8.1) [Computer Software]. Itasca Consulting Group Inc.

Juang, C. H., Wang, L., Liu, Z., Ravichandran, N., Huang, H. & Zhang, J. 2013. Robust geotechnical design of drilled shafts in sand: New design perspective. *Journal of Geotechnical and Geoenvironmental Engineering*, 139(12), 2007–2019.

Kahiel, A., Najjar, S. S. & Sadek, S. 2017. Reliability-based design of spread footings on clays reinforced with aggregate piers. *Georisk: Assessment and Management of Risk for Engineered Systems and Geohazards*, 11(1), 75–89.

Kulhawy, F. H., Prakoso, W. A. & Akbas, S. O. 2005. Evaluation of capacity of rock foundation sockets. In *Proceedings, Alaska Rocks 2005, 40th U.S. Symposium on Rock Mechanics*, American Rock Mechanics Association, Anchorage, AK, 8.

Li, D. Q., Zheng, D., Cao, Z., Tang, X. S. & Phoon, K. K. 2016. Response surface methods for slope reliability analysis: Review and comparison. *Engineering Geology*, 203, 3–14.

Low, B. K. 2021. *Reliability-Based Design in Soil and Rock Engineering: Enhancing Partial Factor Design Approaches*. CRC Press, Boca Raton, Florida.

Meyerhof, G. G. 1957. Penetration tests and bearing capacity of cohesionless soils. *Journal of the Soil Mechanics and Foundation Division*, 82(SM1), 866-1–866-19.

Munfakh, G. A., Arman, J. G., Collin, J. Hung, C. J. & Brouillette, R. P. 2001. *Shallow Foundations Reference Manual. FHWA-NHI-01-023.* Washington, DC: Federal Highway Administration, U.S. Department of Transportation.

Najjar, S. S. & Gilbert, R. B. 2009. Importance of lower-bound capacities in the design of deep foundations. *Journal of Geotechnical and Geoenvironmental Engineering*, 135(7), 890–900.

Najjar, S. S., Saad, G. & Abdallah, Y. 2017. Rational decision framework for designing pile-load test programs. *Geotechnical Testing Journal*, 40(2), 302–316.

Najjar, S. S., Sadek, S. & Farah, Z. 2020. Importance of lower-bound shear strengths in the reliability of spatially random clayey slopes. *Geotechnical and Geological Engineering*, 38(6), 6623–6639.

Paikowsky, S. G. 2004. Load and resistance factor design (LRFD) for deep foundations. *Transportation Research Board*, p.507.

Papaioannou, I. 2012. *Non-Intrusive Finite Element Reliability Analysis Methods*. Technical University of Munich

Phoon, K. K. 2016. *Reliability as a Basis for Geotechnical Design, Reliability of Geotechnical Structures in ISO2394*. CRC Press, London.

Phoon, K. K. 2017. Role of reliability calculations in geotechnical design. *Georisk: Assessment and Management of Risk for Engineered Systems and Geohazards*, 11(1), 4–21.

Phoon, K. K. & Kulhawy, F. H. 2008. Chapter 9. Serviceability limit state reliability-based design. In *Reliability-Based Design in Geotechnical Engineering: Computations and Applications*, 344–383. Abingdon: Taylor & Francis.

Phoon, K. K., Cao, Z., Ji, J., Leung, Y. F., Najjar, S. S., Shuku, T., Tang, C., Yin, Z. Y., Ikumasa, Y. & Ching, J. 2022. Geotechnical uncertainty, modeling, and decision making. *Soils and Foundations*, 62(5), 101189.

Phoon, K. K., Kulhawy, F. H. & Grigoriu, M. D. 1995. *Reliability-Based Design of Foundations for Transmission Line Structures*. Report TR-105000-Palo Alto Electric Power Research Institute.

Phoon, K. K., Kulhawy, F. H. & Grigoriu, M. D. 2003. Multiple resistance factor design for shallow transmission line structure foundations. *Journal of Geotechnical and Geoenvironmental Engineering*, 129(9), 807–818.

Phoon, K. K. & Tang, C. 2019. Characterisation of geotechnical model uncertainty. *Georisk: Assessment and Management of Risk for Engineered Systems and Geohazards*, 13(2), 101–130.

Ravindra, M. K. & Galambos, T. V. 1978. Load and resistance factor design for steel. *Journal of Structural Division, ASCE*, 104(9), 1337–1354.

Ravindra, M. K., Lind, N. C. & Siu, W. 1974. Illustrations of reliability based design. *Journal of Structural Division, ASCE*, 100(9), 1789–1811.

Rocscience Inc. 2021. SLIDE 3 (Version 2021.04) [Computer Software]. Toronto, Canada.

Rosenblueth, E. 1975. Point estimates for probability moments. *Proceedings of the National Academy of Sciences of the United States of America*, 72(10), 3812–3814.

Skempton, A. W. 1951. The bearing capacity of clays. *Proc., Building Research Congress*, 1, 180–189.

Tomlinson, M. J. 1987. *Pile Design and Construction Practice*. Viewpoint Publication.

Transportation Research Board. 2004. *NCHRP Report 507 – Load and Resistance Factor Design (LRFD) for Deep Foundations*. Transportation Research Board, Washington, DC.

Transportation Research Board. 2010. *NCHRP Report 651 – LRFD Design and Construction of Shallow Foundations for Highway Bridge Structures*. Transportation Research Board, Washington, DC.

U.S. Army Corps of Engineers. 1997. *Introduction to Probability and Reliability Methods for Using in Geotechnical Engineering*. ETL 1110-2-547.

Vijayvergiya, V. N. & Focht, J. A. 1971. A new way to predict the capacity of piles in clay. *Proceedings, 4th Annual Offshore Technology Conference*, 2, 865–874.

Wang, Y. & Cao, Z. 2013. Expanded reliability-based design of piles in spatially variable soil using efficient Monte Carlo simulations. *Soils and Foundations*, 53(6), 820–834.

Wolff, T. F. 1994. *Evaluating the Reliability of Existing Levees*. Prepared for the US Army Engineers Waterways Experiment Station Geotechnical Lab, Vicksburg, MS.

Wolff, T. F. 1996. Probabilistic slope stability in theory and practice. In *Proceedings of Uncertainty '96, ASCE Geotechnical Special Publication No. 58*, 419–433.

Yang, Z. & Ching, J. 2020. A novel reliability-based design method based on quantile-based first-order second-moment. *Applied Mathematical Modelling*, 88, 461–473.

Zhang, L. M. & Ng, A. M. Y. 2005. Probabilistic limiting tolerable displacement for serviceability limit state design of foundations. *Geotechnique*, 55(2), 151–161.

10 Probabilistic Inverse Analysis for Geotechnics

Ikumasa Yoshida

ABSTRACT

This chapter discusses the methods for using inverse analysis to estimate model parameters based on observation (measurement, sampling). Starting from the basics of the least-squares method and regularization, their interpretation is given from the viewpoint of probability theory. The regularization term and the weight in the least-squares method are interpreted by prior information in Bayesian updating. Their relations are explained using a simple two-variable numerical example. Recent sample-based methods are introduced and demonstrated with the same two-variable numerical example. The surrogate model approach with active learning is also introduced. The combination of the sample-based method and the surrogate model approach seems a promising tool for probabilistic inverse analysis. Hyperparameter estimation and Laplace approximation of evidence for model selection in inverse analysis are also discussed. The methods introduced in this chapter are general and applicable to inverse problems in many fields, not only geotechnical engineering.

10.1 INTRODUCTION

We have more observation data in geotechnical engineering than in the past, thanks to advances in sampling, sensing, and monitoring technologies. However, the observation information is still scarce for characterizing the real ground, which involves large uncertainties. Estimating model parameters and/or boundary (initial) conditions based on the observation data is called inverse analysis. In geotechnical engineering, inverse analysis usually means parameter identification, and this chapter discusses methods for inverse analysis mainly for the aspect of parameter identification. A brief history of inverse/back analysis in rock mechanics was summarized by Walton and Sinha (2022). An inverse/back analysis can be achieved using an inverse approach or a direct approach. The inverse approach relies on a re-arrangement of the governing equations. The direct approach, in contrast, is based on solutions to the forward problem. This chapter discusses the so-called direct approach.

Deterministic estimates are sometimes used without discussion of their uncertainties; however, the uncertainty quantification of an estimate provides useful and valuable information for enhancing decision-making in practice as stated in the Preface of this book. Inverse analysis or Bayesian updating plays an important role in decision-making in geotechnical engineering. Most of the methodologies for inverse analysis can be consistently derived from Bayes' theorem. The least-squares (LS) method, with regularization such as ridge regression, and the least absolute shrinkage and selection operator (LASSO) can be interpreted from the viewpoint of prior information in Bayes' theorem. The recent sample (particle)-based methods derived from probability theory are introduced and discussed in this chapter, but there is no intent to provide a complete review of available methods for performing inverse analysis.

A brief introduction is given to the LS method with regularization in Section 10.2, and then these methods are interpreted from the viewpoint of probability theory in Section 10.3. Section 10.4 summarizes non-linear inverse problems and recent sample (particle)-based methods for inverse analysis constructed on probability theory. Section 10.5 discusses the surrogate model with active

DOI: 10.1201/9781003333586-12

learning for inverse analysis. Sections 10.6 and 10.7 present the estimation of hyperparameters and the Laplace approximation of evidence for model selection. Finally, optimal observation planning is briefly introduced in Section 10.8 and then Section 10.9 concludes.

10.2 WEIGHTED LEAST-SQUARES METHOD WITH REGULARIZATION

Assume that we have a model to estimate or predict a response. A part of the response is observed, which is denoted as $\boldsymbol{z}$. Let the unknown parameter vector of interest be $\boldsymbol{x}$. Assume that the observation vector $\boldsymbol{z}$ is expressed as the following function of $\boldsymbol{x}$:

$$\boldsymbol{z} = \boldsymbol{h}(\boldsymbol{x}) + \boldsymbol{v} \tag{10.1}$$

where $\boldsymbol{v}$ is the observation error, which includes modeling error. This equation is termed the observation equation. The residual sum of squares between observed and calculated values is

$$J = (\boldsymbol{z} - \boldsymbol{h}(\boldsymbol{x}))^T \boldsymbol{W}(\boldsymbol{z} - \boldsymbol{h}(\boldsymbol{x})) \tag{10.2}$$

where $\boldsymbol{W}$ is the weight matrix. A solution can be obtained such that the objective function J is minimized, namely the weighted least-squares method. In the ordinary least-squares method (LS), an identity matrix is used for the weight matrix. When some of the observation data are more reliable than others, larger values are given to the corresponding diagonal components of $\boldsymbol{W}$. The values of the diagonal component express the reliability of the observation data, the larger the more reliable. When the observation equation is linear, i.e., $\boldsymbol{z}=\boldsymbol{H}\boldsymbol{x}+\boldsymbol{v}$, we have an analytical solution:

$$\hat{\boldsymbol{x}} = \left(\boldsymbol{H}^T \boldsymbol{W} \boldsymbol{H}\right)^{-1} \boldsymbol{H}^T \boldsymbol{W} \boldsymbol{z} \tag{10.3}$$

Most inverse problems in the real world are non-linear, but linear inverse problems provide useful insight for understanding the nature of inverse problems.

NUMERICAL EXAMPLE 10.1

Consider a simple linear inverse problem with three observations and two unknown variables.

$$\boldsymbol{z} = \begin{pmatrix} 160 \\ 170 \\ 100 \end{pmatrix}, \boldsymbol{x} = \begin{pmatrix} x_1 \\ x_2 \end{pmatrix}, \boldsymbol{H} = \begin{bmatrix} 1 & 2 \\ 2 & 1 \\ 1 & 1 \end{bmatrix}, \boldsymbol{W} = \begin{bmatrix} w_{11} & 0 & 0 \\ 0 & w_{22} & 0 \\ 0 & 0 & w_{33} \end{bmatrix} \tag{10.4}$$

The solutions "A", "B", and "C" by weighted LS are shown in unknown variable space in Figure 10.1. The three lines in the figure indicate the observation equation $\boldsymbol{z}=\boldsymbol{H}\boldsymbol{x}$. The solution "A" is obtained when the weight matrix $\boldsymbol{W}$ is the identity matrix, i.e., $w_{11}= w_{22}= w_{33}= 1$. The solution "B" is obtained for the weight matrix with $w_{11}=1$, $w_{22}= w_{33}=9$, in which we consider the observation 160 to be less reliable than the other observation. In the same way, the solution "C" is obtained for the weight matrix with $w_{22}=1$, $w_{11}= w_{33}=9$.

Consider that we have data of only one observation, i.e., there is one line in the figure. It is impossible to determine two variables x1 and x2 from one observation by LS because the inverse matrix of $\boldsymbol{H}^T\boldsymbol{W}\boldsymbol{H}$ cannot be obtained. This problem has infinitely many solutions. Any point on the line satisfies the observation equation. Even in the case of three observation data (three lines), if they are close to parallel, a small observation error will result in a large estimation error. Regularized LS, where the regularization term is considered in the objective function, is often used in this situation. There are several types of regularization, and the simplest one is $\lambda \boldsymbol{x}^T\boldsymbol{x}$, where λ is the weight of the regularization term. The objective function with regularization is

$$J = (\boldsymbol{z} - \boldsymbol{h}(\boldsymbol{x}))^T \boldsymbol{W}(\boldsymbol{z} - \boldsymbol{h}(\boldsymbol{x})) + \lambda \boldsymbol{x}^T \boldsymbol{x} \tag{10.5}$$

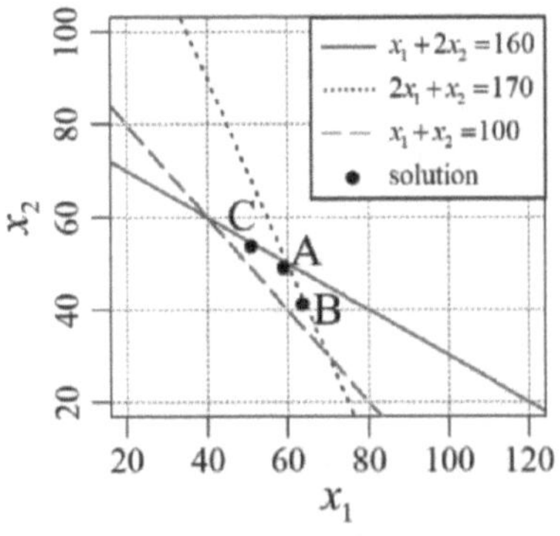

FIGURE 10.1 Solutions of weighted LS, A: $w_{11} = w_{22} = w_{33} = 1$, B: $w_{11}=1$, $w_{22} = w_{33}=9$, C: $w_{22}=1$, $w_{11}= w_{33}=9$.

The first term is the l_2 residual norm in the LS and expresses the fitness to data. The second term is called the regularization term, or penalty, and captures the expected behavior of $\boldsymbol{x}$ through an additional l_2 penalty term. "The expected behavior of $\boldsymbol{x}$" plays an important role for the regularized LS. It will be discussed from the perspective of probability theory in the next section. The regularization parameter λ controls the balance between the two terms. The regularization term is useful to obtain stable solutions. Due to the regularization term, however, the solution is biased. There is the bias-variance trade-off problem in inverse analysis with regularization. The LS with l_2 regularization is called ridge regression or regularized LS (Rasmussen and Williams 2006; Bishop 2006; Hastie et al. 2015).

If the non-linear function $\boldsymbol{h}(\boldsymbol{x})$ is replaced by the linear function $\boldsymbol{Hx}$, the solution can be analytically obtained,

$$\hat{\boldsymbol{x}} = \left(\boldsymbol{H}^T \boldsymbol{W} \boldsymbol{H} + \lambda \boldsymbol{I}\right)^{-1} \boldsymbol{H}^T \boldsymbol{W} \boldsymbol{z} \tag{10.6}$$

where the matrix $\boldsymbol{I}$ is an identity matrix.

Numerical Example 10.2

Consider again the case in which we have data of only one observation, i.e., the new $\boldsymbol{H}$ matrix is [1 2]. Figure 10.2 shows the solutions "a", "b", "c", and "d" when the regularization term λ is 1/4, 1, 4, and 25. When λ is small enough, the solution is almost the same as the one with the generalized inverse matrix. A larger λ results in a solution closer to the origin, namely shrinkage. It introduces bias (shrinkage) into the solution that can reduce variance relative to the ordinary LS solution.

Instead of the l_2 norm, the l_1 norm, i.e., $\lambda|\boldsymbol{x}|$, can be used for the regularization term.

$$J = (\boldsymbol{z} - \boldsymbol{h}(\boldsymbol{x}))^{\mathrm{T}} \boldsymbol{W} (\boldsymbol{z} - \boldsymbol{h}(\boldsymbol{x})) + \lambda|\boldsymbol{x}| \tag{10.7}$$

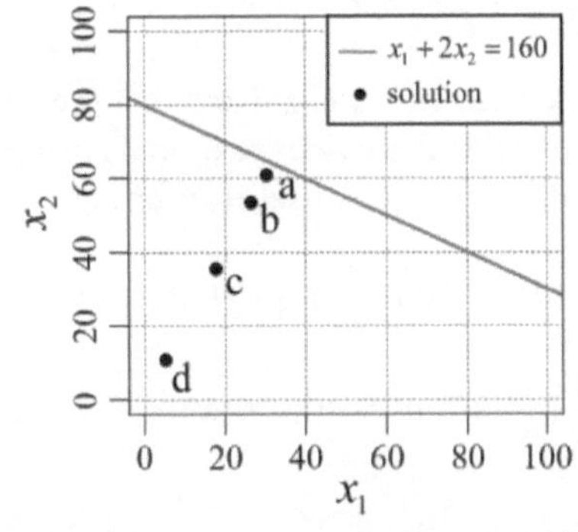

FIGURE 10.2 Solutions of regularized LS, λ values, a: 1/4, b: 1, c: 4, and d: 25.

This formulation is called the least absolute shrinkage and selection operator (LASSO) (Bishop 2006; Hastie et al. 2015), which is one of the most popular sparse modeling approaches. Sparse modeling has attracted attention in many fields, including medical science, astronomy, geophysics, and civil engineering. According to the general principle of sparsity, a phenomenon should be represented by as few variables as possible. Including linear observation equation problems, there is no analytical solution when the l_1 norm is employed. However, effective iterative algorithms have been proposed for obtaining the solution in the case of a linear observation equation, such as the alternating direction method of multipliers (Boyd et al. 2010; Hastie et al. 2015).

10.3 INVERSE ANALYSIS FROM THE VIEWPOINT OF PROBABILITY THEORY

Bayesian model updating provides a useful framework from probability theory for inverse analysis, namely, assimilating the observation data into models, where a prior probability distribution of model parameters is updated to a posterior probability distribution. The theory and perspectives of Bayesian analysis have considerably broader applicability for inverse problems including those in geotechnical engineering.

The theory of Bayesian model updating is next briefly summarized. Engineers can obtain a priori information from either existing physical process data or empirical evidence. Prior knowledge can be updated with observed data. This is represented as an update of the probability distribution from a prior to a posterior using Bayes' theorem:

$$p(\boldsymbol{x} \mid z) = \frac{p(z \mid \boldsymbol{x})p(\boldsymbol{x})}{p(z)} \tag{10.8}$$

where $p(\boldsymbol{x})$ is the prior probability density function (PDF) of the parameter vector of interest $\boldsymbol{x}$, $p(z|\boldsymbol{x})$ is the probabilistic forward model, which is the conditional PDF of z given $\boldsymbol{x}$, and $p(\boldsymbol{x}|z)$ is the posterior PDF of $\boldsymbol{x}$. Typically, prior information is modeled as a Gaussian distribution:

$$\boldsymbol{x} = \bar{\boldsymbol{x}} + \boldsymbol{w} \tag{10.9}$$

where $\bar{\boldsymbol{x}}$ and $\boldsymbol{w}$ are the mean and the probabilistic component of the prior information, respectively. The observation equation, which is the probabilistic forward model for predicting observation z from $\boldsymbol{x}$, is generally expressed as a non-linear function of $\boldsymbol{x}$ contaminated with Gaussian noise $\boldsymbol{v}$, i.e., $z = \boldsymbol{h}(\boldsymbol{x}) + \boldsymbol{v}$, where $\boldsymbol{w}$ and $\boldsymbol{v}$ are Gaussian random variable vectors with zero mean whose covariance matrices are $\boldsymbol{M}$ and $\boldsymbol{R}$, respectively. PDFs $p(\boldsymbol{x})$ and $p(z|\boldsymbol{x})$ are then

$$p(\boldsymbol{x}) = \frac{1}{(2\pi)^{n/2}|\boldsymbol{M}|^{1/2}} \exp\left\{-\frac{1}{2}(\boldsymbol{x}-\bar{\boldsymbol{x}})^T \boldsymbol{M}^{-1}(\boldsymbol{x}-\bar{\boldsymbol{x}})\right\} \tag{10.10}$$

$$p(z \mid \boldsymbol{x}) = \frac{1}{(2\pi)^{m/2}|\boldsymbol{R}|^{1/2}} \exp\left\{-\frac{1}{2}(z-\boldsymbol{h}(\boldsymbol{x}))^T \boldsymbol{R}^{-1}(z-\boldsymbol{h}(\boldsymbol{x}))\right\} \tag{10.11}$$

where n and m are the sizes of the vectors $\boldsymbol{x}$ and z. Note that the denominator $p(z)$ in Eq. (10.8) is just a normalizing constant, often called the evidence or marginalized likelihood (Rasmussen and Williams 2006; Bishop 2006), which provides useful information for model selection. It will be discussed in Section 10.7.

The logarithm of the posterior distribution, $\ln p(\boldsymbol{x}|z)$, is given by

$$\begin{aligned} \ln p(\boldsymbol{x} \mid z) = &-\frac{1}{2}(z-\boldsymbol{h}(\boldsymbol{x}))^T \boldsymbol{R}^{-1}(z-\boldsymbol{h}(\boldsymbol{x})) - \frac{1}{2}(\boldsymbol{x}-\bar{\boldsymbol{x}})^T \boldsymbol{M}^{-1}(\boldsymbol{x}-\bar{\boldsymbol{x}}) \\ &-\frac{1}{2}\ln|\boldsymbol{M}| - \frac{1}{2}\ln|\boldsymbol{R}| - \frac{n+m}{2}\ln(2\pi) - \ln p(z) \end{aligned} \tag{10.12}$$

The vector $\boldsymbol{x}$ which maximizes the posterior PDF $p(\boldsymbol{x}|z)$ is called the most probable value (MPV) or maximum a posteriori (MAP) value (Au 2017). To obtain the MPV or MAP, one can simply minimize the objective function consisting of the first and second terms of $\ln p(\boldsymbol{x}|z)$:

$$J = \frac{1}{2}(z - \boldsymbol{h}(\boldsymbol{x}))^T \boldsymbol{R}^{-1}(z - \boldsymbol{h}(\boldsymbol{x})) + \frac{1}{2}(\boldsymbol{x} - \overline{\boldsymbol{x}})^T \boldsymbol{M}^{-1}(\boldsymbol{x} - \overline{\boldsymbol{x}}) \tag{10.13}$$

where the first term of the summation comes from the observation and the second term comes from the prior. When $\boldsymbol{R}^{-1} = \boldsymbol{W}$, $\overline{\boldsymbol{x}}=0$, $\boldsymbol{M}=(1/\lambda)\boldsymbol{I}$, the objective function derived from the Bayesian framework is equivalent to Eq. (10.5).

Assume that random variable $\boldsymbol{w}$ in Eq. (10.9) follows the (independent) Laplace distribution.

$$p(\boldsymbol{x}) = \prod_{i=1}^{n} \frac{1}{2b} \exp\left(-\frac{|\overline{x}_i|}{b}\right) \tag{10.14}$$

Substituting the Laplace distribution instead of the Gaussian into the prior of Eq. (10.8), we have the following objective function:

$$J = (z - \boldsymbol{h}(\boldsymbol{x}))^T \boldsymbol{R}^{-1}(z - \boldsymbol{h}(\boldsymbol{x})) + \frac{2}{b}\sum_{i=1}^{n} |x_i - \overline{x}_i| \tag{10.15}$$

When $\boldsymbol{R}^{-1} = \boldsymbol{W}$, $\overline{\boldsymbol{x}}=0$, $2/b=\lambda$, the objective function derived from the Bayesian framework with a Laplace prior is equivalent to the objective function of LASSO, Eq. (10.7).

When a Gaussian prior is used and the non-linear function $\boldsymbol{h}(\boldsymbol{x})$ is replaced by the linear function $\boldsymbol{Hx}$, then the solution can be obtained analytically:

$$\hat{\boldsymbol{x}} = \overline{\boldsymbol{x}} + \boldsymbol{P}\boldsymbol{H}^T \boldsymbol{R}^{-1}(z - \boldsymbol{H}\overline{\boldsymbol{x}}) \tag{10.16}$$

$$\boldsymbol{P} = (\boldsymbol{H}^T \boldsymbol{R}^{-1} \boldsymbol{H} + \boldsymbol{M}^{-1})^{-1} \tag{10.17}$$

Because of the choice of a conjugate prior Gaussian distribution, the posterior is also a Gaussian distribution of which the mean and covariance matrix are given by $\hat{\boldsymbol{x}}$ and $\boldsymbol{P}$, respectively. This linear inverse problem is called Bayesian linear regression (Bishop 2006).

The Kalman filter is one of the most widely used algorithms for inverse analysis based on the analytical solution of the posterior probability distribution. The theory consists of observation-updating and time-updating algorithms. In this section, we focus on the observation-updating algorithm because it can be interpreted as a solution to the probabilistic inverse problem. Eqs. (10.16) and (10.17) can be rewritten using the Kalman gain $\boldsymbol{K}_G$ in the Kalman filter algorithm:

$$\hat{\boldsymbol{x}} = \overline{\boldsymbol{x}} + \boldsymbol{K}_G(z - \overline{\boldsymbol{x}}) \tag{10.18}$$

$$\boldsymbol{K}_G = \boldsymbol{M}\boldsymbol{H}^T(\boldsymbol{H}\boldsymbol{M}\boldsymbol{H}^T + \boldsymbol{R})^{-1} \tag{10.19}$$

$$\boldsymbol{P} = \boldsymbol{M} - \boldsymbol{K}_G \boldsymbol{H}\boldsymbol{M} \tag{10.20}$$

These equations are mathematically equivalent to Eqs. (10.16) and (10.17). The Kalman filter is limited to a linear observation equation and Gaussian noise. Many researchers have tried to develop an algorithm for solving non-linear and/or non-Gaussian inverse problems (Ristic et al. 2004). The extended, ensemble, and unscented Kalman filters, and the particle filter (PF) have been proposed to tackle non-linear and/or non-Gaussian problems. The observation update algorithm of PF will be introduced in Section 10.4.

Numerical Example 10.3

The two-variable examples shown in Figures 10.1 and 10.2 are interpreted from a probability perspective. The prior information is not considered when solving the problem in Figure 10.1. Please note that the inverse of observation error covariance matrix $\boldsymbol{R}^{-1}$ corresponds to weight matrix $\boldsymbol{W}$ in weighted LS. When observation error is assumed to be independent and its standard deviations (SDs) of the three observations take a same value, e.g., (30, 30, 30), the posterior mean $\hat{\boldsymbol{x}}$ by Bayesian linear regression is equivalent to the solution "A". When SDs of observation error assumed for the three observation data are (30, 10, 10) and (10, 30, 10), the solutions "B" and "C" are obtained as the posterior means, respectively. When prior information is not considered, the ratio of SD determines the posterior means. For example, in solution "B", the ratio of the inverse of variance is (1/900, 1/100, 1/100), which is equivalent to (1, 9, 9) (=(w_{11}, w_{22}, w_{33})) in numerical example 10.1. Their absolute values do not affect the posterior mean $\hat{\boldsymbol{x}}$, but affect the covariance matrix $\boldsymbol{P}$.

The solutions in Figure 10.2 are obtained by considering prior information on uncorrelated Gaussian distribution. The prior mean and SD are assumed to be 0 and 100, respectively, for both x_1 and x_2. When the SD of observation error is 50, 100, 200, and 500, the solutions "a", "b", "c", and "d" in Figure 10.2 are obtained as the posterior means. Also, the posterior covariance matrix $\boldsymbol{P}$ for each case can be obtained; e.g., the covariance matrix corresponding to "a" is

$$\boldsymbol{P} = \begin{bmatrix} 8095 & -3810 \\ -3810 & 2381 \end{bmatrix}$$

The posterior mean and SD of x_1 are 30 and 90, which are used in the numerical example 10.4.

As stated above, "the expected behavior of $\boldsymbol{x}$" plays an important role for the regularized LS. If this prior information captures the behavior of the problem, the solutions are appropriate. We need to consider "the expected behavior" carefully, e.g., if "the prior mean = 0", "uncorrelated prior" is appropriated.

Gaussian process regression (GPR) (Rasmussen and Williams 2006) or Kriging (Cressie 1991; Christakos 1992) is widely used for the estimation of spatial variability of soil properties. The theory of Kriging, which has focused mostly on two- and three-dimensional spaces, can be interpreted as a special form of a general regression theory GPR, which is also derived as a special case of the probabilistic linear inverse problem. The formulation of GPR is derived and summarized briefly by Yoshida et al. (2021a). The term $\boldsymbol{x}_o$ denotes the soil properties of interest at the observed locations, and $\boldsymbol{x}_u$ denotes the soil properties at the unobserved locations:

$$\boldsymbol{x}^T = \left\{\boldsymbol{x}_o^T, \boldsymbol{x}_u^T\right\} \tag{10.21}$$

Prior covariance matrix $\boldsymbol{M}$ corresponding to $\boldsymbol{x}_o$ and $\boldsymbol{x}_u$ is calculated by one of the covariance functions and is defined as

$$\boldsymbol{M} = \begin{bmatrix} \boldsymbol{M}_{oo} & \boldsymbol{M}_{ou} \\ \boldsymbol{M}_{ou}^T & \boldsymbol{M}_{uu} \end{bmatrix} \tag{10.22}$$

where $\boldsymbol{M}_{ij} = E\left[\left(\boldsymbol{x}_i - \bar{\boldsymbol{x}}_i\right)\left(\boldsymbol{x}_j - \bar{\boldsymbol{x}}_j\right)^T\right]$ (i or j = o or u), and $\bar{\boldsymbol{x}}_o$ and $\bar{\boldsymbol{x}}_u$ are the mean vector of prior information. The observation equation is rewritten as

$$\boldsymbol{z} = \boldsymbol{H}\boldsymbol{x} = \begin{bmatrix} \boldsymbol{I} & \boldsymbol{0} \end{bmatrix} \begin{Bmatrix} \boldsymbol{x}_o \\ \boldsymbol{x}_u \end{Bmatrix} + \boldsymbol{v} \tag{10.23}$$

where $\boldsymbol{I}$ denotes the identify matrix and $\boldsymbol{0}$ denotes a zero matrix. The observation noise is often modeled as white noise with SD σ_n in GPR formulation, namely $\boldsymbol{R}=\sigma_n^2\boldsymbol{I}$. Substituting Eqs. (10.22) and (10.23) into Eqs. (10.18), (10.19), and (10.20) yields the equation for an update by observation $\boldsymbol{z}$:

$$\begin{Bmatrix} \hat{\boldsymbol{x}}_o \\ \hat{\boldsymbol{x}}_u \end{Bmatrix} = \begin{Bmatrix} \bar{\boldsymbol{x}}_o \\ \bar{\boldsymbol{x}}_u \end{Bmatrix} + \begin{bmatrix} \boldsymbol{M}_{oo} \\ \boldsymbol{M}_{ou}^T \end{bmatrix} \left[\boldsymbol{M}_{oo} + \sigma_n^2 \boldsymbol{I} \right]^{-1} (\boldsymbol{z} - \bar{\boldsymbol{x}}_o) \tag{10.24}$$

$$\boldsymbol{P} = \begin{bmatrix} \boldsymbol{P}_{oo} & \boldsymbol{P}_{ou} \\ \boldsymbol{P}_{ou}^T & \boldsymbol{P}_{uu} \end{bmatrix} \tag{10.25}$$

$$\boldsymbol{P}_{ij} = \boldsymbol{M}_{ij} - \boldsymbol{M}_{io}(\boldsymbol{M}_{oo} + \sigma_n^2 \boldsymbol{I})^{-1} \boldsymbol{M}_{oj} \tag{10.26}$$

where $\hat{\boldsymbol{x}}_o$ and $\hat{\boldsymbol{x}}_u$ are the posterior mean vectors of $\boldsymbol{x}_o$ and $\boldsymbol{x}_u$, respectively.

10.4 NON-LINEAR INVERSE PROBLEM AND SAMPLE (PARTICLE)-BASED APPROACH

When the observation equation is non-linear, there is no analytical solution. Gradient-based iterative methods such as the Gauss-Newton, Marquardt, and BFGS methods are often used to obtain the solution that minimizes the objective function. For example, the Gauss-Newton method to minimize the objective function in Eq. (10.13) is shown as follows.

$$\boldsymbol{x}_{i+1} = \boldsymbol{x}_i + \boldsymbol{P}_{xi}\left(\boldsymbol{H}_{xi}^{\mathrm{T}} \boldsymbol{R}^{-1} \left(\boldsymbol{z} - \boldsymbol{h}(\boldsymbol{x}_i) \right) + \boldsymbol{M}^{-1} \left(\bar{\boldsymbol{x}} - \boldsymbol{x}_i \right) \right) \tag{10.27}$$

$$\boldsymbol{P}_{xi} = \left(\boldsymbol{H}_{xi}^{\mathrm{T}} \boldsymbol{R}^{-1} \boldsymbol{H}_{xi} + \boldsymbol{M}^{-1} \right)^{-1} \tag{10.28}$$

where $\boldsymbol{H}_{xi} = \left. \dfrac{d\boldsymbol{h}(\boldsymbol{x})}{d\boldsymbol{x}} \right|_{x=xi}$

The subscript i in the equation is for the iteration to obtain the solution. The posterior PDF can be obtained approximately by linearization of the observation equation with respect to the obtained solution. The converged $\boldsymbol{P}_{xi}$ indicates the approximate posterior PDF. This method is simple and useful; however, it cannot be applied to strongly non-linear problems because accuracy deteriorates. In addition, a non-differential or non-globally identifiable problem cannot be solved by this approach. The inverse problems can be classified into three categories: globally identifiable, locally identifiable, or unidentifiable (Beck and Katafygiotis 1998; Au 2017). A problem is "globally identifiable" if the posterior PDF has a single peak in the parameter space. If there is more than one peak, the problem is said to be "locally identifiable". If there is a continuum of points with the same value as the posterior PDF, the problem is "unidentifiable" (Au 2017). Among linear inverse problems, there are no locally identifiable problems. The problems in Figures 10.1 and 10.2 without a prior are globally identifiable and unidentifiable, respectively. Among non-linear inverse problems, we sometimes encounter locally identifiable problems. Note that the solution of a locally identifiable problem by gradient-based optimization depends on the initial value $\boldsymbol{x}_0$. Boumezerane (2022) reviewed recent publications related to the use of optimization techniques in geotechnical engineering and observed that most of the techniques used recently rely on meta-heuristic methods such as genetic algorithms, particle swarm optimization, and artificial neural networks. These methods are global optimization methods that obtain the highest peak of the posterior PDF, i.e., a representative value; however, they cannot evaluate the uncertainty of the estimated values, i.e., posterior PDF, directly.

Sample (particle)-based approaches have also been developed to solve inverse problems based on probability theory, i.e., to evaluate the posterior probability distribution numerically. Markov chain Monte Carlo (MCMC) is a popular and powerful tool to generate samples from posterior PDFs. However, sampling directly from the target PDF is difficult and generally not practical if the prior and posterior PDFs are significantly different. In order to overcome this, Transitional Markov chain Monte Carlo (TMCMC) was proposed by Ching and Chen (2007). The key idea of TMCMC is to sample from a series of intermediate PDFs which gradually converge to the target PDF. Betz et al. (2016) discussed the properties of TMCMC and proposed modifications to improve efficiency.

Particle filters (PFs) were developed in the 1990s (Gordon et al. 1993, Kitagawa 1996). A PF can be regarded as a Monte Carlo technique for Bayesian updating. Early ideas for the algorithms were proposed in the 1950s and 1960s (Ristic et al. 2004). In a PF algorithm, probability distributions are approximated by their sample realizations, called particles in the PF algorithm. The probabilistic nature of a state vector (uncertain variables of interest) is expressed by these particles instead of by the first and second moments in the Kalman filter. The advantage of a PF-based method is the simplicity of its algorithm, and the implementation is easy.

PFs constitute a state-space approach which has two processes, namely, a time-updating process and an observation-updating process as in a Kalman filter. The example here focuses on the observation-updating process. The algorithm starts by generating random samples (particles) from the prior PDF $p(\boldsymbol{x})$:

$$\boldsymbol{x}^{(j)} \sim p(\boldsymbol{x}), \qquad j = 1,\cdots,n \tag{10.29}$$

where the superscript (j) denotes the j-th particle. The prior PDF $p(\boldsymbol{x})$ then has the following approximate representation, called an empirical PDF:

$$p(\boldsymbol{x}) \cong \frac{1}{n}\sum_{j=1}^{n} \delta(\boldsymbol{x} - \boldsymbol{x}^{(j)}) \tag{10.30}$$

where $\delta(\cdot)$ denotes the Dirac delta function. Substituting Eq. (10.30) into Eq. (10.8) yields

$$p(\boldsymbol{x} \mid z) = \frac{p(z \mid \boldsymbol{x})\frac{1}{n}\sum_{j=1}^{n} \delta(\boldsymbol{x} - \boldsymbol{x}^{(j)})}{\int p(z \mid \boldsymbol{x})\frac{1}{n}\sum_{i=1}^{n} \delta(\boldsymbol{x} - \boldsymbol{x}^{(i)}) d\boldsymbol{x}} \tag{10.31}$$

The integral in the denominator can be evaluated using the properties of the Dirac delta function. The posterior PDF $p(\boldsymbol{x}|z)$ is rewritten as

$$p(\boldsymbol{x} \mid z) = \left(\frac{1}{\sum_{i=1}^{n} p(z \mid \boldsymbol{x}^{(i)})} \right) \sum_{j=1}^{n} p(z \mid \boldsymbol{x}^{(j)}) \delta(\boldsymbol{x} - \boldsymbol{x}^{(j)}) = \sum_{j=1}^{n} a^{(j)} \delta(\boldsymbol{x} - \boldsymbol{x}^{(j)}) \tag{10.32}$$

where

$$a^{(j)} = \frac{p(z \mid \boldsymbol{x}^{(j)})}{\sum_{i=1}^{n} p(z \mid \boldsymbol{x}^{(i)})} \tag{10.33}$$

The term $a^{(j)}$ represents the weight (likelihood ratio) of particle j after updating.

Data assimilation with a PF has attracted attention for updating model parameters and predictions based on the observed data in many research fields such as oceanography, meteorology, and visual processing/tracking (e.g., Ristic et al. 2004). When the SD of observation noise is small, the posterior is significantly different from the prior PDF. The weights tend to concentrate into only a few particles (all other particles degenerate), which causes poor performance in identification and reliability estimation. This is known as degeneracy, and many methods have been proposed to mitigate degeneracy. For example, Nakano et al. (2007) proposed the merging particle filter approach, in which linear combinations of particles, called merging particles, are used to reduce weight variance. Fan et al. (2015) and Manoli et al. (2015) proposed PF algorithms that use iterative schemes and applied them to visual tracking and parameter identification in hydro-geophysical inversion. These methods, however, preserve only the mean and covariance of the particles. They focus only on the most probable states or parameters (representative value) in the updating process and cannot consistently estimate posterior PDFs of the states and parameters of interest.

Numerical Example 10.4

The two-variable example with data of one observation discussed above (Figure 10.2) is used to illustrate the PF approach. The SD of observation error is 50 (solution "a" in Figure 10.2). For comparison, 100 and 1000 samples are generated according to the prior. They are shown in Figure 10.3(a) and (b), respectively. The weight $a^{(j)}$ for particle j, which is calculated by Eq. (10.33), is expressed as the radius of a circle in the figures. The posterior cumulative distribution function (CDF) of $\boldsymbol{x}$ is calculated while considering the weight. The posterior CDFs of x_1 from 100, 300, and 1000 samples are shown in Figure 10.3(c). The true posterior mean and covariance 30 and 90 are obtained analytically as stated in numerical example 10.3. Naturally, the CDFs by 1000 particles have better agreement with the true CDFs than the CDFs by fewer sample cases.

Along an independent line of thought, Straub and Papaioannou (2015) proposed taking a new perspective on the Bayesian updating problem, which gave rise to methods that may be collectively referred to as Bayesian Updating with Structural Reliability (BUS) methods. BUS converts a Bayesian updating problem into an equivalent reliability analysis in the augmented random variable space $(p, \boldsymbol{x})$, where a standard uniform random variable p is added to the space of random variables $\boldsymbol{x}$. The failure domain Ω of this equivalent reliability problem is defined as follows:

$$\Omega = \{p \leq cL(\boldsymbol{x}|z)\} \tag{10.34}$$

where $L(\boldsymbol{x}|z)=p(z|\boldsymbol{x})$ is the likelihood function and c is a positive constant chosen such that $cL(\boldsymbol{x}|z) \leq 1$ holds for all $\boldsymbol{x}$. The advantage of BUS is the ability to incorporate any sophisticated structural reliability methods for efficient calculations. Straub et al. (2016) combined BUS with the first-order reliability method, line sampling, and subset simulation (SuS), and compared their performance. As with a PF, when the SD of observation noise is small, the performance of BUS deteriorates significantly. In order to alleviate this problem, SuS (Au & Beck 2003; Au and Wang 2014) is often

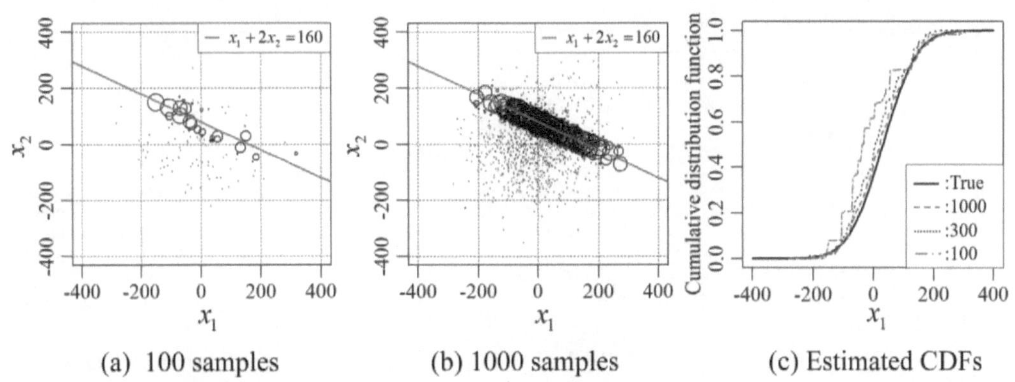

FIGURE 10.3 Weight (likelihood ratio) of generated samples and CDFs estimated by PF with 100, 300, and 1000 samples (a) 100 samples (b) 1000 samples (c) Estimated CDFs.

combined with BUS in order to efficiently generate samples from the posterior distribution. The improvement of BUS with SuS, which addresses the important issue of how to determine constant c in Eq. (10.34), was reported by DiazDelaO et al. (2017) and Betz et al. (2018).

Numerical Example 10.5

The problem shown in Figure 10.3 is solved by BUS with the classical rejection sampling approach. To obtain samples distributed as the posterior PDF, the samples outside the failure domain Ω are rejected. The accepted samples are used for estimating the posterior PDF. Totals of 100 and 1000 samples of variables x_1, x_2, and the additional variable p are generated, where the variable p follows uniform distribution U(0,1). Figures 10.4(a) and (b) show the accepted and rejected samples in x_1, x_2 space. The CDFs of x_1 with 100, 300, and 1000 samples are shown in Figure 10.4(c).

10.5 SURROGATE MODEL WITH ACTIVE LEARNING

An attractive approach is to use a surrogate model (metamodel, response surface) in order to reduce the number of function calls (full model evaluation, e.g., calculation by finite element method), which generally requires a large computational cost in real-world problems. Bichon et al. (2008) and Echard et al. (2011) proposed active learning with the surrogate model by GPR/Kriging with MCS (Monte Carlo Simulation) for reliability analysis. Since then, many modified methods have been proposed for reliability assessment, especially low failure probability. Teixeira et al. (2021) reviewed the implementation of adaptive metamodeling for reliability analysis, including the above papers. Moustapha et al. (2022) also conducted a survey of the recent literature, showing that most of the proposed methods are actually modifications of one or more aspects of those of Bichon et al. (2008) and Echard et al. (2011), and intensively compared the performances of methods proposed by various researchers, using many reliability benchmark problems.

The procedure of AK-MCS (Echard et al. 2011) is briefly summarized here. The terms "particle" and "sample" are used respectively for the vector generated for a probability calculation by MCS and the vector generated for a function call for building a surrogate model to avoid confusion.

Step 1: generate initial samples (model parameter vector) $\boldsymbol{x}_{fc}^{(1)}, ..., \boldsymbol{x}_{fc}^{(j)}$, and calculate the value of the limit state function (LSF). Use these samples to construct a surrogate model of the LSF by Kriging or GPR. Generate a large number of particles $\boldsymbol{x}_{MCS}^{(i)}$, $i = 1, ..., n_{MCS}$ (model parameter vector), for reliability analysis.

Step 2: select the next sample $\boldsymbol{x}_{fc}^{(j+1)}$ among $\boldsymbol{x}_{MCS}^{(i)}$ for the calculation of the LSF value as the one that minimizes the learning function $U(\boldsymbol{x})$ shown below.

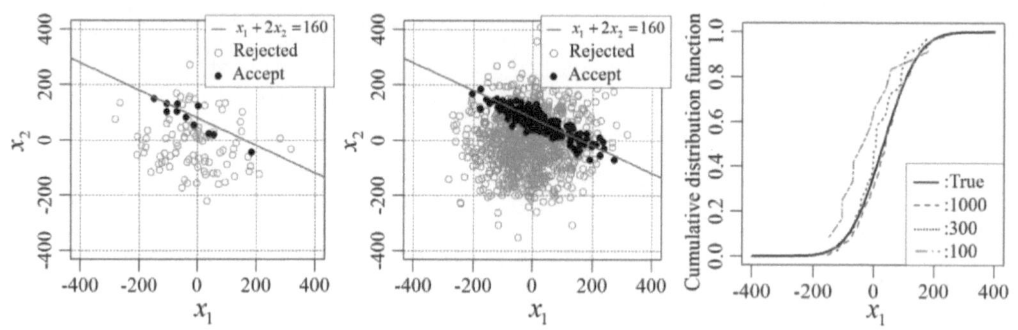

(a) 100 samples (b) 1000 samples (c) Estimated CDFs

FIGURE 10.4 Accepted and rejected samples and CDFs estimated by BUS with rejection sampling with 100, 300, and 1000 samples (a) 100 samples (b) 1000 samples (c) Estimated CDFs.

$$U(\boldsymbol{x}) = \frac{|\mu_g(\boldsymbol{x})|}{\sigma_g(\boldsymbol{x})} \tag{10.35}$$

where $\mu_g(\boldsymbol{x})$ and $\sigma_g(\boldsymbol{x})$ are the mean and SD of the LSF surrogate model.

Step 3: calculate the LSF value at $\boldsymbol{x}_{\text{fc}}^{(j+1)}$. Advance the sample size index j by 1.
Step 4: calculate the failure probability by a large number of particles $\boldsymbol{x}_{\text{MCS}}^{(i)}$ based on the surrogate model constructed by the limited number of samples $\boldsymbol{x}_{\text{fc}}^{(1)}, \ldots, \boldsymbol{x}_{\text{fc}}^{(j)}$.
Step 5: if the stopping criteria are satisfied, then stop. Otherwise, go to Step 2 to select a new sample

One of the most used examples for benchmarking is the one introduced by Echard et al. (2011). The LSF for the benchmarking consists of a series system with four branches.

$$g(x_1, x_2) = \min \left\{ \begin{array}{l} 3 + 0.1(x_1 - x_2)^2 - \dfrac{(x_1 + x_2)}{\sqrt{2}}; \\ 3 + 0.1(x_1 - x_2)^2 + \dfrac{(x_1 + x_2)}{\sqrt{2}}; \\ (x_1 - x_2) + \dfrac{7}{\sqrt{2}}; (x_2 - x_1) + \dfrac{7}{\sqrt{2}} \end{array} \right\} \tag{10.36}$$

The prior information of the random variables $\boldsymbol{x}$ are the standard normal distribution, i.e., $\boldsymbol{x} \sim \text{N}(\boldsymbol{0}, \boldsymbol{I})$. Figure 10.5 shows the process of the evaluation by AK-MCS. The number of initial function calls (initial sample) is 12 in Step 1. The surrogate model of LSF $g(\boldsymbol{x})$ is constructed based on these initial 12 samples, $\boldsymbol{x}_{\text{fc}}^{(1)}, \ldots, \boldsymbol{x}_{\text{fc}}^{(12)}$, as shown in Figure 10.5(a). The estimated LSF (surrogate model) does not agree well with the true one yet. An additional sample for the next function call is determined by the learning function in Step 2. The LSF value is calculated at the additional sample in Step 3. The failure probability is calculated based on the surrogate model updated by the additional sample in Step 4. Until the stopping f criteria are satisfied, these steps are repeated. The samples are increased one by one to enrich the important area. Finally, the accurate surrogate model is obtained

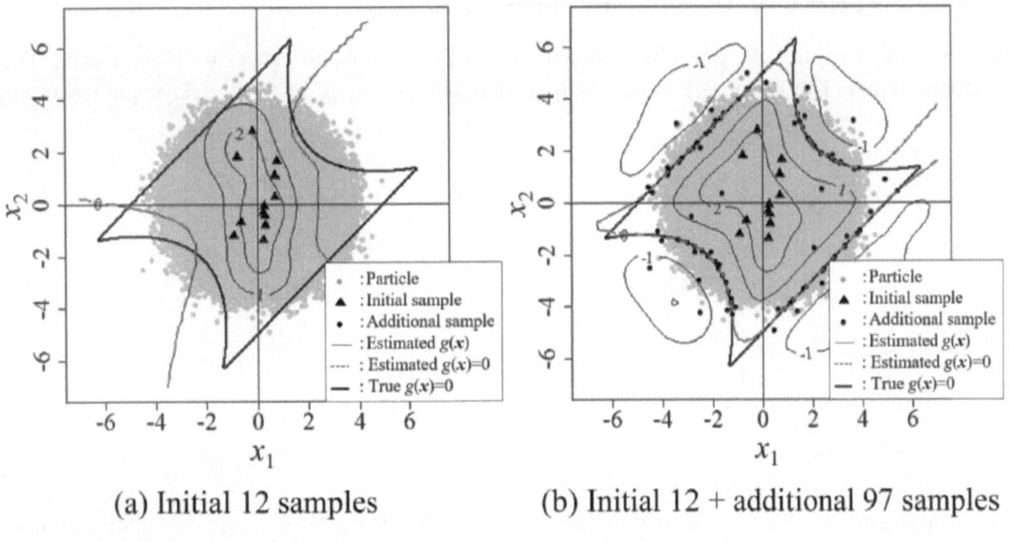

(a) Initial 12 samples (b) Initial 12 + additional 97 samples

FIGURE 10.5 Process of constructing surrogate model LSF by AK-MCS (a) Initial 12 samples (b) Initial 12 + additional 97 samples.

as shown in Figure 10.5(b). The limit state probability can be accurately estimated by the surrogate model constructed with the limited number of function calls.

The surrogate model approach can also be applied to Bayesian updating of model parameters, namely the inverse problem. For example, Wang and Shafieezadeh (2020), Kitahara et al. (2021), and Liu et al. (2022) incorporated adaptive Kriging into BUS with SuS as the surrogate model. Yoshida et al. (2023) directly constructed a surrogate model of the posterior PDF by adaptive GPR and applied the proposed method to a consolidation settlement problem.

Numerical Example 10.6

The two-variable example discussed above is solved by adaptive GPR (Yoshida et al. 2023). A new learning function $U(\boldsymbol{x})$ was developed to serve two basic objectives in active learning for constructing a surrogate model of the posterior PDF: a) finding the location of large values of the posterior PDF, and b) avoiding locations where the posterior PDF has already been calculated (existing function calls).

$$U(\boldsymbol{x}) = r(\boldsymbol{x})(\mu_n(\boldsymbol{x})\text{-min}(\boldsymbol{n}_{\text{cal}})) \tag{10.37}$$

$$r(\boldsymbol{x}) = \sigma_{\text{posterior}}(\boldsymbol{x})/\sigma_{\text{prior}}(\boldsymbol{x}) \tag{10.38}$$

where $\mu_n(\boldsymbol{x})$ is mean of the GPR surrogate of the log-posterior PDF at $\boldsymbol{x}$; $r(\boldsymbol{x})$ is a reduction factor to avoid existing function call locations; $\sigma_{\text{posterior}}(\boldsymbol{x})$ and $\sigma_{\text{prior}}(\boldsymbol{x})$ are the posterior and prior SD at $\boldsymbol{x}$ of the surrogate model by GPR.

Figure 10.6 shows the process of the evaluation of the surrogate model and posterior CDF. The initial surrogate model of the posterior PDF is constructed by GPR based on five initial samples $\boldsymbol{x}_{\text{fc}}$. A new sample $\boldsymbol{x}_{\text{fc}}$ for a function call is selected to maximize the learning function and update the surrogate model. This procedure is repeated until the number of function calls reaches 30. A total of 10,000 particle $\boldsymbol{x}_{\text{MCS}}$ are prepared and used to calculate the posterior CDF based on the surrogate model. Figure 10.6(a) and (b) indicate the distribution of samples for function calls and posterior CDFs estimated by 10, 20, and 30 function calls. Good agreement is obtained in the case of 30 function calls.

10.6 ESTIMATION OF HYPERPARAMETERS AND APPROXIMATION OF THEIR POSTERIOR PDF

Variables which control the probability distribution of the prior and observation error are called hyperparameters. The hyperparameter vector is denoted as $\boldsymbol{\theta}$ here. In a fully Bayesian treatment

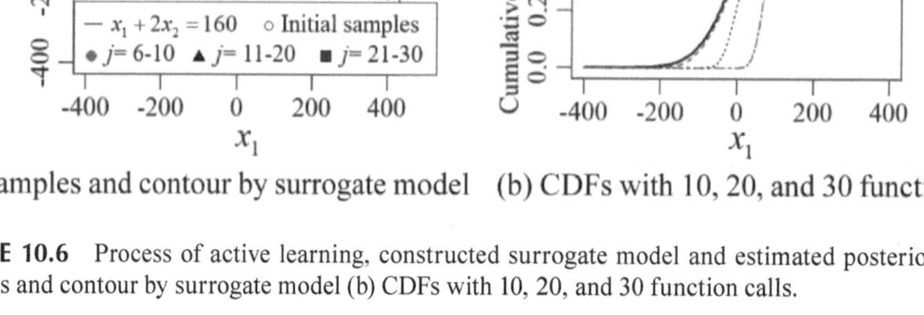

(a) Samples and contour by surrogate model (b) CDFs with 10, 20, and 30 function calls

FIGURE 10.6 Process of active learning, constructed surrogate model and estimated posterior CDFs (a) Samples and contour by surrogate model (b) CDFs with 10, 20, and 30 function calls.

of unknown vector $\boldsymbol{x}$ and $\boldsymbol{\theta}$, the posterior PDF $p(\boldsymbol{x},\boldsymbol{\theta}|z)$ can be formulated if the prior PDF $p(\boldsymbol{x}, \boldsymbol{\theta})$ is defined. However, even in the case of a liner inverse problem, the posterior PDF cannot be analytically obtained in general. As an approximation, instead of evaluating the hyperparameter posterior distribution, one may set the hyperparameters to specific values determined by maximizing the marginal likelihood function with respect to $\boldsymbol{x}$. This approximation is known in the statistics literature as empirical Bayes, type 2 maximum likelihood, or generalized maximum likelihood, and in the machine learning literature this is also called the evidence approximation (Bishop 2006).

The prior distribution reflects one's knowledge about the parameters in the absence of data. "Prior" usually means something before the information in z is used. In this sense, the hyperparameters related to the prior should not be adjusted by observation data z. However, as stated above, the prior information plays the role of a regularization term. The weight of the regularization term is often determined by cross-validation, which requires training and validation data sets. The empirical Bayes approach can determine the weight, i.e., prior information, in a simple way from observation data.

In a linear inverse problem, the marginal likelihood function with respect to $\boldsymbol{x}$ can be obtained analytically. It is assumed that the observation equation is $z=\boldsymbol{Hx}+\boldsymbol{v}$, and the prior information on $\boldsymbol{x}$ and observation noise $\boldsymbol{v}$ follow Gaussian distributions, namely $\boldsymbol{x}\sim N(\bar{\boldsymbol{x}}, \boldsymbol{M})$, $\boldsymbol{v}\sim N(\boldsymbol{0}, \boldsymbol{R})$. Then the marginal likelihood function is obtained by the integration,

$$p(z\,|\,\theta) = \int p(z\,|\,\theta,\boldsymbol{x})p(\boldsymbol{x})d\boldsymbol{x} \tag{10.39}$$

The logarithm of the marginal likelihood by the integration is

$$\ln p\left(z|\theta\right) = \frac{1}{2}\left(-\ln|\boldsymbol{M}| - \ln|\boldsymbol{R}| - (z-\boldsymbol{H}\hat{\boldsymbol{x}})^T\boldsymbol{R}^{-1}(z-\boldsymbol{H}\hat{\boldsymbol{x}}) - (\hat{\boldsymbol{x}}-\bar{\boldsymbol{x}})^T\boldsymbol{M}^{-1}(\hat{\boldsymbol{x}}-\bar{\boldsymbol{x}}) + \ln|\boldsymbol{P}| - m\ln\left(2\pi\right)\right) \tag{10.40}$$

where $\hat{\boldsymbol{x}} = \bar{\boldsymbol{x}} + \boldsymbol{PH}^T\boldsymbol{R}^{-1}(z-\boldsymbol{H}\bar{\boldsymbol{x}})$, $\boldsymbol{P} = (\boldsymbol{H}^T\boldsymbol{R}^{-1}\boldsymbol{H}+\boldsymbol{M}^{-1})^{-1}$ as shown in Section 10.3, $\boldsymbol{\theta}$ is the hyperparameter vector which controls $\boldsymbol{M}$ and $\boldsymbol{R}$ such as SD and scale of fluctuation (correlation length). MAP (or MPV) $\hat{\theta}$ is defined as

$$\hat{\theta} = \underset{\theta}{\arg\max}\left(p(z\,|\,\theta)p(\theta)\right) \tag{10.41}$$

One may use a uniform distribution for $p(\boldsymbol{\theta})$ when one does not want to use prior knowledge. In a globally identifiable case, i.e., when the posterior PDF has a unique maximum, it may be approximated by Gaussian approximation centered on the $\hat{\theta}$ of the posterior distribution of the hyperparameters. Consider $f(z,\boldsymbol{\theta})=\ln(p(z,\boldsymbol{\theta})p(\boldsymbol{\theta}))$, and approximate $f(z,\boldsymbol{\theta})$ by the second-order Taylor series about $\boldsymbol{\theta}$.

$$f(z,\theta) \approx f(z,\hat{\theta}) + \nabla_\theta f(z,\theta)\big|_{\theta=\hat{\theta}}(\theta-\hat{\theta}) + \frac{1}{2}(\theta-\hat{\theta})^T\nabla_\theta^2 f(z,\theta)\big|_{\theta=\hat{\theta}}(\theta-\hat{\theta}) \tag{10.42}$$

where ∇_θ is the gradient operator. The subscript $\boldsymbol{\theta}$ in ∇_θ indicates that the gradient is taken with respect to $\boldsymbol{\theta}$. The posterior PDF of $\boldsymbol{\theta}$ is obtained by $p(\boldsymbol{\theta}|z)=p(z|\boldsymbol{\theta})p(\boldsymbol{\theta})/p(z)$. The posterior PDF is expressed using $f(z,\boldsymbol{\theta})$.

$$p(\theta\,|\,z) \propto \exp\left(f(z,\theta)\right) \tag{10.43}$$

The approximate posterior PDF can be obtained by substituting Eq. (10.42) into Eq. (10.43). It should be noted that the second term of the right-hand side of Eq. (10.42) disappears because $\hat{\theta}$ is a mode of $f(z,\boldsymbol{\theta})$, and a Gaussian distribution has the property that its logarithm is a quadratic function of the variables.

$$p(\theta \mid z) \approx c \exp\left(-\frac{1}{2}(\theta - \hat{\theta})^T P_\theta^{-1}(\theta - \hat{\theta}) \right) \tag{10.44}$$

where c is a normalization coefficient, $P_\theta^{-1} = -\nabla_\theta^2 f(z,\theta)\big|_{\theta=\hat{\theta}}$, P_θ is an approximate covariance matrix of the posterior PDF, and $\nabla_\theta{}^2 f(z,\theta)$ is called the Hessian matrix. This approximation would be useful in situations where the number of data points is relatively large (Bishop 2006). An appropriate transformation of the parameter should be considered, e.g., if $\boldsymbol{\theta}_i$ is a non-negative variable and its SD is large, we can consider a Gaussian approximation with respect to $\ln(\boldsymbol{\theta}_i)$. This will be demonstrated in the numerical example below.

Numerical Example 10.7

For simplicity, we assume $\boldsymbol{M}=(1/\alpha)\boldsymbol{I}$, $\boldsymbol{R}=(1/\beta)\boldsymbol{I}$. A simple example to estimate hyperparameters α and β, i.e., $\boldsymbol{\theta}^{\mathrm{T}}=(\alpha, \beta)$, is shown with the linear regression problem in Figure 10.7. The observation equation is

$$z = \boldsymbol{H}x + v = \begin{bmatrix} 1 & s_1 \\ \vdots & \vdots \\ 1 & s_m \end{bmatrix} \begin{pmatrix} x_1 \\ x_2 \end{pmatrix} + \begin{pmatrix} v_1 \\ \vdots \\ v_m \end{pmatrix} \tag{10.45}$$

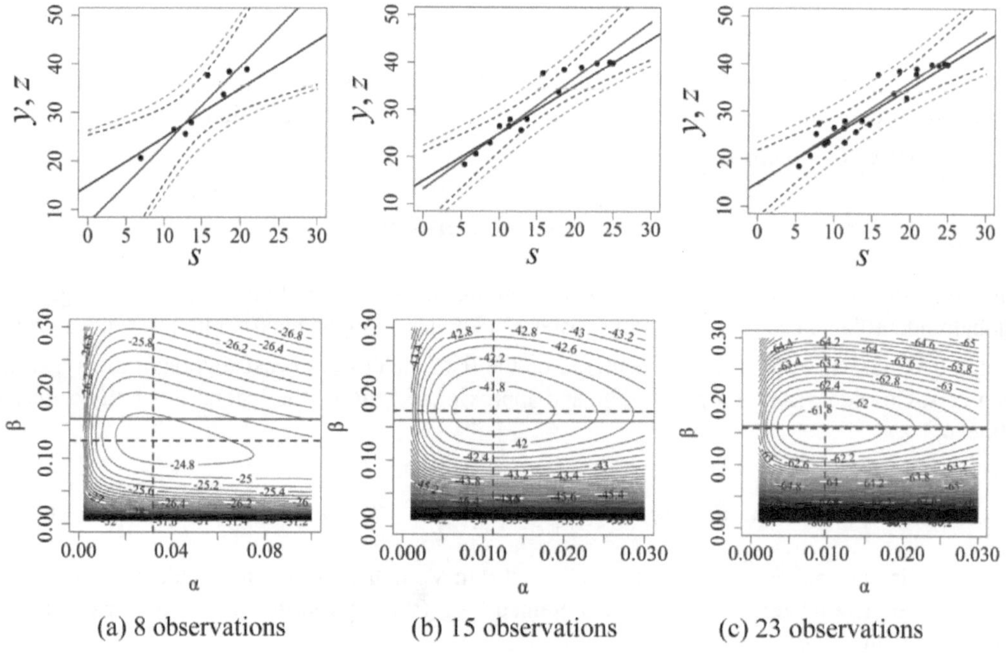

FIGURE 10.7 Estimation of hyperparameters, α, β, with an increase of observation data. Upper panels show observation data, regression lines (solid lines), and a 95% confidence interval; inner broken lines are based on $\boldsymbol{L}^{\mathrm{T}}\boldsymbol{P}\boldsymbol{L}$ and outer broken lines are based on $\boldsymbol{L}^{\mathrm{T}}\boldsymbol{P}\boldsymbol{L}+\boldsymbol{R}$. Lower panels show contour maps of marginal likelihood; solid lines indicate the true value of β (=0.16).

where x_1 and x_2 are the intercept and slope of the regression line, respectively, s_i is the location of the i-th observation data (horizontal axis in upper panels of Figure 10.7). The prediction $\boldsymbol{y}$ is obtained by $\boldsymbol{y}=\boldsymbol{L}\hat{\boldsymbol{x}}$. The matrix $\boldsymbol{L}$ allocates the locations for the output, whereas the matrix $\boldsymbol{H}$ allocates the observation locations. Observation data of z are generated using the Gaussian distribution with variance 6.25 ($=1/\beta$) around the true line of $x_1 = 15$ (intercept), $x_2 = 1$ (slope). β is the inverse of the variance of the observation error. The upper panels in Figure 10.7 show the observation data, regression line ($\boldsymbol{y}=\boldsymbol{L}\hat{\boldsymbol{x}}$), and its 95% confidential interval for numbers of observations m=8, 15, and 23. Two types of confidence interval, a) based on the uncertainty of estimated $\boldsymbol{x}$, i.e., $\boldsymbol{L}^T\boldsymbol{P}\boldsymbol{L}$, and b) observation error in addition to "a)", i.e., $\boldsymbol{L}^T\boldsymbol{P}\boldsymbol{L}+\boldsymbol{R}$, are indicated. The lower panels show the contour maps of the marginal likelihood. The regression lines in the upper panels are calculated using the hyperparameters α and β which maximize the marginal likelihood. As shown in the contour maps, α and β are globally identifiable and therefore gradient-based optimization methods such as BFGS can be used to obtain $\hat{\boldsymbol{\theta}}$. As the number of observations increased, β approaches the true value 0.16 (=1/6.25), while α approaches 0.0088, which is determined based on the difference between the true value and prior mean, $n/(\boldsymbol{x}_t-\bar{\boldsymbol{x}})^T(\boldsymbol{x}_t-\bar{\boldsymbol{x}})=2/(15^2+1^2)=1/113$, where $\boldsymbol{x}_t$ and n are the true value vector and the size of $\boldsymbol{x}$ (Bishop 2006).

The shapes of the contour in the lower panels of Figure 10.7 are distorted, not elliptical. This suggests that the transformation of variables is preferable to implementing Gaussian approximation. Figure 10.8 shows the contour maps with respect to $\ln(\alpha)$ and $\ln(\beta)$. The contour lines are more elliptical compared to Figure 10.7, indicating that Gaussian approximation is more effective. The SDs of the $\ln(\alpha)$ and $\ln(\beta)$, and their correlation can be estimated based on the covariance matrix P_θ in the Gaussian approximation; e.g., the SDs of $\ln(\alpha)$ obtained by P_θ are 1.49, 1.03, and 1.02 when the numbers of observations are 8, 15, and 23, respectively. The correlation coefficients are – 0.3, – 0.02, and – 0.02. The uncertainties in $\ln(\alpha)$ and $\ln(\beta)$ are not considered in the confidence intervals in Figure 10.7. Confidence intervals including the uncertainties in the hyperparameters cannot be obtained analytically. We need a numerical approach to estimate them.

10.7 LAPLACE APPROXIMATION OF EVIDENCE FOR MODEL SELECTION

Model selection is an important issue in inverse analysis because there are many candidates for forward models from simple to complex ones. The most suitable model should be selected based on the given observation data in terms of the trade-off between a data-fit measure and a complexity measure for each model. The evidence (marginal likelihood) $p(z)$ is a useful indicator for the model selection, but the integral to evaluate the evidence is intractable.

$$p(z) = \int p(z \mid \theta) p(\theta) d\theta \tag{10.46}$$

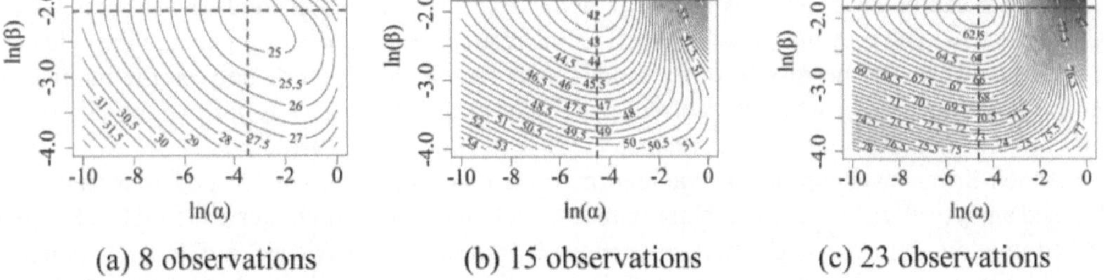

FIGURE 10.8 Contour map of marginal likelihood with respect to $\ln(\alpha)$ and $\ln(\beta)$. Solid lines indicate the true value of $\ln(\beta)$ (=-1.83). (a) 8 observations (b) 15 observations (c) 23 observations.

A method based on MCS is a practical tool for the evaluation of the evidence. Samples of the intermediate distributions by TMCMC (Ching and Chen 2007; Betz et al. 2016) can be used to obtain an estimate of the evidence. Laplace approximation is another practical option for obtaining evidence in approximation. A Laplace approximation is based on a Gaussian approximation centered on the MPV $\hat{\theta}$. Again consider $f(z,\boldsymbol{\theta})=\ln(p(z|\boldsymbol{\theta})p(\boldsymbol{\theta}))$ and approximation of $f(z,\boldsymbol{\theta})$ by the second-order Taylor series about $\boldsymbol{\theta}$ in Eq. (10.42). Substituting the approximation of $f(z,\boldsymbol{\theta})$ into Eq. (10.46) and taking the logarithm, we have

$$\begin{aligned}\ln p(z) &\approx \ln \int \exp\left(f(z,\hat{\theta}) - \frac{1}{2}(\theta-\hat{\theta})^T \boldsymbol{P}_\theta^{-1}(\theta-\hat{\theta}) \right) d\theta \\ &= f(z,\hat{\theta}) + \ln \int \exp\left(-\frac{1}{2}(\theta-\hat{\theta})^T \boldsymbol{P}_\theta^{-1}(\theta-\hat{\theta}) \right) d\theta \\ &= \ln p(z \mid \hat{\theta}) + \ln p(\hat{\theta}) + (n_h/2)\ln(2\pi) - (1/2)\ln\left|\boldsymbol{P}_\theta^{-1}\right| \end{aligned} \tag{10.47}$$

where $\boldsymbol{P}_\theta^{-1} = -\nabla_\theta^2 \ln\left(p(z \mid \theta)p(\theta)\right)\big|_{\theta=\hat{\theta}}$, and n_h is the number of hyperparameters. The matrix $\boldsymbol{P}_\theta$ is the covariance matrix of the MPV $\hat{\theta}$. The first term $\ln p(z \mid \hat{\theta})$ is larger for the model that better fits the data, so this term tends to favor models with more parameters. The 2nd through 4th terms of Eq. (10.47) are called the "Ockham (Occam) factor" (Bishop 2006, Au 2017):

$$\text{Ockham factor} = \ln p(\hat{\theta}) + (n_h/2)\ln(2\pi) + (1/2)\ln\left|\boldsymbol{P}_\theta\right| \tag{10.48}$$

The Ockham factor tends to decrease the number of hyperparameters n_h, namely to penalize model complexity.

Various information criteria such as AIC (Akaike Information Criterion) and BIC (Bayesian Information Criterion) have been proposed that attempt to correct for the bias of maximum likelihood by the addition of a penalty term to compensate for the over-fitting of more complex models (Bishop 2006). AIC and BIC (Akaike 1974; Schwarz 1978; Beck 2010) are defined as follows:

$$\text{AIC} = -2\ln p(z \mid \hat{\theta}) + 2n_h \tag{10.49}$$

$$\text{BIC} = -2\ln p(z \mid \hat{\theta}) + n_h \ln(m) \tag{10.50}$$

In model selection, a smaller AIC or BIC is better, while larger evidence is better. BIC penalizes model complexity more heavily than AIC. Intuitively, the higher n_h (the more complex the model) is, the higher the first term will be; however, the second term will penalize that complexity. The second term can be interpreted as the simplified "Ockham factor". AIC and BIC are simple and easy to calculate, but they may also give misleading results because they approximate the evidence in such a way that the penalty term for model class complexity depends only on the number of uncertain parameters n_h, while the correct penalty term can differ greatly for two model classes with the same number of uncertain parameters (Bishop 2006, Beck 2010).

In a non-linear inverse problem, the marginal likelihood function with respect to $\boldsymbol{x}$ cannot be obtained analytically. An Ockham factor with respect to unknown parameter $\boldsymbol{x}$ should be also considered in addition to one about hyperparameter $\boldsymbol{\theta}$. In information criteria, a likelihood function with respect to $\boldsymbol{x}$ and $\boldsymbol{\theta}$, instead of the marginal likelihood function, is used. Also the number of unknown parameters n in addition to n_h should be considered.

$$\text{AIC} = -2\ln p(z \mid \hat{x},\hat{\theta}) + 2(n+n_h) \tag{10.51}$$

$$\text{BIC} = -2\ln p(z \mid \hat{x},\hat{\theta}) + (n+n_h)\ln(m) \tag{10.52}$$

Numerical Example 10.8

In Figure 10.9, we show the results of fitting polynomials having orders 0, 2, and 3 to the same data set shown in Figure 10.7 (1st-order, 23 observation data). The prior PDF for hyperparameter $p(\boldsymbol{\theta})$ is assumed to follow a uniform distribution. Evidence by numerical integration, Laplace approximation, AIC, and BIC are calculated for the four cases of polynomial order: 0, 1, 2, and 3. Their comparison is shown in Figure 10.10, where AIC and BIC are re-scaled for the comparison (negative half of AIC and BIC, i.e., -AIC/2 = $\ln p(z \mid \hat{\theta}) - n_h$, -BIC/2 = $\ln p(z \mid \hat{\theta}) - n_h \ln(m)/2$). All indices take similar values and maximum values for the 1st-order polynomial which is the true model. This example is easy as a model selection; therefore, all of these four indices select the true model correctly. Only when synthesized data are used, the true model exists. In real data, there is no true model. Evidence and information criteria tell not if the model is true, but if it is suitable for the given data relatively. When the data are limited or the observation error is large, AIC or BIC tends to select a simple model.

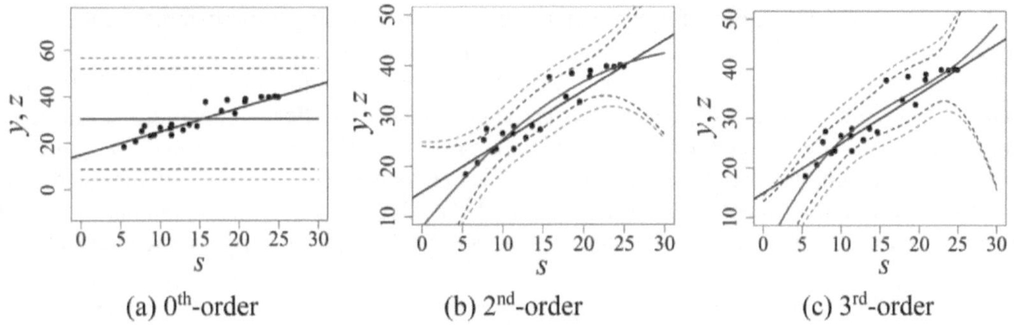

FIGURE 10.9 Polynomial regression with 0th-, 2nd-, and 3rd-order polynomials for 23 observation data. First-order regressions for the same problem are shown in Figure 10.7. (a) 0th-order (b) 2nd-order (c) 3rd-order.

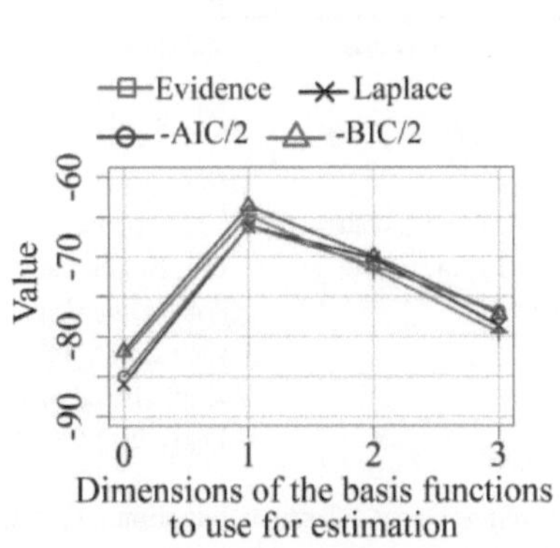

FIGURE 10.10 Model selection for the order of linear regression.

10.8 OPTIMAL OBSERVATION PLANNING

Optimal observation planning including sensor or sampling placement problems is also an important topic in inverse problems. Samples are usually taken at a limited number of locations in geotechnical engineering. This leads to the question of how best to select efficient locations (or timing) for sampling. A probabilistic approach provides useful and valuable information to this question, i.e., the uncertainty of the estimated parameters. The placement can be determined such that the uncertainty is minimized. Several metrics of uncertainty have been used in optimal observation planning. The determinant of the posterior covariance matrix $\boldsymbol{P}$, which is proportional to information entropy in the case of a Gaussian distribution, is used for what is called a D-optimal placement. The metric based on the trace of $\boldsymbol{P}$ is called the A-optimality (Sun 1994).

Metrics of uncertainty, such as D-optimality (=det($\boldsymbol{P}$)) and A-optimality (=trace($\boldsymbol{P}$)), however, do not reflect the significance of uncertainty if the consequence of the uncertainty is not considered. Raiffa and Schlaifer (1961) introduced the theory of Value of Information (VoI), which explicitly considers consequences in decision-making under uncertainty. Ang and Tang (1984) introduced an application of VoI to project decision-making using a decision tree in their textbook. VoI has been used for the optimal placement of soil investigation for slope stability assessment (Jiang et al. 2020; Hu et al. 2021), additional sampling for soil contamination investigation (Yoshida et al. 2022), and so on.

10.9 CONCLUSION

In this chapter, recent methods for parameter identification including hyperparameters are introduced, especially probabilistic sample-based approaches. Table 10.1 summarizes inverse problems from the viewpoint of probability theory. When the observation equation is linear and its noise and the prior information of the parameters are Gaussian, an analytical solution of the unknown parameter exists. We do not need any numerical approach to solve such problems. The posterior mean and covariance matrix of the parameters can be obtained exactly. The Kalman filter and GPR are classified into this category, i.e., the first line of the table. However, hyperparameters cannot be obtained analytically even in this category. The second line indicates the inverse analysis with

TABLE 10.1
Classification of Inverse Problem

Observation Equation	Observation Noise	Regularization Prior Information	Solution	Related Methods
Linear	l_2 norm Gauss	l_2 norm, Gauss	Analytical	Regularized LS Kalman filter, GPR
Linear	l_2 norm Gauss	l_1 norm Laplace	Numerical ADMM, etc.	LASSO (sparse modeling),
Non-linear	l_2 norm Gauss (any)	l_1, l_2 norm Gauss, Laplace (any)	Gradient-based method such as GN, BFGS methods Global Optimization method such as GA, PSO Sample-based method such as PF, BUS, TMCMC	Regularized non-linear LS, surrogate model

ADMM: alternating direction method of multipliers, GA: genetic algorithm, GN: Gauss-Newton, GPR: Gaussian process regression, LS: least-squares, LASSO: least absolute shrinkage and selection operator, PF: particle filter, BUS: Bayesian updating with structural reliability methods, PSO: particle swarm optimization, and TMCMC: transitional Markov chain Monte Carlo

l_1 norm regularization/Laplace prior, which is known as LASSO. There is no analytical solution, but efficient iterative methods have been proposed. The third line in the table indicates the inverse problem with a non-linear observation equation, and/or non-Gaussian noise and/or prior. Most of the real-world inverse problems fall into this category which requires a numerical method to obtain the solutions. Recent methods, especially sample-based methods with/without an adaptive surrogate model, are introduced. PF, BUS, and PF with an adaptive surrogate model are illustrated using a simple two-variable example. The sample-based methods with an adaptive surrogate model seem promising, although they need further investigation to clarify their applicability and limitations. The hyperparameters are not considered in this table. To estimate the hyperparameters, basically, we need a numerical approach even for a linear inverse problem, as shown in numerical example 10.7.

Probability theory provides a comprehensive and unified interpretation of the inverse problem as discussed above. The methods introduced in this chapter can be expected to contribute to the decision-making of problems in geotechnical engineering because they can evaluate the uncertainties as well as representative values, i.e., posterior distributions, of model parameters and their prediction.

REFERENCES

Akaike, H., 1974. A new look at the statistical model identification. *IEEE Trans Autom Control*, 19(6):716–723. https://doi.org/10.1109/TAC.1974.1100705.

Ang, A.H.-S., Tang, W.H., 1984. *Probability Concepts in Engineering Planning and Design, Volume II - Decision, Risk and Reliability*. John Wiley & Sons, New York.

Au, S.K., 2017. *Operational Modal Analysis: Modeling, Bayesian Inference, Uncertainty Laws*. Springer, Singapore. https://doi.org/10.1007/978-981-10-4118-1.

Au, S.K., Beck, J.L., 2003. Subset simulation and its application to seismic risk based on dynamic analysis. *J Eng Mech*, 129(8):901–917. https://doi.org/10.1061/(ASCE)0733-9399(2003)129:8(901).

Au, S.K., Wang, Y., 2014. *Engineering Risk Assessment with Subset Simulation*. John Wiley & Sons, Singapore.

Beck, J.L., 2010. Bayesian system identification based on probability logic. *Struct Control Health Monit*, 17(7):825–847. https://doi.org/10.1002/stc.424.

Beck, J.L., Katafygiotis, L.S., 1998. Updating models and their uncertainties. I: Bayesian statistical framework. *J Eng Mech*, 124(4):455–461.

Betz, W., Papaioannou, I., Beck, J.L., Straub, D., 2018. Bayesian inference with subset simulation: Strategies and improvements. *Comput Methods Appl Mech Eng*, 331:72–93. https://doi.org/10.1016/j.cma.2017.11.021.

Betz, W., Papaioannou, I., Straub, D., 2016. Transitional Markov chain Monte Carlo: Observations and improvements. *J Eng Mech*, 142(5). https://doi.org/10.1061/(ASCE)EM.1943-7889.0001066.

Bichon, B.J., Eldred, M.S., Swiler, L., Mahadevan, S., McFarland, J., 2008. Efficient global reliability analysis for nonlinear implicit performance functions. *AIAA J*, 46(10):2459–2468.

Bishop, C.M., 2006. *Pattern Recognition and Machine Learning*. Springer, New York.

Boumezerane, D., 2022. Recent tendencies in the use of optimization techniques in geotechnics: A review. *Geotechnics*, 2:114–132.

Boyd, S., Parikh, N., Chu, E., Peleato, B., Eckstein, J., 2010. Distributed optimization and statistical learning via alternating direction method of multipliers. *Found Trends Mach Learn*, 3(1):1–122. https://doi.org/10.1561/2200000016.

Ching, J., Chen, Y.C., 2007. Transitional Markov chain Monte Carlo Method for Bayesian model updating, model class selection, and model averaging. *J Eng Mech*, 133(7):816–832. https://doi.org/10.1061/(ASCE)0733-9399(2007)133:7(816).

Christakos, G., 1992. *Random Field Models in Earth Sciences*. Academic Press, New York.

Cressie, N., 1991. *Statistics for Spatial Data*. John Wiley & Sons, New York.

DiazDelaO, F.A., Garbuno-Inigo, A., Au, S.K., Yoshida, I., 2017. Bayesian updating and model class selection with subset simulation. *Comput Methods Appl Mech Eng*, 317:1102–1121. https://doi.org/10.1016/j.cma.2017.01.006.

Echard, B., Gayton, N., Lemaire, M., 2011. AK–MCS: An active learning reliability method combining Kriging and Monte Carlo simulation. *Struct Saf*, 33(2):145–154. https://doi.org/10.1016/j.strusafe.2011.01.002.

Fan, Z., Ji, H., Zhan, Y., 2015. Iterative particle filter for visual tracking. *Signal Process Image Commun*, 36:140–153.

Gordon, N., Salmond, D., Smith, A., 1993. A novel approach to nonlinear/non-Gaussian Bayesian state estimation. *Proceedings of IEEE on Radar and Signal Processing*, 140(2):107–113.

Hastie, T., Tibshirani, R., Wainwright, M., 2015. *Statistical Learning with Sparsity: The Lasso and Generalizations*. Chapman& Hall/CRC, New York. https://doi.org/10.1201/b18401.

Hu, J.Z., Zhang, J., Huang, H.W., Zheng, J.G., 2021. Value of information analysis of site investigation program for slope design. *Comput Geotech*, 131:103938. https://doi.org/10.1016/j.compgeo.2020.103938.

Jiang, S.-H., Papaioannou, I., Straub, D., 2020. Optimization of site-exploration programs for slope-reliability assessment. *ASCE ASME J Risk Uncertainty Eng Syst A*, 6(1):04020004.

Kitagawa, G., 1996. Monte Carlo Filter and smoother for non-Gaussian nonlinear state space models. *J Comp Graph Stat*, 5(1):1–25.

Kitahara, M., Bi, S., Broggi, M., Beer, M., 2021. Bayesian model updating in time domain with metamodel-based reliability method. *ASCE ASME J Risk Uncertainty Eng Syst A*, 7(3):04021030. https://doi.org/10.1061/AJRUA6.0001149.

Liu, L., Li, L., Zhao, S., 2022. Efficient Bayesian updating with two-step adaptive Kriging. *Struct Saf*, 95:102172. https://doi.org/10.1016/j.strusafe.2021.102172.

Manoli, G., Rossi, M., Pasetto, D., Deiana, R., Ferraris, S., Cassiani, G., Putti, M., 2015. An iterative particle filter approach for coupled hydro-geophysical inversion of a controlled infiltration experiment. *J Comp Phys*, 283:35–51.

Moustapha, M., Marelli, S., Sudret, B., 2022. Active learning for structural reliability: Survey, general framework and benchmark. *Struct Saf*, 96:102174.

Nakano, S., Ueno, G., Higuchi, T., 2007. Merging particle filter for sequential data assimilation. *Nonlin Processes Geophys*, 14(4):395–408.

Rasmussen, C.E., Williams, C.K.I., 2006. *Gaussian Processes for Machine Learning*. MIT Press, London.

Raiffa, H., Schlaifer, R., 1961. *Applied Statistical Decision Theory*. MIT Press, Cambridge, MA.

Ristic, B., Arulampalam, S., Gordon, N., 2004. *Beyond the Kalman Filter: Particle Filters for Tracking Applications*. Artech House, Boston.

Schwarz, G., 1978. Estimating the dimension of a model. *Ann Statist*, 6(2):461–464. https://doi.org/10.1214/aos/1176344136.

Straub, D., Papaioannou, I., 2015. Bayesian updating with structural reliability methods. *J Eng Mech*, 141(3):040141134_1–040141134_13.

Straub, D., Papaioannou, I., Betz, W., 2016. Bayesian analysis of rare events. *J Comput Phys J Comp Phys*, 314:538–556. https://doi.org/10.1016/j.jcp.2016.03.018.

Sun, N.-Z., 1994. *Inverse Problems in Groundwater Modelling*. Kluwer Academic Publishers, Boston.

Teixeira, R., Nogal, M., O'connor, A., 2021. Adaptive approaches in metamodel-based reliability analysis: A review. *Struct Saf*, 89:102019.

Walton, G., Sinha, S., 2022. Challenges associated with numerical back analysis in rock mechanics. *J Rock Mech Geotech Eng*, 14:6, 2058-2071.

Wang, Z., Shafieezadeh, A., 2020. Highly efficient Bayesian updating using metamodels: An adaptive Kriging based approach. *Struct Saf*, 84:101915. https://doi.org/10.1016/j.strusafe.2019.101915.

Yoshida, I., Tomizawa, Y., Otake, Y., 2021a. Estimation of trend and random components of conditional random field using Gaussian process regression. *Comput Geotech*, 136:104179. https://doi.org/10.1016/j.compgeo.2021.104179.

Yoshida, I., Tasaki, Y., Tomizawa, Y., 2022. Optimal placement of sampling locations for identification of a two-dimensional space. *Georisk Assess Manag Risk Engineered Syst Geohazards*, 16(1):98–113. https://doi.org/10.1080/17499518.2021.1971255.

Yoshida, I., Nakamura, T., Au, S.-K., 2023. Bayesian updating of model parameters using adaptive Gaussian process regression and particle filter. *Struct Saf*, 102:102328. https://doi.org/10.1016/j.strusafe.2023.102328.

Part 3

Probabilistic software

11 Use of Geotechnical Software for Probabilistic Analysis and Design

Sina Javankhoshdel, Brigid Cami, Terence Ma, Angela Li, Zhanyu Huang, Ellen Yeh, Seok Hyeon Chai, Thamer Yacoub

ABSTRACT

Probabilistic analysis and design in geotechnical problems can easily be accomplished using software packages. This chapter describes the advantages of probabilistic analysis and the convenience of employing commercially available software for such problems. The chapter begins by describing the stochastic analysis for slope stability problems, followed by some discussion on how a similar process can be applied to other geotechnical problems. A few practical case studies are provided to demonstrate its use in slope stability problems as well as other geotechnical applications, in which stochastic analysis procedures are popularly employed.

11.1 WHY USE PROBABILISTIC ANALYSIS?

A deterministic analysis model assumes that the input parameters in an analysis are fixed values. As an example, an engineer may have ten measurements of cohesion for a given slope stability problem, but only one value can be used in the analysis. The cohesion that is selected is often the average value, the worst-case value, or some characteristic value stipulated by regulatory standards. The result of such an analysis is a single result. In the example of slope stability, this is the factor of safety (FOS) value. The engineer must then use this single result to determine the next steps of the problem. However, if the average value was used, the FOS might be unconservative. If the worst-case value was used, the FOS might be too conservative (expensive). An example of a value stipulated by regulatory standards is the Eurocode 7 characteristic value, described as a "cautious estimate of the value affecting the occurrence of the limit state" (BSI, 1995). It is based on judgment or a 5% fractile if statistical methods are used (BSI, 1995).

Rather than forcing multiple values into one representative value, why not consider them all? A probabilistic analysis is a tool that allows the engineer to consider all the available data. Rather than the single input value for cohesion, the engineer can now input a distribution for cohesion. In the case of plentiful data, the distribution can be selected to fit the histogram of the input data. In the case of meager data, the distribution can be defined as a range, or a uniform distribution.

Once the input distribution is defined, it is sampled N times, and the analysis result (e.g., FOS) is computed with each of those samples, N times. In short, all the input data is considered. The result of a probabilistic analysis is not just the deterministic result, but also a summative metric to account for all simulations such as the probability of failure (PF). PF is defined as the percentage of samples that failed, N_f (e.g., samples for which FOS<1), relative to the total number of samples, N, as shown in Eq. 11.1.

$$PF = N_f / N \times 100\% \tag{11.1}$$

DOI: 10.1201/9781003333586-14

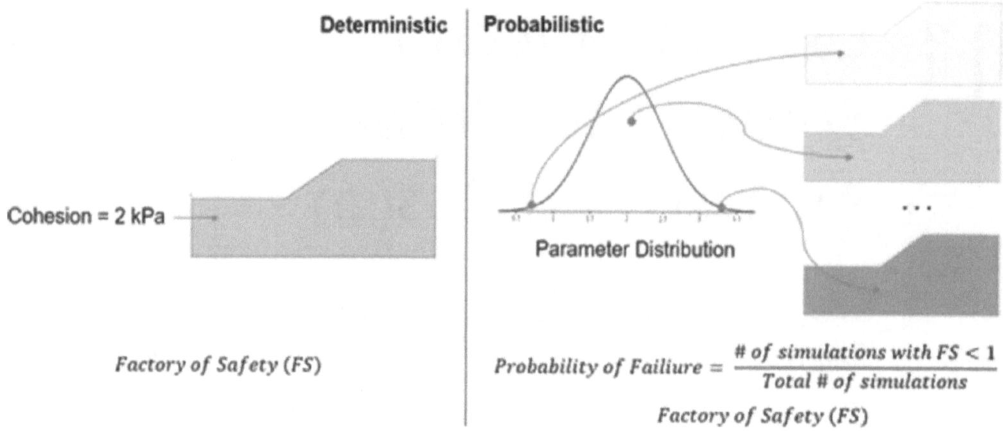

FIGURE 11.1 Deterministic analysis vs probabilistic analysis.

The difference between the deterministic and probabilistic analysis is described pictorially in Figure 11.1.

In summary, the key reasons to use a probabilistic analysis are: 1) probabilistic analysis removes the blind confidence in single parameter values and takes into account the inherent uncertainty in these measurements; 2) probabilistic analysis allows the engineer to consider all the possible cases based on the data, rather than just one case; 3) the PF result that is the output of the analysis is in addition to the deterministic result – the engineer is only gaining knowledge. In addition, it is not well appreciated that an engineer cannot handle multivariate data by judgment alone. For example, the cohesion and friction angles are negatively correlated. It may not be appropriate to reduce both cohesion and friction angles to their worst-case values at the same time.

With the improvements in geotechnical software over recent decades, a probabilistic analysis is also very easy to run.

11.2 ADVANTAGES OF SOFTWARE

Computing technology has made it possible to perform millions of operations per second. Algorithms which would otherwise span years on end for a single person to complete by hand can now be completed in almost no time. Consequently, performing advanced numerical iteration techniques with geotechnical software has become standard practice for many types of geomechanics problems.

A new advantage of geotechnical software is found in the context of stochastic design and analysis, where a user can automatically randomly generate thousands of random simulations without having to reproduce the models each time. It is also practically impossible to select consistent combinations of input parameters for the multivariate case (all realistic problems are multivariate). Simulations involving different spatial distributions of input parameters consistent with measurements are beyond the reach of judgment. Slope stability problems are arguably one of the most notable applications of stochastic analysis in geotechnical engineering practice. As such, the authors describe in detail, and with examples, the process of using a software package primarily for stochastic slope stability problems. However, the process can very similarly be applied to a wide range of other geotechnical problems which require analysis using the same constitutive soil and rock models.

11.3 GENERAL METHODOLOGY

Geotechnical analyses can take many forms. Just to name a few, they can include computations of FOS, settlement analysis, computing the forces and reactions developed in reinforcement, seepage analysis, and often combinations of these. Uncertainty in the strengths and properties of the components in these analyses is very common due to the limited availability and expensive nature of geotechnical information. As such, assuming the mean value of the strength obtained from multiple laboratory samples may not always result in a conservative evaluation of safety. The properties can also vary spatially. For instance, sliding in slopes occurs along the path of least resistance, and in the case where a mean value of the sample strength is used, the strength along the sliding path may be overestimated. As well, there may be uncertainty about the true location of a water table or a fault. To address the uncertainties in the properties of a model, the properties can be represented as random variables in the analysis.

The process of design often entails iterating on its geometrical and material aspects while evaluating the safety and cost-effectiveness of alternative solutions. Where a stochastic analysis is required, a design may be expected to succeed in sustaining the design conditions to an extent where the probability of failure is maintained below a stipulated threshold. For example, a designer may wish to vary the slope angle of an embankment and the amount of reinforcement needed to ensure that a certain probability of failure with respect to slope stability stipulated by design codes is not exceeded. In another instance, one might expect a pile to undergo variable loading conditions and maintain its strength within some margin of the PF.

In each simulation of a probabilistic analysis, some primary analysis procedure is repeated. The primary analysis may already be computationally expensive and require the use of software, such as calculating the FOS or strength reduction factor in an entire slope or determining if an installed barrier is adequate during a rock fall event.

The proceeding sections will study all the aspects of probabilistic analysis and the different options available therein.

11.4 SAMPLING METHOD

Perhaps the most straightforward method for determining the PF would be to randomly sample the variables and input them into the primary analysis procedure and repeat this for N simulations. This approach is commonly referred to as the Monte Carlo method. However, there are more efficient methods which strategically select the samples so that the PF can be estimated faster, such as the Latin Hypercube (LH) and Stochastic Response Surface (SRS) methods.

The Monte Carlo and LH methods are commonplace in the realm of stochastic simulation. However, the most recent advancements in geotechnical software have popularized the SRS method, which is a relatively efficient way to estimate the PF in a stochastic problem compared to the previous methods. It is computationally complex, as its procedure requires the aid of advanced computational processing. First, a subset of initial simulations is generated using a simpler, selective method such as the LH method. This subset is used to fully compute the failure function (e.g., FOS, or some other ratio indicating the strength utilization of the model) for these simulations. The failure function is expressed as a polynomial function of the K input random variables in chaos expansion form. The SRS method fits the results of the initial simulations into the failure function to calibrate its polynomial coefficients. This K^{th}-dimension "response" surface function is then used to estimate the value of the failure function for the desired number of simulations. In this manner, the simulations are still randomly generated but the response surface is used to estimate the PF without having to rerun the computationally expensive primary analysis procedure for each simulation.

In many commercially available software packages for geotechnical problems, the traditional Monte Carlo and LH methods are offered. For example, in slope stability analysis software such as Slide2 RS2 and Slide3 (Rocscience 2023a,b,c), the user can assign random distributions to the material properties and strengths of the reinforcing anchors in the model and select from the Monte Carlo, LH, or SRS methods to compute PF. Similarly, Slope/W offers the Monte Carlo method (GeoStudio, 2022), whereas PLAXIS (Plaxis, 2022) and RS2 offer the Point Estimation Method (PEM) which is an approximate method. The authors of this chapter have published several studies using the commercially available software Slide2 and Slide3 to carry out stochastic slope stability analysis. They have introduced a method called the Random Limit Equilibrium Method (RLEM) which is a combination of the Limit Equilibrium Method (LEM) for slope stability analysis and Global and Local optimization techniques (Mafi et al., 2019). Among these studies are Cami et al. (2018, 2021a, 20221b), Javankhoshdel et al. (2017, 2020, 2021, 2022a, 2022b), and Cylwik et al. (2021, 2023).

Below are a few examples of case studies reported in these publications.

11.5 EXAMPLE ONE: VARIATION OF THE PRIMARY ANALYSIS

As a first example, the authors have chosen to highlight the simple application of stochastic slope stability analysis in the 2D limit equilibrium software, Slide2, and the various primary analysis methods that can be selected. A section of a slope located in the James Bay region in Ontario, Canada is shown in Figure 11.2, adapted from Javankhoshdel et al. (2021) and originally studied by El-Ramly (2001). The dimensions shown in the figure are in meters.

Material Properties

Owing to uncertainties associated with the embankment material and strength properties in the two deepest clay layers, random variables have been assigned to each of these materials accordingly. Specifically, the unit weight and friction angle of the embankment, and the cohesion of the deep clay layers are randomly varied according to the distributions listed in Table 11.1. The random variables in this example are assumed to vary based on normal distributions truncated on both sides at a width equal to three times the standard deviation. A screenshot of the dialog used to specify the random variables for the materials is shown in Figure 11.3.

Deterministic Analysis Results

The deterministic FOS of the slope is computed using a standard non-circular slip surface search in Slide2 via the Particle Swarm search (Kennedy and Eberhart 1995) option. A surface altering (SA)

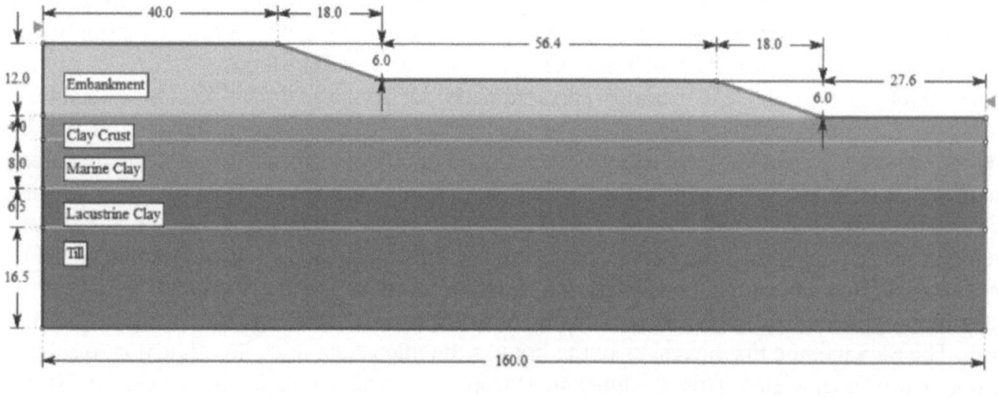

FIGURE 11.2 Example slope for applications in slope stability (adapted from Javankhoshdel et al., 2021).

TABLE 11.1
Material Properties for the James Bay Slope Stability Problem

	Unit Wt. (kN/m³)		c (kPa)		φ (°)	
Layer	**Mean**	**Std Dev.**	**Mean**	**Std Dev.**	**Mean**	**Std Dev.**
Embankment	20.0	1.41	0.0	–	30	2
Clay Crust	18.8	–	43.0	–	0	–
Marine Clay	18.8	–	34.5	8.14	0	–
Lacustrine Clay	20.3	–	31.2	8.65	0	–
Till	20.0	–	∞	–	∞	–

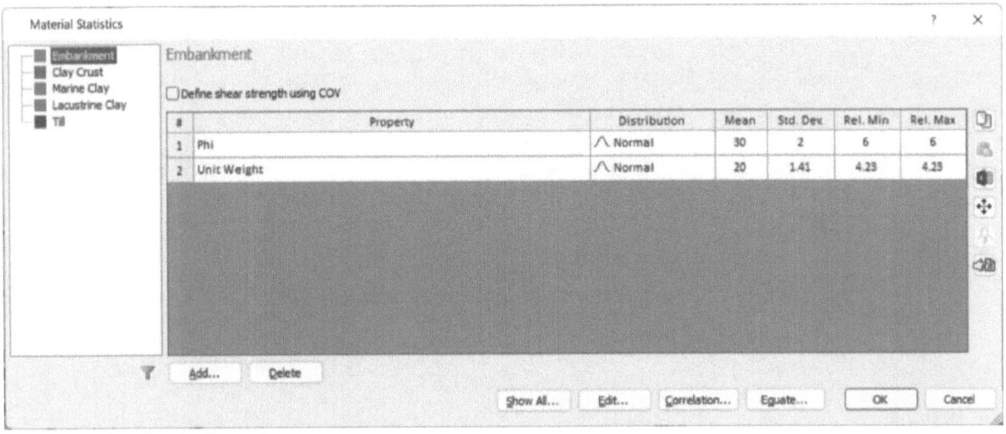

FIGURE 11.3 Example dialog showing the input of random variables for each material in Slide2.

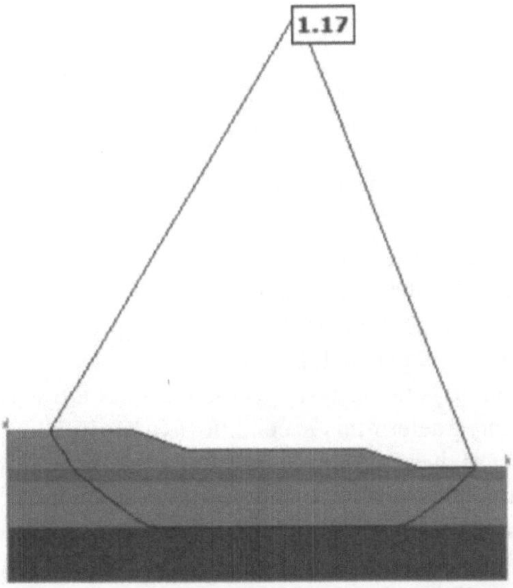

FIGURE 11.4 Deterministic analysis result (Spencer FOS = 1.17) of the James Bay slope stability problem.

scheme is enabled, which further optimizes the slip surface to minimize its FOS (Mafi et al. 2020). The result for the deterministic analysis is a Spencer FOS of 1.17, shown in Figure 11.4.

Primary Analysis Selection

The choice of primary analysis can have a large impact on the time required to complete the probabilistic assessment of the safety of a design. In slope stability analysis, the primary analysis can take several forms. For Slide2, the following options are offered at the time of writing:

- Global Minimum Method.
- Overall Slope Method.

A screenshot of the dialog used to select the sampling method is shown in Figure 11.5. In each case, a representative FOS for the slope is computed as the failure function, and the slope is considered

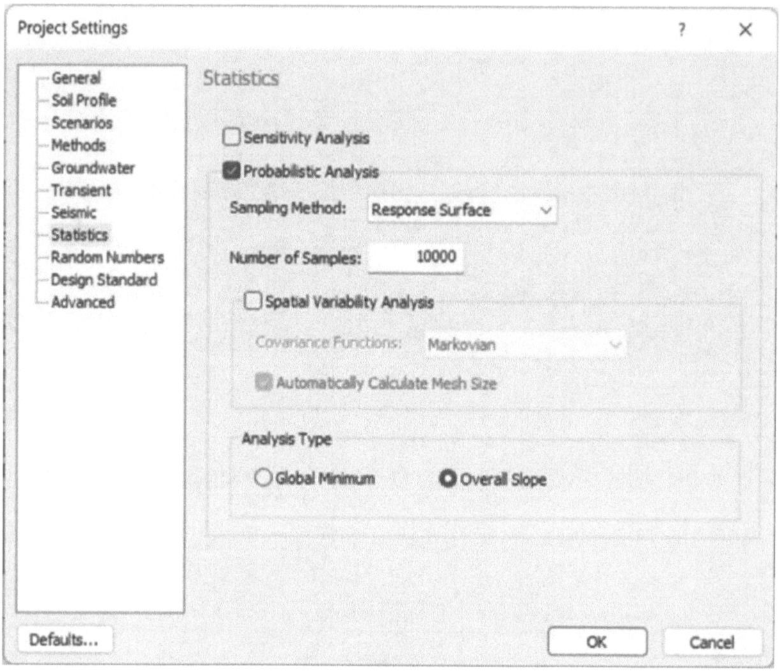

FIGURE 11.5 Example dialog showing the input of sampling methods in Slide2.

to have failed if the computed FOS is below 1.0. In reality, the computed FOS deviates from the actual FOS, because of model idealizations among others. The ratio between the actual FOS and the computed FOS is called the model factor (Phoon and Tang 2019). The mean and coefficient of variation of the model factor are shown in Table 11.2.

The global minimum method is the fastest because the software will perform a holistic search for the critical slip surface under deterministic conditions (where all the random variables are set to their mean values) and then use the same slip surface to calculate the FOS in all of the simulations. As such, only the FOS of the deterministic critical slip surface is calculated for each simulation, which can lead to an unconservative representation of the PF. In many cases, however, the shape and location of the critical slip surface depend highly on the sampled values of the material properties. For this reason, the overall slope method repeats the entire search for the critical slip surface as the primary analysis for every simulation. This method is by far the slowest of the two and in the case of overly complex models, can be facilitated using the more advanced SRS sampling method.

Stochastic Analysis Results

The PF results of selecting the global minimum and overall slope options with respect to the primary analysis are indicated in Table 11.3. The global minimum method uses LH sampling, while the overall slope method was computed using both LH and SRS methods described earlier in this chapter. In all cases, 10,000 simulations are generated. A histogram of the FOS corresponding to the simulations from the overall slope method with SRS is shown in Figure 11.6. Simulations where failure has occurred (FOS < 1) are highlighted.

The overall slope method has a mean FOS of 1.16 for both LH and SRS cases, which is slightly lower than the mean FOS of 1.17 obtained from the global minimum method. The PF obtained by the overall slope method is also higher (23.7–23.9% compared to 21.7%). The reliability index (RI) is the number of standard deviations which separate the mean FOS from the failure FOS of 1.0 - it is lower for the overall slope method as expected. As mentioned previously, although it is faster

TABLE 11.2
Statistics of the Model Factor for the Factor of Safety: Stability of Soil Slope and Base Heave in Excavation

Case	N	Design Method	Mean	COV	Reference
Soil slopes	83	Direct	1.07	0.21	Travis et al. (2011)
	134	Bishop	1	0.2	
	43	Force	0.95	0.2	
	41	Complete	0.97	0.15	
Fill slopes	27	Simplified Bishop	1.11	0.28	Bahsan et al. (2014)
		Spencer	1.19	0.27	
Cut slopes	7	Simplified Bishop	0.89	0.28	
		Spencer	0.9	0.26	
Natural slopes	9	Simplified Bishop	1.41	1	
		Spencer	1.57	0.96	
Base heave	24	Modified Terzaghi	1.02	0.16	Wu et al. (2014)
		Bjerrum-Eide	1.09	0.15	
		Slip circle	1.27	0.22	

(Source: Table A9, Phoon and Tang, 2019)
Note: the model statistics of Travis et al. (2011) and Bahsan et al. (2014) are collected from Dithinde et al. (2016), which are revised from Travis et al. (2011) and Bahsan et al. (2014).

TABLE 11.3
Results of the Probabilistic Analyses of the James Bay Slope Stability Problem. Computer Used for Computation: 8 cores, 2.30GHz speed, 64.0 GB RAM

Analysis	Mean FOS	*PF* (%)	Reliability Index	Computation Time (s)
Global Minimum (LH)	1.17	21.9	0.77	8
Overall Slope (SRS)	1.16	23.7	0.70	77
Overall Slope (LH)	1.16	23.9	0.71	7500

than the overall slope method, the global minimum method typically generates less conservative results because the critical slip surface of the slope is not searched for in each simulation. Figure 11.7 shows all the global minima found by the overall slope throughout the 10,000 simulations. It can be seen that the global minimum method would not have considered the failure mode that runs above the lacustrine clay layer. The computation time is also of note in Table 11.3: the LH and SRS overall slopé methods produce results that are in agreement, but SRS takes about 1% of the computation time that LH takes.

Correlation of Parameters

In the probabilistic analysis described in this example, the variables are sampled independently. That is to say for example that in simulation 1, the friction angle from the embankment material is sampled randomly, and the unit weight of the embankment material is sampled randomly as well, without being impacted by the friction angle sample. However, in many cases, a correlation exists between parameters. For example, cohesion and friction angle have been found to be negatively correlated, meaning that a material with high cohesion generally has a lower friction angle and vice

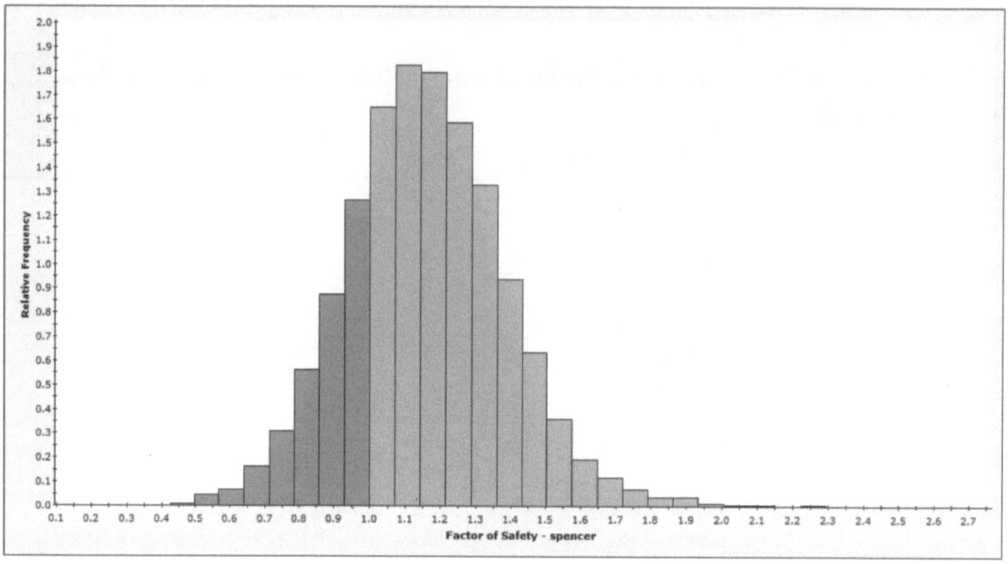

FIGURE 11.6 Histogram distribution of FOS values from the simulations of the overall slope method.

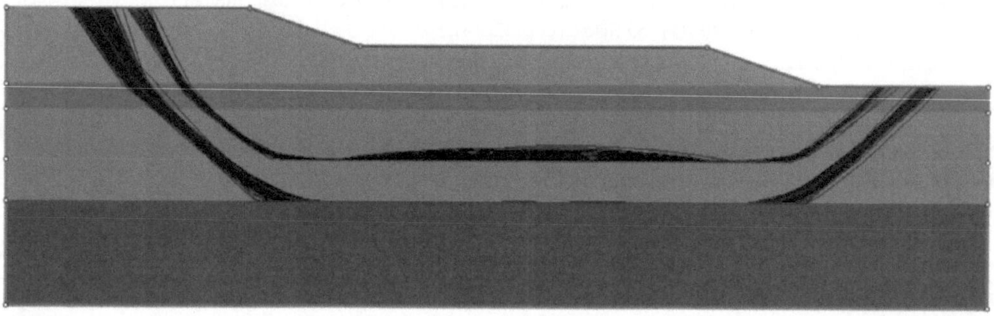

FIGURE 11.7 Global minima surfaces found from all 10,000 simulations using overall slope method. Two failure modes are visible.

versa. Therefore, it would be incorrect to sample them independently; rather, in the same simulation, high samples of one should be paired with lower samples of the other. This is done by applying a correlation coefficient to the correlated parameters. When data is available this correlation can be estimated. Otherwise, typical values can be used from the literature.

11.6 EXAMPLE TWO: SPATIALLY VARIABLE MATERIALS

As described in a previous chapter, the spatial variability of materials can influence the PF in a different way. In fact, for simple slopes, Cylwik et al. (2023) have found that when spatial variability of materials is considered, the deterministic FOS obtained from using the mean values of the material properties can be unconservative because it does not account for the effect of the critical slip surface to optimize through the paths of least resistance in the soils when the materials are spatially varying. To account for this, spatial variability analysis should be considered. In this example, the Random Limit Equilibrium Method (RLEM) is used, whereby a spatial random field is generated for each random variable based on the input distribution and scale of fluctuation. Since the critical slip surface would change for every simulation, only the overall slope method can reasonably be employed for spatial variability models.

This example is a model of an open pit mine located in the Central African Copperbelt obtained from Cylwik et al. (2021). The stratigraphy of the mine contains various complex layers of dolomites, siltstones, shales, and sandstones shown in Figure 11.8. Following the results of a detailed parametric analysis of the available material data, random distributions were assigned to the values of the cohesion and friction angles for each material (Cylwik et al. 2021). These distributions were linked using correlation coefficients since the cohesion and friction angles for each material were shown to be closely correlated. The overall slope method was used to search for the critical slip surface in 1,000 simulations. In each simulation, the material parameters were spatially sampled using a scale of fluctuation estimated using a variogram model (Cylwik et al. 2021).

The results of the Cylwik et al. (2021) study show a deterministic FOS of 1.4 but a mean probabilistic FOS of only 1.12, with $PF = 8.1\%$, shown in the output screenshot from the Slide2 program in Figure 11.9. The red slip surface outline corresponds to the deterministic critical slip surface, and the histogram shows the critical FOS obtained in each simulation. In fact, there are various failure modes along the profile of the slope corresponding to different simulations. The various slip surfaces that can occur along the slope are overlaid using the green and blue lines. Finally, the underlying contours of the spatially variable cohesion are also plotted in the figure for one of the simulation cases, showing localized weak and strong regions of materials. As Cylwik et al. (2021, 2023) report, the introduction of spatially variable materials causes the critical slip surface to optimize through the weaker regions of the model in each simulation, resulting in an analysis more representative of heterogeneous in-situ conditions.

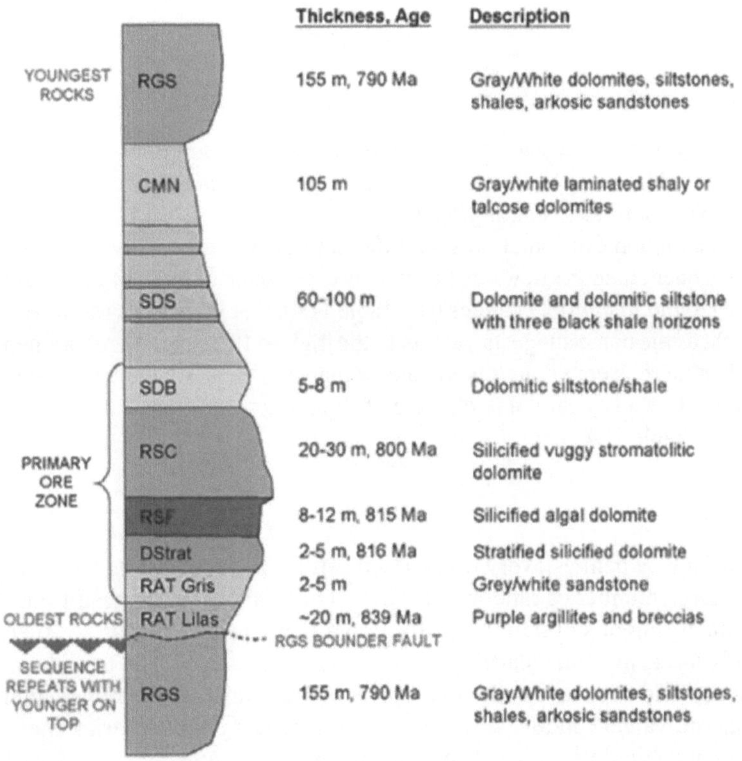

FIGURE 11.8 Stratigraphy of the Central African Copperbelt open pit mine (obtained from Cylwik et al., 2021, with permission).

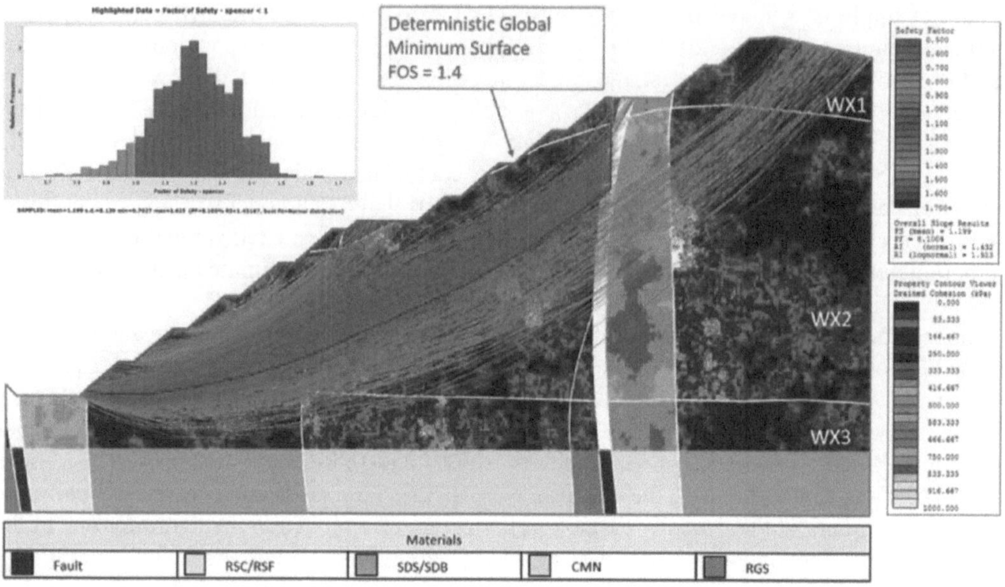

FIGURE 11.9 Results of spatially variable slope analysis a of a section in the Central African Copperbelt open pit mine (obtained from Cylwik et al., 2021, with permission).

It should be noted that if an engineer were to do a deterministic analysis only in this example, the slope may have been considered safe with a FOS of 1.4. After using probabilistic analysis with spatial variability, the results indicate a PF of 8.1%, which is to say an 8% chance of failure.

Worst-Case Scale of Fluctuation

For spatially variable analysis, a scale of fluctuation is always required for each material. The scale of fluctuation is roughly described as the distance (e.g., horizontally or vertically) over which a material exhibits very similar strength properties.

The scale of fluctuation can sometimes be difficult to obtain, especially when field data is lacking. In this case, a phenomenon known as the "worst-case" scale of fluctuation should be used in its stead. It has been found in multiple studies (see Table 11.4), that there exists a "worst-case" scale of fluctuation which, as the name suggests, results in the highest PF value. As recommended by Phoon et al. (2022), when there is not enough field data available to estimate the true scale of fluctuation, the engineer can select a typical value of scale of fluctuation for their soil type and then use the worst-case scale in Table 11.4.

Recent Advances

Since spatial variability defines every discretized cell in the model as its own random variable, this results in a large number of random variables. For this reason, the traditional SRS method would require the fitting of simulation results across many parameters and is therefore infeasible. In recent advances, machine learning methods have been developed to address this problem by simplifying the field of many random variables into a smaller set of representative variables with their own, equivalent random distributions in a process called variational autoencoding (VAE) (Kingma and Welling, 2014). These advances are documented in Javankhoshdel et al. (2022) and greatly improve the speed for conducting spatial variability analysis with the overall slope method.

TABLE 11.4
Examples of "Worst-Case" Scale of Fluctuations Reported in Previous Studies

Study	Problem type	"Worst-case" Definition	Characteristic Length	"Worst-Case" Scale of Fluctuation
Fenton and Griffiths (2003)	Bearing capacity of a footing on a c- φ soil	Mean bearing capacity is minimal	Footing width (B)	1×B
Griffiths et al. (2006)	Bearing capacity of footing(s) on a $\varphi = 0$ soil	Mean bearing capacity is minimal	Footing width (B)	0.5~2×B
Vessia et al. (2009)	Bearing capacity of footing on c-φ soil	Mean bearing capacity is minimal (anisotropic 2D variability)	Footing width (B)	0.3~0.5×B
Fenton and Griffiths (2005)	Differential settlement of footings	Under-design probability is maximal	Footing spacing (S)	1×S
Breysse et al. (2005)	Settlement of a footing system	Footing rotation is maximal	Footing spacing (S)	0.5×S
		Mean different settlement between footings is maximal	Footing spacing (S) Footing width (B)	f(S,B) (no simple equation)
Jaksa et al. (2005)	Settlement of a nine-pad footing system	Under-design probability is maximal	Footing spacing (S)	1×S
Ahmed and Soubra (2014)	Differential settlement of footings	Under-design probability is maximal	Footing spacing (S)	1×S
Stuedlein and Bong (2017)	Differential settlement of footings	Under-design probability is maximal	Footing spacing (S)	1×S
Ali et al. (2014)	Risk of infinite slope	Risk of rainfall-induced slope failure is maximal	Slope height (H)	1×H
Hu and Ching (2015)	Active lateral force for a retaining wall	Mean active lateral force is maximal	Wall height (H)	0.2×H
Fenton et al. (2005)	Active lateral force for a retaining wall	Under-design probability is maximal	Wall height (H)	0.5~1×H
Griffiths et al. (2008)	Passive lateral force for a retaining wall	Under-design probability is maximal	Wall height (H)	0.1 to 0.5×H
Ching and Phoon (2013)	Overall strength of a soil column	Mean strength is Minimal	Column width (W)	1×W (compression) 0×W (simple shear)
Pan et al. (2018)	Stress-strain behavior of cement-treated clay column	Peak global strength	Column diameter (D)	2×D

(Source: Table 4, Phoon et al., 2022).

11.7 EXAMPLE THREE: 3D PROBABILISTIC SLOPE STABILITY

The final slope stability example for this chapter is a 3D slope modeled in Slide3 and shown in Figure 11.10. The limit equilibrium analysis method (LEM) has been extended from 2D to 3D in recent decades for the advantage of modeling the 3D geometrical influences on the true FOS in asymmetrical slopes. LEM in 2D is limited in that it assumes an infinite extrusion of the section perpendicular to the analysis direction and does not always produce a conservative result compared

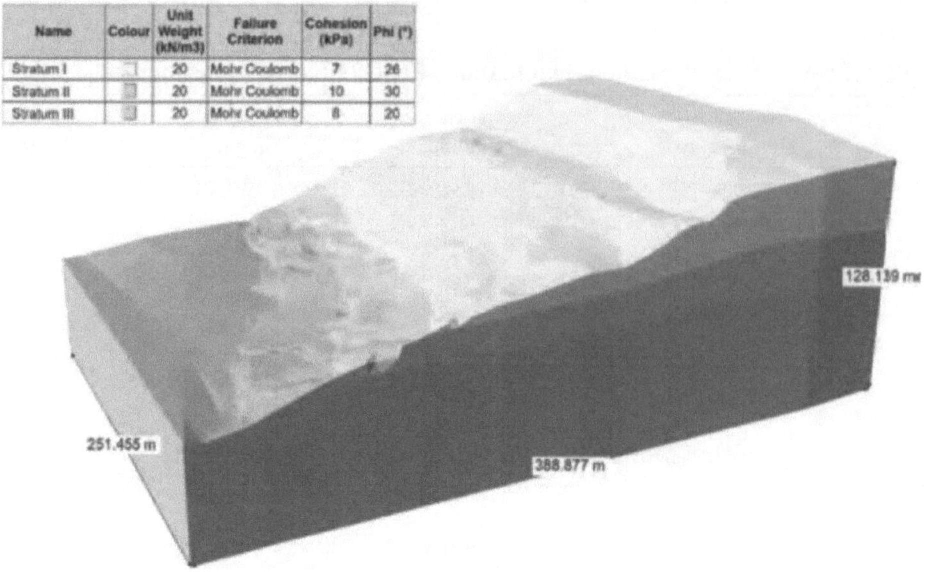

FIGURE 11.10 3D probabilistic slope model obtained from Cami et al. (2021), with permission.

to the 3D FOS (MacRobert et al., 2021; Ma et al., 2022). The FOS of a slip surface can be readily solved in 3D using Slide3 and optimization methods for searching for the critical slip surface like the ones in 2D have been developed for 3D (Ma et al., 2022).

In this example, the cohesion and friction angles are assumed to follow lognormal distributions with mean and standard deviation shown in Table 11.5. An overall slope probabilistic analysis with 1,000 simulations was facilitated using the SRS method, with results showing a mean FOS of 1.23 and PF = 25.8% in Figure 11.11. There are two distinct failure modes in this model, underscoring the importance of searching for more than the single global minimum critical slip surface. To reduce the probability of failure to acceptable levels, reinforcement anchors would be recommended in the locations of both failure modes.

11.8 OTHER APPLICATIONS

In this chapter, the probabilistic analysis of slope stability problems via software programs is presented in detail. However, the same principles of sampling for simulations and running the primary analysis for each simulation can be applied to other geotechnical problems. For instance, two other applications are presented in this section: 1) The simulation of rockfall paths can be accomplished in the RocFall2 and RocFall3 programs, where the failure function can be defined as the probability of rocks in the simulations to impact infrastructure or bounce past a barrier. The primary analysis

TABLE 11.5
Material Properties for the 3D Slope Stability Problem

	c (kPa)		φ (°)	
Layer	**Mean**	**Std Dev.**	**Mean**	**Std Dev.**
Stratum I	7.0	2.8	26.0	5.2
Stratum II	10.0	4.0	30.0	6.0
Stratum III	8.0	3.2	20.0	4.0

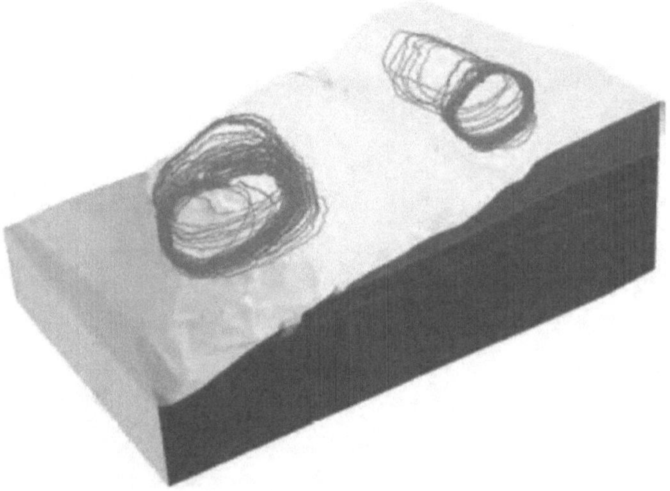

FIGURE 11.11 Differing critical slip surfaces obtained from the overall slope method in the 3D probabilistic slope model obtained from Cami et al. (2021), with permission.

in such a case would take the form of stochastic simulations of the rockfall paths. 2) Statistics in orientation-based analysis in Dips software for the rock slope stability analysis.

11.8.1 Simulation of Rockfall Paths in RocFall2 and RocFall3

A rockfall simulation provides a prediction of the trajectory and energy of the rock as it traverses down the slope rather than a prediction of the FOS. Rockfall trajectories and energies are affected by many factors, including the natural and local variations in the slope geometry, slope material properties, initial conditions, and rock attributes. While these parameters serve as necessary simulation inputs, they are subject to high uncertainty, scale effects, and are difficult to obtain through testing. Accurate prediction of rockfall is practically impossible by virtue of rockfalls being a random phenomenon. For design purposes, deterministic rockfall predictions would not be justifiable due to the high uncertainty in input parameters. Instead, the statistical simulation of rockfalls that computes a large sample of rock trajectories can provide a reasonable distribution of the rock end locations, bounce heights, and energies, which are the quantities important to mitigation design. The interpretation of statistical rockfall simulation would have to be risk-based, as the simulation permits a statistical interpretation of the rockfall hazard while the designer must decide on an acceptable level of consequence.

The following two examples demonstrate the utility of 2D and 3D statistical rockfall simulation in RocFall2 and RocFall3, respectively, for the design of rockfall hazard mitigation.

In the first example using RocFall2 (Rocscience 2022), falling rocks are predicted to reach the roadway infrastructure at a rate of 259 out of 500 simulations (Figure 11.12). The spread in the trajectories is the result of applying statistical distribution to the slope material properties and rock mass, as well as in modeling different rock shapes (i.e., rigid body impact theory). A normal distribution is assigned to the material properties, and a relative minimum and a relative maximum are applied as equal to three times the standard deviation. This assumes a probability distribution bounded by the lowest and highest conceivable values for each parameter, and a probability of 99.7% that the actual material property value falls within the range sampled. In each rockfall simulation, each random variable is sampled using the LH technique, and a different rock shape is adopted, such that there are differences in the input parameters for each of the 500 rockfall simulations. A decision must be made on an acceptable rate of impact on the roadway and whether and how much mitigation is required. Assume a barrier can be installed at location x = 7 m as an option for rock containment at an adequate distance of clearance. To help decide

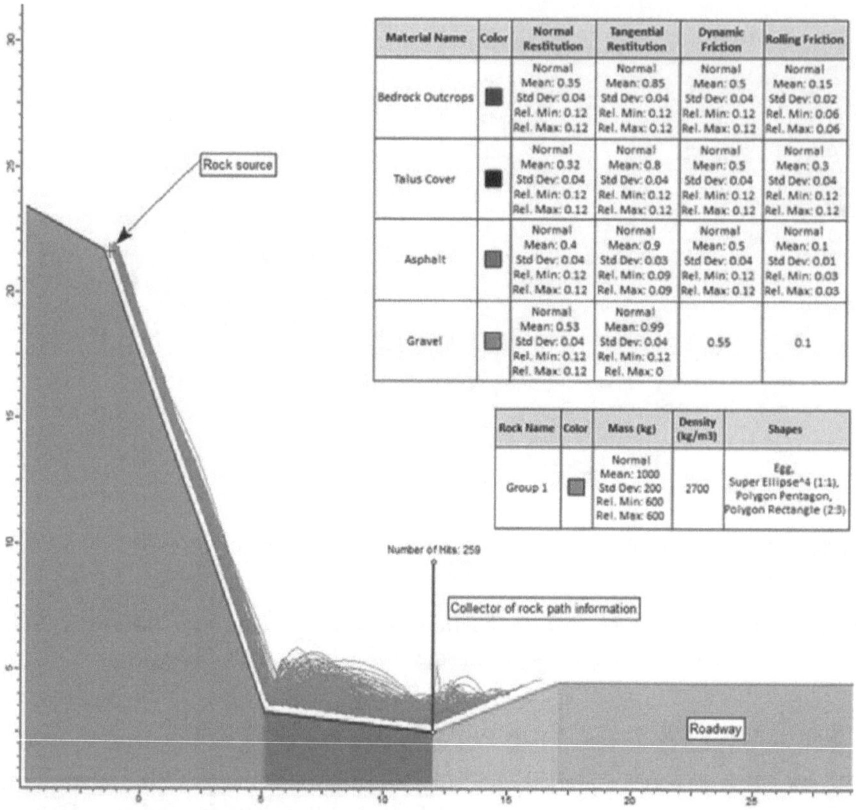

FIGURE 11.12 2D rockfall example model in RocFall2.

on the barrier design, rock path information can be collected at the barrier location, allowing for the probabilistic analysis of rock translational kinetic energies and bounce heights at that location (Figure 11.13). The figure illustrates these quantities plotted as cumulative frequencies, where the designer can view the probabilities of containment according to rock translational kinetic energy and bounce height. The selected percentile (e.g., 95%) would represent the accepted probability of containment. The corresponding values would then be used to select barrier models that have the capacity and height required to mitigate rockfalls at the accepted level of risk.

In the second rockfall example using RocFall3 (Rocscience 2023e), the lateral dispersion of rock trajectories is analyzed as made possible through 3D statistical rockfall simulation. Assuming the formation of a failure surface through the upper benches that can occur with an earthquake, there is potential for rockfall hazard on the downstream benches as well as the open pit floor (Figure 11.14). The random variables in the 3D simulation are the slope material parameters and the initial velocity of the rocks (Tables 11.6 and 11.7). Using the lump mass impact theory, 500 rock paths are simulated, with a resulting lateral dispersion of 133 m from a rock source area of only 20 m (Figure 11.15). The lateral dispersion is the result of a statistical rockfall simulation using 3D geometry, where rock paths are not necessarily confined to a 2D plane but can travel in all directions as influenced by local geometry, randomness in material properties, and rock starting conditions.

A histogram of rock endpoints shows hazards on the benches as well as outside the existing berm, indicating that additional mitigation may be required (Figure 11.16). Using the plane for information collection installed along the berm crest as a reference line for calculating runout distances, about 17% of the 500 simulated rock paths have positive runouts or have fallen outside the berm (Figure 11.17). The designer can then use this information, along with the 3D visualization of the lateral dispersion, to decide on additional mitigation within the open pit space.

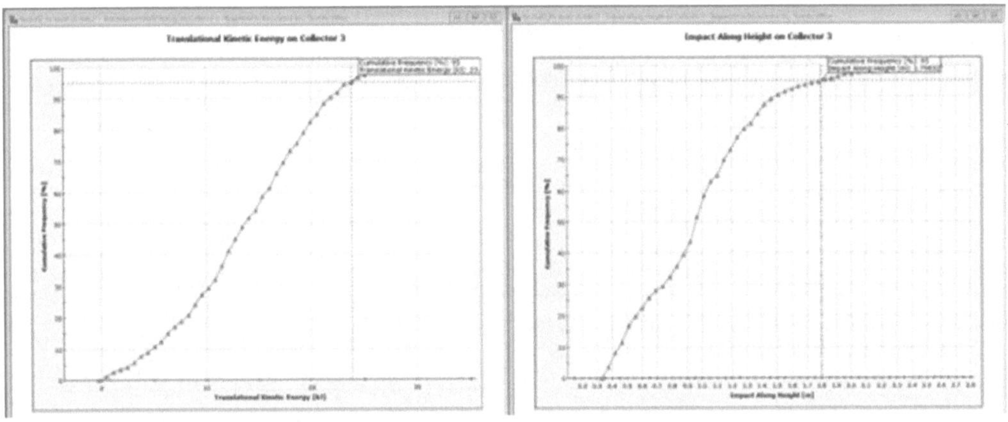

FIGURE 11.13 Cumulative frequency plots of rock translational kinetic energies (left) and bounce heights (right) at barrier location in 2D rockfall example.

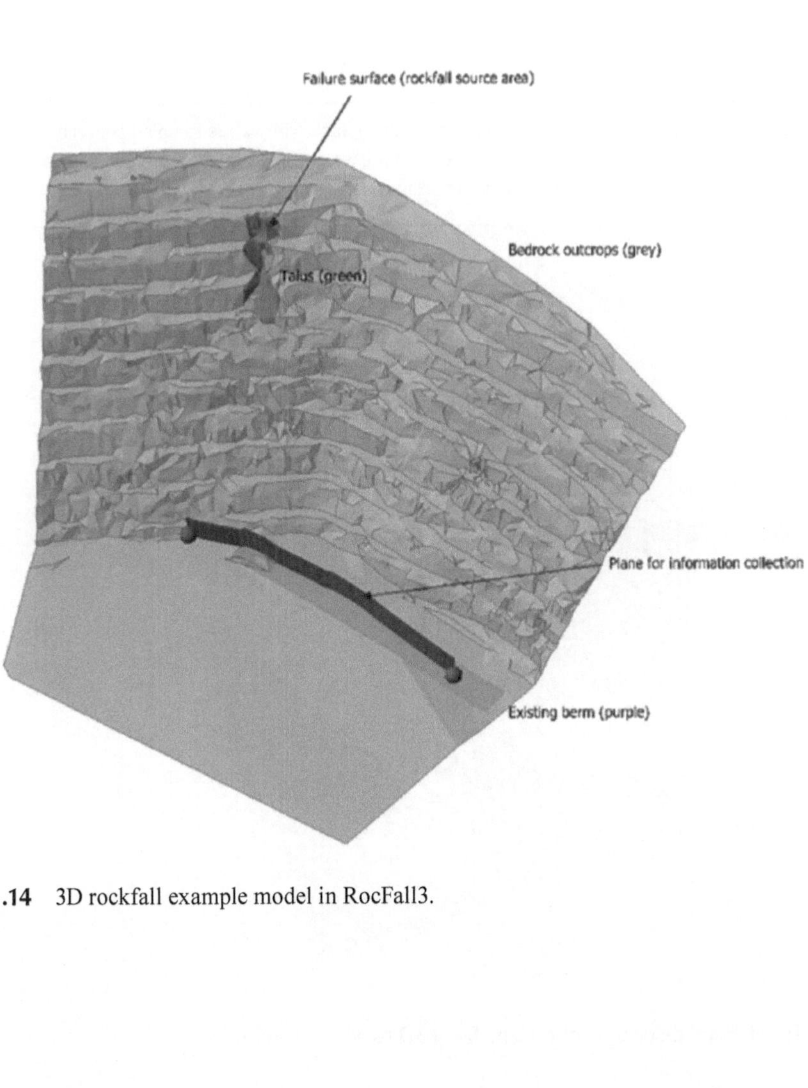

FIGURE 11.14 3D rockfall example model in RocFall3.

TABLE 11.6
Material Properties for the 3D Rockfall Problem

	Normal Restitution		*Tangential Restitution*		*Friction Angle (°)*	
Material	**Mean**	**Std Dev.**	**Mean**	**Std Dev.**	**Mean**	**Std Dev.**
Bedrock Outcrops	0.35	0.04	0.85	0.04	30	0
Talus Cover	0.32	0.04	0.80	0.04	30	0
Berm	0.15	0.04	0.20	0.04	30	0

Note: random variables are normally distributed, with the relative minimum and relative maximum assumed to be three times the standard deviation.

TABLE 11.7
Seeder Properties for the 3D Rockfall Problem

	Mass (kg)				*Initial Translational Velocity (m/s)*		
Seeder	**Mean**	**Std Dev.**	*Density (kg/m³)*	*Number of Rocks*	**Mean**	**Std Dev.**	*Trend of Velocity (deg)*
Property 1	500	100	2700	500	1.0	0.2	180 - 360

Note: the trend of the velocity was assumed to be uniformly distributed with values sampled between 180 to 360 degrees (clockwise from model y-axis); other random variables have normal distributions, with the relative minimum and relative maximum assumed to be three times the standard deviation.

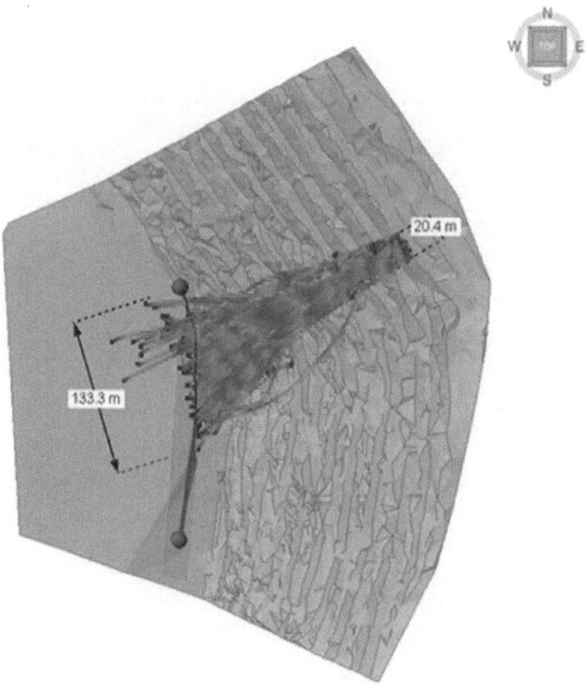

FIGURE 11.15 Lateral dispersion of rock paths in 3D rockfall example.

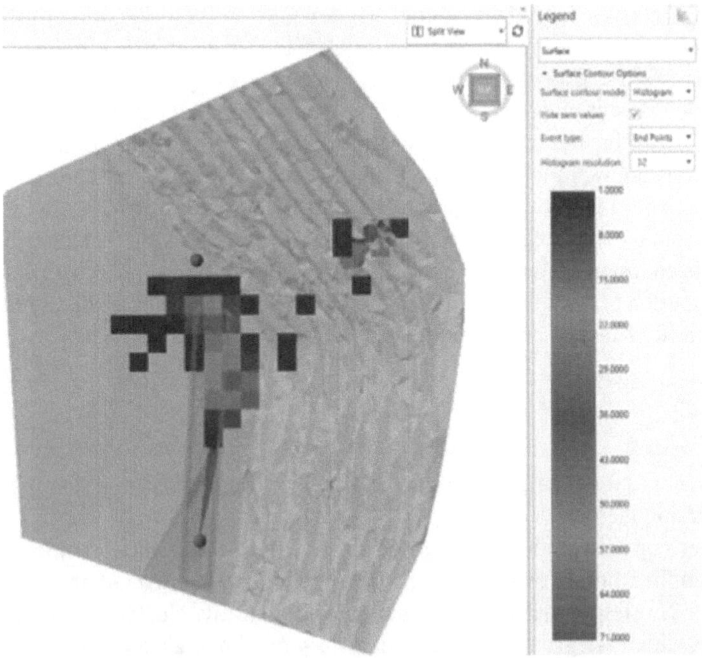

FIGURE 11.16 Histogram of rock endpoints in 3D rockfall example.

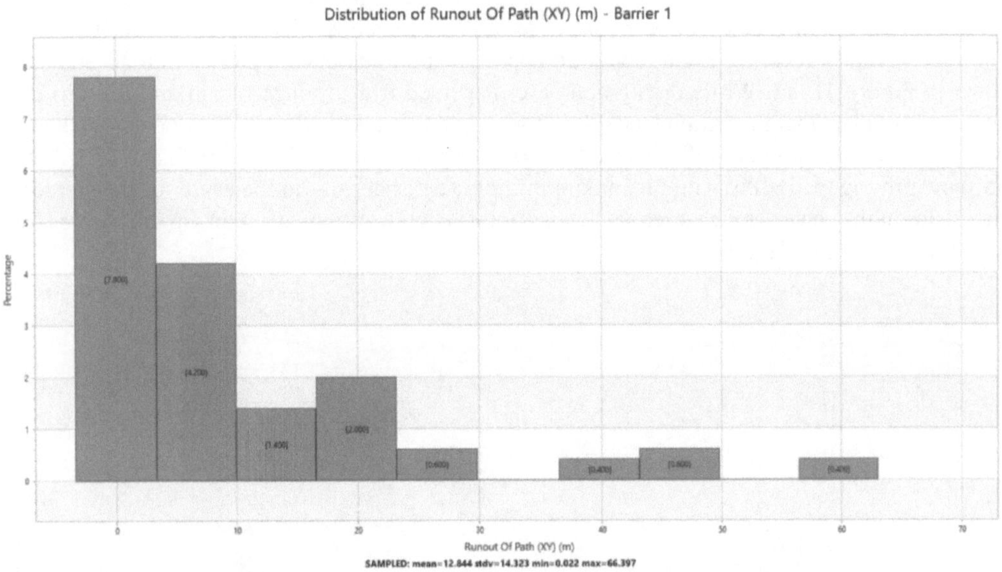

FIGURE 11.17 Percent distribution of rock runout distances calculated from a reference line (i.e., plane for information collection); sum to 17% of 500 simulated rock paths that exceeded the existing berm in the 3D rockfall example.

11.8.2 Statistics in Orientation-Based Analysis

In rock slope stability analysis, the mode of failure is often structurally controlled by planar features such as in the case of planar sliding, wedge sliding, and block toppling. Planar discontinuities include fractures, joints, and faults; they have negligible tensile strength and low shear strength when compared to intact rock and can act as release planes or sliding planes. The likelihood of occurrence of these blocks or wedges of concern depends on the orientation of the slope and the orientation of planes or plane-plane intersections. From purely geometric kinematic criteria, a critical vector must satisfy the following conditions: (1) the vector daylights on the slope; (2) the vector has an unfavorable orientation which dips into or dips steeper than the slope.

Like material strengths, the orientations of these planes can also vary, and these variations can be represented statistically. In a geotechnical context, vectors are often represented by two bearings: dip direction and dip (or trend and plunge, or strike and dip), which are distributed over a spherical range, from 0 to 360 degrees and 0 to 90 degrees, respectively. The distribution of 3D orientation vectors is commonly modeled with the Fisher Distribution (Fisher, 1953), which describes the angular distribution of orientations about a mean orientation vector and is symmetric about the mean. The probability function is expressed as:

$$f(\theta) = \frac{K \sin \theta e^{K \cos \theta}}{e^{K} - e^{-K}}$$

Where θ is the angular deviation from the mean vector (in degrees) and K is the Fisher constant or dispersion factor.

A larger Fisher constant indicates a tighter clustering of orientation vectors, while a smaller Fisher constant implies a more dispersed cluster (analogous to the inverse of variation in a normal distribution). In a 2D stereographic projection or stereonet, this clustering can indicate the existence of joint sets which are groups of joints with the same dip angle and strike angle.

11.8.2.1 Dips Case Study

Dips (Rocscience 2023f) is a software designed for the interactive analysis of orientation-based geological data and will be used for the purpose of this case study. Planar data has been collected over a sedimentary rock slope and input into Dips. The pole vectors are plotted in a 2D stereonet, shown in Figure 11.18a. While orientations are dispersed over a range of dip and dip direction angles, it is evident that the parallel, near horizontal bedding planes dominate, especially when the pole vector density contour is displayed. The contour plot is sampled with the Fisher distribution and shown in Figure 11.18b, with the maximum pole vector density in the center of the stereonet. The cluster in the center of the stereonet forms a joint set with a sampled mean dip of 2 degrees and a dip direction of 124 degrees.

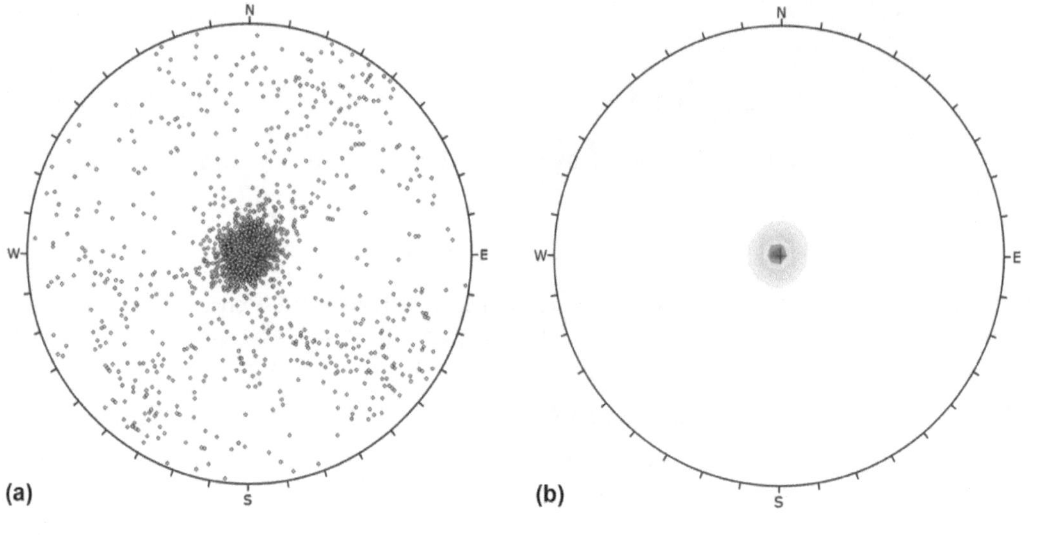

FIGURE 11.18 (a) Pole vectors and (b) pole vector density contours plotted in 2D stereonet.

Another representation of orientation data is in the form of a rosette plot, which is essentially a histogram of strike frequency, plotted radially. In Figure 11.19, the rosette plot shows a tendency in strike angles in the northwest to southeast and in the northeast to southwest directions; this indicates that there may be two additional joint sets which exist.

Due to the concentration of poles which make up the bedding, other orientation data is obscured in the statistical contouring. Using the data filtering option in Dips, the poles in the center of the stereonet belonging to the bedding joint set are removed and the stereonet is re-contoured. Two new regions of high pole vector densities are shown in Figure 11.20, and the potential joint sets indicated in the rosette plot are confirmed. The two additional joint sets have a mean dip and a dip direction of 73 degrees, 307 degrees, 89 degrees, and 32 degrees, respectively. Three orthogonal joint sets are identified, which is typical of a sedimentary rock formation (Figure 11.21).

Now that the key joint structures are identified, the kinematic feasibility of the various modes of slope failure is analyzed with kinematic analysis. For a slope with a slope dip of 65 degrees, slope dip direction of 110 degrees, friction angle of 30 degrees, and lateral limit of 20 degrees on either side of the slope dip direction, the kinematic results are presented for planar sliding, wedge sliding, flexural toppling, and direct toppling failure modes in Figures 11.22–11.25, respectively. Planar sliding is not very kinematically feasible, although still possible (Figure 11.22).

For wedge sliding, the results with all intersections and only the mean set plane intersections considered both suggest that this mode of failure is not likely. However, even though the percent of critical failures considering all intersections is extremely low, it is not zero, as otherwise indicated by only looking at the mean set plane intersections. This reconfirms that the mean is not necessarily a good enough representation of the data and that a range of *possible* values must be considered (Figure 11.23).

For flexural toppling, while the overall plane count does not associate with a high percent of critical toppling vectors, a staggering 50% of the vectors in Set 2 are critical. When it comes to

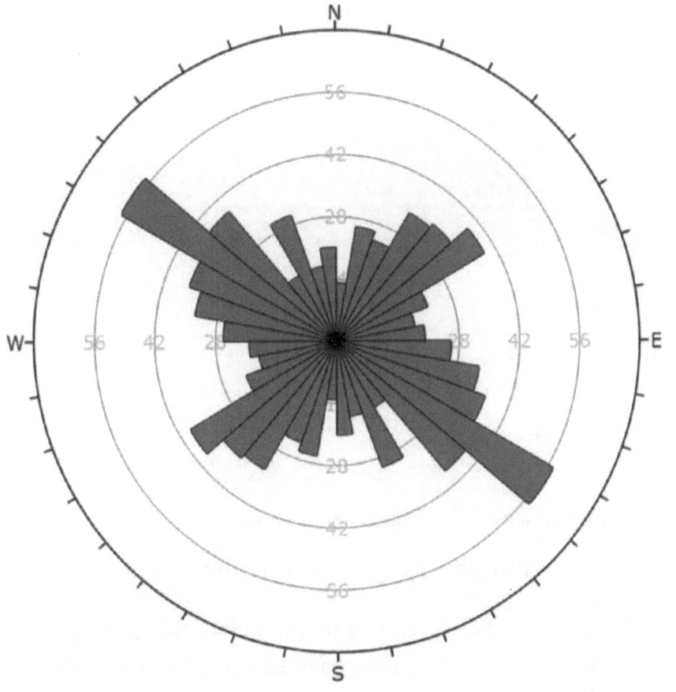

FIGURE 11.19 Rosette plot of strike frequency

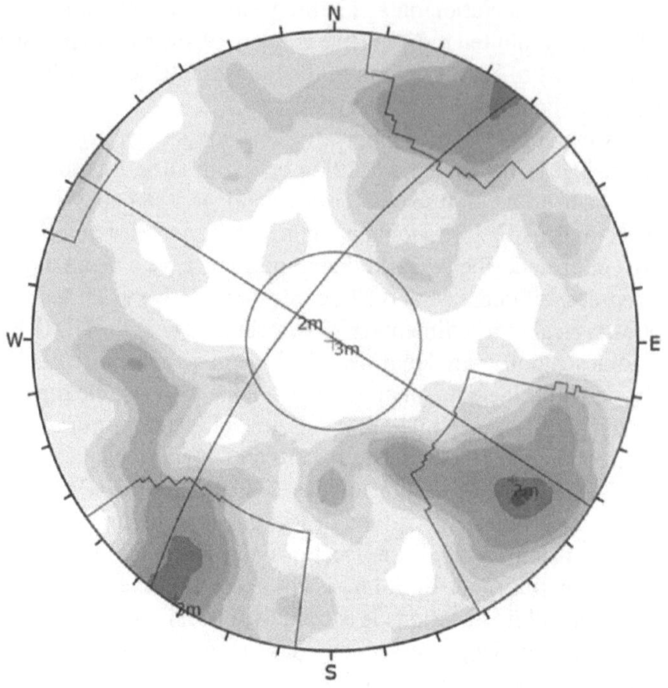

FIGURE 11.20 Pole vector density contours plotted in 2D stereonet with bedding planes removed.

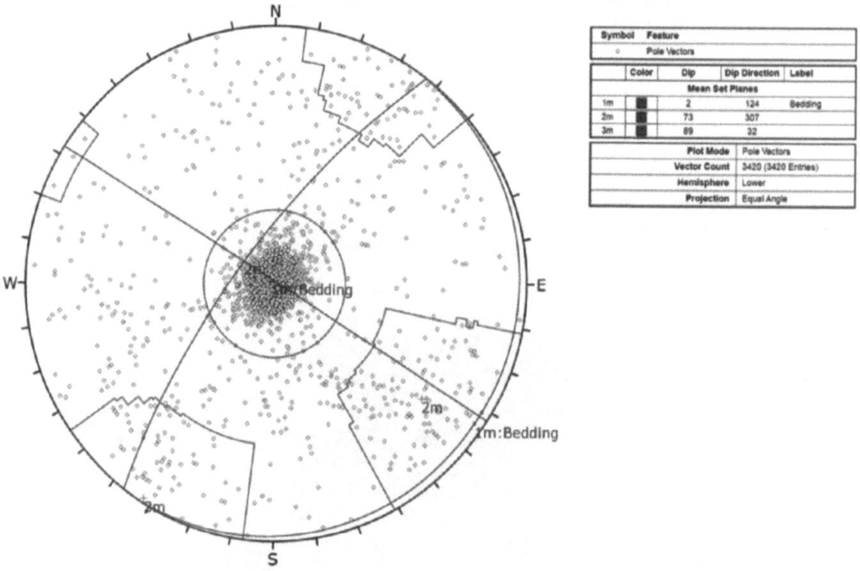

FIGURE 11.21 Pole vector plot with three orthogonal joint sets

sampling, the sampled dataset also plays a big factor in determining what is statistically significant and what is not (Figure 11.24).

Converse to wedge sliding results, the results for direct toppling with all intersections and only the mean set plane intersections considered are contradicting. Even though the percent of critical failures for direct and oblique failure, considering all intersections, is extremely low, the basal planes may be of concern and should be considered (Figure 11.25).

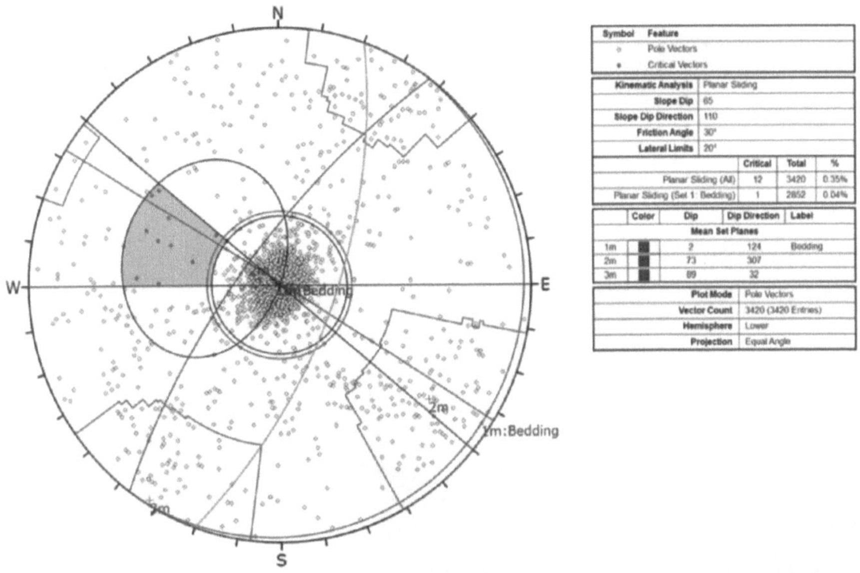

FIGURE 11.22 Kinematic analysis of planar sliding showing critical zone (highlighted) and critical poles (solid markers).

To conclude, when looking at statistics for orientation-based parameters, a different distribution is used to sample and "fit" the data since dip and dip direction are correlated, non-linear, and exist in a spherical space. Mean values or even extreme values may not be enough to capture the significance of the data in terms of feature identification or risk assessment of various failure modes. The significance of the sampled results is determined by the data points which are included or excluded from the analysis.

11.9 RECENT ADVANCES

The practical steps required of an engineer to do a probabilistic analysis from raw Cone Penetration Test (CPT) data would go as follows:

1. Start with a set of raw CPT soundings data.
2. Divide up the observed properties into separate material types.
3. For each material type, obtain the statistical properties (mean, standard deviation, and scale of fluctuation).
4. Create a 3D model with all the soundings (for a 3D analysis) or a 2D model between soundings (for a 2D analysis) by using a typical interpolation method such as kriging.
5. Input all this information into the desired software and compute a probabilistic analysis.

This process can be laborious, especially for the average engineer who isn't a statistical expert. A recent method published by Ching and Phoon (2021) allows for the removal of steps 2–4. The engineer simply needs to input the CPT soundings into the algorithm, and the software automatically completes the intermediate steps. With Bayesian learning the algorithm fills in any missing data in the soundings, obtains the statistical parameters, and builds the 3D model. This is a breakthrough method because engineers would not need to worry about esoteric probabilistic inputs at all. Its future implementation into Settle3 is outlined below.

It is challenging to predict a realistic soil profile and soil layer representation in a given project area as there are a limited number of CPT boreholes that can be collected in the field. The

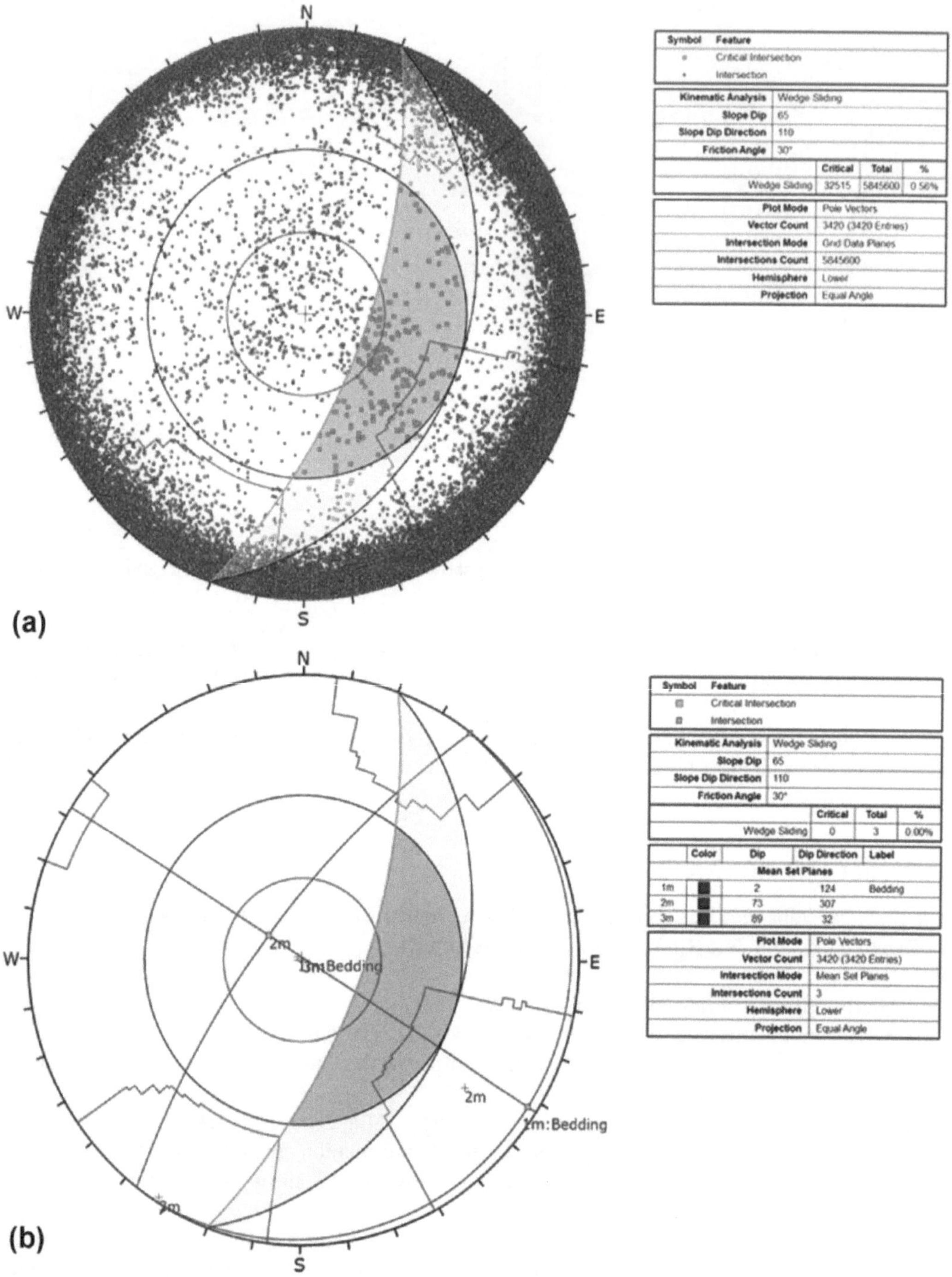

FIGURE 11.23 Kinematic analysis of wedge sliding showing critical zones (area straddled by the slope's great circle and another great circle with dip equal to the friction angle) and intersections for (a) all planar intersections and for (b) mean set plane intersections. Intersections inside the critical zones are considered to be critical vectors which satisfy frictional and kinematic conditions for wedge sliding.

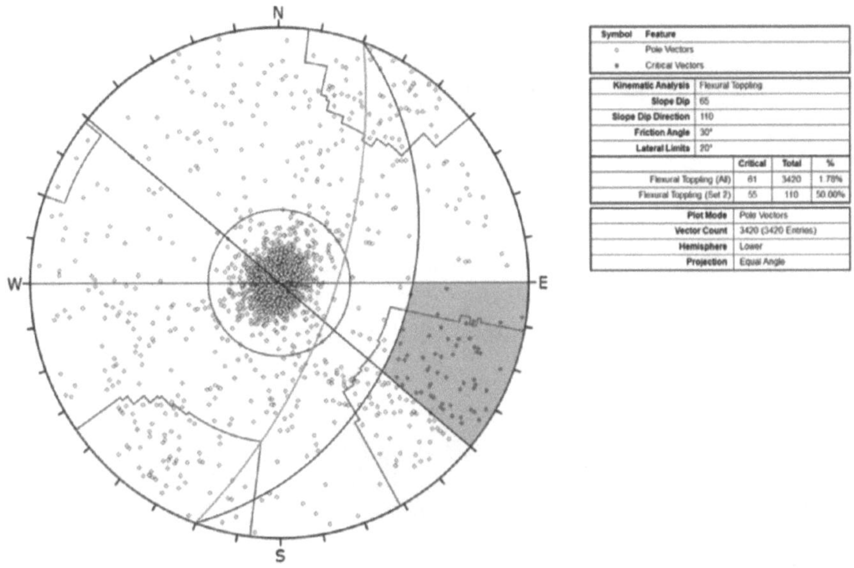

FIGURE 11.24 Flexural toppling.

classification of soil properties with spatial variability under the ground can now be achieved with high confidence. Ching et al. (2021) developed an improved method to address non-lattice data in three-dimensional probabilistic site characterization which determines the missing data points of CPT data based on the surrounding CPT data. Figure 11.26 shows a diagram of the lattice and non-lattice data sets with respect to the CPT holes.

Based on this algorithm, the full 3D surface can then be characterized based on the surrounding known nodes of CPT data points. Figure 11.26b shows the 3D graphical representation of *Ic* values which are soil characteristic values that can be used to categorize soil types (Roberton, 1990, 2010).

The entry number (X/Y) shows the location index of the CPT holes, followed by entry number Z which are the *Ic* values used for the characterization of soil types in the Soil Behavior Type (SBT) chart.

The future development of the Settle3 software (Rocscience 2023g) will include implementing the algorithm presented by Ching et al. (2021). This implementation has two steps: 1) the algorithm will be used to find missing CPT data points (Figure 11.27), and then using the correlations mentioned above and soil characteristics, soil layers can be defined with soil properties that are obtained from these CPT points. 2) Then, the statistical algorithm presented by Ching et al. (2021) can be used to create different realizations of 3D conditional random fields. Using empirical algorithms of settlement in the Settle3 software, the worst, median, and best-case scenarios (with regards to settlement) can be specified, and for those fields, immediate, primary consolidation, and secondary settlement can be computed. Figure 11.28 shows a typical deterministic example of CPT boreholes used to create the soil profile which is then used to carry out the settlement analysis in a 3D domain.

11.10 FINAL NOTE

It is the opinion of the authors that as the technology continues to advance in society and computers become more powerful and accessible, the use of software to simulate probabilistic analyses of geotechnical problems will become widespread and eventually standard for geotechnical design throughout the world.

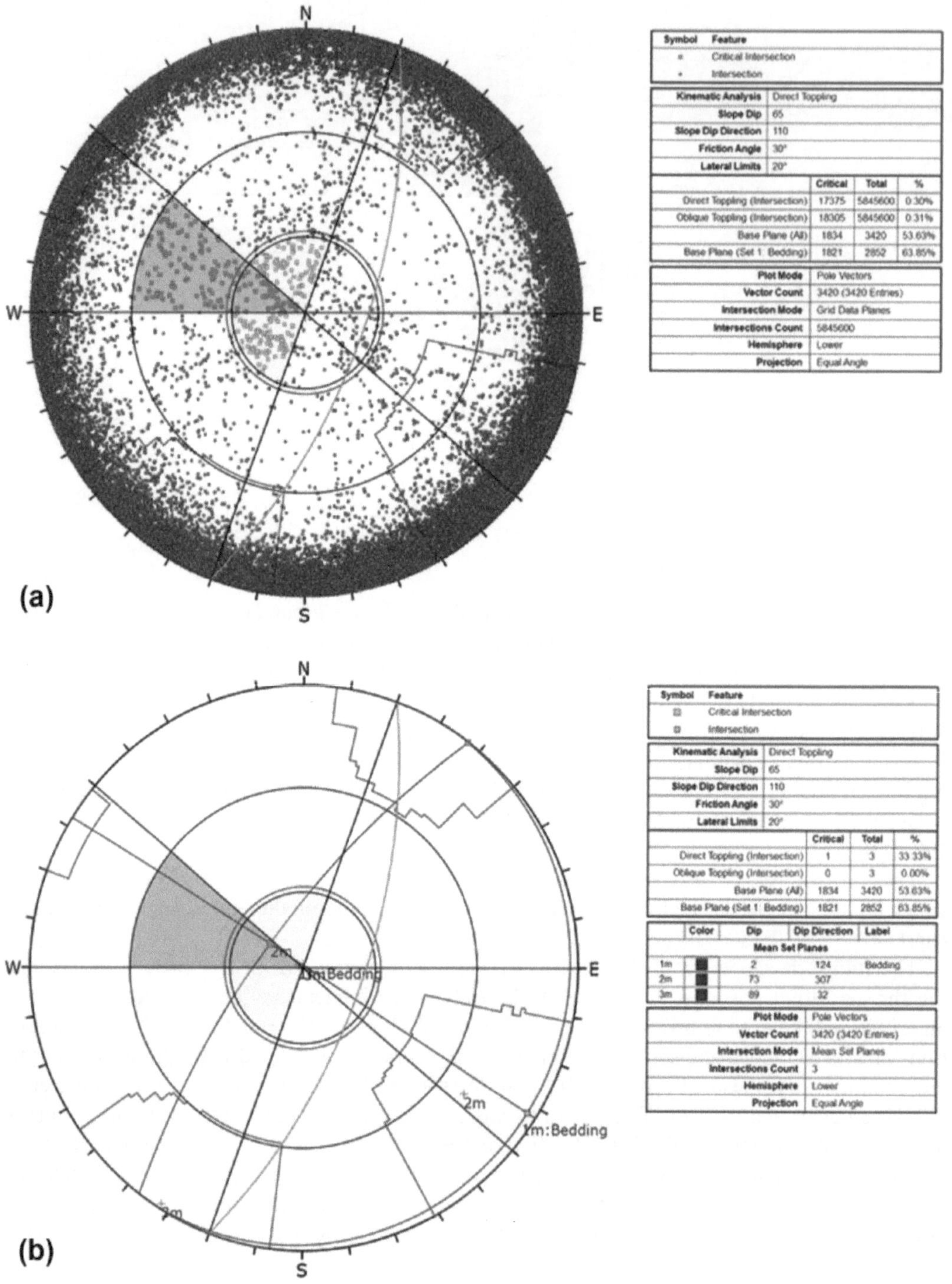

FIGURE 11.25 Direct toppling, (a) all planar intersections and for (b) mean set plane intersections.

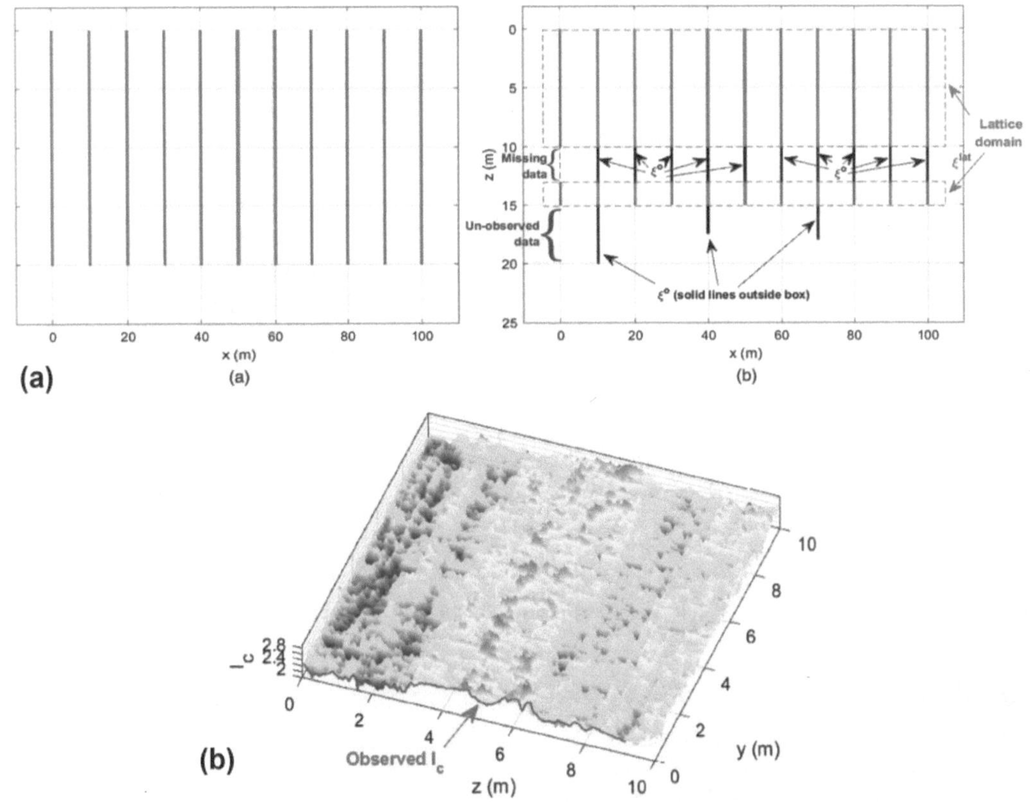

FIGURE 11.26 a) Diagram showing the lattice (left) to non-lattice (right) data, b) 3D view of the site characterization. (Ching et al., 2021)

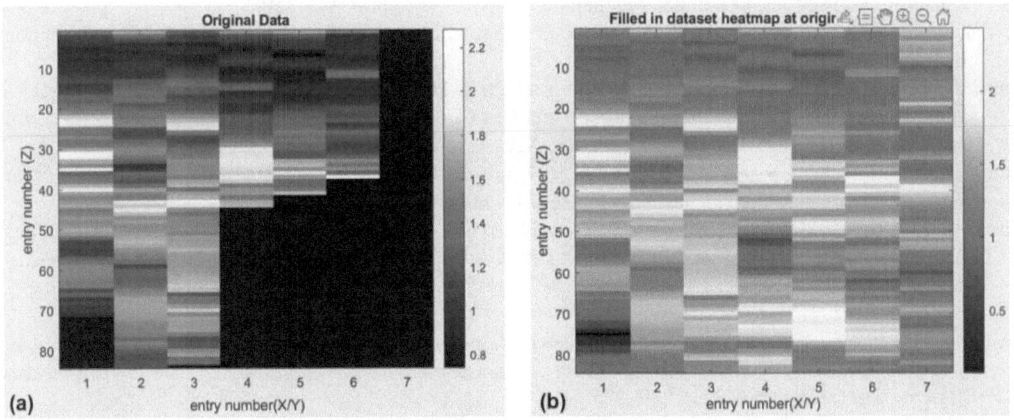

FIGURE 11.27 Finding missing CPT data points using Bayesian algorithm presented by Ching et al., 2021.

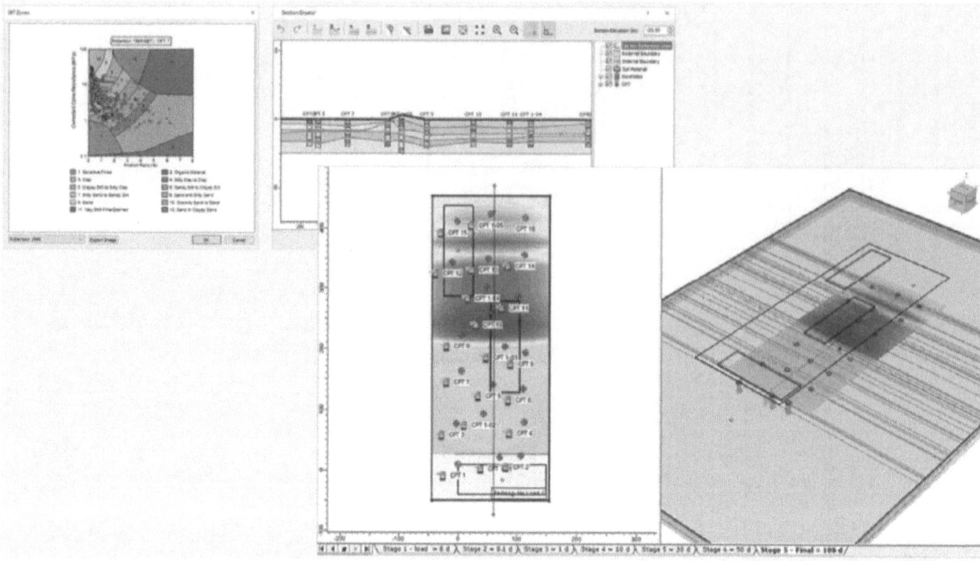

FIGURE 11.28 An example of specifying the soil profile using CPT data and calculating the settlement.

REFERENCES

Ahmed, A. and Soubra, A.H. 2014. Probabilistic analysis at the serviceability limit state of two neighboring strip footings resting on a spatially random soil. *Structural Safety*, 49, 2–9.

Ali, A., Huang, J., Lyamin, A.V., Sloan, S.W., Griffiths, D.V., Cassidy, M.J., and Li, J.H. 2014. Simplified quantitative risk assessment of rainfall-induced landslides modelled by infinite slopes. *Engineering Geology*, 179, 102–116.

Bahsan, E., Liao, H.J., Ching, J.Y., and Lee, S.W. 2014. Statistics for the calculated safety factors of undrained failure slopes. *Engineering Geology*, 172, 85–94.

Breysse, D., Niandou, H., Elachachi, S., and Houy, L. 2005. A generic approach to soil-structure interaction considering the effects of soil heterogeneity. *Géotechnique*, 36(2), 143–150.

British Standards Institution (BSI). 1995. *Eurocode 7. Geotechnical Design. Part 1, General Rules*. London: British Standards Institution.

Cami, B., Javankhoshdel, S., Bathurst, R.J., and Yacoub, T. 2018. Influence of mesh size, number of slices, and number of simulations in probabilistic analysis of slopes considering 2D spatial variability of soil properties. In *IFCEE 2018*, 186–196. Reston, VA: American Society of Civil Engineers.

Cami, B., Javankhoshdel, S., and Yacoub, T. 2021a. Limit equilibrium probabilistic analysis of three-dimensional open pit using the stochastic response surface method. In *IFCEE 2021*, 247–254. ASCE.

Cami, B., Ma, T., Javankhoshdel, S., Yacoub, T., and Corkum, B. 2021b. Considering multiple failure modes: A comparison of probabilistic analysis and multi-modal optimization for a 3D slope stability case study. In *The Evolution of Geotech – 25 Years of Innovation*, RIC 2021, pp. 189–195. Toronto, ON. April 20–21, 2021.

Ching, J. and Phoon, K.K. 2013. Probability distribution for mobilized shear strengths of spatially variable soils under uniform stress states. *Georisk: Assessment and Management of Risk for Engineered Systems and Geohazards*, 7(3), 209–224.

Ching, J., Yang, Z., and Phoon, K.K. 2021. Dealing with nonlattice data in three-dimensional probabilistic site characterization. *Journal of Engineering Mechanics*, 147(5), 06021003.

Cylwik, S.D., Cox, S.B. and Potter, J.J. 2022. Probabilistic analysis of an open pit mine slope in the Centra l African Copperbelt with spatially variable strengths. The Evolution of Geotech, p.558.

Cylwik, S., Javankhoshdel, S., Cami, B., and Ma, T. 2023. Characteristic strength of a slope with spatial variability and cross-correlation. In *Advances in Theory and Innovation and Practice, Geo-Risk 2023*, Arlington VA. July 23–26, 2023.

Dithinde, M., Phoon, K.K., Ching, J.Y., Zhang, L.M., and Retief, J. 2016. Statistical characterization of model uncertainty. In *Reliability of Geotechnical Structures in ISO 2394*, pp. 127–158. Boca Raton, FL: CRC Press.

Dips [Computer Software]. (2023). Toronto, ON: Rocscience Inc.

El-Ramly, H., 2001. Probabilistic analyses of landslide hazards and risks: Bridging theory and practice. PhD dissertation, University of Alberta.

Fenton, G.A. and Griffiths, D.V. 2003. Bearing capacity prediction of spatially random c-φ soils. *Canadian Geotechnical Journal*, 40(1), 54–65.

Fenton, G.A. and Griffiths, D.V. 2005. Three-dimensional probabilistic foundation settlement. *Journal of Geotechnical and Geoenvironmental Engineering, ASCE*, 131(2), 232–239.

Fenton, G.A., Griffiths, D.V., and Williams, M.B. 2005. Reliability of traditional retaining wall design. *Géotechnique*, 55(1), 55–62.

Fisher, R. 1953. Dispersion on a sphere. *Proceedings of the Royal Society of London Series A*, 217(1130), 295–305.

Griffiths, D.V., Fenton, G.A., and Manoharan, N. 2006. Undrained bearing capacity of two-strip footings on spatially random soil. *International Journal of Geomechanics*, 6(6), 421–427.

Griffiths, D.V., Fenton, G.A., and Ziemann, H.R. 2008. Reliability of passive earth pressure. *Georisk: Assessment and Management of Risk for Engineered Systems and Geohazards*, 2(2), 113–121.

Hu, Y.G. and Ching, J. 2015. Impact of spatial variability in undrained shear strength on active lateral force in clay. *Structural Safety*, 52, 121–131.

Jaksa, M.B., Goldsworthy, J.S., Fenton, G.A., Kaggwa, W.S., Griffiths, D.V., Kuo, Y.L., and Poulos, H.G. 2005. Towards reliable and effective site investigations. *Géotechnique*, 55(2), 109–121.

Javankhoshdel, S., Luo, N., and Bathurst, R.J. 2017. Probabilistic analysis of simple slopes with cohesive soil strength using RLEM and RFEM. *Georisk: Assessment and Management of Risk for Engineered Systems and Geohazards*, 11(3), 231–246.

Javankhoshdel, S., Cami, B., Jamshidi Chenari, R., and Dastpak, P. 2020. Probabilistic analysis of slopes with linearly increasing undrained shear strength using RLEM approach. *Transportation Infrastructure Geotechnology*, 8, 114–141.

Javankhoshdel, S., Cami, B., Ma, T., Yacoub, T., and Chenari, R.J. 2021. Probabilistic slope stability analysis of a case study using random limit equilibrium method and surface altering optimization. *Proc., The Evolution of Geotech–25 Years of Innovation*, pp.413–418.

Javankhoshdel, S., Rezvani, M., Fatehi, M., and Jamshidi Chenari, R. 2022a. RLEM versus RFEM in stochastic slope stability analyses in geomechanics. Geo-Congress 2022.

Javankhoshdel, S., Zeger, T., Cami, B., Wahanik, H., and Ma, T. 2022b. A new stochastic response surface method in spatial variability slope stability analysis. In *TVSeminars and Mining One International Conference*, TMIC 2022 (online). December 10–13, 2022.

Kennedy, J. and Eberhart, R. 1995. Particle swarm optimization. In *Proceedings of ICNN'95 – International Conference on Neural Networks*, Perth, WA. November 27–December 1, 1995.

Kingma, D.P. and Welling, M. 2014. Auto-encoding variational Bayes. In *2nd International Conference on Learning Representations*, ICLR 2014, Banff, AB. April 14–16, 2014.

Ma, T., Mafi, R., Cami, B., Javankhoshdel, S., and Gandomi, A.H. 2022. NURBS surface-altering optimization for identifying critical slip surfaces in 3D slopes. *International Journal of Geomechanics*, 22(9), 04022154.

Mafi, R., Javankhoshdel, S., Cami, B., Jamshidi, C.R., and Gandomi, A.H. 2020. Surface altering optimisation in slope stability analysis with non-circular failure for random limit equilibrium method. *Georisk: Assessment and Management of Risk for Engineered Systems and Geohazards*, 15(4), 1–27.

MacRobert, C.J., Mutede, T., and de Koker, N. 2021. Can 2D cross sections be safely extrapolated? In *The Evolution of Geotech –25 Years of Innovation*, RIC 2021, pp. 650–656. Toronto, ON. April 20–21, 2021.

Pan, Y., Liu, Y., Xiao, H., Lee, F.H., and Phoon, K.K. 2018. Effect of spatial variability on short-and long-term behaviour of axially-loaded cement-admixed marine clay column. *Computers and Geotechnics*, 94, 150–168.

Phoon, K.K. and Tang, C. 2019. Characterization of geotechnical model uncertainty. *Georisk: Assessment and Management of Risk for Engineered Systems and Geohazards*, 13(2), 101–130.

Phoon, K.K., Cao, Z., Ji, J., Leung, Y.F., Najjar, S., Shuku, T., Tang, C., Yin, Z.Y., Ikumasa, Y., and Ching, J. 2022. Geotechnical uncertainty, modeling, and decision making. *Soils and Foundations*, 62(5), 101189.

Rocscience 2022. *RocFall2 v.8.020. 2D Statistical rockfall analysis software*. Toronto, On: Canada.

Rocscience 2023d. *RocFall3 v.1.007. 3D Statistical rockfall analysis software*. Toronto, On: Canada.

Rocscience 2023a. *Slide2. 2D Limit Equilibrium Software*. Toronto, On: Canada.

Rocscience 2023b. *RS2. 2D Finite Element Method Software*. Toronto, On: Canada.

Rocscience 2023c. *Slide3 3D Limit Equilibrium Software*. Toronto, On: Canada.

Rocscience 2023d. *Rocfall 2D Rock fall analysis software*. Toronto, On: Canada.

Rocscience 2023e. *Rocfall3 3D Rock fall analysis software*. Toronto, On: Canada.

Rocscience 2023f. *Dips v. Graphical and Statistical Analysis of Orientation Data software*. Toronto, On: Canada.

Rocscience 2023g. *Settle3. Soil Settlement and Consolidation Analysis software*. Toronto, On: Canada.

Stuedlein, A.W. and Bong, T. 2017. Effect of spatial variability on static and liquefaction-induced differential settlements. In *Georisk 2017*, 31–51. Denver, CO: ASCE.

Travis, Q., Schmeeckle, M., and Sebert, D. 2011. Meta-analysis of 301 slope failure calculations. II: Database analysis. *Journal of Geotechnical and Geoenvironmental Engineering*, 137(5), 471–482.

Vessia, G., Cherubini, C., Pieczyńska, J., and Puła, W. 2009. Application of random finite element method to bearing capacity design of strip footing. *Journal of Geology Engineering*, 4, 103–111.

Wu, S.H., Ou, C.Y., and Ching, J.Y. 2014. Calibration of model uncertainties in base heave stability for wide excavations in clay. *Soils and Foundations*, 54(6), 1159–1174.

12 Reliability-based Decision-making with FE Models for Real-life Case Studies

Commend Stéphane, Minini Jocelyn, Groslambert Marc, and Jacot-Descombes Gil

ABSTRACT

In this chapter, probabilistic analyses are applied to non-linear 3D FE models for two real-life case studies: (1) a deep excavation retained by a braced slurry wall and (2) the construction of a tunnel under a sensitive building. The use of accurate meta-models allows sensitivity and reliability analyses to be performed in a reasonable time frame, in parallel with construction works if needed. It is shown that these FE computations have helped design engineers to make the following decisions: (1) eliminating a fourth level of struts for the braced excavation thanks to a live Bayesian analysis; (2) defining the optimal excavation steps for a tunnel excavation beneath a sensitive building, with the help of a reliability analysis considering two different excavation procedures.

12.1 INTRODUCTION

Since the early 2000s, 2D and 3D non-linear finite-element (FE) analyses have become a standard in the design of construction projects involving complex soil-structure interaction, especially when addressing sensitive structures in the vicinity of the project.

The coupling between large FE models and probabilistic approaches is still uncommon in the literature, mainly because of the computational burden. With the evolution of computer power and the use of adequate meta-models approximating the behavior of numerical models, it is now possible to perform FE reliability analyses within a reasonable time frame, which enables engineers to improve their decision-making during the design and execution of construction projects involving complex soil-structure interaction.

In this chapter, we present two real-life problems, addressed with non-linear 3D FE computations: first, a 3D slice representative of an 18-m-deep excavation retained by a braced slurry wall in the Geneva (Switzerland) region. In this case, a Bayesian approach is used during the excavation works to determine whether a fourth level of bracing is necessary to stay within acceptable displacement thresholds. In a second case, a full 3D model of a tunnel excavation for an underground metro line under construction in Lausanne (Switzerland) is presented, where an a priori analysis enables the optimization of the distance between the steel profiles supporting the tunnel excavation. Both the Bayesian approach on the 18-m-deep excavation and the optimization of the distance of the steel profiles are performed with the help of external MATLAB-based software. The software used in this study is UQLab, developed by the Chair of Risk, Safety and Uncertainty Quantification (UQ) at ETH Zurich under the supervision of Prof. B. Sudret and Dr. S. Marelli [1]. UQLab allows the design and development of complex UQ analyses without requiring extensive technical knowledge from the users [1]. UQCloud is also used in this study, which is an OS- and programming language-independent (independent from MATLAB), cloud-based version of UQLab [2].

DOI: 10.1201/9781003333586-15

12.2 COUPLING BETWEEN FE MODELS AND UNCERTAINTY FRAMEWORKS

The development of the finite-element method and its application to civil engineering took place between 1960 and 1970. The first books appearing at this time were Zienkiewicz's (1967), updated several times since [3], and, remarkably, at the time when most contributors confined themselves to linear analysis, Oden's (1971), which anticipated the main ideas that were to be developed over the next 20 years for non-linear analysis [4].

In the early 1980s, the combination of the FEM (finite-element method), plasticity theories applied to soils, appropriate algorithmic techniques, and the appearance of personal computers led to the migration of software that had previously been confined to supercomputers, to PCs, and to the appearance of software dedicated to geomechanics, including ZSoil [5]. ZSoil offers today a unified approach to static and dynamic numerical simulations of soil, rock, and structural mechanics, including excavation and construction stages, soil-structure interaction, underground flow, and heat transfer [6].

Among the different constitutive models present in ZSoil, the Mohr-Coulomb model and the hardening soil model with small strain extension (or HSS) [7] are used in the subsequent analyses, the latter because of its ability to integrate stress-dependent stiffness, as well as a distinction between loading and unloading/reloading moduli.

In parallel, the foundations of probabilistic methods applied to civil engineering problems were also laid in the early 1970s by the works of Benjamin and Cornell [8], Tang and Ang [9], and Vanmarcke [10] which paved the way for a consistent consideration of uncertainties, with the conceptualization of risk and reliability analyses, mainly applied to analytical models. Among remarkable works, Baecher and Christian [11] formalized a reference book on probabilistic methods applied to geomechanical problems.

In 2015, Phoon and Ching edited "Risk and Reliability in Geotechnical Engineering" [12]. Among the notable contributions to this book, two are particularly relevant to this chapter: Sudret proposed the use of chaos polynomials for solving sensitivity or reliability analyses using computationally efficient meta-models, and Straub and Papaioannou defined a clear methodology for updating geotechnical parameters on the basis of in-situ measurements using Bayesian inference.

In the subsequent analyses, uncertainty quantification, sensitivity, reliability, and Bayesian analyses are performed by coupling ZSoil with UQLab, a well-documented probabilistic toolbox [1], and with UQCloud [2].

12.3 IMPORTANCE OF META-MODELING IN PROBABILISTIC ANALYSIS

As explained above, numerical models have become increasingly robust in the last two decades. Nevertheless, this gain in performance is hindered by a major challenge: long computational times. Even though this can be solved by ever-increasing computational power, the solution of a problem solved by the FE method takes longer to obtain compared to analytical methods with a closed-form solution. In the 1990s, Ghanem and Spanos [13] proposed a new approach to approximate the discrete responses of an FE model using a multivariate polynomial series in the input variables. Meanwhile, several authors have proposed techniques to obtain the FE model response in a closed-form [14–16]. These approximation models are also commonly called surrogate models or meta-models. They allow a large number of model evaluations to be run in a reasonable amount of time. Optimization problems, variance analysis, or probabilistic analysis can therefore be performed.

12.3.1 Global Model Behavior – A Sensitivity Indicator

Understanding the global behavior of an FE model is an essential step in numerical modeling. A thorough understanding serves, in particular, to better interpret the results and better understand the model's limits. One of the key tasks the engineer can perform to better his understanding is a parametric study. This technique shows the influence of the variables one-by-one on the model's response, however it does not give any information on the impact of a combination of parameters.

To overcome this limitation, the following paragraph explains why a combination of polynomial regression and sensitivity analysis techniques seems particularly appropriate.

Several studies have shown that polynomial chaos expansions (PCEs) approximate the global behavior of a numerical model well [17–19]. At the stage of a sensitivity analysis, an approximation by PCE is often used to allow the identification of impacting variables [20–23]. Thus, once the model has been approximated by the meta-model, many general sampling-based methods can be performed. Out of these methods, Sobol' (2001) [24] proposed a technique to obtain sensitivity indices based on variance decomposition for independent input variables in 1993. In 2012, Kucherenko (2012) [25] generalized this method to the case of dependent variables. Both methods are general and can be used by any type of meta-model, allowing rapid sampling of the model response, typically by Monte Carlo simulation (MCS). However, in the particular case of an approximation of the original model by PCE, Sudret (2008) [26] proposed a sampling-free method by post-processing the PCE to obtain the Sobol' indices directly from the coefficients of the polynomial basis. In this context, Sobol' indices are given by a closed-form solution, and therefore, no convergence calculation needs to be performed, making the use of the PCE both elegant and efficient.

In its general formulation, the PCE model is expressed as follows:

$$Y \approx \widehat{Y}_{\text{PCE}} = \widehat{\mathcal{M}_{\text{PCE}}}(X) = \sum_{\alpha \in \mathcal{A}} y_\alpha \Psi_\alpha(X) \tag{12.1}$$

12.3.2 Local Model Behavior – A Model Twin

Although regression methods are effective for approximating the overall behavior of a model's response, they may be less effective when an engineer is in need of a precise approximation. In this context, interpolation methods are more suitable and allow the local behavior of the original model to be taken into account. In the 1950s, Krige developed an interpolation technique that was initially used in geostatistics and later applied to the approximation of numerical models by Sacks et al. (1989) [27]. In parallel with regression methods, several authors have proposed improvements to the kriging model over the last two decades. In 2015, by introducing the polynomial chaos Kriging (PCK), Schöbi (2015) [28] proposed a union of the two methods. This new type of meta-model allows the coupling of the global and local behavior of the original model, resulting in an accurate and robust approximation. To the authors' knowledge, PCK provides a reliable approximation for a wide range of geotechnical problems. In this study, the authors advise the use of PCE for sensitivity analysis and the use of PCK for any probabilistic analysis that requires additional precision, such as reliability-based design and Bayesian analysis.

To go further in detail, the global and local behavior is captured by combining a set of orthonormal polynomials $\Psi_\alpha(X)$, which describes the trend of the kriging, and the zero-mean, unit-variance stationary Gaussian process $\sigma^2 Z(X,\omega)$. The formulation of the PCK is then obtained by

$$Y \approx \widehat{Y}_{\text{PCK}} = \widehat{\mathcal{M}_{\text{PCK}}}(X) = \sum_{\alpha \in \mathcal{A}} y_\alpha \Psi_\alpha(X) + \sigma^2 Z(X,\omega) \tag{12.2}$$

With reference to Eq. (12.2), the process of constructing a PCK model involves two fundamental steps: (1) identifying an optimal set of polynomials and (2) determining the kriging parameters. Various combinations of these two steps are available. Two different approaches are implemented in UQLab: sequential PCK and optimal PCK. In sequential PCK, the polynomial set and the kriging are determined sequentially. The multivariate polynomial is then used directly as the trend of the kriging analysis. Its construction follows a process similar to that of standard PCE. In contrast, optimal PCK involves the iterative construction of the PCK model. The objective is to minimize the leave-one-out (LOO) error. This approach guarantees the optimal solution, but it requires a significant amount of computational time to determine the error-minimizing PCK. Especially in cases where the original model is highly non-linear and has many output variables, the computational time required by this approach can be challenging.

12.3.3 Verifying a Surrogate Model

The error estimation of a meta-model is essential for engineers to ensure that their simulations are reliable. This section introduces two methods for this purpose: LOO validation and its graphical variant. On the one hand, the LOO method involves training the meta-model using all but one data point and then numerically testing the accuracy of the model by predicting the omitted point. On the other hand, the graphical variant allows locating the regions of the output domain where the meta-model struggles to accurately approximate the original model response.

12.3.3.1 LOO Error

The LOO approach involves comparing the responses of the original model, built on the full experimental design (ED) $\mathcal{X}$, with those of a meta-model constructed using a reduced ED obtained by excluding a single realization $\mathcal{X}^{(-i)} = \mathcal{X} \setminus \mathcal{X}^{(i)}$. In theory, the computational complexity of calculating the LOO error scales with the size of the ED, which is denoted by $N = \text{card}(\mathcal{X})$. When using polynomial chaos expansion (PCE) or polynomial chaos kriging (PCK) for approximation, the LOO error can be computed analytically without the need to construct N separate models [28–31]. As PCK is a variant of kriging that uses a polynomial trend, the LOO formulation for kriging holds. The LOO error for PCE is given by Equation 3 and that for kriging (and PCK) is given by Equation 4.

$$err_{LOO}^{(\text{PCE})} = \frac{Err_{LOO}^{(\text{PCE})}}{Var[\mathcal{Y}]} = \frac{\sum_{i=1}^{N}\left[\mathcal{Y}^{(i)} - \mathcal{M}_{(-i)}^{(\text{PCE})}\left(\chi^{(i)}\right)\right]^2}{\sum_{i=1}^{N}\left[\mathcal{Y}^{(i)} - \hat{\mu}_{\mathcal{Y}}\right]^2}, \tag{12.3}$$

$$\text{err}_{LOO}^{(\text{K or PCK})} = \frac{Err_{LOO}^{(\text{K or PCK})}}{Var[\mathcal{Y}]} = \frac{\sum_{i=1}^{N}\left[\mathcal{Y}^{(i)} - \mu_{\hat{Y}(-i)}\left(\chi^{(i)}\right)\right]^2}{\sum_{i=1}^{N}\left[\mathcal{Y}^{(i)} - \hat{\mu}_{\mathcal{Y}}\right]^2}, \tag{12.4}$$

where $\hat{\mu}_{\mathcal{Y}} = \frac{1}{N}\sum_{i=1}^{N} \mathcal{M}(\mathcal{X}^{(i)})$ is the estimated mean value of Y and $\mu_{\hat{Y}(-i)}(\mathcal{X}^{(i)})$ is the prediction mean at sample $\mathcal{X}^{(i)}$ by a Kriging meta-model based on the ED $\mathcal{X}^{(-i)}$.

12.3.3.2 Graphical Leave-One-Out Error (GLOO)

Although the LOO technique is a robust approach for estimating the global error of a meta-model, it does not provide any information on areas of the output domain where the approximation model has difficulty fitting the original model response. To address this issue, a graphical variant of LOO validation can be employed, which is used only for qualitative purposes. The procedure is introduced in Figure 12.1. The approach involves several steps, including constructing the initial ED from the original numerical model, dropping one sample from the ED at step i, creating the meta-models from the reduced ED, evaluating the remaining point not used to create the meta-models, and storing the original points and approximations in a matrix of varying size for each i iteration. The boundary of the $\mathcal{H}$ data set is then constructed using an AlphaShape with a maximum shrink factor, and the results are plotted to identify areas where the domain boundary exceeds the set tolerance. Because creating N meta-models can be computationally intensive, a sequential PCK may be sufficient as an initial estimation. Despite this limitation, the method offers a conservative estimate in two ways. First, the AlphaShape technique generates error domains that are larger than the exact error domain. Second, the meta-model based on the full ED guarantees higher accuracy than that obtained with smaller EDs. Thus, this method can be valuable for identifying areas in the output domain where the meta-model is more likely to have errors.

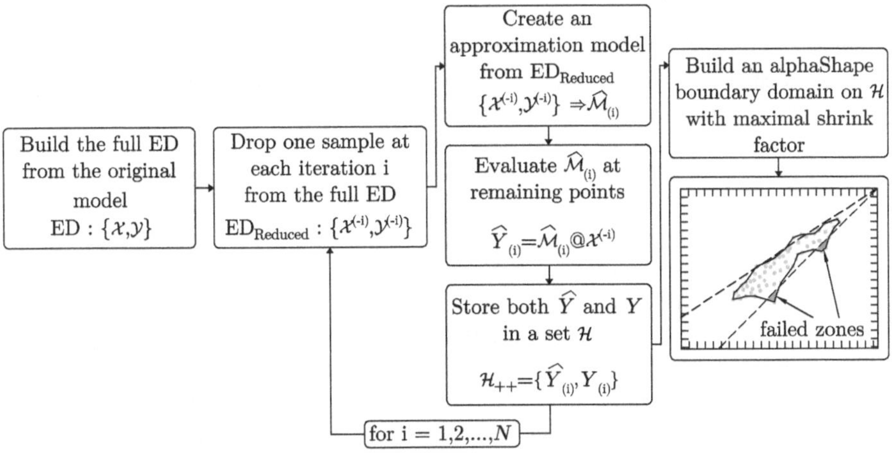

FIGURE 12.1 Flow chart of the GLOO method.

12.4 PREDICTION OF DIAPHRAGM WALL DISPLACEMENTS USING LIVE BAYESIAN MODEL CALIBRATION

To expand its business activities, a banking company has decided to build a cutting-edge edifice in Geneva, Switzerland. The project is set to include four new buildings, including a 90-m-high tower. This development will provide accommodation for over 2,500 workplaces and almost 100 apartments. Completion is scheduled for 2025 [32].

12.4.1 Case Study Description

As shown in Figure 12.2, a deep excavation is required to provide space for the four future basement levels. The 18-m-deep excavation was carried out using a 1 m thick reinforced concrete

FIGURE 12.2 Vertical cross-section of the northern part of the building [33].

"trouser-leg" diaphragm wall. The horizontal stabilization of the wall was achieved by means of three to four levels of circular ROR-type struts. The uncertainty regarding the required number of strut levels to respect displacement limits is the focus of this study. Conventional deterministic partial factor-based FE analyses showed that the three strut levels variant was on the limit of the acceptable thresholds, making it necessary to consider a fourth layer of struts. To take into account the uncertainty of the geomechanical and structural parameters, a probabilistic study including a live update by inverse Bayesian analysis was proposed to the contractor. The probabilistic simulations aimed to show that the three strut level solution would be sufficient to ensure that the ultimate limit state and serviceability thresholds would not be exceeded.

The excavation stages are described in Figure 12.3. A full probabilistic analysis is presented that involves stages S26, S33, and S43 for the Bayesian analyses and S50 for the reliability analysis. Located behind the upper part of the wall, a sensitive water conduit runs along the edge of the excavation (Figure 12.3a), and only small deformations are tolerated in this zone. The wall is built with C30/37 concrete and S500B reinforcing steel with a corresponding $M_{Rd} = 4'100$ kNm/m. The mechanical properties of the horizontal struts are given in Table 12.1.

12.4.2 Geotechnical and Hydrological Conditions

The soils of Geneva are mainly composed of soft and compressible clayey silt deposits sitting on a Wurmian moraine. The interaction between the soil conditions and the water table located three meters below the ground surface leads to distinctive hydromechanical conditions. Commend et al. (2004) [34] showed that these soils exhibit an original behavior, namely, a short-term undrained behavior with a drop in pore water pressure during soil relaxation. A nearby excavation project

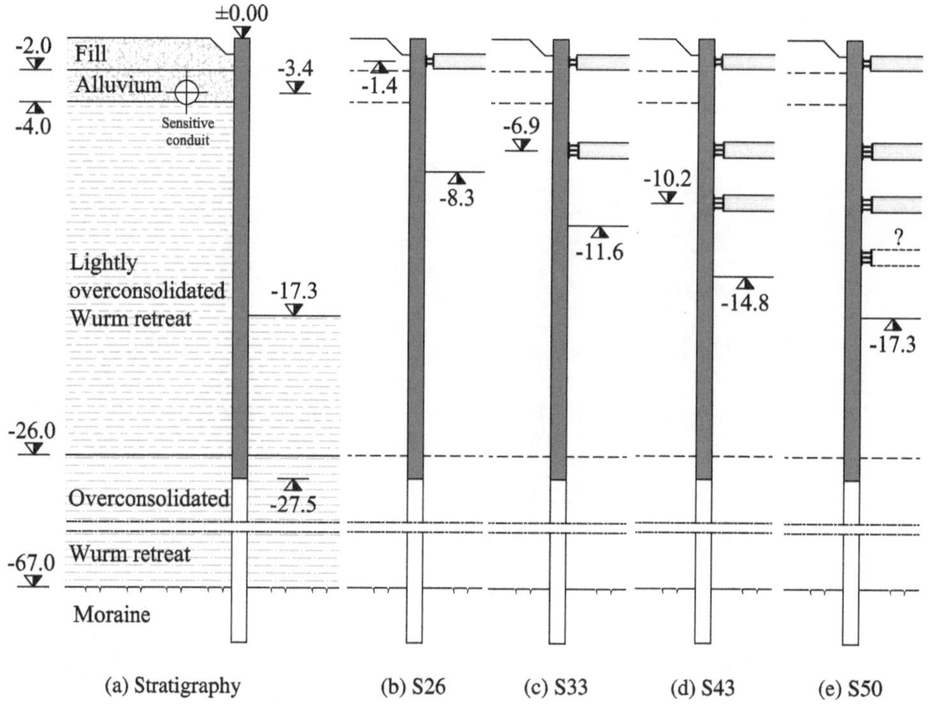

FIGURE 12.3 (a) Diaphragm wall at its final stage with its corresponding stratigraphy. (b–e) Excavation stages with ROR struts. The change in colour of the wall represents the boundary between the continuous and "trouser-legs" parts.

TABLE 12.1
Mechanical Properties of the ROR-Type Struts Used for Horizontal Stabilization

N° and level (m)	Profile type	Steel class	Thickness (mm)	Axial Stiffness (GN)
1) -1.4	ROR ϕ914	S355	17.8	10.5
2) -6.9	ROR ϕ1016	S460	20.0	13.1
3) -10.2	ROR ϕ1016	S355	20.0	13.1

conducted over a year recorded a pressure corresponding to 30% of the reference hydrostatic situation after soil relaxation. A wide range of laboratory and in-situ geotechnical tests performed by De Cérenville Géotechnique SA [35] as well as expert knowledge of the local geotechnical conditions were used to define the hardening soil model parameters listed in Table 12.2.

12.4.3 Numerical Models

According to the considerations made in Section 12.3, this section focuses on the definition of the models used in the current study.

12.4.3.1 Underground Flow and Two-Phase Media

As explained in Section 12.4.2, the hydraulic conditions of the Geneva soils show a strong undrained behavior and make it particularly challenging to create accurate models using simplified approaches. During deterministic analyses, questions arose about the hydromechanical coupling of the model, given the already-known results of Commend et al. (2004) [34]. From a technical and numerical standpoint, it is feasible to apply probabilistic analysis to a FE model including consolidation. However, for this project, it is crucial to consider practical factors, such as :

TABLE 12.2
Hardening Soil Model Mean Parameters

	Specific weight - permeability			Stiffness[a]		Strength		
Formation	γ (kNm^{-3})	k_h (ms^{-1})	k_v (ms^{-1})	$E_{50} = E_{oed}$ (MPa)	E_{ur} (MPa)	φ (°)	ψ (°)	c (kPa)
Fill	20	10^{-4}	10^{-5}	15	60	33	3	1
Alluvium	22	10^{-4}	10^{-5}	50	150	40	10	0
FCR[b]	20	10^{-7}	10^{-9}	7	32	30	0	7
WCR[b]	21	10^{-7}	10^{-9}	18	54	31	1	17
Moraine	22	10^{-7}	10^{-8}	50	150	35.5	5.5	21

Note:
[a] Corresponding stiffness exponent $m = 0.5$ for each soil layer.
[b] (LCR) Lightly Consolidated resp. (OR) Overconsolidated Wurmian Retreat.

(1) If an "as-built" calculation strategy is planned, the engineer must ensure that there is sufficient time to update the FE models and generate the ED as the project progresses. Quick reactions will be difficult to implement with time-consuming calculations.
(2) In the case of a pre-calculation strategy for meta-models, excellent coordination between the company, the surveyor, and the engineer should be planned. If delays or advancements occur during excavation, the consolidation models may no longer be valid, and a new sampling phase would be needed to generate a new ED.

At the beginning of the project, such requirements cannot be ensured. For this reason, a steady-state model is selected for conducting the probabilistic analysis to allow the pre-computation of the meta-models before the beginning of the work. Therefore, the resulting physical model $\mathcal{M}$ is described by the system of differential Eq. (12.5). To improve the model's accuracy and replicate hydrological conditions more closely, a random variable is introduced via a multiplier of the water density noted κ_w (a full probabilistic description of this variable is presented in Section 12.4.4).

$$\mathcal{M}: \begin{cases} (\sigma_{ij} + S\,P\,\delta_{ij})_{,j} + f_i & = 0 \\ v^F_{k,k} - c\dot{p} & = 0 \\ v_i^F = -k_{ij}\,k_r(S)\left(-\dfrac{p}{\gamma_F} + H\right)_{,j} & \end{cases} \qquad \text{Eq. (12.5)}$$

12.4.3.2 FE Model

To ensure accurate simulations, a 3D model based on the hardening soil model with small strain extension [36] is used to account for any out-of-plane effects due to the "trouser-leg" diaphragm wall's discontinuity. The model is created by extruding the 2D mesh represented in Figure 12.4, resulting in a 4.32-m-wide 3D slice. The chosen linearly interpolated B8 FEs are linked by mesh tying to ensure continuity between incompatible meshes, as shown in Figure 12.4. The approximation of the physical model $\mathcal{M}$ is considered a "black box" model $\mathcal{M}_{\text{ZSoil}}$ as an unknown map from the space of input parameters to that of output quantities:

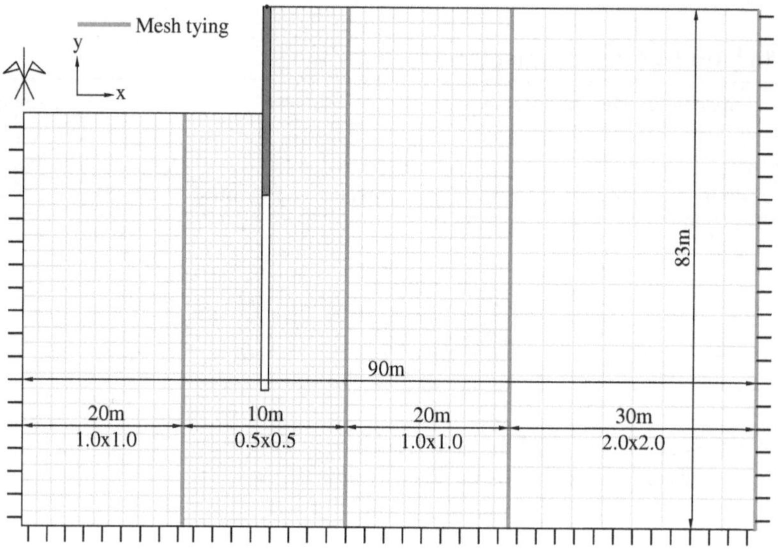

FIGURE 12.4 Mesh strategy of the ZSoil model.

$$Y = \mathcal{M}_{\text{ZSoil}}(X) \qquad \text{Eq. (12.6)}$$

The choice of output quantities, also called quantities of interest (QoIs), is carefully selected based on the specific needs of the project. They are generally determined according to the reliability analyses that are to be conducted. The first output quantity is the horizontal deflection $U_x(y)$, more specifically, the horizontal displacement of the nodes of the shell elements constituting the diaphragm wall. This enabled Bayesian analyses to be performed on multiple measurement values. The maximal deflection $U_{x,\text{Max}}$ is also considered. The second output quantity, maximal bending moment M_{Max}, is necessary to ensure structural safety for ultimate limit state aspects. The third output quantity, the settlement under the sensitive water conduit $U_{y,\text{Pipe}}$, is of great importance to the Geneva sewerage authorities to ensure that the construction of the project does not cause irreversible damage to it.

12.4.3.3 Approximation Model

Regarding the strategy proposed in Section 12.3, the sensitivity analysis is conducted by an approximation model based on a PCE, while the reliability analysis and Bayesian inversions are performed by a PCK model approximation. The separation of the two families of surrogate models requires the creation of two distinct EDs. Each of them is generated stochastically with respect to the statistical distributions of the input variables. Using a stationary model makes it possible to use a pre-calculation strategy. This approach results in a reduction in computation time when processing on-site measurements and conducting live inverse Bayesian analyses. As each sample point is regarded as an independent event, parallel computations can be performed. The simulations are run on a local server allowing 40 parallel calculations, which represent a calculation time of approximately 6.5 hours for generating 100 sample points. PCE and PCK approximation responses are given by Eq. (12.7).

$$Y \approx \hat{Y}_{\text{PCE}} = \widehat{\mathcal{M}}_{\text{PCE}}(X) \quad \text{or} \quad Y \approx \hat{Y}_{\text{PCK}} = \widehat{\mathcal{M}}_{\text{PCK}}(X) \qquad \text{Eq. (12.7)}$$

An ED of 200 sample points is used to construct the PCE approximations, used primarily to identify the global behavior of the model and to allow for sensitivity analysis, as suggested in Section 12.3. In this context, the LOO error values given in Table 12.3 have been accepted. The PCK models are based on an ED of 350 sample points. Before considering the LOO values given in Table 12.3, the graphical evaluation method proposed in Section 12.3 is performed at stage 33 (the worst stage in terms of approximation error). This allows us to identify a point resulting from a combination of unfavorable input variables that had a negative impact on the quality of the PCK results. This sample is marked by a circle in Figure 12.5. Keeping this sample in the ED, the LOO error values at stage 33 correspond to those shown in Table 12.3 with index S33*. As shown by the new LOO values at index S33, neglecting this point allows more accurate PCK approximations to be obtained. The other final LOO errors are listed in the same table.

TABLE 12.3
LOO errors $\times 10^{-4}$ for the QoIs

	$\max(U_x(y))$		$U_{x,\text{Max}}$		$U_{y,\text{Pipe}}$		M_{Max}	
Stage	**PCE**	**PCK**	**PCE**	**PCK**	**PCE**	**PCK**	**PCE**	**PCK**
S26	247	60	67	6.8	86	4.0	9.9	2.3
S33*		68		10		4.3		23
S33	334	61	64	8.9	115	4.0	22	16
S43	331	58	60	5.9	132	3.0	24	4.4
S50	244	38	44	1.2	91	2.2	6.9	0.5

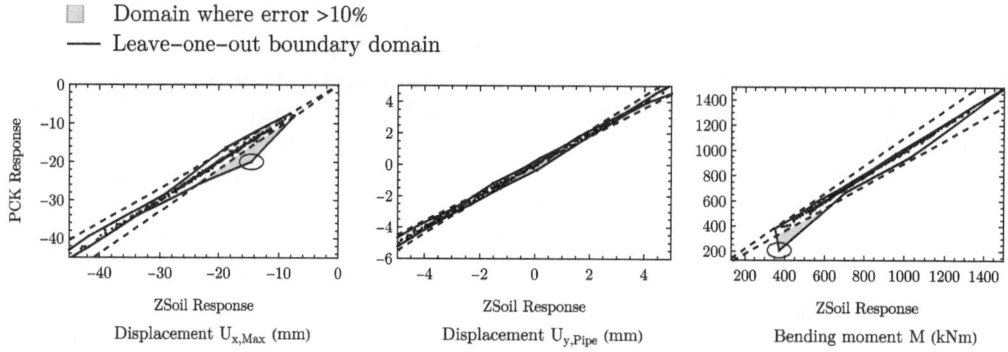

FIGURE 12.5 Graphical LOO cross validation at stage 33.

12.4.4 Quantification of Uncertainties

Statistical characterization of the soil parameters is performed considering that a set of variables can be described by a random vector with a joint probability density function $\mathbf{X} \sim f_{\mathbf{X}}$. The means of the distributions are determined according to the values presented in Table 12.2, while the coefficients of variation and the types of distribution are based on the following references [11, 12, 37, 38]. In addition to the usual variables of a probabilistic analysis (φ, c, E, water table, etc.), two specific parameters are identified as uncertain and therefore treated as random variables. The first is the relationship between the unloading/reloading modulus E_{ur} and the secant modulus E_{50}, which is introduced by a uniformly distributed scalar between 2 and 7. Such values are commonly encountered [39]. As a result, the variable E_{ur} can only take values contained in the specific bin shown in Figure 12.6. The second is the water density multiplier κ_w. As mentioned in Section 12.4.2, this parameter acts as a calibration variable on the water density ($\gamma_w = g\kappa_w$) to account for the effects of pore pressure drop during soil relaxation. Based on experience with the in-situ soils, this parameter is assumed to be uniformly distributed between 35% and 85%. In the first two soil layers, only the stiffness parameters (E_{50} and Γ_{ur}) are statistically considered. In the other layers, two additional random variables are included, the friction angle and the cohesion (φ, c, E_{50} and Γ_{ur}). In addition to these 16 parameters, the height of the water table and κ_w are also taken into account. The result is an initial input consisting of 18 random variables whose characteristics are listed in Table 12.4.

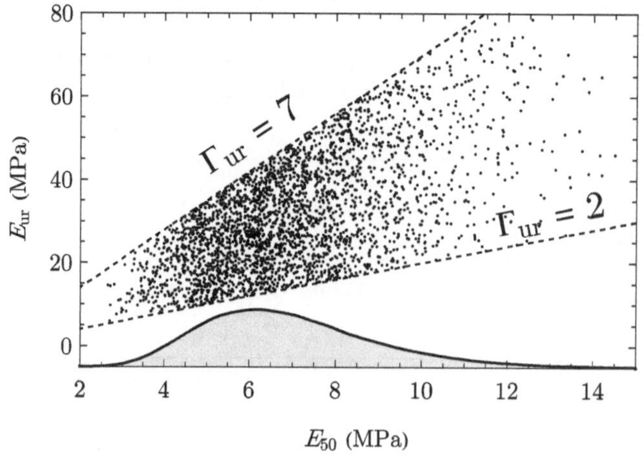

FIGURE 12.6 Relationship between E_{50} and E_{ur} through Γ_{ur}. Dashed lines represent the boundaries of the uniform distribution. The black curve filled in gray represents the lognormal distribution of variable E_{50}.

TABLE 12.4
Marginal Initial Distributions of Random Variables

Parameter	Distribution	f_X	Parameters or CV
Friction angle	Lognormal	$X \sim \mathcal{LN}$	CV = 10%
Cohesion	Lognormal	$X \sim \mathcal{LN}$	CV = 20%
Young's modulus E_{50}	Lognormal	$X \sim \mathcal{LN}$	CV = 30%
Γ_{ur}	Uniform	$X \sim \mathcal{U}$	(2,7)
Water table depth	Gaussian	$X \sim \mathcal{N}$	(3.6,0.05)
Water density multiplier κ_w	Uniform	$X \sim \mathcal{U}$	(0.35,0.85)

Note: No dependency between the variables is considered.

12.4.5 Sensitivity Analysis

Instead of evaluating the sensitivity indices by a sampling-based method, the sensitivity indices are calculated directly based on the post-processing of the PCE proposed by Sudret (2008). The main goal of the sensitivity analysis performed in this study is to reduce the input of 18 variables to gain accuracy and computation time. However, this reduction is allowed only if it is proven that all QoIs are indeed impacted by the same variables. Moreover, as the geomechanical behavior changes with the excavation stage, it must be guaranteed that the impacting variables remain the same throughout the analysis. In summary, the sensitivity analysis is dependent on the excavation stage, the type of QoI, and, in the case of the horizontal wall displacement, also on the depth. Figure 12.7 presents multiple illustrations allowing us to represent the Sobol' indices of the horizontal wall displacement as a function of the depth for the four excavation stages analyzed. Note that similar results to Figure 12.7 are obtained for $U_{x,\max}$, $U_{y,\text{Pipe}}$, and M_{Max}. Regardless of the selected depth, the variables affecting the model response are **X5**: the friction angle of the LC-Wurm retreat; **X7** and **X8**: Young's modulus resp. Γ_{ur} of the LC-Wurm retreat; **X11** and **X12**: Young's modulus resp. Γ_{ur} of the OR-Wurm retreat; and **X18**: water density multiplier. These remaining six parameters are kept as probabilistic variables, while the others are set to their mean value. The remaining six random variables are considered as the prior.

12.4.6 Inverse Bayesian Analysis

Bayesian inversions are applied to excavation steps S26, S33, and S43. Monitoring of the horizontal wall displacements corresponding to these steps is obtained by inclinometer and theodolite measurements. To link the predictions with real in-situ measurements, Equation 12.7 must be extended by an additive discrepancy term ε. Assuming that this discrepancy term is normally distributed with a zero-mean value, the approximation of the real displacements reads

$$Z \approx Y + \varepsilon \approxeq \widehat{\mathcal{M}}_{\text{PCK}}(X) + \varepsilon \quad \text{where } \varepsilon \sim \mathcal{N}(0, \Sigma) \qquad \text{Eq. (12.8)}$$

Considering the additive discrepancy given in Eq. (12.8) and according to Bayes' theorem, the calculation of the distribution of the input *given* one measurement z_i is obtained by

$$f_X(X \mid Z) = \frac{f_X(X)}{\int_{-\infty}^{\infty} \mathcal{L}(x) f_X(x)\,dx} \cdot \prod_{i=1}^{N} \mathcal{N}(z_i \mid \widehat{\mathcal{M}}_{\text{PCK}}(X), \Sigma) \qquad \text{Eq. (12.9)}$$

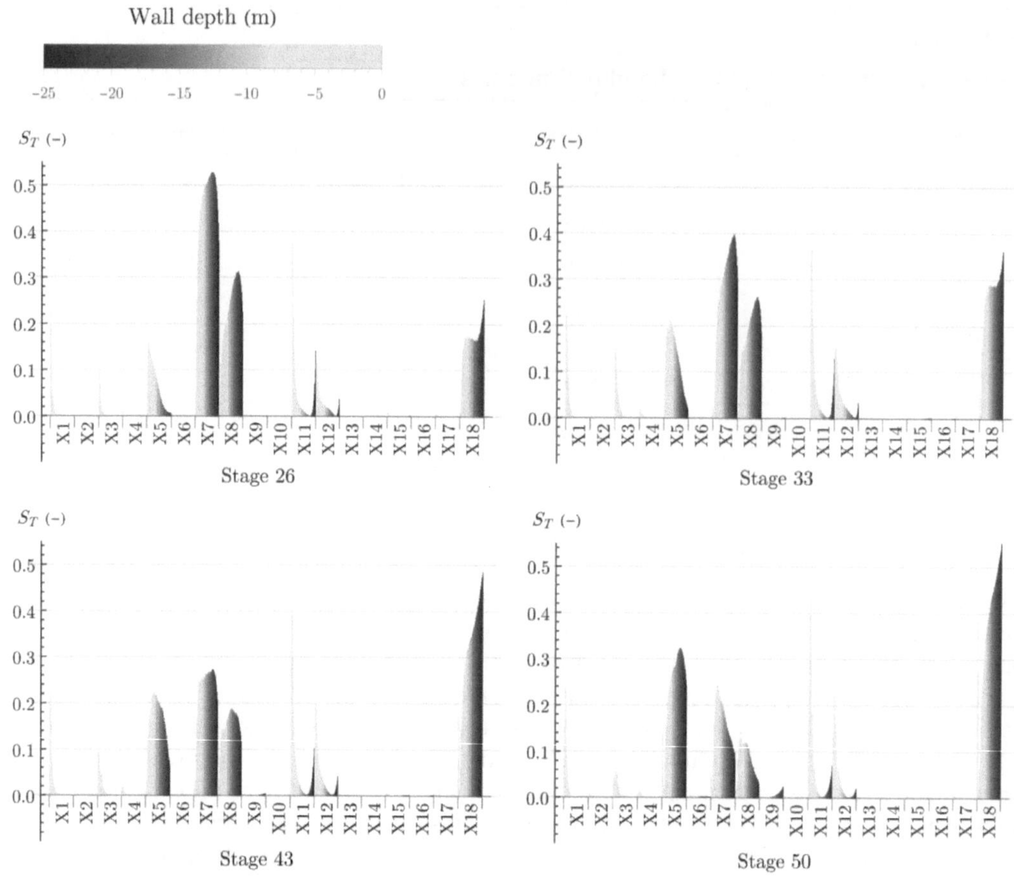

FIGURE 12.7 Total Sobol' indices S_T for output quantity $U_x(y)$.

Referring to Eq. (12.8), the discrepancy is separated into three independent terms: the measurement error $\varepsilon_{\mathcal{Z}}$, the original model error $\varepsilon_{\mathcal{M}}$, and the meta-model error $\varepsilon_{\widehat{\mathcal{M}}}$:

$$\varepsilon = \varepsilon_{\mathcal{Z}} + \varepsilon_{\mathcal{M}} + \varepsilon_{\widehat{\mathcal{M}}} \qquad \text{Eq. (12.10)}$$

$\varepsilon_{\mathcal{Z}}$ quantifies the accuracy obtained from the inclinometer measurements and the error generated by smoothing the noisy measurement data (Figure 12.8) by curve fitting. Considering a confidence interval of 5% resp. 95% on the accuracy of the measurements (given as ± 1 mm), the resulting standard deviation yields $\sigma_{\varepsilon_{\mathcal{Z}}} = 1.0$.

$\varepsilon_{\mathcal{M}}$ is selected based on personal judgment. The error is correlated to the magnitude of the measured settlements with a coefficient of variation $CV_{\varepsilon_{\mathcal{M}}} = 0.20 \rightarrow \sigma_{\varepsilon_{\mathcal{M}}} = CV_{\varepsilon_{\mathcal{M}}} \cdot Z$. This covers the error directly related to the FE model (mesh, constitutive law, etc.).

$\varepsilon_{\widehat{\mathcal{M}}}$ is considered to be zero (low LOO, see Table 12.3).

Eq. (12.9) is solved numerically using the Markov chain Monte Carlo (MCMC) algorithm implemented in UQLab [1]. The sampling method used to construct the Markov chains is provided by the adaptive Metropolis algorithm (AM) [40, 41]. To ensure the convergence of MCMC algorithms, Gelman and Rubin (1992) proposed a quantitative method called Gelman-Rubin diagnostics [42, 43]. This method allows assessing the convergence using a unique scalar noted $\hat{R}_p$. In the context of this method, this scalar yields $\hat{R}_p = 1$ when the algorithm converges perfectly. To the authors'

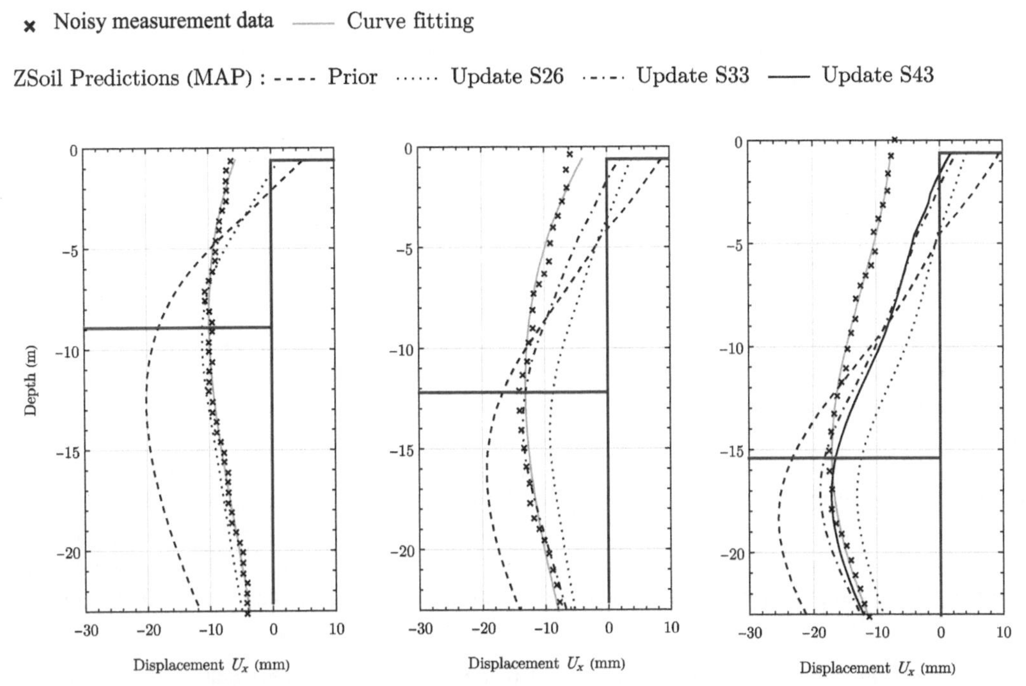

(a) S26 (b) S33 (c) S43

FIGURE 12.8 Horizontal displacements $U_x(y)$ for several stages and updates.

knowledge, a value of $\hat{R}_p < 1.1$ is acceptable and confirms convergence in most practical cases. For this study, Table 12.5 shows the configuration of the AM algorithms and their associated convergence using the Gelman-Rubin diagnostic. As Table 12.5 shows, all three solver configurations meet the convergence criterion.

Figure 12.8 shows the inversions applied to steps S26, S33, and S43. Polynomial fitting is used to smooth the noisy monitoring data for each inversion. The curves shown in Figure 12.8 are obtained by ZSoil simulations where the input parameters are derived from the maximum a posteriori estimation (MAP). In Figure 12.8c, the maximum absolute difference in deflections between Update S33 and Update S43 is 1.8 mm This difference seems small enough to question the need for updating S43. However, the next section shows that Update S43 is indeed necessary for reliable predictions at stage 50. Although the MAP-based ZSoil simulations accurately represented the

TABLE 12.5
Adaptive Metropolis Solver Setup and Convergence Diagnostic

Update N°	Steps[a]	NChains[b]	$\hat{R}_p$
S26	5000	100	1.08
S33	8000	300	1.09
S43	8000	300	1.06

Note:
[a]Number of MCMC iterations per chain.
[b]Number of parallel chains.

displacements at the bottom of the excavation, they cannot readily give satisfactory results in the upper part of the wall. The authors interpret this result as a limitation of the steady-state model. In fact, due to the low permeability of the Wurmian retreat layers, the stress generated by the prestressing of the struts is directly absorbed by the water pressure. This rapid increase in pressure is a reflection of short-term behavior that cannot be captured by a steady-state model.

For ease of reading and comparison, the posterior results of variables X5, X7, and X18 are displayed in a single figure (Figure 12.9). Nevertheless, the three inversions are dependent, i.e., the posterior of an n inversion is the prior of an $n+1$ inversion. The correlation matrices of the posterior output distributions are presented in Figure 12.10. Figure 12.11a shows the evolution of water pressures during construction, with one value missing due to a communication interruption with the cell located at -30 m in the final stage S50. Figure 12.11b shows three linear regressions on the pressure values for the initial scenario before work started, for the first excavation stage S26 and the final stage S50. The slope $m = 9.9$ of the baseline scenario line corresponds to the hydrostatic state's water density. As expected in Geneva's soils, this figure shows a pressure drop during soil relaxation. At the final stage of excavation, the slope of the line is $m = 6.3$, giving a water pressure multiplier of $\kappa_w = 0.63$. The posterior result of the Bayesian analysis of the variable X18 is a Gaussian distribution centered at $\mu_X = 0.61$ with a corresponding standard deviation of $\sigma_X = 0.0434$. Other pressure cells at other locations along the side of the excavation gave similar results with pressures ranging from 60% to 80% of the hydrostatic pressure. The authors interpret

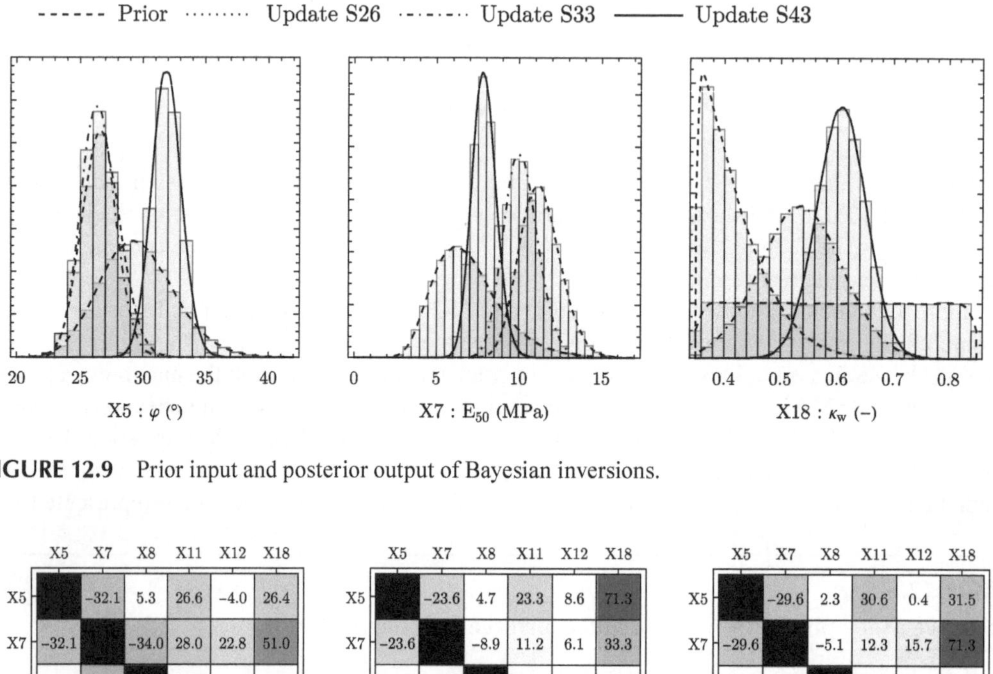

FIGURE 12.9 Prior input and posterior output of Bayesian inversions.

Stage 26	X5	X7	X8	X11	X12	X18
X5		-32.1	5.3	26.6	-4.0	26.4
X7	-32.1		-34.0	28.0	22.8	51.0
X8	5.3	-34.0		5.4	0.4	-0.1
X11	26.6	28.0	5.4		-12.9	14.3
X12	-4.0	22.8	0.4	-12.9		-1.8
X18	26.4	51.0	-0.1	14.3	-1.8	

Stage 26

Stage 33	X5	X7	X8	X11	X12	X18
X5		-23.6	4.7	23.3	8.6	71.3
X7	-23.6		-8.9	11.2	6.1	33.3
X8	4.7	-8.9		5.4	0.7	7.1
X11	23.3	11.2	5.4		-8.6	7.9
X12	8.6	6.1	0.7	-8.6		3.7
X18	71.3	33.3	7.1	7.9	3.7	

Stage 33

Stage 43	X5	X7	X8	X11	X12	X18
X5		-29.6	2.3	30.6	0.4	31.5
X7	-29.6		-5.1	12.3	15.7	71.3
X8	2.3	-5.1		1.1	-0.9	3.8
X11	30.6	12.3	1.1		-22.4	14.0
X12	0.4	15.7	-0.9	-22.4		22.9
X18	31.5	71.3	3.8	14.0	22.9	

Stage 43

FIGURE 12.10 Pearson correlation matrices. Black cells represent a 100% correlation between two parameters.

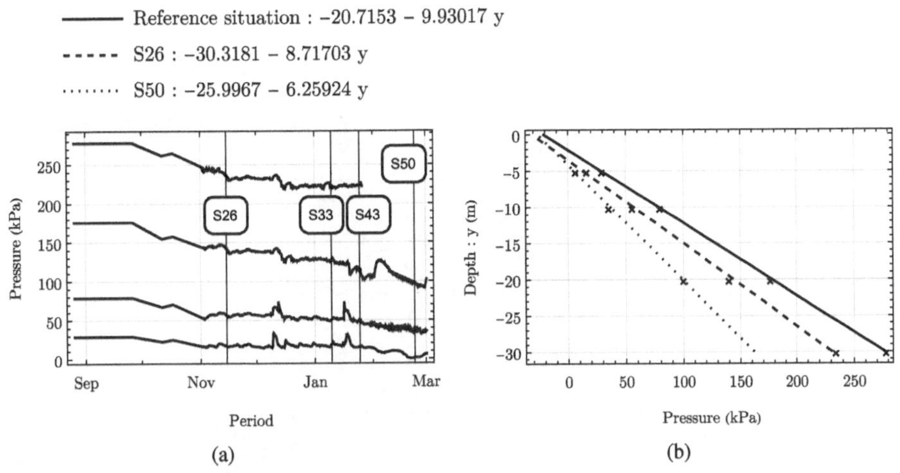

FIGURE 12.11 (a) Pressure values taken from pressure cells (–5 m, –10 m, –20 m and –30 m) with corresponding excavation periods. (b) Linear regressions of pressure values at the start of excavation and at stages S26 and S50.

this result as the model learning to reproduce the compatibility between deformations and water pressures at the back of the wall. Due to the overall stochastic character of the analysis, they also indicate that other study cases in analogous situations should confirm this result.

12.4.7 Final Prediction, Reliability Analysis, and Decision-Making

The successive predictions for the bottom of the excavation are shown in Figure 12.12a. While the first prior-based simulation gave a maximum displacement of $U_{x,\text{Max}} = -44.0$ (mm), the final update reduces this displacement to $U_{x,\text{Max}} = -30.4$ (mm). When the bottom of the excavation is reached, the measured maximum displacement is $z_{U_{x,\text{Max}}} = -30.0\,(\text{mm})$. The corresponding 3D displacement field taken from the FE software is shown in Figure 12.12b. To account for the potential risks associated with three-layered strut support, it is essential to incorporate the uncertainty of the variables reliably and consistently. Simple deterministic MAP-based prediction may be informative but is no longer sufficient for risk assessment. Therefore, a comprehensive reliability analysis is performed to account for the uncertainty of the variables and to finally determine the installation of the fourth row of struts.

In accordance with the input domain $\mathcal{D}_\mathbf{X}$, the failure domain $\mathcal{D}f$ and the safety domain $\mathcal{D}_s$ can be identified by a hypersurface denoted as $g(\mathbf{X})$, which delineates the boundary between safe and unsafe regimes. To evaluate the probability of failure associated with the system, the indicator function of the failure domain

$$\mathbf{1}_{\mathcal{D}_f}(x) = \begin{cases} 1 & \text{if } g(x) \le 0 \\ 0 & \text{if } g(x) > 0 \end{cases}, \; x \in \mathcal{D}_X \qquad \text{Eq. (12.11)}$$

can be used to perform Monte Carlo-driven reliability analyses. The calculation of the probability of failure of a system by MCS is exact when $N \to \infty$. In practice, the calculation of the empirical failure probability usually requires at least 10^{k+2} model evaluations to approximate a probability of magnitude 10^{-k}. This empirical probability of failure is then obtained by dividing the number of realizations of the system that lie within the failure domain by the total number of realizations as follows:

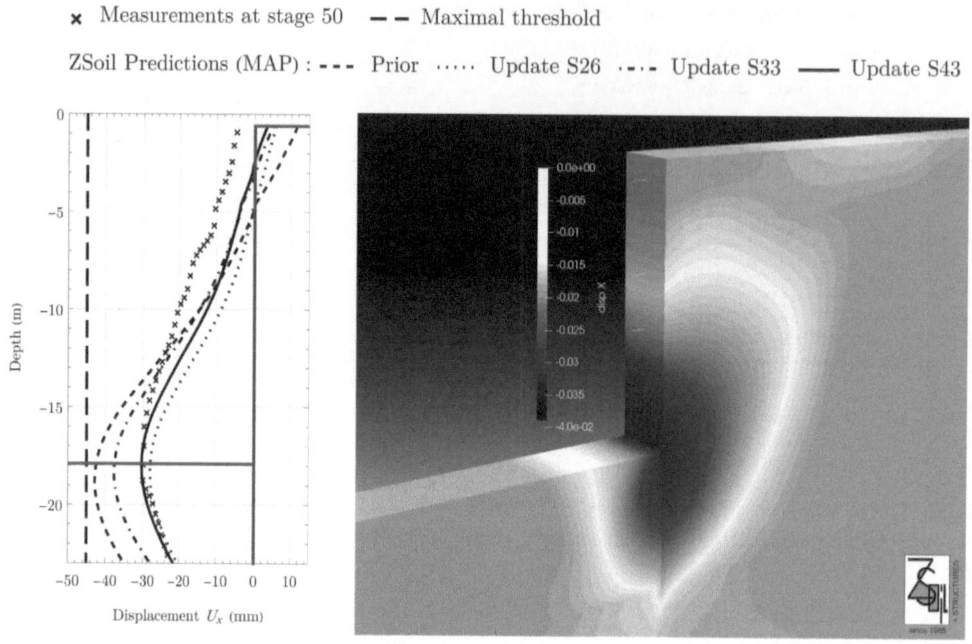

(a) S50 (b) ZSoil model

FIGURE 12.12 (a) Predictions of the horizontal deflection $U_x(y)$ at stage 50 and the last measurement data. (b) Corresponding ZSoil model with U_x displacement field.

$$\widehat{P}_f = \frac{1}{N}\sum_{k=1}^{N} \mathbf{1}_{\mathcal{D}_f}\left(x^{(k)}\right) = \frac{N_{fail}}{N} \qquad \text{Eq. (12.12)}$$

In this study, the $g(\mathbf{X})$ functions are determined through discussions with the contractors. At the time of publication, the European codes state that the minimum reliability requirements are typically established directly by the designer and the contractor, with recommended values in the literature varying between $P_f = 1\%$ and $P_f = 5\%$. The accepted thresholds for different quantities of interest and their associated 10^7 sample-based failure probabilities are listed in Table 12.6. While the prior failure function $g_{M_{\text{Max}}}(\mathbf{X})$ already yields a low and acceptable probability with respect to the pre-established threshold, the probability that the maximum horizontal displacement exceeds the −45 mm threshold is greater than 50%. Additionally, the prior sensitive conduit's displacement is still of concern. Facing such uncertainties, it is not possible to *a priori* determine the necessity

TABLE 12.6
Probabilities of Failure According to $g(X)$

$g(\mathbf{X})$	$45-\lvert U_{x,\text{Max}}\rvert$ (mm)	$20-\lvert U_{y,\text{Pipe}}\rvert$ (mm)	$M_{Rd}-\lvert M_{\text{Max}}\rvert$ (kNm)
Prior	0.5315	7.35×10^{-2}	2.78×10^{-5}
S26	5.12×10^{-4}	4.46×10^{-5}	$<10^{-5}$
S33	1.70×10^{-3}	1.8×10^{-4}	$<10^{-5}$
S43	$<10^{-5}$	$<10^{-5}$	$<10^{-5}$

of the fourth row of struts. Thanks to Bayesian analyses, the latest update significantly reduces the prior probability to a posterior probability well below the selected thresholds. For graphical and informative purposes, the distributions of the different uncertainty propagations along the Bayesian inversions are shown in Figure 12.13, where the reduction of uncertainties as a function of updates is easily recognizable.

In agreement with the contractor and based on the latest calculated probability of exceeding the displacement thresholds, the need for a fourth row of struts is ruled out. Measurements made during concreting of the bottom slab showed that the horizontal displacements had stabilized at −30 mm. Figure 12.14 shows the concreting of the bottom slab after the excavation with three strut layers.

12.4.8 Case Study Conclusion

This case study presents a probabilistic analysis of an 18-m-deep excavation with live Bayesian inversions. Given the geotechnical uncertainties, the objective is to determine the feasibility of a

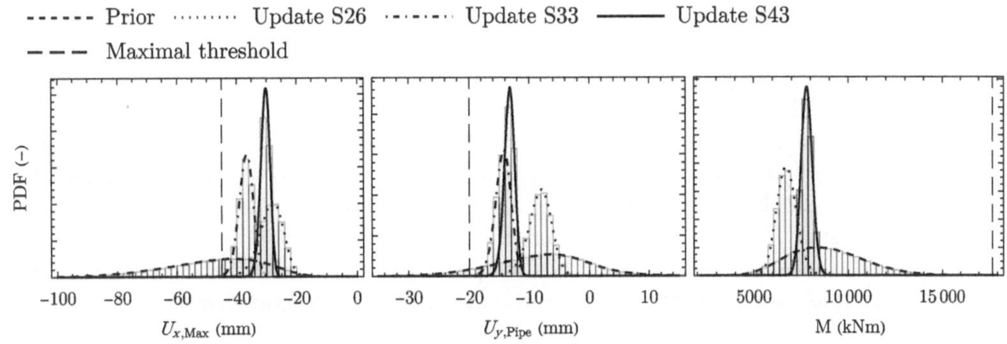

FIGURE 12.13 Probability density functions for QoIs $U_{x,\text{Max}}$, $U_{y,\text{Pipe}}$ and M_{Max}.

FIGURE 12.14 In-situ final excavation stage with three strut levels.

three-level support option and ensure that pre-set thresholds are not exceeded. After identifying the variables that have the most influence on the results using a PCE-based sensitivity analysis, three PCK-based Bayesian inversions are sequentially performed using measurements from on-site inclinometers. The inverse solution is then used to obtain MAP-based deterministic predictions as well as Monte Carlo-driven probabilistic predictions in the form of probability density functions and associated failure probabilities. Decision-making is finally formulated based on these posterior failure probabilities.

In light of the results of this study, the principal findings and interpretations are listed as follows:

(1) The influencing geotechnical variables are located in the Wurmian retreats, which consist of clayey silts. Water pressure, introduced as a multiplier of water density, also plays an important role in the model response. These results are in line with the designers' expectations at the beginning of the study.
(2) Using a steady-state model to represent partially drained soil behavior is achieved by introducing the water pressure as a random variable expressed as a multiplier of its density. The steady-state formulation allows the model to be independent of the excavation time and possible delays and advances in the work. Therefore, a pre-computation strategy is established to prepare time-consuming EDs before the start of the work. This strategy increases the efficiency of the calculations and decreased the reaction time for decision-making during the excavation process.
(3) The calibrated model reproduces the horizontal displacements at the bottom of the excavation with high accuracy but has difficulty reproducing the displacements at the top of the wall. This problem is interpreted as a limitation of the steady-state model.
(4) In this case study, Bayesian analyses enable an optimized variant of the diaphragm wall support system to be proposed. Based on the posterior failure probabilities, the decision for a three-level strut support system is made during the construction phase.

12.5 3D PROBABILISTIC FE MODEL FOR A TUNNEL EXCAVATION

The second example of the use of probabilistic methods associated with reliability computations to help decision-making concerns the construction of a third underground metro line in Lausanne, Switzerland [44]. The tunnels are to be excavated mostly in rock composed of marl and sandstone, with a conventional excavation method using steel profiles and shotcrete for the temporary support chosen by engineers. Due to the adjacent buildings and shallow cover, a detailed examination of construction options is necessary. This investigation includes considerations such as whether to use a full-face or calotte-stross excavation, the implementation of a pipe umbrella, nailing of the tunnel face, and determining the optimal spacing between the steel profiles of the temporary support.

The present example concerns the evaluation of the probability of exceeding the admissible settlement threshold of a sensitive building according to the spacing between the steel profiles: a small spacing limits the settlements but is more expensive and takes more time to execute than a larger spacing. Two variants of steel profile spacing are considered: one meter and two meters. During the contemplation process for the development of the associated document addressing the implementation of reliability analyses in the new Eurocodes [38], the serviceability limit state (SLS) threshold is deemed subject to the judgment of both design engineers and contractors. Here, the decision on the spacing is therefore based on the probability of exceeding a pre-set threshold. For one of the variants to be selected and according to literature [45], the probability must be less than $P_f = 6.7\%$. Additionally, the resisting bending moment of the steel profiles for ultimate state verification is an essential requirement, albeit outside the scope of this study.

12.5.1 FE Model Description

To investigate the passage of the tunnel beneath an existing six-story building (see Figure 12.15), a 3D FE model is constructed using ZSoil [6] (see Figure 12.16). In this sector, the tunnel crown is located in marlstone and requires the installation of a pipe umbrella, as well as fiberglass bolts in the front of the tunnel (see Figure 12.17). In accordance with the contractor, the settlement of the building must not exceed 20 mm. The construction sequence for a span of 14 m is as follows:

- Installation of the pipe umbrella made of ROR 159/10 steel profiles.
- Installation of the front nailing with fiberglass bolts.
- Excavation meter by meter with the installation of a steel profile every meter or every two meters for both the excavation step and the steel profile construction (see Figure 12.17), both accompanied by a 30 cm layer of shotcrete.

The characteristics of the FE model, whose dimensions cover 7 times the 14 m pipe umbrella span, are as follows:

- For rocks: 400,000 bilinear 3D elements with 8 nodes, with volumetric locking treatment by means of adequate methods (enhanced assumed strains formulation).
- For support: 4-node thin shell elements, taking into account the equivalent stiffness and area of the steel profiles and shotcrete.

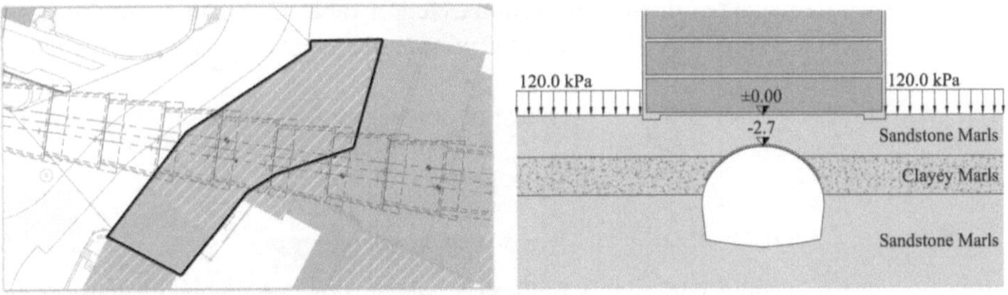

(a) Situation (b) Cross section

FIGURE 12.15 Excavation of the tunnel below a sensitive existing building.

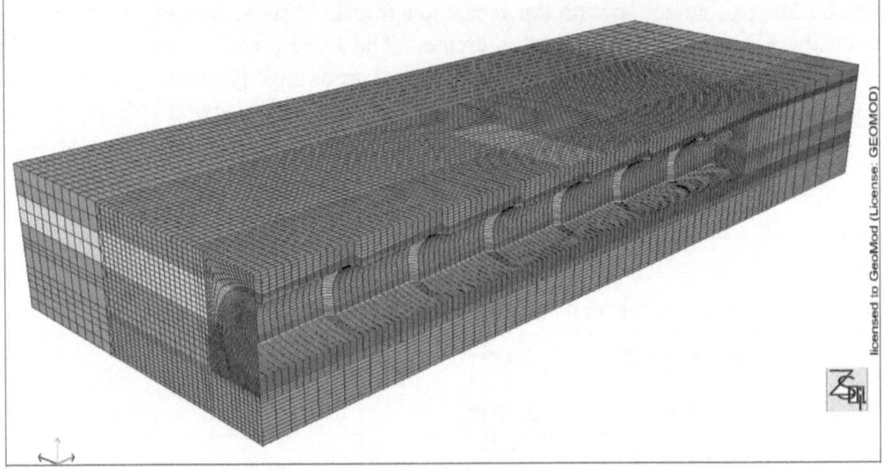

FIGURE 12.16 Complete 3D FE model. Size = 135 m × 50 m × 20 m.

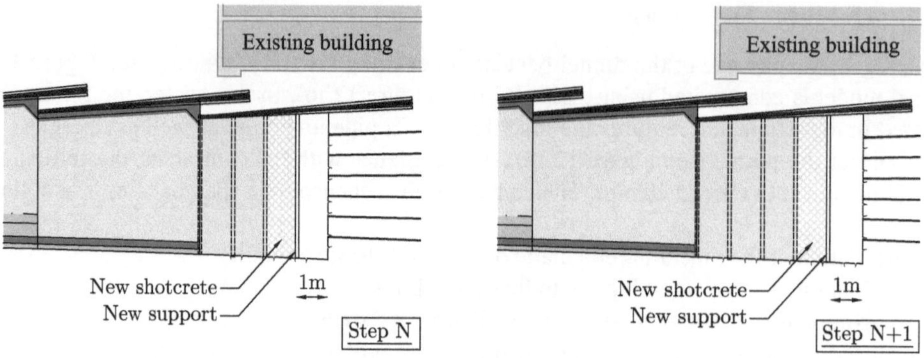

FIGURE 12.17 Excavation steps.

- For the pipe umbrella and the front nailing: nail elements.
- Interface elements allow the correct consideration of the interaction between the rock and the temporary support.
- The foundation of the existing building is modeled using thin shell elements with a 170 kPa surface load applied.
- The soil on either side of the building (6 m high) is modeled by a 120 kPa surface load.
- The rock layers are modeled using an elastic-perfectly plastic constitutive law (Mohr-Coulomb), whose mean parameters are summarized in Table 12.7. The use of this model, which is simpler than the HSS law used in the previous example, is justified as the excavation is made in rock.
- The geotechnical studies show no presence of a water table and, therefore, water is not taken into account in the computations.

12.5.2 Reliability Analysis and Decision-Making

The following uncertain input parameters are considered in this analysis, with their associated probability density function described in Table 12.8: elastic modulus, cohesion, and friction angle for both rock layers, as well as the surface load applied on the building foundation. Then, the ED is created, composing of 100 samples drawn with the Latin Hypercube Sampling method which involves 100 ZSoil model evaluations. The quantity of interest in this model is the maximal settlement of the building's foundation with the acceptable threshold set at 20mm.

Based on the ED, a PCK meta-model is created. The LOO error is used to assess if the ED size is sufficient to achieve an accurate meta-model. Approximately 50 samples are required to achieve convergence (LOO error below 0.5%). Such a level of error is regarded as sufficiently low to consider the PCK model as accurate.

TABLE 12.7
Mohr-Coulomb Model Parameters

	Specific Weight	Stiffness	Strength			Stress State
Formation	γ (kN.m^{-3})	E (MPa)	φ (°)	ψ (°)	c (kPa)	K_0 (-)
Clayey marl	23	50	22	4	150	1
Sandstone marl	25	900	35	5	1300	1

TABLE 12.8
Probabilistic Input Parameters

Parameter	Distribution	f_X	CV
Friction angle	Lognormal	$X \sim \mathcal{LN}$	CV = 10%
Cohesion	Lognormal	$X \sim \mathcal{LN}$	CV = 20%
Young's modulus	Lognormal	$X \sim \mathcal{LN}$	CV = 30%
Building loads	Normal	$X \sim \mathcal{N}$	CV = 10%

Using a Monte Carlo simulation with a sample size of 10^6 and the PCK meta-model, the probability of failure is computed: $P_f(u_y > 20\,mm) = 2.65\%$. The target reliability index for the serviceability limit state, according to [45], is 1.5 which corresponds to $P_f = 6.7\%$. The excavation process can therefore be optimized as we are $\sim 4\%$ below the target probability of failure. In the original project, the excavation is carried out meter by meter, with a steel profile placed every meter as well (see Figure 12.17). The scrutinized optimization aims to define if a sequence of 2 meters for both the excavation step and the steel profile installation is possible. To assess the reliability of the optimized excavation process, the same workflow is applied:

- A new FE model is created based on the former model. The only difference is in the excavation process and the spacing between steel profiles (2 m).
- A new study is conducted using the same ED to create 100 input files using the new model. These input files are then analyzed using ZSoil.

Using the Monte Carlo simulation on the PCK model with a LOO error of 3.3×10^{-3}, the probability of failure is computed as $P_f(2m, u_y > 20mm) = 8.40\%$. A comparison between the histograms of the probability distribution function for 1 m and 2 m is presented in Figure 12.19.

The optimized excavation process does therefore not fulfill the requirements, and the 1 m spacing is maintained as the construction method.

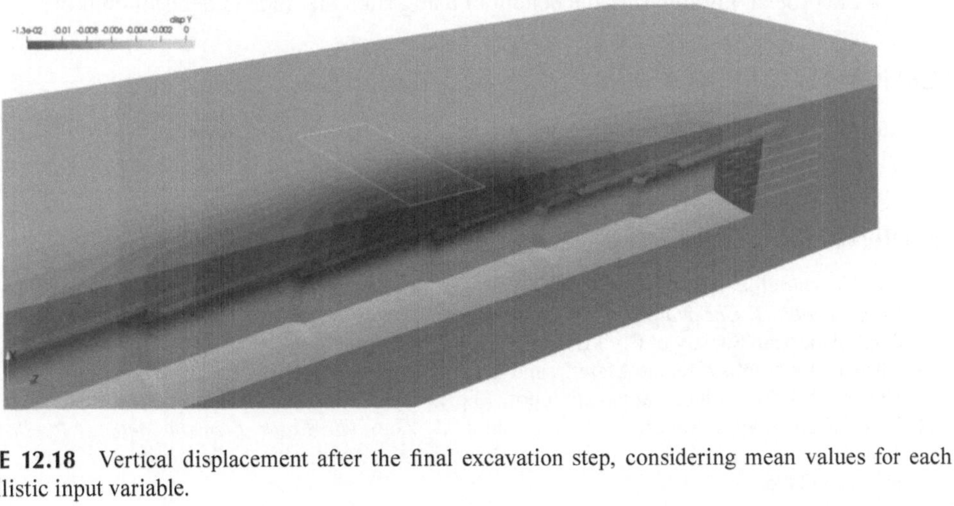

FIGURE 12.18 Vertical displacement after the final excavation step, considering mean values for each probabilistic input variable.

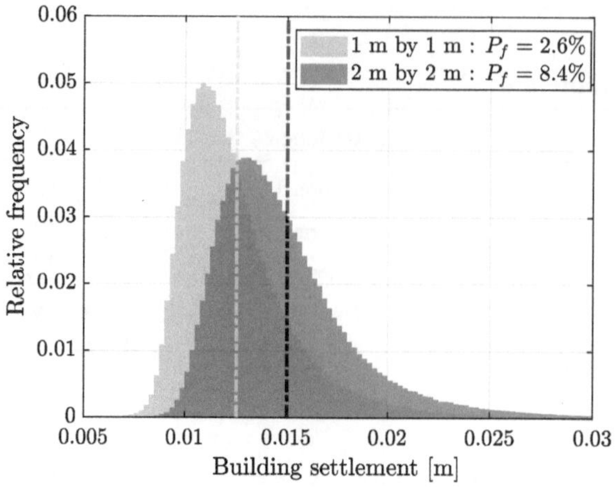

FIGURE 12.19 PDF of the settlement, with the original excavation process (1 m × 1 m) and the optimized process (2 m × 2 m). Dot-dashed lines represent the mean of the distributions.

12.6 CONCLUSION

In this chapter, probabilistic analyses are applied to non-linear 3D FE models of two real-life case studies: a deep excavation retained by a braced slurry wall, and the construction of a tunnel under a sensitive building. By coupling the FE software ZSoil and the framework for uncertainty quantification UQLab, sensitivity, reliability, and inverse Bayesian analyses are performed whilst meeting time and cost requirements imposed in daily practice. The methodology applied in the presented case studies is feasible for optimizing the design of future construction projects. In the authors' view, two final remarks are particularly relevant :

1. Sensitivity analysis is a fundamental tool to aid the decision-making process in construction projects. It allows the engineer's intuitions to be confirmed or challenged by analyzing the real behavior and sensitivity of their model. Furthermore, it enables a better understanding of the results of Bayesian analyses when the posterior distributions become geotechnically or physically meaningless.
2. Once predictions are updated by Bayesian and reliability analyses, optimized solutions can be proposed to the contractor. This can result in cost reductions and material savings for the project, highlighting the benefit of using such methods in decision-making.

ACKNOWLEDGMENTS

This study has been financed by the "Ingénierie et Architecture" department of HES-SO, University of Applied Sciences Western Switzerland. The financial support is gratefully acknowledged.

REFERENCES

1. Stefano Marelli and Bruno Sudret. UQLab: A framework for uncertainty quantification in MATLAB. In *Vulnerability, Uncertainty, and Risk*, edited by Michael Beer, Siu-Kui Au, Jim W. Hall, pages 2554–2563, American Society of Civil Engineers, Liverpool, June 2014.
2. Christos Lataniotis, Stefano Marelli, and Bruno Sudret. Uncertainty Quantification in the cloud with UQCloud, 2021. Medium: application/pdf, 10 p. accepted version Publisher: ETH Zurich.
3. O. C. Zienkiewicz, Robert L. Taylor, and J. Z. Zhu. *The Finite Element Method: Its Basis and Fundamentals.* Elsevier, Butterworth-Heinemann, Amsterdam, 7th edition, 2013. OCLC: ocn852808496.

4. J. Tinsley Oden. *Finite Elements of Nonlinear Continua*. Dover Publications, Mineola, New York, Dover edition, 2006.
5. T. Zimmermann, C. Rodriguez, and B. Dendrou. Z-SOIL.PC: A program for solving soil mechanics problems on a personal computer using plasticity theory. In *Proceedings of the Sixth International Conference on Numerical Methods in Geomechanics, Innsbruck*, pages 2121–2126, Balkema, Innsbruck, April 1988.
6. S. Commend, S. Kivell, R. Obrzud, K. Podles, A. Truty, and T. Zimmermann. *Computational Geomechanics: Getting Started with ZSOIL.PC*. Rossolis; Zace Services, [Bussigny], Préverenges, 7th edition, 2022. OCLC: 1038708800.
7. T. Schanz, P. A. Vermeer, and P. G. Bonnier. The hardening soil model: Formulation and verification. In Ronald B. J. Brinkgreve, editor, *Beyond 2000 in Computational Geotechnics*, pages 281–296. Routledge, 1st edition, January 2019.
8. Jack R. Benjamin and C. Allin Cornell. *Probability, Statistics, and Decision for Civil Engineers*. Dover Publications, Inc, Mineola, New York, Dover edition, 2014.
9. W. H. Tang and A. H.-S. Ang. *Modeling, Analysis and Updating of Uncertainties*. American Society of Civil Engineers, 1973.
10. Erik H. Vanmarcke. Probabilistic modeling of soil profiles. *Journal of the Geotechnical Engineering Division*, 103(11):1227–1246, November 1977.
11. Gregory B. Baecher and John T. Christian. *Reliability and Statistics in Geotechnical Engineering*. J. Wiley, Chichester, West Sussex, Hoboken, NJ, 2003.
12. Kok-Kwang Phoon and Jianye Ching, editors. *Risk and Reliability in Geotechnical Engineering*. CRC Press, Boca Raton, October 2018.
13. Roger G. Ghanem and Pol D. Spanos. *Stochastic Finite Elements: A Spectral Approach*. Springer, New York, 1991.
14. Trevor Hastie, Robert Tibshirani, and J. H. Friedman. *The Elements of Statistical Learning: Data Mining, Inference, and Prediction*. Springer Series in Statistics. Springer, New York, 2nd edition, 2009.
15. Curtis B. Storlie, Laura P. Swiler, Jon C. Helton, and Cedric J. Sallaberry. Implementation and evaluation of nonparametric regression procedures for sensitivity analysis of computationally demanding models. *Reliability Engineering & System Safety*, 94(11):1735–1763, November 2009.
16. Wengang Zhang, Runhong Zhang, Chongzhi Wu, Anthony Teck Chee Goh, Suzanne Lacasse, Zhongqiang Liu, and Hanlong Liu. State-of-the-art review of soft computing applications in underground excavations. *Geoscience Frontiers*, 11(4):1095–1106, July 2020.
17. Zheng-Wei Li, Qiu-Jing Pan, and Rui-Zheng Fei. Probabilistic evaluation of three-dimensional seismic active earth pressures using sparse polynomial chaos expansions. *Computers & Geotechnics*, 129:103869, January 2021.
18. Jian Zhang, Weijie Gong, Xinxin Yue, Maolin Shi, and Lei Chen. Efficient reliability analysis using prediction-oriented active sparse polynomial chaos expansion. *Reliability Engineering & System Safety*, 228:108749, December 2022.
19. Xujia Zhu, Marco Broccardo, and Bruno Sudret. Seismic fragility analysis using stochastic polynomial chaos expansions. *Probabilistic Engineering Mechanics*, 72:103413, April 2023.
20. Pramudita Satria Palar, Lavi Rizki Zuhal, Koji Shimoyama, and Takeshi Tsuchiya. Global sensitivity analysis via multi-fidelity polynomial chaos expansion. *Reliability Engineering & System Safety*, 170:175–190, February 2018.
21. Max Ehre, Iason Papaioannou, and Daniel Straub. Global sensitivity analysis in high dimensions with PLS-PCE. *Reliability Engineering & System Safety*, 198:106861, June 2020.
22. Thierry A. Mara and William E. Becker. Polynomial chaos expansion for sensitivity analysis of model output with dependent inputs. *Reliability Engineering & System Safety*, 214:107795, October 2021.
23. Mishal Thapa and Samy Missoum. Uncertainty quantification and global sensitivity analysis of composite wind turbine blades. *Reliability Engineering & System Safety*, 222:108354, June 2022.
24. I. M. Sobol. Global sensitivity indices for non-linear mathematical models and their Monte Carlo estimates. *Mathematics & Computers in Simulation*, 55(1–3):271–280, February 2001.
25. S. Kucherenko, S. Tarantola, and P. Annoni. Estimation of global sensitivity indices for models with dependent variables. *Computer Physics Communications*, 183(4):937–946, April 2012.
26. Bruno Sudret. Global sensitivity analysis using polynomial chaos expansions. *Reliability Engineering & System Safety*, 93(7):964–979, July 2008.
27. Jerome Sacks, Susannah B. Schiller, and William J. Welch. Designs for computer experiments. *Technometrics*, 31(1):41–47, February 1989.

28. R. Schöbi, B. Sudret, and J. Wiart. Polynomial-chaos-based kriging, 1(2), 2015. Publisher: arXiv Version Number: 1.
29. Olivier Dubrule. Cross validation of kriging in a unique neighborhood. *Journal of the International Association for Mathematical Geology*, 15(6):687–699, December 1983.
30. Gilbert Saporta. *Probabilités, Analyse des Données et Statistique*. Éd. Technip, Paris, 3e éd. révisée edition, 2011.
31. Géraud Blatman and Bruno Sudret. Sparse polynomial chaos expansions and adaptive stochastic finite elements using a regression approach. *Comptes Rendus Mécanique*, 336(6):518–523, June 2008.
32. Pictet Group. Campus Pictet de Rochemont - New building construction in Praille Acacias Vernets in Geneva, Switzerland, 2023.
33. dl-a, Designlab-architecture sa. *Extension du siège de la banque Pictet & Cie à Genève, Suisse*, 2023.
34. S. Commend, F. Geiser, and J. Crisinel. Numerical simulation of earthworks and retaining system for a large excavation. *Advances in Engineering Software*, 35(10):669–678, 2004.
35. De Cérenville Géotechnique sa. *Extension du siège de la banque Pictet & Cie à Genève, Suisse*, 2023.
36. Marcin Cudny and Andrzej Truty. Refinement of the Hardening Soil model within the small strain range. *Acta Geotechnica*, 15(8):2031–2051, August 2020.
37. Kok-Kwang Phoon, Zi-Jun Cao, Jian Ji, Yat Fai Leung, Shadi Najjar, Takayuki Shuku, Chong Tang, Zhen-Yu Yin, Yoshida Ikumasa, and Jianye Ching. Geotechnical uncertainty, modeling, and decision making. *Soils & Foundations*, 62(5):101189, October 2022.
38. TG-C3. Reliability-based methods for geotechnical design and assessment. Guideline document for the next-generation Eurocodes. Technical Report CEN/TC250/SC7/TG-C3, 2023.
39. Rafal Obrzud and Andrzej Truty. The hardening soil model-A practical guidebook Z_soil. Technical Report PC100701, 2010.
40. Heikki Haario, Eero Saksman, and Johanna Tamminen. An adaptive metropolis algorithm. *Bernoulli*, 7(2):223, April 2001.
41. P.-R. Wagner, J. Nagel, S. Marelli, and B. Sudret. UQLab user manual – Bayesian inversion for model calibration and validation. Technical report, Chair of Risk, Safety and Uncertainty Quantification, ETH Zurich, Switzerland, 2021.
42. Andrew Gelman and Donald B. Rubin. Inference from iterative simulation using multiple sequences. *Statistical Science*, 7(4), November 1992.
43. Steve Brooks, Andrew Gelman, Galin Jones, and Xiao-Li Meng, editors. *Handbook of Markov Chain Monte Carlo*. Chapman and Hall/CRC, New York, May 2011.
44. Etat de Vaud. Métros - Construction du métro m3 et transformation du m2, 2023. https://www.vd.ch/themes/mobilite/loffre-de-mobilite-a-votre-disposition/metros.
45. H. Gulvanessian, J.-A. Calgaro, and M. Holicky. *Designer' Guide to Eurocode: Basis of Structural Design EN 1990*, ICE Publishing, 2nd edition, 2012.

Part 4

Probabilistic applications

13 Soil-Structure Interaction in Spatially Variable Ground

Andy Y.F. Leung

ABSTRACT

The variability and uncertainty in soil properties propagate into uncertainty in the response of engineering systems involving soil-structure interactions. An important feature of geomaterials lies in the significant spatial variability in their properties, where the existence of strong and weak spots within the soil mass would lead to failure mechanisms that deviate from theoretical predictions based on assumptions of homogenized media. While spatial variability of material properties may be characterized by the scale of fluctuation, previous studies have found that many soil-structure interaction problems entail a certain 'worst-case' scale of fluctuation in soil properties, which leads to the largest discrepancy from nominal estimates of system responses by deterministic or homogenization approaches. This section presents some recent findings related to these phenomena and the key considerations in various geotechnical problems, including shallow and deep foundations, slope stability, and also excavation and retaining structures.

13.1 INTRODUCTION

Due to natural variations that arise from the weathering process forming the soil materials, together with various subsequent transport and depositional processes, soils are observed to display features of spatial variability. The often limited geotechnical investigation data also lead to significant uncertainty both in the statistics (e.g., mean value and standard deviation) of the soil properties and their spatial distributions. The variable and uncertain properties of soils and rocks manifest themselves as uncertainty in the estimates of the performance of slopes, foundations, excavations, or other problems involving soil-structure interactions. In the past, to investigate the propagation of uncertainty in material properties to that of the system performance, a logical step beyond the deterministic approach involves probabilistic analyses that model the properties of "homogenized" material layers as random variables. These assume that materials properties are uniform across the layers with the associated parameters being uncertain. The uncertainty in material parameters of the homogenized layers can be considered through multiple simulations either in the Monte Carlo fashion or via reliability approaches such as the first order second moment (FOSM) method or first order reliability method (FORM) discussed in earlier chapters. Some earlier works adopting this approach include Alonso (1976), Vanmarcke (1977), Li and Lumb (1987), Christian et al. (1994), Duncan (2000), etc., while some of them also pointed out the potential limitations of the homogenization approach.

For example, Christian et al. (1994) and Duncan (2000) discussed the applications of first-order approximations to reliability estimates of slopes, consolidations, and other soil-structure interaction problems. The former considered the effects of spatial variability (or spatial autocorrelation) via the reduction ratio of variance; while the latter illustrated the evaluation of the mean values and standard deviations of factors of safety for these geotechnical problems with soil layers being homogenized with properties represented as random variables. In these early studies where computational power might be a bottleneck for extensive random field simulations, the homogenization approach and consideration of variance reduction factors provide practitioners with a handy tool for

DOI: 10.1201/9781003333586-17

quick approximations on the uncertainty in system responses arising from variable material parameters and loading. However, spatially variable ground properties may lead to failure mechanisms that cannot be captured by simulations assuming uniform or homogenized geomaterials properties.

At the element testing level, Kim and Santamarina (2008) showed that spatial variability features such as correlation lengths and anisotropy in void ratios of the material affect the phenomena of strain localization, local drainage, and hence the shear strength development. The distinctive characteristics of spatial variability would manifest themselves in a variety of ways in soil-structure interaction problems with different stress paths. As a case in point, considering a building foundation with symmetric geometry and load conditions founded on/embedded in a soil layer, deterministic or probabilistic analyses with a homogenized soil layer would yield a deformation profile that is also symmetrical. In reality, soil properties vary spatially within the same layer, and tilting of foundations or differential settlements between two footings may arise due to this phenomenon. A homogenized soil layer may be described as a case of perfect spatial correlation with the scale of fluctuation being infinity; the other extreme would involve uncorrelated material properties with the scale of fluctuation being zero. As described by Christian and Baecher (2011), the intermediate conditions (or values of the scale of fluctuation), instead of the two extremes, often constitute the most critical case for geotechnical problems. These phenomena have been investigated extensively by Fenton and Griffiths (2008) and others in recent studies, and some of their key findings are summarized in terms of the worst-case scale of fluctuation (Table 13.1) in Vessia et al. (2021) and Phoon et al. (2022).

The existence of the worst-case scale of fluctuation generally refers to the phenomenon where the complex deformation patterns or failure mechanisms arising from specific correlation patterns are associated with the largest deviations, often unconservative ones, from conventional 'nominal' responses based on assumptions of uniform or homogenized layers. Such worst-case scales of fluctuation are usually represented as some multiples of the characteristic length of the geotechnical structure. Following the previous example of building foundations on spatially variable soils, the differential settlements between two footings may be largest when the scale of fluctuation is close to a certain multiple of the footing spacing, as this may imply the existence of strong and weak spots of soils at two adjacent footings. This concept of the worst-case scale of fluctuation is valuable in the context of reliability-based designs because in the absence of a precise determination of site-specific spatial variability features, assuming the worst-case scenarios usually leads to a conservative design (Vessia et al. 2021).

This chapter describes various aspects of soil-structure interaction problems that can be revealed by random field models of spatially variable ground and summarizes recent findings associated with the worst-case scale of fluctuation for these problems. Recent research has contributed to the development of general recommendations and guidelines pertinent to spatially variable geomaterials, while a number of commercial software packages (described in Chapter 11) now include built-in functionality for probabilistic analyses. As "command-driven" user interfaces or compatibility with Python scripting are becoming common for finite element and finite difference packages (e.g., PLAXIS, FLAC), random field modeling is also within reach for practitioners who wish to apply these techniques to their project-specific conditions.

13.2 SHALLOW FOOTINGS

Using random field modeling techniques, Griffiths et al. (2002) and Fenton and Griffiths (2003) investigated the bearing capacities of strip footings founded on spatially variable Tresca soils and c-ϕ soils, respectively. Considering the ultimate limit state, they revealed that, on average, the bearing capacity would be generally smaller than the deterministic Prandtl solution based on the mean values of shear strength parameters. Griffiths et al. (2002) explained that the weak soil elements, rather than strong soil elements, tend to dominate the overall performance of the footing. As would be expected, the failure surfaces also deviate from the classical log spiral curve as the soil fails along the "weakest path", a phenomenon which was also demonstrated by Li et al. (2015b) through the developments of multiple asymmetric shear planes in spatially variable soils

TABLE 13.1
Examples of "Worst-Case" Scale of Fluctuations Reported in Previous Studies

Study	Problem type	"Worst-case" Definition	Characteristic Length	"Worst-Case" Scale of Fluctuation
Fenton and Griffiths (2003)	Bearing capacity of a footing on a $c-\phi$ soil	Mean bearing capacity is minimal	Footing width (B)	$1 \times B$
Griffiths et al. (2006)	Bearing capacity of footing(s) on a $\phi = 0$ soil	Mean bearing capacity is minimal	Footing width (B)	$0.5\sim2 \times B$
Vessia et al. (2009)	Bearing capacity of footing on $c-\phi$ soil	Mean bearing capacity is minimal (anisotropic 2D variability)	Footing width (B)	$0.3\sim0.5 \times B$
Soubra et al. (2008)	Bearing capacity under punching failure mode for a strip footing	Mean bearing capacity is minimal	Footing width (B)	$1 \times B$ for isotropic $5 \times B$ for anisotropic
Pula et al. (2017)	Bearing capacity of a strip footing on $c-\phi$ soil	Under-design probability is maximal	Footing width (B)	$8 \times B$ for isotropic case, with 2D random fields of *c*
Luo and Bathurst (2017)	Bearing-capacity factor of a strip footing on a cohesive slope	Mean bearing-capacity factor is minimal	Footing width (B)	$0.25\sim1 \times B$
Tabarroki (2020)	Mobilized shear strength of a strip footing	Mean mobilized shear strength is minimal	Footing width (B)	$0.1–0.5 \times B$
Fenton and Griffiths (2005)	Differential settlement of footings	Under-design probability is maximal	Footing spacing (S)	$1 \times S$
Breysse et al. (2005)	Settlement of a footing system	Footing rotation is Maximal	Footing spacing (S)	$0.5 \times S$
		Mean different settlement between footings is maximal	Footing spacing (S) Footing width (B)	f(S,B) (no simple equation)
Jaksa et al. (2005)	Settlement of a nine-pad footing system	Under-design probability is maximal	Footing spacing (S)	$1 \times S$
Ahmed and Soubra (2014)	Differential settlement of footings	Under-design probability is maximal	Footing spacing (S)	$1 \times S$

(*Continued*)

TABLE 13.1 (CONTINUED)
Examples of "Worst-Case" Scale of Fluctuations Reported in Previous Studies

Study	Problem type	"Worst-case" Definition	Characteristic Length	"Worst-Case" Scale of Fluctuation
Stuedlein and Bong (2017)	Differential settlement of footings	Under-design probability is maximal	Footing spacing (S)	1 × S
Naghibi et al. (2016)	Differential settlement of a two-pile system	Mean differential settlement is maximal	Pile spacing (S)	1 × S
Leung and Lo (2018)	Differential settlement of a pile group	Standard deviation of differential settlement is maximal	Width of pile group (B)	0.5~1 × B
Tabarroki et al. (2013)	FS of a cohesive slope	Mean FS is minimal	Slope height (H)	0.1~1 × H
Ali et al. (2014)	Risk of infinite slope	Risk of rainfall-induced slope failure is maximal	Slope height (H)	1 × H
Hu and Ching (2015)	Active lateral force for a retaining wall	Mean active lateral force is maximal	Wall height (H)	0.2 × H
Fenton et al. (2005)	Active lateral force for a retaining wall	Under-design probability is maximal	Wall height (H)	0.5~1 × H
Griffiths et al. (2008)	Passive lateral force for a retaining wall	Under-design probability is maximal	Wall height (H)	0.1–0.5 × H
Ching et al. (2017)	FS for basal heave of an excavation in clay	Mean FS is minimal	Wall penetration depth (H_p)	0.1~0.4 × H_p
Tabarroki (2020)	Mobilized shear strength of a retaining wall	Mean mobilized shear strength is minimal	Wall height (H)	0.05 × H
Tabarroki (2020)	Mobilized shear strength of a deep excavation in clay	Mean mobilized shear strength is minimal	Penetration depth (H_p)	0.2 × H_p
Ching and Phoon (2013)	Overall strength of a soil column	Mean strength is minimal	Column width (W)	1 × W (compression) 0 × W (simple shear)
Tabarroki (2020)	Mobilized shear strength of a soil column under compression	Mean mobilized shear strength is minimal	Column width (W)	0.5 × W
Pan et al. (2018)	Compressive strength of cement-treated clay column	Mean strength is minimal	Column diameter (D)	1 × D

(updated from Ching et al. 2017 and Vessia et al. 2021)

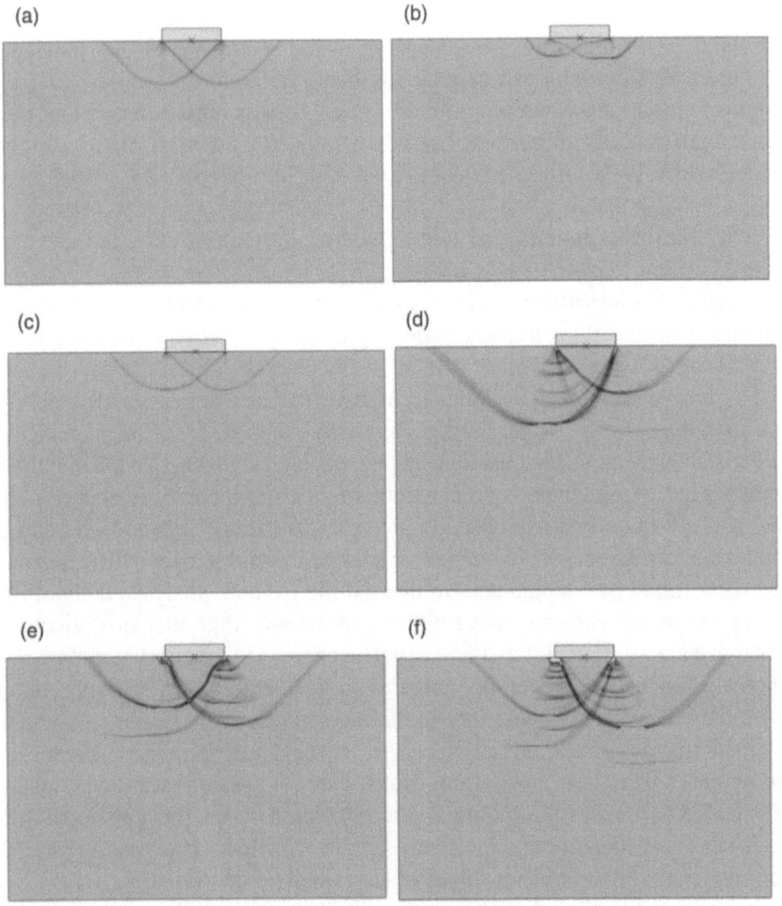

FIGURE 13.1 Different shear planes from realizations of shallow footings on spatially variable soils (reproduced from Li et al. 2015b, with permission).

(Figure 13.1). The reduction in the expected bearing capacity and the asymmetric failure path are generally more evident with a high coefficient of variation (COV) in soil properties, and vary also with the scale of fluctuation (Griffiths et al. 2002; Kasama and Whittle 2011). In addition, the cross-correlation between shear strength parameters in c-ϕ soils also affects the probabilistic estimates of bearing capacities in footings, and neglecting the negative cross-correlation between c and ϕ is deemed to be more conservative (Fenton and Griffiths 2003; Cho and Park 2010).

Fenton and Griffiths (2003) hypothesized that the greatest reduction in bearing capacity would be associated with some intermediate correlation lengths approximately equal to the footing width. For isotropic spatial variability patterns, they also recommended a practical approach to predict the statistics of the bearing capacity, by utilizing Prandtl's classical solution together with the geometric average of soil properties over the domain of plastically deformed failure region. The results are close-form approximations to the mean values and standard deviations of bearing capacity factors. On the other hand, consideration of anisotropy in spatial variability patterns is found to result in lower mean bearing capacity and higher probability of failure (Puła and Chwała 2018), while the non-stationarity of random fields and its influence on bearing capacity was studied by Li et al. (2015a). In particular, the gradient of the increasing trend was found to be a dominant factor affecting the shape of failure envelopes and the statistics of bearing capacity, while the size of failure envelopes is mainly influenced by the scale of fluctuation (Wu et al. 2019; Shen et al. 2023). Recently, Tabarroki et al. (2022a) calibrated a model called the weakest-path model using the results of random finite element simulations. Their approach is based on the observation that the mobilized shear strength is approximately equal to the spatial average along a critical slip curve (Ching and Phoon 2013), which emerges according to the spatial variability patterns to seek the weak zones. Therefore, the model can capture the behavior of footing failure along the weakest path. For offshore foundations where combined vertical, horizontal, and moment (V-H-M) loading conditions are particularly important, Cassidy et al. (2013) demonstrated the possibilities to establish probabilistic V-H-M failure envelopes that took into account spatial variability features of the founding soil. With recent advances in three-dimensional numerical methods, the statistics of bearing capacity for three-dimensional footing geometries can be obtained and compared with plane-strain models (Simões et al. 2014; Chwała 2019).

The overall and differential settlements correspond to the serviceability limit states of the footings, and these phenomena on spatially variable ground are topics that receive significant research interest. Some of the early contributions were made by Fenton and Griffiths (2002, 2005) where single footing and two footing systems were studied in plane-strain and then three-dimensional models with spatially variable elastic modulus. For the single footing case, they suggested that the statistics of the settlement could be satisfactorily approximated by using the geometric average of the modulus underneath the footing, which is in line with the recommendation for bearing capacity evaluation. Tabarroki et al. (2022b) further showed that using the "pseudo incremental energy" method, the numerical values of footing settlements can be reasonably approximated by a weighted geometric average, where the weights can be determined by a single deterministic finite element analysis. It should be noted that approximation in numerical values is more difficult to achieve than that in statistics because two random variables can have identical statistics and yet be completely uncorrelated. Meanwhile, Lo and Leung (2017) conducted a series of conditional random field simulations to explore the influence of soil sampling on the reduction of uncertainty in footing settlement predictions, and quantified these effects for different spatial correlation lengths (in a vertical direction) and different depths of soil sample. Based on the random field simulation results, they suggested that in most cases, a sampling depth (*D*) between 0.5 to 1 times the footing width (*B*) would be the most effective in reducing the uncertainty in footing settlement predictions, an example of which is shown in Figure 13.2 considering footings on soils with different mean values of undrained shear strengths (μ_{cu}).

Another important implication of spatial variability lies on the differential settlements between two footings or among multiple footings, which would not be captured by simulations adopting homogenized layers ($\delta \rightarrow \infty$) or uncorrelated soil properties ($\delta \rightarrow 0$). Various researchers (e.g., Jaksa et al. 2005; Ahmed and Soubra 2014) have arrived at similar conclusions, that the worst condition would entail the scale of fluctuation being close to the horizontal spacing between footings.

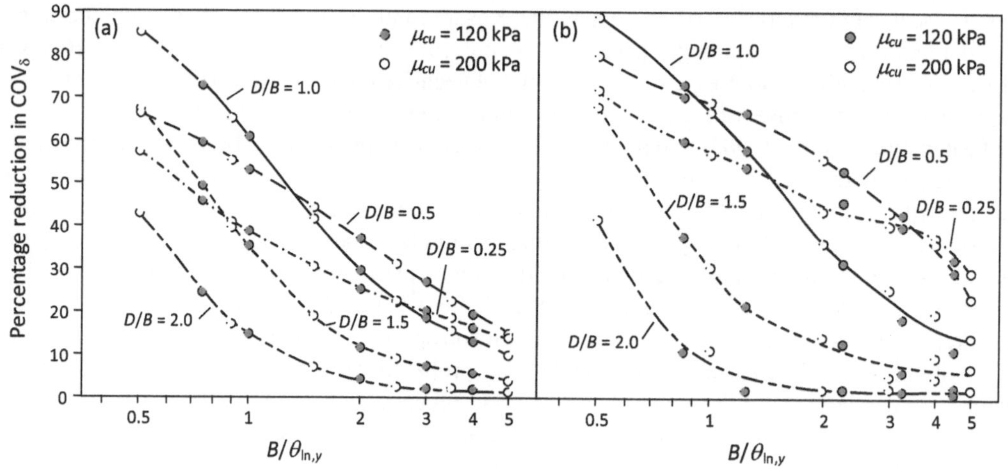

FIGURE 13.2 Reduction in uncertainty (COV) of footing settlement predictions due to sampling, for soils with different COV in undrained shear strength: (a) COVcu = 0.15; (b) COVcu = 0.4 (modified from Lo and Leung 2017).

13.3 PILED FOUNDATIONS

The evaluation of bearing capacity and deformations of piled foundations is often associated with significant model uncertainty (Tang and Phoon 2021), but this section focuses mainly on the influence of spatial variability on their performance. In the case of end-bearing piles, their response is mostly dominated by properties of the stiff stratum on which the piles are founded. On the other hand, floating piles develop much of their resistance from friction between the piles and the surrounding soils. Therefore, the spatial variability patterns can have substantial impacts on the foundation performance. Since geomaterials often entail much higher variability than concrete or structural steel (Chapter 2), the properties of pile materials are often assumed to be deterministic in many numerical studies. Early attempts to understand pile performance in spatially variable soils include Phoon et al. (1990) and Quek et al. (1992), who adopted the first-order and second-order reliability methods to establish design charts of reliability indices for settlements of single piles embedded in soils with variable elastic moduli. Alternatively, as soil-pile interactions are often modeled with different spring stiffness and strength (or "t-z" curves) along the pile, it would be a natural extension to vary the spring properties along the pile based on the random field theory, as described by Fenton and Griffiths (2008). Later, Fan et al. (2014) studied the serviceability limit state of drilled shafts in spatially variable soils subjected to axial and lateral loading, in which the t-z method (including q-w curves) and p-y curves were used to model axial and lateral soil-pile interactions, respectively. They found that the failure probability in the vertical movement limit state and the system failure probability would be sensitive to the variations in the scale of fluctuation, but the failure probabilities for the lateral deflection limit state and the angular distortion limit state were not. Using the random finite difference method (RFDM), Haldar and Babu (2008) investigated the allowable capacity of a laterally loaded pile embedded in undrained clays with spatial variations in strength properties, and the worst-case scale of fluctuation that leads to minimum allowable load was observed to be half of the pile length. Zhou et al. (2022) reached similar conclusions for a laterally loaded offshore monopile case, analyzed by the random finite element method (RFEM), and they also demonstrated that the failure plane exhibited an unsymmetrical pattern and an irregular shape due to soil spatial variability.

For large pile groups embedded in soils, the deformation analyses involve complex interaction effects among the piles and are generally computationally demanding if three-dimensional numerical analyses are performed using the finite element or finite difference methods. The problem is exacerbated for probabilistic assessments that often require a large number of analyses in the typical Monte Carlo fashion. On a similar basis to the work of Phoon et al. (1990), Quek et al. (1991) studied the reliability of pile groups with a rigid pile cap and up to 25 piles, where the pile interaction effects were considered through the soil flexibility matrix. Breysse et al. (2005) and Niandou and Breysse (2007) investigated the structural response of a piled raft foundation in soils with horizontal spatial variability, considering various relative slab to pile stiffness ratios, and revealed that the worst-case horizontal scales of fluctuation for the maximum bending moment and the load on central pile would be close to the span and half span of the foundation, respectively. Later, Fenton and Naghibi (2011) developed a mathematical model to estimate the probability of failure and the resistance factors for the ultimate limit state design of deep foundations in spatially variable sandy soils, while Naghibi et al. (2014, 2016) studied the serviceability limit states of single and two-pile systems in spatially variable ground, by establishing theoretical solutions of the statistics of total and differential settlements which were validated by RFEM. Later, Leung and Lo (2018) extended the approach by Quek et al. (1991) to incorporate three-dimensional random fields, flexible pile caps, and elastoplastic pile response, with probabilistic analyses conducted through response surface method and Monte Carlo simulations and applied to re-analyses of foundation case studies (Figure 13.3). Adopting the soil flexibility matrix for pile-soil interaction effects and the response surface method reduces the computational demands, allowing the approach to be utilized for re-analyses of several foundation cases in Europe, and the ensuing probabilistic analyses revealed the risks of foundation tilting due to spatially variable soil properties. Based on a series of parametric studies, they further

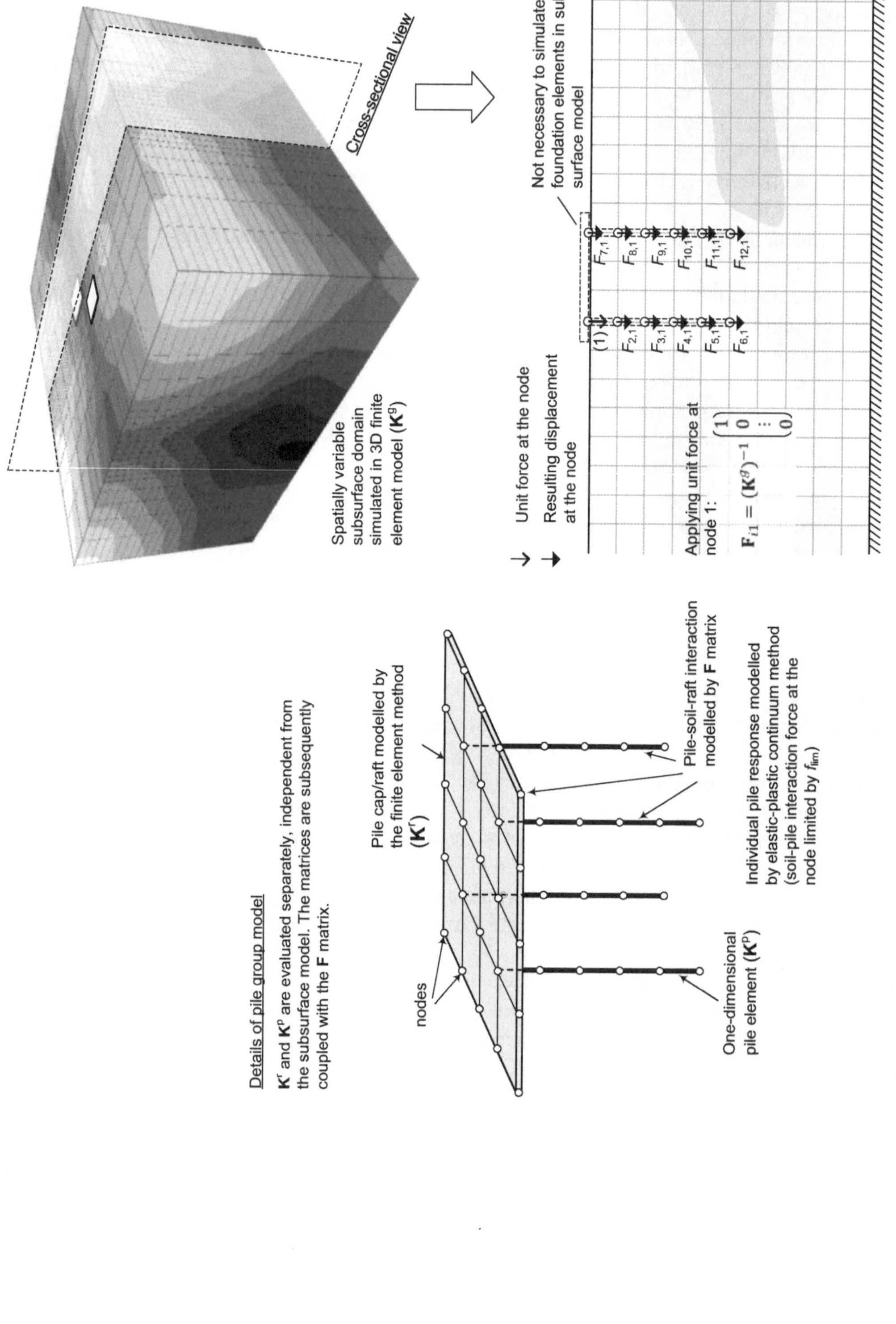

FIGURE 13.3 Probabilistic analysis approach for pile groups (modified from Leung and Lo 2018).

suggested that the worst-case scenarios for differential settlements in pile groups would involve the horizontal scale of fluctuation being close to the width of the pile group or piled raft.

13.4 SLOPES

Landslide hazards pose a significant threat in many major cities around the world, and the reliability of man-made and natural slopes are important topics for researchers, engineers, and government authorities. Conventional deterministic analysis of slopes usually focuses on the evaluation of the factor of safety (FOS), while probabilistic analysis evaluates the reliability index of the slope, or the probability of failure which is often defined, in the context of the Monte Carlo method, as the proportion of the simulations associated with FOS below unity. It may be tempting to assume that the critical slip surface obtained by the deterministic approach (with minimum FOS) would also result in the highest probability of failure in a probabilistic analysis. This is, however, generally not the case unless the stability of the slope is dominated by particularly weak seams or planes constituting the slip surfaces. The differences between critical surfaces in deterministic and probabilistic analyses have been discussed in the literature (e.g., Li and Lumb 1987; Hassan and Wolff 1999). Conceptually, as the mean values of the properties would differ from those at the design point (closest to the limit state function), the critical slip surfaces should be different, even when assumptions of homogenized layers are adopted.

Examples of early works incorporating the random field theory with finite element or limit equilibrium methods in slope reliability assessments include those by El-Ramly et al. (2002), Griffiths and Fenton (2004), Bubu and Mukesh (2004), Cho (2007), and Griffiths et al. (2009a). In particular, Griffiths and Fenton (2004) showed that modeling the slope with homogenized material properties as single random variables could lead to unconservative results when the coefficient of variation of the shear strength parameter is high or factors of safety (based on mean strength) is low. Furthermore, Griffiths et al. (2009a) introduced a "critical" COV value, beyond which the single-random-variable assumption yields unconservative results. They revealed that the critical value would generally be lower for steeper slopes and stronger c-ϕ cross-correlation. Similar to the discussions above on bearing capacities of shallow footings, soils fail along the weakest path within the slopes, which do not necessarily resemble a circular arc or other smooth mathematical functions in cases of significant spatial variability in the shear strength properties (Tabarroki et al. 2013). The "irregular" slip surfaces are naturally catered for by strength reduction methods in finite element and finite difference analyses, while some modern software packages for the limit equilibrium method also include functionality to search for irregular critical failure planes. Recent advances in numerical techniques for large-deformation geotechnical problems facilitate the evaluation of landslide runout distances during the post-failure process. For example, Wang et al. (2016) combined the material point method with random field theory to investigate the runout distance of landslides, while Wang et al. (2019) and Mori et al. (2020) adopted the smoothed particle hydrodynamics method for large-deformation analyses of slope failure incorporating random fields of soil properties. Together with Liu et al. (2019), they showed that the spatial distributions of soil properties have a considerable influence on the evolution of slope failure modes and the slope failure consequence.

While recent research explores the worst-case scale of fluctuation associated with slope stability, there has not been a clear consensus or hard and fast rules on its value. This may be attributed to the fact that different researchers focused on different slope geometries, e.g., infinite slopes in Ali et al. (2014) versus 2D slopes in Tabarroki et al. (2013) and Javankhoshd et al. (2017), or different slope angles, slope heights, and depths to hard stratum (Griffiths and Yu 2015). There can also be different definitions of the "worstcase", such as minimum FOS versus maximum probability of failure. Moreover, when the mean FOS is above 1.0, a higher standard deviation leads to a higher probability of failure, but the probability of failure reduces with the increasing standard deviation in FOS when the mean is below 1.0. Nonetheless, the worst-case scale of fluctuation is found to be in a similar order as the slope height (or soil layer thickness for infinite slopes).

The significance of anisotropic random fields, in particular those with rotated anisotropy features, is gradually being recognized in slope reliability assessments. In general, a higher probability of failure is associated with the case where the dip direction of the slope is aligned with that of the rotated anisotropy in the spatial variation of soil strength. Meanwhile, the three-dimensional effects of spatial variability affect the failure mechanisms, size of the failure mass, and the validity of 2D plane-strain assumptions (e.g., Griffiths et al. 2009b, Hicks et al. 2014, Huang and Leung 2021), and these can provide useful indications in the design of long embankments. In particular, Hicks et al. (2014) discussed three categories of failure modes depending on the horizontal scale of fluctuation θ_h (along the longitudinal direction). When θ_h is larger than the slope height but smaller than half of the slope length along the longitudinal direction, there is a tendency for discrete failures to occur (Figure 13.4) and the reliability reduces as the slope length increases. These phenomena cannot be easily approximated by 2D simulations, and they also highlight the unique features of spatial variability that cannot be captured by homogenized layers or probabilistic analyses adopting single random variables, indicating the influence of spatial variability features (e.g., θ_h) on the lengths and volumes of landslide failure masses.

Considering the influence of soil spatial variability on slope stability and reliability, and the fact that only limited site-specific geotechnical data can be obtained in most projects, there are recent studies that aim to explore the optimal sampling location for slopes. To this end, Yang et al. (2019) conducted a series of numerical simulations on slopes with undrained shear strength increasing with depth. Based on the simulation results, they found that when only one CPT sounding is available, the slope crest seemed to be the optimal sampling location to reduce the probabilities of making the wrong decisions, i.e., suggesting that a slope is safe when it is actually unsafe (type I error), or vice versa (type II error). Lo and Leung (2018) developed an approach to quantify the effectiveness of soil sampling in uncertainty reduction of the factor of safety of slopes in spatially variable soils. They then established a series of design charts on optimal sampling locations (depth of sampling point and horizontal distance from slope toe), catering for various scenarios of slope angles, coefficients of variation of undrained shear strengths, or friction angles. Their design charts also enable quick approximations of the failure probability of the slope, depending on the sampled value of soil strength and its deviation from the adopted mean values.

13.5 LATERAL EARTH PRESSURE AND EXCAVATIONS

Spatial variability in ground properties influences the analyses of retaining structures through the magnitudes of lateral earth pressures, as classical solutions of active and passive pressures do not account for the tendency for soils to fail along the weakest paths in heterogeneous media. The impacts of spatial variability on the active and passive thrusts on a retaining wall have been investigated by various researchers through random field simulations. Based on the results of random finite element analyses of walls in cohesions soils, Fenton et al. (2005) and Griffiths et al. (2008) suggested that

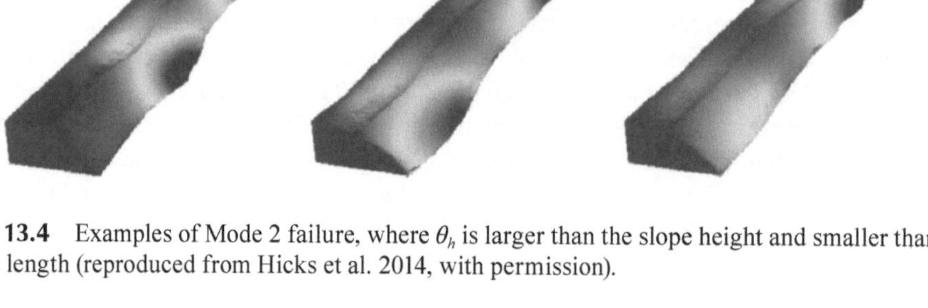

FIGURE 13.4 Examples of Mode 2 failure, where θ_h is larger than the slope height and smaller than half of the slope length (reproduced from Hicks et al. 2014, with permission).

the worst-case correlation lengths that lead to the maximum probability of under-design (compared with Rankine's theory) are of the order of the height of the retaining wall, being 0.5 to 1 times of wall height for active lateral force, and 0.1 to 0.5 times of the wall height for passive lateral force. Hu and Ching (2015) studied the impact of spatial variability in clayey soils on the active lateral force and found that a critical scale of fluctuation exists, at around 0.2 times the wall height, that corresponds to the maximum active lateral force acting on the retaining wall. Tabarroki et al. (2022a) showed that, however, the weakest path seeking for retaining structures is not as significant as that for shallow footings, because the potential failure surfaces for retaining structures are more constrained.

For deep excavations, the design considerations extend beyond the active and passive earth pressures, as basal heave, wall deflections, and structural forces in various elements often govern the system's performance. Ching et al. (2017) studied the basal heave factor of safety for deep excavations in clays and derived that the worst-scale (vertical) scale of fluctuation varies from 0.1 to 0.4 times the wall penetration depth, depending on the horizontal scale of fluctuation. Meanwhile, Luo et al. (2018, 2020) demonstrated the effects of spatially variable ground properties on the wall and ground displacements behind the deep excavation supported by strutted sheet pile wall, and explored the risks of different failure modes such as excessive bending moments or shear forces in the wall and buckling of struts. Lo and Leung (2019) illustrated the effects of 3D spatial variability on the wall displacements, especially the non-uniform displacements along the longitudinal direction of the excavation. These are realistic failure modes or risks in excavation projects which are often observed in the field but cannot be predicted by conventional analyses that do not explicitly consider spatial variability of the ground. In the work by Lo and Leung (2019), they further proposed the use of the Bayesian framework to update the knowledge of spatial variability patterns based on measured responses at the early stages of the excavation. The approach allows for continuous refinements on the estimates of retaining wall response, which help to support the decision-making process as the construction progresses in deep excavation projects.

REFERENCES

Ahmed, A. and Soubra, A. H. 2014. Probabilistic analysis at the serviceability limit state of two neighboring strip footings resting on a spatially random soil. *Structural Safety*, 49, 2–9.

Ali, A., Huang, J., Lyamin, A. V., Sloan, S. W., Griffiths, D. V., Cassidy, M. J. and Li, J. H. 2014. Simplified quantitative risk assessment of rainfall-induced landslides modelled by infinite slopes. *Engineering Geology*, 179, 102–116.

Alonso, E. E. 1976. Risk analysis of slopes and its application to slopes in Canadian sensitive clays. *Géotechique*, 26(3), 453–472.

Babu, G. L. S. and Mukesh, M. D. 2004. Effect of soil variability on reliability of soil slopes. *Géotechnique*, 54(5), 335–337.

Breysse, D., Niandou, H., Elachachi, S. and Houy, L. 2005. A generic approach to soil–structure interaction considering the effects of soil heterogeneity. *Géotechnique*, 55(2), 143–150.

Cassidy, M. J., Uzielli, M. and Tian, Y. 2013. Probabilistic combined loading failure envelopes of a strip footing on spatially variable soil. *Computers and Geotechnics*, 49, 191–205.

Ching, J. and Phoon, K. K. 2013. Mobilized shear strength of spatially variable soils under simple stress states. *Structural Safety*, 41, 20–28.

Ching, J., Phoon, K. K. and Sung, S. P. 2017. Worst case scale of fluctuation in basal heave analysis involving spatially variable clays. *Structural Safety*, 68, 28–42.

Cho, S. E. 2007. Effects of spatial variability of soil properties on slope stability. *Engineering Geology*, 92, 97–109.

Cho, S. E. and Park, H. C. 2010. Effect of spatial variability of cross-correlated soil properties on bearing capacity of strip footing. *International Journal for Numerical and Analytical Methods in Geomechanics*, 34(1), 1–26.

Christian, J. T. and Baecher, G. B. 2011. Unresolved problems in geotechnical risk and reliability. *Georisk*, 2011, 50–63.

Christian, J. T., Ladd, C. C. and Baecher, G. B. 1994. Reliability applied to slope stability analysis. *Journal of Geotechnical Engineering*, 120(12), 2180–2207.

Chwała, M. 2019. Undrained bearing capacity of spatially random soil for rectangular footings. *Soils and Foundations*, 59(5), 1508–1521.

Duncan, J. M. 2000. Factors of safety and reliability in geotechnical engineering. *Journal of Geotechnical and Geoenvironmental Engineering*, 126(4), 307–316.

El-Ramly, H., Morgenstern, N. R. and Cruden, D. M. 2002. Probabilistic slope stability analysis for practice. *Canadian Geotechnical Journal*, 39(3), 665–683.

Fan, H., Huang, Q. and Liang, R. 2014. Reliability analysis of piles in spatially varying soils considering multiple failure modes. *Computers and Geotechnics*, 57, 97–104.

Fenton, G. A. and Griffiths, D. V. 2002. Probabilistic foundation settlement on spatially random soil. *Journal of Geotechnical and Geoenvironmental Engineering*, 128(5), 381–390.

Fenton, G. A. and Griffiths, D. V. 2003. Bearing-capacity prediction of spatially random c-ϕ soils. *Canadian Geotechnical Journal*, 40(1), 54–65.

Fenton, G. A. and Griffiths, D. V. 2005. Three-dimensional probabilistic foundation settlement. *Journal of Geotechnical and Geoenvironmental Engineering*, 131(2), 232–239.

Fenton, G. A. and Griffiths, D. V. 2008. *Risk Assessment in Geotechnical Engineering*. New York: J. Wiley.

Fenton, G. A., Griffiths, D. V. and Williams, M. B. 2005. Reliability of traditional retaining wall design. *Géotechnique*, 55(1), 55–62.

Fenton, G. A. and Naghibi, M. 2011. Geotechnical resistance factors for ultimate limit state design of deep foundations in frictional soils. *Canadian Geotechnical Journal*, 48(11), 1742–1756.

Griffiths, D. V. and Fenton, G. A. 2004. Probabilistic slope stability analysis by finite elements. *Journal of Geotechnical and Geoenvironmental Engineering*, 130(5), 507–518.

Griffiths, D. V., Fenton, G. A. and Manoharan, N. 2002. Bearing capacity of rough rigid strip footing on cohesive soil: Probabilistic study. *Journal of Geotechnical and Geoenvironmental Engineering*, 128(9), 743–755.

Griffiths, D. V., Fenton, G. A. and Manoharan, N. 2006. Undrained bearing capacity of two-strip footings on spatially random soil. *ASCE International Journal of Geomechanics*, 6(6), 421–427.

Griffiths, D. V., Fenton, G. A. and Ziemann, H. R. 2008. Reliability of passive earth pressure. *Georisk: Assessment and Management of Risk for Engineered Systems and Geohazards*, 2(2), 113–121.

Griffiths, D. V., Huang, J. and Fenton, G. A. 2009a. Influence of spatial variability on slope reliability using 2-D random fields. *Journal of Geotechnical and Geoenvironmental Engineering*, 135(10), 1367–1378.

Griffiths, D. V., Huang, J. and Fenton, G. A. 2009b. On the reliability of earth slopes in three dimensions. *Proceedings of the Royal Society Series A*, 465(2110), 3145–3164.

Griffiths, D. V. and Yu, X. 2015. Another look at the stability of slopes with linearly increasing undrained strength. *Géotechnique*, 65(10), 824–830.

Haldar, S. and Babu, G. S. 2008. Effect of soil spatial variability on the response of laterally loaded pile in undrained clay. *Computers and Geotechnics*, 35(4), 537–547.

Hassan, A. M. and Wolff, T. F. 1999. Search algorithm for minimum reliability index of earth slopes. *Journal of Geotechnical and Geoenvironmental Engineering*, 125(4), 301–308.

Hicks, M. A., Nuttall, J. D. and Chen, J. 2014. Influence of heterogeneity on 3D slope reliability and failure consequence. *Computers and Geotechnics*, 61, 198–208.

Hu, Y. G. and Ching, J. 2015. Impact of spatial variability in undrained shear strength on active lateral force in clay. *Structural Safety*, 52, 121–131.

Huang, L. and Leung, Y. F. 2021. Reliability assessment of slopes with three-dimensional rotated transverse anisotropy in soil properties. *Canadian Geotechnical Journal*, 58(9), 1365–1378.

Jaksa, M. B., Goldsworthy, J. S., Fenton, G. A., Kaggwa, W. S., Griffiths, D. V., Kuo, Y. L. and Poulos, H. G. 2005. Towards reliable and effective site investigations. *Géotechnique*, 55(2), 109–121.

Javankhoshdel, S., Luo, N. and Bathurst, R. J. 2017. Probabilistic analysis of simple slopes with cohesive soil strength using RLEM and RFEM. *Georisk: Assessment and Management of Risk for Engineered Systems and Geohazards*, 11(3), 231–246.

Kasama, K. and Whittle, A. J. 2011. Bearing capacity of spatially random cohesive soil using numerical limit analyses. *Journal of Geotechnical and Geoenvironmental Engineering*, 137(11), 989–996.

Kim, H. K. and Santamarina, J. C. 2008. Spatial variability: Drained and undrained deviatoric load response. *Géotechnique*, 58(8), 805–814.

Leung, Y. F. and Lo, M. K. 2018. Probabilistic assessment of pile group response considering superstructure stiffness and three-dimensional soil spatial variability. *Computers and Geotechnics*, 103, 193–200.

Li, D. Q., Qi, X. H., Cao, Z. J., Tang, X. S., Zhou, W., Phoon, K. K. and Zhou, C. B. 2015a. Reliability analysis of strip footing considering spatially variable undrained shear strength that linearly increases with depth. *Soils and Foundations*, 55(4), 866–880.

Li, J., Tian, Y. and Cassidy, M. J. 2015b. Failure mechanism and bearing capacity of footings buried at various depths in spatially random soil. *Journal of Geotechnical and Geoenvironmental Engineering*, 141(2), 04014099.

Li, K. S. and Lumb, P. 1987. Probabilistic design of slopes. *Canadian Geotechnical Journal*, 24(4), 520–535.

Liu, X., Wang, Y. and Li, D. Q. 2019. Investigation of slope failure mode evolution during large deformation in spatially variable soils by random limit equilibrium and material point methods. *Computers and Geotechnics*, 111, 301–312.

Lo, M. K. and Leung, Y. F. 2017. Probabilistic analyses of slopes and footings with spatially variable soils considering cross-correlation and conditioned random field. *Journal of Geotechnical and Geoenvironmental Engineering*, 143(9), 04017044.

Lo, M. K. and Leung, Y. F. 2018. Reliability assessment of slopes considering sampling influence and spatial variability by Sobol' sensitivity index. *Journal of Geotechnical and Geoenvironmental Engineering*, 144(4), 04018010.

Lo, M. K. and Leung, Y. F. 2019. Bayesian updating of subsurface spatial variability for improved prediction of braced excavation response. *Canadian Geotechnical Journal*, 56(8), 1169–1183.

Luo, N. and Bathurst, R. J. 2017. Reliability bearing capacity analysis of footings on cohesive soil slopes using RFEM. *Computers and Geotechnics*, 89, 203–212.

Luo, Z., Di, H., Kamalzare, M. and Li, Y. 2020. Effects of soil spatial variability on structural reliability assessment in excavations. *Underground Space*, 5(1), 71–83.

Luo, Z., Hu, B., Wang, Y. and Di, H. 2018. Effect of spatial variability of soft clays on geotechnical design of braced excavations: A case study of Formosa excavation. *Computers and Geotechnics*, 103, 242–253.

Mori, H., Chen, X., Leung, Y. F., Shimokawa, D. and Lo, M. K. 2020. Landslide hazard assessment by smoothed particle hydrodynamics with spatially variable soil properties and statistical rainfall distribution. *Canadian Geotechnical Journal*, 57(12), 1953–1969.

Naghibi, F., Fenton, G. A. and Griffiths, D. V. 2014. Serviceability limit state design of deep foundations. *Géotechnique*, 64(10), 787–799.

Naghibi, F., Fenton, G. A. and Griffiths, D. V. 2016. Probabilistic considerations for the design of deep foundations against excessive differential settlement. *Canadian Geotechnical Journal*, 53(7), 1167–1175.

Niandou, H. and Breysse, D. 2007. Reliability analysis of a piled raft accounting for soil horizontal variability. *Computers and Geotechnics*, 34(2), 71–80.

Pan, Y., Liu, Y., Fook, H. X., Lee, H. and Phoon, K. K. 2018. Effect of spatial variability on short-and long-term behaviour of axially-loaded cement-admixed marine clay column. *Computers and Geotechnics*, 94, 150–168.

Phoon, K. K., Cao, Z., Ji, J., Leung, Y. F., Najjar, S., Shuku, T., Tang, C., Yin, Z. Y., Yoshida, I. and Ching, J. 2022. Geotechnical uncertainty, modeling, and decision making. *Soils and Foundations*, 62(5), 101189.

Phoon, K. K., Quek, S. T., Chow, Y. K. and Lee, S. L. 1990. Reliability analysis of pile settlement. *Journal of Geotechnical Engineering*, 116(11), 1717–1734.

Puła, W., Pieczyńska-Kozłowska, J. M. and Chwała, M. 2017. Search for the worst-case correlation length in the bearing capacity probability of failure analyses. Geo-risk 2017 in ASCE geotechnical special publication, GSP 283, 534–544.

Puła, W. and Chwała, M. 2018. Random bearing capacity evaluation of shallow foundations for asymmetrical failure mechanisms with spatial averaging and inclusion of soil self-weight. *Computers and Geotechnics*, 101, 176–195.

Quek, S. T., Phoon, K. K. and Chow, Y. K. 1991. Pile group settlement: A probabilistic approach. *International Journal for Numerical and Analytical Methods in Geomechanics*, 15(11), 817–832.

Quek, S. T., Chow, Y. K. and Phoon, K. K. 1992. Further contributions to reliability-based pile-settlement analysis. *Journal of Geotechnical Engineering*, 118(5), 726–741.

Shen, Z., Pan, Q., Chian, S. C., Gourvenec, S. and Tian, Y. 2023. Probabilistic failure envelopes of strip foundations on soils with non-stationary characteristics of undrained shear strength. *Géotechnique*, 73(8), 716–735.

Simões, J. T., Neves, L. C., Antão, A. N. and Guerra, N. M. 2014. Probabilistic analysis of bearing capacity of shallow foundations using three-dimensional limit analyses. *International Journal of Computational Methods*, 11(2), 1342008.

Soubra, A. H., Massih, D. S. Y. A. and Kalfa, M. 2008. Bearing capacity of foundations resting on a spatially random soil. *ASCE Geotechnical Special Publication*, 178, 66–73.

Stuedlein, A. W. and Bong, T. 2017. Effect of spatial variability on static and liquefaction-induced differential settlements. ASCE geotechnical special publication, GSP, 282, 31–51.

Tang, C. and Phoon, K. K. 2021. *Model Uncertainties in Foundation Design*. CRC Press, London.

Tabarroki, M. 2020. PhD Dissertation, Dept of Civil Engineering, National Taiwan University.

Tabarroki, M., Ahmad, F., Banaki, R., Jha, S. K. and Ching, J. 2013. Determining the factors of safety of spatially variable slopes modeled by random fields. *Journal of Geotechnical and Geoenvironmental Engineering*, 139(12), 2082–2095.

Tabarroki, M., Ching, J., Phoon, K. K. and Chen, Y. Z. 2022a. Mobilisation-based characteristic value of shear strength for ultimate limit states. *Georisk: Assessment and Management of Risk for Engineered Systems and Geohazards*, 16(3), 413–434.

Tabarroki, M., Ching, J., Lin, C. P., Liou, J. J. and Phoon, K. K. 2022b. Homogenizing spatially variable young modulus using pseudo-incremental energy method. *Structural Safety*, 97, 102226.

Vanmarcke, E. H. 1977. Reliability of earth slopes. *Journal of the Geotechnical Engineering Division*, 103(11), 1247–1265.

Vessia, G., Cherubini, C., Pieczyńska, J. and Puła, W. 2009. Application of random finite element method to bearing capacity design of strip footing. *Journal of GeoEngineering*, 4, 103–111.

Vessia, G., Zhou, Y. G., Leung, A., Pula, W., Di Curzio, D., Tabarroki, M. and Ching, J. 2021. Numerical evidences for worst-case scale of fluctuation. In TC304 State-of-the-art review of inherent variability and uncertainty, 204–215.

Wang, B., Hicks, M. A. and Vardon, P. J. 2016. Slope failure analysis using the random material point method. *Géotechnique Letters*, 6(2), 113–118.

Wang, Y., Qin, Z., Liu, X. and Li, L. 2019. Probabilistic analysis of post-failure behavior of soil slopes using random smoothed particle hydrodynamics. *Engineering Geology*, 261, 105266.

Wu, Y., Zhou, X., Gao, Y., Zhang, L. and Yang, J. 2019. Effect of soil variability on bearing capacity accounting for non-stationary characteristics of undrained shear strength. *Computers and Geotechnics*, 110, 199–210.

Yang, R., Huang, J., Griffiths, D. V., Li, J. and Sheng, D. 2019. Importance of soil property sampling location in slope stability assessment. *Canadian Geotechnical Journal*, 56(3), 335–346.

Zhou, Z., Li, X. and Zhao, H. 2022. Probabilistic study of offshore monopile foundations considering soil spatial variability. *ASCE-ASME Journal of Risk and Uncertainty in Engineering Systems, Part A: Civil Engineering*, 8(3), 04022035.

14 Reduction of Uncertainty through Piling Data within the Same Site in the Press-in Piling Method

Naoki Suzuki

ABSTRACT

This chapter provides an overview of the current use of piling data, focusing on the application of piling data to improve the quality of geotechnical structures. After summarizing previous studies on piling data and introducing the variation of bearing capacity estimation, potential issues are explored based on a pile load test database using multiple piling data within the same site, such as within- and inter-site errors in estimation. Two examples of the practical applications of press-in piling data are presented. In the first example, simplified load tests were conducted using a press-in piling machine due to the uncertain geotechnical performance of the mudstone. At the relatively large site in the second example, variations of the bearing layer were observed by piling data. The high spatial correlation of the piling data allows us to reduce the individual estimation error and to identify trends within the site.

14.1 INTRODUCTION

The construction industry has made significant advancements with the integration of building information modeling (BIM) and digital twins. Some technology is used for construction management and reducing costs where measuring instruments are installed in construction machinery, such as intelligent compaction which facilitate real-time compaction monitoring and timely adjustments. However, despite these advances, the use of piling data has remained relatively unchanged, especially in small projects.

Geotechnical and geological uncertainties have been identified for piled structures. For the former, the uncertainties have been reported and clarified by many years of research (e.g., Evangelista et al., 1978; Vanmarcke, 1977; Phoon and Kulhawy, 1999a; Chen and Zhang, 2013). As for the latter, reports of sensitivity analysis for pile-bearing capacity and cantilever retaining walls are beginning to show (Fei et al., 2022; Suzuki et al., 2021). It is still difficult to completely remove both uncertainties from the preliminary ground investigation alone. The design based on safety margins leaves room for rationalization.

Pile-bearing capacity estimation methods have been developed for each piling method. The coefficients of variation (COV) of these estimation methods are smaller than those estimated from soil information alone, as will be explained later. The author believes that developing a method by which piling data can be used to improve the quality of geotechnical structures will contribute to the rationalization (e.g., Lacasse et al., 1989; Otake et al., 2021) and, in turn, make the piling data more valuable.

However, with pre-fabricated piles, it is naturally difficult to modify the pile design after obtaining piling data, which makes it difficult to make decisions based on the piling data (Van Baars

DOI: 10.1201/9781003333586-18

and Vrijling, 2005). In response to this problem, the author proposed a design and construction framework based on piling data and contingency plans, and discussed its applicability (Suzuki and Ishihara, 2019a). The author determined that the value of the pile driving data would increase dramatically if the variation in bearing capacity estimated from the piling data was less than it is now.

Piling data reduce spatial variability because they are in-situ data. On the other hand, the variability in estimation can be divided into inter-site and within-site variability based on analysis of variance (ANOVA). Multiple data within the same site should reduce this within-site variability, and the characteristics of these variations and how multiple data are used are important.

This chapter will present an overview of how piling data improves the quality of the pile foundations and the retaining wall. After the current methods of estimating bearing capacity from piling data are summarized, the within- and inter-site errors are reported based on a database. Press-in piling method and its simplified load test are introduced, which records the continuous penetration resistance. Finally, two examples of the application of within-site press-in piling data are presented for cases related to geotechnical uncertainty and geological uncertainty, respectively.

14.2 PILING DATA

14.2.1 Estimation of Bearing Capacity from Piling Data

Data related to piling includes one during pile penetration and subsequent load tests. Reliability analysis has been conducted by Zhang et al. (2006), Zhang (2004) and McVay et al. (2000), and partial coefficients for different types and numbers of load tests have been presented in the standards (AASHTO, 2012; EN1997-1, 2004). In this chapter, piling data is defined as the former, which can be easily obtained by construction machinery with little preparation or procedures.

Table 14.1 summarizes the previous studies on statistics of bearing capacity estimation with pre-fabricated piles based on piling data, comparing with model errors in the calculation based on geotechnical investigation. The former includes dynamic formulas (Hiley, 1925) and wave equations such as Case Pile Wave Analysis Program (CAPWAP, Rausche et al., 1972) for driven piles, and torque correlation for helical piles.

Impact driving is the oldest piling method, and the relationship between the hammer energy and the pile displacement after a blow has long been used as a reference for construction management. In Japan, it was reported to have been used for the construction management of the foundation of the Tokyo Station (Kanai, 1915). Ng and Sritharan (2016) reported that the COV in CAPWAP is 19%, including the setup effect. Although the dynamic formulas have more variations, they are still used in practice. Reddy and Stuedlein (2013) reported COVs of 25–28% with some estimation formulas. Mizutani and Matsumura (2016) also reported that the Hiley formula had a COV of 25% within the same site.

For vibratory-driven piles, formulas have been proposed to estimate the dynamic bearing capacity of a pile using the current and voltage values of the motor and the penetration rate, though Holeyman (2002) warned about using it with extreme caution because of the significant uncertainties involved. In Japan, the relationship with the ultimate bearing capacity was reported (JRA, 2020), and a method using the wave equation has also been developed.

Screw piles (helical piles) are becoming widespread for medium-scale piles (300 to 800mm diameter) (Lehane, 2005). The International Building Code (IBC, 2009) presented the torque-correlation as a method to evaluate the ultimate bearing capacity. Ohki et al. (2005) reported the depths of the bearing layer estimated from piling torque at 41 sites and emphasized the usefulness of piling management based on the data.

Press-in piling, also known as pile jacking, is one of the oldest pile installation methods, along with impact driving. Its mechanism is easy to understand due to the similarity between pile installation and pile loading. Since press-in piling has difficulty penetrating piles into stiff soil, some auxiliary methods are increasingly being used (IPA, 2021). In the rotary cutting press-in piling, Ishihara et al. (2015) proposed separating the base and shaft resistance based on the relationship between speed and resistance in the rotational and insertion directions, and estimating the N-value

TABLE 14.1
Summary of Statistics of Bearing Capacity Estimation of Pile Foundations

Remarks		n	Mean*	COV	References
Model errors in the calculation based on geotechnical investigation					
JSHB method	Driven piles	129	0.93	41%	Nanazawa et al., 2019
	Bored piles	124	0.97	30%	*ditto*
β-Method	Sand	20	0.85	62%	Paikowsky, 2004
α-API	Clay	19	1.27	54%	*ditto*
ICP-05	Aged filtered	70	0.94	30%	Yang et al., 2015
UWA-05	Aged filtered	70	1.05	35%	*ditto*
Recurrent NN (CPT-based)	Driven piles and drilled shafts, s_{max}=10%D	7**	1.01	8%	Shahin, 2016
NN (SPT-based)	Jacked-in reinforced concrete piles at Ha Nam, Vietnam	About 690 (1)**	1.01	11%	Pham et al., 2020
Random forest (*ditto*)	*ditto*		1.01	9%	*ditto*
PMT	All piles	174	0.96	25%	Burlon et al., 2014
Estimation errors from piling data					
<u>Dynamic formula</u>					
Janbu	Recalibrated, EOD vs. SLT	202	1.00	25%	Reddy and Stuedlein, 2013
Danish	Recalibrated, EOD vs. SLT	202	1.00	25%	*ditto*
FHWA Gates	Recalibrated, EOD vs. SLT	202	1.02	27%	*ditto*
HKCA	Based on Davisson's failure	31	0.85	20%	Zhang et al., 2006
Measured energy	Based on Davisson's failure	31	1.02	17%	*ditto*
Hiley	Based on Davisson's failure	31	0.64	28%	*ditto*
	EOD vs. CAPWAP	31 (1)	1.36	30%	Mizutani and Matsumura, 2016
Proposed	vs. SLT	14	1.04	26%	Salgado et al., 2017
<u>Wave theory</u>					
WEAP	EOD	99	0.60	72%	Paikowsky, 2004
	EOD with setup consideration	30	0.97	19%	Ng and Sritharan, 2016
PDA	Driven piles	48	0.74	33%	McVay et al., 2000
CAPWAP	Based on Davisson's failure	30	0.82	17%	Zhang et al., 2006
<u>Machine learning through driving record</u>					
NN		35**	1.02	23%	Goh, 1995
NN	vs. CAPWAP	34**	1.00	12%	Chan et al., 1995
GEP and NN with SPT	Based on Davisson's failure	6**	0.95	14%	Alkroosh and Nikraz, 2014
<u>Other piling methods</u>					
Empirical formula	Vibratory-driven piles	6	1.12	32%	JRA, 2020
Torque correlation	Helical piles (Round-Shape, Single-helix), Clay	75	1.09	26%	Tang and Phoon, 2018
<u>Extrapolation of load-settlement curve</u>					
Chin's formula	s_{max}= 5%D, CFA	10	—	15%	Galbraith et al., 2014
	s_{max}=10%D, CFA	10	—	10%	*ditto*

Notes: Some statics are calculated based on the figure shown in the reference. *The prediction normalized by the measurement. If the statistics of measurement normalized by the prediction were presented in references, the mean of the inverse and the COV were shown as it is. **Only validation data. n = no. of piles or (number of sites or piers), s_{max} = maximum settlement, BOR = beginning of redriving, CAPWAP = Case Pile Wave Analysis Program, CFA = continuous flight auger piles, D = pile diameter, EOD = end of driving, FHWA = Federal Highway Administration, GEP = gene expression programming, HKCA = Hong Kong Construction Association, JSHB = Japanese Specifications for Highway Bridges, NN = neural network, PDA = case method using the pile driving analyzer, PMT = pressure-meter tests, SLT = static load test, WEAP = Wave Equation Analysis Program.

of the standard penetration test (SPT). Ishihara (2023) also proposed updating the bearing capacity based on UWA-05 method (Lehane, 2005) with the piling data of pressing and rotational resistance.

In addition, the pile capacity changes during piling and after curing, which is called the setup effect. For the effect, many estimation formulas have been proposed for each soil type (Tang and Phoon, 2021). Yu et al. (2012) proposed a method to extrapolate the setup effect by re-pressing the pile with a time delay.

Estimation methods using neural networks (NN) over piling data have also been proposed relatively early (Goh, 1995; Chan et al., 1995). In addition, models considering correlation with SPT have also been proposed and have yielded better results (Alkroosh and Nikraz, 2014).

As described above, piling data is used to determine the bearing layer and to estimate the subsoil information and the bearing capacity; however, some guidelines advise caution in using them because of the variation in the estimation. The COV in the ultimate bearing capacity estimation using the piling data is smaller, approximately 20–30%, while those estimated from the ground investigation are 30–40%.

14.2.2 Rotary Cutting Press-in Piling and Simplified Load Tests

A walk-on type press-in piling machine uses installed piles as a reaction force and does not require additional dead weights (Figure 14.1). The machinery units can be set on top of installed piles, which does not require large working yards. This advantage and the low noise and vibration procedure have led to increased adoptions of the press-in piling in urban or mountainous areas where construction is difficult.

Another feature is the simplified load test, which can be performed as easily as a normal pile penetration by the press-in piling machine. After the pile was installed, static push-in can be done directly with the same machine. It can be conducted for all piles, providing valuable insights into individual pile performance. A detailed test and an example of measured settlement at a service load will be presented in the second half of this chapter.

Piles are often required to be embedded into hard ground, and some assistance methods have been developed. In the rotary cutting press-in piling, a pile with cutting bits attached to the pile tip is rotationally pushed into the ground. This allows penetration into relatively hard bedrock and reinforced concrete of existing structures. But in most cases, the ultimate bearing capacity of a pile exceeds the pressing capacity of the machine (e.g., 2000 kN for a machine 1000 mm pile diameter). Thus, the load-settlement curve needs to be extrapolated to estimate the ultimate bearing capacity.

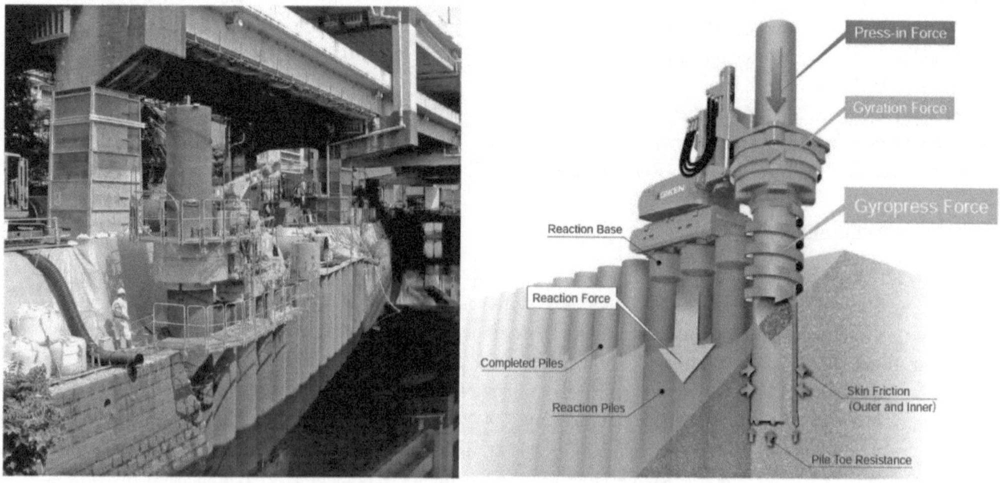

FIGURE 14.1 Machinery for rotary cutting press-in piling (walk-on pile type) (Giken Ltd., 2017).

Numerous formulas to extrapolate/interpolate load-settlement curves have been proposed by Chin (1972), Van Der Veen (1957), Uto et al. (1982), and Zhang and Zhang (2012), and were compared and discussed by researchers such as Paikowsky and Tolosko (1999) and Fellenius and Rahman (2019).

Suzuki and Ishihara (2019b) compared the loading and the simplified load tests at three sites with rotary cutting press-in piles. The results showed that they agree well and that Chin's extrapolation underestimated the measured ultimate bearing capacity. Suzuki (2022a) also found that the maximum pressing force could be the lower bound of the estimated variability in the extrapolated ultimate bearing capacity from the load-settlement curves. This can reduce the estimated variation.

14.3 WITHIN- AND INTER-SITE VARIABILITY IN ESTIMATION FROM PILING DATA

14.3.1 Overview

Piles are not often used individually; in most cases, multiple piles are installed to construct foundations and walls within a single site. Based on ANOVA, estimation variability includes within- and inter-site errors, also referred to as intra- and between-site errors, respectively. Zhang et al. (2020) reported these errors from the geotechnical investigation; similar errors occur in the estimation from piling data. This section will analyze the relationship between the within- and inter-site errors in CAPWAP and Chin's extrapolation based on a database.

The Federal Highway Administration (FHWA) database (Kalavar and Ealy, 2000) with load test results for multiple piles at the same site was used. The bearing capacity at 10% displacement of the pile diameter measured by the static load test was taken as the true value, and the estimated variations by CAPWAP and Chin's extrapolation method were checked (Table 14.2). In Chin's extrapolation, the ultimate bearing capacity was estimated using the load–settlement curves from the static load test at settlements up to 2–5% of the pile diameter, *D*.

14.3.2 Results

Figure 14.2 plots an example of the estimation errors for each site, which indicates estimation normalized by measurement value. Figure 14.2 a is for 3% *D* for the same load tests. Sub-figures b and c are for the same pile with static load tests and CAPWAP at beginning of redriving (BOR).

To separate the estimation error from the inter- and within-site error, Markov chain Monte Carlo methods (MCMC) are implemented. The errors were examined in four combinations using a normal or lognormal distribution with within-site and between-site errors common to all sites, respectively. The model is selected considering widely available information criterion (WAIC, Watanabe, 2010); it is assumed that both errors follow a lognormal distribution (Figure 14.3). The convergence of MCMC is confirmed for each trace plot and Rhat values of less than 1.05 (Gelman et al., 2013).

Figure 14.4 shows the posterior distributions of inter-site mean and inter-/within-site standard deviations by MCMC. In Chin's extrapolations, both inter- and within-site standard deviations

TABLE 14.2
Overview of the Databases

Case	Piling stage	Estimation	Material	Number Piles	Sites	Miscellaneous Piles
A	BOR	CAPWAP	Steel	26	10	23
B	BOR	CAPWAP	Concrete	23	7	10
C	BOR	Chin's extrapolation	All	163	37	—

Note: BOR = beginning of redriving

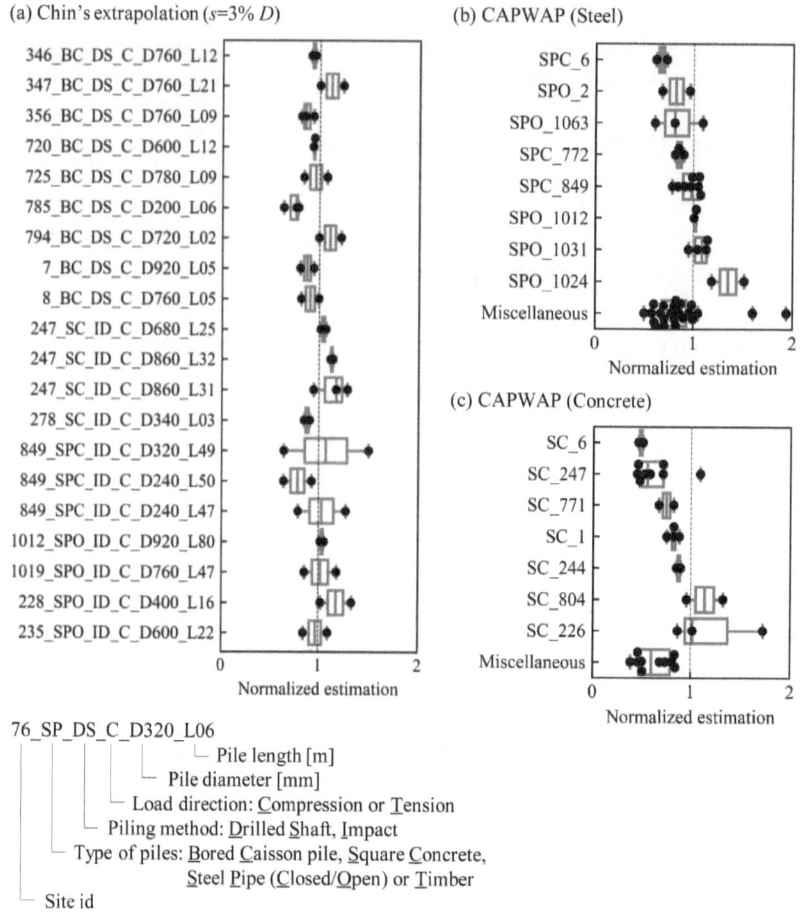

FIGURE 14.2 Estimation errors at each site (Source: FHWA database).

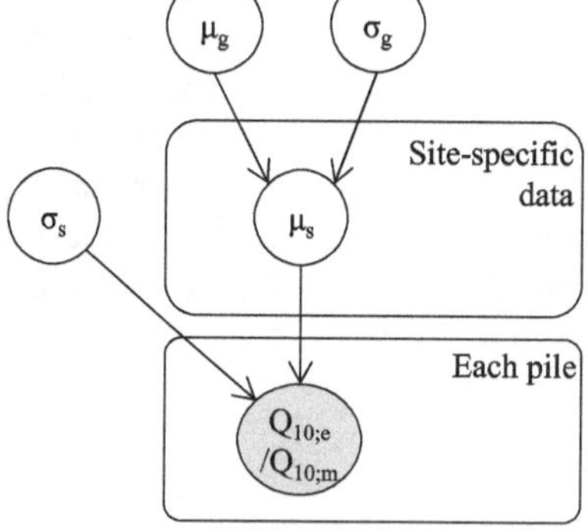

FIGURE 14.3 Graphical model for MCMC.

decrease with increasing settlement. The inter-site error was greater than the within-site error in CAPWAP and vice versa in Chin's extrapolation of load-settlement curves. This may be due to the fact that Chin's extrapolation is derived from the same load test data when getting the ultimate capacity and extrapolating it, whereas it is different in CAPWAP. This can lead to not only measurement errors and human errors (Fellenius, 1988), but also differences in setup effect.

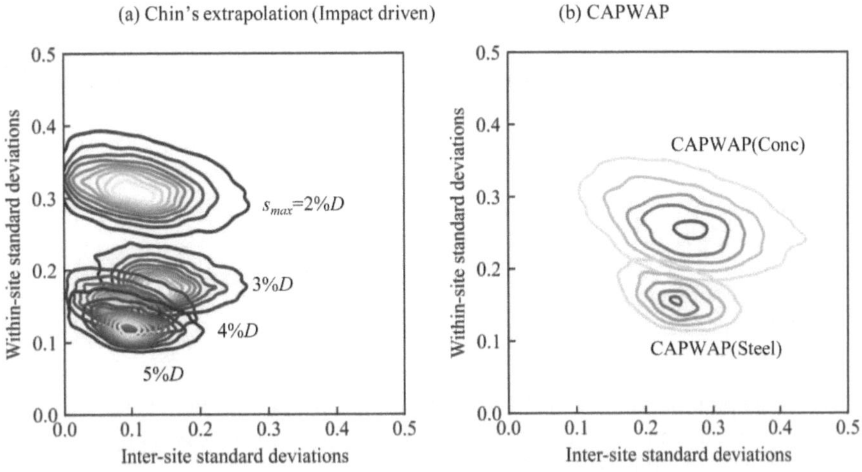

FIGURE 14.4 Posterior distribution with Kernel density estimation obtained by MCMC: inter-site means and inter-/within-site standard deviations.

14.3.3 Using Piling Data for Quality Control of Pile Foundations

There are two main ways to improve these estimation errors in the field: reducing within-site estimation error and inter-site estimation error.

A site-specific database could improve estimations of bearing capacity at each pile. A hierarchical Bayesian method (e.g., Bozorgzadeh et al., 2019) and random field modeling (e.g., Wang and Cao, 2013; El Haj et al., 2019) are possible options. These methods can reduce within-site uncertainty. Another option is to compensate for the bias by getting true capacity by more reliable methods such as load test, which is recommended in various guidelines (ICC-ES, 2007) and will reduce the inter-site variance.

This can be illustrated using the within-site and inter-site variances (Figure 14.5). In cases where the x and y axes are the inter-site and within-site standard deviations (SD), respectively, the estimation SD is the distance from the origin. In the two previous methods, the variations to be

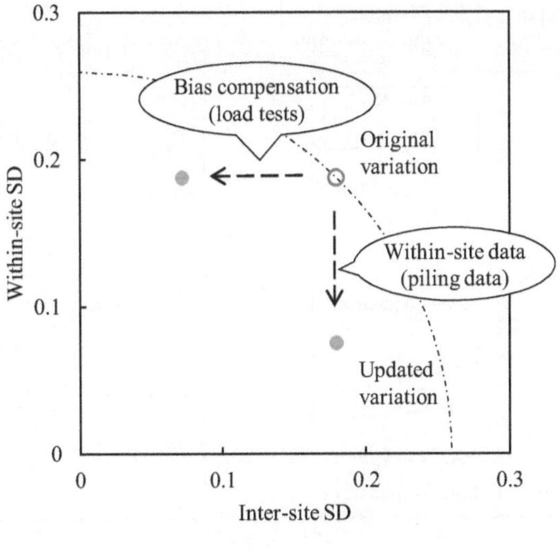

FIGURE 14.5 Illustrative image of decreasing uncertainty.

reduced are completely different, and the appropriate way to reduce the variation depends on the type of the piling data. For estimation methods with large within-site variation, such as extrapolating load-settlement curves as shown in Figure 14.4, collecting within-site data can be effective.

14.4 APPLICATION OF PILING DATA

Two field examples are presented in this section. The purposes of each piling data were quality control of pile foundations and checking the bearing layer for river levees.

14.4.1 River Walls and Foundations

14.4.1.1 Overview

This project was the pile foundation works for a 30-m-long bridge abutment substructure and overhanging walkway (Figure 14.6): 4 abutment support piles and 24 overhanging walkway foundations for a bus stop (D = 1000 mm; L = 10.5 m). After completion of the pile penetration, simplified load tests were conducted to check the integrity of the piles. In addition, a dynamic load test was conducted on one pile as a representative.

After penetrating 5.2 m of unreinforced concrete revetment, 4.0 m of the pile was installed in the mudstone, which is the bearing layer (Figure 14.7a). Geotechnical investigations were conducted on

FIGURE 14.6 Photos of the projects (a: during press-in piling, b: after piling)

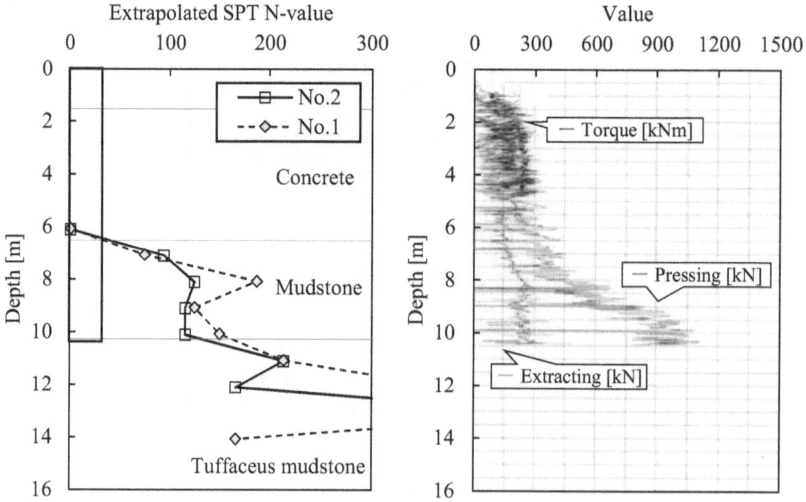

FIGURE 14.7 Pile profile and SPT-N-value a) and overview of press-in piling data (pile W11) b).

both sides of the bridge, and the ground was found to be horizontally homogeneous (No. 2 SPT is on the side of the pile work, and No. 1 is on the other side of the river).

Figure 14.7b shows an example of press-in piling data, which represents the envelope of the maximum value of data at each depth. The mudstone layer was observed to be approximately 1 m shallower than designed, as judged from the trend of increasing pressing force and torque, and the piles were installed into the designed depth.

14.4.1.2 Simplified Load Tests

The test was conducted using a staged loading method, where each load was held for one minute at the beginning and was last held at 1150 kN or 1550 kN of the service loads for ten minutes. The load was applied by the hydraulic pressure of the press-in machine, and the displacement was measured by a dial gauge fixed by the adjasent pile (Figure 14.8).

Figure 14.9 shows a load-settlement curve. Note that the weight of the chuck of the press-in machine and the weight of the follower (approximately 130 kN) were omitted. Static results from the dynamic load test which was conducted two days after the simplified test were added with a red cross in the figure. The results of the last pressing in the simplified test and the first blow in the dynamic load test were roughly close.

Figure 14.10 shows the spatial distribution of the settlement at the service loads. The elastic deformation of the pile, assuming the load distribution during the dynamic load test, is shown by the dotted line.

Although the settlements seem spatially correlated, some settlements are unusually small; those of piles W16 and W19 are about 1/2 smaller than those of the adjacent piles. Some settlements are smaller than the estimated elastic deformation of the piles.

Details are omitted in this paper, and no clear correlation could be found between the following conditions and the amount of a settlement: average water discharge in the last 2 m, press-in duration in the last 2.5 m, press-in force at the last penetration, and curing period. This will indicate that the directly measured settlements themselves are more important.

Finally, Gaussian process (GP) (Rasmussen and Williams, 2006) is used to analyze the spatial characteristics of the settlement at the time of pile failure. The settlement is assumed to follow a lognormal distribution, and a Matérn function is used as the kernel.

FIGURE 14.8 Overview of simplified load tests. The lower right photo was reversed left to right for readability.

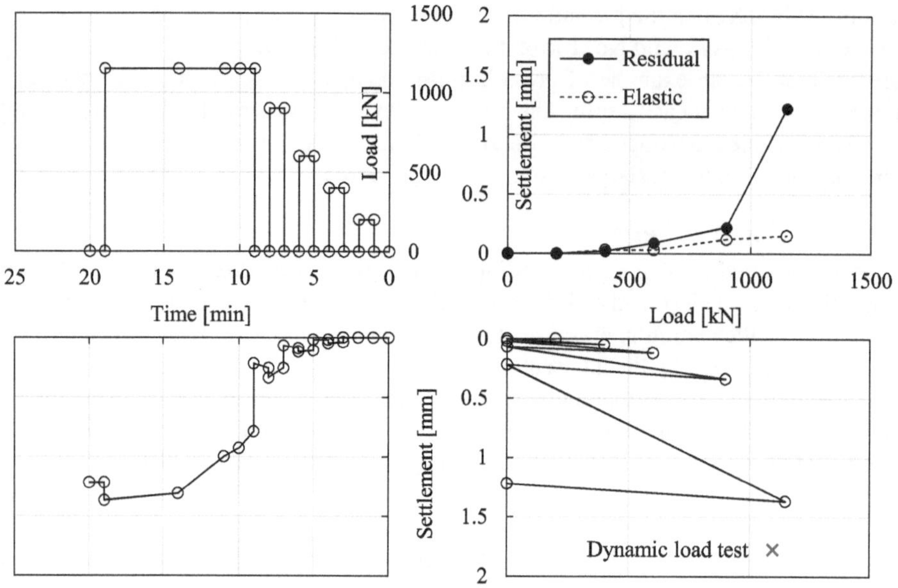

FIGURE 14.9 Load-Time-Settlement diagram by the simplified load test (pile W11).

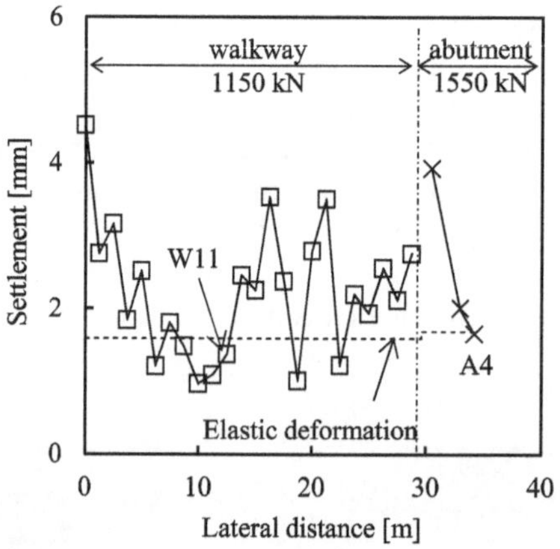

FIGURE 14.10 Relationship between settlement and horizontal position.

$$\rho(\Delta x) = \frac{2^{1-\nu}}{\Gamma(\nu)}\left(\frac{\sqrt{2\nu}\Delta x}{\theta}\right)^{\nu} K_{\nu}\left(\frac{\sqrt{2\nu}\Delta x}{\theta}\right) \cdots (\text{Matérn model}) \tag{14.1}$$

where K_ν is the modified Bessel function of the second kind.

The smoothness parameter is, $\nu = 1.6$, and the length scale, $\theta = 7$ m. The scale of fluctuation (SOF) becomes 16 m, which is smaller than the SOF reported by Phoon and Kulhawy (1999b). The hyperparameters are within the general range, and the estimated results seemed reasonable. Figure 14.11a shows the median value and the range of two standard deviations based on GPR. The dotted lines in Figure 14.11b is fitted with a lognormal distribution. The estimated values generally fit the observations except for two points.

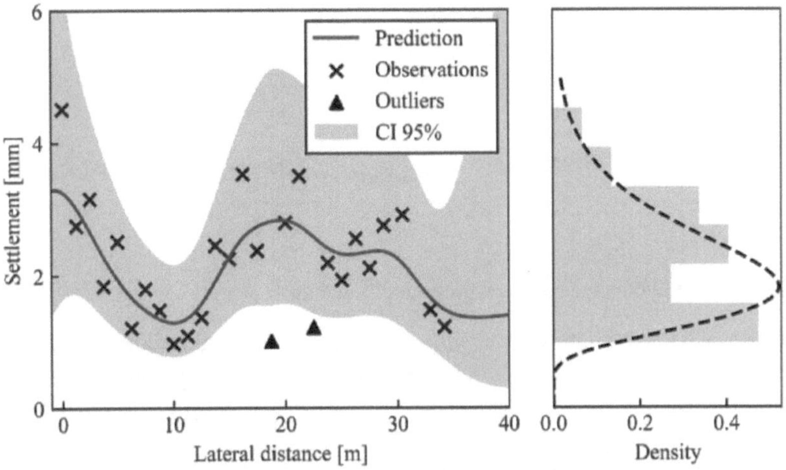

FIGURE 14.11 Spatial distribution of observed settlements and GPR estimations (a: observed and posterior distribution, b: probability distribution of observed).

14.4.2 Raising of River Levees (Suzuki, 2022b)

14.4.2.1 Overview

The second project attempted to raise river levees, and 160 steel pipe piles of 800 mm diameter were installed in a straight line for approximately 450 m. The boundary layer is depicted in Figure 14.12 using 13 borehole logs and SPTs. The piles were embedded into tuff layer with a depth of approximately 5–20 m. The depth at which the SPT N-value exceeded 50 was 1 m deeper than the bearing layer.

14.4.2.2 Estimations from Piling Data

Figure 14.13 depicts the geotechnical map with estimated N-values based on the piling data and Ishihara et al. (2015). Note that while they have recommended using incremental filling ratio (IFR) or plug length ratio (PLR) based on the measurement results, PLR was assumed to be constant at 0.5 due to the lack of data. This may be why the N-value is underestimated overall.

When comparing with the estimated boundary from the investigation, those estimated from piling data are in general agreement, but there are significant differences, especially in pile Nos. 110 to 130. Within the range, the estimated depth varied by up to 3 m, and it is believed that the variation was difficult to explore using only geotechnical surveys before piling.

The estimated boundary was fitted with GP using kernels of the Gaussian function (a.k.a. squared-exponential, radial basis function: RBF) (Figure 14.14). The horizontal SOF of the

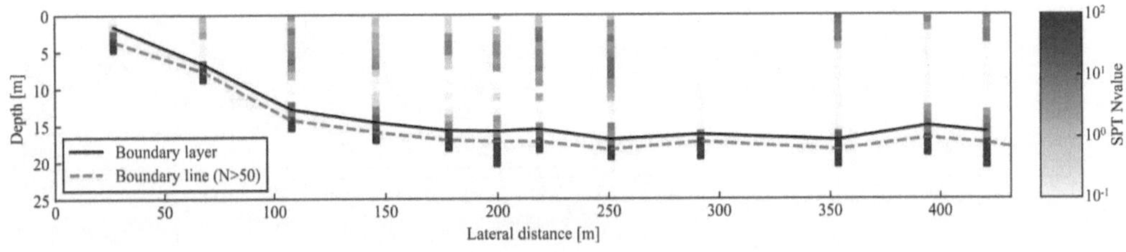

FIGURE 14.12 Longitudinal profiles: boundary layers based on borehole log (solid black line) and SPT-N-value (red dotted line).

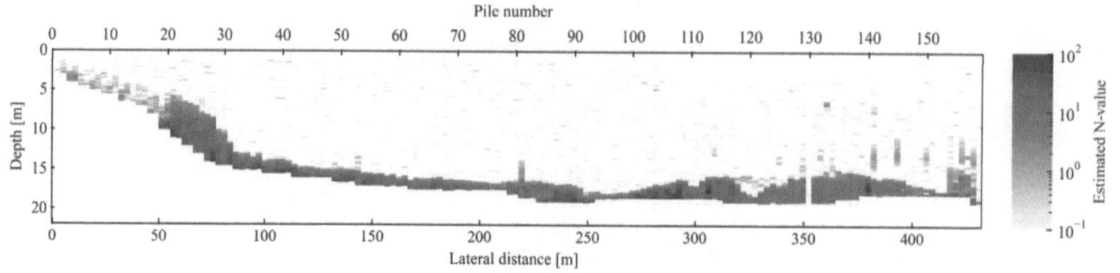

FIGURE 14.13 Estimated N-value based on press-in piling data.

short-term trend was approximately 24 m, which was near the N-value. The residual had a normal distribution with a standard deviation of 0.3 m, which was small enough and practical.

14.4.3 Discussion

The first application involved direct measurement of pile settlement, while the second application focused on indirect estimation of the bearing layer using pile data. I know it is difficult to make a clear distinction between direct and indirect measurements, because the former includes changes due to time effects and measurement errors, while the latter is also close to direct data, because the data itself is processed from the vertical resistance of the pile in the press-in method. However, if they are ignored and simplified, what kind of uncertainty is included in the confidence interval given in Figures 14.11 and 13?

In Figure 14.11, the confidence interval considers variability related to pile-specific random errors, while in Figure 14.13, the confidence interval accounts for within-site estimation errors. Since piles are man-made structures, they may exhibit variability due to piling workmanship, which cannot be known before piling. Therefore, piling data is crucial for comprehending and evaluating this type of error.

The layer is considered spatially continuous in the absence of fault displacement, and interpolating and extrapolating adjacent values help accurate estimation of the bearing layer. Indirect estimations need to account for bias, which can be achieved by comparing the results with other reliable surveys. These comparisons are summarized in Table 14.3.

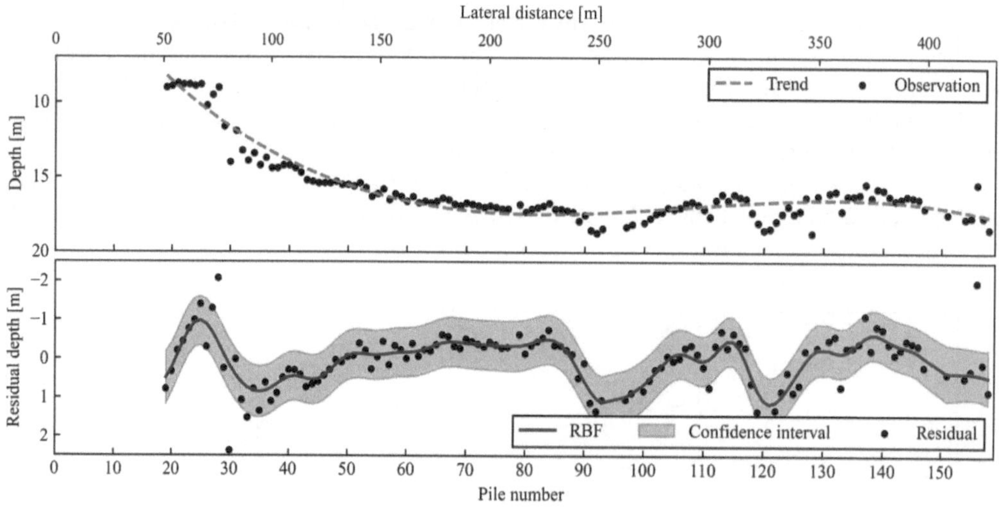

FIGURE 14.14 Boundary depth and estimated error.

TABLE 14.3
Type of Uncertainties in the Applications

Target	Method	Type	Uncertainty	Comparison
Pile bearing capacity	Chin's extrapolation and dynamic formula	Indirect	Within-site estimation error	Load test
Settlement	Simpified load test	Direct	Random noise due to piling workmanship	Dynamic load test
Depth of bearing layer	Estimation of SPT N-value	Indirect	Within-site estimation error	Bore-hole and SPT

14.5 SUMMARY

The design based on geotechnical investigations requires a margin of safety against both geotechnical and geological uncertainties. The author believes that the use of piling data can streamline the project, while piling data is now used primarily for anomaly detection only. This chapter reviewed the status of the usage of piling data.

First, previous studies on piling data were summarized, and the variation of bearing capacity estimation was introduced. The COV estimated from ground investigation is 30–40%, while the COV estimated from piling data is 20–30%, with no significant difference between piling methods. Furthermore, the previous studies were focused on the total error of estimation.

This variability can be divided into within- and inter-site errors. This distinction is important when using multiple pile data within the same site. The ratio of inter-site to within-site error differs depending on the estimation method used. When the inter-site error is large, it is effective to correct the estimation bias by loading the test piles separately. In this study based on a database of pile load tests, the inter-site error was greater than the within-site error in CAPWAP and vice versa in Chin's extrapolation of load–settlement curves. When the within-site error is large, the error is expected to be reduced by the within-site correlation, such as spatial correlation.

Section 14.4 presented two examples of the application of press-in piling data; the first is related to geotechnical uncertainty, and the second to geological uncertainty. The first site was an example of simplified load tests of piles with uncertain performance due to unfamiliar mudstone: walkway and bridge abutment foundation. The tests were aimed at confirming the elastic conditions of the pile foundation. Though high spatial correlation of settlement was observed over 30 m, COV in random noise was around 20% and individual measurement would be important.

The second site is an example of a relatively large site with geologic variations that were confirmed by piling data: raising river levees with a continuous steel pipe pile wall. The support depths estimated from the piling data were generally consistent with those point-by-point from the geotechnical investigations.

The estimation bias (i.e., inter-site variation) based on piling data was observed to be small by the dynamic load test and the borehole investigations, which may be because estimation techniques were based on static press-in resistance. Moreover, the high spatial correlation of the piling data allowed us to identify trends within the site, which could lead to a reduction in within-site errors.

Piling techniques have a history of development led by specialized contractors. For piling data to be actively used in the future, as in other fields, the usage must be beneficial to both contractors and clients. The author believes it will ultimately promote development, thus improving the quality of geotechnical structures.

ACKNOWLEDGMENTS

I am grateful to Kakuto Co., Ltd. and Giken Seko Co., Ltd. for providing their valuable construction data, and FHWA for opening the database.

REFERENCES

AASHTO, 2012. AASHTO LRFD bridge design specifications, customary U.S. Units. American Association of State Highway and Transportation Officials (AASHTO).

Alkroosh, I., Nikraz, H., 2014. Predicting pile dynamic capacity via application of an evolutionary algorithm. *Soils Found.* 54(2), 232–242.

Bozorgzadeh, N., Harrison, J.P., Escobar, M.D., 2019. Hierarchical Bayesian modelling of geotechnical data: Application to rock strength. *Géotechnique* 69(12), 1056–1070.

Burlon, S., Frank, R., Baguelin, F., Habert, J., Legrand, S., 2014. Model factor for the bearing capacity of piles from pressuremeter test results – Eurocode 7 approach. *Géotechnique* 64(7), 513–525.

Chan, W.T., Chow, Y.K., Liu, L.F., 1995. Neural network: An alternative to pile driving formulas. *Comput. Geotech.* 17(2), 135–156.

Chen, J.J., Zhang, L., 2013. Effect of spatial correlation of cone tip resistance on the bearing capacity of piles. *J. Geotech. Geoenviron. Eng.* 139(3), 494–500.

Chin, F., 1972. The inverse slope as a prediction of ultimate bearing capacity of piles. In *Proceedings of the 3rd South-East Asian Conference on Soil Engineering*, Libra Press, Hong Kong, pp. 83–91.

El Haj, A.K., Soubra, A.H., Fajoui, J., 2019. Probabilistic analysis of an offshore monopile foundation taking into account the soil spatial variability. *Comput. Geotech.* 106, 205–216.

EN1997-1, 2004. Eurocode 7: Geotechnical design-part 1: General rules.

Evangelista, A., Pellegrino, A., Viggiani, C., 1978. Variability among piles of the same foundation. In *9th International Conference on Soil Mechanics and Foundation Engineering (Tokyo)*, pp. 493–500.

Fei, C., Jianfeng, X., Fang-Bao, T., Kevin, D., Ken, G., 2022. The influence of a thin weak clay layer on the close-ended pile behaviors in sand. In *Proceedings of the 8th International Symposium for Geotechnical Safety & Risk*, 14–16 December 2022, Newcastle, pp. 363–368.

Fellenius, B.H., 1988. Variation of CAPWAP results as a function of the operator. In *Proceedings of the 3rd International Conference on the Application of Stress-Wave Theory to Piles*, pp. 814–825.

Fellenius, B.H., Rahman, M.M., 2019. Load-movement response by t-z and q-z functions. *Geotech. Eng. Journal of the SEAGS & AGSSEA Journal* 50, 11–19.

Galbraith, A.P., Farrell, E.R., Byrne, J.J., 2014. Uncertainty in pile resistance from static load tests database. *Proc. Inst. Civ. Eng. Geotech. Eng.* 167(5), 431–446.

Gelman, A., Carlin, J.B., Stern, H.S., Dunson, D.B., Vehtari, A., Rubin, D.B., 2013. *Bayesian Data Analysis.* Third Edition. CRC press, New York.

Giken Ltd., 2017. Product catalog: Silent piling technologies, Available: https://www.giken.com/ja/wp-content/uploads/2017/06/press-in_method_variations.pdf

Goh, A.T.C., 1995. Back-propagation neural networks for modeling complex systems. *Artif. Intell. Eng.* 9(3), 143–151.

Hiley, A., 1925. Rational pile-driving formula and its application in piling practice explained. *Engineering* 657, p721.

Holeyman, A.E., 2002. Soil behavior under vibratory driving. In *Vibratory Pile Driving and Deep Soil Compaction – TRANSVIB2002*, CRC Press, pp. 3–19.

International Building Code (IBC), 2009. Chapter 18 Soils and Foundations, Section 1810, International Code Council, Washington, D.C.

International Code Council's Evaluation services (ICC-ES), 2007. ACC358 Acceptance criteria, for helical foundation systems and devices. Washington, D.C., 21p.

International Press-in Association (IPA), 2021. Press-in retaining structures: a handbook (2nd edition 2021). IPA, Tokyo.

Ishihara Y., 2023. Use of press-in piling data for estimating subsurface information and pile performance. Dissertation of University of Tokyo.

Ishihara, Y., Haigh, S., Bolton, M., 2015. Estimating base resistance and N value in rotary press-in. *Soils Found.* 55(4), 788–797.

Japan Road Association (JRA), 2020. *Pile Foundation Design Handbook (2020 Revised Edition).* JRA, Tokyo.

Kalavar, S., Ealy, C., 2000. FHWA deep foundation load test database. In *Proceedings of Sessions of Geo-Denver 2000 – New Technological and Design Developments in Deep Foundations, GSP 100.* American Society of Civil Engineers, pp. 192–206.

Kanai, H., 1915. Tokyo station building report. *Doboku Gakkai Shi* 1, 49–76 (in Japanese).

Lacasse, S., Guttormsen, T.R., Goulois, A., 1989. Bayesian updating of axial capacity of single pile.

Lehane, B.M., 2005. Technical session 2h: Pile foundations (II): Installation, quality control, performance, and case histories. In *Proceedings of the 16th International Conference on Soil Mechanics and Geotechnical Engineering: Geotechnology in Harmony with the Global Environment*, IOS Press, pp. 3089–3096.

Lehane, B.M., Schneider, J.A., Xu, X., 2005. The UWA-05 method for prediction of axial capacity of driven piles in sand. In: *Proceedings of the 1st International Symposium on Frontiers in Offshore Geotechnics*, pp. 683–689.

McVay, M.C., Birgisson, B., Zhang, L., Perez, A., Putcha, S., 2000. Load and resistance factor design (LRFD) for driven piles using dynamic methods – A Florida perspective. *Geotech. Test. J.* 23(1), 55–66.

Mizutani, T., Matsumura, S., 2016. Correction of Hiley's equation and its applicability base on dynamic loading tests at Mizushima port. *J. Jpn Soc. Civ. Eng. Ser. B3 Ocean Eng.* 72, I_396-I_401 (in Japanese).

Nanazawa, T., Kouno, T., Sakashita, G., Oshiro, K., 2019. Development of partial factor design method on bearing capacity of pile foundations for Japanese Specifications for Highway Bridges. *Georisk Assess. Manag. Risk Eng. Syst. Geohazards* 13(3), 166–175.

Ng, K.W., Sritharan, S., 2016. A procedure for incorporating setup into load and resistance factor design of driven piles. *Acta Geotech.* 11(2), 347–358.

Ohki, H., Nagata, M., Saeki, E., Kuwabara, H., 2005. Fluctuation on bearing strata levels of piles (part 2). In *Proceedings of the 40th Technical Report of the Annual Meeting of the Japan Geotechnical Society, Japan*, pp. 1549–1550 (in Japanese).

Otake, Y., Watanabe, S., Mizutani, T., 2021. Improvement of side resistance prediction for pile foundation using construction information. *Can. Geotech. J.* 58(4), 496–513.

Paikowsky, G., 2004. *Load and Resistance Factor Design (LRFD) for Deep Foundations, Load and Resistance Factor Design (LRFD) for Deep Foundations.* Transportation Research Board, Washington, D.C.

Paikowsky, S.G., Tolosko, T.A., 1999. Extrapolation of pile capacity from non-failed load tests. FHWA-RD-99-107, Federal Highway Administration, p169.

Pham, T.A., Ly, H.B., Tran, V.Q., Giap, L. Van, Vu, H.L.T., Duong, H.A.T., 2020. Prediction of pile axial bearing capacity using artificial neural network and random forest. *Appl. Sci.* 10(5), 1871.

Phoon, K.K., Kulhawy, F.H., 1999a. Characterization of geotechnical variability. *Can. Geotech. J.* 36(4), 612–624.

Phoon, K.K., Kulhawy, F.H., 1999b. Evaluation of geotechnical property variability. *Can. Geotech. J.* 36(4), 625–639.

Rasmussen, C. E., Williams, C. K. I., 2006. *Gaussian processes for machine learning*, The MIT Press, Cambridge, Massachusetts.

Rausche, F., Moses, F., Goble, G.G., 1972. Soil resistance predictions from pile dynamics. *J. Soil Mech. Found. Div.* 98(9), 917–937.

Reddy, S.C., Stuedlein, A.W., 2013. Accuracy and reliability-based region-specific recalibration of dynamic pile formulas. *Georisk* 7(3), 163–183.

Salgado, R., Zhang, Y., Abou-Jaoude, G., Loukidis, D., Bisht, V., 2017. Pile driving formulas based on pile wave equation analyses. *Comput. Geotech.* 81, 307–321.

Shahin, M.A., 2016. State-of-the-art review of some artificial intelligence applications in pile foundations. *Geosci. Front.* 7(1), 33–44.

Suzuki, N., 2022a. Three-parameter lognormal distribution to estimate ultimate bearing capacity of pile foundations with extrapolation of load-settlement curves. In *Proceedings of the 8th International Symposium for Geotechnical Safety & Risk*, 14–16 December 2022, Newcastle, pp. 138–143.

Suzuki, N., 2022b. Geotechnical mapping using press-in piling data to estimate bearing layer. In *The 11th International Symposium on Field Monitoring in Geomechanics*, London.

Suzuki, N., Ishihara, Y., 2019. Case study on the application of press-in piling data to design and construction of pile foundations for reducing the expected total cost. *International Conference on Case Histories & Soil Properties, Singapore,* No.157.

Suzuki, N., Ishihara, Y., 2019b. Discussion on the method of estimating the ultimate pile capacity from the load-displacement curve at the end of pressing. In *2019 JSCE Annual Meeting*, 74 (in Japanese).

Suzuki, N., Nagai, K., Sanagawa, T., 2021. Reliability analysis on cantilever retaining walls embedded into stiff ground (Part 1: contribution of major uncertainties in the elasto-plastic subgrade reaction method). In *Proceedings of the Second International Conference on Press-in Engineering 2021*, Kochi, pp. 306–314.

Tang, C., Phoon, K.-K., 2021. Model uncertainties in foundation design, CRC press, p.588.

Tang, C., Phoon, K.-K., 2018. Statistics of model factors and consideration in reliability-based design of axially loaded helical piles. *J. Geotech. Geoenviron. Eng.* 144(8), 04018050.

Uto, K., Fuyuki, M., Sakurai, M., 1982. How to organize the results of pile loading tests. *Kisoko* 10, 21–30 (in Japanese).

Van Baars, S., Vrijling, J.K., 2005. Geotechnical applications and conditions of the observational method. *Heron* 50, 155–172.

Van Der Veen, C., 1957. The bearing capacity of a pile. *Proceedings of the 4th ICSMFE* 2, 72–75.

Vanmarcke, E.H., 1977. Probabilistic modeling of soil profiles. *ASCE J. Geotech. Eng. Div.* 103(11), 1227–1246.

Wang, Y., Cao, Z.J., 2013. Expanded reliability-based design of piles in spatially variable soil using efficient Monte Carlo simulations. *Soils Found.* 53(6), 820–834.

Watanabe, S., 2010. Asymptotic equivalence of Bayes cross validation and widely applicable information criterion in singular learning theory. *J. Mach. Learn. Res.* 11, 3571–3594.

Yang, Z., Jardine, R., Guo, W., Chow, F., 2015. *A Comprehensive Database of Tests on Axially Loaded Piles Driven in Sand, A Comprehensive Database of Tests on Axially Loaded Piles Driven in Sand.* Academic Press, Cambridge.

Yu, F., Kou, H., Liu, J., Yang, Y., 2012. Jacking installation of displacement piles: From empiricism toward scientism. *Electron. J. Geotech. Eng.* 17J, 1581–1590.

Zhang, J., Hu, J., Li, X., Li, J., 2020. Bayesian network based machine learning for design of pile foundations. *Autom. Constr.* 118, 103295.

Zhang, L., 2004. Reliability verification using proof pile load tests. *J. Geotech. Geoenviron. Eng.* 130(11), 1203–1213.

Zhang, L., Shek, L.M.P., Pang, H.W., Pang, C.F., 2006. Knowledge-based design and construction of driven piles. *Proc. Inst. Civ. Eng. Geotech. Eng.* 159(3), 177–185.

Zhang, L.M., Li, D.Q., Tang, W.H., 2006. Level of construction control and safety of driven piles. *Soils Found*, 46(4):415–25.

Zhang, Q., Zhang, Z., 2012. A simplified nonlinear approach for single pile settlement analysis. *Can. Geotech. J.* 49(11), 1256–1266.

15 Slope Reliability Assessments for Linear Infrastructures

Michael A. Hicks

ABSTRACT

The random finite element method (RFEM) is applied to the modeling of slope stability problems for engineered structures, with a particular focus on linear infrastructures such as dykes, cuttings, and embankments. RFEM uses random field theory to model the spatial variability of soils and the finite element method to model structure response, and it takes account of the uncertainty arising from incomplete knowledge of the spatial variability by carrying out a Monte Carlo analysis. The output from an RFEM analysis is a distribution of possible responses from which the reliability of a structure may be determined. Through a review of simple illustrative examples and a case history involving a regional dyke in the Netherlands, this chapter provides insight into how geotechnical assessments are affected by soil spatial variability and illustrates why a probabilistic (reliability-based) approach in general is beneficial in geotechnical practice.

15.1 INTRODUCTION

This chapter considers the application of the so-called random finite element method (RFEM) (Fenton and Griffiths 2008) to the modeling of slope stability problems for engineered structures. In particular, it focuses on linear infrastructures, which include flood defense structures such as dykes and transport infrastructures such as cuttings and embankments. RFEM uses random field theory to model the spatial variability of soils and the finite element method to model structure response, and it takes account of the uncertainty arising from incomplete knowledge of the spatial variability by carrying out a Monte Carlo analysis. Hence, an RFEM analysis is composed of multiple realizations of the same problem, with each realization involving the prediction of the spatial variability (using a random field generator) and the analysis of the problem domain using finite elements. The output from an RFEM analysis is a distribution of possible responses from which the reliability of a structure may be determined.

The main purpose of this chapter is to simply illustrate the benefits of using a reliability-based approach in geotechnical practice. Although RFEM is a form of probabilistic analysis that is relatively high level, and currently perceived by practitioners as being too complex for everyday use, it is useful as an illustrative tool for providing insight into how geotechnical assessments are affected by soil spatial variability. For this purpose, the methodology and example analyses are herein introduced in a straightforward manner, and the chapter concludes by emphasizing why a probabilistic (reliability-based) approach in general is beneficial.

The chapter starts with a simple introduction to the methodology. This is followed by analyses of idealized 2D and 3D embankment slope stability problems, which give insight into the influence of spatial variability of soil properties on embankment performance as well as into the propagation of uncertainty from the material level to the structure level. The concept of spatial variability of soil properties is then used to provide an interpretation of characteristic values in Eurocode 7. Finally, the benefits of a reliability-based approach in practice are discussed and a brief overview is given of the practical use of RFEM for a dyke safety assessment in the Netherlands. This case history is also used to introduce alternative, simpler forms of probabilistic analysis.

DOI: 10.1201/9781003333586-19

15.2 OUTLINE OF METHODOLOGY

15.2.1 Spatial Variability of Soil Properties

It is well known that, even in so-called uniform layers of soil, there exists spatial variability of property values (Phoon and Kulhawy 1999; Hicks 2007). The conventional (deterministic) approach is to assign a single value for each material property within the layer, such as the mean value or some other representative value. Hence, for the assessment of the stability of a geotechnical structure, for example, this leads to the calculation of a single factor of safety for which there is no information regarding the probability of failure.

In contrast, a stochastic approach makes use of all available data for a soil layer and represents each property value by a statistical distribution characterized by its mean, μ, and standard deviation, σ. These point statistics are often combined to give the coefficient of variation, $V = \sigma / \mu$, and either or both of these point statistics may be functions of depth below the ground surface. Figure 15.1 shows a simple illustration for a soil property X in a so-called uniform layer. In this case, the mean value of X is constant with depth [Figure 15.1 (left)], while the distribution of properties about the mean is represented by a normal distribution [Figure 15.1 (right)].

In addition to the point statistics, the spatial correlation of soil properties may be represented by an autocorrelation function characterized by the scale of fluctuation, θ, which is a measure of the distance over which soil property values are significantly correlated (Vanmarcke 1983). This is simply illustrated in Figure 15.1 (left) as being a function of the distance between adjacent strong (or weak) zones in the vertical direction. However, Figure 15.2 more clearly illustrates the influence

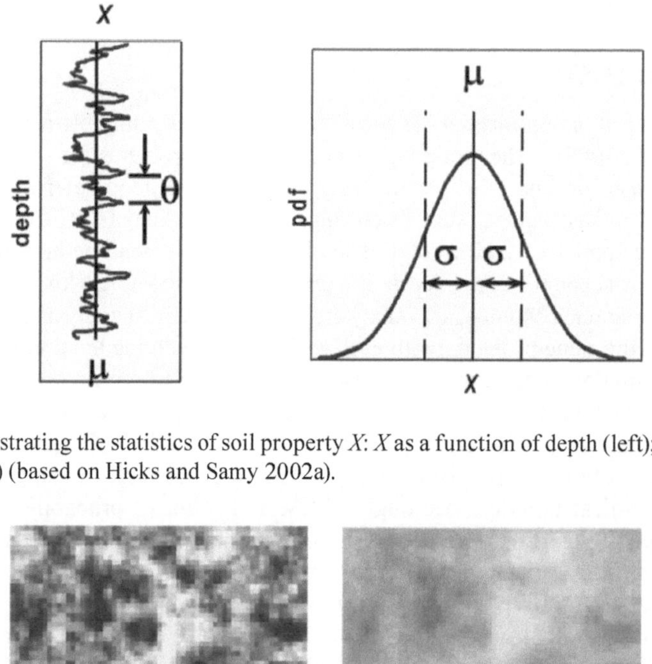

FIGURE 15.1 Illustrating the statistics of soil property X: X as a function of depth (left); probability density function of X (right) (based on Hicks and Samy 2002a).

FIGURE 15.2 Two-dimensional random fields of X: $\theta/D = 0.1$ (left); $\theta/D = 1.0$ (right) (based on Hicks and Samy 2002b).

of θ in two dimensions for the variation in X over a square domain of size D, in which the dark colors indicate high values of X and the light colors indicate low values of X. Figure 15.2 illustrates the influence of the value of θ/D on the characteristics of the spatial correlation for the case of isotropic spatial variability, which is when the value of θ is the same in all directions. However, due to the natural process of soil deposition, the value of θ will generally be much smaller in the vertical direction than in the horizontal plane (Hicks and Samy 2002a, 2002b), and it may also change with direction in the horizontal plane (Varkey et al. 2023).

A consequence of spatial variability within soil layers is that it can affect material behavior and structure response (Hicks and Onisiphorou 2005; de Gast et al. 2021b). It also means that we are never quite sure about the ground conditions at a site, although, of course, we may carry out in-situ tests to gain knowledge and reduce the uncertainty. For this purpose, the cone penetration test (CPT) is an ideal candidate, in that each CPT profile provides a continuous data source that may be used in deriving soil property distributions [Figure 15.1 (right)] as well as scales of fluctuation in the vertical direction. Multiple CPTs at a site may also be used in deriving scales of fluctuation in the horizontal plane, although this is more difficult to achieve accurately (even for closely spaced CPTs) due to the generally sparse nature of the data (Lloret-Cabot et al. 2014; de Gast et al. 2021a). Hence, while point statistics and vertical scales of fluctuation are readily obtained from CPT data, conservative values for the horizontal scale of fluctuation (in terms of their impact on structure response) are often chosen instead (Varkey et al. 2023).

15.2.2 Quantification and Propagation of Uncertainty

The uncertainty associated with spatial variability within a soil layer arises from incomplete knowledge about the layer due to limitations in testing frequency, although there will be additional uncertainties in soil property values due to measurement error, transformation error, and so on. Focusing only on the uncertainty due to the spatial variability itself (for the purpose of this chapter), numerical predictions of the spatial variability throughout a layer may be generated by using the point and spatial statistics derived from, for example, CPT data collected at discrete locations within the layer. These numerical predictions are known as random fields, although they are not random at all, but spatially correlated distributions characterized by the input statistics. For a given set of input statistics, there are theoretically an infinite number of possible random fields. They will all look similar, because they will all have been generated using the same set of statistics, but they will differ with respect to the spatial distribution of strong and weak zones. Figure 15.3 shows an example of four random fields that have been generated using the same set of input statistics. The

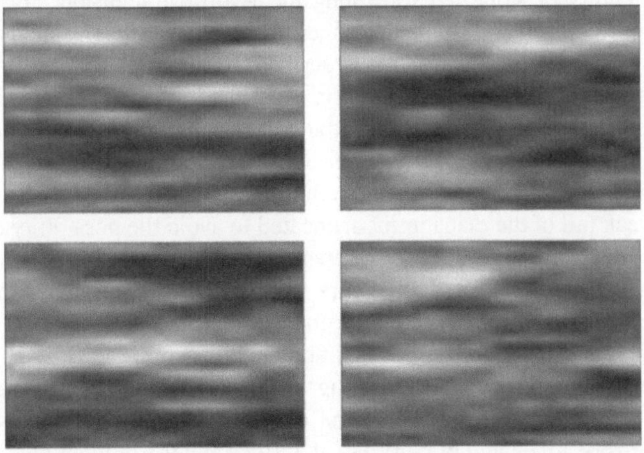

FIGURE 15.3 Four random fields of X based on the same input statistics (based on Hicks and Samy 2002b).

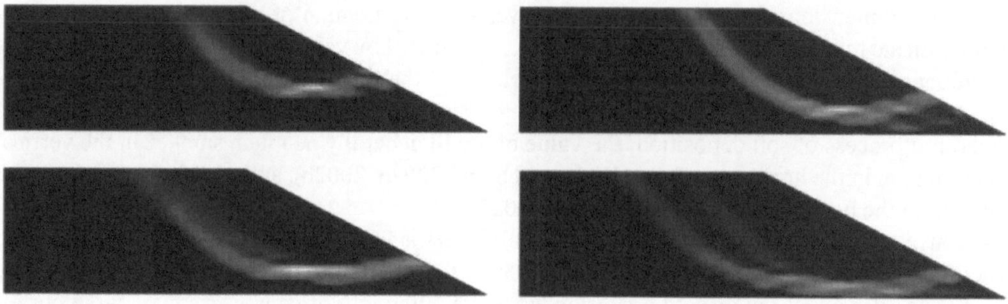

FIGURE 15.4 Four slope failure mechanisms based on the same input statistics (based on Hicks and Samy 2002b).

range of possible random fields represents the uncertainty associated with the spatial variability in the soil layer, although this uncertainty can be reduced by conditioning the random fields so that they match known soil profiles at the CPT locations (Lloret-Cabot et al. 2012; Li et al. 2016). When this is done, the range of possible random fields represents the uncertainty in the spatial variability between the CPT locations.

RFEM is a numerical approach for quantifying the propagation of uncertainty from the material level to the structure response level, by linking finite elements with random fields as described in Section 15.1. In each realization of the Monte Carlo simulation, the generated random field of property values is mapped onto the finite element mesh, and the structure response is then computed using the finite element method. Hence, each realization involves a standard finite element analysis, but with every element (or integration point) in the finite element mesh having a different property value. As illustrated in Figure 15.4 for a simple slope stability investigation, in every realization the computed response of the structure will be different because of the different spatial distribution of material properties, so that the output from an RFEM analysis is a distribution of possible responses (e.g., possible factors of safety) from which the probability of failure can be computed. For this purpose, the Monte Carlo simulation is run until the statistics of the performance distribution have converged to an acceptable level; for example, there is convergence of both the mean and standard deviation of the computed factor of safety from all the realizations.

15.3 INFLUENCE OF SOIL SPATIAL VARIABILITY ON SLOPE RELIABILITY

This section uses RFEM to investigate the influence of spatial variability of undrained shear strength on the reliability of slopes that are long in the third dimension. For this purpose, it is assumed that the slope comprises a single heterogeneous soil layer that rests on an underlying firm stratum. Firstly, a simple 2D investigation is described, to illustrate some of the basic characteristics of analyses involving soil spatial variability. Next, the influence of soil spatial variability on slope failure in three dimensions is investigated, and the implications for dykes and embankments are discussed. Finally, the influence of spatial variability on failure consequence is considered. For each numerical example, the undrained shear strength is simply represented by a normal distribution, but with the weak tail of the distribution truncated to avoid the possibility of negative values. This is a reasonable distribution for this soil property, due to the coefficient of variation generally lying in the range $0.0 < V < 0.3$; that is, the very small proportion of values that will need to be truncated will have a negligible influence on the analysis (Hicks and Samy 2002a). However, other types of distribution (e.g., the lognormal distribution) are also commonly used.

The RFEM analyses have been conducted using "in-house" computer software [e.g., as described in Hicks and Spencer (2010)]. The finite element codes have been developed within the general finite element framework advocated by Smith et al. (2013), and the random field generator is based on Local Average Subdivision (Fenton and Vanmarcke 1990).

15.3.1 RFEM Analysis of 2D Slope Reliability

Figure 15.5 shows the finite element mesh used to model a 1:2 slope characterized by a spatially varying undrained shear strength, c_u (Hicks and Samy 2002b). The mesh is composed of 8-node quadrilateral elements, with each element using 2 × 2 Gaussian integration. The height of the slope is 10 m, the volumetric weight of the soil is $\gamma = 20$ kN/m^3, and the statistics of c_u are a mean that increases linearly with depth, from 10 kPa at the top boundary to 50 kPa at the bottom boundary, and a constant coefficient of variation of 0.3. The vertical scale of fluctuation is $\theta_v = 1.0$ m, whereas various horizontal scales of fluctuation have been considered (Hicks and Samy 2002b). Figure 15.6 shows a typical random field of c_u for $\xi = \theta_h/\theta_v = 12$, in which the darker zones indicate higher values of c_u. The soil has been modeled by a Tresca failure criterion and the following elastic properties: Young's modulus, $E = 100{,}000$ kPa, and Poisson's ratio, $\nu = 0.3$.

The slope has been analyzed using the strength reduction method. Hence, for each realization in the Monte Carlo (RFEM) simulation, i.e., for each random field of c_u, gravitational loading is applied to generate the in-situ stresses in the slope and the crest settlement, Δ, due to the soil self-weight, is recorded. The slope is then repeatedly analyzed for progressively weaker soil profiles (i.e., by scaling down the property values in the original random field) until the slope fails, as indicated by a sudden increase in the crest settlement. For each re-analysis, the random field of c_u is the original random field scaled down by a factor F. The scaling factor that causes failure is the factor of safety of the slope (based on the original random field).

Figure 15.7 shows that, when no heterogeneity is considered, the factor of safety is close to the analytical solution (Hunter and Schuster 1968) of $F = 1.6$. However, Figure 15.8 shows that, when heterogeneity is considered (in this case, for $\xi = \infty$), there is a wide range of possible solutions. Moreover, the mean factor of safety is significantly less than the deterministic solution based on the underlying depth-dependent mean. This is because failure mechanisms follow the path of least resistance; i.e., they are attracted to the weaker zones and try to avoid (where possible) the stronger zones. Figure 15.8 shows the results of only 30 realizations. However, when sufficient realizations are performed to provide convergence of the mean and standard deviation of the factor of safety, a probability distribution of the factor of safety can be plotted, from which the probability of failure can be determined (i.e., the area under the distribution for which the factor of safety is less than unity).

Hicks and Samy (2002b) considered several values of ξ for this boundary value problem and showed that the distribution of the factor of safety tended to converge for values of ξ that were

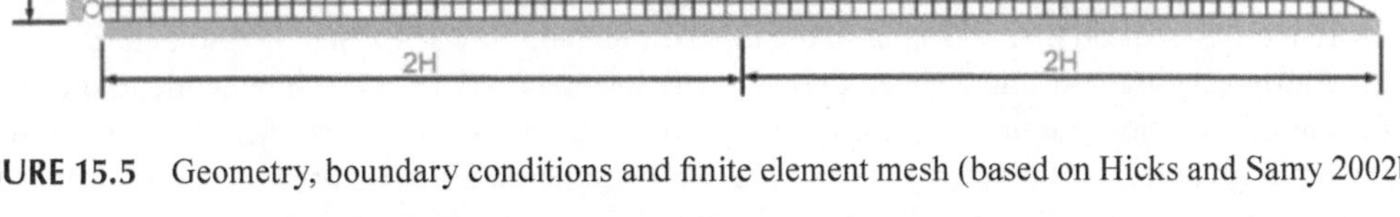

FIGURE 15.5 Geometry, boundary conditions and finite element mesh (based on Hicks and Samy 2002b).

FIGURE 15.6 Typical random field for $\xi = 12$ (based on Hicks and Samy 2002b).

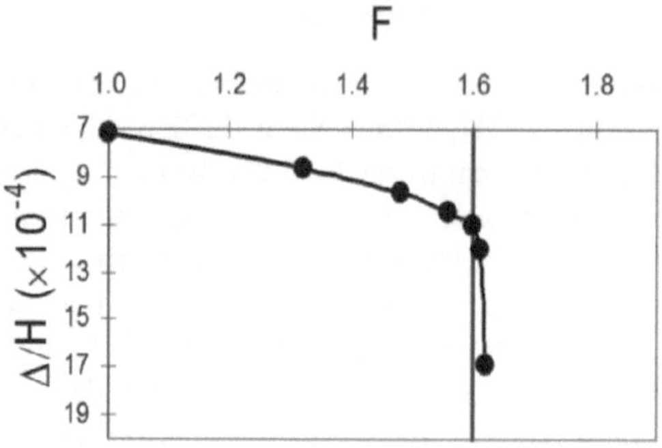

FIGURE 15.7 Mobilized safety factor versus crest settlement (deterministic solution) (based on Hicks and Samy 2002b).

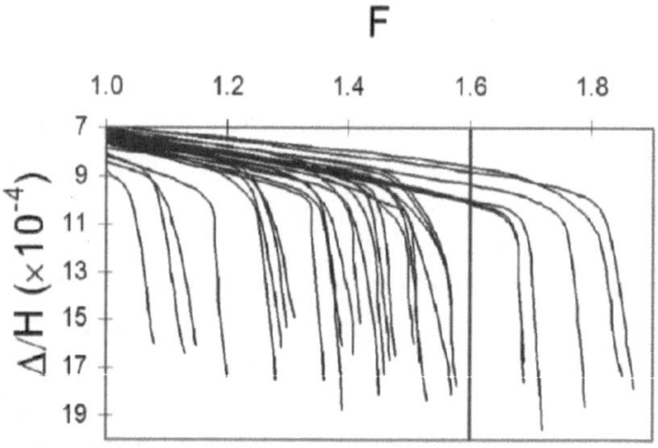

FIGURE 15.8 Mobilized safety factor versus crest settlement (stochastic solution, $\xi = \infty$) (based on Hicks and Samy 2002b).

lower than those likely on site. This implied that θ_h need not always be accurately known since a conservative solution could be found by assuming $\xi = \infty$. However, more recent analyses in 3D have suggested that a more accurate knowledge of θ_h may be important, as will be illustrated in Section 15.3.2. Meanwhile, Hicks and Samy (2002a) carried out a detailed 2D investigation to illustrate the importance of accounting for both the anisotropy of the heterogeneity ($\xi > 1$) and the depth-dependency of the underlying mean of c_u, whereas Hicks and Samy (2004) investigated the influence of slope angle.

15.3.2 RFEM Analysis of 3D Slope Reliability

Spencer and Hicks (2007) and Hicks and Spencer (2010) carried out 3D RFEM analyses for a slope of constant cross-section that was long in the third dimension. It was formed in a soil characterized by a spatially varying undrained shear strength with a constant (i.e., depth-independent) mean and coefficient of variation. Figure 15.9 shows the problem geometry and mesh details. The 1:1 slope is $H = 5$ m high and $L = 100$ m long and is modeled using 8000 20-node brick elements with $2 \times 2 \times 2$ Gaussian integration. The soil has again been modeled by a Tresca failure criterion and by the elastic properties $E = 100{,}000$ kPa and $\nu = 0.3$.

As in Section 15.3.1, the slope has been loaded by applying gravity loading to generate the in-situ stresses. However, rather than analyzing the slope for a given (problem-specific) set of statistics and finding the distribution of the factor of safety using the strength reduction method, this investigation focuses on finding the relationship between the reliability of the slope and the factor of safety based only on the mean c_u. The process starts with calculating $\mu_{F=1.0}$, which is the mean c_u at which the slope would just start to fail if there were no spatial variation in c_u. In other words, it is the value

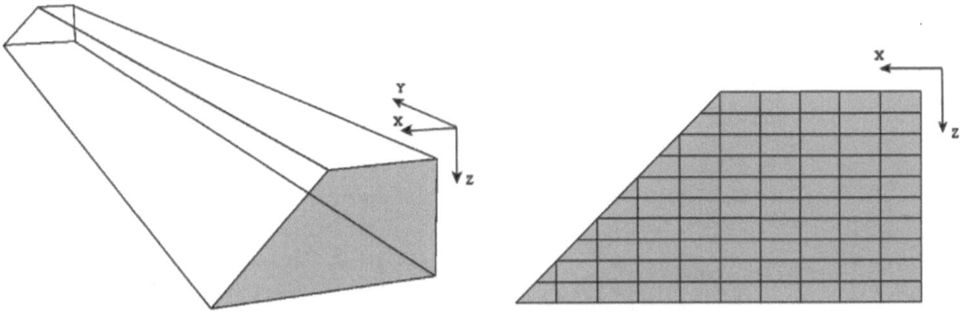

FIGURE 15.9 Isometric projection of 3D slope and cross-section through mesh (based on Spencer and Hicks 2007, Hicks and Spencer 2010).

of c_u corresponding to $F = 1.0$. The process then involves finding the reliability of the slope for different values of F (where F is the factor of safety based only on the mean c_u, i.e., not accounting for heterogeneity). For a given value of F and ξ, the point and spatial statistics of c_u are calculated as: $\mu = \mu_{F=1.0} \times F$; $\sigma = \mu \times V$; and $\theta_h = \theta_v \times \xi$, where, for this investigation, $\mu_{F=1.0} = 16.1$ kPa, $V = 0.3$ and $\theta_v = 1.0$ m. These statistics are used to generate N random fields of c_u for the chosen value of F, and, for each realization, the slope is analyzed by finite elements. The percentage reliability is then given by $R = (1 - (N_f/N)) \times 100$, where N_f is the number of realizations in which the slope fails under its self-weight. Hence, N–N_f is the number of realizations in which the slope remains stable.

For comparative purposes, Figure 15.10 shows the relationship between reliability and factor of safety (based on the mean c_u) for a 2D (i.e., plane strain) analysis of the same problem. At a factor of safety of 1.0 (based on the mean c_u), $R < 50\%$ for all values of ξ, due to failure being attracted to the weaker zones. It is clear that, although the solution is dependent on the horizontal scale of fluctuation, the solution has converged for $\xi > 6$ in this example.

Figure 15.11 shows the results of the equivalent 3D analyses. In evaluating these results, Spencer and Hicks (2007) and Hicks and Spencer (2010) identified three categories of failure mode, which are illustrated by the typical deformed meshes in Figure 15.12. The failure mode categories are as follows:

- Mode 1: for $\theta_h < H$, the scale of fluctuation is relatively small in all directions, making it harder for failure to propagate through semi-continuous weaker zones. In particular, for very small values of θ_h the failure mechanism passes through strong and weak zones in almost equal measure. Hence, there is considerable averaging of soil properties over

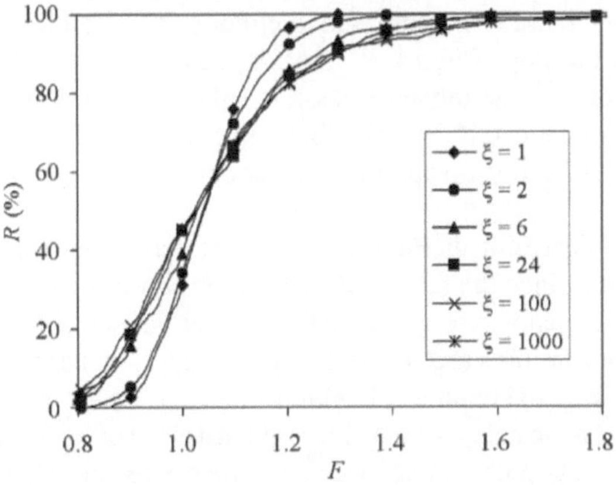

FIGURE 15.10 Influence of degree of anisotropy of the heterogeneity on reliability versus global factor of safety (2D analysis) (based on Spencer and Hicks 2007, Hicks and Spencer 2010).

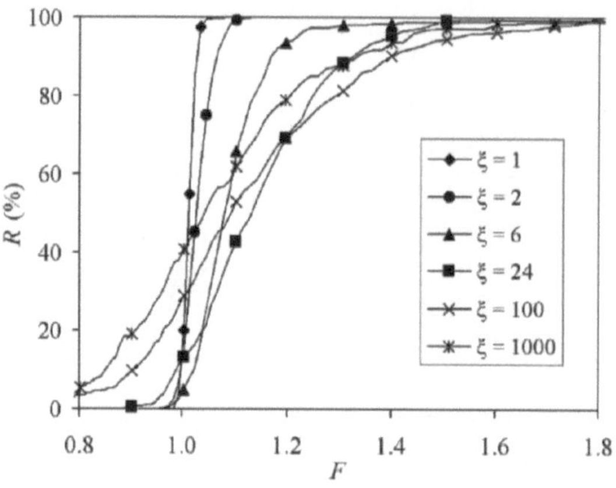

FIGURE 15.11 Influence of degree of anisotropy of the heterogeneity on reliability versus global factor of safety (3D analysis) (based on Spencer and Hicks 2007, Hicks and Spencer 2010).

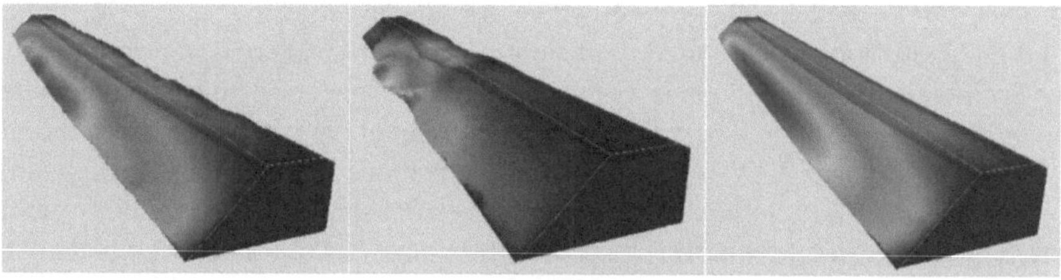

FIGURE 15.12 Typical deformed meshes and contours of out-of-face displacement for various 3D mechanisms (left to right: Mode 1, Mode 2, Mode 3) (based on Spencer and Hicks 2007, Hicks and Spencer 2010).

potential failure planes and the soil layer behaves like a homogeneous soil characterized by the mean c_u. This explanation is supported by R increasing from 0–100% as F passes through 1.0 (Figure 15.11) and by failure originating from the slope toe and extending along the entire length of the slope [Figure 15.12 (left)].

- Mode 2: for $H < \theta_h < L/2$, it is possible for failure to propagate through semi-continuous weaker zones, which results in discrete (3D) failure mechanisms [Figure 15.12 (center)]. In this case, R is a function of slope length, since, as the slope becomes longer, there is an increased possibility of encountering a zone weak enough to trigger failure.
- Mode 3: for $\theta_h > L/2$, there is an increased likelihood of the failure mechanism extending along the length of the slope [Figure 15.12 (right)], as in Mode 1. However, in contrast to Mode 1, there is now a large range of possible solutions, due to the depth of the failure mechanism being influenced by the distribution of strong and weak "sub-layers". In this case, the R versus F relationship approaches that obtained for the 2D stochastic analysis.

A practical implication of the results in Figure 15.11 is that, for Modes 1 and 3, the solution is independent of the slope length, since the failure mechanism is two-dimensional. In contrast, the solution for Mode 2 is length-dependent because the failure mechanism is three-dimensional. This has practical implications since failures are generally three-dimensional and it is clearly impractical to analyze very long slopes (e.g., dykes and embankments) in 3D. However, Hicks and Spencer (2010) carried out a detailed stochastic analysis of a "representative" 50 m long section of an embankment and then combined the results with simple probability theory to successfully predict the behavior of longer embankment sections that had also been analyzed using 3D RFEM. This approach was also followed by Hicks and Li (2018), who looked at embankment sections up to 1 km in length.

15.3.3 Influence of Heterogeneity on Failure Consequence

The investigation in Section 15.3.2 was extended by Hicks et al. (2014), who implemented a simple numerical scheme for automatically computing slide geometries in 3D RFEM simulations [based on Hicks et al. (2008)]. Figure 15.13 shows the results for a similar slope to that analyzed in Hicks and Spencer (2010), except that, in this case, $V = 0.2$ and a Von Mises failure criterion was adopted. The figure shows the influence of the horizontal scale of fluctuation on reliability versus factor of safety (based on the mean property value), as well as on failure volumes and lengths for individual realizations (which have been expressed as percentages of the total mesh volume and length, respectively). Because the same slope geometry and θ_v as used in Hicks and Spencer (2010) were adopted, the results for the values of ξ shown in Figure 15.13 are directly comparable with those in Figure 15.11.

Figure 15.13(a) shows that, when $\theta_h = H/5$, the slide volumes and lengths are consistent with those obtained when the slope fails along its entire length (indicating Mode 1 failure). In contrast, Figure 15.13(b) shows that, when $\theta_h = 1.2H$, there is a wide range of slide geometries (indicating Mode 2 failure). Similarly, Figures 15.13(c) and 15.13(d), corresponding to $\theta_h \approx L/8$ and $\theta_h \approx L/2$,

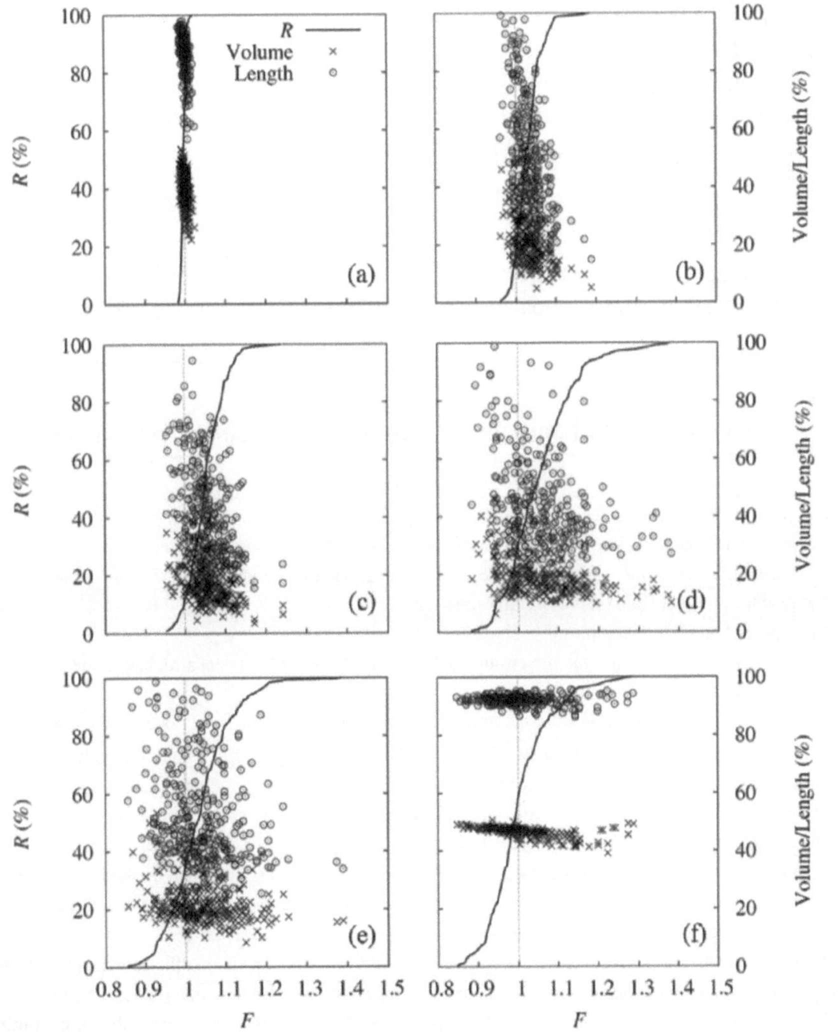

FIGURE 15.13 Influence of ξ on slide volume and length for a 3D slope; (a) $\xi = 1$, (b) $\xi = 6$, (c) $\xi = 12$, (d) $\xi = 48$, (e) $\xi = 100$, (f) $\xi = \infty$ (based on Hicks et al. 2014).

respectively, indicate Mode 2 failure. Although Figure 15.13(e) does reveal an increase in the number of larger slides for $\theta_h = L$, suggesting some Mode 3 failures, it is apparent that most slides are still Mode 2. Indeed, although Figure 15.13(f) shows mainly Mode 3 failures for $\theta_h = \infty$, Hicks et al. (2014) demonstrated just how difficult it is to compute 2D slope failures in a soil that is heterogeneous. They also presented results for a slope with a foundation layer which showed an even greater tendency for Mode 2 failures.

The ability to automatically compute slide volumes is an important first step towards benchmarking simpler semi-analytical and probabilistic models used in design (Li et al. 2015; Hicks and Li 2018; Varkey et al. 2019). This is because there is a need to quantify slide geometries when comparing with simpler methods based on predefined (e.g., cylindrical) failure mechanisms (Vanmarcke 1977; Calle 1985). The computation of slide volumes is also important in quantifying the risk posed by potential slope failures, especially for problems involving retrogressive failure mechanisms as in liquefiable sands and sensitive clays (Wang et al. 2016; Remmerswaal et al. 2021).

15.4 STOCHASTIC EXPLANATION OF CHARACTERISTIC VALUES

This section considers the issue of soil spatial variability within the context of characteristic soil property values advocated in Eurocode 7 (EC7) (CEN 2004). It is shown that stochastic analysis may be used as an aid to understanding the philosophy and nature of characteristic values, as well as providing a framework for deriving reliability-based characteristic values in line with EC7 (Hicks and Samy 2002b; Hicks 2012; Hicks and Nuttall 2012).

15.4.1 Extracts from Eurocode 7

The importance of accounting for the variability of soils is highlighted in Section 2.4.5.2 of EC7, "Characteristic values of geotechnical parameters" (CEN 2004). Table 15.1 lists some of the

TABLE 15.1
Extracts from Section 2.4.5.2 of Eurocode 7 (based on CEN 2004, Hicks and Nuttall 2012)

No.	Clause
(4)P	The selection of characteristic values for geotechnical parameters shall take account of the following: • geological and other background information, such as data from previous projects. • the variability of measured property values and other relevant information, e.g. from existing knowledge. • the extent of the field and laboratory investigation. • the type and number of samples. • the extent of the zone of ground governing the behavior of the geotechnical structure at the limit state being considered. • the ability of the geotechnical structure to transfer loads from weak to strong zones in the ground.
(7)	The zone of ground governing the behavior of a geotechnical structure at a limit state is usually much larger than a test sample or the zone of ground affected in an in-situ test. Consequently, the value of the governing parameter is often the mean of the range of values covering a large surface or volume of the ground. The characteristic value should be a cautious estimate of this mean value.
(8)	If the behavior of the geotechnical structure at the limit state considered is governed by the lowest or highest value of the ground property, the characteristic value should be a cautious estimate of the lowest or highest value occurring in the zone governing the behavior.
(11)	If statistical methods are used, the characteristic value should be derived such that the calculated probability of a worse value governing the occurrence of the limit state under consideration is not greater than 5%. NOTE: In this respect, a cautious estimate of the mean value is a selection of the mean value of the limited set of geotechnical parameter values, with a confidence level of 95%; where local failure is concerned, a cautious estimate of the low value is a 5% fractile.

main clauses, including Clause (4)P, which highlights the spatial nature of soil variability, the uncertainty this causes, and the problem-dependency of characteristic values; Clause (7), which emphasizes the importance of the mean property value over the domain of influence; Clause (8), which considers the special case of local failure; and Clause (11), which considers the use of statistical methods.

Hicks (2012) gave a detailed review of Section 2.4.5.2 by explaining selected clauses, clarifying the relationship between clauses, and addressing areas of potential confusion. In particular, the paper focused on the statistical definition of a characteristic value given in Clause (11) and explained how it is, despite first appearances, completely consistent with Section 2.4.5.2 as a whole, including Clauses (7) and (8) and the footnote to Clause (11).

Clause (11) states that "the characteristic value should be derived such that the calculated probability of a worse value governing the occurrence of the limit state under consideration is not greater than 5%". This implies a minimum level of reliability of 95% regarding the response of the structure (before the application of partial safety factors), and appears to contradict Clauses (7) and (8) and the footnote to Clause (11) which focus on property values rather than structure response. However, Hicks (2012) used Figure 15.14 to demonstrate that the latter are merely special cases of Clause (11).

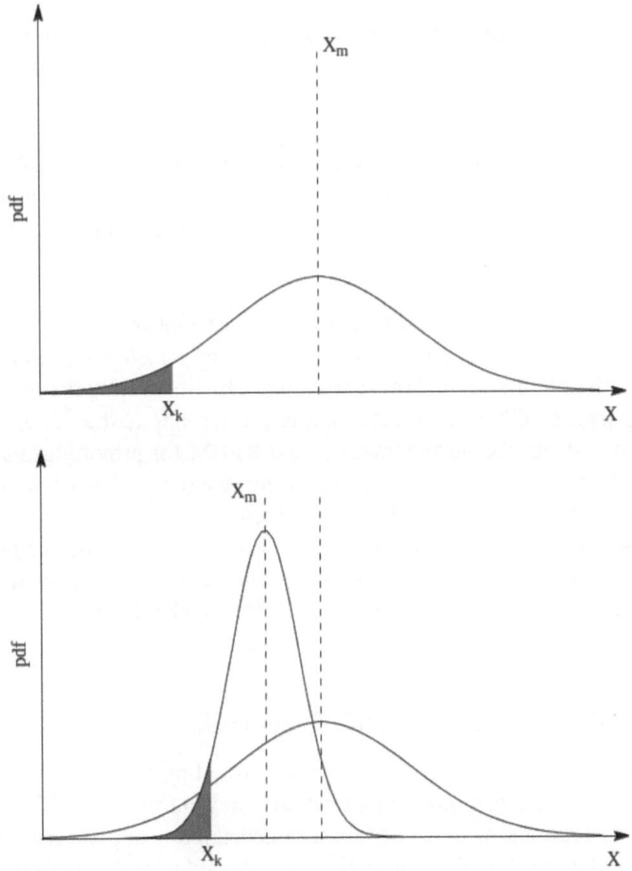

FIGURE 15.14 Derivation of characteristic property values satisfying Eurocode 7; basic definition of X_k (top), general definition of X_k (bottom) (based on Hicks 2012).

15.4.2 Reliability-Based Characteristic Values

Figure 15.14 (top) shows the probability density function of a material property X, which, to simplify the illustration, is assumed to be normal with a mean value X_m. The simplest way to derive a reliability-based characteristic value X_k is to proportion the area under the distribution as indicated. However, this is not consistent with Clause (11), as it merely defines a value of X_k for which there is a 95% probability of a larger value.

Figure 15.14 (bottom) gives a more general derivation of X_k that is consistent with Clause (11) and, thereby, with all other clauses in Section 2.4.5.2. This involves proportioning the area under a modified distribution of X that has been back-figured from the geotechnical response of the structure itself. The modified distribution is narrower than the underlying property distribution due to the averaging of property values over potential failure surfaces. It is also shifted to the left, due to the tendency for failure to propagate through weaker zones. Hence, although it may be reasonable to take a conservative estimate of the mean property value over a potential failure surface as the characteristic value for that mechanism, this mean will generally be smaller than the mean of the underlying property distribution. Variance reduction methods may therefore give an unsafe solution if no account is taken of the reduction in the mean.

Hicks (2012) explained how the modified property distribution is a function of the underlying property distribution, the spatial correlation of property values, the problem being analyzed, and the quality and extent of site investigation data. Moreover, the modified distribution has two limits:

- When the spatial scale of fluctuation is very small relative to the problem domain there is much averaging of soil properties, so that the standard deviation approaches zero and the mean tends to the mean of the underlying distribution. In this case, a cautious estimate of the mean is appropriate, as advocated by Clause (7) and the first part of the footnote to Clause (11).
- When the spatial scale of fluctuation is very large relative to the problem domain there is a very large range of possible solutions, so that the modified distribution approaches the underlying distribution. In this case, Clause (8) and the second part of the footnote to Clause (11) are relevant.

Hicks (2012) also explained how the modified property distribution in Figure 15.14 (bottom) may be derived for general problems, based on earlier work using RFEM by Hicks and Samy (2002b), while Hicks and Nuttall (2012) extended this earlier work by illustrating the process for a 3D slope. However, the main purpose of these publications was not to promote the use of RFEM for deriving characteristic values. Instead, the purpose was to use RFEM for providing insight into the influence of spatial variability of material properties on the response of geotechnical structures, and thereby to provide insight and guidance into what influences reliability-based characteristic values of material properties. Indeed, the notion of using RFEM to routinely derive problem-dependent, reliability-based characteristic values is counter-intuitive, since the reason for using reliability-based characteristic values in deterministic analysis is to avoid having to use probabilistic methods – but more will be said on this in Section 15.5.

15.5 APPLICATION TO ENGINEERING PRACTICE

The previous sections have used RFEM as a tool for providing insight into the influence of soil spatial variability on structure response. In particular, they introduced RFEM (and probabilistic methods in general) as a rational way to account for uncertainties in geotechnical assessments and designs. The traditional deterministic approach uses a global factor of safety to account for uncertainties implicitly, but the result is a single factor of safety for which there is no information regarding the probability of failure. In contrast, a probabilistic approach (such as RFEM) quantifies the uncertainty at the material level and then propagates the uncertainty to the structure level in order to quantify the structure response in terms of reliability, i.e., the probability that failure will not occur. The advantage of a probabilistic approach is that it quantifies the uncertainty when assessing stability and provides a rational framework whereby the uncertainty can be reduced through additional information (Vardon et al. 2016; Liu et al. 2018). This additional information can come in the form of additional laboratory or field test data for a site, or it could come in the form of monitoring (e.g., pore pressure or settlement) data during the lifetime of a structure.

A common misconception about the use of probabilistic methods is that a lot of data are needed to make it worthwhile, and this deters their wider use in practice. While more data are indeed desirable, this is true for both deterministic and probabilistic assessments. Deterministic methods compensate for a lack of data by using a larger factor of safety, but, as there is no information regarding the probability of failure, if more data do become available there is no rational framework for guiding by how much the factor of safety may be changed. In contrast, probabilistic methods account for a lack of data by assigning larger coefficients of variation to the input parameters, which in turn leads to a larger coefficient of variation of the predicted structure response. In this case, if more data become available, this is reflected by a reduction in the coefficients of variation of the input parameters and thereby a reduction in the coefficient of variation of the structure response. Of course, a reduction in the coefficient of variation of the computed structure response is no guarantee that the probability of failure will reduce, and it may in certain instances increase; however, it is more usual that the probability of failure reduces and this then leads to a more efficient engineering solution. Take, for example, the illustration of characteristic property values in Figure 15.14 (bottom) as an analogy to the computed response distribution for a structure. The modified distribution is narrower than the underlying distribution because of the additional available data (in this instance, the scales of fluctuation and data relating to the problem being analyzed), indicating a reduced level of uncertainty. The modified distribution lies completely within the range of values defined by the underlying distribution, but where it is located within that range depends on the relative means of the distributions. For the case in Figure 15.14 (bottom), it is seen that a reduced coefficient of variation of X leads to a value of X_k that is larger than that derived from the underlying distribution (even though the mean of the distribution is smaller). It can also be inferred from the figure that this will generally be the case, except for those relatively few occasions when the mean of the modified distribution is much smaller than that of the underlying distribution.

The following section gives a brief outline of the use of RFEM for a regional dyke in the Netherlands and includes a discussion on the use of simpler forms of probabilistic methods. This practical application is typical of issues facing Dutch engineers. In the Netherlands there are hundreds of kilometers of regional dykes that do not meet the latest safety guidelines, prompting the re-design and upgrading of some dyke sections, even if they have never experienced a failure in their lifetime. Similarly, there are hundreds of kilometers of railway embankments not meeting the latest safety guidelines. Such structures need to be reassessed, and upgraded where necessary, to satisfy the guidelines in the light of climate change and the expected increases in the frequency of, and loads imposed by, rail traffic.

15.5.1 Case History

Figure 15.15 shows a typical cross-section through a regional dyke located at Starnmeer in the Netherlands. This ring dyke is 13 km in length and managed by the water board Hoogheemraadschap Hollands Noorderkwartier (HHNK), who initiated a stability assessment of representative cross-sections of the dyke. These revealed factors of safety for some sections as low as 0.5, and prompted a re-design of some sections of the dyke even though the structure had remained stable for hundreds of years. This re-design involved relocating the ditch in the polder further from the dyke, so that a stabilizing berm could be constructed on the shoulder of the dyke (Hicks et al. 2019).

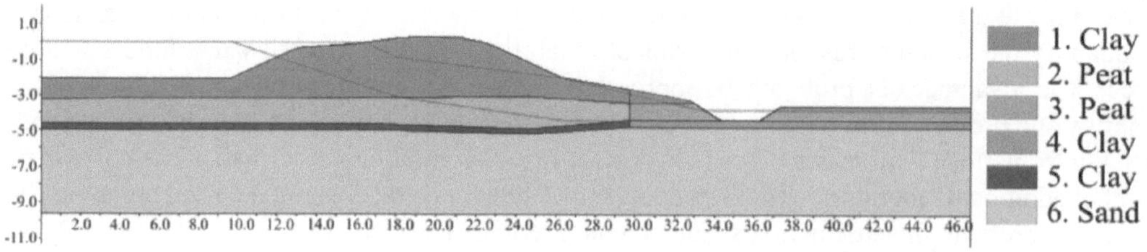

FIGURE 15.15 Starnmeer dyke cross-section (dimensions in meters) (based on Hicks et al. 2019)

As part of the re-design process, Hicks et al. (2019) and Varkey et al. (2020) conducted RFEM analyses of the cross-section shown in Figure 15.15. This involved defining the point and spatial statistics of effective cohesion and friction angle for each material zone in the cross-section, and, based on these sets of statistics, each realization in the RFEM simulation involved the generation of separate random fields for each zone. It was shown that, by taking account of the spatial nature of soil variability in the RFEM simulations and targeting acceptable reliability levels, the required size of the berm was significantly reduced, thereby reducing the cost of remedial works and the level of intrusion on the land adjacent to the dyke.

Varkey et al. (2020) extended the investigation to consider alternative, simpler forms of probabilistic analysis in analyzing the cross-section shown in Figure 15.15. On the one hand, RFEM may be

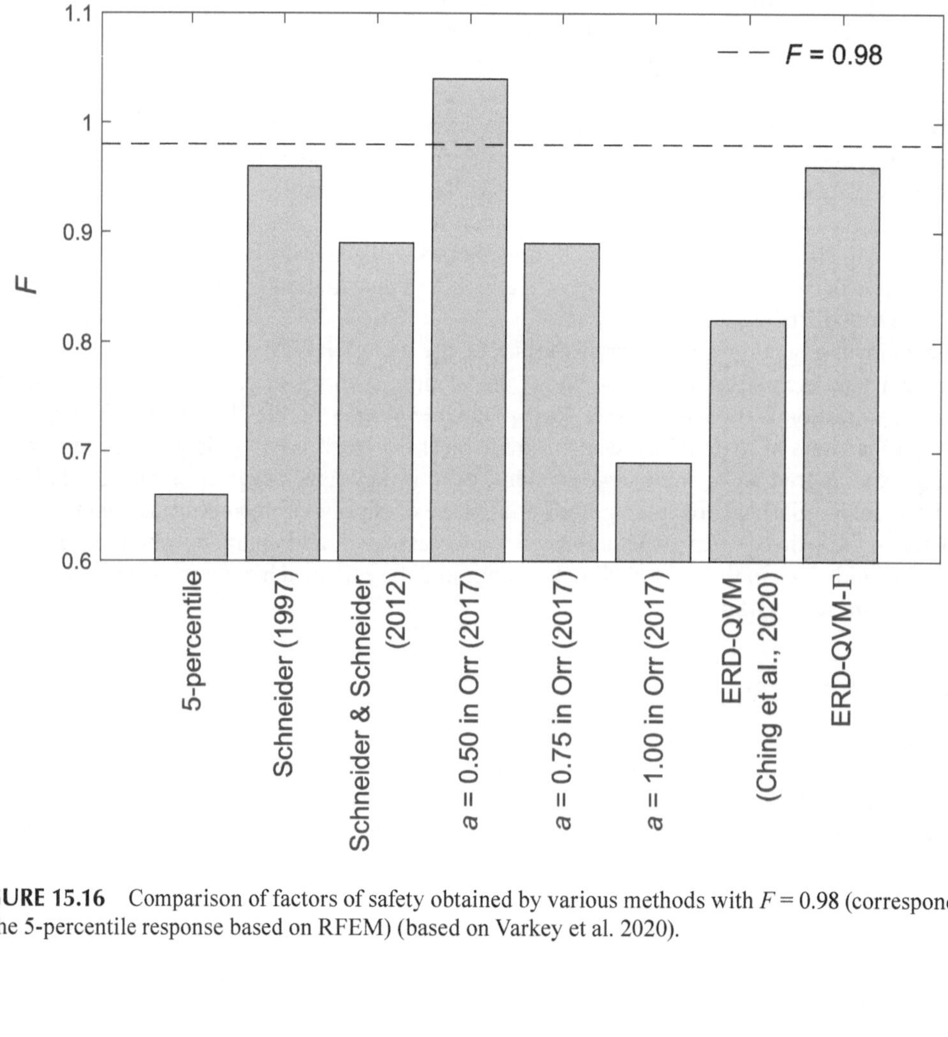

FIGURE 15.16 Comparison of factors of safety obtained by various methods with $F = 0.98$ (corresponding to the 5-percentile response based on RFEM) (based on Varkey et al. 2020).

considered a fully probabilistic approach, in that it achieves a target reliability level for a structure directly, without having to resort to the use of characteristic values or partial factors. (This is why it is counter-intuitive to use RFEM to derive characteristic values, as was alluded to at the end of Section 15.4.2.) In contrast, semi-probabilistic approaches involve the determination of reliability-based characteristic values, which, when factored down by partial factors and used in deterministic analysis, are designed to satisfy some target reliability level indirectly. Figure 15.16 shows a comparison between the reliability-based factor of safety computed using RFEM (indicated by the broken horizontal line) (Hicks et al. 2019) and various semi-probabilistic approaches by Schneider (1997), Schneider and Schneider (2012), Orr (2017) and Ching et al. (2020), as well a combination of two of these methods (designated as ERD-QVM-Γ) proposed by Varkey et al. (2020) and the value of F based on the 5-percentile values of the underlying property distributions. It is seen that although simpler approaches can work well by accounting for variance reduction due to spatial averaging, which method is the best may be problem-dependent. Varkey et al. (2022) also used RFEM to reassess six regional dyke cross-sections from other locations in the Netherlands, thereby demonstrating that RFEM can be useful as a tool for proposing partial factors, which, when used in conventional semi-probabilistic assessments, achieve the desired reliability levels required by practice.

REFERENCES

Calle, E. O. 1985. Probabilistic analysis of stability of earth slopes. In *Proceedings 11th International Conference on Soil Mechanics and Foundation Engineering*, San Francisco, 809–812.

CEN. 2004. *Eurocode 7: Geotechnical Design. Part 1: General Rules*. EN 1997-1. European Committee for Standardization.

Ching, J., K. K. Phoon, K. F. Chen, T. L. L. Orr, and H. R. Schneider. 2020. Statistical determination of multivariate characteristic values for Eurocode 7. *Structural Safety* 82:101893.

de Gast, T., P. J. Vardon, and M. A. Hicks. 2021a. Assessment of soil spatial variability for linear infrastructure using cone penetration tests. *Géotechnique* 71(11):999–1013.

de Gast, T., M. A. Hicks, A. P. van den Eijnden, and P. J. Vardon. 2021b. On the reliability assessment of a controlled dyke failure. *Géotechnique* 71(11):1028–1043.

Fenton, G. A., and D. V. Griffiths. 2008. *Risk Assessment in Geotechnical Engineering*. Hoboken, NJ: John Wiley & Sons.

Fenton, G. A., and E. H. Vanmarcke. 1990. Simulation of random fields via local average subdivision. *Journal of Engineering Mechanics, ASCE* 116(8):1733–1749.

Hicks, M. A. (ed.). 2007. *Risk and Variability in Geotechnical Engineering*. London: Thomas Telford.

Hicks, M. A. 2012. An explanation of characteristic values of soil properties in Eurocode 7. In *Modern Geotechnical Design Codes of Practice: Development, Calibration and Experiences*, ed. P. G. Arnold, G. A. Fenton, M. A. Hicks, T. Schweckendiek, and B. Simpson, 36–45. Amsterdam: IOS Press.

Hicks, M. A., and Y. Li. 2018. Influence of length effect on embankment slope reliability in 3D. *International Journal for Numerical and Analytical Methods in Geomechanics* 42(7):891–915.

Hicks, M. A., and J. D. Nuttall. 2012. Influence of soil heterogeneity on geotechnical performance and uncertainty: A stochastic view on EC7. In *Proceedings 10th International Probabilistic Workshop*, Stuttgart, Germany, 215–227.

Hicks, M. A., and C. Onisiphorou. 2005. Stochastic evaluation of static liquefaction in a predominantly dilative sand fill. *Géotechnique* 55(2):123–133.

Hicks, M. A., and K. Samy. 2002a. Influence of heterogeneity on undrained clay slope stability. *Quarterly Journal of Engineering Geology and Hydrogeology* 35(1):1–49.

Hicks, M. A., and K. Samy. 2002b. Reliability-based characteristic values: A stochastic approach to Eurocode 7. *Ground Engineering* 35(12):30–34.

Hicks, M. A., and K. Samy. 2004. Stochastic evaluation of heterogeneous slope stability. *Italian Geotechnical Journal* 38:54–66.

Hicks, M. A., and W. A. Spencer. 2010. Influence of heterogeneity on the reliability and failure of a long 3D slope. *Computers and Geotechnics* 37(7–8):948–955.

Hicks, M. A., J. Chen, and W. A. Spencer. 2008. Influence of spatial variability on 3D slope failures. In *Proceedings 6th International Conference on Computer Simulation in Risk Analysis and Hazard Mitigation*. Cephalonia, Greece, 335–342.

Hicks, M. A., J. D. Nuttall, and J. Chen. 2014. Influence of heterogeneity on 3D slope reliability and failure consequence. *Computers and Geotechnics* 61:198–208.

Hicks, M. A., D. Varkey, A. P. van den Eijnden, T. de Gast, and P. J. Vardon. 2019. On characteristic values and the reliability-based assessment of dykes. *Georisk: Assessment and Management of Risk for Engineered Systems and Geohazards* 13(4):313–319.

Hunter, J. H., and R. L. Schuster. 1968. Stability of simple cuttings in normally consolidated clays. *Géotechnique* 18(3):372–378.

Li, Y. J., M. A. Hicks, and J. D. Nuttall. 2015. Comparative analyses of slope reliability in 3D. *Engineering Geology* 196:12–23.

Li, Y. J., M. A. Hicks, and P. J. Vardon. 2016. Uncertainty reduction and sampling efficiency in slope designs using 3D conditional random fields. *Computers and Geotechnics* 79:159–172.

Liu, K., P. J. Vardon, and M. A. Hicks. 2018. Sequential reduction of slope stability uncertainty based on temporal hydraulic measurements via the ensemble Kalman filter. *Computers and Geotechnics* 95:147–161.

Lloret-Cabot, M., G. A. Fenton, and M. A. Hicks. 2014. On the estimation of scale of fluctuation in geostatistics. *Georisk: Assessment and Management of Risk for Engineered Systems and Geohazards* 8(2):129–140.

Lloret-Cabot, M., M. A. Hicks, and A. P. van den Eijnden. 2012. Investigation of the reduction in uncertainty due to soil variability when conditioning a random field using Kriging. *Géotechnique Letters* 2(3):123–127.

Orr, T. L. L. 2017. Defining and selecting characteristic values of geotechnical parameters for designs to Eurocode 7. *Georisk: Assessment and Management of Risk for Engineered Systems and Geohazards* 11(1):103–115.

Phoon, K. K., and F. H. Kulhawy. 1999. Characterization of geotechnical variability. *Canadian Geotechnical Journal* 36(4):612–624.

Remmerswaal, G., P. J. Vardon, and M. A. Hicks. 2021. Evaluating residual dyke resistance using the Random Material Point Method. *Computers and Geotechnics* 133:104034.

Schneider, H. R. 1997. Definition and characterization of soil properties. In *Proceedings 14th International Conference on Soil Mechanics and Geotechnical Engineering*, Hamburg, Germany, 273–281.

Schneider, H. R., and M. A. Schneider. 2012. Dealing with uncertainties in EC7 with emphasis on determination of characteristic soil properties. In *Modern Geotechnical Design Codes of Practice: Development, Calibration and Experiences*, ed. P. G. Arnold, G. A. Fenton, M. A. Hicks, T. Schweckendiek, and B. Simpson, 87–101. Amsterdam: IOS Press.

Smith, I. M., D. V. Griffiths, and L. Margetts. 2013. *Programming the Finite Element Method* (5th edition). Wiley.

Spencer, W. A., and M. A. Hicks. 2007. A 3D finite element study of slope reliability. In *Proceedings 10th International Symposium on Numerical Models in Geomechanics*, Rhodes, Greece, 539–543.

Vanmarcke, E. H. 1977. Reliability of earth slopes. *Journal of Geotechnical Engineering Division, ASCE* 103(11):1247–1265.

Vanmarcke, E. H. 1983. *Random Fields: Analysis and Synthesis*. Cambridge, MA: The MIT Press.

Vardon, P. J., K. Liu, and M. A. Hicks. 2016. Reduction of slope stability uncertainty based on hydraulic measurement via inverse analysis. *Georisk: Assessment and Management of Risk for Engineered Systems and Geohazards* 10(3):223–240.

Varkey, D., M. A. Hicks, and P. J. Vardon. 2019. An improved semi-analytical method for 3D slope reliability assessments. *Computers and Geotechnics* 111:181–190.

Varkey, D., M. A. Hicks, A. P. van den Eijnden, and P. J. Vardon. 2020. On characteristic values for calculating factors of safety for dyke stability. *Géotechnique Letters* 10(2):353–359.

Varkey, D., M. A. Hicks, and A. P. van den Eijnden. 2022. Reliability-based partial factors considering spatial variability of strength parameters. In *Proceedings 16th International Conference of the International Association for Computer Methods and Advances in Geomechanics*, Turin, Italy, 299–304.

Varkey, D., M. A. Hicks, and P. J. Vardon. 2023. Effect of uncertainties in geometry, inter-layer boundary and shear strength properties on the probabilistic stability of a 3D embankment slope. *Georisk: Assessment and Management of Risk for Engineered Systems and Geohazards* 17(2):262–276.

Wang, B., M. A. Hicks, and P. J. Vardon. 2016. Slope failure analysis using the random material point method. *Géotechnique Letters* 6(2):113–118.

16 Uncertainty, Modeling, and Decision-Making for Ground Improvement

Yutao Pan and Rui Tao

ABSTRACT

Ground improvement has been extensively used in unfavorable ground conditions to improve mechanical features of the ground, such as strength, stiffness, and watertightness. Uncertainties related to a random fluctuation of improved ground property and construction errors may greatly impact the global performance of the improved ground. This chapter classifies the two types of uncertainties and provides statistical summaries of them. Risk assessment methods and results are then provided to shed some light on the impact of these two types of uncertainties. Finally, limitations and future works are recommended at the end of the chapter.

16.1 INTRODUCTION

Most civil infrastructures such as residential buildings, roads, railways, and airports have to be safely constructed on stable ground. In unfavorable geological conditions and/or challenging infrastructural contexts, ground improvement must be used to improve the mechanical or hydraulic properties of various types of natural soils, such that the sustained infrastructures satisfy safety and serviceability requirements.

People's desire for safer and more comfortable infrastructure drives a steady increase in the ground improvement market. The global soil stabilization market has been steadily increasing and is estimated to reach USD 27.80 billion in 2022.

During ground improvement, the soil is chemically or physically treated to improve its engineering properties such as strength, stiffness, and permeability. This distinguishes itself from the notion of "soil stabilization" in earth environment sectors, which typically involves the stabilization of heavy metals or other contaminants in soils. Ground improvement has been applied in the soil rehabilitation of building foundations, the seepage mitigation of dam foundations, and the contamination control of landfills. Various kinds of ground improvement approaches have been developed and applied (e.g., deep mixing, jet grouting, pre-loading, compaction, and minipile). The two most widely-used ground improvement methods are deep mixing and jet grouting. We will focus on these two methods in this chapter.

Deep cement mixing involves in-situ mixing of binder (usually in the form of wet cement/lime slurry or dry cement/lime powders) by rotating blades. Jet grouting cuts and partially replaces in-situ soils with high-momentum cement slurry. Whatever method is used, the mixing is commonly done in a columnar fashion so that the treated ground mass comprises contiguous columns.

Although the ground improvement work has been conducted with due discretion, significant uncertainties in the improved ground have been reported, and the uncertainties can be divided into two categories: random fluctuation and construction error. The former stems from the fact that the final mixture after installation is still heterogenous due to uneven mixture and natural variation of soil properties. The binder contents in some parts are higher than other points. This results in an uneven spatial distribution

DOI: 10.1201/9781003333586-20

of engineering properties (e.g., strength, modulus, and permeability). The construction error is mainly caused by limited accuracy in workmanship and equipment. Typical examples of construction errors are random inclinations of column axis and column diameter deviation from ideal assumption.

This chapter aims at identifying and summarizing the sources and statistics of uncertainties in cement-based ground improvement. Related modeling methods and risk assessment approaches are then introduced. The preliminary framework can be used to shed some light on design and decision-making in the design and construction phases.

16.2 SOURCES OF UNCERTAINTIES IN GROUND IMPROVEMENT

As mentioned above, there are two major types of uncertainties in cement-based ground improvement, i.e. 1) random fluctuation and 2) construction error.

1) Random Fluctuation

Many factors could contribute to the random fluctuations. The two most important factors are the uneven distribution of binders and the natural variation of soil properties. They will determine the local mix ratios (e.g., cement content, water content) of the final mixture, which further determines the strength and stiffness (Xiao et al. 2014). Given that the binder and soil properties both variate spatially, the strength and stiffness also fluctuate spatially.

Chen et al. (2016) evaluated the spatial distribution of binder (in the form of cement slurry) concentration using centrifuge modeling, and showed that the uniformity is largely influenced by the binder density and blade rotation number. It is difficult to directly derive the mechanical properties (strength, stiffness) of improved ground solely from binder and soil properties because many factors (in-situ conditions: soil mineralogy, water content, effective stress, temperature; binder: cement content, water content) would affect the result. A more practical and effective approach for quality control is direct strength measurement through coring or in-situ testing. Larsson et al. (2005) used hand-operated CPT to directly evaluate the strength of cement/lime-treated soils. Unconfined Compression Tests (UCT) are also widely used for coring samples in Singapore and Japan (Honjo 1982; Liu et al. 2017, 2019). Statistical approaches were used to characterize the uncertainties of cement-treated ground. Random field theory is widely used in the characterization and replication of spatial variability for cement-treated ground, in which the spatial variability is considered by probability distribution functions and correlation structures. The three most important statistical features, namely mean value, coefficient of variation (COV, standard deviation divided by mean value), and scale of fluctuation (SOF, similar definitions of autocorrelation length), are used. The first two describe the average level and spread of the properties at a point, providing rough probabilistic distribution of the marginal distribution. Whereas the SOF provides extra information in the spatial domain, or in other words, a measure of the texture. Table 16.1 summarizes the statistical features of cement-treated soils. The mean unconfined compressive strength (UCS or q_u) ranges from 600–4000 kPa. Depending on the workmanship and test scheme, the COV ranges from 0.2–0.6 in most cases. The SOF measurement is more complex and varies greatly from study to study. Some studies (Hedman and Kuokkanen 2003; Larsson et al. 2005; Al-Naqshabandy et al. 2012) show a weakly correlated structure in cement-treated soils, with SOF at the scale of 0.3–1.0 m. In contrast, Navin and Filz (2005) show very long SOFs in horizontal directions, at the scale of 20 m. It has been recently explained by Liu et al. (2019) that the difference lies in the source of spatial variability, namely intra-column variation and inter-column variation. The former is mainly accredited to the uneven distribution of binder due to insufficient mixing. The latter is due to the natural variation of in-situ water content, which according to Phoon and Kulhawy (1999) ranges from 10 to 100 m.

2) Construction Errors

Construction errors are the result of the limited accuracy of workmanship and machines. The improved columns are installed underground using elongated steel rods. For deep mixing, the

TABLE 16.1
Statistical Characteristics of Cement-Admixed Soil

References	Test (Result)	Mean Value	COV	Scale of fluctuation* (m)		Skewness	Kurtosis	Marginal Distribution
				Vertical	Horizontal			
Honjo (1982)	Unconfined Compressive Test (UCS)	0.6–8.0 MPa	0.21–0.36(clay) 0.32– 0.4(sandy soils)	0.8–8.0	—	—1.19 ~ 2.55	2.7–4.4	Normal
Babasaki et al. (1996)	Unconfined Compressive Test (UCS)	—	0.22– 0.27	—	—	—	—	—
Hedman and Kuokkanen (2003)	Hand-operated penetrometer test (c_u)	—	—	0.38– 1.12	0.07–0.33‡			
Navin and Filz (2005)	Unconfined Compressive Test (UCS)	1.0–4.7 MPa	0.34–0.79	—	Approximate 24.0	—	—	Lognormal
Larsson et al.(2005) ‡	Hand-operated penetrometer test (c_u)	—	<0.60	—	Radial:<0.13 Orthogonal:<0.32‡	—	—	—
Larsson and Nilsson (2009)	Cone penetration test (Tip resistance)	—	0.20–0.60	—	1.8–3.6	—	—	—
Chen et al. (2011) (MFBC)	Unconfined Compressive Test (UCS)	2.0–2.7 MPa	0.29–0.46	—	—	0.48 ~ 1.34	—	-
Chen et al. (2011) (NCHS)	Unconfined Compressive Test (UCS)	3.2–4.5 MPa	0.29	—	—	—1.4 ~ —0.7	—	-
Al-Naqshabandy et al. (2012)	Cone penetration test (Tip resistance)	—	0.22–0.67	0.2– 0.7	2.0–3.0	—	—	
Bergman et al. (2013)	Cone penetration test (Tip resistance)			0.08– 0.77 m	<3.5 m			
Namikawa and Koseki (2013)	Unconfined compressive test (UCS)	1.7 MPa	0.2–0.4	—	—	—	—	Normal
Bruce et al. (2013)	Unconfined Compressive test (UCS)	0.7–2.1 MPa	0.34–0.79	—	—		—	—
Chen et al. (2016)	Binder concentration	29%	0.19	—	—	—	—	Normal
Liu et al. (2017)† (MFBC)	Unconfined Compressive Test (UCS)	1.7 MPa	0.42	—	—	1.10	4.67	Beta-distribution
Liu et al. (2017)† (Marina One)	Unconfined Compressive Test (UCS)	2.1 MPa	0.44	—	—	1.31	4.78	Beta-distribution
Liu et al. (2019)	Centrifuge test (Binder Concentration)			1.0–3.33	**Small Scale[c] SOF:** Intra—column: Radial: 0.12D–0.28D Circumferential: 67°–133° Intercolumn: 0.12D–0.28D **Large** Scale[d] **SOF:** 25 m Explanations of c and d are missing			

Notes:
*The concept "auto-correlation distance" used in some studies (e.g., Namikawa and Koseki 2013) is converted to "scale of fluctuation" Vanmarcke (1983) by multiplying 2.0;
†Liu et al. (2017) normalized the strength to 28-day equivalent strength to eliminate the effect of the curing period.
‡SOF within the column cross-section
c due to uneven distribution of binder during mixing
d due to effect of in-situ water content

major construction error is the positioning error at the production level. It is mainly the deviation of a machine at ground level and the random orientation of rods that contribute to this error. Jet grouting has an extra source of construction errors, namely diameter variation along the column. The diameter of solidified columns is influenced by several factors, such as the in-situ soil properties (strength and permeability), and the variation of supplied pressure for jet-grouted fluid. Intuitively, the column diameter diminishes with low-permeability high strength soil layers (e.g., stiff clay). As a result, the diameter of jet-grout columns will vary along the column axis. Figure 16.1 illustrates the two sources of construction error.

Specifically, the random orientations consist of two independent parts, i.e. azimuth (α) and inclination angle (β). The azimuth (α) is assumed to follow the uniform distribution within $[0, 2\pi]$, while the inclination angle (β) is assumed as a Gaussian distribution with a mean value of 0, and a standard deviation based on the workmanship and execution standard. Table 16.2 shows the standard deviation of the inclination angle (β) is around 0.1–0.2 degrees, while the COV of the

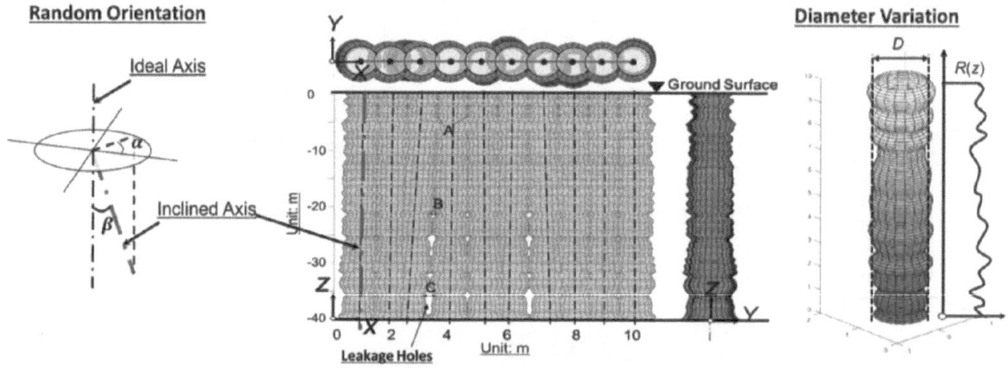

FIGURE 16.1 Illustration of construction errors for jet-grouted cut-off walls (JGCOW).

TABLE 16.2
Statistical Characteristics of Geometric Imperfections for Jet-Grout Column

No.	Type of natural soil	S.D. (β)**	Average of diameter *D* (m)	COV(*D*)*	Remarks	References
1	Clay-Silt			0.02–0.05		Croce et al.
	Sands			0.02–0.10		
	Gravels			0.05–0.25		
2	Sandy Clay		1.1	0.06–0.19	Derived from field data of diameter at different depths, horizontal column	Langhorst et al.
3	Silty Sand	0.07°	0.71–1.11	0.06	Vertical columns in Vesuvius site	Croce et al. (2007)
	Sandy Gravel		1.06–1.20	0.19	Vertical columns in Polcevera site	
4	Silty Sand	0.16°	2.5		Vertical columns in Barcelona site	Eramo et al. [6]
5	Sandy Clay		0.38	0.13	Vertical column with lower water content	Arroyo et al. [1]
			0.48		Vertical column with lower water content	
		0.17°	0.75	0.17	(Sub) Horizontal column	

TABLE 16.3
Verticality requirements in different standards

Source	Type of technique used	Maximum Deviation from Verticality	Corresponding S.D. (β) *
ASCE Jet Grouting Guideline (2009)	Jet grouting	1/100	0.3°
Ryan and Jasperse (1989)	Deep mixing	1/100	0.3°
Singapore Standard (2003)	Jet grouting	1/75	0.4°
Amos et al. (2008)	Drilled concrete piles	1/200**	0.15°

*the S.D. (β) is evaluated by assuming that only 5% of the cases exceed that off-verticality

diameter is around 0.02–0.2. The average diameter depends highly on the ground conditions and the contractors' configuration for jet grouting's operation parameters (e.g., penetration pressure, rotation number, etc.). It can be determined by site mock-up or empirical approaches. Table 16.3 summarizes the execution standard for verticality of deep mixing and jet grouting. In most execution standards, the tolerance is 1/100, which corresponds to around 0.3 degrees of standard deviation (Pan et al. 2019b).

16.3 MODELING OF GLOBAL PERFORMANCE

The global performances (global strength, stiffness, and watertightness) of the improved ground are significantly influenced by the two types of uncertainties. It is therefore important to quantitatively evaluate the global performance with the given uncertainties. Since improved grounds are used to improve the mechanical (strength, stiffness) and hydrological (watertightness) performances, this subsection consists of two parts, 1) global stress-strain behavior, and 2) watertightness.

16.3.1 Global Stress-Strain Behavior

Cement-treated columns are usually installed in an overlapping way such that it forms a stiff continuum (slabs or berms) that can transfer the load with less deformation. Figure 16.2 shows a typical application. The columns are installed deep underground before the excavation phase using deep mixing or jet grouting. After the injected materials harden and the retaining walls are in position, the excavation starts. The lateral earth pressure behind the retaining walls will be gradually transferred to the improved ground, which is usually in the form of slabs, berms, or strips. This form of underground strut is advantageous over temporary internal struts because it can be cast before excavation, which is effective in reducing the inward displacement of retaining walls. The laboratory test shows that the average unconfined compressive strength is around 1–4 MPa, depending on the mix ratio and curing conditions. However, the design strength was assumed to be as low as 600 kPa, due largely to the random spatial variability and the brittle behavior of cement-treated soils. It is therefore of great interest to factually quantify the effect of these two factors on the global performance of cement-treated ground and to provide a rational guideline on the reduction from the mean value. This potentially saves costs and reduces carbon emissions.

Reliability analysis is usually used to quantify uncertainties and their impacts. It involves the establishment of analytical performance functions and the propagation of uncertainties from random input variables and performance indicators. Honjo (1982) carried out a systematic analysis with a wide spectrum of work, including statistical analyses on UCS of coring samples, reporting the mean value, and COV and SOF of UCS. He further proposed a bundle model to consider the effect of strength variability (COV) on the reduction ratio of the design value from the average UCS of the coring samples. The effect of SOF was implicitly considered in the number of elements.

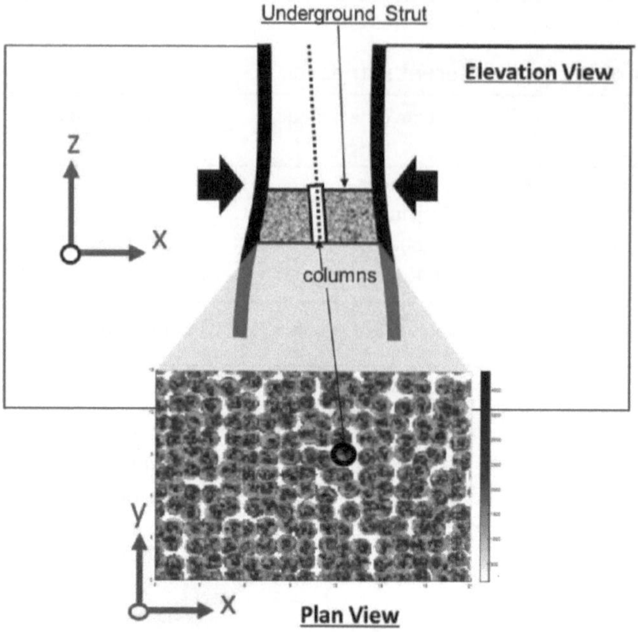

FIGURE 16.2 Illustration of deep-mixed underground strut.

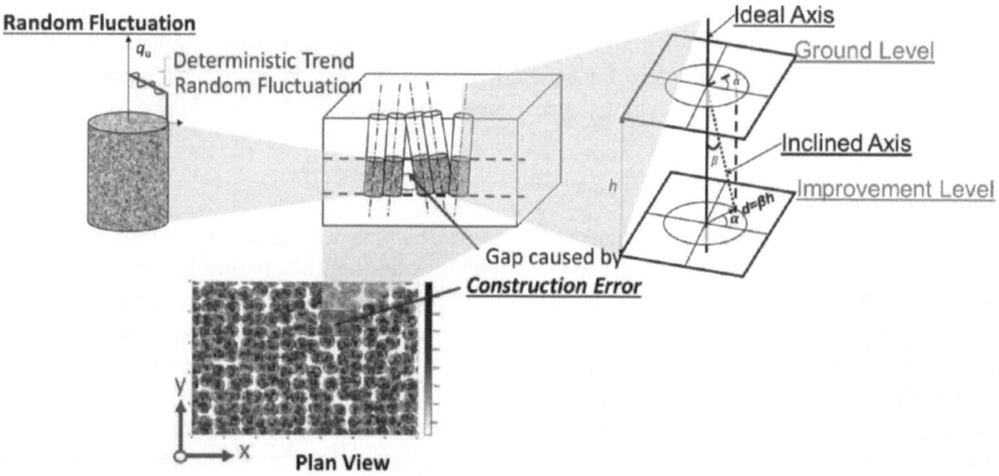

FIGURE 16.3 Illustration of source of random fluctuation and construction errors of deep-mixed underground strut.

Omine et al. (2005) used a combined method of weakest link and bundle model to further evaluate the effect of both variability and scale of the problem.

When combined with previous methods, numerical approaches provide solutions for more complex boundary value problems. The most representative methods are Random Numerical Limit Analysis (RNLA) and Random Finite Element Method (RFEM). Both methods involve replications of spatial variability using a random field and running a large number of numerical analyses to obtain the distribution of relevant global performance indicators (Monte Carlo Simulations). With the help of powerful software packages, these approaches can consider more complex boundary value problems and output more information than just a few indicators. The RNLA provides solutions for the upper and lower bound of global stability, while the RFEM directly provides information about both stability and deformation.

Kasama et al. (2010, 2011, 2012) performed numerical limit analysis to investigate the stability of cement-treated ground with spatial variability. The undrained shear strength of cement-treated soils was assumed to be lognormally distributed and a random field was generated using a Cholesky decomposition technique. Then lower and upper bound numerical limit analyses were performed to obtain a solution domain that brackets the true solution.

Random Finite Element Method (RFEM) is a powerful method that combines random field theory and finite element method. The RFEM can factually consider the effect of both spatial variability and complex constitutive behaviors in boundary value questions. Although the computation cost of RFEM is usually high and currently difficult to be introduced to the industry, the computational cost becomes less of a problem with the ever-upgrading hardware and software.

Namikawa and Koseki (2013) used the RFEM to simulate the effect of inherent random fluctuation of strength on the compressive strength of a single cement-treated sand column. These works were based on a systematical study ranging from basic cell tests, constitutive modeling, numerical modeling, and finally risk assessment via RFEM. Liu et al. (2015) simulated three-dimensional random spatial variability considering both the random distribution of binder and random deviation of column positioning using RFEM. The simple elastic-plastic Tresca criteria were used to simulate undrained behavior. Pan et al. (2018, 2019a, 2021a) further investigated the combined effect of both spatial variability and softening behavior on the global performance of cement-treated ground. These works were based on the earlier contributions of basic triaxial tests (Xiao et al. 2014), constitutive behavior (Xiao et al. 2017; Pan et al. 2016), efficient random field generation method (Liu et al. 2014), and quantification of random fluctuation (Chen et al. 2016; Liu et al. 2017). These works were then integrated using the random finite element method (Pan et al. 2018). This is further used in the reliability-based design of cement-treated slabs (Pan et al. 2019a), and tunneling with cement-treated surroundings (Pan et al. 2021a). The results show that the positioning error has a greater impact on the global strength and stiffness of cement-treated soil slab than the random fluctuation of strength since the random fluctuation has far smaller dimensions than the untreated gaps caused by positioning errors.

In this chapter, a demonstration example is used to evaluate the effect of random fluctuation and construction error on the global stress and stiffness of a cement-treated slab in an undrained condition. Figure 16.4 shows the flowchart of the random finite element method application. More details can be found in Pan et al. (2019a).

16.3.2 Watertightness

Cut-off walls are usually used to prevent water infiltration in excavations, dams, and landfills. Diaphragm walls, secant pile walls, and jet-grouted cut-off walls (JGCOW) are three representative types of cut-off walls. In cases where the ground is unfavorable for heavy machine deployment (e.g., ground is very soft), or when the construction clearance is too small for deployment of installation machines, JGCOW is a good alternative. The JGCOW consists of one or more rows of overlapping columns. An overlap between two adjacent columns of 200–300 mm is often used in practice. However, with the random orientation and diameter variation, it is possible that some continuous seepage passages may occur, see Figure 16.1. Existence of such defects may significantly shorten the seepage pathway by providing a direct flow passage across the wall, leading to an increase in flow rate in the vicinity of the untreated zones (Shirlaw 1996; Amos et al. 2008), erosion of dams (Richardson et al. 2015), or contamination of adjacent bodies (Inazumi et al. 2009).

Croce and Modoni (2007) examined the effect of geometric imperfections on the watertightness of cement-treated ground with deep mixing. Simplified probabilistic frameworks were then established to evaluate the size of geometric defects based on a wide range of statistical field data (Croce et al. 2004; Croce and Modoni 2007; Eramo et al. 2011; Arroyo et al. 2012, Modoni et al. 2016).

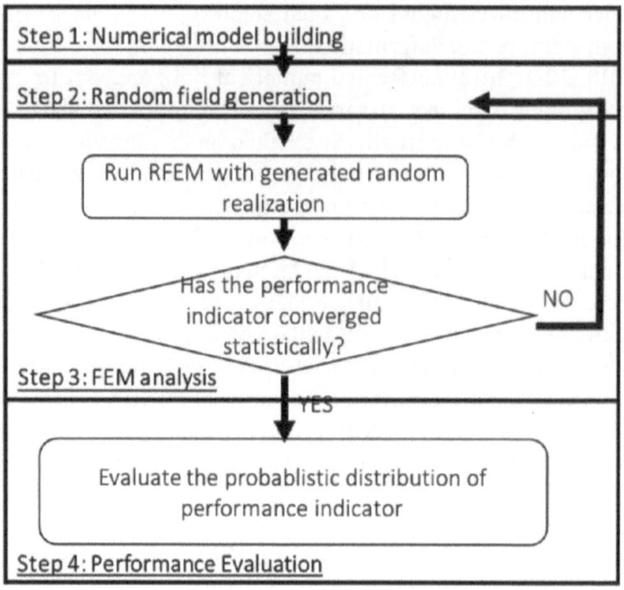

FIGURE 16.4 Flowchart for RFEM.

Although RFEM can be used, it involves a huge calculation expense because the mesh size must be very small to capture the small defectives and Monte Carlo simulations are involved (Pan et al. 2020). Pan et al. (2017, 2019bc, 2021b) proposed a three-dimensional discretized algorithm (TDA) to quantify the flow rate through defective cut-off structures. The flow chart and detailed illustrations of TDA are shown in Figure 16.5 and Figure 16.6, respectively.

This algorithm starts with the generation of random configuration parameters (inclination angle, azimuth, and diameter variation). The inclination angle and azimuth are simulated as independent and identically distributed random variables with given marginal distributions summarized in Table 16.2. For special cases where multi-shaft installation is used, higher correlation among the same flight can be considered by generating correlated random numbers. The diameter variation is continuous along the shaft and can be simulated using a one-dimensional random field, with a given mean value, coefficient of variation, and scale of fluctuation. The scale of fluctuation was assumed to be close to that of the natural soil.

Then the treated ground is discretized. Two kinds of planes are used as references to facilitate further analysis. As shown in Figure 16.6(b), vertical planes that are parallel to the wall plane are called slices and horizontal planes are called layers. These planes are used to discretize the space with uniform grids.

The space between two adjacent columns are places where most leakage holes occur. Therefore, a penetration check is conducted for each cell to examine the existence of penetrating zones. A cell is penetrated when 1) untreated zones exist in every slice and 2) any arbitrarily chosen pair of adjacent slices overlap, as shown in Figure 16.6(c). If none of the cells are penetrated, then the cut-off wall is not penetrated. If any cell is penetrated, then the leakage discharge will be evaluated.

For each penetrated passage, the discharge rate is proportional to the harmonic average of the cross-sectional area along the continuous passage, detailed procedures of derivation are provided by Pan et al. (2017). This is intuitive because the harmonic average is low-value sensitive, that is, it approaches zero when a very small cross-section (bottle-neck) presides. The cross-sectional area on each slice can be evaluated by counting the number of untreated nodes on that specific slice.

The TDA has shown the capability to provide a sufficiently accurate and remarkably quick evaluation of the leakage discharge rate through defective jet-grouted cut-off walls. This algorithm does not involve a solution of massive linear equation series, as the finite element methods did. It requires

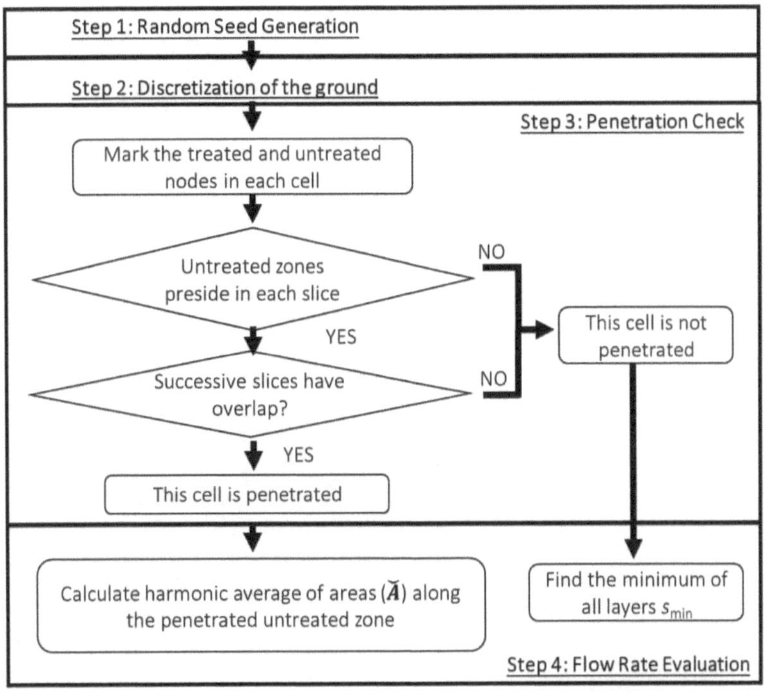

FIGURE 16.5 Flow chart for TDA.

remarkably less computational cost. A direct validation was made between the TDA and Finite Element Method (FEM) (Pan et al. 2020). The same random realizations of defective cut-off walls were generated for both TDA and a full three-dimensional FEM analysis, and the discharge rates of TDA were plotted against the corresponding FEM results, showing a remarkably good linear correlation. The TDA results show a consistent augmentation factor of 10–30%, because the corrugated inner boundary along the irregular drainage passage may lengthen the seepage passage, which cannot be simulated in TDA. The calculation expense of TDA was estimated to be only 1/10000 to 1/1000 that of FEM analysis. The main reason for the difference was accredited to the fact that the TDA does not involve the time-consuming solutions of large linear equation functions.

16.4 RISK ASSESSMENT AND DECISION-MAKING

It is interesting to know the connection between the input random variables and the global performance, such that the probabilistic distribution of the global performance of the improved ground can be quantified. This section provides two simple case studies that apply to the design of cement-treated slabs as an underground strut, and jet-grouted cut-off walls.

16.4.1 Global Stiffness and Strength of Cement-Treated Slabs

As mentioned earlier, random fluctuation and construction errors were considered in this study.

Random Fluctuation

The random fluctuation is characterized by the scale of fluctuation and marginal distribution. In this section, the marginal beta distribution is used with the dimensionless parameters COV, with skewness and kurtosis being calibrated by Liu et al. (2015). Based on the statistical analysis in Table 16.1, a COV of 0.4 was assumed and the SOF along radial, and the

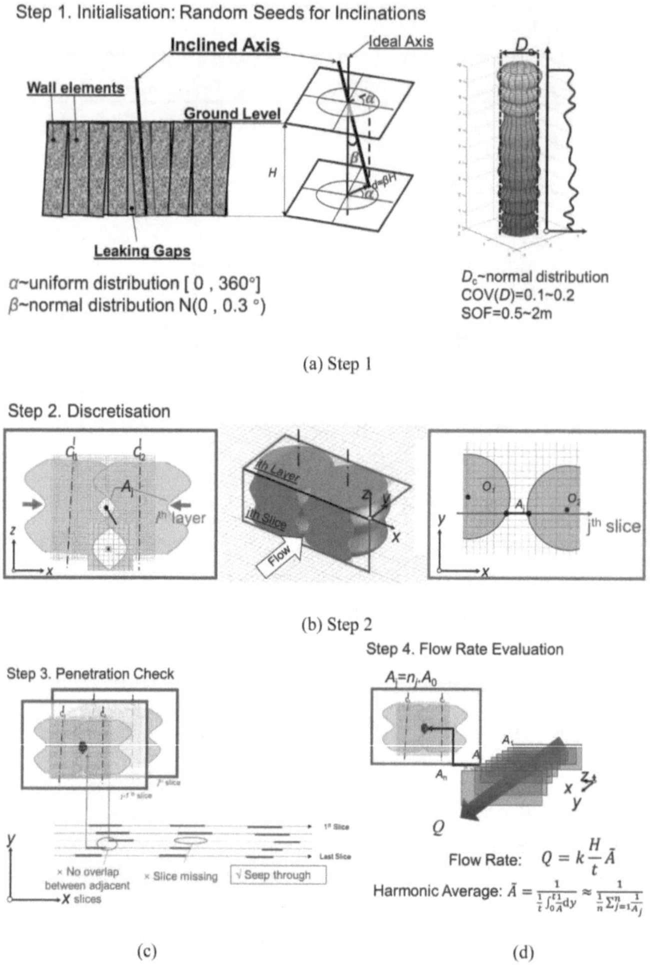

FIGURE 16.6 Detailed illustrations for each step.

circumferential and vertical directions used were R/3, π/4, and 5*R*. Beta distribution was used to characterize the probabilistic distribution UCS. The shape factors of Probability Density Function (PDF) of unconfined compressive strength are obtained according to the Marina Bay Financial Center (MBFC) project (Liu et al. 2015). The modified linear estimation method (Liu et al. 2014) is employed to generate this random field with a marginal beta field.

Positioning Error

The deviation distance and deviation angle follow a two-dimension bi-variate random field with marginal PDF of uniform distribution. The maximum radial deviation is D/4 (D is 1.5 m in this study). This corresponds to a case where the embedment depth is about 20 m and the maximum allowable off-verticality is 1:75 (Singapore Standard 2003).

Typical Realizations

The unconfined compressive strength for a random realization is shown in Figure 16.3 (darker color refers to higher strength, white color indicates untreated gaps). A typical spread of global stress-strain curves of treated slab is shown in Figure 16.7. The hollow dots indicate the global peak strength, which may not be achieved at the same global strain level due to the random distribution of material properties. Similarly, the global moduli (E) of all random cases are also evaluated. The global moduli are evaluated by fitting the initial

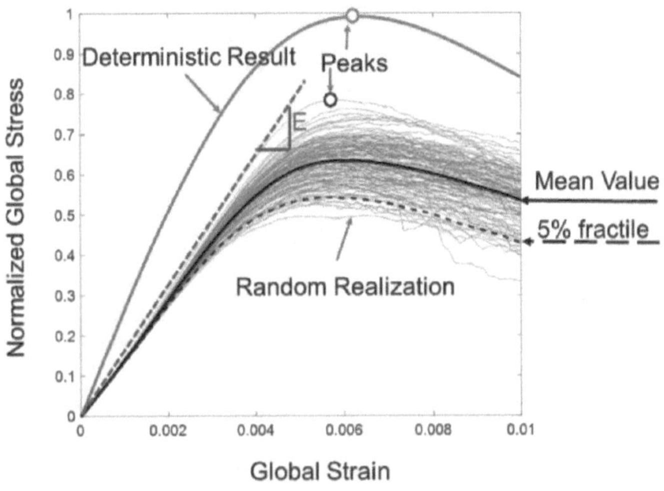

FIGURE 16.7 Global stress-strain curves.

linear portion of each curve. Normal distributions are assumed for further probabilistic analyses. This is not unreasonable because the COV of global behavior is basically low.

Effect of Positioning Error

The positioning error in this study varies between 0 and 0.35 times the column diameter. Figure 16.8(a) shows the average and 5th percentile (5% fractile) of the normalized global strength (global strength of random realizations divided by the deterministic global strength) in an undrained scenario. It is clear that both the mean and 5th percentile reduce with increasing positioning error. When the positioning error is zero, the randomness arises from the random fluctuation of material only. In such a case, the normalized global strength is very close to the 1 in Figure 16.8a, showing a limited influence of inherent random variability (that arises from the uneven binder distribution) on the global behavior. This is reasonable because cement-treated soil is regarded as a weakly correlated material if only intra-column variability is taken into account, and this leads to a much smaller horizontal scale of fluctuation (around 1/3 of radius) than the dimension of the slab. As a result, the impact of such spatial variability is diminished by a strong local averaging effect. In contrast, the positioning error may occasionally lead to connected gaps (as shown in Figure 16.3). These gaps are larger in size and are essentially untreated or weakly cemented. The output COV increases with positioning error, indicating more variability in results. A similar trend is observed for normalized global stiffness (Figure 16.8b but with less reduction).

Implementation

Figure 16.9 shows the flowchart of implementation. One should determine the initial boundary setup. Relevant factors like depth, action standard for positioning error, and exceeding probability of reliability index should be determined. With these factors, the normalized positioning error can be determined and then design factors (R_1 and R_2 in Figure 16.10) can be evaluated. It helps by doing site mock-up and coring to further determine the mean value of unconfined compressive strength (q_{u_ave}). The ratio of elastic modulus to the unconfined compressive strength can also be roughly estimated from the UCT curves (Lee et al. 2005). If no site data are available, data from adjacent areas are suggested.

Depending on the design target, one should choose to calculate a higher or lower estimate. If the design target is the underground strut itself, then it is conservative to find a reasonably low value in both strength and stiffness. If the design target is the internal force of the retaining wall that is supported by the strut, then it is conservative to assume a

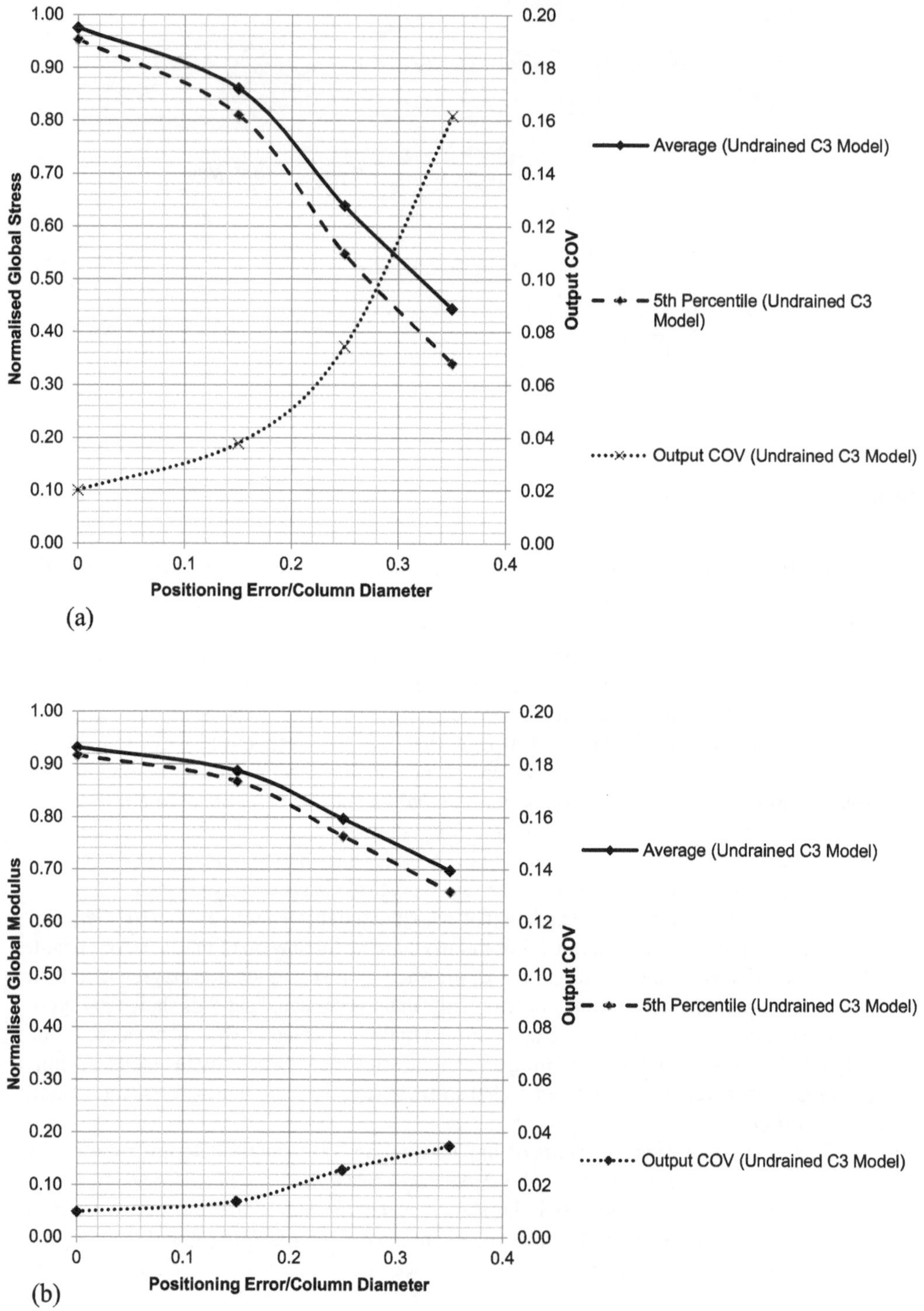

FIGURE 16.8 Effect of positioning error on the undrained global peak strength (a) and global stiffness (b).

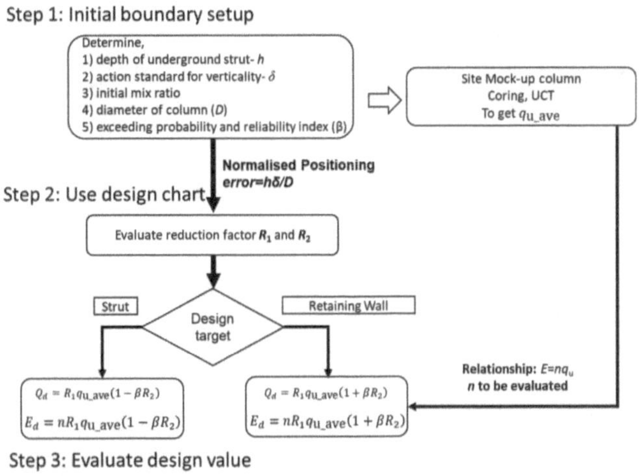

FIGURE 16.9 Flowchart for reliability-based design.

stiffer and stronger strut. This is because a stiffer and stronger strut would lead to a higher moment in the retaining wall, and in this way, we can avoid under-design of retaining walls. With the rational design values, one can then determine the dimension (thickness of the strut) according to the depth and mix ratio of the strut. Optimization is possible if one restarts the process with a different assumption (e.g., mix ratio, depth of strut).

16.4.2 Watertightness of Jet-Grouted Cut-Off Walls

Watertightness is evaluated using TDA as mentioned in 16.3.2. Direct Monte Carlo simulation is used to evaluate the probabilistic distribution of flow rate through defective cut-off walls. Table 16.4 shows the configurations of parametric studies. The effect of design parameters (number of rows and column spacing) and construction errors (standard deviation of inclination angle and COV of

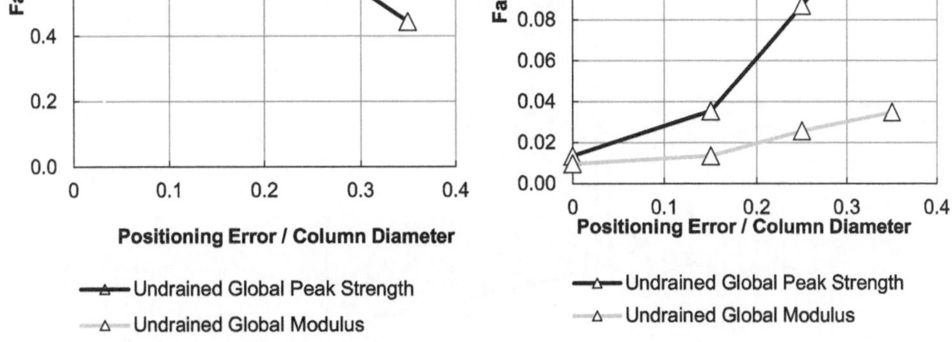

FIGURE 16.10 Design chart for underground strut (undrained).

TABLE 16.4
Configurations of Model for JGCOW

Remarks	Number of Rows n_{row}	Diameter D (m)	Spacing s_x (m)	Spacing s_y (m)	Excavation Depth Depth (m)	COV (D)	S.D. (β)
Reference Case	**2**	**1.0**	**0.8**	**0.5**	**10**	**0.1**	**0.3°**
Effect of n_{row}	1	1.0	0.8	0.5	10	0.1	0.3°
	2	**1.0**	**0.8**	**0.5**	**10**	**0.1**	**0.3°**
	3	1.0	0.8	0.5	10	0.1	0.3°
Effect of s_x	2	1.0	0.6	0.5	10	0.1	0.3°
	2	1.0	0.7	0.5	10	0.1	0.3°
	2	**1.0**	**0.8**	**0.5**	**10**	**0.1**	**0.3°**
Effect of COV of	2	1.0	0.8	0.5	10	0.05	0.3°
Diameter	**2**	**1.0**	**0.8**	**0.5**	**10**	**0.1**	**0.3°**
	2	1.0	0.8	0.5	10	0.2	0.3°
Effect of Standard	2	1.0	0.8	0.5	10	0.1	0.1°
Deviation of	**2**	**1.0**	**0.8**	**0.5**	**10**	**0.1**	**0.3°**
Inclination	2	1.0	0.8	0.5	10	0.1	0.5°

diameter) are evaluated. A reference case is set as a standard basis for calculation. The physical meanings of the design parameters are illustrated in Figure 16.11.

A wide range of parametric studies were done to provide the basis for a reliability-based design, see Figure 16.12. It links the global performance (normalized flow rate) with the design parameters (number of rows, spacing, and depth of the cut-off wall) and construction error level (inclination

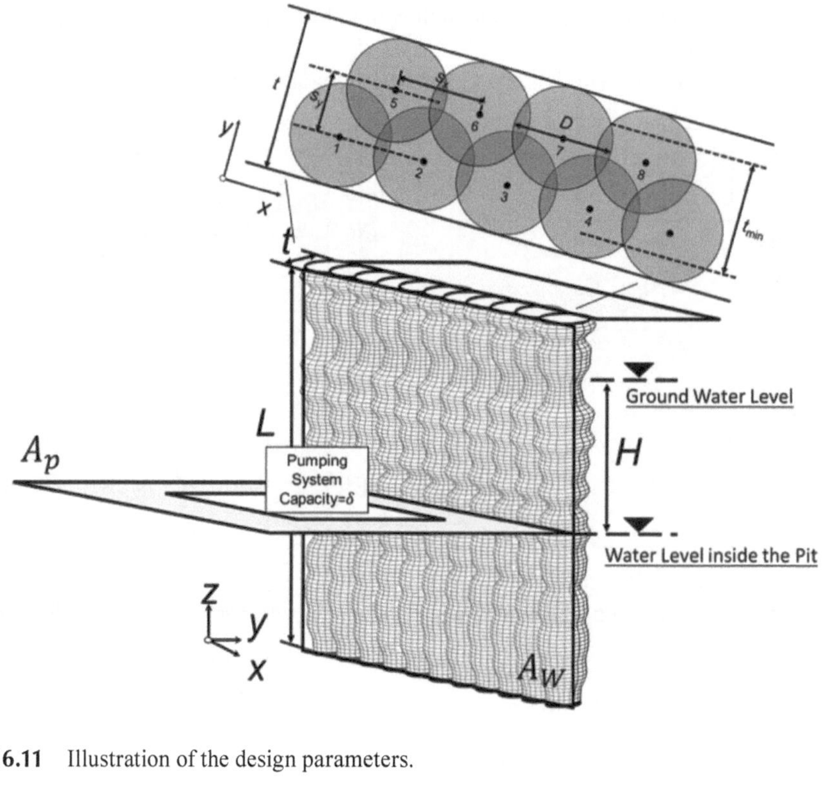

FIGURE 16.11 Illustration of the design parameters.

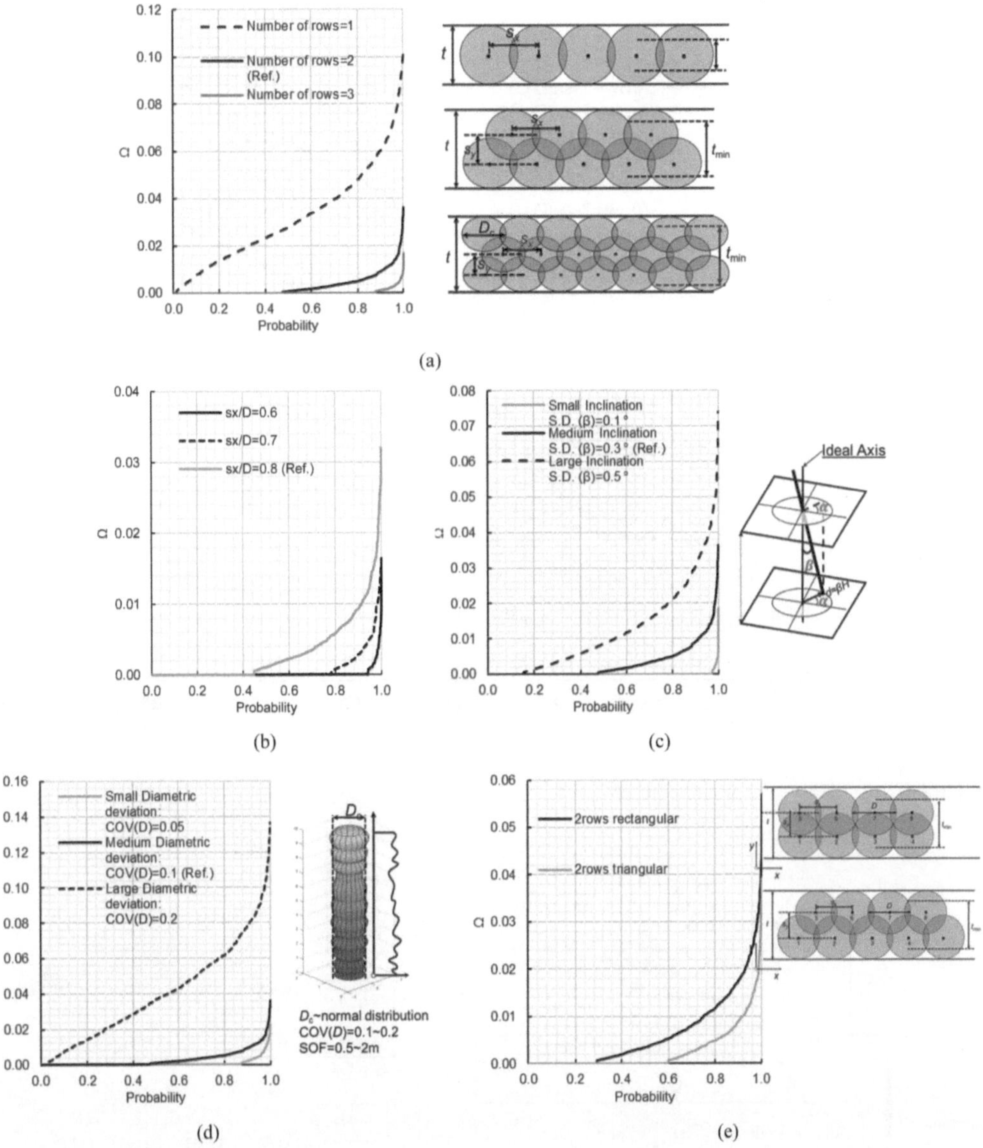

FIGURE 16.12 Cumulative distribution of normalized flow rate with different configurations.

and diametric variation). As expected, the normalized flow rate diminishes with more rows of jet-grouted columns, closer spacing, smaller inclination, smaller diametric variation, and usage of triangular layout (Figure 16.12a-e).

A rationally conservative estimate of cut-off performance (flow rate) is 95% fractile, as a high flow rate is considered an unfavorable condition. A design is regarded as reasonably safe if the estimated flow rate is higher than the drainage capacity. Based on this assumption, the equivalent drainage capacity corresponding to 95% fractile of the normalized flow rate is calculated. The design chart is shown in Figure 16.13.

Implementation

Figure 16.14 shows the flowchart. It starts with initializing boundary setup. Relevant information like the length of the cut-off wall, water head difference on both sides of the cut-off wall, drainage

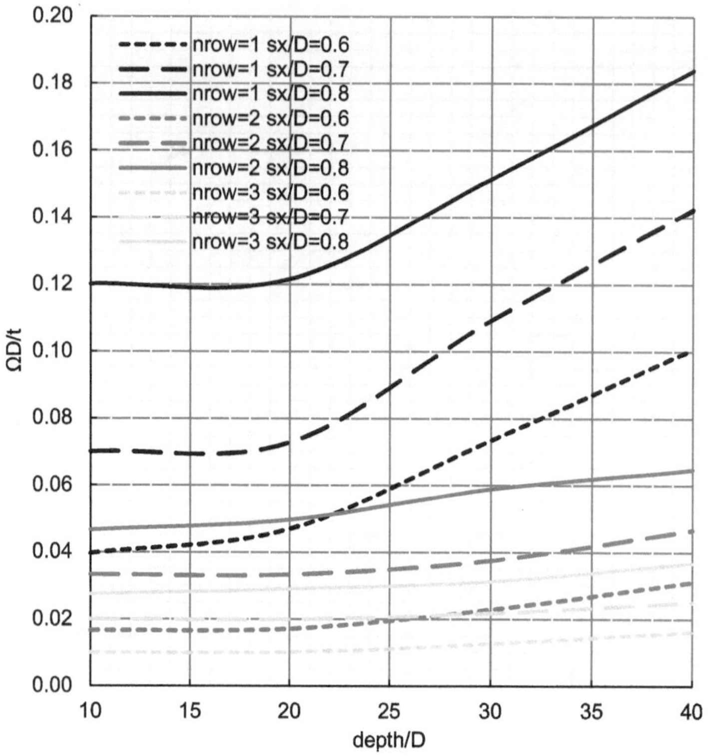

FIGURE 16.13 Design chart for JGCOW.

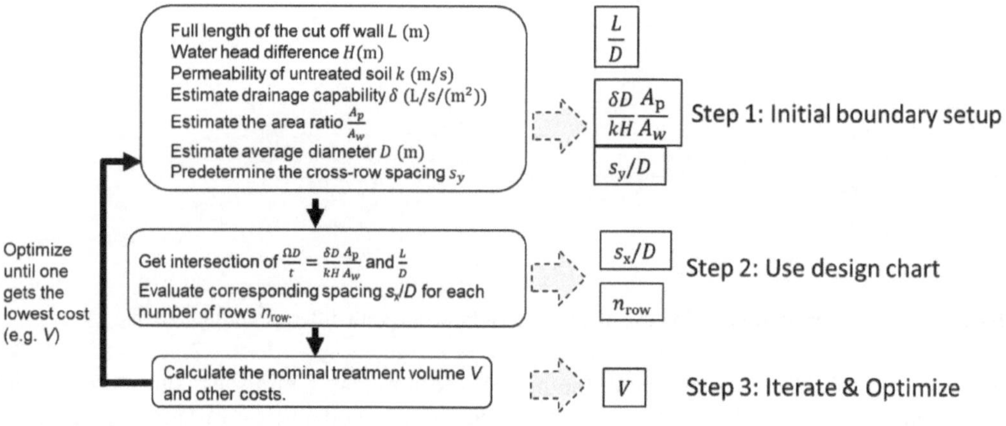

FIGURE 16.14 Flow chart for design of JGCOW.

capacity, and permeability coefficient of untreated soil should be determined. Draw a vertical line with known normalized wall length L/D, then intersect the line with equivalent drainage capacity $\delta D/kH.A_p/A_w$. With the intersection point, one can determine the normalized spacing (Sx/D). One may have more than one choice, as one can choose either fewer rows with smaller normalized spacing (bigger overlap), or more rows with bigger spacing (smaller overlap). An optimized design can be achieved when one can find the one with the lowest cost. In this illustration example, it is assumed that the design with the lowest volume is the most optimized. However, there may be other factors, too. Manpower and carbon emissions may also be considered for the final performance evaluation. The given method provides a quantification platform to strike a balance among rivaling factors.

16.5 LIMITATIONS AND FUTURE WORK

This chapter aims at showing the risk assessment framework for cement-treated ground used as underground struts and cut-off walls. The RFEM and TDA are effective risk assessment tools for accurate quantification of global performance.

The following future works are required to fulfill more demands in the future:

1) It can be foreseen that automation in construction will be more and more popular and necessary in the future, as the worldwide shrinkage of the labor force is more and more pronounced. In such a background, we would expect necessities in on-site monitoring and risk assessment. This would give rise to more demanding requirements on both calculation accuracy and speed. Therefore, a data-driven approach should be used to provide instant mapping of the monitoring data and performance spectrum. An aiding solution toolbox should also be in place to help the operators or machines to automatically adjust or mitigate the consequence of existing construction errors.
2) The current version of the TDA algorithm does not distinguish independent penetrating passages inside one cell. Therefore, a more accurate algorithm is needed to reduce the errors caused by this inaccuracy.

REFERENCES

Al-Naqshabandy, M. S., Bergman, N. S., & Larsson, S. (2012). Strength variability in lime-cement columns based on CPT data. *Proceedings of the Institution of Civil Engineers – Ground Improvement*, 165(1), 15–30.

Amos, P. D., Bruce, D. A., Lucchi, M., Watkins, N., & Wharmby, N. (2008). Design and construction of deep secant pile seepage cut-off walls under the Arapuni dam in New Zealand. In *USSD 2008 Conference*, Portland, OR, April.

Arroyo, M., Gens, A., Croce, P., & Modoni, G. (2012, September). Design of jet-grouting for tunnel waterproofing. In *Proceedings of the 7th International Symposium on the Geotechnical Aspects of Underground Construction in Soft Ground: TC28-IS Rome*. London: Taylor & Francis Group, pp. 181–188.

ASCE Geo-Institute Grouting Committee and Jet Grouting Task Force. (2009). Jet grouting guideline. Reston, VA, USA: American Society of Civil Engineers.

Babasaki, R., Terashi, M., Suzuki, T., Maekawa, A., Kawamura, M., & Fukazawa, E. (1996, May). JGS TC report: Factors influencing the strength of improved soil. In *Proceedings of the 2nd International Conference on Ground Improvement Geosystems, Grouting and Deep Mixing*, Vol. 2. London: Taylor & Francis, pp. 913–918.

Bergman, N., Al-Naqshabandy, M. S., & Larsson, S. (2013). Variability of strength and deformation properties in lime–cement columns evaluated from CPT and KPS measurements. *Georisk: Assessment and Management of Risk for Engineered Systems and Geohazards*, 7(1), 21–36.

Bruce, M. E. C., Berg, R. R., Collin, J. G., Filz, G. M., Terashi, M., & Yang, D. S. (2013). Federal highway administration design manual: Deep mixing for embankment and foundation support (no. FHWA-HRT-13-046).

Chen, J., Lee, F. H., & Ng, C. C. (2011). Statistical analysis for strength variation of deep mixing columns in Singapore. In Han, J. and Alzamora, D.E. (eds.), *Geo-Frontiers 2011 Advances in Geotechnical Engineering. Dallas*. Reston, VA: American Society of Civil Engineering, pp. 13–16.

Chen, E. J., Liu, Y., & Lee, F. H. (2016). A statistical model for the unconfined compressive strength of deep-mixed columns. *Géotechnique*, 66(5), 351–365.

Croce, P., & Modoni, G. (2007). Design of jet-grouting cut-offs. *Proceedings of the Institution of Civil Engineers – Ground Improvement*, 11(1), 11–20.

Croce, P., Flora, A., & Modoni, G. (2004). Jet Grouting: Tecnica, Progetto e Controllo. Hevelius (In Italian).

Eramo, N., Modoni, G., & Arroyo, M. (2011, May). Design control and monitoring of a jet grouted excavation bottom plug. In *7th International Symposium on Geotechnical Aspects of Underground Construction in Soft Ground.*

Hedman, P., & Kuokkanen, M. (2003). Strength distribution in lime-cement columns – Field tests at Strängnäs. MSc Thesis 03/06, Royal Institute of Technology. Stockholm, pp. 47–59 (in Swedish).

Honjo, Y. (1982). A probabilistic approach to evaluate shear strength of heterogeneous stabilized ground by deep mixing method. *Soils and Foundations*, 22(1), 23–38.

Inazumi, S., Ohtsu, H., Otake, Y., Kimura, M., & Kamon, M. (2009). Evaluation of environmental feasibility of steel pipe sheet pile cutoff wall at coastal landfill sites. *Journal of Material Cycles and Waste Management*, 11, 55–64.

Kasama, K., & Whittle, A. J. (2011). Bearing capacity of spatially random cohesive soil using numerical limit analyses. *Journal of Geotechnical and Geoenvironmental Engineering*, 137(11), 989–996.

Kasama, K., Whittle, A. J., & Zen, K. (2012). Effect of spatial variability on the bearing capacity of cement-treated ground. *Soils and Foundations*, 52(4), 600–619.

Kasama, K., Zen, K., & Whittle, A. J. (2010, December). Effects of spatial variability of cement-treated soil on undrained bearing capacity. In *Numerical Modelling of Construction Processes in Geotechnical Engineering for Urban Environment: Proceedings of the International Conference on Numerical Simulation of Construction Processes in Geotechnical Engineering for Urban Environment*, 23–24 March 2006, Bochum, Germany (p. 305). Taylor & Francis.

Larsson, S., & Nilsson, A. (2009). Horizontal strength variability in lime-cement columns. In *Deep Mixing 2009 Okinawa Symposium, DM′ 09. Okinawa, Japan*, 19–21 May 2009 (pp. 629–634). SANWA CO., LTD.

Larsson, S., Stille, H., & Olsson, L. (2005). On horizontal variability in lime-cement columns in deep mixing. *Geotechnique*, 55(1), 33–44.

Lee, F. H., Lee, Y., Chew, S. H., & Yong, K. Y. (2005). Strength and modulus of marine clay-cement mixes. Journal of geotechnical and geoenvironmental engineering, 131(2), 178–186.

Liu, Y., He, L. Q., Jiang, Y. J., Sun, M. M., Chen, E. J., & Lee, F. H. (2019). Effect of in situ water content variation on the spatial variation of strength of deep cement-mixed clay. *Géotechnique*, 69(5), 391–405.

Liu, Y., Jiang, Y. J., Xiao, H., & Lee, F. H. (2017). Determination of representative strength of deep cement-mixed clay from core strength data. *Géotechnique*, 67(4), 350–364.

Liu, Y., Lee, F. H., Quek, S. T., & Beer, M. (2014). Modified linear estimation method for generating multi-dimensional multi-variate Gaussian field in modelling material properties. *Probabilistic Engineering Mechanics*, 38, 42–53.

Liu, Y., Lee, F. H., Quek, S. T., Chen, E. J., & Yi, J. T. (2015). Effect of spatial variation of strength and modulus on the lateral compression response of cement-admixed clay slab. *Géotechnique*, 65(10), 851–865.

Liu, Y., Pan, Y., Sun, M., Hu, J., & Yao, K. (2017). Lateral compression response of overlapping jet-grout columns with geometric imperfections in radius and position. *Canadian Geotechnical Journal*. https://doi.org/10.1139/cgj-2017-0280.

Modoni, G., Flora, A., Lirer, S., Ochmański, M., & Croce, P. (2016). Design of jet grouted excavation bottom plugs. *Journal of Geotechnical and Geoenvironmental Engineering*, 04016018.

Namikawa, T., & Koseki, J. (2013). Effects of spatial correlation on the compression behaviour of a cement-treated column. *Journal of Geotechnical and Geoenvironmental Engineering*, 139(8), 1346–1359.

Navin, M. P., & Filz, G. M. (2005). Statistical analysis of strength data from ground improved with DMM columns. Deep Mixing '05: International Conference on Deep Mixing Best Practice and Recent Advances.

Omine, K., Ochiai, H., & Yasufuku, N. (2005). Evaluation of scale effect on strength of cement-treated soils based on a probabilistic failure model. *Soils and Foundations*, 45(3), 125–134.

Pan, Y. T., Xiao, H. W., Lee, F. H., & Phoon, K. K. (2016). Modified isotropic compression relationship for cement-admixed marine clay at low confining stress. *Geotechnical Testing Journal*, 39(4), 695–702.

Pan, Y., Liu, Y., Hu, J., Sun, M., & Wang, W. (2017). Probabilistic investigations on the water-tightness of jet-grouted ground considering geometric imperfections in diameter and position. *Canadian Geotechnical Journal*, 54(10), 1447–1459.

Pan, Y., Liu, Y., Xiao, H., Lee, F. H., & Phoon, K. K. (2018). Effect of spatial variability on short-and long-term behaviour of axially-loaded cement-admixed marine clay column. *Computers and Geotechnics*, 94, 150–168.

Pan, Y., Liu, Y., Lee, F. H., & Phoon, K. K. (2019a). Analysis of cement-treated soil slab for deep excavation support–a rational approach. *Géotechnique*, 69(10), 888–905.

Pan, Y., Liu, Y., & Chen, E. J. (2019b). Probabilistic investigation on defective jet-grouted cut-off wall with random geometric imperfections. *Géotechnique*, 69(5), 420–433.

Pan, Y., Yi, J., Goh, S. H., Hu, J., Wang, W., & Liu, Y. (2019c). A three-dimensional algorithm for estimating water-tightness of cement-treated ground with geometric imperfections. *Computers and Geotechnics*, 115, 103176.

Pan, Y., Liu, Y., Tyagi, A., Lee, F. H., & Li, D. Q. (2021a). Model-independent strength-reduction factor for effect of spatial variability on tunnel with improved soil surrounds. *Géotechnique*, 71(5), 406–422.

Pan, Y., Hicks, M. A., & Broere, W. (2021b). An efficient transient-state algorithm for evaluation of leakage through defective cutoff walls. *International Journal for Numerical and Analytical Methods in Geomechanics*, 45(1), 108–131.

Phoon, K. K., & Kulhawy, F. H. (1999). Characterization of geotechnical variability. *Canadian Geotechnical Journal*, 36(4), 612–624.

Richardson, M., Murray, T., & George, A. (2015). Seepage management control for a dam rehabilitation project using deep cut-off wall construction and jet grouting. In *Canadian Dam Association 2015 Annual Conference*, October 5–8, 2015, Mississauga, ON.

Ryan, C., & Jasperse, B. H. (1989). Deep soil mixing at the Jackson lake dam. In F. H. Kulhawy (ed.), *Foundation Engineering: Current Principles and Practices, GSP 22*, vol. 5. New York: American Society of Civil Engineers, pp. 25–29.

Shirlaw, J. N. (1996). Ground treatment for bored tunnels. In R. Mair and R. Taylor (eds), *Geotechnical Aspects of Underground Construction in Soft Ground*. London: Balkema, pp. 19–25.

Singapore Standard. (2003). *CP4:2003: Code of Practice for Foundations*. Singapore: Spring.

Vanmarcke, E., (1983 March). Random fields. *Random Fields, by Erik Vanmarcke*. ISBN 0-262-72045-0. Cambridge, MA: The MIT Press, p. 372.

Xiao, H., Lee, F. H., & Chin, K. G. (2014). Yielding of cement-treated marine clay. *Soils and Foundations*, 54(3), 488–501.

Xiao, H., Lee, F. H., & Liu, Y. (2017). Bounding surface cam-clay model with cohesion for cement-admixed clay. *International Journal of Geomechanics*, 17(1), 04016026.

Index

Note: Locators in *italics* represent figures and **bold** indicate tables in the text.

Accuracy, 167
Adaptive Metropolis algorithm (AM), 412
Adaptive Monte Carlo simulation (AMCS), 229–230
AI, *see* Artificial intelligence
AIC, *see* Akaike information criterion
Akaike information criterion (AIC), 155
Allowable stress design (ASD), 26, 27, 182
Alternate point estimate methods (APEM), 241
AM, *see* Adaptive Metropolis algorithm
AMCS, *see* Adaptive Monte Carlo simulation
Analysis of variance (ANOVA), 442
ANICREEP model, 117, 128, *128*
Anisotropic random field, 74, 436
ANN, *see* Artificial neural network
ANOVA, *see* Analysis of variance
APEM, *see* Alternate point estimate methods
Approximation model, 409
Artificial intelligence (AI), 17
Artificial neural network (ANN), 119, *119*, *133*
ASD, *see* Allowable stress design
Autocorrelation models, common
in geotechnical engineering, **66–67**
variance reduction function for, **69–71**

Back-propagation neural network (BPNN), 281, 284–285, *284*, 293, *295*
Bayesian analysis, 234
Bayesian compressive sampling/sensing (BCS), 150
Bayesian inference, 118
Bayesian information criterion (BIC), 155
Bayesian linear regression, 355
Bayesian machine learning, 11
Bayesian methods, 65
Bayesian model calibration, 354, 405
case study description, 405–406
geotechnical and hydrological conditions, 406–407
numerical models, 407
approximation model, 409
case study conclusion, 417–418
final prediction, reliability analysis, and decision-making, 415–417
finite-element (FE) model, 408–409
inverse Bayesian analysis, 411–415
quantification of uncertainties, 410
sensitivity analysis, 411
underground flow and two-phase media, 407–408
Bayesian neural network (BNN), *119*, 119–120
Bayesian parameter identification, framework of, 123–126
Bayesian perspective, data-driven site characterization from, 145–146
Bayesian Updating with Structural Reliability (BUS) methods, 359
Bayes' theorem, 411
BCS, *see* Bayesian compressive sampling/sensing
Bearing capacity estimation from piling data, 442–444
Benchmarking, 164
example of, 169–171
performance metrics, 166–169
training and validation set, 166
virtual ground, 164–165, **164**
Bernstein–von Mises theorem, 14
BIC, *see* Bayesian information criterion
BID, *see* Big indirect data
Big indirect data (BID), 10, 105, 143, 186
BIM, *see* Building information modeling
Bivariate dataset, 83, 85, 89, 97, 98, 105, *105*
BNN, *see* Bayesian neural network
BPNN, *see* Back-propagation neural network
Building information modeling (BIM), 18, 105
Burland Triangle, 1, 7–10, 105
data-centric geotechnics, 16–18
expanded, *10*
numerical modeling, role of, 1–4
prediction exercises, 4–7
probabilistic applications, 28–29
probabilistic methods, 26–27
probabilistic software, 27–28
statistical characterization, 24–26
uncertainty quantification, 11–16
BUS methods, *see* Bayesian Updating with Structural Reliability methods

Canadian Highway Bridge Design Code, 325
CAPWAP, *see* Case Pile Wave Analysis Program
Case Pile Wave Analysis Program (CAPWAP), 442
Cautious estimate, 48
CDF, *see* Cumulative distribution function
Cement-admixed soils, statistical characteristics of, **63–64**
Characteristic value, 42, 43
stochastic explanation of, 466
Eurocode 7, extracts from, 466–467
reliability-based characteristic values, 468
Cholesky decomposition, 228
Class A protocol, 7
Clay databases, multivariate probability models for, 88
CLAY/5/345 database, 93–96, **94**, *95*, **96**, *97*
CLAY/6/535 database, 89–93, **92–93**
CLAY/10/7490 database, 88–89, **88**, **89**, *90*, **91**, *94*, *95*, *96*, *97*
CMC model, *see* Coupled Markov chain model
Coefficient of variation (COV), 40–42, 48, 52–64, 106–107, 228, 431, 474
of index parameters, *58*
of laboratory parameters, *59*
for spatial variability, **55**, *61*, *62*
Collaborative stochastic finite element method (CSFEM), 281, 287, 300
computational issues in, 291–292
non-intrusive stochastic finite element method and, *297*

response conditioning method, target analysis using, 289–290
for slope stability problem, *288*
subset simulation, preliminary analysis using, 288–292
Common autocorrelation models
in geotechnical engineering, **66–67**
variance reduction function for, **69–71**
Comparable experience, 28
Compression index, 122
Computational complexity, 169
Conditional distribution, 98, 99–101, *100, 101*
Cone Penetration Test (CPT) data, 28, 144, 393, 395, *398*, 459
Confidence interval, width of, 167
"Conservatively assessed mean" value, 49
Constitutive model, data-driven, **129**
Constitutive models, uncertainty in, 117
correlating soil properties considering uncertainty
Bayesian neural network, 119–120
examples and validation, 122–123
framework of ML-based uncertainty model, 121–122
literature survey, 118
Monte Carlo dropout, 120–121
variational inference (VI), 120
identifying soil model parameters considering uncertainty
examples and validation, 128
framework of Bayesian parameter identification, 123–126
literature survey, 123
parallel computing DE-TMCMC, 126–128, *127*
proposition of enhanced DE-TMCMC, 125–126
integrating uncertainty into the framework of constitutive modeling
examples and validation, 134
incorporating uncertainty, 133
literature survey, 128–130
multi-fidelity residual modeling, 132–133
physics-informed neural network, 130–132
Constrained optimization algorithms, 221–222
Construction errors, 474–477
Conventional parametric studies, 13
Cornell's reliability index, 220, *220*
Cosine Whittle–Matérn (CosWM) model, 68
CosWM model, *see* Cosine Whittle–Matérn model
Coupled Markov chain (CMC) model, 159–161, *160, 161, 163*
Coverage probability, 167
CPT data, *see* Cone Penetration Test data
CSFEM, *see* Collaborative stochastic finite element method
Cumulative distribution function (CDF), 359, *359, 360, 362*
Cut-off walls, 479

Database assessment of prediction variability, 182–183
geotechnical databases, 186
model factor approach, 183–186
statistics of model factor and implications, 186–191
Data-centric geotechnics, 16–18
Data centricity, 17
Data-driven constitutive model, 129, **129**, 135
Data-driven model, 129–130, 134, 135, *178*, 179, 180
Data-driven site characterization (DDSC), 18, 40, 104–105, 143
from Bayesian perspective, 145–146
benchmarking, 164
example of, 169–171
performance metrics, 166–169
training and validation set, 166
virtual ground, 164–165, **164**
DDSC-C1 Methods, **149**
geologic uncertainty, categorical data, 158
coupled Markov chain simulation, 159–161
DDSC-C3 methods, 163
Markov random field (MRF) model, 161–163
geologic uncertainty, numerical data, 153
Gaussian mixture model (GMM), 155, *155*
kernel methods, 156, *156*
K-means clustering, 153, *153*
geotechnical and geologic uncertainty, numerical data, 156–158
geotechnical uncertainty, numerical data, 146
Gaussian process regression (GPR), 151–152, *152*
sparse modeling, 150–151
numerical and categorical data in, *144*
tentative literature scan of, **41**
2×2 matrix to organize, *147*
types of data and corresponding uncertainty, 144–145
Data ground truths, 16
Data infrastructure, 24
Data science, 179
DDSC, *see* Data-driven site characterization
DDSC-C3 methods, 163
Decision-making, 23, 30, 41, 415–417
reliability-based, *see* Reliability-based decision-making
risk assessment and, *see* Risk assessment and decision-making
Deep neural networks (DNNs), 146
DEs, *see* Design experiments
Design experiments (DEs), 258
Design value, 43
Deterministic analysis, 12
vs probabilistic analysis, *374*
Deterministic approach, 23
Deterministic methods, 123
Deterministic slope stability analysis, 281–282
Diaphragm wall displacements prediction using live Bayesian model calibration, 405
case study description, 405–406
geotechnical and hydrological conditions, 406–407
numerical models, 407
approximation model, 409
case study conclusion, 417–418
final prediction, reliability analysis, and decision-making, 415–417
finite-element (FE) model, 408–409
inverse Bayesian analysis, 411–415
quantification of uncertainties, 410
sensitivity analysis, 411
underground flow and two-phase media, 407–408
Digital twin, 18, 105
Direct toppling, 392, *396*
Ditlevsen bounds for system reliability analysis, 227–228
DMD, *see* Dynamic mode decomposition

DMD-C, *see* Dynamic mode decomposition with control
DNNs, *see* Deep neural networks
Dynamic mode decomposition (DMD), 248, 250–253, 256, 257
Dynamic mode decomposition with control (DMD-C), 270–271

EC7, *see* Eurocode 7
Effective friction angle, 55
Effective property, 52
Effective random dimension (ERD), 327
Effective Young's modulus, 80
Elasto-plastic (EP) theoretical concept, 130
EM algorithm, *see* Expectations and maximization algorithm
EN1990:2002, 320
Enhanced DE-TMCMC, proposition of, 125–126
EP theoretical concept, *see* Elasto-plastic theoretical concept
ERD, *see* Effective random dimension
Eurocode 7 (EC7), 43, 466–467, **466**
Exact DMD, 253
Expectations and maximization (EM) algorithm, 150

Factor of safety (FOS), 320, 373, 378, 435
Failure and reliability index, target probability of, 320–321
Failure consequence, influence of heterogeneity on, 465–466
FDM, *see* Finite difference method
Federal Highway Administration (FHWA) database, 438
FEM, *see* Finite element method
FE model, *see* Finite-element model
FHWA database, *see* Federal Highway Administration database
Finite difference method (FDM), 280
Finite-element (FE) model, 40, 402, 408–409
Finite element method (FEM), 260, 280, 281, 481
First-order reliability method (FORM), 258, 280, 332, 427
  constrained optimization algorithms, 221–222
  Hasofer–Lind (HL) reliability index for uncorrelated normal variables, 221
  implicit performance function, practical form for
    Hasofer–Lind–Rackwitz–Fiessler (HLRF) recursive algorithms, 223–227
    response surface method, 222–223
First-order second moment (FOSM) method, 39, 241–243, 258, 280, 347, 427
  experimental points and reliability results of the, **346**
  RBD using, 335
  reliability-based design of strip footing using, 345–346
Fisher Distribution, 390
FISH functions, 330
Flexural toppling, 391, *395*
Forcing load operator, 268
FORM, *see* First-order reliability method
FOS, *see* Factor of safety
FOSM method, *see* First-order second moment method
Foundation load test databases, **22–23**
Fourier series expansion, 255
Frequentist constants, 14

Gaussian distribution, 354
Gaussian mixture model (GMM), 155, *155*
Gaussian models, mixture of, 155
Gaussian process (GP), 449
Gaussian process regression (GPR), 151–152, *152*, 356
Generalized model factor approach, 200
Generalized subset simulations (GSS), 229, 233
Geologic uncertainty, 25
  categorical data, 158
    coupled Markov chain simulation, 159–161
    DDSC-C3 methods, 163
    Markov random field (MRF) model, 161–163
  numerical data, 153
    Gaussian mixture model (GMM), 155, *155*
    kernel methods, 156, *156*
    K-means clustering, 153, *153*
Geotechnical and geologic uncertainty, numerical data, 156–158
Geotechnical reliability analysis
  for soil spatial variability, 240–241
  via stand-alone numerical codes, 241–243
Geotechnical resistance factors, **326–327**
Geotechnical uncertainty, numerical data, 146
  Gaussian process regression (GPR), 151–152, *152*
  sparse modeling, 150–151
Glasso, 157
Glasso-BFs, 150, *151*, 157
Global model behavior, 397–398
Global stiffness and strength of cement-treated slabs, 481
  implementation, 483–485
  positioning error, 482
  positioning error, effect of, 483
  random fluctuation, 481–489
  typical realizations, 482–483
Global stress-strain behavior, 477–479
GLOO method, *see* Graphical leave-one-out error method
GMM, *see* Gaussian mixture model
Goodness-of-fit function, 124
GP, *see* Gaussian process
GPotts model, *168*
GPR, *see* Gaussian process regression
GPUs, *see* Graphics processing units
Graphical leave-one-out error (GLOO) method, 404, *405*
Graphics processing units (GPUs), 127
Ground improvement, 473–474
  limitations and future work, 489
  modeling of global performance, 477
    global stress-strain behavior, 477–479
    watertightness, 479–481
  risk assessment and decision-making, 481
    global stiffness and strength of cement-treated slabs, 481–485
    watertightness of jet-grouted cut-off walls, 485–489
  sources of uncertainties in, 474
    construction errors, 474–477
    random fluctuation, 474
GSS, *see* Generalized subset simulations

Hard assignment, 155
Hasofer–Lind (HL) reliability index, 221, *221*
  for uncorrelated normal variables, 221
Hasofer–Lind–Rackwitz–Fiessler (HLRF) recursive algorithms, 223
  reliability analysis of strut with complex supports, 224–227

HBM, *see* Hierarchical Bayesian model
HCV, *see* Highest conceivable value
Hermite polynomial chaos expansion (HPCE), 281, 283
HF data, *see* High-fidelity data
Hierarchical Bayesian model (HBM), 98
Highest conceivable value (HCV), 58
High-fidelity (HF) data, 129
High-performance computing (HPC) techniques, 126
HL reliability index, *see* Hasofer–Lind reliability index
HLRF recursive algorithms, *see* Hasofer–Lind–Rackwitz–Fiessler recursive algorithms
HPCE, *see* Hermite polynomial chaos expansion
HPC techniques, *see* High-performance computing techniques
Hydraulic conductivity, 81
Hyperparameters, estimation of, 362–365, *364*

IBC, *see* International Building Code
Identification rate (IR), 167
IFR, *see* Incremental filling ratio
Implicit performance function, practical form for
  HLRF recursive algorithms, 223–227
  response surface method, 222–223
Importance sampling, 229, 232
Incremental filling ratio (IFR), 451
Information entropy, 167
Information technology (IT), 17
International Building Code (IBC), 442
Inverse analysis from the viewpoint of probability theory, 354–357
Inverse Bayesian analysis, 411–415
Inverse problem, classification of, **368**
Inverse reliability method, RBD by
  multiple design parameters, 237–238
  single design parameters, 236–237
IR, *see* Identification rate
ISO 2394:2015, 320
IT, *see* Information technology

Jet-grouted cut-off walls (JGCOW), *476*, 479
  configurations of model for, **486**
  design chart for, *488*
  flow chart for design of, *488*
  watertightness of, 485–489
JGCOW, *see* Jet-grouted cut-off walls
Jinping I left abutment slope, *305*, *306*, 310
Johnson distribution, 86, *87*

Kalman filter, 355
Karhunen–Loève expansion (KLE), 240, 250, 283
Kernel methods, 156, *156*
Kernel trick, 156
KL divergence, *see* Kullback–Leibler divergence
KLE, *see* Karhunen–Loève expansion
K-means clustering, 153, *153*, 156
Kullback–Leibler (KL) divergence, 120

Laplace approximation of evidence for model selection, 365–367
Large-diameter open-ended pile (LDOEP), 189
LASSO, *see* Least absolute shrinkage and selection operator
Lateral earth pressure and excavations, 436–437
Latin Hypercube (LH), 375
Latin hypercube sampling (LHS) technique, 229, 230, 284
LCV, *see* Lowest conceivable value
LDOEP, *see* Large-diameter open-ended pile
Least absolute shrinkage and selection operator (LASSO), 351
Least-square estimation (LSE), 148
Least-square method (LSM), 148
Least-squares (LS) method, 351
Leave-one-out (LOO) error, 403–404
LEM, *see* Limit equilibrium method
LH, *see* Latin Hypercube
LHS technique, *see* Latin hypercube sampling technique
LI, *see* Liquidity index
Limit equilibrium method (LEM), 281–282, 375, 383
Limit state function (LSF), 219
Limit state surface (LSS), 221
Linear regression, model selection for the order of, *367*
Line sampling (LS), 229, 230–232
Liquidity index (LI), 16, 55
Load and resistance factor design (LRFD), 40, 200, 320, 325, 329
Load-movement curve, 16
Local failure, 48
Local model behavior, 403
LOO error, *see* Leave-one-out error
Lowest conceivable value (LCV), 58
LRFD, *see* Load and resistance factor design
LS, *see* Line sampling
LSE, *see* Least-square estimation
LSF, *see* Limit state function
LSM, *see* Least-square method
LS method, *see* Least-squares method
LSS, *see* Limit state surface

Machine learning (ML) algorithms, 25, 26, 143
  ML-based constitutive models of soils, **130**
  ML-based uncertainty model, 121
MAP value, *see* Maximum a posteriori value
Markov chain, 159–161
Markov chain Monte Carlo (MCMC) methods, 150, 357, 412, 438
  graphical model for, *439*
  posterior distribution with Kernel density estimation obtained by, *447*
Markov random field (MRF) model, 145, 161–163, *163*
MATLAB, 126
Maximum a posteriori (MAP) value, 118, 351, 413
Maximum likelihood estimation (MLE), 146
MCD, *see* Monte Carlo dropout
MCMC methods, *see* Markov chain Monte Carlo methods
MC method, *see* Monte Carlo method
MCS, *see* Monte Carlo simulation
Mean absolute error, 166
Meta-modeling in probabilistic analysis, 402
  global model behavior, 402–403
  local model behavior, 403
  surrogate model, verifying, 404
    graphical leave-one-out error (GLOO), 404
    leave-one-out (LOO) error, 404, **409**
ML algorithms, *see* Machine learning algorithms
MLE, *see* Maximum likelihood estimation
Mobilized value, 51–52, 108
  statistics of, 74
    hydraulic conductivity, 81

serviceability limit state (SLS), 80–81
ultimate limit state, 77–80
Model calibration, 196
Model factor, *17*
classification based on bias and imprecision, **104**
defined, 4, 40, 101
Modélisation des Ouvrages et Mesures In Situ (MOMIS), *see* MOMIS database
Model uncertainties, 14
Mohr–Coulomb failure criterion, 298
MOMIS database, 197–198
Monte Carlo (MC) method, 118
Monte Carlo analysis, 457
Monte Carlo dropout (MCD), 118, *119*, 120–121
Monte Carlo simulation (MCS), 12, 13, 26, 228, 280, *342*, 403
adaptive, 229–230
advanced simulation techniques
adaptive MCS, 229–230
generalized subset simulation, 233
importance sampling, 232
latin hypercube sampling, 230
line sampling, 230–232
subset simulation, 232–233
weighted uniform simulations (WUS), 233
-based full probabilistic design, 238–239
design point, approximation of, 239–240
RBD using, 329–330
reliability-based design of strip footing using, 337–338
sampling-based sensitivity analysis, 234
sensitivity to probability distributions of random variables, 234–235
sensitivity to statistical moment values of random variables, 234
and theoretical lognormal, *345*
Most probable failure point (MPP), 221
Most probable value (MPV), 355
MPP, *see* Most probable failure point
MPV, *see* Most probable value
MRFD method, *see* Multiple Load and Resistance Factor Design method
MRF model, *see* Markov random field model
MR framework, *see* Multi-fidelity residual framework
Multi-fidelity residual (MR) framework, 130, 132–133
Multiple design parameters, 237–238
Multiple Load and Resistance Factor Design (MRFD) method, 325, 329
Multivariate probability model, 84–86
for clay databases, 88
CLAY/5/345 database, 93–96, **94**, *95*, **96**, *97*
CLAY/6/535 database, 89–93, **92–93**
CLAY/10/7490 database, 88–89, **88**, **89**, *90*, **91**, *94*, *95*, *96*, *97*
conditioning from, 98–101
Multivariate soil databases, incomplete versus complete, 84
MUSIC-3X, 18, 107, 108
MUSIC-X, 180, 186

National Cooperative Highway Research Program (NCHRP) study, 61
NCHRP study, *see* National Cooperative Highway Research Program study
Negative log-likelihood (NLL), 121
Neural networks (NN), 444
Newton–Raphson iterations, 238
NIGPA, *see* Non-Intrusive Geotechnical Probabilistic Analysis
NISFEM, *see* Non-intrusive stochastic finite element method
NLL, *see* Negative log-likelihood
NN, *see* Neural networks
Non-Intrusive Geotechnical Probabilistic Analysis (NIGPA), 312–314
framework of, *314*
major interfaces of, *315*
Non-intrusive stochastic finite element method (NISFEM), 281
and collaborative stochastic finite element method, *297*
deterministic slope stability analysis, 281–282
implementation flowchart of, *287*
slope failure
estimation of probability of, 285
quantitative risk assessment of, 285–287
surrogate model, construction of, 282
back-propagation neural network (BPNN), 284–285, *284*, 293, *295*
Hermite polynomial chaos expansion (HPCE), 283
quadratic polynomial, 282–283
Non-linear inverse problem and sample (particle)-based approach, 357–360
Non-local failure, 48
N-sigma rule, 40
Numerical modeling, role of, 1–4

Ockham (Occam) factor, 366
One-dimensional (1D) consolidation equation
analytical model applying, *254*
analytical solution of, *256*
Optimal observation planning, 368
Orientation-based analysis, statistics in, 389–393

Parallel computing DE-TMCMC, 126–128, *127*
Particle filters (PFs), 355, 358
PCEs, *see* Polynomial chaos expansions
PCK, *see* Polynomial chaos kriging
PDF, *see* Probability density function; Probability distribution function
Pearson correlation matrices, *414*
PEM, *see* Point Estimate Method
Performance metrics, 166–169
PF, *see* Particle filters; Probability of failure
PhILMs, *see* Physics-informed learning machines
Physics-informed and data-driven characterization, 200–201
Physics-informed learning machines (PhILMs), 180
Physics-informed machine learning methodology, 25–26
Physics-informed neural network, 130–132
PI, *see* Plasticity index
PIE method, *see* Pseudo incremental energy method
Piezocone test, 81, 84
Pile-bearing capacity estimation methods, 441
Piled foundations, 433, 435
Piling data, 442
application of, 448
discussion, 452
river levees, raising of, 451–452

river walls and foundations, 448–450
bearing capacity estimation from, 442–444
rotary cutting press-in piling and simplified load tests, 444–445
within- and inter-site variability in estimation from, 438
for quality control of pile foundations, 447–448
results, 438
Plasticity index (PI), 55
PLAXIS, 330
PLAXIS 2D, 338, 340, 345
deterministic design using, 336
PLR, *see* Plug length ratio
Plug length ratio (PLR), 451
POD, *see* Proper orthogonal decomposition
Point Estimate Method (PEM), 347, 375
experimental points and reliability results of the, **344**
RBD using, 333–335
reliability-based design of strip footing using, 343–345
Point value, 46–48
Polynomial chaos expansions (PCEs), 403, 404
Polynomial chaos kriging (PCK), 403, 404
Positioning error, 482, 483
Potts model, 161, *162*
Practical reliability analysis, Monte Carlo simulations for, 228
advanced simulation techniques
adaptive MCS, 229–230
generalized subset simulation, 233
importance sampling, 232
latin hypercube sampling, 230
line sampling, 230–232
subset simulation, 232–233
weighted uniform simulations (WUS), 233
sampling-based sensitivity analysis, 234
sensitivity to probability distributions of random variables, 234–235
sensitivity to statistical moment values of random variables, 234
Practical stochastic finite element analysis software, 312–314
Precision construction, 3
Preconsolidation stress, 16
Prediction events in geotechnics, **5–7**
Prediction exercises, 4–7
dispersion exhibited in, 26
in geotechnical engineering, 1
Prediction method categories, 181–182
Prediction variability, 176, 177
background, 176–180
classification of, *199*
database assessment of, 182
geotechnical databases, 186
model factor approach, 183–186
statistics of model factor and implications, 186–191
international prediction events, lessons from, 191–195
numerical analyses versus in situ measurements, 195
MOMIS database, 197–198
results and implications, 198–200
objective and scope, 180–181
physics-informed and data-driven characterization, 200–201
Primary analysis, variation of (example)
correlation of parameters, 379–380
deterministic analysis results, 376–382
material properties, 376
primary analysis selection, 377–378
stochastic analysis results, 378–379
Principal component analysis (PCA), 250, 258
Prior information-based neural network (PiNet), 130, *131*
Probabilistic analysis, 12, 373
applications, 384
orientation-based analysis, statistics in, 389–393
rockfall simulation, 385–389
vs deterministic analysis, *374*
general methodology, 374–375
need to use, 373–374
primary analysis, variation of (example)
correlation of parameters, 379–380
deterministic analysis results, 376–382
material properties, 376
primary analysis selection, 377–378
stochastic analysis results, 378–379
recent advances, 393–395
sampling method, 375–376
software, advantages of, 374
spatially variable materials (example), 380
recent advances, 382
worst-case scale of fluctuation, 382
3D probabilistic slope stability (example), 383–384, *384*
Probabilistic applications, 28–29
Probabilistic inverse analysis for geotechnics, 351
hyperparameters, estimation of, 362–365, *364*
inverse analysis from the viewpoint of probability theory, 354–357
Laplace approximation of evidence for model selection, 365–367
non-linear inverse problem and sample (particle)-based approach, 357–360
optimal observation planning, 368
surrogate model with active learning, 360–362
weighted least-squares method with regularization, 352–354
Probabilistic methods, 26–27, 123
Probabilistic software, 27–28
Probability density function (PDF), 16, 83, 219, 354, 362–365, 482
Probability distribution, 12
Probability distribution function (PDF), 11, 47–48
Probability function, 390
Probability of failure (PF), 373
and reliability index, 220
Probability of state, 160, 161, 167
Problem-specific RBD, 329
Projected DMD, 253
Proper orthogonal decomposition (POD), 247–250, 256, 257, *256*
Pseudo incremental energy (PIE) method, 52, 80

QoI, *see* Quantity of interest
Quadratic polynomial, 282–283
Quantification of uncertainties, 410
Quantile Value Method (QVM), 325, 327
Quantity of interest (QoI), 4
Quasi-site-specific model, 25
QVM, *see* Quantile Value Method

Radial basis function (RBF), 149
Random field theory, 474
Random finite difference method (RFDM), 433
Random finite element method (RFEM), 52, 77, 78, 80, 433, 457, 460, 478, 479
  flowchart for, *480*
  3D slope reliability, RFEM analysis of, 462–465
  2D slope reliability, RFEM analysis of, 461–462
Random fluctuation, 474, 481–489
Random Limit Equilibrium Method (RLEM), 375, 380
Random Numerical Limit Analysis (RNLA), 478
Random variables
  sensitivity to probability distributions of, 234–235
  sensitivity to statistical moment values of, 234
Random vector model, 82
RBD, *see* Reliability-based design
RBF, *see* Radial basis function
Recurrent neural networks (RNN), 129
Reduced order model (ROM), 247, 248
  analysis, 258
  application in the information-oriented construction control of retaining walls
    dynamic mode decomposition with control (DMD-C), 270–271
    extending the model with uncertainty, 273–275
    modeling of deformation time evolution, 271–272
    results and considerations, 275
  characterization of
    problem setting, 254–255
    result and consideration, 255–257
  dynamic mode decomposition (DMD), 250–253
  initial data processing, 258
  mode decomposition of numerical analysis results, 261
  proper orthogonal decomposition (POD), 247–250, *256*
  reliability analysis using POD ROM, 263–267
  target structure and basic analysis method, 260–261
Relevant vector machine (RVM), 150
Reliability analysis, 26, 219
  background, 219
  and decision-making, 420–421
  Ditlevsen bounds for system reliability analysis, 227–228
  first-order reliability method (FORM)
    constrained optimization algorithms, 221–222
    Hasofer–Lind (HL) reliability index, 221
    Hasofer–Lind–Rackwitz–Fiessler (HLRF) recursive algorithms, 223–227
    response surface method, 222–223
  Monte Carlo simulations (MCS), 228
    adaptive MCS, 229–230
    generalized subset simulation, 233
    importance sampling, 232
    latin hypercube sampling, 230
    line sampling, 230–232
    sampling-based sensitivity analysis, 234–235
    subset simulation, 232–233
    weighted uniform simulations (WUS), 233
  probability of failure and reliability index, 220
  with reduced order model, *see* Reduced order model
  reliability-based design (RBD), 236
    by inverse reliability method, 236–238
    by Monte Carlo simulations, 238–240
  for soil spatial variability, 240–241
  via stand-alone numerical codes, 241–243
Reliability-based characteristic values, 52, 468
Reliability-based decision-making, 401
  diaphragm wall displacements prediction, 405
    approximation model, 409
    case study conclusion, 417–418
    case study description, 405–406
    FE model, 408–409
    final prediction, reliability analysis, and decision-making, 415–417
    geotechnical and hydrological conditions, 406–407
    inverse Bayesian analysis, 411–415
    quantification of uncertainties, 410
    sensitivity analysis, 411
    underground flow and two-phase media, 407–408
  finite-element (FE) models and uncertainty frameworks, coupling between, 402
  meta-modeling in probabilistic analysis, 402
    global model behavior, 402–403
    graphical leave-one-out error (GLOO), 404
    leave-one-out (LOO) error, 404, **409**
    local model behavior, 403
    surrogate model, verifying, 403–404
  tunnel excavation, 3D probabilistic FE model for, 418
    FE model description, 419–420
    reliability analysis and decision-making, 420–421
Reliability-based design (RBD), 26, 42, 180, 182, 236, 319
  in design codes, 321–329
  in geotechnical LRFD, 321–328
  as implemented in design codes to numerical modeling, 328–329
  by inverse reliability method
    multiple design parameters, 237–238
    single design parameters, 236–237
  by Monte Carlo simulations
    approximation of design point, 239–240
    MCS-based full probabilistic design, 238–239
  optimization, 224
  PLAXIS 2D, deterministic design using, 336
  problem definition, 336
  of strip footing
    using first-order second moment (FOSM), 345–346
    using Monte Carlo Simulation (MCS), 337–338
    using Point Estimate Method (PEM), 343–345
    using response surfaces, 338–343
  in structural design codes, 321
  target probability of failure and reliability index, 320–321
  using first-order second moment method (FOSM), 335
  using numerical models
    Monte Carlo Simulation (MCS), 329–330
    problem-specific RBD, 329
  using Point Estimate Method (PEM), 333–335
  using response surface methods, 331–333
Reliability software with geotechnical features, **10**
Resistance factors, 323, **324**
Response conditioning method, target analysis using, 289–290
Response surface method (RSM), 222–223, *342*
  RBD using, 331–333
  reliability-based design of strip footing using, 338–343
RFDM, *see* Random finite difference method
RFEM, *see* Random finite element method
Risk assessment and decision-making, 481

global stiffness and strength of cement-treated slabs, 481
effect of positioning error, 483
implementation, 483–485
positioning error, 482
random fluctuation, 481–489
typical realizations, 482–483
watertightness of jet-grouted cut-off walls, 485
implementation, 487–489
River levees, raising of, 451
piling data, estimations from, 451–452
River walls and foundations, 448
simplified load tests, 449–450
RLEM, *see* Random Limit Equilibrium Method
RMSE, *see* Root mean square error
RNLA, *see* Random Numerical Limit Analysis
RNN, *see* Recurrent neural networks
Rockfall simulation, 385–389
ROM, *see* Reduced order model
Root mean square error (RMSE), 166, *170*
Rotary cutting press-in piling and simplified load tests, 444–445
RSM, *see* Response surface method
RVM, *see* Relevant vector machine

Sampling-based sensitivity analysis, 234
random variables
sensitivity to probability distributions of, 234–235
sensitivity to statistical moment values of, 234
SA scheme, *see* Surface altering scheme
SBL, *see* Sparse Bayesian learning
SBT index, *see* Soil behavior type index
Scale of fluctuation (SOF), 68, 474
"Select one value" approach, 43
Sensitivity analysis, 13, 15, 411
Serviceability limit state (SLS), 52, 74, 80–81, 337
SFEM, *see* Stochastic finite element methods
Shallow footings, 428–432
Shannon entropy, *see* Information entropy
Shear strength parameters, 310, **312**
Shear strength reduction method (SSRM), 281–282
Simplified load tests, 449–450
Simpson's paradox, 15
Simulation techniques, advanced
adaptive Monte Carlo simulations, 229–230
generalized subset simulation, 233
importance sampling, 232
Latin hypercube sampling (LHS), 230
line sampling, 230–232
subset simulation, 232–233
weighted uniform simulations (WUS), 233
Single design parameters, 236–237
Single program/multiple data (SPMD) technique, 123, 126
Site-specific correlations, 96–97
Site-specific spatial variability, **56–57**
Site uniqueness, 18
Slope failure
estimation of probability of, 285
quantitative risk assessment of, 285–287
Slope reliability, 457
case history, 469–471
characteristic values, stochastic explanation of, 466
Eurocode 7, extracts from, 466–467
reliability-based characteristic values, 468
engineering practice, application to, 468–469
quantification and propagation of uncertainty, 459–460
soil spatial variability on, 460
failure consequence, influence of heterogeneity on, 465–466
3D slope reliability, RFEM analysis of, 462–465
2D slope reliability, RFEM analysis of, 461–462
spatial variability of soil properties, 458–459
Slopes, 435–436
Slope stability analysis and risk assessment, *see* Stochastic finite element methods
SLS, *see* Serviceability limit state
SOF, *see* Scale of fluctuation
Soft assignments, 155
Software, geotechnical, 374
Soil and rock parametric uncertainties, 39
coefficient of variation (COV), 40–42, 48, 52–64
cross-correlations, 81
clay databases, multivariate probability models for, 88–96
incomplete versus complete multivariate soil databases, 84
multivariate probability model, 84–86
site-specific correlations, 96–97
data-driven site characterization (DDSC), **41**, 104–105
mobilized value, statistics of, 74
hydraulic conductivity, 81
serviceability limit state (SLS), 80–81
ultimate limit state, 77–80
spatial average, statistics of, 64–74
transformation model, 97
multivariate probability model, conditioning from, 98–101
transformation bias and uncertainty, 101–103
Soil behavior type (SBT) index, 144, *165*
Soil model parameters identification considering uncertainty
Bayesian parameter identification, framework of, 123–126
enhanced DE-TMCMC, proposition of, 125–126
examples and validation, 128
literature survey, 123
parallel computing DE-TMCMC, 126–128, *127*
Soil parameter, characteristic value of, 42
mobilized value, 51–52
point value, 46–48
spatial average, 48–51
Soil properties correlation considering uncertainty
Bayesian neural network, 119–120
examples and validation, 122–123
literature survey, 118
machine learning (ML)-based uncertainty model, 120
Monte Carlo dropout, 120–121
variational inference (VI), 120
Soil/rock property databases, **19–20**
Soil-structure interaction in spatially variable ground, 427
lateral earth pressure and excavations, 436–437
piled foundations, 433, 435
shallow footings, 428–432
slopes, 435–436
Sparse Bayesian learning (SBL), 150–151
Sparse modeling, 150–151
Spatial average, 48–51

statistics of, 64–74
Spatial correlation, 108
Spatially variable materials (example), 380
recent advances, 382
worst-case scale of fluctuation, 382
Spatial variability
coefficient of variation for, **55**, *61*, *62*
site-specific, **56–57**
SPMD technique, *see* Single program/multiple data technique
SPT, *see* Standard penetration test
SRS methods, *see* Stochastic Response Surface methods
SS, *see* Subset simulations
SSRM, *see* Shear strength reduction method
Standard penetration test (SPT), 144, 181, 442
Statistical characterization, 24–26
Statistical uncertainties, 13, 14
Statistics-based characteristic values, 50
Stochastic finite element methods (SFEM), 280
collaborative stochastic finite element method (CSFEM), 287
computational issues in CSFEM, 291–292
preliminary analysis using subset simulation, 288–292
for slope stability problem, *288*
target analysis using response conditioning method, 289–290
non-intrusive stochastic finite element method, 281
back-propagation neural network (BPNN), 284–285, *284*, 293, *295*
deterministic slope stability analysis, 281–282
estimation of probability of slope failure, 285
Hermite polynomial chaos expansion (HPCE), 283
quadratic polynomial, 282–283
quantitative risk assessment of slope failure, 285–287
surrogate model, construction of, 282
practical stochastic finite element analysis software, 312–314
3D large-scale engineered slope (example)
deterministic stability analysis of rock slope, 305–310
reliability analysis results of rock slope, 310–312
slope overview, 304–305
3D long slope in spatially varying soils (example)
benchmark analysis, 298–301
3D spatial variability, effect of, 302–304, *303*, *304*
two-layered cohesive slope (example), 292
reliability analysis results, 293–294
risk analysis results, 294–296
Stochastic Response Surface (SRS) methods, 375
Strip footing, reliability-based design of
using first-order second moment (FOSM), 345–346
using Monte Carlo Simulation (MCS), 337–338
using Point Estimate Method (PEM), 343–345
using response surfaces, 338–343
"Structured" variability, 145
Subset simulation, 232–233
preliminary analysis using, 288–292
Subset simulations (SS), 229
Subsurface condition, 177–178
Surface altering (SA) scheme, 376–382
Surrogate model
with active learning, 360–362
construction of, 282
back-propagation neural network, 284–285
Hermite polynomial chaos expansion (HPCE), 283
quadratic polynomial, 282–283
verifying, 404
graphical leave-one-out error (GLOO), 404
leave-one-out (LOO) error, 404, **409**
Swedish weight sounding (SWS) test, 144
SWS test, *see* Swedish weight sounding test
System reliability analysis, Ditlevsen bounds for, 227–228

Target probability of failure and reliability index, 320–321
Target reliability index, 320, **323**
TDA, *see* Three-dimensional discretized algorithm
Terzaghi's consolidation equation, 254
Three-dimensional discretized algorithm (TDA), 480
flow chart for, *481*
3D large-scale engineered slope (example)
deterministic stability analysis of rock slope, 305–310
reliability analysis results of rock slope, 310–312
slope overview, 304–305
3D long slope in spatially varying soils (example)
3D spatial variability, effect of, 302–304, *303*, *304*
benchmark analysis, 298–301
3D probabilistic FE model for tunnel excavation, 418
FE model description, 419–420
reliability analysis and decision-making, 420–421
3D probabilistic slope stability (example), 383–384, *384*
3D slope reliability, RFEM analysis of, 462–465
3D stratification mapping, 25
TMCMC approach, *see* Transitional Markov chain Monte Carlo approach
Toppling
direct, 392, *396*
flexural, 391, *395*
Total uncertainty analyses, 53, *53*, **54**
Training and validation set, 166
Transformation model, 97
multivariate probability model, conditioning from, 98–101
transformation bias and uncertainty, 101–103
Transitional Markov chain Monte Carlo (TMCMC) approach, 123, 125, 126, 357–358
Transition probabilities, 159
Tunnel excavation, 3D probabilistic FE model for, 418
FE model description, 419–420
reliability analysis and decision-making, 420–421
2D slope reliability, RFEM analysis of, 461–462
Two-layered cohesive slope (example), 292
reliability analysis results, 293–294
risk analysis results, 294–296
Typical realizations, 482–483

UCT, *see* Unconfined Compression Tests
"Ugly data" challenge, 18, 104
ULS, *see* Ultimate limit state
Ultimate limit state (ULS), 77–80
Uncertainty in constitutive models, 117
correlating soil properties considering uncertainty
Bayesian neural network, 119–120
examples and validation, 122–123
framework of ML-based uncertainty model, 121–122

literature survey, 118
Monte Carlo dropout, 120–121
variational inference (VI), 120
identifying soil model parameters considering uncertainty
examples and validation, 128
framework of Bayesian parameter identification, 123–126
literature survey, 123
parallel computing DE-TMCMC, 126–128, *127*
proposition of enhanced DE-TMCMC, 125–126
integrating uncertainty into the framework of constitutive modeling
examples and validation, 134
incorporating uncertainty, 133
literature survey, 128–130
multi-fidelity residual modeling, 132–133
physics-informed neural network, 130–132
Uncertainty in data-driven site characterization, 143
Bayesian perspective, DDSC from, 145–146
benchmarking, 164
example of benchmarking, 169–171
performance metrics, 166–169
training and validation set, 166
virtual ground, 164–165, **164**
geologic uncertainty, categorical data, 158
coupled Markov chain simulation, 159–161
DDSC-C3 methods, 163
Markov random field (MRF) model, 161–163
geologic uncertainty, numerical data, 153
Gaussian mixture model (GMM), 155, *155*
kernel methods, 156, *156*
K-means clustering, 153, *153*
geotechnical and geologic uncertainty, numerical data, 156–158
geotechnical uncertainty, numerical data, 146
Gaussian process regression (GPR), 152, *152*
sparse modeling, 150–151
types of data and corresponding uncertainty, 144–145
Uncertainty-informed Burland Triangle, 1, *25*
data-centric geotechnics, 16–18
numerical modeling, role of, 1–4
prediction exercises, 4–7
probabilistic applications, 28–29
probabilistic methods, 26–27
probabilistic software, 27–28
statistical characterization, 24–26
uncertainty quantification, 11–16
Uncertainty quantification, 11–16
Unconfined Compression Tests (UCT), 474
Undrained shear-MODE, 261
"Unstructured" variability, 145
Uplift resistance factors, **328**

VAE, *see* Variational autoencoding
Vane shear tests (VSTs), 84
Vanmarcke's spatial averaging, 77–78, 80
Vanmarcke's variance reduction function, 107
Variable redundancy, 328
Variance reduction function, 65, 68, 74, 107, 328
for common autocorrelation models, **69–71**
for an inclined line, **75–76**
Variational autoencoding (VAE), 382
Variational inference (VI), 120
VI, *see* Variational inference
Virtual ground, 164–165, **164**
VSTs, *see* Vane shear tests

Watertightness, 479–481
Watertightness of jet-grouted cut-off walls, 485–489
Weighted least-squares method with regularization, 352–354
Weighted uniform simulations (WUS), 229, 233
WUS, *see* Weighted uniform simulations

Young's modulus, 80

ZSoil model, 402, *408*